# Calculus

## A New Horizon

# Calculus

## A NEW HORIZON

SIXTH EDITION

**HOWARD ANTON**
Drexel University

JOHN WILEY & SONS, INC.
New York  Chichester  Brisbane  Toronto  Singapore

**Mathematics Editor:** Barbara Holland
**Associate Editor:** Sharon Smith
**Senior Developmental Editor:** Madalyn Stone
**Senior Photo Editor:** Hilary Newman
**Senior Marketing Manager:** Leslie Hines
**Copy Editor:** Lilian Brady
**Production and Text Design:** HRS Electronic Text Management
**Electronic Illustration:** Techsetters, Inc.
**Typesetting:** Techsetters, Inc.
**Cover Design:** Madelyn Lesure
**Cover Photo:** © Dann Coffey/The Image Bank

This book was set in Times Roman by Techsetters, Inc., and printed and bound by Von Hoffmann Press, Inc. The cover was printed by The Phoenix Color Corp.

Recognizing the importance of preserving what has been written, it is a policy of John Wiley & Sons, Inc. to have books of enduring value published in the United States printed on acid-free paper, and we exert our best efforts to that end.

The paper on this book was manufactured by a mill whose forest management programs include sustained yield harvesting of its timberlands. Sustained yield harvesting principles ensure that the numbers of trees cut each year does not exceed the amount of new growth.

*Derive* is a registered trademark of Soft Warehouse, Inc.
*Maple* is a registered trademark of Waterloo Maple Software, Inc.
*Mathematica* is a registered trademark of Wolfram Research, Inc.

ISBN     0-471-24331-0

Printed in the United States of America

10   9   8   7   6   5   4   3   2   1

**H**oward Anton obtained his B.A. from Lehigh University, his M.A. from the University of Illinois, and his Ph.D. from the Polytechnic University of Brooklyn, all in mathematics. In the early 1960s he worked for Burroughs Corporation and Avco Corporation at Cape Canaveral, Florida, where he was involved with missile tracking problems for the manned space program. In 1968 he joined the Mathematics Department at Drexel University, where he taught full time until 1983. Since that time he has been an adjunct professor at Drexel and has devoted the majority of his time to textbook writing and activities for mathematical associations. Dr. Anton was President of the EPADEL Section of the Mathematical Association of America (MAA), served on the board of Governors of that organization, and guided the creation of the Student Chapters of the MAA. He has published numerous research papers in Functional Analysis, Approximation Theory, and Topology, as well as pedagogical papers on applications of mathematics. He is best known for his textbooks in mathematics, which are among the most widely used in the world. There are currently more than ninety versions of his books, including translations into Spanish, Arabic, Portuguese, Italian, Indonesian, French, Japanese, Chinese, Hebrew, and German. Dr. Anton has an avid interest in computer technology as it relates to mathematical education and publishing. He has developed pedagogical software for teaching calculus and linear algebra as well as various software programs for the publishing industry that automate the production of four-color mathematical text and art. For relaxation he enjoys traveling and photography.

To
**My Wife Pat**
**My Children Brian, David, and Lauren**

In Memory of
**My Mother Shirley**
**Stephen Girard (1750–1831)—Benefactor**
**Albert Herr—Esteemed Colleague and Contributor**

# A NOTE FROM THE AUTHOR

When I began writing the first edition of this calculus text almost 25 years ago, the task, though daunting, was straightforward in that the content and organization of a standard calculus course was nearly universal—the challenge for me at that time was to present the material in a livelier style and with greater clarity than my predecessors. Since this calculus text is still among the most widely used in the world, I take comfort that the goals I set for myself as a young writer and mathematician have been achieved.

However, times are changing, and the era of a standard and universal calculus course seems destined for the repository of slide rules and three-cent stamps. We are witnessing a lot of experimentation with the content, organization, and goals of calculus—some of which has been successful and some of which has not. Thus, my challenge in writing the sixth edition has been to create a text that has all of the strengths of the earlier editions, yet incorporates those new ideas that are clearly important and have withstood the objective scrutiny of skilled and thoughtful teachers.

In preparing for this edition, I sought advice from outstanding teachers at a wide variety of institutions. Needless to say, I received a diversity of opinions—some reviewers advised against any major changes, arguing that the book was already clearly written and working well in the classroom, while others felt that major changes were required to incorporate technology and make the book more contemporary. I listened carefully, and the lively discussions that followed helped me formulate my philosophy for the new edition. Many of the specific changes are itemized in the preface, but here are some of the general goals:

- Add graphing calculator and CAS materials to the text in a way that will allow students who have those tools to use them but that will not prevent the text from being used by those students who do not have access to that technology.
- Place more emphasis on mathematical modeling and applications.
- Incorporate new examples and exercises that will be meaningful to today's students and will more accurately convey the role of calculus in the real world.
- Widen the variety of exercises to focus more on *conceptual understanding* through conjecture, multistep analysis, expository writing, and what-if analysis.

In addition, I wanted to provide some optional innovative materials that would capture the student's interest and provide the kind of problem-solving experience that he or she might find in a research or industrial setting. This gave birth to an exciting set of modules that we have called *Expanding the Calculus Horizon*. These modules appear at the ends of selected chapters and each has an optional Internet component that we hope will grow dynamically over time with input from teachers and students.

In developing this edition I have stood firm on two principles that were adhered to in earlier editions:

- The text material is presented at a mathematical level that is suitable for students who will embark on careers in engineering and science.
- It remains a primary goal of the text to teach the student clear, logical, mathematical thinking. Informal discussions play an important motivational role in the exposition and are used extensively, but eventually I want the student to be able to read and understand the language of mathematics.

Although this edition has many changes and new features, they have been implemented in a *flexible* way that will accommodate a wide variety of teaching philosophies. Thus, I am confident that professors who have had positive experiences with earlier editions will be comfortable with this revision, and I am hopeful that those professors who are looking for a contemporary text with an established history of success in the classroom will be pleased with the innovations in this new edition.

Sincerely,

Howard Anton

Howard Anton

At times the words of a complete stranger are difficult to accept. That is why I am about to take this first opportunity to introduce myself. Hopefully by revealing a bit about myself and how I relate to this textbook may help you find these words more compelling.

Hello, my name is Ajay Arora and I am an Electrical Engineering student at McMaster University in Hamilton, Canada. I too was in your place when I began my entry into the much dreaded field of *CALCULUS*. The vast amounts of rate of change and antiderivative problems were overwhelming. With a little struggle and hard work, I successfully completed that course only to be faced with three more advanced level calculus courses. What I am about to write is the unbiased truth on how you can be successful in calculus and how this textbook will assist you on your journey.

I have been a member of the Student Advisory Board for this textbook for over a year now. The committee came together as a venture from the authors and publisher to get more student input in the development stages instead of simply focusing on feedback when the book was published. After a chapter was completed by the author, each student committee member evaluated, commented, and in some cases, recommended alternative approaches. These tasks involved lots of special deliveries, E-mails, faxes, telephone calls, conference calls, and of course, a whole lot of calculus! But in the end it was a total rewarding experience.

How many times have you asked yourself, "Is math really useful?" Or how about, "Will I ever use calculus in the real world?" I know I have! This textbook will definitely help you answer some of these questions with true applications of the theories you learn. The modules entitled Expanding the Calculus Horizon have been included for precisely that purpose. Every module has been critiqued extensively by the Student Advisory Board, and I encourage you to try them. Not only will these applications of calculus surprise you, but they may actually help give you direction in a field that you might want to pursue after college.

I wish you success in this course, as well as the many others you will face during your college career. Good Luck!

Sincerely,

Ajay Arora
McMaster University
A.Arora@ieee.org

## Best Wishes for Success from the Student Advisory Board

Dan Arndt, *University of Texas at Dallas*
Ajay Arora, *McMaster University*
Scott E. Barnett, *Wayne State University*

Fatenah Issa, *Loyola University of Chicago*
Laurie Haskell Messina, *University of Oklahoma*
Steven E. Pav, *Alfred University*

## ABOUT THIS EDITION

This is a major revision. In keeping with current trends in calculus, the goal for this edition is to focus more on ***conceptual understanding*** and ***applicability*** of the subject matter. In designing this edition, we worked closely with a talented team of reviewers to ensure that the book is sufficiently ***flexible*** that it will continue to meet the needs of those using the last edition and at the same time provide a fresh approach for those instructors who are taking their calculus course in a new direction. Some of the more significant changes are as follows:

**Technology** This edition provides extensive materials for instructors who want to use graphing calculators or computer algebra systems. However, these materials are implemented in a way that allows the text to be used in courses where technology is used heavily, moderately, or not at all. To provide a sound foundation for the technology material, I have added a new section entitled Graphing Functions on Calculators and Computers; Computer Algebra Systems (Section 1.3).

**Horizon Modules** Selected chapters end with modules called Expanding the Calculus Horizon. As the name implies, these modules are intended to take the student a step beyond the traditional calculus text. The modules, all of which are optional, can be assigned either as individual or group projects and can be used by instructors to tailor the calculus course to meet their specific needs and teaching philosophies. For example, there are modules that touch on iteration and dynamical systems, modeling from experimental data by curve fitting, applications, expository report writing, and so forth.

**Mathematical Modeling** Mathematical modeling plays a more prominent role in this edition. The concept of a mathematical model is introduced in Section 1.5, and the terminology of modeling is used extensively thereafter. The optional Horizon module for Chapter 5 discusses how to obtain various kinds of mathematical models from experimental data, and Chapter 10 discusses mathematical modeling using differential equations.

**Applicability of Calculus** One of the goals in this edition is to link calculus more closely to the real world and to the student's own experience. This theme starts with the Introduction and is carried through in the modules, examples, and exercises. Applications appearing in exercises and examples are carefully chosen to be sufficiently simple that they do not divert time from learning important mathematical fundamentals. More extensive applications appear in various Horizon modules.

**Earlier Differential Equations** Basic ideas about differential equations, initial-value problems, direction fields, and integral curves are introduced concurrently with integration and then revisited in more detail in Chapter 10.

**Quicker Entry to Functions** Chapter 1 begins immediately with functions, and the precalculus material that formed the first chapter in earlier editions has been moved to the appendix.

**For the Reader** This element is new. At various points in the exposition the student is assigned a brief task. Some tasks are appropriate for all readers, while others are appropriate only for readers who have a graphing calculator or a CAS. The tasks for all readers are designed to immerse the student more deeply into the text by asking them to think about an idea and reach some conclusion; the tasks for students using technology are designed to familiarize them with the procedures for using that technology by asking them to read their documentation and perform some text-related computation. Some instructors may want to make these tasks part of their assignments.

**Earlier Logarithms and Exponentials**   Logarithmic and exponential functions are introduced in Chapter 4 from the exponent point of view and then revisited in Section 7.9 from the integral point of view. This provides a richer variety of functions to work with earlier in the text, fits in better with the discussions of modeling, and makes for a less fragmented presentation of the analysis of graphs. However, for instructors who prefer a later presentation of logarithmic and exponential functions, there is an instructor's guide that explains how to move this material to Chapter 7 and provides a reference list of those exercises in Chapters 4, 5, and 6 that involve logarithmic and exponential functions.

**Early Parametric Option**   There is a new option for introducing parametric curves in Section 1.7 of Chapter 1 and revisiting the material in Chapter 13, where calculus-related issues are discussed. Instructors who prefer the traditional late discussion of parametric equations will have no problem teaching Section 1.7 as part of Chapter 13, as in earlier editions.

**More Variety in Exercises**   The exercise sets have been revised extensively to create a richer variety—there are many more exercises that include conjecture, exploration, multistep analysis, and expository writing. The goal has been to put more focus on *conceptual understanding*. There are also many new exercises that are intended to be solved using a graphing calculator or a CAS. These are marked with icons for easy identification.

**Analysis of Functions**   The old "curve-sketching" material has been replaced by Sections 5.1–5.3 on the Analysis of Functions. The name change reflects a more contemporary approach to the material—there is more emphasis on the interplay between technology and calculus and more focus on the problem of finding a *complete graph*, that is, a graph that contains all of the significant features of concern.

**Principles of Integral Evaluation**   The old "Techniques of Integration" has been renamed Principles of Integral Evaluation to reflect its more contemporary approach to the material. The chapter has been condensed and there is now more emphasis on general methods and less on tricks for evaluating complicated or obscure integrals. The section entitled Using Integral Tables and Computer Algebra Systems has been expanded and rewritten extensively.

**Supplementary Exercises**   Supplementary exercises have been added at the ends of chapters.

**New Appendix on Solving Polynomial Equations**   Appendix F, entitled Solving Polynomial Equations, is new. It reviews the Factor Theorem, the Remainder Theorem, and procedures for finding rational roots. Many students are weak on this material, yet it plays an important role in determining whether a polynomial graph generated on a calculator or computer is complete.

**Rule of Four**   The "rule of four" refers to the presentation of material from the verbal, algebraic, visual, and numerical points of view. It is used more extensively in this edition, where appropriate.

**Internet**  An internet site http://www.wiley.com/college/anton has been established to complement the text. This site contains additional Horizon modules and technology materials. The site is experimental, but we expect it to grow dynamically over time.

## OTHER FEATURES

**Flexibility**  This edition has a built-in flexibility that is designed to serve a broad spectrum of calculus philosophies, ranging from traditional to reform. Graphing technology can be used heavily, moderately, or not at all; and the order of presentation of many sections can be permuted to accommodate specific course needs.

**Trigonometry Review**  Deficiencies in trigonometry plague many students, so I have included a substantial trigonometry review in Appendix E.

**Historical Notes**  The biographies and historical notes have been a hallmark of this text from its first edition and have been maintained in this edition. All of the biographical materials have been distilled from standard sources with the goal of capturing the personalities of the great mathematicians and bringing them to life for the student.

**Graded Exercise Sets**  Section Exercise Sets are graded to begin with routine problems and progress gradually toward problems of greater difficulty. However, in the Supplementary Exercises I have opted not to grade the exercises by level of difficulty to avoid giving the student a predisposition about the level of effort required.

**Rigor**  The challenge of writing a good calculus book is to strike the right balance between rigor and clarity. My goal is to present precise mathematics to the fullest extent possible for the freshman audience, but where clarity and rigor conflict I choose clarity. However, I believe it to be essential that the student understand the difference between a careful proof and an informal argument, so I try to make it clear to the reader when arguments are informal. Theory involving $\delta$-$\epsilon$ arguments appear in separate sections, so they can be bypassed if desired.

**Mathematical Level**  This book is written at a mathematical level that is suitable for students planning on careers in engineering or science.

**Student Review**  A Student Advisory Board was actively involved in the development process of this edition to provide information on pedagogical clarity and to advise on the development of examples, exercises, and modules that students would find interesting and relevant.

**RESOURCES FOR THE
STUDENT**

**Student Resource and Survival CD**                          0-471-24622-0

This CD for IBM compatibles or Macintosh platforms provides students with an electronic form of detailed solutions to odd-numbered exercises, multiple choice and true–false sample tests for each section and chapter of the text, precalculus review material, and a brief introduction to those aspects of linear algebra that are of immediate concern to the calculus student. Two demonstration modules from the Windows-based multimedia calculus program *Calculus Connections, A Multimedia Adventure* are also available on this CD.

**Student Resource Manual**                                  0-471-24616-6

This manual provides students with detailed solutions to odd-numbered exercises and multiple choice and true–false sample tests for each section and chapter of the text.

**RESOURCES FOR THE
INSTRUCTOR**

Hard copy and electronic resources are available for the instructor. These can be obtained by sending a request on your institutional letterhead to Leslie Hines, Senior Marketing Manager, John Wiley & Sons, Inc., 605 Third Avenue, New York, NY 10158-0012, or by requesting them from your local Wiley representative.

# ACKNOWLEDGMENTS

It has been my good fortune to have the advice and guidance of many talented people whose knowledge and skills have enhanced this book in many ways. For their valuable help I thank:

## Reviewers and Contributors to Earlier Editions

Edith Ainsworth, *University of Alabama*

Loren Argabright, *Drexel University*

David Armacost, *Amherst College*

John Bailey, *Clark State Community College*

Robert C. Banash, *St. Ambrose University*

George R. Barnes, *University of Louisville*

Larry Bates, *University of Calgary*

John P. Beckwith, *Michigan Technological University*

Joan E. Bell, *Northeastern Oklahoma State University*

Irl C. Bivens, *Davidson College*

Harry N. Bixler, *Bernard M. Baruch College, CUNY*

Marilyn Blockus, *San Jose State University*

Ray Boersma, *Front Range Community College*

Barbara Bohannon, *Hofstra University*

David Bolen, *Virginia Military Institute*

Daniel Bonar, *Denison University*

George W. Booth, *Brooklyn College*

Phyllis Boutilier, *Michigan Technological University*

Mark Bridger, *Northeastern University*

John Brothers, *Indiana University*

Stephen L. Brown, *Olivet Nazarene University*

Virginia Buchanan, *Hiram College*

Robert C. Bueker, *Western Kentucky University*

Robert Bumcrot, *Hofstra University*

Christopher Butler, *Case Western Reserve University*

Carlos E. Caballero, *Winthrop University*

James Caristi, *Valparaiso University*

Stan R. Chadick, *Northwestern State University*

Hongwei Chen, *Christopher Newport University*

Chris Christensen, *Northern Kentucky University*

Robert D. Cismowski, *San Bernardino Valley College*

Patricia Clark, *Rochester Institute of Technology*

Hannah Clavner, *Drexel University*

David Clydesdale, *Sauk Valley Community College*

David Cohen, *University of California, Los Angeles*

Michael Cohen, *Hofstra University*

Pasquale Condo, *University of Lowell*

Robert Conley, *Precision Visuals*

Cecil J. Coone, *State Technical Institute at Memphis*

Norman Cornish, *University of Detroit*

Terrance Cremeans, *Oakland Community College*

Lawrence Cusick, *California State University–Fresno*

Michael Dagg, *Numerical Solutions, Inc.*

Stephen L. Davis, *Davidson College*

A. L. Deal, *Virginia Military Institute*

Charles Denlinger, *Millersville University*

William H. Dent, *Maryville College*

Blaise DeSesa, *Drexel University*

Dennis DeTurck, *University of Pennsylvania*

Jacqueline Dewar, *Loyola Marymount University*

Preston Dinkins, *Southern University*

Irving Drooyan, *Los Angeles Pierce College*

Tom Drouet, *East Los Angeles College*

Clyde Dubbs, *New Mexico Institute of Mining and Technology*

Della Duncan, *California State University–Fresno*

Ken Dunn, *Dalhousie University*

Sheldon Dyck, *Waterloo Maple Software*

Hugh B. Easler, *College of William and Mary*

Scott Eckert, *Cuyamaca College*

Joseph M. Egar, *Cleveland State University*

Judith Elkins, *Sweet Briar College*

Brett Elliott, *Southeastern Oklahoma State University*

Garret J. Etgen, *University of Houston*

Benny Evans, *Oklahoma State University*

James H. Fife, *Educational Testing Service*

Dorothy M. Fitzgerald, *Golden West College*

Barbara Flajnik, *Virginia Military Institute*

Daniel Flath, *University of South Alabama*

Ernesto Franco, *California State University–Fresno*

Nicholas E. Frangos, *Hofstra University*

Katherine Franklin, *Los Angeles Pierce College*

Marc Frantz, *Indiana University–Purdue University at Indianapolis*

Michael Frantz, *University of La Verne*

Susan L. Friedman, *Bernard M. Baruch College, CUNY*

William R. Fuller, *Purdue University*

Daniel B. Gallup, *Pasadena City College*

Mahmood Ghamsary, *Long Beach City College*

G. S. Gill, *Brigham Young University*

Michael Gilpin, *Michigan Technological University*

Kaplana Godbole, *Michigan Technological Institute*

S. B. Gokhale, *Western Illinois University*

Morton Goldberg, *Broome Community College*

Mardechai Goodman, *Rosary College*

Sid Graham, *Michigan Technological University*

Raymond Greenwell, *Hofstra University*

Gary Grimes, *Mt. Hood Community College*

Jane Grossman, *University of Lowell*

Michael Grossman, *University of Lowell*

Diane Hagglund, *Waterloo Maple Software*

Douglas W. Hall, *Michigan State University*

Nancy A. Harrington, *University of Lowell*

Kent Harris, *Western Illinois University*

Jim Hefferson, *St. Michael College*

Albert Herr, *Drexel University*

Peter Herron, *Suffolk County Community College*

Warland R. Hersey, *North Shore Community College*

Konrad J. Heuvers, *Michigan Technological University*

Robert Higgins, *Quantics Corporation*

Rebecca Hill, *Rochester Institute of Technology*

Edwin Hoefer, *Rochester Institute of Technology*

Louis F. Hoelzle, *Bucks County Community College*

Robert Homolka, *Kansas State University–Salina*

Jerry Johnson, *University of Nevada–Reno*

John M. Johnson, *George Fox College*

Wells R. Johnson, *Bowdoin College*

Herbert Kasube, *Bradley University*

Phil Kavanaugh, *Mesa State College*

Maureen Kelley, *Northern Essex Community College*

Harvey B. Keynes, *University of Minnesota*

Richard Krikorian, *Westchester Community College*

Paul Kumpel, *SUNY, Stony Brook*

Fat C. Lam, *Gallaudet University*

Leo Lampone, *Quantics Corporation*

James F. Lanahan, *University of Detroit–Mercy*

Bruce Landman, *University of North Carolina at Greensboro*

Kuen Hung Lee, *Los Angeles Trade–Technology College*

Marshall J. Leitman, *Case Western Reserve University*

Benjamin Levy, *Lexington H.S., Lexington, Mass.*

Darryl A. Linde, *Northeastern Oklahoma State University*

Phil Locke, *University of Maine, Orono*

Leland E. Long, *Muscatine Community College*

John Lucas, *University of Wisconsin–Oshkosh*

Stanley M. Lukawecki, *Clemson University*

Nicholas Macri, *Temple University*

Melvin J. Maron, *University of Louisville*

Mauricio Marroquin, *Los Angeles Valley College*

Majid Masso, *Brookdale Community College*

Larry Matthews, *Concordia College*

Thomas McElligott, *University of Lowell*

Phillip McGill, *Illinois Central College*

Judith McKinney, *California State Polytechnic University, Pomona*

Joseph Meier, *Millersville University*

Aileen Michaels, *Hofstra University*

Janet S. Milton, *Radford University*

Robert Mitchell, *Rowan College of New Jersey*

Marilyn Molloy, *Our Lady of the Lake University*

Ron Moore, *Ryerson Polytechnical Institute*

Barbara Moses, *Bowling Green State University*

David Nash, *VP Research, Autofacts, Inc.*

Kylene Norman, *Clark State Community College*

Roxie Novak, *Radford University*

Richard Nowakowski, *Dalhousie University*

Stanley Ocken, *City College–CUNY*

Donald Passman, *University of Wisconsin*

David Patterson, *West Texas A & M*

Walter M. Patterson, *Lander University*

Edward Peifer, *Ulster County Community College*

Robert Phillips, *University of South Carolina at Aiken*

Mark A. Pinsky, *Northeastern University*

Catherine H. Pirri, *Northern Essex Community College*

Father Bernard Portz, *Creighton University*

David Randall, *Oakland Community College*

Richard Remzowski, *Broome Community College*

Guanshen Ren, *College of Saint Scholastica*

William H. Richardson, *Wichita State University*

David Rollins, *University of Central Florida*

Naomi Rose, *Mercer County Community College*

Sharon Ross, *DeKalb College*

David Ryeburn, *Simon Fraser University*

David Sandell, *U.S. Coast Guard Academy*
Ned W. Schillow, *Lehigh County Community College*
Dennis Schneider, *Knox College*
Dan Seth, *Morehead State University*
George Shapiro, *Brooklyn College*
Parashu R. Sharma, *Grambling State University*
Donald R. Sherbert, *University of Illinois*
Howard Sherwood, *University of Central Florida*
Bhagat Singh, *University of Wisconsin Centers*
Martha Sklar, *Los Angeles City College*

John L. Smith, *Rancho Santiago Community College*
Wolfe Snow, *Brooklyn College*
Ian Spatz, *Brooklyn College*
Jean Springer, *Mount Royal College*
Norton Starr, *Amherst College*
Richard B. Thompson, *The University of Arizona*
William F. Trench, *Trinity University*
Walter W. Turner, *Western Michigan University*
Richard C. Vile, *Eastern Michigan University*
David Voss, *Western Illinois University*
Shirley Wakin, *University of New Haven*

James Warner, *Precision Visuals*
Peter Waterman, *Northern Illinois University*
Evelyn Weinstock, *Glassboro State College*
Candice A. Weston, *University of Lowell*
Bruce F. White, *Lander University*
Gary L. Wood, *Azusa Pacific University*
Yihren Wu, *Hofstra University*
Richard Yuskaitis, *Precision Visuals*
Michael Zeidler, *Milwaukee Area Technical College*
Michael L. Zwilling, *Mount Union College*

## Development Team for the Sixth Edition

The following people critiqued and reviewed various parts of the manuscript and suggested many of the ideas that found their way into this new edition:

Judith Broadwin, *Jericho High School*
Christopher D. Butler, *Case Western Reserve University*
Larry Cusick, *California State University–Fresno*
Philip Farmer, *Diablo Valley College*
Sally E. Fischbeck, *Rochester Institute of Technology*
J. Derrick Head, *University of Minnesota–Morris*
Tommie Ann Hill-Natter, *Prairie View A&M University*
Holly Hirst, *Appalachian State University*
Dan Kemp, *South Dakota State University*
Holly A. Kresch, *Diablo Valley College*
Marshall Leitman, *Case Western Reserve University*
Thomas W. Mason, *Florida A&M University*

Gary L. Peterson, *James Madison University*
Douglas Quinney, *University of Keele*
William H. Richardson, *Wichita State University*
Lila F. Roberts, *Georgia Southern University*
Avinash Sathaye, *University of Kentucky*
Mary Margaret Shoaf-Grubbs, *College of New Rochelle*
Mark Stevenson, *Oakland Community College*
John A. Suvak, *Memorial University of Newfoundland*
Skip Thompson, *Radford University*
Bruce R. Wenner, *University of Missouri–Kansas City*

The following people participated in phone surveys that helped to answer important questions about organization, philosophy, technology, and content:

Linda Bridge, *Long Beach City College*
Ted Clinkenbeard, *Des Moines Area Community College*
Victor Feser, *University of Maryland*
David Gross, *University of Connecticut*
Dennis Hadah, *Saddleback Community College*
Henry Horton, *University of West Florida*
Emmett Johnson, *Grambling State University*

Bill Kavanagh, *Mesa College*
Phil Locke, *University of Maine*
Thomas W. Mason, *Florida A&M University*
Ralph Okojie, *Elizabeth City State University*
David Robbins, *Trinity College*
Skip Thompson, *Radford University*
Paul Vesce, *University of Missouri–Kansas City*
Ronald Wagoner, *California State University–Fresno*

The following people read the sixth edition at various stages for mathematical and pedagogical accuracy and/or assisted with the critically important job of preparing answers to exercises:

Larry Cusick, *California State University–Fresno*
Stephen Davis, *Davidson College*
Blaise DeSesa, *Allentown College of St. Francis de Sales*
Thomas Vanden Eynden, *Thomas More College*
Susan Gerstein
Konrad Heuvers, *Michigan Technological University*

Majid Masso, *University of Delaware*
Kylene Norman, *Clark State Community College*
Irwin Pressman, *Carleton University*
David Ryeburn, *Simon Fraser University*
Shirley Wakin, *University of New Haven*
Neil Wigley, *University of Windsor*

The following students are members of the Student Advisory Board that critiqued the manuscript for clarity and provided valuable advice on making material interesting and relevant to today's students:

Dan Arndt, *University of Texas at Dallas*
Ajay Arora, *McMaster University*
Scott E. Barnett, *Wayne State University*

Fatenah Issa, *Loyola University of Chicago*
Laurie Haskell Messina, *University of Oklahoma*
Steven E. Pav, *Alfred University*

The following people created materials for tests and other supplements:

William H. Barker, *Bowdoin College*
Henry Smith, *Southeastern Louisiana University*

James E. Ward, *Bowdoin College*
Neil Wigley, *University of Windsor*

I would also like to thank Gary S. Stoudt of the *University of Indiana of Pennsylvania* for his assistance in locating various obscure historical materials.

## Content Contributions

The following people assisted in the creation of modules and exercises or contributed valuable ideas that helped in the development of those materials:

Mary Ann Connors, *U.S. Military Academy at West Point*

Art Davis, *San Jose State University*

Gloria S. Dion, *Educational Testing Service*

Iris Brann Feta, *Clemson University*

Dixie Griffin, Jr., *Louisiana Tech University*

Hugh E. Huntley, *University of Michigan*

Lynn Kiaer, *Rose-Hulman Institute of Technology*

Cecilia Knoll, *Florida Institute of Technology*

Michael Magill, *Purdue University*

Robert Meitz, *Arizona State University*

John Rickert, *Rose-Hulman Institute of Technology*

Michael D. Shaw, *Florida Institute of Technology*

P. Narayana Swamy, *Southern Illinois University*

Josef Torok, *Rochester Institute of Technology*

I gratefully acknowledge the permission to adapt various exercises, examples, and text materials from the following publications:

*Physics*, 3rd ed., Cutnell and Johnson, John Wiley & Sons, Inc., 1995.

*Fundamentals of Physics*, 4th ed., Halliday, Resnick, and Walker, John Wiley & Sons, Inc., 1993.

*Applications of Calculus*, Philip Straffin, Ed., MAA Notes Number 29, The Mathematical Association of America, 1993.

*Calculus Problems for a New Century*, Robert Fraga, Ed., MAA Notes Number 28, Vol. 2, The Mathematical Association of America, 1993.

*Engineering Mechanics*, Meriam and Kraige, 3rd ed., Vol. 2, John Wiley & Sons, Inc., 1992.

## Special Contributions

A special debt of gratitude to:

Barbara Holland, my editor, for sharing a vision and having unwavering faith in my work.

Madalyn Stone, for her infectious enthusiasm.

Ann Berlin, Pam Kennedy, and Charlotte Hyland of the Wiley Production Department for a scheduling miracle.

Sharon Smith, for again successfully coordinating a complex set of supplements.

Maddy Lesure, for capturing the "horizon" theme so beautifully in the cover.

Hilary Newman, for unearthing my obscure photographic requests.

Lilian Brady, for her unerring eye for aesthetics and typography.

The group at HRS, for a beautiful design and their patience with an (occasionally) cranky author.

The group at Techsetters, for superb composition and illustration and their devotion to excellence.

Lynn Kiaer, John Rickert, and Mary Ann Connors for their creative contributions to the modules.

Neil Wigley, for enjoyable E-mail humor and an outstanding job in writing the solutions manuals.

David Ryeburn, for his attention to detail and valued advice on applications of calculus.

My assistant, Dolores Morgan, whose superb organizational skills played a major role in keeping me on schedule.

# Contents

# CHAPTER 4. LOGARITHMIC AND EXPONENTIAL FUNCTIONS    225

# CHAPTER 5. ANALYSIS OF FUNCTIONS AND THEIR GRAPHS    289

# CHAPTER 6. APPLICATIONS OF THE DERIVATIVE    329

# CHAPTER 7. INTEGRATION    377

## VOLUME 2

## VOLUME 3

## BRIEF EDITION

## COMPLETE EDITION

# FOR THE STUDENT

**C**alculus is a compilation of ideas that provides a way of viewing and analyzing the physical world. As with all mathematics courses, calculus involves equations and formulas. However, if you successfully learn to use all of the formulas and solve all of the problems in this text but don't master the underlying ideas, you will have missed the most important part of calculus. Keep in mind that every single problem in this text has already been solved by somebody, so your ability to solve those problems gives you nothing unique. However, if you master the ideas of calculus, then you will have the tools to go beyond what other people have done, limited only by your own talents and creativity.

Before starting your studies, you may find it helpful to leaf through this text to get a general feeling for its different parts.

▶ At the beginning of each chapter you will find a page that gives an overview of the chapter, and at the beginning of each section you will find an introduction that gives an overview of that section. To help you locate specific information, sections are divided into topics described by headings in the margin.

▶ Each section ends with a set of exercises. The answers to most odd-numbered exercises appear in the back of the book. Worked-out solutions to the odd-numbered exercises are given in the *Student Resource Manual* and on a CD, which are available as supplements to the text.

▶ Some of the exercises are tagged with icons to indicate that some kind of technology is required for their solution. If your calculus course does not incorporate the use of technology, then your instructor will probably not assign these. Those exercises tagged with the icon ⊡ require graphing technology, which might be either a graphing calculator or a computer program that produces graphs from equations. Those exercises tagged with the icon [c] require a computer algebra system (called a CAS), which is a program that can perform symbolic as well as numerical calculations. The most common CAS programs are *Mathematica*, *Maple*, and *Derive*. Some of the newer calculators incorporate CAS capabilities.

▶ Each chapter ends with a set of supplementary exercises, many of which involve a combination of ideas from various sections within the chapter.

▶ Near the end of the text you will find seven appendices. Appendices A–F review some precalculus material, including trigonometry, and Appendix G contains some proofs that may or may not be part of your course.

▶ There is also reference material on the endpapers that are inside the front and back covers of the text.

▶ Illustrations in the exposition are referenced using a triple-number system. For example, Figure 1.6.3 is the third figure in Section 1.6, and Figure 7.2.5 is the fifth figure in Section 7.2. The same numbering system is used for theorems and definitions. Illustrations in the exercises are identified by the exercise number with which they are associated. For example, in a particular exercise set, Figure Ex-7 would be associated with Exercise 7.

▶ The ideas in this text were created by real people with interesting personalities and backgrounds. Pictures and signatures of many of these people appear on the opening pages of the chapters, and biographical sketches of various mathematicians appear throughout the text as footnotes.

▶ At various places in the text you will see elements labeled "For the Reader," which are designed to reinforce ideas in the text. Some of these ask you to think about an idea, some ask you to perform a computation, and (for students using technology) some ask you to read your reference manual and then use the technology to perform a computation or to generate a graph.

As you read through this book, you will find some ideas that you understand immediately, others that you don't understand until you have read them several times, and others that you do not understand, even after numerous readings. Don't become discouraged—some calculus ideas take time to "percolate," and you may well find that the idea suddenly becomes clear later when you least expect it.

If you find that your answer to an exercise does not match that in the back of the book, do not presume immediately that your answer is incorrect—there may be more than one way to express the answer. For example, if your answer is $\sqrt{3}/3$ and the text answer is $1/\sqrt{3}$, then both are correct, since your answer can be obtained by rationalizing the text answer. In general, if your answer does not match that in the text, then your best first step is to look for an algebraic manipulation or a trigonometric identity that relates the two answers. In cases where the answer is a decimal approximation, your answer may differ from that in the text because of different choices in the number of decimal places used in the computations.

Some exercises require a verbal answer. Express those answers in complete, correctly punctuated, logical sentences—not fragmented phrases and formulas.

It is *not* essential to have graphing technology to read and use this text. Exercises requiring technology have been tagged with icons precisely so they can be omitted if necessary. Text elements requiring technology are relegated to the "For the Reader," so they can be omitted as well. If you have graphing technology, then you may want to use it as you read the text or to check your work in exercises that are not tagged with icons. However, it is not essential.

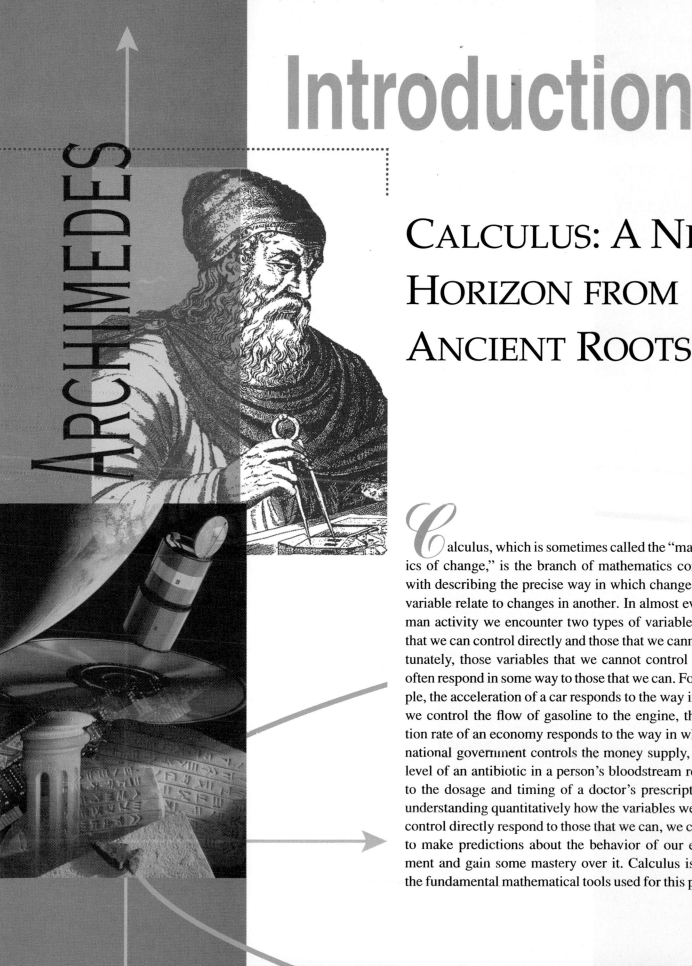

# Introduction

## CALCULUS: A NEW HORIZON FROM ANCIENT ROOTS

*C*alculus, which is sometimes called the "mathematics of change," is the branch of mathematics concerned with describing the precise way in which changes in one variable relate to changes in another. In almost every human activity we encounter two types of variables: those that we can control directly and those that we cannot. Fortunately, those variables that we cannot control directly often respond in some way to those that we can. For example, the acceleration of a car responds to the way in which we control the flow of gasoline to the engine, the inflation rate of an economy responds to the way in which the national government controls the money supply, and the level of an antibiotic in a person's bloodstream responds to the dosage and timing of a doctor's prescription. By understanding quantitatively how the variables we cannot control directly respond to those that we can, we can hope to make predictions about the behavior of our environment and gain some mastery over it. Calculus is one of the fundamental mathematical tools used for this purpose.

Calculus has an enormous, but often unnoticed, impact on our daily lives. To provide some sense of how you and I are being affected by calculus, I have selected a few of its applications to fields of contemporary research. All of these applications involve other branches of science and mathematics, but they all use calculus in some essential way. The first three applications are based on a new and exciting area of mathematics called the theory of *wavelets*. Wavelets make it possible to capture and store mathematical representations of images and signals using much less data than previously possible. As a result, the current research literature is literally exploding with new applications of wavelets to such diverse fields as astronomy, acoustics, nuclear engineering, image processing, neurophysiology, music, medicine, speech synthesization, earthquake prediction, and pure mathematics, to name only a few.

**FBI Fingerprint Compression** — The U.S. Federal Bureau of Investigation began collecting fingerprints and handprints in 1924 and now has more than 30 million such prints in its files, all of which are being digitized for storage on computer. It takes about 0.6 megabyte of storage space to record a fingerprint and 6 megabytes to record a pair of handprints, so that digitizing the current FBI archive would result in about $200 \times 10^{12}$ bytes of data to be stored, which is the capacity of roughly 138 million floppy disks. At today's prices for computer equipment, storage media, and labor, this would cost roughly 200 million dollars. To reduce this cost, the FBI's Criminal Justice Information Service Division began working in 1993 with the National Institute of Standards, the Los Alamos National Laboratory, and several other groups to devise compression methods for reducing the storage space. These methods, which are based on wavelets, are proving to be highly successful. Figure 1 is a good example—the image on the left is an original thumbprint and the one on the right is a mathematical reconstruction from a 26:1 data compression.

**Music** — Researchers with the Numerical Algorithms Research Group at Yale University have investigated the application of wavelets to sound synthesis (musical and voice). To approximate the sound of a musical instrument or voice, samples are taken and decomposed mathematically into numbers called *wavelet packet coefficients*. These coefficients can be stored on a computer and later the sound can be reconstructed (synthesized) from the computer data. This area of research makes it possible to reproduce complex sounds from a small amount of data and to transmit those data electronically in a highly compressed form. This research may eventually speed up the transmission of sound over the Internet, for example.

**Removing Noise from Data** — In fields ranging from planetary science to molecular spectroscopy, scientists are faced with the problem of recovering a true signal from incomplete or noisy data. For example, weak signals from deep space probes are often so overwhelmed with background noise that the signal itself is barely detectable, yet the signal must be used to produce a photograph or provide other information. Researchers at Stanford University and elsewhere have been working for several years on using wavelet methods to filter out such noise. For example, Figure 2 shows a signal from a medical imaging signal that has been cleaned up (de-noised) using wavelets.

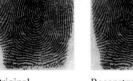

Original    Reconstruction

Figure 1

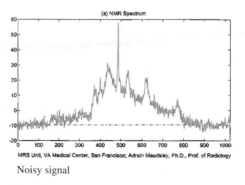

Noisy signal

De-noised signal

Figure 2

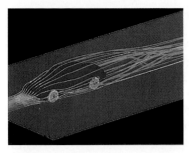

Airflow past a Saturn SL2

Figure 3

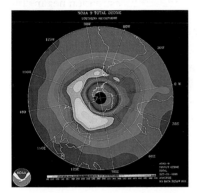

Ozone hole in the Southern Hemisphere

Figure 4

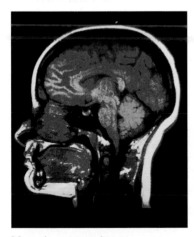

Magnetic resonance image

Figure 5

Figure 6

**Airflow Past an Automobile** — Problems involving fluid flow (air, water, and blood, for example) are a major focus of scientific research. The Army High Performance Computing Research Center (AHPCRC) sponsors numerous unclassified research projects that involve teams of researchers from various science and engineering disciplines. One such project deals with airflow past an automobile (they use a General Motors Saturn SL2). The problem is quite complex since it takes into account the body contours, the wheels, the recessed headlights, and the spoiler. Figure 3 shows a simulation of airflow past an automobile that was produced using state-of-the-art mathematical methods and a Cray T3D supercomputer.

**Weather Prediction** — Modern meteorology is a marriage between mathematics and physics. Today's meteorologists are concerned with much more than predicting daily weather changes—their research delves into such areas as global warming, holes in the ozone layer (Figure 4), and weather patterns on other planets. In 1904 the Norwegian meteorologist Vilhelm Bjerknes (1862–1951) proposed that the state of the atmosphere at any future time can be determined by measuring appropriate variables at a single instant of time and then solving certain hydrodynamic equations. Although Bjerknes' idea is true in principle, it is difficult to apply because of uncertainties in measured variables, the enormous amounts of data to be processed, and technical complications involved with solving the equations. However, new mathematical discoveries have dramatically improved meteorological predictions and spawned enormous economic benefits. For example, it costs about 50 million dollars to prepare for a hurricane over 300 nautical miles of coastline, even if the hurricane does not hit the area. On the other hand, if the hurricane hits without adequate preparation, then the added costs can mount to billions of dollars (let alone the loss of life). Thus, each new mathematical breakthrough that produces more accurate hurricane prediction translates into enormous economic savings and preservation of human life.

**Medical Imaging and DNA Structure** — Advances in *nuclear magnetic resonance* (NMR) have made it possible to determine the structure of biological macromolecules, study DNA replication, and determine how proteins act as enzymes and antibodies. Related advances in *magnetic resonance imaging* (MRI) have made it possible to view internal human tissue without invasive surgery and to provide real-time images during surgical procedures (Figure 5). High-quality NMR and MRI would not be possible without mathematical discoveries that have occurred within the last decade.

**Controlling Chaotic Behavior in the Human Heart** — Chaos theory, which is one of the most exciting new branches of mathematics, is concerned with identifying regularities in phenomena that on the surface seem random and unpredictable (Figure 6). Today's research literature abounds with applications of chaos theory to almost every imaginable branch of science. Recently, researchers at the Applied Chaos Laboratory at Georgia Tech University collaborated with physicians at the Emory University Medical Center in applying chaos theory to control the chaotic behavior of heart tissue that is undergoing ventricular fibrillation (cardiac arrest). The research, though experimental, is already showing promising results.

**The World Model of the Future** — In anticipation of the 1992 United Nations Earth Summit, researchers at the Institute for Economic Analysis (IEA) at New York University were commissioned by a number of world leaders with the daunting task of creating a model that would predict the economic and environmental future of the world. They started with the World Model and World Database developed by Nobel laureate Wassily Leontief and his colleagues at Harvard in the 1970s, but they expanded on the model by incorporating such environmental factors as the cost of controlling pollutant emissions (from mining, energy creation, and automobiles, for example). They also accounted for the effect of population growth rates on the added demand for energy and other natural resources. Models such as this require a team effort by government, academic, and industrial experts in a variety of fields and play an important role in guiding the decisions of governmental agencies.

**Deep Space Exploration** — Alexander Wolszczan of Penn State University may go down in history as the first scientist to identify a planetary system beyond our own. While searching the radio sky, Professor Wolszczan discovered a new pulsar, PSR1257+12, that seemed to wobble as it traveled through space. As a result of an extensive mathematical analysis, many scientists are now convinced that the wobble is caused by two or three planets orbiting PSR1257+12. Although scientists have been able to detect pulsars for some time by searching for faint periodic radio signals from outer space, it is only recently that the mathematical techniques have been developed to analyze the data in a way that stands up to scientific scrutiny. Wolszczan predicts that the planets orbiting PSR1257+12 are barren and inhospitable because of stellar winds, but his methods open the possibility of discovering new planetary systems that may sustain intelligent life.

**THE ROOTS OF CALCULUS**

Today's exciting applications of calculus have roots that can be traced to the work of the Greek mathematician Archimedes, but the actual discovery of the fundamental principles of calculus was made independently by Isaac Newton and Gottfried Leibniz in the late seventeenth century. The work of Newton and Leibniz was motivated by four major classes of scientific and mathematical problems of the time:

- Find the tangent line to a general curve at a given point.
- Find the area of a general region, the length of a general curve, and the volume of a general solid.
- Find the maximum or minimum value of a quantity—for example, the maximum and minimum distances of a planet from the Sun, or the maximum range attainable for a projectile by varying its angle of fire.
- Given a formula for the distance traveled by a body in any specified amount of time, find the velocity and acceleration of the body at any instant. Conversely, given a formula that specifies the acceleration of velocity at any instant, find the distance traveled by the body in a specified period of time.

**INFINITE PROCESSES**

Even though these problems may seem diverse and unrelated, we will see later that they are all closely linked by the fundamental principles of calculus and that all of them involve *infinite processes* in some way. These same principles and processes underlie the contemporary applications that we discussed at the beginning of this section.

There is something very satisfying about starting a task and bringing it to a step-by-step conclusion. However, the real world is replete with processes that by their very nature cannot be completed in finitely many steps, and hence must be left unfinished in some sense. For example, whereas the complete decimal expansion of the fraction $1/8$ can be obtained in three steps by long division,

$$1/8 = .125$$

the complete decimal expansion of $\sqrt{2}$ cannot be obtained in a finite number of steps by *any* procedure. Although there are numerous *algorithms* (i.e., step-by-step procedures) for approximating $\sqrt{2}$ to any desired degree of accuracy, none of them produces the exact value in finitely many steps. One such algorithm, called the *mechanic's rule*, is based on the formula

$$y_0 = 1, \quad y_{n+1} = \frac{1}{2}\left(y_n + \frac{2}{y_n}\right) \tag{1}$$

These equations can be used to generate an *infinite sequence* of approximations

$$y_0, y_1, y_2, y_3, \ldots \tag{2}$$

that get closer and closer to $\sqrt{2}$, achieving an arbitrary degree of accuracy in finitely many steps. This is done by first setting $y_0 = 1$ and then using the second part of Formula (1) to generate each new approximation $y_{n+1}$ from the preceding approximation $y_n$. For example, Table 1 shows the first six approximations of $\sqrt{2}$ produced by the mechanic's rule. The fractions in the table were obtained using a computer program, called a *Computer Algebra*

*System* (CAS),[*] that is capable of performing algebraic operations exactly. We used the same program to convert the fractions to decimal approximations with 12 digits, but we could also have used a calculator. At $n = 4$ the decimal approximations began to repeat because we had reached the accuracy limit of a 12-digit display.

**Table 1**

| $n$ | $y_0 = 1, \quad y_{n+1} = \frac{1}{2}\left(y_n + \frac{2}{y_n}\right)$ | DECIMAL APPROXIMATION |
|---|---|---|
| | $y_0 = 1$ (Starting value) | 1.00000000000 |
| 0 | $y_1 = \frac{1}{2}\left[1 + \frac{2}{1}\right] = \frac{3}{2}$ | 1.50000000000 |
| 1 | $y_2 = \frac{1}{2}\left[\frac{3}{2} + \frac{2}{3/2}\right] = \frac{17}{2}$ | 1.41666666667 |
| 2 | $y_3 = \frac{1}{2}\left[\frac{17}{2} + \frac{2}{17/2}\right] = \frac{577}{408}$ | 1.41421568627 |
| 3 | $y_4 = \frac{1}{2}\left[\frac{577}{408} + \frac{2}{577/408}\right] = \frac{665,857}{470,832}$ | 1.41421356237 |
| 4 | $y_5 = \frac{1}{2}\left[\frac{665,857}{470,832} + \frac{2}{665,857/470,832}\right] = \frac{886,731,088,897}{627,013,566,048}$ | 1.41421356237 |

**INFINITE SERIES**

We learn in elementary arithmetic that the decimal expansion of $\frac{1}{3}$ is

$$\frac{1}{3} = .33333\ldots$$

(all decimal digits being 3). We can rewrite this equation as

$$\frac{1}{3} = .3 + .03 + .003 + .0003 + .00003 + \cdots$$

or alternatively, as

$$\frac{1}{3} = \frac{3}{10} + \frac{3}{10^2} + \frac{3}{10^3} + \frac{3}{10^4} + \frac{3}{10^5} + \cdots \tag{3}$$

This formula expresses the number $\frac{1}{3}$ as an unending sum with infinitely many terms; such sums are called *infinite series*. An infinite series denotes an addition process that cannot be completed in finitely many steps—one can add the first 10 terms, the first 100 terms, or even the first 10,000 terms (with the help of a computer), but one cannot add *all* of the terms in the usual sense because there are infinitely many of them. However, if we start at the beginning of the series and add terms one by one, then at each step the sum gets closer and closer to $\frac{1}{3}$. For example,

$$\frac{1}{3} \approx \frac{3}{10} = .3$$

$$\frac{1}{3} \approx \frac{3}{10} + \frac{3}{10^2} = .33$$

$$\frac{1}{3} \approx \frac{3}{10} + \frac{3}{10^2} + \frac{3}{10^3} = .333$$

$$\frac{1}{3} \approx \frac{3}{10} + \frac{3}{10^2} + \frac{3}{10^3} + \frac{3}{10^4} = .3333$$

---

[*]The most widely used CAS programs are *Mathematica*, by Wolfram Research, Inc.; *Maple*, by Waterloo Maple Software, Inc.; and *Derive*, by Soft Warehouse, Inc.

Thus, Formula (3) is interpreted to mean that $\frac{1}{3}$ can be approximated to any desired degree of accuracy by adding sufficiently many terms from the beginning of the series.

**CALCULUS AND THE SEARCH FOR $\pi$**

The need for an accurate approximation of $\pi$ dates back to the surveyors of the early Babylonian and Egyptian civilizations. The Greek mathematician Archimedes calculated $\pi$ to two decimal places by geometric means and later mathematicians obtained greater accuracy by improving his geometric methods. With the advent of calculus, various infinite series for $\pi$ were discovered. The first such series was

$$\pi = 4\left(1 - \tfrac{1}{3} + \tfrac{1}{5} - \tfrac{1}{7} + \tfrac{1}{9} - \cdots\right)$$

Although this series is quite beautiful in its simplicity, it has little practical value because enormous numbers of terms are required to achieve good approximations. For example, a computer computation of the sum of the first 5000 terms in this series yields the approximation

$$\pi \approx 3.14139$$

which is correct to only three decimal places in spite of the large amount of computation. (Compare this to the value of $\pi$ that your calculator produces.) Later in this text you will encounter infinite series that are better suited for approximating $\pi$ because they produce better accuracy using fewer terms. The activity of finding algorithms that produce better accuracy with less computation is an area of current mathematical research in a branch of mathematics called ***numerical analysis***.

**AREA**

Formulas for the areas of plane regions with straight-line boundaries (squares, rectangles, triangles, trapezoids, etc.) were well known in many early civilizations. However, obtaining formulas for regions with curvilinear boundaries (a circle being the simplest case) caused problems for early mathematicians. An idea for computing the area of a circle to an arbitrary degree of accuracy was suggested around 430 B.C. by the Greek scholar Antiphon and was later systematized by the Greek mathematician Eudoxus into an algorithm called the ***method of exhaustion***. That method, when applied to a circle of radius $r$, consists of inscribing a succession of regular polygons in the circle and allowing the number of sides $n$ to increase indefinitely (Figure 7). As $n$ increases, the polygons tend to "exhaust" the region inside the circle, and the areas of those polygons become better and better approximations to the exact area of the circle.

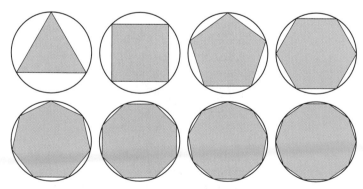

Figure 7

To see how this works numerically, let $p(n)$ denote the area of a regular $n$-sided polygon inscribed in a circle of radius $r$. We can find a formula for $p(n)$ by subdividing the polygon into $n$ congruent triangles (Figure 8$a$) and adding the areas of those triangles to obtain the area of the entire polygon. Each triangle is isosceles, since two of its sides are radii of the circle; and the angle at the apex of each triangle is $2\pi/n$, since the triangles divide the

**Table 2**

| $n$ | $p(n)$ |
|---|---|
| 100 | $3.13952597647\ r^2$ |
| 200 | $3.14107590781\ r^2$ |
| 300 | $3.14136298250\ r^2$ |
| 400 | $3.14146346236\ r^2$ |
| 500 | $3.14150997084\ r^2$ |
| 600 | $3.14153523487\ r^2$ |
| 700 | $3.14155046835\ r^2$ |
| 800 | $3.14156035548\ r^2$ |
| 900 | $3.14156713408\ r^2$ |
| 1000 | $3.14157198278\ r^2$ |
| 2000 | $3.14158748588\ r^2$ |
| 3000 | $3.14159035683\ r^2$ |
| 4000 | $3.14159136166\ r^2$ |
| 5000 | $3.14159182676\ r^2$ |
| 6000 | $3.14159207940\ r^2$ |
| 7000 | $3.14159223174\ r^2$ |
| 8000 | $3.14159233061\ r^2$ |
| 9000 | $3.14159239839\ r^2$ |
| 10000 | $3.14159244688\ r^2$ |

central angle of the circle into $n$ equal parts. Thus, with the help of some basic trigonometry (Figure 8$b$), we deduce that the area of each triangle is

$$\text{area} = \tfrac{1}{2} \cdot \text{base} \cdot \text{height}$$
$$= \tfrac{1}{2} \cdot 2(r \sin \pi/n)(r \cos \pi/n) = r^2 \sin(\pi/n) \cos(\pi/n)$$

from which it follows that the area of $n$ triangles is

$$p(n) = nr^2 \sin(\pi/n) \cos(\pi/n) \tag{4}$$

As $n$ increases, this formula should produce better and better approximations to the exact area of the circle. To see that this is so, we used a calculator set to the radian mode to generate Table 2. Later, using the tools of calculus, we will show definitively that $p(n)$ converges to the limit $\pi r^2$ as $n$ increases.

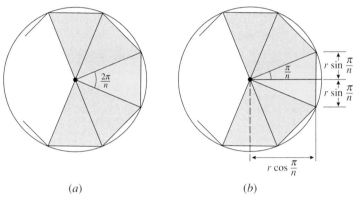

$(a)$ $\qquad\qquad\qquad$ $(b)$

Figure 8

Figure 9 shows a variation of the method of exhaustion that appeared in a seventeenth century Japanese manuscript. In that manuscript the area of the circle is approximated by inscribed rectangles, rather than polygons.

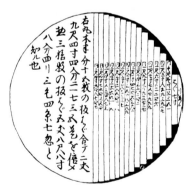

Figure 9

A form of calculus called *yenri* (circle principle) was developed in seventeenth century Japan by the mathematician Seki Kōwa and his pupils. This illustration, dating to 1670, shows an approximation of the area of a circle with inscribed rectangles.

**TANGENT LINES**

Tangent lines to general curves were of great interest to the mathematicians and scientists of the seventeenth century because of their application to the design of lenses. To determine how a ray of light passes through a lens using the laws of optics, one must know the angle at which the ray strikes the lens. This angle is measured between the ray and the normal line to the lens surface, the normal line being perpendicular to the tangent line (Figure 10). Thus, the study of various lens shapes led to the mathematical problem of finding the tangent line at a point on a general curve.

For circles, the concept of a tangent line is simple—a line is tangent to the circle if it meets the circle at precisely one point. However, this does not work for other kinds of curves (Figure 11).

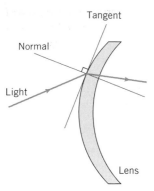

Figure 10

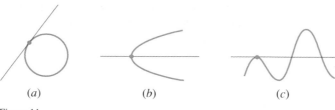

(a)          (b)          (c)

Figure 11

In order to apply the concept of a tangent line to curves other than circles, we must view tangent lines another way. For this purpose, suppose that we are interested in the tangent line at a point $P$ on a circle, and let $Q$ be any point on the circle different from $P$ (Figure 12a). If we draw the secant line through $P$ and $Q$, and then allow $Q$ to move along the circle toward $P$, then intuition suggests that the secant line will rotate toward a "limiting position" that coincides with the tangent line at $P$. This viewpoint about tangent lines is important because it can be applied to more general curves (Figure 12b). Thus, the geometric problem of finding a tangent line leads to a problem involving an infinite process—finding the limiting position of secant lines.

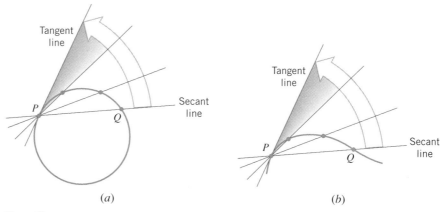

(a)                              (b)

Figure 12

### CALCULUS AND THE MYSTERY OF CONTINUOUS MOTION

One of the early triumphs of calculus was its use in clarifying and quantifying continuous motion. The ancient Greeks had two schools of thought on the nature of space and time: the *discrete* and the *continuous*. From the discrete viewpoint, space and time are composed of small *indivisible* units (points and instants) and motion is a succession of small discrete jumps that gives the illusion of smoothness to the eye (like a movie). From the continuous viewpoint, every unit of space and time, no matter how small, can be further subdivided, and motion is a smooth continuous process.

The Greek philosopher Zeno (born c. 490 B.C.) raised perplexing questions about both theories of motion with some paradoxes. (A ***paradox*** is an argument that appears to be logically correct but that leads to a contradiction or reaches a conclusion that flies in the face of common sense.) Zeno questioned the discrete theory of motion with his *Arrow Paradox* and the continuous theory of motion with his *Paradox of Achilles and the Tortoise*:

> ***Zeno's Arrow Paradox:***  *If time and space are discrete, then an arrow cannot move through the air. For at each instant of time the arrow is at a definite point and hence is at rest at that instant. Thus, the arrow is always at rest.*

> ***Zeno's Paradox of Achilles and the Tortoise:***  *If time and space are continuous, and if a tortoise is given the slightest head start in a race with Achilles, then Achilles will never catch*

*the tortoise. For when Achilles reaches the tortoise's starting point, the tortoise will have moved ahead to a point B. When Achilles reaches the point B the tortoise will have moved ahead to a point C—ad infinitum. Thus, the tortoise will always be ahead, even if by a hair* (Figure 13).

Figure 13

Even today the paradoxes of Zeno raise bothersome philosophical issues about the nature of motion. However, in the early fourteenth century the emphasis shifted away from the philosophical issues toward the quantitative study of speed and acceleration. The difficulty faced by mathematicians and scientists of that period was the lack of a precise *definition* of speed that could be used as a starting point for quantitative analysis—and that turned out to be a nontrivial matter.

To understand the difficulty, suppose that a car travels 75 miles in a 3-hour period. We say that the *average* speed of the car is 25 miles per hour (75/3 = 25 mi/h). More generally, the *average speed* of an object during a specified time interval is defined as

$$\text{average speed} \; = \; \frac{\text{distance traveled}}{\text{time elapsed}}$$

However, it is important to recognize that this is just an average—a car with an average speed of 25 mi/h on a trip need not travel at a constant speed of 25 mi/h—it may speed up and it may slow down. Moreover, average speed is not a very useful quantity in certain situations. For example, if the car happens to hit a tree, then the resulting damage will not be determined by the average speed up to the time of impact, but rather by the *instantaneous* speed at the precise moment of impact.

But exactly what do we mean by "instantaneous speed" and how do we compute it? We cannot simply carry over the process for computing average speed, since in any given instant the distance traveled is 0 and the time elapsed is 0, so the distance traveled divided by the time elapsed is 0/0, which is meaningless. Thus, although instantaneous speed is a physical reality, there is difficulty computing it for lack of a precise definition.

Mathematicians and scientists ultimately resolved this difficulty by using the well-defined notion of average speed together with an infinite process to define the concept of instantaneous speed. The idea is as follows: Suppose that we are interested in the instantaneous speed of an object at some time $t$. Intuition suggests that over a *small* time interval the speed of the object should not change very much. Thus, if $t + h$ is a point in time slightly later than $t$, then the average speed over the time interval from $t$ to $t + h$ should be very close to the instantaneous speed at time $t$. Moreover, the closer $t + h$ is to $t$, the better we should expect the approximation to be. This suggests that the instantaneous speed at time $t$ be *defined* as the limiting value of the average speed computed over smaller and smaller time intervals starting at time $t$.

What is fascinating about this is the link between precise mathematical definition and real-world applications—once physicists were armed with the right definition of instantaneous velocity, they were ultimately able to find equations for the motion of the planets and develop fundamental theories about gravitational attraction.

**THE ROLE OF RIGOR AND PROOF IN CALCULUS**

Although the concept of deductive proof dates back to Euclid, most of the developments in mathematics from about 200 B.C. to 1870 were based on intuition and empirical discovery—the idea of proving new mathematical results rigorously was largely ignored. However, as calculus developed, concepts related to infinite processes began to challenge the reliability of intuition and eventually an emphasis on precise definitions and careful proof was reestablished.

To illustrate how intuition can fail when dealing with infinite processes, consider the infinite series

$$1 - 1 + 1 - 1 + 1 - 1 + \cdots$$

We might be tempted to conclude that the sum of this series is zero by grouping the terms as

$$(1 - 1) + (1 - 1) + (1 - 1) + \cdots = 0 + 0 + 0 + \cdots$$

However, we can also group the terms as

$$1 + (-1 + 1) + (-1 + 1) + (-1 + 1) + \cdots = 1 + 0 + 0 + 0 + \cdots$$

which suggests that the sum is 1. Something has to be wrong! (Later, we will see that neither conclusion is correct.) The difficulty is our lack of mathematical precision; we have not established a precise definition of what we mean by the sum of an infinite series, and we have assumed without justification that the rules of grouping for finitely many terms also apply to infinite series.

In this text many ideas will be introduced informally at first to develop our intuition, but eventually we will take great care to define terms precisely and state exact conditions under which results are valid. The preceding example should convince you that this is not an idle mathematical exercise but rather an essential part of avoiding serious mathematical errors.

## THE DISCOVERY OF CALCULUS

The development of calculus was an evolutionary process that culminated in the discovery of a fundamental relationship between the problem of finding areas and the problem of finding tangent lines. The discovery of that result, which was made independently by Sir Isaac Newton (English) and Gottfried Wilhelm Leibniz (German), is considered to be the "discovery" of calculus. Newton made the discovery 10 years before Leibniz but did not publish his work until 20 years after Leibniz published his work. This situation led to a stormy debate over the rightful discoverer of calculus that engulfed Europe for half a century, with the scientists of the Continent supporting Leibniz and those from England supporting Newton. The conflict was extremely unfortunate because Newton's inferior notation badly hampered scientific development in England, and the Continent in turn lost the benefit of Newton's discoveries in astronomy and physics for nearly 50 years. In spite of it all, Newton and Leibniz were sincere admirers of each other's work.

### ISAAC NEWTON (1642–1727)

Newton was born in the village of Woolsthorpe, England. His father died before he was born and his mother raised him on the family farm. As a youth he showed little evidence of his later brilliance, except for an unusual talent with mechanical devices—he apparently built a working water clock and a toy flour mill powered by a mouse. In 1661 he entered Trinity College in Cambridge with a deficiency in geometry. Fortunately, Newton caught the eye of Isaac Barrow, a gifted mathematician and teacher. Under Barrow's guidance Newton immersed himself in mathematics and science, but he graduated without any special distinction. Because the Plague was spreading rapidly through London, Newton returned to his home in Woolsthorpe and stayed there during the years of 1665 and 1666. In those two momentous years the entire framework of modern science was miraculously created in Newton's mind—he discovered calculus, recognized the underlying principles of planetary motion and gravity, and determined that "white" sunlight was composed of all colors, red to violet. For some reasons he kept his discoveries to himself. In 1667 he returned to Cambridge to obtain his Master's degree and upon graduation became a teacher at Trinity. Then in 1669

Newton succeeded his teacher, Isaac Barrow, to the Lucasian chair of mathematics at Trinity, one of the most honored chairs of mathematics in the world. Thereafter, brilliant discoveries flowed from Newton steadily. He formulated the law of gravitation and used it to explain the motion of the Moon, the planets, and the tides; he formulated basic theories of light, thermodynamics, and hydrodynamics; and he devised and constructed the first modern reflecting telescope.

Throughout his life Newton was hesitant to publish his major discoveries, revealing them only to a select circle of friends, perhaps because of a fear of criticism or controversy. In 1687, only after intense coaxing by the astronomer, Edmond Halley (Halley's comet), did Newton publish his masterpiece, *Philosophiae Naturalis Principia Mathematica* (The Mathematical Principles of Natural Philosophy). This work is generally considered to be the most important and influential scientific book ever written. In it Newton explained the workings of the solar system and formulated the basic laws of motion which to this day are fundamental in engineering and physics. However, not even the pleas of his friends could convince Newton to publish his discovery of calculus. Only after Leibniz published his results did Newton relent and publish his own work on calculus.

After 25 years as a professor, Newton suffered depression and a nervous breakdown. He gave up research in 1695 to accept a position as warden and later master of the London mint. During the 25 years that he worked at the mint, he did virtually no scientific or mathematical work. He was knighted in 1705 and on his death was buried in Westminster Abbey with all the honors his country could bestow. It is interesting to note that Newton was a learned theologian who viewed the primary value of his work to be its support of the existence of God. Throughout his life he worked passionately to date biblical events by relating them to astronomical phenomena. He was so consumed with this passion that he spent years searching the Book of Daniel for clues to the end of the world and the geography of hell.

Newton described his brilliant accomplishments as follows: "I seem to have been only like a boy playing on the seashore and diverting myself in now and then finding a smoother pebble or prettier shell than ordinary, whilst the great ocean of truth lay all undiscovered before me."

## GOTTFRIED WILHELM LEIBNIZ (1646–1716)

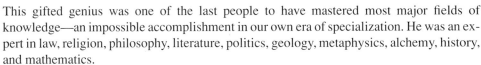

This gifted genius was one of the last people to have mastered most major fields of knowledge—an impossible accomplishment in our own era of specialization. He was an expert in law, religion, philosophy, literature, politics, geology, metaphysics, alchemy, history, and mathematics.

Leibniz was born in Leipzig, Germany. His father, a professor of moral philosophy at the University of Leipzig, died when Leibniz was six years old. The precocious boy then gained access to his father's library and began reading voraciously on a wide range of subjects, a habit that he maintained throughout his life. At age 15 he entered the University of Leipzig as a law student and by the age of 20 received a doctorate from the University of Altdorf. Subsequently, Leibniz followed a career in law and international politics, serving as counsel to kings and princes.

During his numerous foreign missions, Leibniz came in contact with outstanding mathematicians and scientists who stimulated his interest in mathematics—most notably, the physicist Christian Huygens. In mathematics Leibniz was self-taught, learning the subject by reading papers and journals. As a result of this fragmented mathematical education, Leibniz often rediscovered the results of others, and this helped to fuel the debate over the discovery of calculus.

Leibniz never married. He was moderate in his habits, quick-tempered, but easily appeased, and charitable in his judgment of other people's work. In spite of his great achievements, Leibniz never received the honors showered on Newton, and he spent his final years as a lonely embittered man. At his funeral there was one mourner, his secretary. An eyewitness stated, "He was buried more like a robber than what he really was—an ornament of his country."

**EXERCISE SET FOR INTRODUCTION** ☐ Graphing Calculator  ☐c CAS

1. The repeating decimal $0.137137137\ldots$ can be expressed as a ratio of integers by writing

$$x = 0.137137137\ldots$$
$$1000x = 137.137137137\ldots$$

and subtracting to obtain $999x = 137$ or $x = \frac{137}{999}$. Use this idea, where needed, to express the following decimals as ratios of integers.
   (a) $0.123123123\ldots$      (b) $12.7777\ldots$
   (c) $38.07818181\ldots$      (d) $0.4296000\ldots(=0.4296)$

> All decimals fall into two categories: ***repeating decimals*** and ***nonrepeating decimals***. In a repeating decimal there is some point after which a fixed block of integers repeats over and over. For example, all of the decimals in Exercise 1 are repeating. In a nonrepeating decimal there is no *fixed block* of digits that repeats over and over. For example, although the decimal $0.101001000100001\ldots$ has a definite pattern, it is nonrepeating. Those real numbers whose decimals are repeating are called ***rational numbers*** and those whose decimals are nonrepeating are called ***irrational numbers***. It can be proved that the rational numbers are precisely those real numbers that can be expressed as the ratio of two integers (as in Exercise 1). Some familiar irrational numbers are $\pi$, $\sqrt{2}$, $\sqrt{3}$, $\sqrt{5}$, $\ldots$. Such numbers cannot be expressed as the ratio of two integers.

☐c 2. (a) Use the preceding discussion to explain in one sentence why $\pi$ cannot be equal to $\frac{22}{7}$.
   (b) The accompanying figure shows $\pi$ to 500 decimal places. Use it to determine whether $\frac{22}{7}$ is greater than or less than $\pi$.
   (c) Use a CAS to duplicate the results in the accompanying figure.

3.1415926535897932384626433832795028841971693993751058209749445923078164062862089986280348253421170679821480865132823066470938446095505822317253594081284811174502841027019385211055596446229489549303819644288109756659334461284756482337867831652712019091456485669234603486104543266482133936072602491412737245870066063155881748815209209628292540917153643678925903600113305305488204665213841469519415116094330572703657595919530921861173819326117931051185480744623799627495673518857527248912279381830119491

Figure Ex-2

3. The following are all famous approximations of $\pi$:

$$\frac{22}{7}, \quad \frac{223}{71}, \quad \frac{333}{106}, \quad \frac{355}{113}, \quad \frac{63}{25}\left(\frac{17+15\sqrt{5}}{7+15\sqrt{5}}\right)$$

   (a) Use a calculating device to order these approximations according to size.
   (b) Which of these approximations is closest to but larger than $\pi$?
   (c) Which of these approximations is closest to but smaller than $\pi$?
   (d) Which of these approximations is most accurate?
   (e) The last approximation is due to a famous self-taught Indian mathematician, named Ramanujan (1887–1920). Do some reading about his fascinating but tragic life.

4. The Rhind Papyrus, which is a fragment of Egyptian mathematical writing from about 1650 B.C., is one of the oldest known examples of written mathematics. It is stated in the papyrus that the area $A$ of a circle is related to its diameter $D$ by

$$A = \left(\tfrac{8}{9}D\right)^2$$

   (a) What approximation of $\pi$ were the Egyptians using?
   (b) Use a calculating device to determine if this approximation is better or worse than the approximation $\frac{22}{7}$.

5. In this section we stated that $\pi$ can be expressed as the infinite series

$$\pi = 4\left(1 - \tfrac{1}{3} + \tfrac{1}{5} - \tfrac{1}{7} + \tfrac{1}{9} - \cdots\right)$$

However, this series is of little practical value because it *converges* too slowly; that is, too many terms are required to obtain a good approximation. A more practical approach is based on the following formula, discovered in 1706 by the English astronomer John Machin (1680–1752):

$$\pi = 16\left(\frac{1}{5} - \frac{1}{3\cdot 5^3} + \frac{1}{5\cdot 5^5} - \cdots\right)$$
$$-4\left(\frac{1}{239} - \frac{1}{3\cdot 239^3} + \frac{1}{5\cdot 239^5} - \cdots\right)$$

Machin's formula was used in 1949 on the ENIAC computer at the Ballistic Research Laboratories to produce the first computer calculation of $\pi$ (2037 decimal places). Show that the terms shown in Machin's formula give a more accurate approximation of $\pi$ than the sum of the first 10 terms of the first series. [*Suggestion:* Compare your calculated values to the approximation of $\pi$ in Figure Ex-2.]

6. In each part, use a calculating device to find the decimal expansion of the fraction, and then use that expansion to express the fraction as an infinite series. (Show at least the first six terms of the series.)

   (a) $\frac{1}{9}$        (b) $\frac{5}{27}$        (c) $\frac{14}{45}$

7. Repeat the directions of Exercise 6 for

   (a) $\frac{7}{11}$        (b) $\frac{8}{33}$        (c) $\frac{5}{12}$.

---

In Formula (1) we gave an algorithm, called the *mechanic's rule*, for approximating $\sqrt{2}$ to any degree of accuracy. That algorithm is a special case of the following more general algorithm for approximating the square root of any positive number $p$ to any degree of accuracy:

$$y_0 = 1, \quad y_{n+1} = \frac{1}{2}\left(y_n + \frac{p}{y_n}\right)$$

Use this result in Exercises 8 and 9.

---

8. In each part use the algorithm stated above to approximate the square root to four decimal places.

   (a) $\sqrt{3}$        (b) $\sqrt{5}$

9. Repeat the directions of Exercise 8 for

   (a) $\sqrt{7}$        (b) $\sqrt{50}$.

10. If $a$ and $b$ are distinct real numbers, say $a < b$, then it can be proved that there must be real numbers between $a$ and $b$. One such number is the arithmetic average $\frac{1}{2}(a+b)$.

    (a) Explain why there must be infinitely many real numbers between any two distinct real numbers.

    (b) Do you think it is true that
    $$0.9999999\ldots < 1.0000000\ldots?$$
    Explain your reasoning.

    (c) Find the decimal representation of the arithmetic average of $0.9999999\ldots$ and $1.0000000\ldots$. Is this result consistent with your answer in part (b)? Explain.

    (d) Use the method of Exercise 1 to express the decimal $0.9999999\ldots$ as a ratio of two integers. Is this result consistent with your answer in part (b)? Explain.

Rene' Descartes

# 1

# FUNCTIONS

$\mathcal{O}$ ne of the important themes in calculus is the analysis of relationships between physical or mathematical quantities. Such relationships can be described in terms of graphs, formulas, numerical data, or words. In this chapter we will develop the concept of a *function*, which is the basic idea that underlies almost all mathematical and physical relationships, regardless of the form in which they are expressed. We will study properties of some of the most basic functions that occur in calculus, and we will examine some familiar ideas involving lines, polynomials, and trigonometric functions from viewpoints that may be new. We will also discuss ideas relating to the use of graphing utilities such as graphing calculators and graphing software for computers. Before you start reading, you may want to scan through the appendices, since they contain various kinds of precalculus material that may be helpful if you need to review some of those ideas.

# 1.1 FUNCTIONS AND THE ANALYSIS OF GRAPHICAL INFORMATION

---

*In this section we will define and develop the concept of a function. Functions are used by mathematicians and scientists to describe the relationships between variable quantities and hence play a central role in calculus and its applications.*

---

**SCATTER PLOTS AND TABULAR DATA**

Many scientific laws are discovered by collecting, organizing, and analyzing experimental data. Since graphs play a major role in studying data, we will begin by discussing the kinds of information that a graph can convey.

To start, we will focus on paired data. For example, Table 1.1.1 shows the top qualifying speed by year in the Indianapolis 500 auto race from 1975 to 1994. This table pairs up each year $t$ between 1975 and 1994 with the top qualifying speed $S$ for that year. This paired data can be represented graphically in a number of ways:

- One possibility is to plot the paired data points in a rectangular $tS$-coordinate system ($t$ horizontal and $S$ vertical), in which case we obtain a **scatter plot** of $S$ versus $t$ (Figure 1.1.1*a*).

- A second possibility is to enhance the scatter plot visually by joining successive points with straight-line segments, in which case we obtain a **line graph** (Figure 1.1.1*b*).

- A third possibility is to represent the paired data by a **bar graph** (Figure 1.1.1*c*).

All three graphical representations reveal an upward trend in the data, as one would expect with improvements in automotive technology.

**Table 1.1.1**

INDIANAPOLIS 500
QUALIFYING SPEEDS

| YEAR $t$ | SPEED $S$ (mi/h) |
|---|---|
| 1975 | 193.976 |
| 1976 | 188.957 |
| 1977 | 198.884 |
| 1978 | 202.156 |
| 1979 | 193.736 |
| 1980 | 192.256 |
| 1981 | 200.546 |
| 1982 | 207.004 |
| 1983 | 207.395 |
| 1984 | 210.029 |
| 1985 | 212.583 |
| 1986 | 216.828 |
| 1987 | 215.390 |
| 1988 | 219.198 |
| 1989 | 223.885 |
| 1990 | 225.301 |
| 1991 | 224.113 |
| 1992 | 232.482 |
| 1993 | 223.967 |
| 1994 | 228.011 |

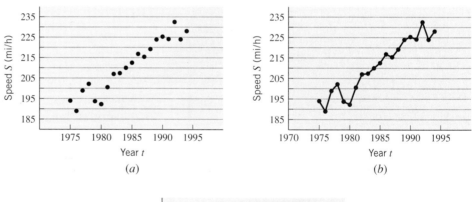

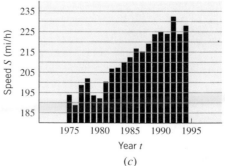

Figure 1.1.1

**EXTRACTING INFORMATION FROM GRAPHS**

One of the first books to use graphs for representing numerical data was *The Commercial and Political Atlas*, published in 1786 by the Scottish political economist William Playfair (1759–1823). Figure 1.1.2*a* shows an engraving from that work that compares exports and imports by England to Denmark and Norway (combined). In spite of its antiquity, the engraving is modern in spirit and provides a wealth of information. You should be able to extract the following information from Playfair's graphs:

- In the year 1700 imports were valued at about 70,000 pounds and exports at about 35,000 pounds.

- During the period from 1700 to about 1754 imports exceeded exports (a trade deficit for England).

- In the year 1754 the imports and exports were equal (a trade balance in today's economic terminology).

- From 1754 to 1780 exports exceeded imports (a trade surplus for England). The greatest surplus occurred in 1780, at which time exports exceeded imports by about 95,000 pounds.

- During the period from 1700 to 1725 imports were rising. They peaked in 1725, and then slowly fell until about 1760, at which time they bottomed out and began to rise again slowly until 1780.

- During the period from 1760 to 1780 exports and imports were both rising, but exports were rising more rapidly than imports, resulting in an ever-widening trade surplus for England.

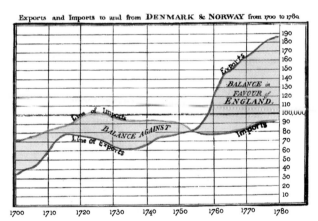

*Playfair's Graph of 1786:* The horizontal scale is in years from 1700 to 1780 and the vertical scale is in units of 1,000 pounds sterling from 0 to 200.

(*a*)

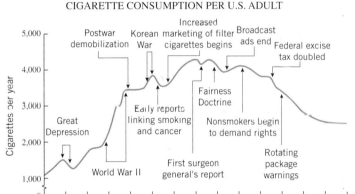

*Source:* U.S. Department of Health and Human Services.

(*b*)

Figure 1.1.2

Figure 1.1.2*b* is a more contemporary graph; it describes the per capita consumption of cigarettes in the United States between 1925 and 1995.

FOR THE READER.  Use the graph in Figure 1.1.2*b* to provide reasonable answers to the following questions:

- When did the maximum annual cigarette consumption per adult occur and how many were consumed?

- What factors are likely to cause sharp decreases in cigarette consumption?

- What factors are likely to cause sharp increases in cigarette consumption?

- What were the long- and short-term effects of the first surgeon general's report on the health risks of smoking?

Graphs can be used to describe mathematical equations as well as physical data. For example, consider the equation

$$y = x\sqrt{9 - x^2} \tag{1}$$

For each value of $x$ in the interval $-3 \leq x \leq 3$, this equation produces a corresponding real value of $y$, which is obtained by substituting the value of $x$ into the right side of the equation. Some typical values are shown in Table 1.1.2.

**Table 1.1.2**

| $x$ | $-3$ | $-2$ | $-1$ | $0$ | $1$ | $2$ | $3$ |
|---|---|---|---|---|---|---|---|
| $y$ | $0$ | $-2\sqrt{5} \approx -4.47214$ | $-2\sqrt{2} \approx -2.82843$ | $0$ | $2\sqrt{2} \approx 2.82843$ | $2\sqrt{5} \approx 4.47214$ | $0$ |

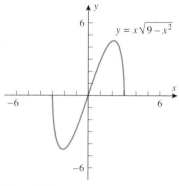

Figure 1.1.3

The set of *all* points in the $xy$-plane whose coordinates satisfy an equation in $x$ and $y$ is called the **graph** of that equation in the $xy$-plane. Figure 1.1.3 shows the graph of Equation (1) in the $xy$-plane. Notice that the graph extends only over the interval $[-3, 3]$. This is because values of $x$ outside of this interval produce complex values of $y$, and in these cases the ordered pairs $(x, y)$ do not correspond to points in the $xy$-plane. For example, if $x = 8$, then the corresponding value of $y$ is $y = 8\sqrt{-55} = 8\sqrt{55}\,i$, and the ordered pair $(8, 8\sqrt{55}\,i)$ is not a point in the $xy$-plane.

### Example 1

Figure 1.1.4 shows the graph of an unspecified equation that was used to obtain the values that appear in the shaded parts of the accompanying tables. Examine the graph and confirm that the values in the tables are reasonable approximations. ◀

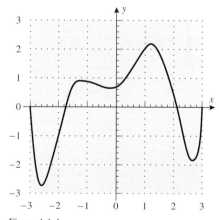

| $x$ | $y$ |
|---|---|
| $-3$ | $0$ |
| $-2$ | $-1$ |
| $-1$ | $0.9$ |
| $0$ | $0.7$ |
| $1$ | $2$ |
| $2$ | $0.4$ |
| $3$ | $0$ |

| $x$ | $y$ |
|---|---|
| None | $-3$ |
| $-2.3, -2.8$ | $-2$ |
| $-2, -2.9, 2.4, 2.9$ | $-1$ |
| $-3, -1.7, 2.1, 3$ | $0$ |
| $0.3, 1.8$ | $1$ |
| $1, 1.4$ | $2$ |
| None | $3$ |

Figure 1.1.4

Tables, graphs, and equations provide three methods for describing how one quantity depends on another—numerical, visual, and algebraic. The fundamental importance of this idea was recognized by Leibniz in 1673 when he coined the term *function* to describe the dependence of one quantity on another. The following examples illustrate how this term is used:

• The area $A$ of a circle depends on its radius $r$ by the equation $A = \pi r^2$, so we say that *A is a function of r*.

- The velocity $v$ of a ball falling freely in the Earth's gravitational field increases with time $t$ until it hits the ground, so we say that $v$ *is a function of* $t$.
- In a bacteria culture, the number $n$ of bacteria present after 1 hour of growth depends on the number $n_0$ of bacteria present initially, so we say that $n$ *is a function of* $n_0$.

This idea is captured in the following definition.

> **1.1.1**   DEFINITION.   If a variable $y$ depends on a variable $x$ in such a way that each value of $x$ determines exactly one value of $y$, then we say that **$y$ is a function of $x$**.

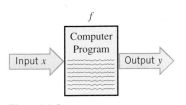

Figure 1.1.5

In the mid-eighteenth century the Swiss mathematician Leonhard Euler[*] (pronounced "oiler") conceived the idea of denoting functions by letters of the alphabet, thereby making it possible to describe functions without stating specific formulas, graphs, or tables. To understand Euler's idea, think of a function as a computer program that takes an *input x*, operates on it in some way, and produces exactly one *output y*. The computer program is an object in its own right, so we can give it a name, say $f$. Thus, the function $f$ (the computer program) associates a unique output $y$ with each input $x$ (Figure 1.1.5). This suggests the following definition.

> **1.1.2**   DEFINITION.   A **function $f$** is a rule that associates a unique output with each input. If the input is denoted by $x$, then the output is denoted by $f(x)$ (read "$f$ of $x$").

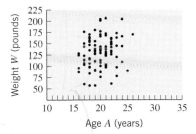

Figure 1.1.6

REMARK.   In this definition the term *unique* means "exactly one." Thus, a function cannot assign two different outputs to the same input. For example, Figure 1.1.6 shows a scatter plot of weight versus age for a random sample of 100 college students. This scatter plot does not describe the weight $W$ as a function of the age $A$ because there are some values of $A$ with more than one corresponding value of $W$. This is to be expected, since two people with the same age need not have the same weight. In contrast, Table 1.1.1 describes $S$ as a function of $t$ because there is only one top qualifying speed in a given year; similarly, Equation (1) describes $y$ as a function of $x$ because each input $x$ in the interval $-3 \leq x \leq 3$ produces exactly one output $y = x\sqrt{9 - x^2}$.

---

[*] LEONHARD EULER (1707–1783). Euler was probably the most prolific mathematician who ever lived. It has been said that "Euler wrote mathematics as effortlessly as most men breathe." He was born in Basel, Switzerland, and was the son of a Protestant minister who had himself studied mathematics. Euler's genius developed early. He attended the University of Basel, where by age 16 he obtained both a Bachelor of Arts degree and a Master's degree in philosophy. While at Basel, Euler had the good fortune to be tutored one day a week in mathematics by a distinguished mathematician, Johann Bernoulli. At the urging of his father, Euler then began to study theology. The lure of mathematics was too great, however, and by age 18 Euler had begun to do mathematical research. Nevertheless, the influence of his father and his theological studies remained, and throughout his life Euler was a deeply religious, unaffected person. At various times Euler taught at St. Petersburg Academy of Sciences (in Russia), the University of Basel, and the Berlin Academy of Sciences. Euler's energy and capacity for work were virtually boundless. His collected works form more than 100 quarto sized volumes and it is believed that much of his work has been lost. What is particularly astonishing is that Euler was blind for the last 17 years of his life, and this was one of his most productive periods! Euler's flawless memory was phenomenal. Early in his life he memorized the entire Aeneid by Virgil and at age 70 could not only recite the entire work, but could also state the first and last sentence on each page of the book from which he memorized the work. His ability to solve problems in his head was beyond belief. He worked out in his head major problems of lunar motion that baffled Isaac Newton and once did a complicated calculation in his head to settle an argument between two students whose computations differed in the fiftieth decimal place.

Following the development of calculus by Leibniz and Newton, results in mathematics developed rapidly in a disorganized way. Euler's genius gave coherence to the mathematical landscape. He was the first mathematician to bring the full power of calculus to bear on problems from physics. He made major contributions to virtually every branch of mathematics as well as to the theory of optics, planetary motion, electricity, magnetism, and general mechanics.

Functions can be represented in four basic ways:

- Numerically by tables
- Geometrically by graphs
- Algebraically by formulas
- Verbally

The method of representation often depends on how the function arises. For example:

- Table 1.1.1 is a numerical representation of $S$ as a function of $t$. This is the natural way in which data of this type are recorded.
- Figure 1.1.7 shows a seismic graph of an earthquake's intensity $H$ as a function of the elapsed time $t$. In this case the function originates as a graph.
- Some of the most familiar examples of functions arise as formulas; for example, the formula $C = 2\pi r$ expresses the circumference $C$ of a circle as a function of its radius $r$.
- Sometimes functions are described in words. For example, Isaac Newton's Universal Law of Gravitation is often stated as follows: The gravitational force of attraction between two bodies in the Universe is directly proportional to the product of their masses and inversely proportional to the square of the distance between them. This is the verbal description of the formula

$$F = G\frac{m_1 m_2}{r^2} \tag{2}$$

in which $F$ is the force of attraction, $m_1$ and $m_2$ are the masses, $r$ is the distance between them, and $G$ is a constant.

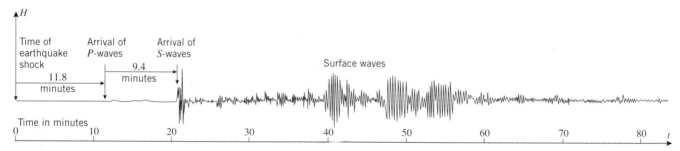

Figure 1.1.7

**Table 1.1.3**

U.S. POPULATION

| YEAR $t$ | POPULATION $P$ (millions) |
|---|---|
| 1790 | 3.9 |
| 1800 | 5.3 |
| 1810 | 7.2 |
| 1820 | 9.6 |
| 1830 | 12 |
| 1840 | 17 |
| 1850 | 23 |

*Source: The World Almanac.*

Sometimes it is desirable to convert one representation of a function into another. For example, in Figure 1.1.1 we converted the numerical relationship between $S$ and $t$ into a graphical relationship, and in writing Formula (2) we converted the verbal representation of the Universal Law of Gravitation into an algebraic relationship.

The problem of converting numerical representations of functions into algebraic formulas often requires special techniques known as ***curve fitting***. For example, Table 1.1.3 gives the U.S. population at 10-year intervals from 1790 to 1850. This table is a numerical representation of the function $P = f(t)$ that relates the U.S. population $P$ to the year $t$. If we plot $P$ versus $t$, we obtain the scatter plot in Figure 1.1.8a, and if we use curve-fitting methods that will be discussed later, we can obtain the approximation

$$P \approx 3.94(1.03)^{t-1790}$$

Figure 1.1.8b shows the graph of this equation imposed on the scatter plot.

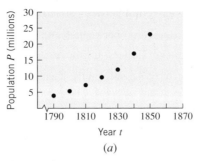
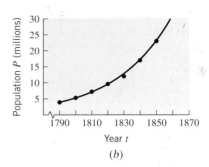

Figure 1.1.8

**DISCRETE VERSUS CONTINUOUS DATA**

Engineers and physicists distinguish between ***continuous data*** and ***discrete data***. Continuous data have values that vary *continuously* over an interval, whereas discrete data have values that make *discrete* jumps. For example, for the seismic data in Figure 1.1.7 both the time and intensity vary continuously, whereas in Table 1.1.3 and Figure 1.1.8*a* both the year and population make discrete jumps. As a rule, continuous data lead to graphs that are continuous, unbroken curves, whereas discrete data lead to scatter plots consisting of isolated points. Sometimes, as in Figure 1.1.8*b*, it is desirable to approximate a scatter plot by a continuous curve. This is useful for making conjectures about the values of the quantities between the recorded data points.

**GRAPHS AS PROBLEM-SOLVING TOOLS**

Sometimes a function is buried in the statement of a problem, and it is up to the problem solver to uncover it and use it in an appropriate way to solve the problem. Here is an example that illustrates the power of graphical representations of functions as a problem-solving tool.

### Example 2

Figure 1.1.9*a* shows an offshore oil well located at a point $W$ that is 5 km from the closest point $A$ on a straight shoreline. Oil is to be piped from $W$ to a shore point $B$ that is 8 km from $A$. It costs \$1,000,000/km to lay pipe under water and \$500,000/km over land. In your role as project manager you receive three proposals for piping the oil from $W$ to $B$. Proposal 1 claims that it is cheapest to pipe directly from $W$ to $B$, since the shortest distance between two points is a straight line. Proposal 2 claims that it is cheapest to pipe directly to point $A$ and then along the shoreline to $B$, thereby using the least amount of expensive underwater pipe. Proposal 3 claims that it is cheapest to compromise by piping under water to some well-chosen point between $A$ and $B$, and then piping along the shoreline to $B$. Which proposal is correct?

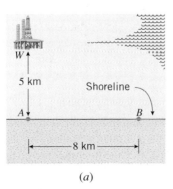
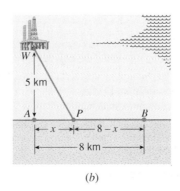
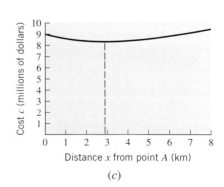

Figure 1.1.9

*Solution.* Let $P$ be any point between $A$ and $B$ (Figure 1.1.9$b$), and let

$x =$ distance (in kilometers) between $A$ and $P$

$c =$ cost (in millions of dollars) for the entire pipeline

Proposal 1 claims that $x = 8$ results in the least cost, Proposal 2 claims that it is $x = 0$, and Proposal 3 claims it is some value of $x$ between 0 and 8. From Figure 1.1.9$b$ the length of pipe along the shore is

$$8 - x \tag{3}$$

and from the Theorem of Pythagoras, the length of pipe under water is

$$\sqrt{x^2 + 25} \tag{4}$$

Thus, from (3) and (4) the total cost $c$ (in millions of dollars) for the pipeline is

$$c = 1\left(\sqrt{x^2 + 25}\right) + 0.5(8 - x) = \sqrt{x^2 + 25} + 0.5(8 - x) \tag{5}$$

where $0 \leq x \leq 8$. The graph of Equation (5), shown in Figure 1.1.9$c$, makes it clear that Proposal 3 is correct—the most cost-effective strategy is to pipe to a point a little less than 3 km from point $A$. ◄

---

**EXERCISE SET 1.1**  ⌇ Graphing Calculator   [C] CAS

· · · · · · · · · · · · · · · · · · · · · · · · · · · · · · · · · · · · · · · · · · · · · · · · · · · · · · · · · · · · · · · · · · · · · · · · · · · · · · · · · ·

1. Use the cigarette consumption graph in Figure 1.1.2$b$ to answer the following questions, making reasonable approximations where needed.
   (a) When did the annual cigarette consumption reach 3000 per adult for the first time?
   (b) When did the annual cigarette consumption per adult reach its peak, and what was the peak value?
   (c) Can you tell from the graph how many cigarettes were consumed in a given year? If not, what additional information would you need to make that determination?
   (d) What factors are likely to cause a sharp increase in annual cigarette consumption per adult?
   (e) What factors are likely to cause a sharp decline in annual cigarette consumption per adult?

2. The accompanying graph shows the median income in U.S. households (adjusted for inflation) between 1975 and 1995. Use the graph to answer the following questions, making reasonable approximations where needed.
   (a) When did the median income reach its maximum value, and what was the median income when that occurred?
   (b) When did the median income reach its minimum value, and what was the median income when that occurred?
   (c) The median income was declining during the 4-year period between 1989 and 1993. Was it declining more rapidly during the first 2 years or the second 2 years of that period? Explain your reasoning.

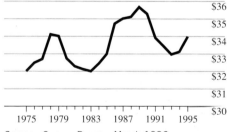

MEDIAN U.S. HOUSEHOLD INCOME IN
THOUSANDS OF CONSTANT 1995 DOLLARS

*Source:* Census Bureau, March 1996
[1996 measures 1995 income].

Figure Ex-2

3. Use the accompanying graph to answer the following questions, making reasonable approximations were needed.
   (a) For what values of $x$ is $y = 1$?
   (b) For what values of $x$ is $y = 3$?
   (c) For what values of $y$ is $x = 3$?
   (d) For what values of $x$ is $y \leq 0$?
   (e) What are the maximum and minimum values of $y$ and for what values of $x$ do they occur?

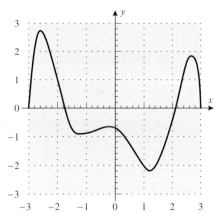

Figure Ex-3

**4.** Use the table in the accompanying figure to answer the questions posed in Exercise 3.

| x | −2 | −1 | 0 | 2 | 3 | 4 | 5 | 6 |
|---|---|---|---|---|---|---|---|---|
| y | 5 | 1 | −2 | 7 | −1 | 1 | 0 | 9 |

Figure Ex-4

**5.** Use the equation $y = x^2 - 6x + 8$ to answer the following questions.
(a) For what values of $x$ is $y = 0$?
(b) For what values of $x$ is $y = -10$?
(c) For what values of $x$ is $y \geq 0$?
(d) Does $y$ have a minimum value? A maximum value? If so, find them.

**6.** Use the equation $y = 1 + \sqrt{x}$ to answer the following questions.
(a) For what values of $x$ is $y = 4$?
(b) For what values of $x$ is $y = 0$?
(c) For what values of $x$ is $y \geq 6$?
(d) Does $y$ have a minimum value? A maximum value? If so, find them.

**7.** (a) If you had a device that could record the Earth's population continuously, would you expect the graph of population versus time to be a continuous (unbroken) curve? Explain what might cause breaks in the curve.
(b) Suppose that a hospital patient receives an injection of an antibiotic every 8 hours and that between injections the concentration $C$ of the antibiotic in the bloodstream decreases as the antibiotic is absorbed by the tissues. What might the graph of $C$ versus the elapsed time $t$ look like?

**8.** (a) If you had a device that could record the temperature of a room continuously over a 24-hour period, would you expect the graph of temperature versus time to be a continuous (unbroken) curve? Explain your reasoning.
(b) If you had a computer that could track the number of boxes of cereal on the shelf of a market continuously over a 1-week period, would you expect the graph of the number of boxes on the shelf versus time to be a continuous (unbroken) curve? Explain your reasoning.

**9.** A construction company wants to build a rectangular enclosure with an area of 1000 square feet by fencing in three sides and using its office building as the fourth side. Your objective as supervising engineer is to design the enclosure so that it uses the least amount of fencing. Proceed as follows.
(a) Let $x$ and $y$ be the dimensions of the enclosure, and let $L$ be the length of fencing required for those dimensions. Since the area must be 1000 square feet, we must have $xy = 1000$. Find a formula for $L$ in terms of $x$ and $y$, and then express $L$ in terms of $x$ alone by using the area equation.
(b) Are there any restrictions on the value of $x$? Explain.
(c) Make a graph of $L$ versus $x$ over a reasonable interval, and use the graph to estimate the value of $x$ that results in the smallest value of $L$.
(d) Estimate the smallest value of $L$.

**10.** A manufacturer constructs open boxes from sheets of cardboard that are 6 inches square by cutting small squares from the corners and folding up the sides (as shown in the accompanying figure). The Research and Development Department asks you to determine the size of the square that produces a box of greatest volume. Proceed as follows.
(a) Let $x$ be the length of a side of the square to be cut, and let $V$ be the volume of the resulting box. Show that $V = x(6 - 2x)^2$.
(b) Are there any restrictions on the value of $x$? Explain.
(c) Make a graph of $V$ versus $x$ over an appropriate interval, and use the graph to estimate the value of $x$ that results in the largest volume.
(d) Estimate the largest volume.

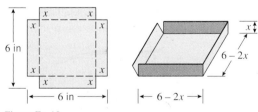

Figure Ex-10

**11.** A soup company wants to manufacture a can in the shape of a right circular cylinder that will hold 500 cm³ of liquid. The material for the top and bottom costs 0.02 cent/cm², and the material for the sides costs 0.01 cent/cm².
(a) Use the method of Exercises 9 and 10 to estimate the radius $r$ and height $h$ of the can that costs the least to manufacture. [*Suggestion:* Express the cost $C$ in terms of $r$.]
(b) Suppose that the tops and bottoms of radius $r$ are punched out from square sheets with sides of length $2r$ and the scraps are waste. If you allow for the cost of

the waste, would you expect the can of least cost to be taller or shorter than the one in part (a)? Explain.

(c) Estimate the radius, height, and cost of the can in part (b), and determine whether your conjecture was correct.

12. The designer of a sports facility wants to put a quarter-mile (1320 ft) running track around a football field, oriented as in the accompanying figure. The football field is 360 ft long (including the end zones) and 160 ft wide. The track consists of two straightaways and two semicircles.

(a) Show that it is possible to construct a quarter-mile track around the football field. [*Suggestion:* Find the shortest track that can be constructed around the field.]

(b) Let $L$ be the length of a straightaway (in feet), and let $x$ be the distance (in feet) between a sideline of the football field and a straightaway. Make a graph of $L$ versus $x$.

(c) Use the graph to estimate the value of $x$ that produces the shortest straightaways, and then find this value of $x$ exactly.

(d) Use the graph to estimate the length of the longest possible straightaways, and then find that length exactly.

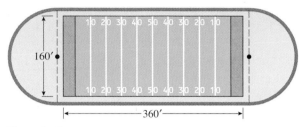

Figure Ex-12

## 1.2 PROPERTIES OF FUNCTIONS

*In this section we will explore properties of functions in more detail. We will assume that you are familiar with the standard notation for intervals and the basic properties of absolute value. Reviews of these topics are provided in Appendices A and B.*

**INDEPENDENT AND DEPENDENT VARIABLES**

Recall from the last section that a function $f$ is a rule that associates a unique output $f(x)$ with each input $x$. This output is sometimes called the *value* of $f$ at $x$ or the *image* of $x$ under $f$. Sometimes we will want to denote the output by a single letter, say $y$, and write

$$y = f(x)$$

This equation expresses $y$ as a function of $x$; the variable $x$ is called the **independent variable** (or **argument**) of $f$, and the variable $y$ is called the **dependent variable** of $f$. This terminology is intended to suggest that $x$ is free to vary, but that once $x$ has a specific value a corresponding value of $y$ is determined. For now we will only consider functions in which the independent and dependent variables are real numbers, in which case we say that $f$ is a **real-valued function of a real variable**. Later, we will consider other kinds of functions as well.

Table 1.2.1 can be viewed as a numerical representation of a function of $f$. For this function we have

$f(0) = 3$    <small>$f$ associates $y = 3$ with $x = 0$.</small>

$f(1) = 4$    <small>$f$ associates $y = 4$ with $x = 1$.</small>

$f(2) = -1$    <small>$f$ associates $y = -1$ with $x = 2$.</small>

$f(3) = 6$    <small>$f$ associates $y = 6$ with $x = 3$.</small>

To illustrate how functions can be defined by equations, consider

$$y = 3x^2 - 4x + 2 \tag{1}$$

This equation has the form $y = f(x)$, where

$$f(x) = 3x^2 - 4x + 2 \tag{2}$$

The outputs of $f$ (the $y$-values) are obtained by substituting numerical values for $x$ in this formula. For example,

$f(0) = 3(0)^2 - 4(0) + 2 = 2$    <small>$f$ associates $y = 2$ with $x = 0$.</small>

$f(-1.7) = 3(-1.7)^2 - 4(-1.7) + 2 = 17.47$    <small>$f$ associates $y = 17.47$ with $x = -1.7$.</small>

$f(\sqrt{2}) = 3(\sqrt{2})^2 - 4\sqrt{2} + 2 = 8 - 4\sqrt{2}$    <small>$f$ associates $y = 8 - 4\sqrt{2}$ with $x = \sqrt{2}$.</small>

**Table 1.2.1**

| $x$ | 0 | 1 | 2 | 3 |
|-----|---|---|----|---|
| $y$ | 3 | 4 | -1 | 6 |

REMARK.   Although $f$, $x$, and $y$ are the most common notations for functions and variables, any letters can be used. For example, to indicate that the area $A$ of a circle is a function of the radius $r$, it would be more natural to write $A = f(r)$ [where $f(r) = \pi r^2$]. Similarly, to indicate that the circumference $C$ of a circle is a function of the radius $r$, we might write $C = g(r)$ [where $g(r) = 2\pi r$]. The area function and the circumference function are different, which is why we denoted them by different letters, $f$ and $g$.

**DOMAIN AND RANGE**

If $y = f(x)$, then the set of all possible inputs ($x$-values) is called the ***domain*** of $f$, and the set of outputs ($y$-values) that result when $x$ varies over the domain is called the ***range*** of $f$. For example, consider the equations

$$y = x^2 \quad \text{and} \quad y = x^2, \quad x \geq 2$$

In the first equation there is no restriction on $x$, so we may assume that any real value of $x$ is an allowable input. Thus, the equation defines a function $f(x) = x^2$ with domain $-\infty < x < +\infty$. In the second equation, the inequality $x \geq 2$ restricts the allowable inputs to be greater than or equal to 2, so the equation defines a function $g(x) = x^2$, $x \geq 2$ with domain $2 \leq x < +\infty$.

As $x$ varies over the domain of the function $f(x) = x^2$, the values of $y = x^2$ vary over the interval $0 \leq y < +\infty$, so this is the range of $f$. By comparison, as $x$ varies over the domain of the function $g(x) = x^2$, $x \geq 2$, the values of $y = x^2$, $x \geq 2$ vary over the interval $4 \leq y < +\infty$, so this is the range of $g$.

It is important to understand here that even though $f(x) = x^2$ and $g(x) = x^2$, $x \geq 2$ involve the same formula, we regard them to be different functions because they have different domains. In short, *to fully describe a function you must not only specify the rule that relates the inputs and outputs, but you must also specify the domain, that is, the set of allowable inputs.*

**GRAPHS OF FUNCTIONS**

If $f$ is a real-valued function of a real variable, then the ***graph*** of $f$ in the $xy$-plane is defined to be the graph of the equation $y = f(x)$. For example, the graph of the function $f(x) = x$ is the graph of the equation $y = x$, shown in Figure 1.2.1. That figure also shows the graphs

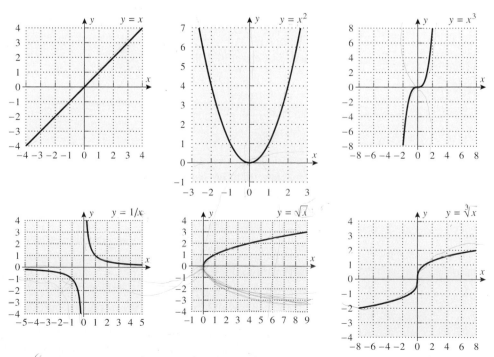

Figure 1.2.1

of some other basic functions that may already be familiar to you. Later in this chapter we will discuss techniques for graphing functions using graphing calculators and computers.

Graphs can provide useful visual information about a function. For example, because the graph of a function $f$ in the $xy$-plane consists of all points whose coordinates satisfy the equation $y = f(x)$, the points on the graph of $f$ are of the form $(x, f(x))$; hence each $y$-coordinate is the value of $f$ at the $x$-coordinate (Figure 1.2.2$a$). Pictures of the domain and range of $f$ can be obtained by projecting the graph of $f$ onto the coordinate axes (Figure 1.2.2$b$). The values of $x$ for which $f(x) = 0$ are the $x$-coordinates of the points where the graph of $f$ intersects the $x$-axis (Figure 1.2.2$c$); these values of $x$ are called the *zeros* of $f$, the *roots* of $f(x) = 0$, or the *x-intercepts* of $y = f(x)$.

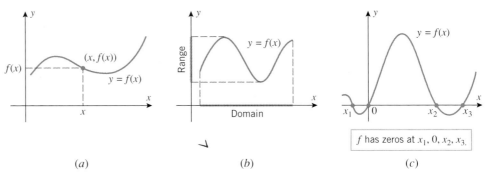

(a)          (b)          (c)

Figure 1.2.2

**THE VERTICAL LINE TEST**

Not every curve in the $xy$-plane is the graph of a function. For example, consider the curve in Figure 1.2.3, which is cut at two distinct points, $(a, b)$ and $(a, c)$, by a vertical line. This curve cannot be the graph of $y = f(x)$ for any function $f$; otherwise, we would have

$$f(a) = b \quad \text{and} \quad f(a) = c$$

which is impossible, since $f$ cannot assign two different values to $a$. Thus, there is no function $f$ whose graph is the given curve. This illustrates the following general result, which we will call the *vertical line test*.

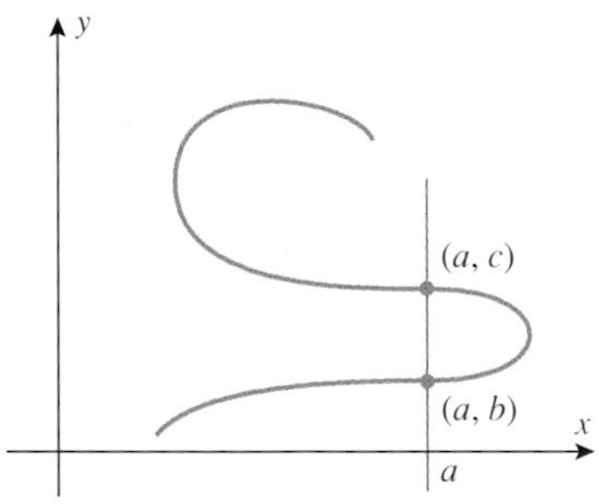

> **1.2.1**   THE VERTICAL LINE TEST.   *A curve in the xy-plane is the graph of some function f if and only if no vertical line intersects the curve more than once.*

Figure 1.2.3

### Example 1

The graph of the equation

$$x^2 + y^2 = 25 \tag{3}$$

is a circle of radius 5, centered at the origin (see Appendix D for a review of circles), and hence there are vertical lines that cut the graph more than once. This can also be seen algebraically by solving (3) for $y$ in terms of $x$:

$$y = \pm\sqrt{25 - x^2}$$

This equation does not define $y$ as a function of $x$ because the right side is "multiple valued" in the sense that values of $x$ in the interval $(-5, 5)$ produce two corresponding values of $y$. For example, if $x = 4$, then $y = \pm 3$, and hence $(4, 3)$ and $(4, -3)$ are two points on the circle that lie on the same vertical line (Figure 1.2.4$a$). However, we can regard the circle as the union of two semicircles:

$$y = \sqrt{25 - x^2} \quad \text{and} \quad y = -\sqrt{25 - x^2}$$

(Figure 1.2.4$b$), each of which defines $y$ as a function of $x$.   ◄

Figure 1.2.4

## THE ABSOLUTE VALUE FUNCTION

Recall that the ***absolute value*** or ***magnitude*** of a real number $x$ is defined by

$$|x| = \begin{cases} x, & x \geq 0 \\ -x, & x < 0 \end{cases}$$

The effect of taking the absolute value of a number is to strip away the minus sign if the number is negative and to leave the number unchanged if it is nonnegative. Thus,

$$|5| = 5, \quad \left|-\tfrac{4}{7}\right| = \tfrac{4}{7}, \quad |0| = 0$$

A more detailed discussion of the properties of absolute value is given in Appendix B. However, for convenience we provide the following summary of its algebraic properties.

---

**1.2.2** PROPERTIES OF ABSOLUTE VALUE. *If a and b are real numbers, then*

(a) $|-a| = |a|$      A number and its negative have the same absolute value.

(b) $|ab| = |a|\,|b|$      The absolute value of a product is the product of the absolute values.

(c) $|a/b| = |a|/|b|$      The absolute value of a ratio is the ratio of the absolute values.

(d) $|a+b| \leq |a| + |b|$      The ***triangle inequality***

---

REMARK. Symbols such as $+x$ and $-x$ are deceptive, since it is tempting to conclude that $+x$ is positive and $-x$ is negative. However, this need not be so, since $x$ itself can be positive or negative. For example, if $x$ is negative, say $x = -3$, then $-x = 3$ is positive and $+x = -3$ is negative.

The graph of the function $f(x) = |x|$ can be obtained by graphing the two parts of the equation

$$y = \begin{cases} x, & x \geq 0 \\ -x, & x < 0 \end{cases}$$

separately. For $x \geq 0$, the graph of $y = x$ is a ray of slope 1 with its endpoint at the origin, and for $x < 0$, the graph of $y = -x$ is a ray of slope $-1$ with its endpoint at the origin. Combining the two parts produces the V-shaped graph in Figure 1.2.5.

Absolute values have important relationships to square roots. To see why this is so, recall from algebra that every positive real number $x$ has two square roots, one positive and one negative. By definition, the symbol $\sqrt{x}$ denotes the *positive* square root of $x$. To denote the negative square root you must write $-\sqrt{x}$. For example, the positive square root of 9 is $\sqrt{9} = 3$, and the negative square root is $-\sqrt{9} = -3$. (Do not make the mistake of writing $\sqrt{9} = \pm 3$.)

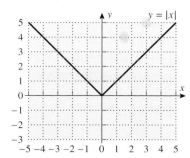

Figure 1.2.5

Care must be exercised in simplifying expressions of the form $\sqrt{x^2}$, since it is *not* always true that $\sqrt{x^2} = x$. This equation is correct if $x$ is nonnegative, but it is false for negative $x$. For example, if $x = -4$, then

$$\sqrt{x^2} = \sqrt{(-4)^2} = \sqrt{16} = 4 \neq x$$

A statement that is correct for all real values of $x$ is

$$\sqrt{x^2} = |x|$$

FOR THE READER.    Verify this relationship by using a graphing utility to show that the equations $y = \sqrt{x^2}$ and $y = |x|$ have the same graph.

**FUNCTIONS DEFINED PIECEWISE**

The absolute value function $f(x) = |x|$ is an example of a function that is defined *piecewise* in the sense that the formula for $f$ changes, depending on the value of $x$.

### Example 2

Sketch the graph of the function defined piecewise by the formula

$$f(x) = \begin{cases} 0, & x \leq -1 \\ \sqrt{1 - x^2}, & -1 < x < 1 \\ x, & x \geq 1 \end{cases}$$

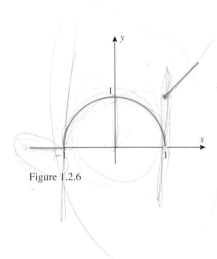

Figure 1.2.6

*Solution.* The formula for $f$ changes at the points $x = -1$ and $x = 1$. (We call these the **breakpoints** for the formula.) A good procedure for graphing functions defined piecewise is to graph the function separately over the open intervals determined by the breakpoints, and then graph $f$ at the breakpoints themselves. For the function $f$ in this example the graph is the horizontal line segment $y = 0$ on the interval $(-\infty, -1)$, it is the semicircle $y = \sqrt{1 - x^2}$ on the interval $(-1, 1)$, and it is the line segment $y = x$ on the interval $(1, +\infty)$. The formula for $f$ specifies that the equation $y = 0$ applies at the breakpoint $-1$ [so $y = f(-1) = 0$], and it specifies that the equation $y = x$ applies at the breakpoint 1 [so $y = f(1) = 1$]. The graph of $f$ is shown in Figure 1.2.6. ◀

REMARK.    In Figure 1.2.6 the solid dot and open circle at the breakpoint $x = 1$ serve to emphasize that the point on the graph lies on the line segment and not the semicircle. There is no ambiguity at the breakpoint $x = -1$ because the two parts of the graph join together continuously there.

### Example 3

Increasing the speed at which air moves over a person's skin increases the rate of moisture evaporation and makes the person feel cooler. (This is why we fan ourselves in hot weather.) The **windchill index** is the temperature at a wind speed of 4 mi/h that would produce the same sensation on exposed skin as the current temperature and wind speed combination. An empirical formula (i.e., a formula based on experimental data) for the windchill index $W$ at $32°F$ for a wind speed of $v$ mi/h is

$$W = \begin{cases} 32, & 0 \leq v \leq 4 \\ 91.4 + 59.4(0.0203v - 0.304\sqrt{v} - 0.474), & 4 < v < 45 \\ -3.6, & v \geq 45 \end{cases}$$

A computer-generated graph of $W(v)$ is shown in Figure 1.2.7. ◀

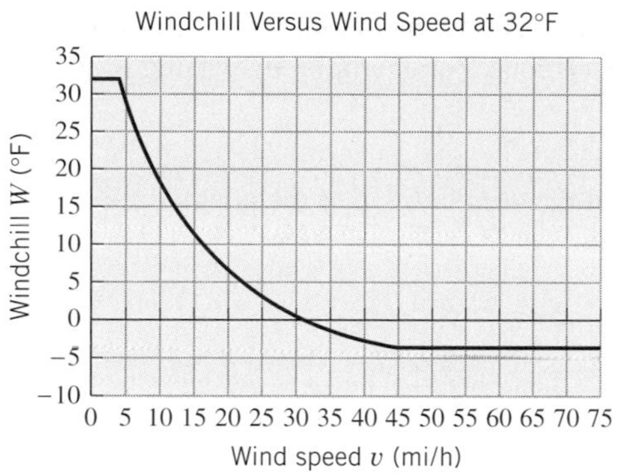
Windchill Versus Wind Speed at 32°F
Windchill W (°F)
35
30
25
20
15
10
5
0
−5
−10
0 5 10 15 20 25 30 35 40 45 50 55 60 65 70 75
Wind speed v (mi/h)

Figure 1.2.7

## THE NATURAL DOMAIN

Sometimes, restrictions on the allowable values of an independent variable result from a mathematical formula that defines the function. For example, if $f(x) = 1/x$, then $x = 0$ must be excluded from the domain to avoid division by zero, and if $f(x) = \sqrt{x}$, then negative values of $x$ must be excluded from the domain, since we are only considering real-valued functions of a real variable for now. We make the following definition.

> **1.2.3** DEFINITION. If a real-valued function of a real variable is defined by a formula, and if no domain is stated explicitly, then it is to be understood that the domain consists of all real numbers for which the formula yields a real value. This is called the ***natural domain*** of the function.

### Example 4

Find the natural domain of

(a) $f(x) = x^3$ (b) $f(x) = 1/(x-1)(x-3)$

(c) $f(x) = \tan x$ (d) $f(x) = \sqrt{x^2 - 5x + 6}$

*Solution (a).* The function $f$ has real values for all real $x$, so its natural domain is the interval $(-\infty, +\infty)$.

*Solution (b).* The function $f$ has real values for all real $x$, except $x = 1$ and $x = 3$, where divisions by zero occur. Thus, the natural domain is

$$\{x : x \neq 1 \text{ and } x \neq 3\} = (-\infty, 1) \cup (1, 3) \cup (3, +\infty)$$

*Solution (c).* Since $f(x) = \tan x = \sin x / \cos x$, the function $f$ has real values except where $\cos x = 0$, and this occurs when $x$ is an odd integer multiple of $\pi/2$. Thus, the natural domain consists of all real numbers except

$$x = \pm\frac{\pi}{2}, \pm\frac{3\pi}{2}, \pm\frac{5\pi}{2}, \ldots$$

*Solution (d).* The function $f$ has real values, except when the expression inside the radical is negative. Thus the natural domain consists of all real numbers $x$ such that

$$x^2 - 5x + 6 = (x - 3)(x - 2) \geq 0$$

This inequality is satisfied if $x \leq 2$ or $x \geq 3$ (verify), so the natural domain of $f$ is

$$(-\infty, 2] \cup [3, +\infty)$$ ◀

REMARK. In some problems we will want to limit the domain of a function by imposing specific restrictions. For example, by writing

$$f(x) = x^2, \quad x \geq 0$$

we can limit the domain of $f$ to the positive $x$-axis (Figure 1.2.8).

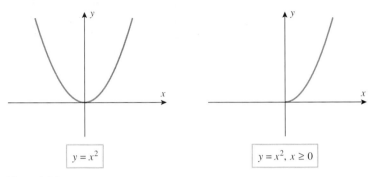

Figure 1.2.8

**THE EFFECT OF ALGEBRAIC OPERATIONS ON THE DOMAIN**

Algebraic expressions are frequently simplified by canceling common factors in the numerator and denominator. However, care must be exercised when simplifying formulas for functions in this way, since this process can alter the domain.

### Example 5

The natural domain of the function

$$f(x) = \frac{x^2 - 4}{x - 2}$$

consists of all real $x$ except $x = 2$. However, if we factor the numerator and then cancel the common factor in the numerator and denominator, we obtain

$$f(x) = \frac{(x - 2)(x + 2)}{x - 2} = x + 2$$

which *is* defined at $x = 2$ [since $f(2) = 4$ for the altered function $f$]. Thus, the algebraic simplification has altered the domain of the function. Geometrically, the graph of $y = x + 2$ is a line of slope 1 and $y$-intercept 2, whereas the graph of $y = (x^2 - 4)/(x - 2)$ is the same line, but with a hole in it at $x = 2$, since $y$ is undefined there (Figure 1.2.9). Thus, the geometric effect of the algebraic cancellation is to eliminate the hole in the original graph. In some situations such minor alterations in the domain are irrelevant to the problem under consideration and can be ignored. However, if we wanted to preserve the domain in this example, then we would express the simplified form of the function as

$$f(x) = x + 2, \quad x \neq 2 \qquad \blacktriangleleft$$

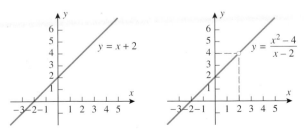

Figure 1.2.9

**Example 6**

Find the domain and range of

(a)  $f(x) = 2 + \sqrt{x - 1}$     (b)  $f(x) = (x + 1)/(x - 1)$

*Solution (a).*  Since no domain is stated explicitly, the domain of $f$ is the natural domain $[1, +\infty)$. To determine the range, it will be convenient to introduce a dependent variable $y = 2 + \sqrt{x - 1}$. As $x$ varies over the interval $[1, +\infty)$, the value of $\sqrt{x - 1}$ varies over the interval $[0, +\infty)$, so the value of $y = 2 + \sqrt{x - 1}$ varies over the interval $[2, +\infty)$, which is the range of $f$. The domain and range are shown graphically in Figure 1.2.10*a*.

*Solution (b).*  The given function $f$ is defined for all real $x$, except $x = 1$, so the natural domain of $f$ is

$$\{x : x \neq 1\} = (-\infty, 1) \cup (1, +\infty)$$

As in the preceding part of this example, it will be convenient to introduce a dependent variable

$$y = \frac{x + 1}{x - 1} \tag{4}$$

Although the set of possible $y$-values is not immediately evident from this equation, the graph of (4), which is shown in Figure 1.2.10*b*, suggests that the range of $f$ consists of all $y$, except $y = 1$. To see that this is so, we solve (4) for $x$ in terms of $y$:

$$(x - 1)y = x + 1$$
$$xy - y = x + 1$$
$$xy - x = y + 1$$
$$x(y - 1) = y + 1$$
$$x = \frac{y + 1}{y - 1}$$

It is now evident from the right side of this equation that $y = 1$ is not in the range; otherwise we would have a division by zero. No other values of $y$ are excluded by this equation, so the range of the function $f$ is $\{y : y \neq 1\} = (-\infty, 1) \cup (1, +\infty)$, which agrees with the result obtained graphically.  ◀

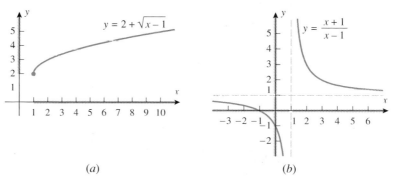

(a)                    (b)

Figure 1.2.10

**DOMAIN AND RANGE IN APPLIED PROBLEMS**

In applications, physical considerations often impose restrictions on the domain and range of a function.

### Example 7

An open box is to be made from a 16 in by 30 in piece of cardboard by cutting out squares of equal size from the four corners and bending up the sides (Figure 1.2.11*a*).

(a)   Let $V$ be the volume of the box that results when the squares have sides of length $x$. Find a formula for $V$ as a function of $x$.

(b)   Find the domain of $V$.

(c)   Use the graph of $V$ given in Figure 1.2.11*c* to estimate the range of $V$.

(d)   Describe in words what the graph tells you about the volume.

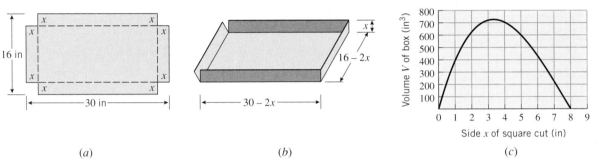

(a)                                     (b)                                     (c)

Figure 1.2.11

*Solution* (*a*). As shown in Figure 1.2.11*b*, the resulting box has dimensions $16 - 2x$ by $30 - 2x$ by $x$, so the volume $V(x)$ is given by

$$V(x) = (16 - 2x)(30 - 2x)x = 480x - 92x^2 + 4x^3$$

*Solution* (*b*). The domain is the set of $x$-values and the range is the set of $V$-values. Because $x$ is a length, it must be nonnegative, and because we cannot cut out squares whose sides are more than 8 in long (why?), the $x$-values in the domain must satisfy

$$0 \le x \le 8$$

*Solution* (*c*). From the graph of $V$ versus $x$ in Figure 1.2.11*c* we estimate that the $V$-values in the range satisfy

$$0 \le V \le 725$$

Note that this is an approximation. Later we will show how to find the range exactly.

*Solution* (*d*). The graph tells us that the box of maximum volume occurs for a value of $x$ that is between 3 and 4 and that the maximum volume is approximately 725 in$^3$. Moreover, the volume decreases toward zero as $x$ gets closer to 0 or 8.   ◀

In applications involving time, formulas for functions are often expressed in terms of a variable $t$ whose starting value is taken to be $t = 0$.

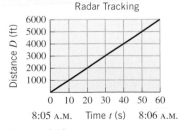

Figure 1.2.12

### Example 8

At 8:05 A.M. a car is clocked at 100 ft/s by a radar detector that is positioned at the edge of a straight highway. Assuming that the car maintains a constant speed between 8:05 A.M. and 8:06 A.M., find a function $D(t)$ that expresses the distance traveled by the car during that time interval as a function of the time $t$.

*Solution.* It would be clumsy to use clock time for the variable $t$, so let us agree to measure the elapsed time in seconds, starting with $t = 0$ at 8:05 A.M. and ending with $t = 60$ at

8:06 A.M. At each instant, the distance traveled (in ft) is equal to the speed of the car (in ft/s) multiplied by the elapsed time (in s). Thus,

$$D(t) = 100t, \quad 0 \le t \le 60$$

The graph of $D$ versus $t$ is shown in Figure 1.2.12. ◀

**ISSUES OF SCALE AND UNITS**

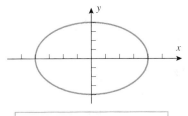

The circle is squashed because 1 unit on the $y$-axis has a smaller length than 1 unit on the $x$-axis.

Figure 1.2.13

In geometric problems where you want to preserve the "true" shape of a graph, you must use units of equal length on both axes. For example, if you graph a circle in a coordinate system in which 1 unit in the $y$-direction is smaller than 1 unit in the $x$-direction, then the circle will be squashed vertically into an elliptical shape (Figure 1.2.13). You must also use units of equal length when you want to apply the distance formula

$$d = \sqrt{(x_2 - x_1)^2 + (y_2 - y_1)^2}$$

to calculate the distance between two points $(x_1, y_1)$ and $(x_2, y_2)$ in the $xy$-plane.

However, sometimes it is inconvenient or impossible to display a graph using units of equal length. For example, consider the equation

$$y = x^2$$

If we want to show the portion of the graph over the interval $-3 \le x \le 3$, then there is no problem using units of equal length, since $y$ only varies from 0 to 9 over that interval. However, if we want to show the portion of the graph over the interval $-10 \le x \le 10$, then there is a problem keeping the units equal in length, since the value of $y$ varies between 0 and 100. In this case the only reasonable way to show all of the graph that occurs over the interval $-10 \le x \le 10$ is to compress the unit of length along the $y$-axis, as illustrated in Figure 1.2.14.

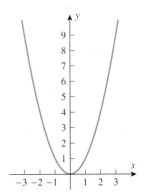
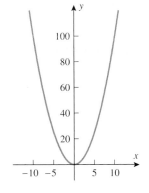

Figure 1.2.14

REMARK.   In applications where the variables on the two axes have unrelated units (say, centimeters on the $y$-axis and seconds on the $x$-axis), then nothing is gained by requiring the units to have equal lengths; choose the lengths to make the graph as clear as possible.

**EXERCISE SET 1.2**  ⊠ Graphing Calculator  ⊏c⊐ CAS

**1.** Find $f(0)$, $f(2)$, $f(-2)$, $f(3)$, $f(\sqrt{2})$, and $f(3t)$.

 (a) $f(x) = 3x^2 - 2$

 (b) $f(x) = \begin{cases} \dfrac{1}{x}, & x > 3 \\ 2x, & x \le 3 \end{cases}$

**2.** Find $g(3)$, $g(-1)$, $g(\pi)$, $g(-1.1)$, and $g(t^2 - 1)$.

 (a) $g(x) = \dfrac{x+1}{x-1}$

 (b) $g(x) = \begin{cases} \sqrt{x+1}, & x \ge 1 \\ 3, & x < 1 \end{cases}$

In Exercises 3–6, find the natural domain of the function algebraically, and confirm that your result is consistent with the graph produced by your graphing utility. [*Note:* Set your graphing utility to the radian mode when graphing trigonometric functions.]

**3.** (a) $f(x) = \dfrac{1}{x-3}$      (b) $g(x) = \sqrt{x^2 - 3}$

    (c) $G(x) = \sqrt{x^2 - 2x + 5}$      (d) $f(x) = \dfrac{x}{|x|}$

    (e) $h(x) = \dfrac{1}{1 - \sin x}$

**4.** (a) $f(x) = \dfrac{1}{5x + 7}$      (b) $h(x) = \sqrt{x - 3x^2}$

    (c) $G(x) = \sqrt{\dfrac{x^2 - 4}{x - 4}}$      (d) $f(x) = \dfrac{x^2 - 1}{x + 1}$

    (e) $h(x) = \dfrac{3}{2 - \cos x}$

**5.** (a) $f(x) = \sqrt{3 - x}$      (b) $g(x) = \sqrt{4 - x^2}$

    (c) $h(x) = 3 + \sqrt{x}$      (d) $G(x) = x^3 + 2$

    (e) $H(x) = 3 \sin x$

**6.** (a) $f(x) = \sqrt{3x - 2}$      (b) $g(x) = \sqrt{9 - 4x^2}$

    (c) $h(x) = \dfrac{1}{3 + \sqrt{x}}$      (d) $G(x) = \dfrac{3}{x}$

    (e) $H(x) = \sin^2 \sqrt{x}$

**7.** In each part of the accompanying figure, determine whether the graph defines $y$ as a function of $x$.

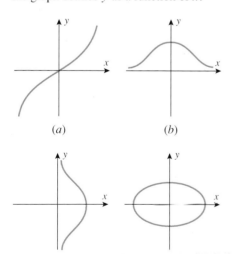

       (*a*)              (*b*)

       (*c*)              (*d*)

Figure Ex-7

**8.** Express the length $L$ of a chord of a circle with radius 10 cm as a function of the central angle $\theta$ (see the accompanying figure).

**9.** As shown in the accompanying figure, a pendulum of constant length $L$ makes an angle $\theta$ with its vertical position. Express the height $h$ as a function of the angle $\theta$.

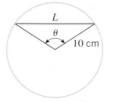
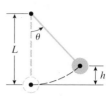

Figure Ex-8            Figure Ex-9

**10.** A cup of hot coffee sits on a table. You pour in some cool milk and let it sit for an hour. Sketch a rough graph of the temperature of the coffee as a function of time.

**11.** A boat is bobbing up and down on some gentle waves. Suddenly it gets hit by a large wave and sinks. Sketch a rough graph of the height of the boat above the ocean floor as a function of time.

**12.** Make a rough sketch of your weight as a function of time from birth to the present.

In Exercises 13 and 14, express the function in piecewise form without using absolute values. [*Suggestion:* It may help to generate the graph of the function.]

**13.** (a) $f(x) = |x| + 3x + 1$      (b) $g(x) = |x| + |x - 1|$

**14.** (a) $f(x) = 3 + |2x - 5|$      (b) $g(x) = 3|x - 2| - |x + 1|$

**15.** As shown in the accompanying figure, an open box is to be constructed from a rectangular sheet of metal, 8 inches by 15 inches, by cutting out squares with sides of length $x$ from each corner and bending up the sides.

    (a) Express the volume $V$ as a function of $x$.

    (b) Find the natural domain and the range of the function, ignoring any physical restrictions on the values of the variables.

    (c) Modify the domain and range appropriately to account for the physical restrictions on the values of $V$ and $x$.

    (d) In words, describe how the volume $V$ of the box varies with $x$, and discuss how one might construct boxes of maximum volume and minimum volume.

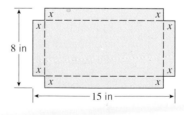

Figure Ex-15

**16.** As shown in the accompanying figure, a camera is mounted at a point 3000 ft from the base of a rocket launching pad.

The shuttle rises vertically when launched, and the camera's elevation angle is constantly adjusted to follow the bottom of the rocket.

(a) Choose letters to represent the height of the rocket and the elevation angle of the camera, and express the height as a function of the elevation angle.

(b) Find the natural domain and the range of the function, ignoring any physical restrictions on the values of the variables.

(c) Modify the domain and range appropriately to account for the physical restrictions on the values of the variables.

(d) Generate the graph of height versus the elevation on a graphing utility, and use it to estimate the height of the rocket when the elevation angle is $\pi/4 \approx 0.7854$ radian. Compare this estimate to the exact height. [*Suggestion:* If you are using a graphing calculator, the trace and zoom features will be helpful here.

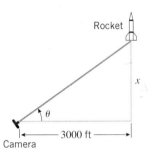

Figure Ex-16

In Exercises 17 and 18: (i) Explain why the function $f$ has one or more holes in its graph, and state the $x$-values at which those holes occur. (ii) Find a function $g$ whose graph is identical to that of $f$, but without the holes.

**17.** $f(x) = \dfrac{(x+2)(x^2-1)}{(x+2)(x-1)}$    **18.** $f(x) = \dfrac{x+\sqrt{x}}{\sqrt{x}}$

**19.** For a given outside temperature $T$ and wind speed $v$, the windchill index (WCI) is the equivalent temperature that exposed skin would feel with a wind speed of 4 mi/h. An empirical formula for the WCI (based on experience and observation) is

$$\text{WCI} = \begin{cases} T, & 0 \le v \le 4 \\ 91.4 + (91.4 - T)(0.0203v - 0.304\sqrt{v} - 0.474), & 4 < v < 45 \\ 1.6T - 55, & v \ge 45 \end{cases}$$

where $T$ is the air temperature in $^\circ$F, $v$ is the wind speed in mi/h, and WCI is the equivalent temperature in $^\circ$F. Find the WCI to the nearest degree if the air temperature is $25\,^\circ$F and

(a) $v = 3$ mi/h    (b) $v = 15$ mi/h
(c) $v = 46$ mi/h.

[Adapted from UMAP Module 658, *Windchill*, W. Bosch and L. Cobb, COMAP, Arlington, MA.]

In Exercises 20–22, use the formula for the windchill index described in Exercise 19.

**20.** Find the air temperature to the nearest degree if the WCI is reported as $-60\,^\circ$F with a wind speed of 48 mi/h.

**21.** Find the air temperature to the nearest degree if the WCI is reported as $-10\,^\circ$F with a wind speed of 8 mi/h.

**22.** Find the wind speed to the nearest mile per hour if the WCI is reported as $-15\,^\circ$F with an air temperature of $20\,^\circ$F.

**23.** At 9:23 A.M. a lunar lander that is 1000 ft above the Moon's surface begins a vertical descent, touching down at 10:13 A.M. Assuming that the lander maintains a constant speed, find a function $D(t)$ that expresses the altitude of the lander above the Moon's surface as a function of $t$.

## 1.3 GRAPHING FUNCTIONS ON CALCULATORS AND COMPUTERS; COMPUTER ALGEBRA SYSTEMS

*In this section we will discuss issues that relate to generating graphs of equations and functions with graphing utilities (graphing calculators and computers). Because graphing utilities vary widely, it is difficult to make general statements about them. Therefore, at various places in this section we will ask you to refer to the documentation for your own graphing utility for specific details about the way it operates.*

**GRAPHING CALCULATORS AND COMPUTER ALGEBRA SYSTEMS**

The development of new technology has significantly changed how and where mathematicians, engineers, and scientists perform their work, as well as their approach to problem solving. Not only have portable computers and handheld calculators with graphing capabilities become standard tools in the scientific community, but there have been major new innovations in computer software. Among the most significant of these innovations are programs called **Computer Algebra Systems** (abbreviated CAS), the most common being *Mathematica*, *Maple*, and *Derive*.[*] Computer algebra systems not only have powerful graphing capabilities, but, as their name suggests, they can perform many of the symbolic computations that occur in algebra, calculus, and branches of higher mathematics. For example, it is a trivial task for a CAS to perform the factorization

$$x^6 + 23x^5 + 147x^4 - 139x^3 - 3464x^2 - 2112x + 23040 = (x + 5)(x - 3)^2(x + 8)^3$$

or the exact numerical computation

$$\left(\frac{63456}{3177295} - \frac{43907}{22854377}\right)^3 = \frac{225191245716420829125932020230122866923}{3828959558193692044495659453692037646883755}$$

Technology has also made it possible to generate graphs of equations and functions in seconds that in the past might have taken hours to produce. Graphing technology includes handheld graphing calculators, computer algebra systems, and software designed for that purpose. Figure 1.3.1 shows the graphs of the function $f(x) = x^4 - x^3 - 2x^2$ produced with various graphing utilities; the first two were generated with the CAS programs, *Mathematica* and *Maple*, and the third with a graphing calculator. Graphing calculators produce coarser graphs than most computer programs but have the advantage of being compact and portable.

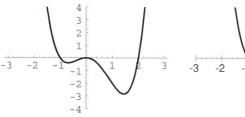
*Generated by Mathematica*

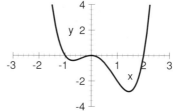
*Generated by Maple*

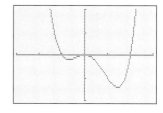
*Generated by a graphing calculator*

Figure 1.3.1

**VIEWING WINDOWS**

Graphing utilities can only show a portion of the $xy$-plane in the viewing screen, so the first step in graphing an equation is to determine which rectangular portion of the $xy$-plane you want to display. This region is called the **viewing window** (or **viewing rectangle**). For example, in Figure 1.3.1 the viewing window extends over the interval $[-3, 3]$ in the $x$-direction and over the interval $[-4, 4]$ in the $y$-direction, so we say that the viewing window is $[-3, 3] \times [-4, 4]$ (read "$[-3, 3]$ by $[-4, 4]$"). In general, if the viewing window is $[a, b] \times [c, d]$, then the window extends between $x = a$ and $x = b$ in the $x$-direction and between $y = c$ and $y = d$ in the $y$-direction. We will call $[a, b]$ the **$x$-interval** for the window and $[c, d]$ the **$y$-interval** for the window (Figure 1.3.2).

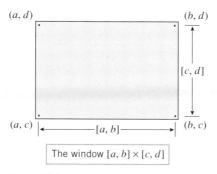

The window $[a, b] \times [c, d]$

Figure 1.3.2

---

[*]*Mathematica* is a product of Wolfram Research, Inc.; *Maple* is a product of Waterloo Maple Software, Inc.; and *Derive* is a product of Soft Warehouse, Inc.

Different graphing utilities designate viewing windows in different ways. For example, the first two graphs in Figure 1.3.1 were produced by the commands

```
Plot[x^4 - x^3 -2*x^2, {x, -3, 3}, PlotRange->{-4, 4}]
```
(*Mathematica*)

```
plot( x^4 - x^3 -2*x^2, x = -3..3, y = -4..4);
```
(*Maple*)

and the last graph was produced on a graphing calculator by pressing the GRAPH button after setting the following values for the variables that determine the $x$-interval and $y$-intervals:

$$x\text{Min} = -3, \quad x\text{Max} = 3, \quad y\text{Min} = -4, \quad y\text{Max} = 4$$

FOR THE READER.    Use your own graphing utility to generate the graph of the function $f(x) = x^4 - x^3 - 2x^2$ in the window $[-3, 3] \times [-4, 4]$.

**TICK MARKS AND GRID LINES**

To help locate points in a viewing window visually, graphing utilities provide methods for drawing **tick marks** (also called **scale marks**) on the coordinate axes or at other locations in the viewing window. With computer programs such as *Mathematica* and *Maple*, there are specific commands for designating the spacing between tick marks, but if the user does not specify the spacing, then the programs make certain *default* choices. For example, in the first two parts of Figure 1.3.1, the tick marks shown were the default choices.

On graphing calculators the spacing between tick marks is determined by two **scale variables** (also called **scale factors**), which we will denote by

$$x\text{Scl} \quad \text{and} \quad y\text{Scl}$$

(The notation varies among calculators.) These variables specify the spacing between the tick marks in the $x$- and $y$-directions, respectively. For example, in the third part of Figure 1.3.1 the window and tick marks were designated by the settings

$$x\text{Min} = -3 \qquad x\text{Max} = 3$$
$$y\text{Min} = -4 \qquad y\text{Max} = 4$$
$$x\text{Scl} = 1 \qquad y\text{Scl} = 1$$

Most graphing utilities allow for variations in the design and positioning of tick marks. For example, Figure 1.3.3 shows two variations of the graphs in Figure 1.3.1; the first was generated on a computer using an option for placing the ticks and numbers on the edges of a box, and the second was generated on a graphing calculator using an option for drawing grid lines to simulate graph paper.

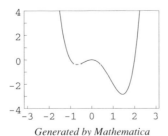

Generated by Mathematica        Generated by a graphing calculator

Figure 1.3.3

## Example 1

Figure 1.3.4*a* shows the window $[-5, 5] \times [-5, 5]$ with the tick marks spaced .5 unit apart in the $x$-direction and 10 units apart in the $y$-direction. Note that no tick marks are actually visible in the $y$-direction because the tick mark at the origin is covered by the $x$-axis, and all other tick marks in the $y$-direction fall outside of the viewing window. ◀

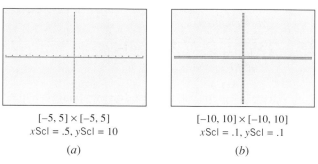

$[-5, 5] \times [-5, 5]$
$x\text{Scl} = .5, y\text{Scl} = 10$

$(a)$

$[-10, 10] \times [-10, 10]$
$x\text{Scl} = .1, y\text{Scl} = .1$

$(b)$

Figure 1.3.4

### Example 2

Figure 1.3.4$b$ shows the window $[-10, 10] \times [-10, 10]$ with the tick marks spaced .1 unit apart in the $x$- and $y$-directions. In this case the tick marks are so close together that they create the effect of thick lines on the coordinate axes. When this occurs you will usually want to increase the scale factors to reduce the number of tick marks and make them legible. ◀

FOR THE READER. Graphing calculators provide a way of clearing all settings and returning them to *default values*. For example, on the author's calculator the default window is $[-10, 10] \times [-10, 10]$ and the default scale factors are $x\text{Scl} = 1$ and $y\text{Scl} = 1$. Check your documentation to determine the default values for your calculator and how to reset the calculator to its default configuration. If you are using a computer program, check your documentation to determine the commands for specifying the spacing between tick marks.

**CHOOSING A VIEWING WINDOW**

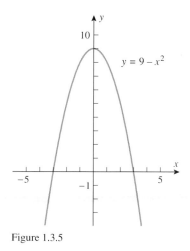

Figure 1.3.5

When the graph of a function extends indefinitely in some direction, no single viewing window can show the entire graph. In such cases the choice of the viewing window can drastically affect one's perception of how the graph looks. For example, Figure 1.3.5 shows a computer-generated graph of $y = 9 - x^2$, and Figure 1.3.6 shows four views of this graph generated on the author's calculator:

- In part $(a)$ the graph falls completely outside of the window, so the window is blank (except for the ticks and axes).

- In part $(b)$ the graph is broken into two pieces because it passes in and out of the window.

- In part $(c)$ the graph appears to be a straight line because we have zoomed in on such a small segment of the curve.

- In part $(d)$ we have a more complete picture of the graph shape because the window encompasses all of the important points, namely the high point on the graph and the intersections with the $x$-axis.

For a function whose graph does not extend indefinitely in either the $x$- or $y$-directions, the domain and range of the function can be used to obtain a viewing window that contains the entire graph.

### Example 3

Use the domain and range of the function $f(x) = \sqrt{12 - 3x^2}$ to determine a viewing window that contains the entire graph.

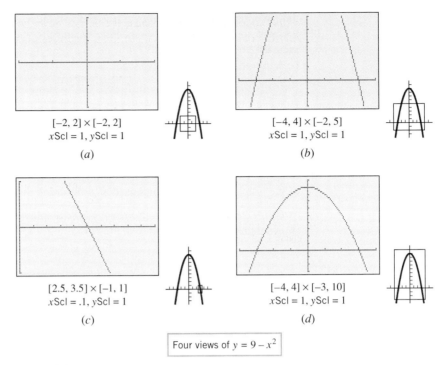

$[-2, 2] \times [-2, 2]$
$x\text{Scl} = 1, y\text{Scl} = 1$

(a)

$[-4, 4] \times [-2, 5]$
$x\text{Scl} = 1, y\text{Scl} = 1$

(b)

$[2.5, 3.5] \times [-1, 1]$
$x\text{Scl} = .1, y\text{Scl} = 1$

(c)

$[-4, 4] \times [-3, 10]$
$x\text{Scl} = 1, y\text{Scl} = 1$

(d)

Four views of $y = 9 - x^2$

Figure 1.3.6

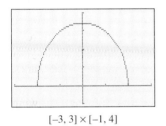

$[-3, 3] \times [-1, 4]$
$x\text{Scl} = 1, y\text{Scl} = 1$

Figure 1.3.7

*Solution.* The natural domain of $f$ is $[-2, 2]$ and the range is $[0, \sqrt{12}]$ (verify), so the entire graph will be contained in the viewing window $[-2, 2] \times [0, \sqrt{12}]$. For clarity, it is desirable to use a slightly larger window to avoid having the graph too close to the ends of the screen. For example, taking the viewing window to be $[-3, 3] \times [-1, 4]$ yields the graph in Figure 1.3.7. ◀

If the graph of $f$ extends indefinitely in either the $x$- or $y$-direction, then it will not be possible to show the entire graph in any one viewing window. In such cases one tries to choose the window to show all of the important features for the problem at hand. (Of course, what is important in one problem may not be important in another, so the choice of the viewing window will often depend on the objectives in the problem.)

### Example 4

Graph the equation $y = x^3 - 12x^2 + 18$ in the following windows and discuss the advantages and disadvantages of each window.

(a)  $[-10, 10] \times [-10, 10]$ with $x\text{Scl} = 1, y\text{Scl} = 1$
(b)  $[-20, 20] \times [-20, 20]$ with $x\text{Scl} = 1, y\text{Scl} = 1$
(c)  $[-20, 20] \times [-300, 20]$ with $x\text{Scl} = 1, y\text{Scl} = 20$
(d)  $[-5, 15] \times [-300, 20]$ with $x\text{Scl} = 1, y\text{Scl} = 20$
(e)  $[1, 2] \times [-1, 1]$ with $x\text{Scl} = .1, y\text{Scl} = .1$

*Solution (a).* The window in Figure 1.3.8a has chopped off the portion of the graph that intersects the $y$-axis, and it shows only two of three possible real roots for the given cubic polynomial. To remedy these problems we need to widen the window in both the $x$- and $y$-directions.

*Solution (b).* The window in Figure 1.3.8b shows the intersection of the graph with the $y$-axis and the three real roots, but it has chopped off the portion of the graph between the two positive roots. Moreover, the ticks in the *y*-direction are nearly illegible because they are so close together. We need to extend the window in the negative *y*-direction and increase *y*Scl. We do not know how far to extend the window, so some experimentation will be required to obtain what we want.

*Solution* (*c*). The window in Figure 1.3.8*c* shows all of the main features of the graph. However, we have some wasted space in the *x*-direction. We can improve the picture by shortening the window in the *x*-direction appropriately.

*Solution* (*d*). The window in Figure 1.3.8*d* shows all of the main features of the graph without a lot of wasted space. However, the window does not provide a clear view of the roots. To get a closer view of the roots we must forget about showing all of the main features of the graph and choose windows that zoom in on the roots themselves.

*Solution* (*e*). The window in Figure 1.3.8*e* displays very little of the graph, but it clearly shows that the root in the interval [1, 2] is slightly less than 1.3, say $x \approx 1.29$. ◀

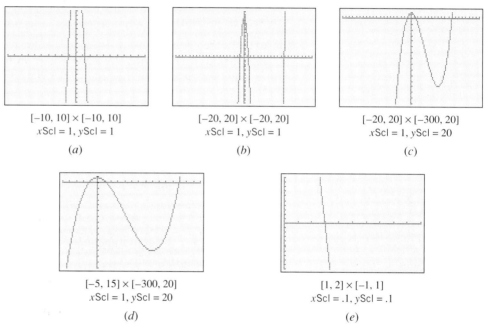

$[-10, 10] \times [-10, 10]$
$x$Scl = 1, $y$Scl = 1

(*a*)

$[-20, 20] \times [-20, 20]$
$x$Scl = 1, $y$Scl = 1

(*b*)

$[-20, 20] \times [-300, 20]$
$x$Scl = 1, $y$Scl = 20

(*c*)

$[-5, 15] \times [-300, 20]$
$x$Scl = 1, $y$Scl = 20

(*d*)

$[1, 2] \times [-1, 1]$
$x$Scl = .1, $y$Scl = .1

(*e*)

Figure 1.3.8

FOR THE READER.    Sometimes you will want to determine the viewing window by choosing the *x*-interval for the window and allowing the graphing utility to determine a *y*-interval that encompasses the maximum and minimum values of the function over the *x*-interval. Most graphing utilities provide some method for doing this, so check your documentation to determine how to use this feature. Allowing the graphing utility to determine the *y*-interval of the window takes some of the guesswork out of problems like that in part (b) of the preceding example.

**ZOOMING**

The process of enlarging or reducing the size of a viewing window is called *zooming*. If you reduce the size of the window, you see less of the graph as a whole, but more detail of the part shown; this is called *zooming in*. In contrast, if you enlarge the size of the window, you see more of the graph as a whole, but less detail of the part shown; this is called *zooming out*. Most graphing calculators provide menu items for zooming in or zooming out by fixed factors. For example, on the author's calculator the amount of enlargement or reduction is controlled by setting values for two **zoom factors**, $x$Fact and $y$Fact. If

$$x\text{Fact} = 10 \quad \text{and} \quad y\text{Fact} = 5$$

then each time a zoom command is executed the viewing window is enlarged or reduced by a factor of 10 in the $x$-direction and a factor of 5 in the $y$-direction. With computer programs such as *Mathematica* and *Maple*, zooming is controlled by adjusting the $x$-interval and $y$-interval directly; however, there are ways to automate this by programming.

FOR THE READER.   If you are using a graphing calculator, read your documentation to determine how to use the zooming feature.

**COMPRESSION**

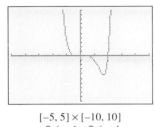

$[-5, 5] \times [-1000, 1000]$
$x$Scl $= 1$, $y$Scl $= 500$

(a)

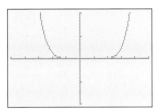

$[-5, 5] \times [-10, 10]$
$x$Scl $= 1$, $y$Scl $= 1$

(b)

Figure 1.3.9

Enlarging the viewing window for a graph has the geometric effect of compressing the graph, since more of the graph is packed into the calculator screen. If the compression is sufficiently great, then some of the detail in the graph may be lost. Thus, the choice of the viewing window frequently depends on whether you want to see more of the graph or more of the detail. Figure 1.3.9 shows two views of the equation

$$y = x^5(x - 2)$$

In part (a) of the figure the $y$-interval is very large, resulting in a vertical compression that obscures the detail in the vicinity of the $x$-axis. In part (b) the $y$-interval is smaller, and consequently we see more of the detail in the vicinity of the $x$-axis but less of the graph in the $y$-direction.

### Example 5

Describe the graph of the function $f(x) = x + 0.01 \sin 50\pi x$; then graph the function in the following windows and explain why the graphs do or do not differ from your description.

(a) $[-10, 10] \times [-10, 10]$      (b) $[-1, 1] \times [-1, 1]$

(c) $[-.1, .1] \times [-.1, .1]$      (d) $[-.01, .01] \times [-.01, .01]$

*Solution.*   The formula for $f$ is the sum of the function $x$ (whose graph is a straight line) and the function $0.01 \sin 50\pi x$ (whose graph is a sinusoidal curve with an amplitude of $0.01$ and a period of $2\pi/50\pi = 0.04$). Intuitively, this suggests that the graph of $f$ will follow the general path of the line $y = x$ but will have small bumps resulting from the contributions of the sinusoidal oscillations.

To generate the four graphs, we first set the calculator to the radian mode.* Because the windows in successive parts of this example are decreasing in size by a factor of 10, it will be convenient to use the zoom in feature of the calculator with the zoom factors set to 10 in the $x$- and $y$-directions. In Figure 1.3.10$a$ the graph appears to be a straight line

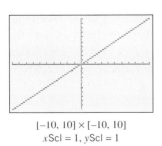

$[-10, 10] \times [-10, 10]$
$x$Scl $= 1$, $y$Scl $= 1$

(a)

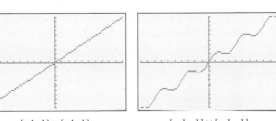

$[-1, 1] \times [-1, 1]$
$x$Scl $= .1$, $y$Scl $= .1$

(b)

$[-.1, .1] \times [-.1, .1]$
$x$Scl $= .01$, $y$Scl $= .01$

(c)

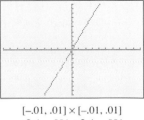

$[-.01, .01] \times [-.01, .01]$
$x$Scl $= .001$, $y$Scl $= .001$

(d)

Figure 1.3.10

---

*In this text we follow the convention that angles are measured in radians unless degree measure is specified.

because compression has hidden the small sinusoidal oscillations. (Keep in mind that the amplitude of the sinusoidal portion of the function is only 0.01.) In part (*b*) the oscillations have begun to appear since the *y*-interval has been reduced, and in part (*c*) the oscillations have become very clear because the vertical scale is more in keeping with the amplitude of the oscillations. In part (*d*) the graph appears to be a line segment because we have zoomed in on such a small portion of the curve.   ◄

**ASPECT RATIO DISTORTION**

Figure 1.3.11*a* shows a circle of radius 5 and two perpendicular lines graphed in the window $[-10, 10] \times [-10, 10]$ with $x\text{Scl} = 1$ and $y\text{Scl} = 1$. However, the circle is distorted and the lines do not appear perpendicular because the calculator has not used the same length for 1 unit on the *x*-axis and 1 unit on the *y*-axis. (Compare the spacing between the ticks on the axes.) This is called **aspect ratio distortion**. Many calculators provide a menu item for automatically correcting the distortion by adjusting the viewing window appropriately. For example, the author's calculator makes this correction to the viewing window $[-10, 10] \times [-10, 10]$ by changing it to

$$[-16.9970674487, 16.9970674487] \times [-10, 10]$$

(Figure 1.3.11*b*). With computer programs such as *Mathematica* and *Maple*, aspect ratio distortion is controlled with adjustments to the physical dimensions of the viewing window on the computer screen, rather than altering the *x*- and *y*-intervals of the viewing window.

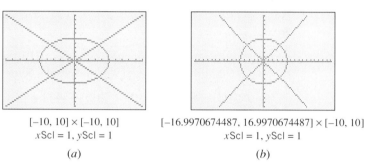

$[-10, 10] \times [-10, 10]$
$x\text{Scl} = 1, y\text{Scl} = 1$

(*a*)

$[-16.9970674487, 16.9970674487] \times [-10, 10]$
$x\text{Scl} = 1, y\text{Scl} = 1$

(*b*)

Figure 1.3.11

FOR THE READER.   Read the documentation for your graphing utility to determine how to control aspect ratio distortion.

**PIXELS AND RESOLUTION**

Sometimes graphing utilities produce unexpected results. For example, Figure 1.3.12 shows the graph of $y = \cos(10\pi x)$ generated on the author's graphing calculator in four different windows. (Your own calculator may produce different results.) The first graph has the correct shape, but the remaining three do not. To explain what is happening here we need to understand more precisely how graphing utilities generate graphs.

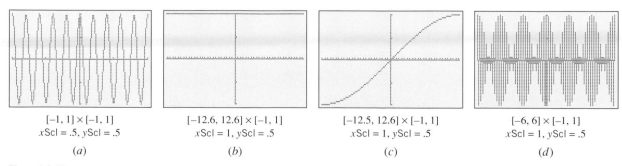

$[-1, 1] \times [-1, 1]$
$x\text{Scl} = .5, y\text{Scl} = .5$

(*a*)

$[-12.6, 12.6] \times [-1, 1]$
$x\text{Scl} = 1, y\text{Scl} = .5$

(*b*)

$[-12.5, 12.6] \times [-1, 1]$
$x\text{Scl} = 1, y\text{Scl} = .5$

(*c*)

$[-6, 6] \times [-1, 1]$
$x\text{Scl} = 1, y\text{Scl} = .5$

(*d*)

Figure 1.3.12

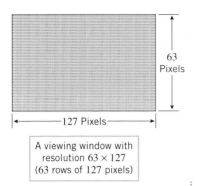

63
Pixels

|←———127 Pixels———→|

A viewing window with
resolution 63 × 127
(63 rows of 127 pixels)

Figure 1.3.13

Screen displays for graphing utilities are divided into rows and columns of rectangular blocks, called *pixels*. For black-and-white displays each pixel has two possible states—an activated (or dark) state and a deactivated (or light) state. Since graphical elements are produced by activating pixels, the more pixels that a screen has to work with, the greater the amount of detail it can show. For example, the author's calculator has a *resolution* of $63 \times 127$, meaning that there are 63 rows with 127 pixels per row (Figure 1.3.13). In contrast, the author's computer screen has a resolution of $1024 \times 1280$ (1024 rows with 1280 pixels per row), so the computer screen is capable of displaying much smoother graphs than the calculator.

FOR THE READER.    If you are using a graphing calculator, check the documentation to determine its resolution.

## SAMPLING ERROR

The procedure that a graphing utility follows to generate a graph is similar to the procedure for plotting points by hand. When a viewing window is selected and an equation is entered, the graphing utility determines the $x$-coordinates of certain pixels on the $x$-axis and computes the corresponding points $(x, y)$ on the graph. It then activates the pixels whose coordinates most closely match those of the calculated points and uses some built-in algorithm to activate additional intermediate pixels to create the curve shape. The point to keep in mind here is that *changing the window changes the points plotted by the graphing utility*. Thus, it is possible that a particular window will produce a false impression about the graph shape because significant characteristics of the graph occur *between* the plotted pixels. This is called *sampling error*. This is exactly what occurred in Figure 1.3.12 when we graphed $y = \cos(10\pi x)$. In part (*b*) of the figure the plotted pixels happened to fall at the peaks of the cosine curve, giving the false impression that the graph is a horizontal line at $y = 1$. In part (*c*) the plotted pixels fell at successively higher points along the graph, and in part (*d*) the plotted pixels fell in a strange way that created yet another misleading impression of the graph shape.

REMARK.    Figure 1.3.12 suggests that for trigonometric graphs with rapid oscillations, restricting the $x$-interval to a few periods is likely to produce a more accurate representation about the graph shape.

## FALSE GAPS

Sometimes graphs that are continuous appear to have gaps when they are generated on a calculator. These *false gaps* typically occur where the graph rises so rapidly that vertical space is opened up between successive pixels.

### Example 6

Figure 1.3.14 shows the graph of the semicircle $y = \sqrt{9 - x^2}$ in two viewing windows. Although this semicircle has $x$-intercepts at the points $x = \pm 3$, part (*a*) of the figure shows false gaps at those points because there are no pixels with $x$-coordinates $\pm 3$ in the window

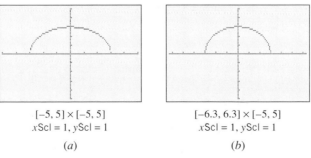

$[-5, 5] \times [-5, 5]$
$x\text{Scl} = 1, y\text{Scl} = 1$

(*a*)

$[-6.3, 6.3] \times [-5, 5]$
$x\text{Scl} = 1, y\text{Scl} = 1$

(*b*)

Figure 1.3.14

selected. In part (*b*) no gaps occur because there are pixels with *x*-coordinates $x = \pm 3$ in the window being used.   ◄

In addition to creating false gaps in continuous graphs, calculators can err in the opposite direction by placing *false line segments* in the gaps of discontinuous curves.

### Example 7

Figure 1.3.15*a* shows the graph of $y = 1/(x - 1)$ in the default window on the author's calculator. Although the graph appears to contain vertical line segments near $x = 1$, they should not be there. There is actually a gap in the curve at $x = 1$, since a division by zero occurs at that point (Figure 1.3.15*b*).   ◄

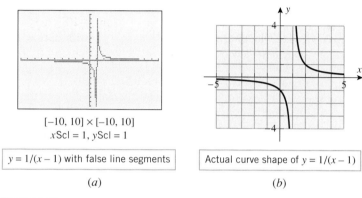

$[-10, 10] \times [-10, 10]$
$x$Scl $= 1$, $y$Scl $= 1$

| $y = 1/(x - 1)$ with false line segments | | Actual curve shape of $y = 1/(x - 1)$ |

(*a*)                                    (*b*)

Figure 1.3.15

Most graphing utilities use logarithms to evaluate functions with fractional exponents such as $f(x) = x^{2/3} = \sqrt[3]{x^2}$. However, because logarithms are only defined for positive numbers, many (but not all) graphing utilities will omit portions of the graphs of functions with fractional exponents. For example, the author's calculator graphs $y = x^{2/3}$ as in Figure 1.3.16*a*, whereas the actual graph is as in Figure 1.3.16*b*. (See Exercise 29 for a discussion of how to circumvent this problem.)

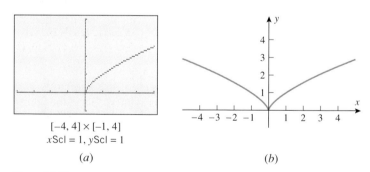

$[-4, 4] \times [-1, 4]$
$x$Scl $= 1$, $y$Scl $= 1$

(*a*)                                    (*b*)

Figure 1.3.16

FOR THE READER.   Determine whether your graphing utility produces the complete graph of $y = x^{2/3}$.

Although graphing utilities are powerful tools for generating graphs quickly, they can produce misleading graphs as a result of compression, sampling error, false gaps, and false line segments. In short, *graphing utilities can suggest graph shapes*, *but they cannot establish them with certainty*. Thus, the more you know about the functions you are graphing, the easier it will be to choose good viewing windows, and the better you will be able to judge the reasonableness of the results produced by your graphing utility.

........................................

**MORE INFORMATION ON GRAPHING AND CALCULATING UTILITIES**

The main source of information about your graphing utility is its own documentation, and from time to time we will suggest that you refer to that documentation to learn some particular technique.

## EXERCISE SET 1.3
........................................................................

1. Use a graphing utility to generate the graph of the function $f(x) = x^4 - x^2$ in the given viewing windows, and specify the window that you think gives the best view of the graph.
   (a) $-50 \le x \le 50$, $-50 \le y \le 50$
   (b) $-5 \le x \le 5$, $-5 \le y \le 5$
   (c) $-2 \le x \le 2$, $-2 \le y \le 2$
   (d) $-2 \le x \le 2$, $-1 \le y \le 1$
   (e) $-1.5 \le x \le 1.5$, $-0.5 \le y \le 0.5$

2. Use a graphing utility to generate the graph of the function $f(x) = x^5 - x^3$ in the given viewing windows, and specify the window that you think gives the best view of the graph.
   (a) $-50 \le x \le 50$, $-50 \le y \le 50$
   (b) $-5 \le x \le 5$, $-5 \le y \le 5$
   (c) $-2 \le x \le 2$, $-2 \le y \le 2$
   (d) $-2 \le x \le 2$, $-1 \le y \le 1$
   (e) $-1.5 \le x \le 1.5$, $-0.5 \le y \le 0.5$

3. Use a graphing utility to generate the graph of the function $f(x) = x^2 + 12$ in the given viewing windows, and specify the window that you think gives the best view of the graph.
   (a) $-1 \le x \le 1$, $13 \le y \le 15$
   (b) $-2 \le x \le 2$, $11 \le y \le 15$
   (c) $-4 \le x \le 4$, $10 \le y \le 28$
   (d) A window of your choice

4. Use a graphing utility to generate the graph of the function $f(x) = -12 - x^2$ in the given viewing windows, and specify the window that you think gives the best view of the graph.
   (a) $-1 \le x \le 1$, $-15 \le y \le -13$
   (b) $-2 \le x \le 2$, $-15 \le y \le -11$
   (c) $-4 \le x \le 4$, $-28 \le y \le -10$
   (d) A window of your choice

In Exercises 5 and 6, use the domain and range of $f$ to determine a viewing window that contains the entire graph, and generate the graph in that window.

5. $f(x) = \sqrt{16 - 2x^2}$  6. $f(x) = \sqrt{3 - 2x - x^2}$

7. Graph the function $f(x) = x^3 - 15x^2 - 3x + 45$ using the stated windows and tick spacing, and discuss the advantages and disadvantages of each window.
   (a) $-10 \le x \le 10$, $-10 \le y \le 10$
   with $x$Scl $= 1$ and $y$Scl $= 1$

   (b) $-20 \le x \le 20$, $-20 \le y \le 20$
   with $x$Scl $= 1$ and $y$Scl $= 1$
   (c) $-5 \le x \le 20$, $-500 \le y \le 50$
   with $x$Scl $= 5$ and $y$Scl $= 50$
   (d) $-2 \le x \le -1$, $-1 \le y \le 1$
   with $x$Scl $= 0.1$ and $y$Scl $= 0.1$
   (e) $9 \le x \le 11$, $-486 \le y \le -484$
   with $x$Scl $= 0.1$ and $y$Scl $= 0.1$

8. Graph the function $f(x) = -x^3 - 12x^2 + 4x + 48$ using the stated windows and tick spacing, and discuss the advantages and disadvantages of each window.
   (a) $-10 \le x \le 10$, $-10 \le y \le 10$
   with $x$Scl $= 1$ and $y$Scl $= 1$
   (b) $-20 \le x \le 20$, $-20 \le y \le 20$
   with $x$Scl $= 1$ and $y$Scl $= 1$
   (c) $-16 \le x \le 4$, $250 \le y \le 50$
   with $x$Scl $= 2$ and $y$Scl $= 25$
   (d) $-3 \le x \le -1$, $-1 \le y \le 1$
   with $x$Scl $= 0.1$ and $y$Scl $= 0.1$
   (e) $-9 \le x \le -7$, $-241 \le y \le -239$
   with $x$Scl $= 0.1$ and $y$Scl $= 0.1$

In Exercises 9–16, generate the graph of $f$ in a viewing window that you think is appropriate.

9. $f(x) = x^2 - 9x - 36$  10. $f(x) = \dfrac{x + 7}{x - 9}$

11. $f(x) = 2 \cos 80x$  12. $f(x) = 12 \sin(x/80)$

13. $f(x) = 300 - 10x^2 + 0.01x^3$

14. $f(x) = x(30 - 2x)(25 - 2x)$

15. $f(x) = x^2 + \dfrac{1}{x}$  16. $f(x) = \sqrt{11x - 18}$

In Exercises 17 and 18, generate the graph of $f$ and determine whether your graphs contain false line segments. Sketch the actual graph and see if you can make the false line segments disappear by changing the viewing window.

17. $f(x) = \dfrac{x}{x^2 - 1}$  18. $f(x) = \dfrac{x^2}{4 - x^2}$

**19.** The graph of the equation $x^2 + y^2 = 16$ is a circle of radius 4 centered at the origin.

(a) Find a function whose graph is the upper semicircle and graph it.

(b) Find a function whose graph is the lower semicircle and graph it.

(c) Graph the upper and lower semicircles together. If the combined graphs do not appear circular, see if you can adjust the viewing window to eliminate the aspect ratio distortion.

(d) Graph the portion of the circle in the first quadrant.

(e) Is there a function whose graph is the right half of the circle? Explain.

**20.** In each part, graph the equation by solving for $y$ in terms of $x$ and graphing the resulting functions together.

(a) $x^2/4 + y^2/9 = 1$     (b) $y^2 - x^2 = 1$

**21.** Read the documentation for your graphing utility to determine how to graph functions involving absolute values, and graph the given equation.

(a) $y = |x|$     (b) $y = |x - 1|$

(c) $y = |x| - 1$     (d) $y = |\sin x|$

(e) $y = \sin |x|$     (f) $y = |x| - |x + 1|$

**22.** Based on your knowledge of the absolute value function, sketch the graph of $f(x) = |x|/x$. Check your result using a graphing utility.

**23.** Make a conjecture about the relationship between the graph of $y = f(x)$ and the graph of $y = |f(x)|$; check your conjecture with some specific functions.

**24.** Make a conjecture about the relationship between the graph of $y = f(x)$ and the graph of $y = f(|x|)$; check your conjecture with some specific functions.

**25.** (a) Based on your knowledge of the absolute value function, sketch the graph of $y = |x - a|$, where $a$ is a constant. Check your result using a graphing utility and some specific values of $a$.

(b) Sketch the graph of $y = |x - 1| + |x - 2|$; check your result with a graphing utility.

**26.** How are the graphs of $y = |x|$ and $y = \sqrt{x^2}$ related? Check your answer with a graphing utility.

Most graphing utilities provide some way of graphing functions that are defined piecewise; read the documentation for your graphing utility to find out how to do this. However, if your goal is just to find the general shape of the graph, you can graph each portion of the function separately and combine the pieces with a hand-drawn sketch. Use this method in Exercises 27 and 28.

**27.** Draw the graph of

$$f(x) = \begin{cases} \sqrt[3]{x - 2}, & x \le 2 \\ x^3 - 2x - 4, & x > 2 \end{cases}$$

**28.** Draw the graph of

$$f(x) = \begin{cases} x^3 - x^2, & x \le 1 \\ \dfrac{1}{1 - x}, & 1 < x < 4 \\ x^2 \cos \sqrt{x}, & 4 \le x \end{cases}$$

We noted in the text that for functions involving fractional exponents (or radicals), graphing utilities sometimes omit portions of the graph. If $f(x) = x^{p/q}$, where $p/q$ is a positive fraction in *lowest terms*, then you can circumvent this problem as follows:

- If $p$ is even and $q$ is odd, then graph $g(x) = |x|^{p/q}$ instead of $f(x)$.
- If $p$ is odd and $q$ is odd, then graph $g(x) = (|x|/x)|x|^{p/q}$ instead of $f(x)$.

We will explain why this works in the exercises of the next section.

**29.** (a) Generate the graphs of $f(x) = x^{2/5}$ and $g(x) = |x|^{2/5}$, and determine whether your graphing utility missed part of the graph of $f$.

(b) Generate the graphs of the functions $f(x) = x^{1/5}$ and $g(x) = (|x|/x)|x|^{1/5}$, and then determine whether your graphing utility missed part of the graph of $f$.

(c) Generate a complete graph of the equation

$$y = (x - 1)^{4/5}$$

(d) Generate a complete graph of the equation

$$y = (x + 1)^{3/4}$$

**30.** The graphs of $y = (x^2 - 4)^{2/3}$ and $y = [(x^2 - 4)^2]^{1/3}$ should be the same. Does your graphing utility produce the same graph for both equations? If not, what do you think is happening?

**31.** In each part, graph the function for various values of $c$, and write a paragraph or two that describes how changes in $c$ affect the graph in each case.

(a) $y = cx^2$     (b) $y = x^2 + cx$

(c) $y = x^2 + x + c$

**32.** The graph of an equation of the form $y^2 = x(x - a)(x - b)$ (where $0 < a < b$) is called a ***bipartite cubic***. The accompanying figure shows a typical graph of this type.

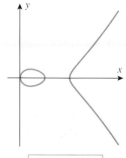

Bipartite cubic

Figure Ex-32

(a) Graph the bipartite cubic $y^2 = x(x-1)(x-2)$ by solving for $y$ in terms of $x$ and graphing the two resulting functions.

(b) Find the $x$-intercepts of the bipartite cubic

$$y^2 = x(x-a)(x-b)$$

and make a conjecture about how changes in the values of $a$ and $b$ would affect the graph. Test your conjecture by graphing the bipartite cubic for various values of $a$ and $b$.

**33.** Based on your knowledge of the graphs of $y = x$ and $y = \sin x$, make a sketch of the graph of $y = x\sin x$. Check your conclusion using a graphing utility.

**34.** What do you think the graph of $y = \sin(1/x)$ looks like? Test your conclusion using a graphing utility. [*Suggestion:* Examine the graph on a succession of smaller and smaller intervals centered at $x = 0$.]

## 1.4 NEW FUNCTIONS FROM OLD

*Just as numbers can be added, subtracted, multiplied, and divided to produce other numbers, so functions can be added, subtracted, multiplied, and divided to produce other functions. In this section we will discuss these operations and some others that have no analogs in ordinary arithmetic.*

**ARITHMETIC OPERATIONS ON FUNCTIONS**

Two functions, $f$ and $g$, can be added, subtracted, multiplied, and divided in a natural way to form new functions $f + g$, $f - g$, $fg$, and $f/g$. For example, $f + g$ is defined by the formula

$$(f + g)(x) = f(x) + g(x) \tag{1}$$

which states that for each input the value of $f + g$ is obtained by adding the values of $f$ and $g$. For example, if

$$f(x) = x \quad \text{and} \quad g(x) = x^2$$

then

$$(f + g)(x) = f(x) + g(x) = x + x^2$$

Equation (1) provides a formula for $f + g$ but does not say anything about the domain of $f + g$. However, for the right side of this equation to be defined, $x$ must lie in the domain of $f$ *and* in the domain of $g$, so we define the domain of $f + g$ to be the intersection of those two domains. More generally, we make the following definition:

**1.4.1 DEFINITION.** Given functions $f$ and $g$, we define

$$(f + g)(x) = f(x) + g(x)$$
$$(f - g)(x) = f(x) - g(x)$$
$$(fg)(x) = f(x)g(x)$$
$$(f/g)(x) = f(x)/g(x)$$

For the functions $f + g$, $f - g$, and $fg$ we define the domain to be the intersection of the domains of $f$ and $g$, and for the function $f/g$ we define the domain to be the intersection of the domains of $f$ and $g$ but with the points where $g(x) = 0$ excluded (to avoid division by zero).

REMARK. If $f$ is a constant function, say $f(x) = c$ for all $x$, then the product of $f$ and $g$ is $cg$, so multiplying a function by a constant is a special case of multiplying two functions.

## Example 1

Let

$$f(x) = 1 + \sqrt{x - 2} \quad \text{and} \quad g(x) = x - 3$$

Find $(f + g)(x)$, $(f - g)(x)$, $(fg)(x)$, $(f/g)(x)$, and $(7f)(x)$; state the domains of $f + g$, $f - g$, $fg$, $f/g$, and $7f$.

*Solution.* First, we will find formulas for the functions and then the domains. The formulas are

$$(f + g)(x) = f(x) + g(x) = (1 + \sqrt{x - 2}) + (x - 3) = x - 2 + \sqrt{x - 2} \qquad (2)$$

$$(f - g)(x) = f(x) - g(x) = (1 + \sqrt{x - 2}) - (x - 3) = 4 - x + \sqrt{x - 2} \qquad (3)$$

$$(fg)(x) = f(x)g(x) \quad = (1 + \sqrt{x - 2})(x - 3) \qquad (4)$$

$$(f/g)(x) = f(x)/g(x) \quad = \frac{1 + \sqrt{x - 2}}{x - 3} \qquad (5)$$

$$(7f)(x) = 7f(x) \quad = 7 + 7\sqrt{x - 2} \qquad (6)$$

In all five cases the natural domain determined by the formula is the same as the domain specified in Definition 1.4.1, so there is no need to state the domain explicitly in any of these cases. For example, the domain of $f$ is $[2, +\infty)$, the domain of $g$ is $(-\infty, +\infty)$, and the natural domain for $f(x) + g(x)$ determined by Formula (2) is $[2, +\infty)$, which is precisely the intersection of the domains of $f$ and $g$. ◄

REMARK. There are situations in which the natural domain associated with the formula resulting from an operation on two functions is not the correct domain for the new function. For example, if $f(x) = \sqrt{x}$ and $g(x) = \sqrt{x}$, then according to Definition 1.4.1 the domain of $fg$ should be $[0, +\infty) \cap [0, +\infty) = [0, +\infty)$. However, $(fg)(x) = \sqrt{x}\sqrt{x} = x$, which has a natural domain of $(-\infty, +\infty)$. Thus, to be precise in describing the formula for $fg$, we must write $(fg)(x) = x$, $x \geq 0$.

### STRETCHES AND COMPRESSIONS

Multiplying a function $f$ by a *nonnegative* constant $c$ has the geometric effect of stretching or compressing the graph of $f$ vertically. For example, examine the graphs of $y = f(x)$, $y = 2f(x)$, and $y = \frac{1}{2}f(x)$ shown in Figure 1.4.1a. Multiplying by 2 doubles each $y$-coordinate, thereby stretching the graph, and multiplying by $\frac{1}{2}$ cuts each $y$-coordinate in

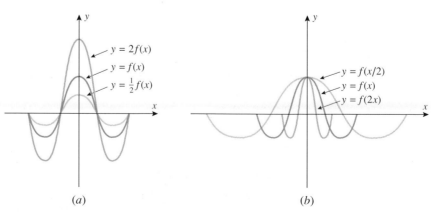

(a)                    (b)

Figure 1.4.1

half, thereby compressing the graph. In general, if $c > 0$, then the graph of $y = cf(x)$ can be obtained from the graph of $y = f(x)$ by compressing the graph of $y = f(x)$ vertically by a factor of $1/c$ if $0 < c < 1$, or stretching it by a factor of $c$ if $c > 1$.

Analogously, multiplying the independent variable of a function $f$ by a **nonnegative** constant $c$ has the geometric effect of stretching or compressing the graph of $f$ horizontally. For example, examine the graphs of $y = f(x)$, $y = f(2x)$, and $y = f(x/2)$ shown in Figure 1.4.1$b$. Multiplying $x$ by 2 compresses the graph by a factor of 2 and multiplying $x$ by $\frac{1}{2}$ stretches the graph by a factor of 2. [This is a little confusing, but think of it this way: The value of $2x$ changes twice as fast as the value of $x$, so a point moving along the $x$-axis will only have to move half as far from the origin for $y = f(2x)$ to have the same value as $y = f(x)$.] In general, if $c > 0$, then the graph of $y = f(cx)$ can be obtained from the graph of $y = f(x)$ by stretching the graph of $y = f(x)$ horizontally by a factor of $c$ if $0 < c < 1$, or compressing it by a factor of $c$ if $c > 1$.

**SUMS OF FUNCTIONS**

Adding two functions can be accomplished geometrically by adding the corresponding $y$-coordinates of their graphs. For example, Figure 1.4.2 shows line graphs of yearly new car sales $N(t)$ and used car sales $U(t)$ in the United States between 1985 and 1995. The sum of these functions, $T(t) = N(t) + U(t)$, represents the yearly total car sales for that period. As illustrated in the figure, the graph of $T(t)$ can be obtained by adding the values of $N(t)$ and $U(t)$ together at each time $t$ and plotting the resulting value.

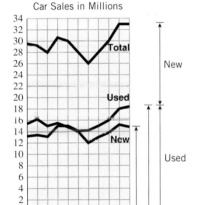

Car Sales in Millions

Source: NADA.

Figure 1.4.2

### Example 2

Referring to Figure 1.2.2 for the graphs of $y = \sqrt{x}$ and $y = 1/x$, make a sketch that shows the general shape of the graph of $y = \sqrt{x} + 1/x$ for $x \geq 0$.

*Solution.* To add the corresponding $y$-values of $y = \sqrt{x}$ and $y = 1/x$ graphically, just imagine them to be "stacked" on top of one another. This yields the sketch in Figure 1.4.3. ◄

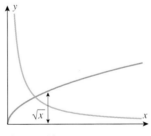
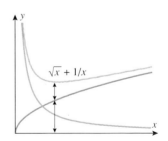

Figure 1.4.3

**COMPOSITION OF FUNCTIONS**

We now consider an operation on functions, called *composition*, which has no direct analog in ordinary arithmetic. Informally stated, the operation of composition is performed by substituting some function for the independent variable of another function. For example, suppose that

$$f(x) = x^2 \quad \text{and} \quad g(x) = x + 1$$

If we substitute $g(x)$ for $x$ in the formula for $f$, we obtain a new function

$$f(g(x)) = (g(x))^2 = (x + 1)^2$$

which we denote by $f \circ g$. Thus,

$$(f \circ g)(x) = f(g(x)) = (g(x))^2 = (x + 1)^2$$

In general, we make the following definition.

---

**1.4.2**   DEFINITION.   Given functions $f$ and $g$, the **composition** of $f$ with $g$, denoted by $f \circ g$, is the function defined by

$$(f \circ g)(x) = f(g(x))$$

The domain of $f \circ g$ is defined to consist of all $x$ in the domain of $g$ for which $g(x)$ is in the domain of $f$.

---

REMARK.   Although the domain of $f \circ g$ may seem complicated at first glance, it makes sense intuitively: To compute $f(g(x))$ one needs $x$ in the domain of $g$ to compute $g(x)$, then one needs $g(x)$ in the domain of $f$ to compute $f(g(x))$.

**COMPOSITIONS VIEWED AS COMPUTER PROGRAMS**

In Section 1.1 we noted that a function $f$ can be viewed as a computer program that takes an input $x$, operates on it, and produces an output $f(x)$. From this viewpoint composition can be viewed as two programs, $g$ and $f$, operating in succession: An input $x$ is fed first to a program $g$, which produces the output $g(x)$; then this output is fed as input to a program $f$, which produces the output $f(g(x))$ (Figure 1.4.4). However, rather than have two separate programs operating in succession, we could create a **single** program that takes the input $x$ and directly produces the output $f(g(x))$. This program is the composition $f \circ g$ since $(f \circ g)(x) = f(g(x))$.

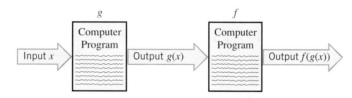

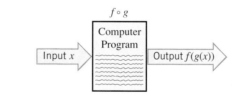

Figure 1.4.4

## Example 3

Let $f(x) = x^2 + 3$ and $g(x) = \sqrt{x}$. Find

(a)  $(f \circ g)(x)$       (b)  $(g \circ f)(x)$

*Solution (a).* The formula for $f(g(x))$ is

$$f(g(x)) = [g(x)]^2 + 3 = (\sqrt{x})^2 + 3 = x + 3$$

Since the domain of $g$ is $[0, +\infty)$ and the domain of $f$ is $(-\infty, +\infty)$, the domain of $f \circ g$ consists of all $x$ in $[0, +\infty)$ such that $g(x) = \sqrt{x}$ lies in $(-\infty, +\infty)$; thus, the domain of $f \circ g$ is $[0, +\infty)$. Therefore,

$$(f \circ g)(x) = x + 3, \quad x \geq 0$$

*Solution (b).* The formula for $g(f(x))$ is

$$g(f(x)) = \sqrt{f(x)} = \sqrt{x^2 + 3}$$

Since the domain of $f$ is $(-\infty, +\infty)$ and the domain of $g$ is $[0, +\infty)$, the domain of $g \circ f$ consists of all $x$ in $(-\infty, +\infty)$ such that $f(x) = x^2 + 3$ lies in $[0, +\infty)$. Thus, the domain of

$g \circ f$ is $(-\infty, +\infty)$. Therefore,

$$(g \circ f)(x) = \sqrt{x^2 + 3}$$

There is no need to indicate that the domain is $(-\infty, +\infty)$, since this is the natural domain of $\sqrt{x^2 + 3}$.  ◄

REMARK.    Note that the functions $f \circ g$ and $g \circ f$ in the preceding example are not the same. Thus, the order in which functions are composed can (and usually will) make a difference in the end result.

Compositions can also be defined for three or more functions; for example, $(f \circ g \circ h)(x)$ is computed as

$$(f \circ g \circ h)(x) = f(g(h(x)))$$

In other words, first find $h(x)$, then find $g(h(x))$, and then find $f(g(h(x)))$.

### Example 4

Find $(f \circ g \circ h)(x)$ if

$$f(x) = \sqrt{x}, \quad g(x) = 1/x, \quad h(x) = x^3$$

*Solution.*

$$(f \circ g \circ h)(x) = f(g(h(x))) = f(g(x^3)) = f(1/x^3) = \sqrt{1/x^3} = 1/x^{3/2}$$  ◄

**EXPRESSING A FUNCTION AS A COMPOSITION**

Many problems in mathematics are attacked by "decomposing" functions into compositions of simpler functions. For example, consider the function $h$ given by

$$h(x) = (x + 1)^2$$

To evaluate $h(x)$ for a given value of $x$, we would first compute $x + 1$ and then square the result. These two operations are performed by the functions

$$g(x) = x + 1 \quad \text{and} \quad f(x) = x^2$$

We can express $h$ in terms of $f$ and $g$ by writing

$$h(x) = (x + 1)^2 = [g(x)]^2 = f(g(x))$$

so we have succeeded in expressing $h$ as the composition $h = f \circ g$.

The thought process in this example suggests a general procedure for decomposing a function $h$ into a composition $h = f \circ g$:

- Think about how you would evaluate $h(x)$ for a specific value of $x$, trying to break the evaluation into two steps performed in succession.

- The first operation in the evaluation will determine a function $g$ and the second a function $f$.

- The formula for $h$ can then be written as $h(x) = f(g(x))$.

For descriptive purposes, we will refer to $g$ as the "inside function" and $f$ as the "outside function" in the expression $f(g(x))$. The inside function performs the first operation and the outside function performs the second.

### Example 5

Express $h(x) = (x - 4)^5$ as a composition of two functions.

*Solution.*   To evaluate $h(x)$ for a given value of $x$ we would first compute $x - 4$ and then raise the result to the fifth power. Therefore, the inside function (first operation) is

$$g(x) = x - 4$$

and the outside function (second operation) is

$$f(x) = x^5$$

so $h(x) = f(g(x))$. As a check,

$$f(g(x)) = [g(x)]^5 = (x - 4)^5 = h(x)$$   ◀

### Example 6

Express $\sin(x^3)$ as a composition of two functions.

*Solution.*   To evaluate $\sin(x^3)$, we would first compute $x^3$ and then take the sine, so $g(x) = x^3$ is the inside function and $f(x) = \sin x$ the outside function. Therefore,

$$\sin(x^3) = f(g(x)) \qquad \boxed{g(x) = x^3 \text{ and } f(x) = \sin x}$$   ◀

### Example 7

Table 1.4.1 gives some more examples of decomposing functions into compositions.

**Table 1.4.1**

| FUNCTION | $g(x)$ INSIDE | $f(x)$ OUTSIDE | COMPOSITION |
|---|---|---|---|
| $(x^2 + 1)^{10}$ | $x^2 + 1$ | $x^{10}$ | $(x^2 + 1)^{10} = f(g(x))$ |
| $\sin^3 x$ | $\sin x$ | $x^3$ | $\sin^3 x = f(g(x))$ |
| $\tan(x^5)$ | $x^5$ | $\tan x$ | $\tan(x^5) = f(g(x))$ |
| $\sqrt{4 - 3x}$ | $4 - 3x$ | $\sqrt{x}$ | $\sqrt{4 - 3x} = f(g(x))$ |
| $8 + \sqrt{x}$ | $\sqrt{x}$ | $8 + x$ | $8 + \sqrt{x} = f(g(x))$ |
| $\dfrac{1}{x + 1}$ | $x + 1$ | $\dfrac{1}{x}$ | $\dfrac{1}{x + 1} = f(g(x))$ |

◀

REMARK.   It should be noted that there is always more than one way to express a function as a composition. For example, here are two ways to express $(x^2 + 1)^{10}$ as a composition that differ from that in Table 1.4.1:

$$(x^2 + 1)^{10} = \left[(x^2 + 1)^2\right]^5 = f(g(x)) \qquad \boxed{g(x) = (x^2 + 1)^2 \text{ and } f(x) = x^5}$$

$$(x^2 + 1)^{10} = \left[(x^2 + 1)^3\right]^{10/3} = f(g(x)) \qquad \boxed{g(x) = (x^2 + 1)^3 \text{ and } f(x) = x^{10/3}}$$

**SYMMETRY**

Figure 1.4.5 shows the graphs of three curves that have certain obvious symmetries. The graph in part (*a*) is **symmetric about the x-axis** in the sense that for each point $(x, y)$ on the graph the point $(x, -y)$ is also on the graph; the graph in part (*b*) is **symmetric about the y-axis** in the sense that for each point $(x, y)$ on the graph the point $(-x, y)$ is also on the graph; and the graph in part (*c*) is **symmetric about the origin** in the sense that for each point $(x, y)$ on the graph the point $(-x, -y)$ is also on the graph. Geometrically, symmetry about the origin occurs if rotating the graph $180°$ about the origin leaves the graph unchanged.

Symmetries can often be detected from the equation of a curve. For example, the graph of

$$y = x^3 \tag{7}$$

must be symmetric about the origin because for any point $(x, y)$ whose coordinates satisfy (7), the coordinates of the point $(-x, -y)$ also satisfy (7), since substituting these

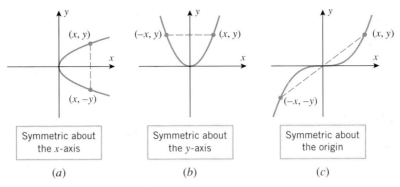

Figure 1.4.5

coordinates in (7) yields

$$-y = (-x)^3$$

which simplifies to (7). This suggests the following symmetry tests (Figure 1.4.6).

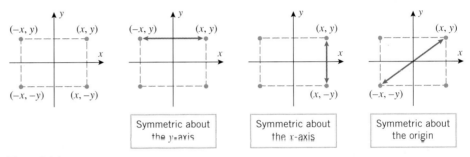

Figure 1.4.6

---

**1.4.3** THEOREM (*Symmetry Tests*).

(a) *A plane curve is symmetric about the y-axis if and only if replacing x by −x in its equation produces an equivalent equation.*

(b) *A plane curve is symmetric about the x-axis if and only if replacing y by −y in its equation produces an equivalent equation.*

(c) *A plane curve is symmetric about the origin if and only if replacing both x by −x and y by −y in its equation produces an equivalent equation.*

---

**EVEN AND ODD FUNCTIONS**

For the graph of a function $f$ to be symmetric about the $y$-axis, the equations $y = f(x)$ and $y = f(-x)$ must be equivalent; for this to happen we must have

$$f(x) = f(-x)$$

A function with this property is called an ***even function***. Some examples are $x^2, x^4, x^6$, and $\cos x$. Similarly, for the graph of a function $f$ to be symmetric about the origin, the equations $y = f(x)$ and $-y = f(-x)$ must be equivalent; for this to happen we must have

$$f(x) = -f(-x)$$

A function with this property is called an ***odd function***. Some examples are $x, x^3, x^5$, and $\sin x$.

FOR THE READER.   Explain why the graph of a nonzero function cannot by symmetric about the $x$-axis.

**TRANSLATIONS**

Once you know the graph of an equation $y = f(x)$, there are some techniques that can be used to help visualize the graphs of the equations

$$y = f(x) + c, \quad y = f(x) - c, \quad y = f(x + c), \quad y = f(x - c)$$

where $c$ is any positive constant.

If a positive constant is added to or subtracted from $f(x)$, the geometric effect is to translate the graph of $y = f(x)$ parallel to the $y$-axis; addition translates the graph in the positive direction and subtraction translates it in the negative direction. This is illustrated in Table 1.4.2. Similarly, if a positive constant is added to or subtracted from the independent variable $x$, the geometric effect is to translate the graph of the function parallel to the $x$-axis; subtraction translates the graph in the positive direction, and addition translates it in the negative direction. This is also illustrated in Table 1.4.2.

**Table 1.4.2**

| OPERATION ON $y = f(x)$ | Add a positive constant $c$ to $f(x)$ | Subtract a positive constant $c$ from $f(x)$ | Add a positive constant $c$ to $x$ | Subtract a positive constant $c$ from $x$ |
|---|---|---|---|---|
| NEW EQUATION | $y = f(x) + c$ | $y = f(x) - c$ | $y = f(x + c)$ | $y = f(x - c)$ |
| GEOMETRIC EFFECT | Translates the graph of $y = f(x)$ up $c$ units | Translates the graph of $y = f(x)$ down $c$ units | Translates the graph of $y = f(x)$ left $c$ units | Translates the graph of $y = f(x)$ right $c$ units |
| EXAMPLE | $y = x^2 + 2$ $y = x^2$ | $y = x^2$ $y = x^2 - 2$ | $y = (x + 2)^2$ $y = x^2$ | $y = (x - 2)^2$ $y = x^2$ |

Before proceeding to the following examples, it will be helpful to review the graphs in Figure 1.2.1.

### Example 8

Sketch the graph of

    (a) $y = \sqrt{x - 3}$    (b) $y = \sqrt{x + 3}$

*Solution.* The graph of the equation $y = \sqrt{x - 3}$ can be obtained by translating the graph of $y = \sqrt{x}$ right 3 units, and the graph of $y = \sqrt{x + 3}$ by translating the graph of $y = \sqrt{x}$ left 3 units (Figure 1.4.7).   ◄

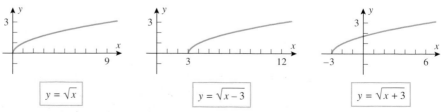

$y = \sqrt{x}$        $y = \sqrt{x - 3}$        $y = \sqrt{x + 3}$

Figure 1.4.7

## Example 9

Sketch the graph of $y = |x - 3| + 2$.

*Solution.*  The graph can be obtained by two translations: first translate the graph of $y = |x|$ right 3 units to obtain the graph of $y = |x - 3|$, then translate this graph up 2 units to obtain the graph of $y = |x - 3| + 2$ (Figure 1.4.8).  ◄

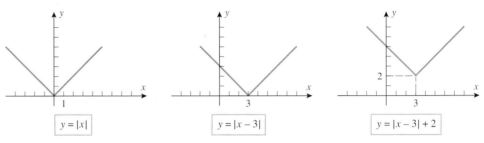

$y = |x|$    $y = |x - 3|$    $y = |x - 3| + 2$

Figure 1.4.8

REMARK.  The graph in the preceding example could also have been obtained by performing the translations in the opposite order: first translating the graph of $y = |x|$ up 2 units to obtain the graph of $y = |x| + 2$, then translating this graph right 3 units to obtain the graph of $y = |x - 3| + 2$.

## Example 10

Sketch the graph of $y = x^2 - 4x + 5$.

*Solution.*  Completing the square on the first two terms yields

$$y = (x^2 - 4x + 4) - 4 + 5 = (x - 2)^2 + 1$$

(see Appendix D for a review of this technique). In this form we see that the graph can be obtained by translating the graph of $y = x^2$ right 2 units because of the $x - 2$, and up 1 unit because of the $+1$ (Figure 1.4.9).  ◄

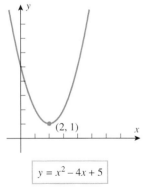

$(2, 1)$

$y = x^2 - 4x + 5$

Figure 1.4.9

## Example 11

By completing the square, an equation of the form $y = ax^2 + bx + c$ with $a \neq 0$ can be expressed as

$$y = a(x - h)^2 + k \tag{8}$$

Sketch the graph of this equation.

*Solution.*  We can build up Equation (8) in three steps from the equation $y = x^2$. First, we can multiply by $a$ to obtain $y = ax^2$. If $a > 0$, this operation has the geometric effect of stretching or compressing the graph of $y = x^2$; and if $a < 0$, it has the geometric effect of reflecting the graph about the $x$-axis, in addition to stretching or compressing it. Since stretching or compressing does not alter the general parabolic shape of the original curve, the graph of $y = ax^2$ looks roughly like one of those in Figure 1.4.10a. Next, we can subtract $h$ from $x$ to obtain the equation $y = a(x - h)^2$, and then we can add $k$ to obtain $y = a(x - h)^2 + k$. Subtracting $h$ causes a horizontal translation (right or left, depending on the sign of $h$), and adding $k$ causes a vertical translation (up or down, depending on the sign of $k$). Thus, the graph of (8) looks roughly like one of those in Figure 1.4.10b, which are shown with $h > 0$ and $k > 0$ for simplicity.  ◄

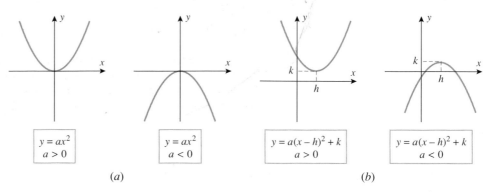

Figure 1.4.10

**REFLECTIONS**

The graph of $y = f(-x)$ is the reflection of the graph of $y = f(x)$ about the $y$-axis, and the graph of $y = -f(x)$ [or equivalently, $-y = f(x)$] is the reflection of the graph of $y = f(x)$ about the $x$-axis. Thus, if you know what the graph of $y = f(x)$ looks like, you can obtain the graphs of $y = f(-x)$ and $y = -f(x)$ by making appropriate reflections. This is illustrated in Table 1.4.3.

**Table 1.4.3**

| OPERATION ON $y = f(x)$ | Replace $x$ by $-x$ | Multiply $f(x)$ by $-1$ |
|---|---|---|
| **NEW EQUATION** | $y = f(-x)$ | $y = -f(x)$ |
| **GEOMETRIC EFFECT** | Reflects the graph of $y = f(x)$ about the $y$-axis | Reflects the graph of $y = f(x)$ about the $x$-axis |
| **EXAMPLE** | $y = \sqrt{-x}$    $y = \sqrt{x}$ | $y = \sqrt{x}$    $y = -\sqrt{x}$ |

### Example 12

Sketch the graph of $y = \sqrt[3]{2 - x}$.

*Solution.* The graph can be obtained by a reflection and a translation: first reflect the graph of $y = \sqrt[3]{x}$ about the $y$-axis to obtain the graph of $y = \sqrt[3]{-x}$, then translate this graph right 2 units to obtain the graph of the equation $y = \sqrt[3]{-(x - 2)} = \sqrt[3]{2 - x}$ (Figure 1.4.11). ◄

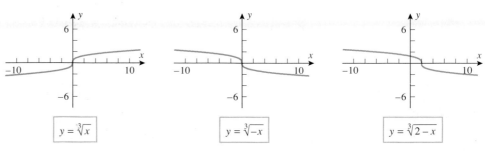

$y = \sqrt[3]{x}$      $y = \sqrt[3]{-x}$      $y = \sqrt[3]{2 - x}$

Figure 1.4.11

**Example 13**

Sketch the graph of $y = 4 - |x - 2|$.

*Solution.* The graph can be obtained by a reflection and two translations: first translate the graph of $y = |x|$ right 2 units to obtain the graph of $y = |x - 2|$; then reflect this graph about the $x$-axis to obtain the graph of $y = -|x - 2|$; and then translate this graph up 4 units to obtain the graph of the equation $y = -|x - 2| + 4 = 4 - |x - 2|$ (Figure 1.4.12). ◀

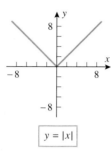

$$y = |x|$$

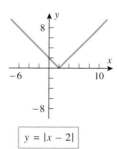

$$y = |x - 2|$$

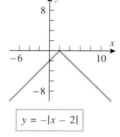

$$y = -|x - 2|$$

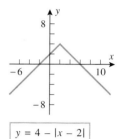

$$y = 4 - |x - 2|$$

Figure 1.4.12

---

## EXERCISE SET 1.4  ⊠ Graphing Calculator  □c CAS

**1.** The graph of a function $f$ is shown in the accompanying figure. Sketch the graphs of the following equations.
  (a) $y = f(x) - 1$      (b) $y = f(x - 1)$
  (c) $y = \frac{1}{2}f(x)$        (d) $y = f\left(-\frac{1}{2}x\right)$

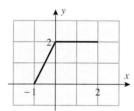

Figure Ex-1

**2.** Use the graph in Exercise 1.4.1 to sketch the graphs of the following equations.
  (a) $y = -f(-x)$      (b) $y = f(2 - x)$
  (c) $y = 1 - f(2 - x)$    (d) $y = \frac{1}{2}f(2x)$

**3.** The graph of a function $f$ is shown in the accompanying figure. Sketch the graphs of the following equations.
  (a) $y = f(x + 1)$      (b) $y = f(2x)$
  (c) $y = |f(x)|$         (d) $y = 1 - |f(x)|$

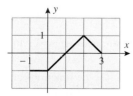

Figure Ex-3

**4.** Use the graph in Exercise 1.4.3 to sketch the graph of the equation $y = f(|x|)$.

In Exercises 5–12, sketch the graph of the equation by translating, reflecting, compressing, and stretching the graph of $y = x^2$ appropriately, and then use a graphing utility to confirm that your sketch is correct.

⊠ **5.** $y = 1 + (x - 2)^2$    ⊠ **6.** $y = 2 - (x + 1)^2$
⊠ **7.** $y = -2(x + 1)^2 - 3$    ⊠ **8.** $y = \frac{1}{2}(x - 3)^2 + 2$
⊠ **9.** $y = x^2 + 6x$        ⊠ **10.** $y = x^2 + 6x - 10$
⊠ **11.** $y = 1 + 2x - x^2$    ⊠ **12.** $y = \frac{1}{2}(x^2 - 2x + 3)$

In Exercises 13–16, sketch the graph of the equation by translating, reflecting, compressing, and stretching the graph of $y = \sqrt{x}$ appropriately, and then use a graphing utility to confirm that your sketch is correct.

⊠ **13.** $y = 3 - \sqrt{x + 1}$    ⊠ **14.** $y = 1 + \sqrt{x - 4}$
⊠ **15.** $y = \frac{1}{2}\sqrt{x} + 1$      ⊠ **16.** $y = -\sqrt{3x}$

In Exercises 17–20, sketch the graph of the equation by translating, reflecting, compressing, and stretching the graph of $y = 1/x$ appropriately, and then use a graphing utility to confirm that your sketch is correct.

⊠ **17.** $y = \dfrac{1}{x - 3}$       ⊠ **18.** $y = \dfrac{1}{1 - x}$

**19.** $y = 2 - \dfrac{1}{x+1}$          **20.** $y = \dfrac{x-1}{x}$

In Exercises 21–24, sketch the graph of the equation by translating, reflecting, compressing, and stretching the graph of $y = |x|$ appropriately, and then use a graphing utility to confirm that your sketch is correct.

**21.** $y = |x+2| - 2$          **22.** $y = 1 - |x-3|$

**23.** $y = |2x-1| + 1$          **24.** $y = \sqrt{x^2 - 4x + 4}$

In Exercises 25–28, sketch the graph of the equation by translating, reflecting, compressing, and stretching the graph of $y = \sqrt[3]{x}$ appropriately, and then use a graphing utility to confirm that your sketch is correct.

**25.** $y = 1 - 2\sqrt[3]{x}$          **26.** $y = \sqrt[3]{x-2} - 3$

**27.** $y = 2 + \sqrt[3]{x+1}$          **28.** $y + \sqrt[3]{x-2} = 0$

**29. (a)** Sketch the graph of $y = x + |x|$ by adding the corresponding $y$-coordinates on the graphs of $y = x$ and $y = |x|$.

   **(b)** Express the equation $y = x + |x|$ in piecewise form with no absolute values, and confirm that the graph you obtained in part (a) is consistent with this equation.

**30.** Sketch the graph of $y = x + (1/x)$ by adding corresponding $y$-coordinates on the graphs of $y = x$ and $y = 1/x$. Use a graphing utility to confirm that your sketch is correct.

In Exercises 31–34, find formulas for $f + g$, $f - g$, $fg$, and $f/g$, and state the domains of the functions.

**31.** $f(x) = 2x$, $g(x) = x^2 + 1$

**32.** $f(x) = 3x - 2$, $g(x) = |x|$

**33.** $f(x) = 2\sqrt{x-1}$, $g(x) = \sqrt{x-1}$

**34.** $f(x) = \dfrac{x}{1+x^2}$, $g(x) = \dfrac{1}{x}$

**35.** Let $f(x) = \sqrt{x}$ and $g(x) = x^3 + 1$. Find
   **(a)** $f(g(2))$          **(b)** $g(f(4))$
   **(c)** $f(f(16))$          **(d)** $g(g(0))$.

**36.** Let $g(x) = \pi - x^2$ and $h(x) = \cos x$. Find
   **(a)** $g(h(0))$          **(b)** $h(g(\sqrt{\pi/2}\,))$
   **(c)** $g(g(1))$          **(d)** $h(h(\pi/2))$.

**37.** Let $f(x) = x^2 + 1$. Find
   **(a)** $f(t^2)$          **(b)** $f(t+2)$          **(c)** $f(x+2)$
   **(d)** $f\left(\dfrac{1}{x}\right)$          **(e)** $f(x+h)$          **(f)** $f(-x)$
   **(g)** $f(\sqrt{x}\,)$          **(h)** $f(3x)$.

**38.** Let $g(x) = \sqrt{x}$. Find
   **(a)** $g(5s+2)$          **(b)** $g(\sqrt{x}+2)$          **(c)** $3g(5x)$
   **(d)** $\dfrac{1}{g(x)}$          **(e)** $g(g(x))$          **(f)** $(g(x))^2 - g(x^2)$
   **(g)** $g(1/\sqrt{x}\,)$          **(h)** $g((x-1)^2)$.

In Exercises 39–44, find formulas for $f \circ g$ and $g \circ f$, and state the domains of the functions.

**39.** $f(x) = 2x + 1$, $g(x) = x^2 - x$

**40.** $f(x) = 2 - x^2$, $g(x) = x^3$

**41.** $f(x) = x^2$, $g(x) = \sqrt{1-x}$

**42.** $f(x) = \sqrt{x-3}$, $g(x) = \sqrt{x^2 + 3}$

**43.** $f(x) = \dfrac{1+x}{1-x}$, $g(x) = \dfrac{x}{1-x}$

**44.** $f(x) = \dfrac{x}{1+x^2}$, $g(x) = \dfrac{1}{x}$

In Exercises 45 and 46, find a formula for $f \circ g \circ h$.

**45.** $f(x) = x^2 + 1$, $g(x) = \dfrac{1}{x}$, $h(x) = x^3$

**46.** $f(x) = \dfrac{1}{1+x}$, $g(x) = \sqrt[3]{x}$, $h(x) = \dfrac{1}{x^3}$

In Exercises 47–50, express $f$ as a composition of two functions; that is, find $g$ and $h$ such that $f = g \circ h$. [*Note:* Each exercise has more than one solution.]

**47. (a)** $f(x) = \sqrt{x+2}$          **(b)** $f(x) = |x^2 - 3x + 5|$

**48. (a)** $f(x) = x^2 + 1$          **(b)** $f(x) = \dfrac{1}{x-3}$

**49. (a)** $f(x) = \sin^2 x$          **(b)** $f(x) = \dfrac{3}{5 + \cos x}$

**50. (a)** $f(x) = 3\sin(x^2)$          **(b)** $f(x) = 3\sin^2 x + 4\sin x$

In Exercises 51 and 52, express $F$ as a composition of three functions; that is, find $f$, $g$, and $h$ such that $F = f \circ g \circ h$. [*Note:* Each exercise has more than one solution.]

**51. (a)** $F(x) = \left(1 + \sin(x^2)\right)^3$          **(b)** $F(x) = \sqrt{1 - \sqrt[3]{x}}$

**52. (a)** $F(x) = \dfrac{1}{1-x^2}$          **(b)** $F(x) = |5 + 2x|$

**53.** Use the table in the accompanying figure to make a scatter plot of $y = f(g(x))$.

| $x$ | $-3$ | $-2$ | $-1$ | $0$ | $1$ | $2$ | $3$ |
|------|------|------|------|------|------|------|------|
| $f(x)$ | $-4$ | $-3$ | $-2$ | $-1$ | $0$ | $1$ | $2$ |
| $g(x)$ | $-1$ | $0$ | $1$ | $2$ | $3$ | $-2$ | $-3$ |

Figure Ex-53

**54.** Find the domain of $g \circ f$ for the functions $f$ and $g$ in Exercise 53.

**55.** Sketch the graph of $y = f(g(x))$ for the functions graphed in the accompanying figure.

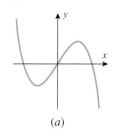

Figure Ex-55

**56.** Sketch the graph of $y = g(f(x))$ for the functions graphed in Exercise 55.

**57.** Use the graphs of $f$ and $g$ in Exercise 55 to estimate the solutions of the equations $f(g(x)) = 0$ and $g(f(x)) = 0$.

**58.** Use the table in Exercise 53 to solve the equations $f(g(x)) = 0$ and $g(f(x)) = 0$.

In Exercises 59–62, find
$$\frac{f(x + h) - f(x)}{h}$$
and simplify as much as possible.

**59.** $f(x) = 3x^2 - 5$      **60.** $f(x) = x^2 + 6x$

**61.** $f(x) = 1/x$      **62.** $f(x) = 1/x^2$

**63.** In each part of the accompanying figure determine whether the graph is symmetric about the $x$-axis, the $y$-axis, the origin, or none of the preceding.

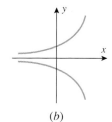

    (a)                (b)

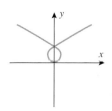

    (c)                (d)

Figure Ex-63

**64.** The accompanying figure shows a portion of a graph. Complete the graph so that the entire graph is symmetric about
(a) the $x$-axis      (b) the $y$-axis      (c) the origin.

Figure Ex-64

**65.** Complete the table in the accompanying figure so that the graph of $y = f(x)$ (which is a scatter plot) is symmetric about
(a) the $y$-axis                (b) the origin.

| $x$ | −3 | −2 | −1 | 0 | 1 | 2 | 3 |
|-----|----|----|----|----|----|----|----|
| $f(x)$ | 1 | | −1 | 0 | | −5 | |

Figure Ex-65

**66.** The accompanying figure shows a portion of the graph of a function $f$. Complete the graph assuming that
(a) $f$ is an even function      (b) $f$ is an odd function.

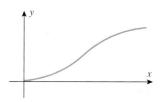

Figure Ex-66

**67.** Classify the functions graphed in the accompanying figure as even, odd, or neither.

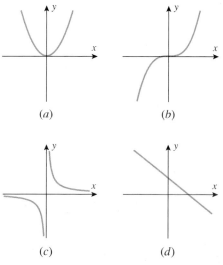

Figure Ex-67

**68.** Classify the functions whose values are given in the following table as even, odd, or neither.

| $x$ | $-3$ | $-2$ | $-1$ | $0$ | $1$ | $2$ | $3$ |
|-----|------|------|------|-----|-----|-----|-----|
| $f(x)$ | 5 | 3 | 2 | 3 | 1 | $-3$ | 5 |
| $g(x)$ | 4 | 1 | $-2$ | 0 | 2 | $-1$ | $-4$ |
| $h(x)$ | 2 | $-5$ | 8 | $-2$ | 8 | $-5$ | 2 |

**69.** In each part, classify the function as even, odd, or neither.
(a) $f(x) = x^2$
(b) $f(x) = x^3$
(c) $f(x) = |x|$
(d) $f(x) = x + 1$
(e) $f(x) = \dfrac{x^5 - x}{1 + x^2}$
(f) $f(x) = 2$

In Exercises 70 and 71, use Theorem 1.4.3 to determine whether the graph has symmetries about the $x$-axis, the $y$-axis, or the origin.

**70.** (a) $x = 5y^2 + 9$
(b) $x^2 - 2y^2 = 3$
(c) $xy = 5$

**71.** (a) $x^4 = 2y^3 + y$
(b) $y = \dfrac{x}{3 + x^2}$
(c) $y^2 = |x| - 5$

In Exercises 72 and 73: (i) Use a graphing utility to graph the equation in the first quadrant. [*Note:* To do this you will have to solve the equation for $y$ in terms of $x$.] (ii) Use symmetry to make a hand-drawn sketch of the entire graph. (iii) Confirm your work by generating the graph of the equation in the remaining three quadrants.

**72.** $9x^2 + 4y^2 = 36$
**73.** $4x^2 + 16y^2 = 16$

**74.** The graph of the equation $x^{2/3} + y^{2/3} = 1$, which is shown in the accompanying figure, is called a ***four-cusped hypocycloid***.
(a) Use Theorem 1.4.3 to confirm that this graph is symmetric about the $x$-axis, the $y$-axis, and the origin.
(b) Find a function $f$ whose graph in the first quadrant coincides with the four-cusped hypocycloid, and use a graphing utility to confirm your work.
(c) Repeat part (b) for the remaining three quadrants.

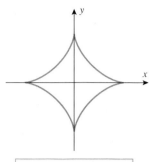

Four-cusped hypocycloid

Figure Ex-74

**75.** The equation $y = |f(x)|$ can be written as
$$y = \begin{cases} f(x), & f(x) \geq 0 \\ -f(x), & f(x) < 0 \end{cases}$$
which shows that the graph of $y = |f(x)|$ can be obtained from the graph of $y = f(x)$ by retaining the portion that lies on or above the $x$-axis and reflecting about the $x$-axis the portion that lies below the $x$-axis. Use this method to obtain the graph of $y = |2x - 3|$ from the graph of $y = 2x - 3$.

In Exercises 76 and 77, use the method described in Exercise 75.

**76.** Sketch the graph of $y = |1 - x^2|$.

**77.** Sketch the graph of
(a) $f(x) = |\cos x|$
(b) $f(x) = \cos x + |\cos x|$.

**78.** The ***greatest integer function***, $[x]$, is defined to be the greatest integer that is less than or equal to $x$. For example, $[2.7] = 2$, $[-2.3] = -3$, and $[4] = 4$. Sketch the graph of
(a) $f(x) = [x]$
(b) $f(x) = [x^2]$
(c) $f(x) = [x]^2$
(d) $f(x) = [\sin x]$.

**79.** Is it ever true that $f \circ g = g \circ f$ if $f$ and $g$ are nonconstant functions? If not, prove it; if so, give some examples for which it is true.

**80.** In the discussion preceding Exercise 29 of Section 1.3, we gave a procedure for generating a complete graph of $f(x) = x^{p/q}$ in which we suggested graphing the function $g(x) = |x|^{p/q}$ instead of $f(x)$ when $p$ is even and $q$ is odd and graphing $g(x) = (|x|/x)|x|^{p/q}$ if $p$ is odd and $q$ is odd. Show that in both cases $f(x) = g(x)$ if $x > 0$ or $x < 0$. [*Hint:* Show that $f(x)$ is an even function if $p$ is even and $q$ is odd and is an odd function if $p$ is odd and $q$ is odd.]

## **1.5** MATHEMATICAL MODELS; LINEAR MODELS

*In this section we will discuss mathematical modeling, which is the process that is used to express scientific laws in mathematical form. We will also review some results about lines and apply those results to mathematical modeling.*

This section includes a quick review of precalculus material on lines. Readers who want to review this material in more depth are referred to Appendix C.

**MATHEMATICAL MODELS**

A *mathematical model* of a physical law is a description of that law in the language of mathematics. The process of constructing a mathematical model is called *mathematical modeling*. For example, suppose that two variables, $x$ and $y$, are related by some physical law that we would like to describe by a mathematical model. Models can be expressed in terms of graphs, tables, or equations, ranging from simple to extremely complicated. However, many important mathematical models are simply equations of the form

$$y = f(x)$$

that relate $x$ and $y$. For such models the fundamental problem is to find a function $f$ that accurately describes the physical relationship between the variables. Sometimes an appropriate function $f$ might be suggested by experimental data, in which case we say that the model is obtained *inductively*, and sometimes it might be derived from some general theory proposed by a researcher, in which case we say that the model is obtained *deductively*.

The more factors one takes into account when creating a mathematical model, the more complicated the model tends to become, so there is always a balance to be struck between keeping a model mathematically simple and accounting for all of the physical factors that might affect the relationship between the variables. For example, if a meteorologist were trying to model the relationship between the speed of a raindrop when it hits the ground and the height of the cloud in which it was formed, then he or she would certainly have to take air resistance into account. However, the meteorologist would likely ignore the gravitational pull of the planet Pluto since its effect is so small.

Once a mathematical model of a physical law is obtained, it may be possible to use mathematical methods to deduce results about the physical world that are not self-evident or have never been observed. For example, the possibility of placing a satellite in orbit around the Earth was deduced mathematically from Isaac Newton's model of mechanics nearly 200 years before the launching of *Sputnik*, and Albert Einstein's relativistic model of mechanics in 1915 explained a precession (position shift) in the perihelion of the planet Mercury that was not confirmed by physical measurement until 1967.

A good mathematical model is one that produces results that are consistent with observations in the physical world. If a time comes when the mathematical results produced by the model do not agree with real-world observations, then the model must be abandoned in favor of a new model that does. This is the nature of the scientific method—old models constantly being replaced by new models that more accurately describe the real world.

**A QUICK REVIEW OF LINES**

An equation that is expressible in the form

$$Ax + By + C = 0 \tag{1}$$

where $A$ and $B$ are not both zero, is called a *first-degree equation* or a *linear equation* in $x$ and $y$. It is shown in precalculus that every first-degree equation in $x$ and $y$ has a straight line as its graph and, conversely, every straight line can be represented by a first-degree equation in $x$ and $y$. For this reason (1) is sometimes called the *general equation* of a line.

Recall that equations of lines can be written in several different forms:

$$y = mx + b \qquad \boxed{\text{Slope-intercept form}} \qquad (2)$$

$$y - y_1 = m(x - x_1) \qquad \boxed{\text{Point-slope form}} \qquad (3)$$

$$\frac{x}{a} + \frac{y}{b} = 1 \qquad \boxed{\text{Double-intercept form}} \qquad (4)$$

In these equations $m$ is the slope of the line, $a$ is the $x$-intercept, $b$ is the $y$-intercept, and $(x_1, y_1)$ is any point on the line (Figure 1.5.1). Keep in mind that these equations do not apply to vertical lines. For vertical lines the slope is *undefined*, or stated informally, a vertical line has **infinite slope**. Vertical and horizontal lines have particularly simple equations:

$$x = a \qquad \boxed{\text{The vertical line with } x\text{-intercept } a} \qquad (5)$$

$$y = b \qquad \boxed{\text{The horizontal line with } y\text{-intercept } b} \qquad (6)$$

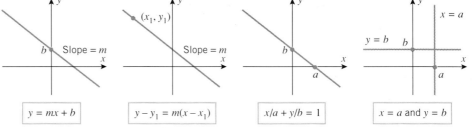

Figure 1.5.1

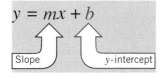

Figure 1.5.2

Equation (2) is especially useful because the slope and the $y$-intercept of the line can be determined by inspection: the slope is the coefficient of $x$, and the $y$-intercept is the constant term (Figure 1.5.2). This equation expresses $y$ as a function of $x$, the function being $f(x) = mx + b$. A function of this form is called a **linear function** of $x$.

**INTERPRETATIONS OF SLOPE**

The slope $m$ of a nonvertical line $y = mx + b$ has two important interpretations (which are related but different in viewpoint):

- $m$ is a measure of the *steepness* of the line.
- $m$ is the rate of change of $y$ with respect to $x$.

The steepness interpretation has an analog in surveying. Surveyors measure the grade or slope of a hill as the ratio of its rise over its run (Figure 1.5.3$a$). The same idea applies to lines. Consider a particle that moves left to right along a nonvertical line from a point

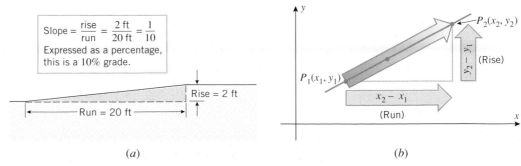

$(a)$

$(b)$

Figure 1.5.3

$P_1(x_1, y_1)$ to a point $P_2(x_2, y_2)$. In the course of its travel the point moves $y_2 - y_1$ units vertically as it travels $x_2 - x_1$ units horizontally (Figure 1.5.3b). The vertical change, which is denoted by $\Delta y = y_2 - y_1$, is called the **rise**, and the horizontal change, which is denoted by $\Delta x = x_2 - x_1$, is called the **run**. The ratio of the rise over the run is always equal to the slope, regardless of where the points $P_1$ and $P_2$ are located on the line; that is,

$$m = \frac{\Delta y}{\Delta x} = \frac{y_2 - y_1}{x_2 - x_1} \tag{7}$$

REMARK.  The symbols $\Delta x$ and $\Delta y$ should not be interpreted as products; rather, $\Delta x$ should be viewed as a single entity representing the *change* in the value of $x$, and $\Delta y$ as a single entity representing the *change* in the value of $y$. In general, if $v$ is any variable whose value changes from an initial value of $v_1$ to a final value of $v_2$, then we call $\Delta v = v_2 - v_1$ (final value minus initial value) an **increment** in $v$. Increments can be positive or negative, depending on whether the final value is larger or smaller than the initial value.

## ANGLE OF INCLINATION

The slope of a nonvertical line $L$ is related to the angle that $L$ makes with the positive $x$-axis. If $\phi$ is the smallest positive angle measured counterclockwise from the $x$-axis to $L$, then the slope of the line can be expressed as

$$m = \tan \phi \tag{8}$$

(Figure 1.5.4a). The angle $\phi$, which is called the **angle of inclination** of the line, satisfies $0° \leq \phi < 180°$ in degree measure (or, equivalently, $0 \leq \phi < \pi$ in radian measure). If $\phi$ is an acute angle, then $m = \tan \phi$ is positive and the line slopes up to the right, and if $\phi$ is an obtuse angle, then $m = \tan \phi$ is negative and the line slopes down to the right. For example, a line whose angle of inclination is $45°$ has slope $m = \tan 45° = 1$, and a line whose angle of inclination is $135°$ has a slope of $m = \tan 135° = -1$ (Figure 1.5.4b). Figure 1.5.5 shows a convenient way of using the line $x = 1$ as a "ruler" for visualizing the relationship between lines of various slopes.

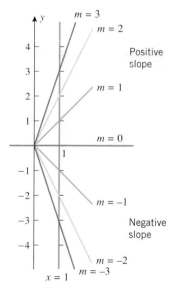

Figure 1.5.5

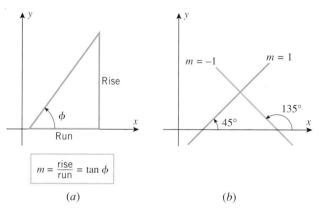

$$m = \frac{\text{rise}}{\text{run}} = \tan \phi$$

(a)                              (b)

Figure 1.5.4

## SLOPES OF LINES IN APPLIED PROBLEMS

In applied problems, changing the units of measurement can change the slope of a line, so it is essential to include the units when calculating the slope. The following example illustrates this.

### Example 1

Suppose that a uniform rod of length 40 cm (= 0.4 m) is thermally insulated around the lateral surface and that the exposed ends of the rod are held at constant temperatures of $25°C$ and $5°C$, respectively (Figure 1.5.6a). It is shown in physics that under appropriate

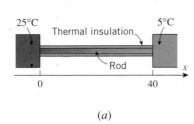 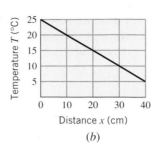 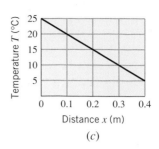

(a)                    (b)                    (c)

Figure 1.5.6

conditions the graph of the temperature $T$ versus the distance $x$ from the left-hand end of the rod will be a straight line. Parts (b) and (c) of Figure 1.5.6 show two such graphs: one in which $x$ is measured in centimeters and one in which it is measured in meters. The slopes in the two cases are

$$m = \frac{5 - 25}{40 - 0} = \frac{-20}{40} = -0.5°\text{C/cm} \tag{9}$$

$$m = \frac{5 - 25}{0.4 - 0} = \frac{-20}{0.4} = -50°\text{C/m} \tag{10}$$

The slope in (9) implies that the temperature *decreases* at a rate of 0.5°C per centimeter of distance from the left end of the rod, and the slope in (10) implies that the temperature decreases at a rate of 50°C per meter of distance from the left end of the rod. The two statements are equivalent physically, even though the slopes differ. ◀

### Example 2

Find the slope-intercept form of the equation of the temperature distribution in the preceding example if the temperature $T$ is measured in degrees Celsius (°C) and the distance $x$ is measured in (a) centimeters and (b) meters.

*Solution (a).* The slope is $m = -0.5°\text{C/cm}$ and the intercept on the $T$-axis is 25°, so

$$T = -0.5x + 25, \quad 0 \le x \le 40$$

where the restriction on $x$ is required because the rod is 40 cm in length. The graph of this equation with the restriction is a line segment rather than a line.

*Solution (b).* The slope is $m = -50°\text{C/m}$, the intercept on the $T$-axis is 25°, and the restriction on $x$ is $0 \le x \le 0.4$. Thus, the equation is

$$T = -50x + 25, \quad 0 \le x \le 0.4 \qquad ◀$$

························································

**LINEAR MATHEMATICAL MODELS**

If $y$ is a linear function of $x$, say $y = mx + b$, then it follows from (7) that

$$\Delta y = m \Delta x$$

Thus, a 1-unit increase in $x$ ($\Delta x = 1$) produces an $m$-unit change in $y$ ($\Delta y = m$). Moreover, this is true at every point on the line (Figure 1.5.7), so we say that $y$ changes at a *constant rate* with respect to $x$, and we call $m$ the **rate of change of $y$ with respect to $x$**. This idea can be summarized as follows.

> **1.5.1** LINEAR MATHEMATICAL MODELS. *If a variable $y$ is related to a variable $x$ in such a way that the rate of change of $y$ with respect to $x$ is constant, say $m$, then $y$ is a linear function of $x$ of the form*
>
> $$y = mx + b$$
>
> *and we say that $y$ is related to $x$ by a **linear mathematical model**. Conversely, if $y$ is a linear function of $x$ whose graph has slope $m$, then the rate of change of $y$ with respect to $x$ is constant and equal to $m$.*

A 1-unit increase in $x$ always produces an $m$-unit change in $y$.

Figure 1.5.7

It follows from this that linear models are appropriate whenever experimentation or theory suggests that the rate of change of $y$ with respect to $x$ is constant.

**UNIFORM RECTILINEAR MOTION**

One of the important themes in calculus is the study of motion. To describe the motion of an object completely, one must specify its *speed* (how fast it is going) and the direction in which it is moving. The speed and the direction of motion together comprise what is called the *velocity* of the object. For example, knowing that the speed of an aircraft is 500 mi/h tells us how fast it is going, but not which way it is moving. In contrast, knowing that the velocity of the aircraft is 500 mi/h *due south* pins down the speed and the direction of motion.

Later, we will study the motion of particles that move along curves in two- or three-dimensional space, but for now we will focus on motion along a line; this is called *rectilinear motion*. In general rectilinear motion, a particle can move back and forth along the line (as with a piston moving up and down in a cylinder); however, for now we will only consider the simple case in which the particle moves in just *one direction* along a line (as with a car traveling on a straight road).

For simplicity, we will assume that the motion is along a coordinate line, such as an $x$-axis or $y$-axis, and that the particle is moving in the positive direction. In general discussions we will usually name the coordinate line the $s$-axis to avoid being specific. A graphical description of rectilinear motion along an $s$-axis can be obtained by making a plot of the $s$-coordinate of the particle versus the elapsed time $t$. This is called the *position versus time curve* for the particle. Figure 1.5.8a shows a typical position versus time curve for a car moving in the positive direction along an $s$-axis.

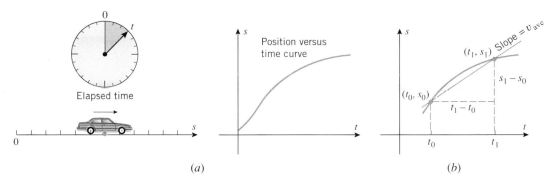

(a)    (b)

Figure 1.5.8

FOR THE READER. How can you tell from the position versus time curve in Figure 1.5.8a that the car does not reverse direction?

Because we are assuming that the particle is moving in the positive direction of the $s$-axis, there is no ambiguity about the direction of motion, and hence the terms "speed" and "velocity" can be used interchangeably. However, later, when we consider general rectilinear motion or motion along a curved path, it will be necessary to distinguish between these terms, since the direction of motion may vary.

For a particle in rectilinear motion along a coordinate axis, we define the *average velocity* $v_{\text{ave}}$ of the particle during the time interval from $t_0$ to $t_1$ to be

$$v_{\text{ave}} = \frac{s_1 - s_0}{t_1 - t_0} = \frac{\Delta s}{\Delta t} \tag{11}$$

where $s_0$ and $s_1$ are the $s$-coordinates of the particle at times $t_0$ and $t_1$, respectively. Geometrically, this is the slope of the secant line connecting the points $(t_0, s_0)$ and $(t_1, s_1)$ on the position versus time curve (Figure 1.5.8b). The quantity $\Delta s = s_1 - s_0$ is called

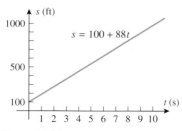

Figure 1.5.9

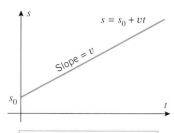

Position versus time curve for a particle with coordinate $s_0$ at time $t = 0$ and moving with constant velocity $v$

Figure 1.5.10

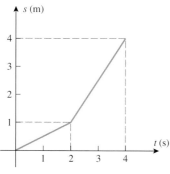

Figure 1.5.11

...................................................

**CONSTANT ACCELERATION**

the **displacement** or **change in position** of the particle during the time interval from $t_0$ to $t_1$. With this terminology, Formula (11) states that for a particle in rectilinear motion *the average velocity over a time interval is the displacement during the time interval divided by the length of the time interval.* For example, if a car moving in one direction along a straight road travels 75 miles in 3 hours, then its average velocity is $75/3 = 25$ mi/h.

In the special case where the average velocity of a particle in rectilinear motion is the same over every time interval, the particle is said to have **constant velocity** and **uniform rectilinear motion**. If the average velocity over every time interval is $v$, then we will refer to $v$ as the **velocity** of the particle (dropping the adjective "average").

For a particle with uniform rectilinear motion the displacement over *any* time interval is given by the formula

$$\text{displacement} = \text{velocity} \times \text{elapsed time} \tag{12}$$

### Example 3

Suppose that a car moves with a constant velocity of 88 ft/s in the positive direction of an $s$-axis. Given that the $s$-coordinate of the car at time $t = 0$ is $s = 100$, find an equation for $s$ as a function of $t$, and graph the position versus time curve.

*Solution.* It follows from (12) that in a period of $t$ seconds, the car will move $88t$ feet from its starting point, so its coordinate $s$ at time $t$ will be

$$s = 100 + 88t$$

The graph of this equation is the line in Figure 1.5.9. ◀

It is not accidental that the position versus time curve turned out to be a line in the last example; this will always be the case for uniform rectilinear motion. To see why this is so, suppose that a particle moves with constant velocity $v$ in the positive direction along an $s$-axis, starting at the point $s_0$ at time $t = 0$. It follows from (12) that in $t$ units of time the particle will move $vt$ units from its starting point $s_0$, so its coordinate $s$ at time $t$ will be

$$s = s_0 + vt$$

which is a line with $s$-intercept $s_0$ and slope $v$ (Figure 1.5.10). It follows from this equation and 1.5.1 that we can view the velocity $v$ as the rate of change of $s$ with respect to $t$, that is, the rate of change of position with respect to time.

### Example 4

Figure 1.5.11 shows the position versus time curve for a particle moving along an $s$-axis. Describe the motion of the particle in words.

*Solution.* At time $t = 0$ the particle is at the origin. From time $t = 0$ to $t = 2$ the slope of the line segment is $\frac{1}{2}$, so the particle is moving with a constant velocity of $\frac{1}{2} = 0.5$ m/s. At time $t = 2$ the particle is at the point $s = 1$ (i.e., 1 meter from the origin). From time $t = 2$ to $t = 4$ the slope of the line segment is $\frac{3}{2}$, so the particle is moving with a constant velocity of $\frac{3}{2} = 1.5$ m/s. At time $t = 4$ it is at the point $s = 4$. ◀

In everyday language we say that an object is "accelerating" if it is speeding up and "decelerating" if it is slowing down. Mathematically, the **acceleration** of a particle in rectilinear motion is defined to be the *rate of change of velocity with respect to time*, where the acceleration is positive if the velocity is increasing and negative if it is decreasing. Thus, for a particle that moves in the positive direction of an $s$-axis, negative acceleration means the particle is "decelerating" in everyday language. Acceleration, like velocity, can be variable or constant. For example, by pressing the gas pedal of a car toward the floor smoothly, the driver can make the car's velocity increase at a constant rate (a constant acceleration); however, if the driver suddenly slams the pedal to the floor, the car will lurch forward, reflecting a nonconstant acceleration. Later in the text we will study acceleration in more depth, but for now we will only consider the case in which acceleration is constant.

REMARK.  The units of acceleration are units of velocity divided by units of time. For example, if the velocity of a particle is increasing at a rate of 3 feet per second each second, then its acceleration is 3 ft/s/s (velocity in ft/s divided by time in s); this is usually written as 3 ft/s$^2$ (read "3 feet per second per second" or "3 feet per second squared"). Similarly, if the velocity of a particle is decreasing at a rate of 3 feet per second each second, then it has an acceleration of $-3$ ft/s$^2$.

Graphical information about the acceleration of a particle can be obtained from the graph of velocity versus time; this is called the ***velocity versus time curve***. In the case where the particle has constant acceleration, the velocity versus time curve will be linear, and its slope, which is the rate of change of velocity with time, will be the acceleration (Figure 1.5.12).

## Example 5

Suppose that a car moves in the positive direction of an $s$-axis in such a way that its velocity $v$ increases at a constant rate of 2 ft/s$^2$.

(a)  Assuming that the velocity of the car is 88 ft/s at time $t = 0$, find an equation for $v$ as a function of $t$.

(b)  Make a graph of velocity versus time, and mark the point on the graph at which the car attains a velocity of 100 ft/s.

*Solution* (*a*).  Since the rate of change of $v$ with respect to $t$ is 2 ft/s$^2$, and since $v = 88$ ft/s if $t = 0$, the equation for velocity as a function of time is

$$v = 88 + 2t \tag{13}$$

*Solution* (*b*).  To find the time it takes for the car to reach a velocity of 100 ft/s, we substitute $v = 100$ in (13) and solve for $t$. This yields $t = 6$. The graph of (13) and the point at which the velocity reaches 100 ft/s is shown in Figure 1.5.13.  ◀

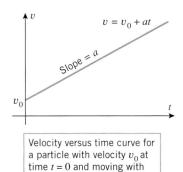

Velocity versus time curve for a particle with velocity $v_0$ at time $t = 0$ and moving with constant acceleration $a$

Figure 1.5.12

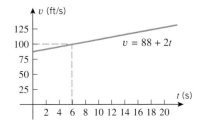

Velocity versus time curve for a particle with a velocity of 88 ft/s at time $t = 0$ and moving with a constant acceleration of 2 ft/s$^2$

Figure 1.5.13

**LINEAR MODELS FROM DIRECT PROPORTION**

Recall that a variable $y$ is said to be ***directly proportional*** to a variable $x$ if there is a positive constant $k$, called the ***constant of proportionality***, such that

$$y = kx \tag{14}$$

The graph of this equation is a line through the origin whose slope $k$ is the constant of proportionality. Thus, linear models are appropriate in physical problems where one variable is directly proportional to another.

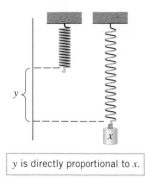

y is directly proportional to x.

Figure 1.5.14

Hooke's law[*] in physics provides a nice example of direct proportion. It follows from this law that if a weight of $x$ units is suspended from a spring, then the spring will be stretched by an amount $y$ that is directly proportional to $x$, that is, $y = kx$ (Figure 1.5.14). The constant $k$ depends on the stiffness of the spring: the stiffer the spring, the smaller the value of $k$ (why?).

### Example 6

Figure 1.5.15 shows an old-fashioned spring scale that is calibrated in pounds.

(a)  Given that the pound scale marks are 0.5 in apart, find an equation that expresses the length $y$ that the spring is stretched (in inches) in terms of the suspended weight $x$ in pounds).

(b)  Graph the equation obtained in part (a).

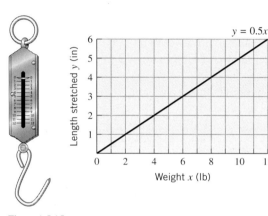

Figure 1.5.15

*Solution (a).* It follows from Hooke's law that $y$ is related to $x$ by an equation of the form $y = kx$. To find $k$ we rewrite this equation as $k = y/x$, and use the fact that a weight of $x = 1$ lb stretches the spring $y = 0.5$ in. Thus,

$$k = \frac{y}{x} = \frac{0.5}{1} = 0.5 \qquad \text{and hence} \qquad y = 0.5x$$

*Solution (b).* The graph of the equation $y = 0.5x$ is shown in Figure 1.5.15. ◀

**LINEAR MODELS FROM GRAPHICAL DATA**

Sometimes linear models are suggested by graphical data. For example, Figure 1.5.16a shows a graph of temperature versus altitude that was transmitted by the *Magellan* spacecraft when it entered the atmosphere of Venus in October 1991. The graph strongly suggests that there is a linear relationship between temperature and altitude for altitudes between 35 km and 60 km.

### Example 7

(a)  Use the graph transmitted by the *Magellan* spacecraft to find a linear model of temperature versus altitude in the Venusian atmosphere that is valid for altitudes between 35 km and 60 km.

(b)  Use the model to estimate the temperature at the surface of Venus, and discuss the assumptions you are making in obtaining the estimate.

---

[*] Hooke's law, named for the English physicist Robert Hooke (1635–1703), applies only for small displacements that do not stretch the spring to the point of permanently distorting it.

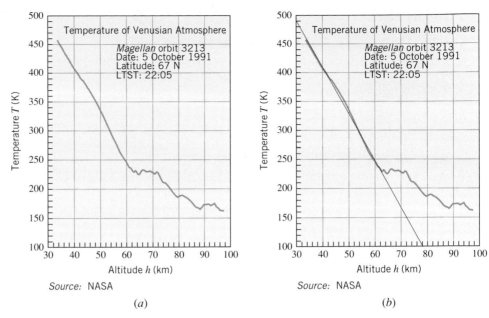

Figure 1.5.16

*Solution (a).* Let $T$ be the temperature in kelvins and $h$ the altitude in kilometers. We will first estimate the slope $m$ of the linear portion of the graph, then estimate the coordinates of a data point $(h_1, T_1)$ on that portion of the graph, and then use the point-slope form of a line

$$T - T_1 = m(h - h_1) \tag{15}$$

The graph nearly passes through the point $(60, 250)$, so we will take $h_1 \approx 60$ and $T_1 \approx 250$. In Figure 1.5.16b we have sketched a line that closely approximates the linear portion of the data. Using the intersections of that line with the edges of the grid box, we estimate the slope to be

$$m \approx \frac{100 - 490}{78 - 30} = -\frac{390}{48} \approx -8.125 \text{ K/km}$$

Substituting our estimates of $h_1$, $T_1$, and $m$ into (15) yields the equation

$$T - 250 = -8.125(h - 60)$$

or equivalently,

$$T = -8.125h + 737.5 \tag{16}$$

*Solution (b).* The *Magellan* spacecraft stopped transmitting data at an altitude of approximately 35 km, so we cannot be certain that the linear model applies at lower altitudes. However, since we have no other data to work with, let us *assume* that the model is valid at all lower altitudes, in which case we can approximate the temperature at the surface of Venus by setting $h = 0$ in (16). We obtain $T \approx 737.5$ K.   ◀

REMARK.   The method of the preceding example is crude, at best, since it relies on extracting rough estimates of numerical data from a graph. Nevertheless, the final result is quite good, since the most recent information from NASA places the surface temperature of Venus at about 740 K (hot enough to melt lead).

**LINEAR MODELS FROM NUMERICAL DATA**

One method for determining whether $n$ points

$$(x_1, y_1), (x_2, y_2), \ldots, (x_n, y_n)$$

lie on a line is to compare the slopes of the line segments joining successive points. The points lie on a line if and only if those slopes are equal (Figure 1.5.17).

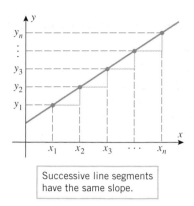

Successive line segments have the same slope.

Figure 1.5.17

**Table 1.5.1**

| x | y |
|-----|-----|
| 1.5 | 0.3 |
| 2.5 | 1.1 |
| 3.5 | 1.9 |
| 5.5 | 3.5 |
| 9.5 | 6.7 |

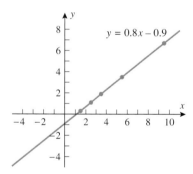

Figure 1.5.18

### Example 8

Consider the data in Table 1.5.1.

(a) Explain why a linear model is appropriate for the data in the table.

(b) Find a linear equation that relates $x$ and $y$, and graph the equation and the data together.

*Solution (a).* The five data points lie on a line, since each 1-unit increase in $x$ produces a corresponding 0.8-unit increase in $y$. Thus, the slope of the line segment joining any two successive data points is

$$m = \frac{\Delta y}{\Delta x} = \frac{0.8}{1} = 0.8$$

*Solution (b).* A linear equation relating $x$ and $y$ can be obtained from the point-slope form of the line using the slope $m = 0.8$ calculated in part (a) and any one of the five data points. If we use the first data point, $(1.5, 0.3)$, we obtain

$$y - 0.3 = 0.8(x - 1.5)$$

or in slope-intercept form,

$$y = 0.8x - 0.9$$

The graph of this equation together with the given data are shown in Figure 1.5.18.  ◀

REMARK.    Sometimes, data points that should theoretically lie on a line do not because of experimental error and other factors. In such cases curve-fitting techniques are used to find a line that most closely fits the data. Such techniques will be discussed later in the text.

### OTHER APPLICATIONS OF LINEAR FUNCTIONS

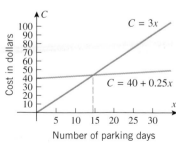

Figure 1.5.19

Linear functions arise in a variety of practical problems. Here is a typical example.

### Example 9

A university parking lot charges $3.00 per day but offers a $40.00 monthly sticker with which the student pays only $0.25 per day.

(a) Find equations for the cost $C$ of parking for $x$ days per month under both payment methods, and graph the equations for $0 \le x \le 30$. (Treat $C$ as a continuous function of $x$, even though $x$ only assumes integer values.)

(b) Find the value of $x$ for which the graphs intersect, and discuss the significance of this value.

*Solution (a).* The cost in dollars of parking for $x$ days at $3.00 per day is $C = 3x$, and the cost for the $40.00 sticker plus $x$ days at $0.25 per day is $C = 40 + 0.25x$ (Figure 1.5.19).

*Solution (b).* The graphs intersect at the point where

$$3x = 40 + 0.25x$$

which is $x = 40/2.75 \approx 14.5$. This value of $x$ is not an option for the student, since $x$ must be an integer. However, it is the dividing point at which the monthly sticker method becomes less expensive than the daily payment method; that is, for $x \geq 15$ it is cheaper to buy the monthly sticker and for $x \leq 14$ it is cheaper to pay the daily rate.   ◄

**EXERCISE SET 1.5**   ∼ Graphing Calculator   ⃞c CAS

Exercises 1–26 involve the basic properties of lines and slope. In some of these exercises you will need to use slopes to determine whether two lines are parallel or perpendicular. If you have forgotten how to do this, review Appendix C.

**1.** (a) Find the slopes of the sides of the triangle with vertices $(0, 3)$, $(2, 0)$, and $\left(6, \frac{8}{3}\right)$.
   (b) Is this a right triangle? Explain.

**2.** (a) Find the slopes of the sides of the quadrilateral with vertices $(-3, -1)$, $(5, -1)$, $(7, 3)$, and $(-1, 3)$.
   (b) Is this a parallelogram? Explain.

**3.** List the lines in the accompanying figure in the order of increasing slope.

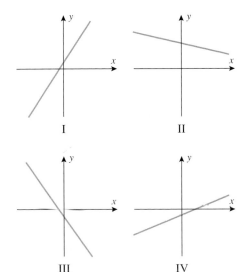

Figure Ex-3

**4.** List the lines in the accompanying figure in the order of increasing slope.

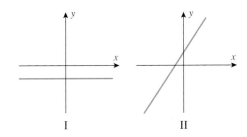

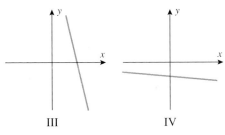

Figure Ex-4

**5.** Use slopes to determine whether the given points lie on the same line.
   (a) $(1, 1)$, $(-2, -5)$, and $(0, -1)$
   (b) $(-2, 4)$, $(0, 2)$, and $(1, 5)$

**6.** A particle, initially at $(7, 5)$, moves along a line of slope $m = -2$ to a new position $(x, y)$.
   (a) Find $y$ if $x = 9$.
   (b) Find $x$ if $y = 12$.

**7.** A particle, initially at $(1, 2)$, moves along a line of slope $m = 3$ to a new position $(x, y)$.
   (a) Find $y$ if $x = 5$.
   (b) Find $x$ if $y = -2$.

**8.** Find $x$ and $y$ if the line through $(0, 0)$ and $(x, y)$ has slope $\frac{1}{2}$, and the line through $(x, y)$ and $(7, 5)$ has slope 2.

**9.** Find $x$ if the slope of the line through $(1, 2)$ and $(x, 0)$ is the negative of the slope of the line through $(4, 5)$ and $(x, 0)$.

In Exercises 10 and 11, find the angle of inclination of the line with slope $m$ to the nearest degree. Use a calculating utility, where needed.

**10.** (a) $m = \frac{1}{2}$      (b) $m = -1$
     (c) $m = 2$      (d) $m = -57$

**11.** (a) $m = -\frac{1}{2}$      (b) $m = 1$
     (c) $m = -2$      (d) $m = 57$

In Exercises 12 and 13, find the angle of inclination of the line to the nearest degree. Use a calculating utility, where needed.

**12.** (a) $3y = 2 - \sqrt{3}x$      (b) $y - 4x + 7 = 0$

**13.** (a) $y = \sqrt{3}x + 2$      (b) $y + 2x + 5 = 0$

**14.** Find equations for the $x$- and $y$-axes.

In Exercises 15–22, find the slope-intercept form of the equation of the line satisfying the stated conditions, and check your answer using a graphing utility.

**15.** Slope $= -2$, $y$-intercept $= 4$

**16.** $m = 5$, $b = -3$

**17.** The line is parallel to $y = 4x - 2$ and its $y$-intercept is 7.

**18.** The line is parallel to $3x + 2y = 5$ and passes through $(-1, 2)$.

**19.** The line is perpendicular to the equation $y = 5x + 9$ and has $y$-intercept 6.

**20.** The line is perpendicular to $x - 4y = 7$ and passes through $(3, -4)$.

**21.** The line passes through $(2, 4)$ and $(1, -7)$.

**22.** The line passes through $(-3, 6)$ and $(-2, 1)$.

**23.** In each part, classify the lines as parallel, perpendicular, or neither.
(a) $y = 4x - 7$ and $y = 4x + 9$
(b) $y = 2x - 3$ and $y = 7 - \frac{1}{2}x$
(c) $5x - 3y + 6 = 0$ and $10x - 6y + 7 = 0$
(d) $Ax + By + C = 0$ and $Bx - Ay + D = 0$
(e) $y - 2 = 4(x - 3)$ and $y - 7 = \frac{1}{4}(x - 3)$

**24.** In each part, classify the lines as parallel, perpendicular, or neither.
(a) $y = -5x + 1$ and $y = 3 - 5x$
(b) $y - 1 = 2(x - 3)$ and $y - 4 = -\frac{1}{2}(x + 7)$
(c) $4x + 5y + 7 = 0$ and $5x - 4y + 9 = 0$
(d) $Ax + By + C = 0$ and $Ax + By + D = 0$
(e) $y = \frac{1}{2}x$ and $x = \frac{1}{2}y$

In Exercises 25 and 26, use the graph to find the equation of the line in slope-intercept form, and then check your result by using a graphing utility to graph the equation.

**25.**

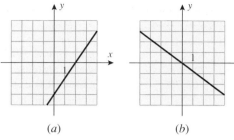

(a)            (b)

Figure Ex-25

**26.**

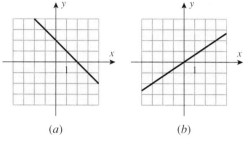

(a)            (b)

Figure Ex-26

**27.** The accompanying figure shows the position versus time curve for a particle moving along an $x$-axis.
(a) What is the velocity of the particle?
(b) What is the $x$-coordinate of the particle at time $t = 0$?
(c) What is the $x$-coordinate of the particle at time $t = 2$?
(d) At what time does the particle have an $x$-coordinate of $x = 4$?

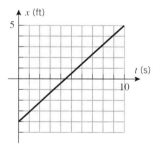

Figure Ex-27

**28.** A particle moving along an $x$-axis with constant velocity is at the point $x = 1$ when $t = 2$ and is at the point $x = 5$ when $t = 4$.
(a) Find the velocity of the particle if $x$ is in meters and $t$ is in seconds.
(b) Find an equation that expresses $x$ as a function of $t$.
(c) What is the coordinate of the particle at time $t = 0$?

**29.** A particle moving along an $x$-axis with constant acceleration has velocity $v = 3$ ft/s at time $t = 1$ and velocity $v = -1$ ft/s at time $t = 4$.
(a) Find the acceleration of the particle.
(b) Find an equation that expresses $v$ as a function of $t$.
(c) What is the velocity of the particle at time $t = 0$?

**30.** The accompanying figure shows the velocity versus time curve for a particle moving along the *x*-axis.
  (a) What is the acceleration of the particle?
  (b) What is the velocity of the particle at time $t = 0$?
  (c) What is the velocity of the particle at time $t = 2$?
  (d) At what time does the particle have a velocity of $v = 3$ ft/s?

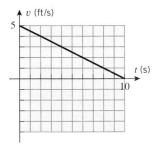

Figure Ex-30

**31.** The accompanying figure shows the position versus time curve for a particle moving along an *x*-axis.
  (a) Describe the motion of the particle in words.
  (b) Find the average velocity of the particle from $t = 0$ to $t = 10$.
  (c) Find the average speed of the particle from $t = 0$ to $t = 10$.

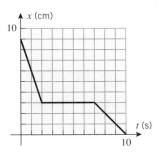

Figure Ex-31

**32.** The accompanying figure shows the velocity versus time curve for a particle moving along an *x*-axis. Describe the motion of the particle in words.

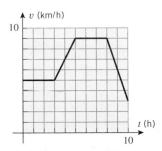

Figure Ex-32

**33.** A locomotive travels on a straight track at a constant speed of 40 mi/h, then reverses direction and returns to its starting point, traveling at a constant speed of 60 mi/h.
  (a) What is the average velocity for the round-trip?
  (b) What is the average speed for the round-trip?
  (c) What is the total distance traveled by the train if the total trip took 5 h?

**34.** A ball is tossed straight up at time $t = 0$ with an initial velocity of 64 ft/s. We will show later using basic principles of physics that the velocity of the ball as a function of time is $v = 64 - 32t$.
  (a) What direction is the ball traveling 3 s after it is released? Explain your reasoning.
  (b) At what time does the ball reach its maximum height above the ground? Explain your reasoning.
  (c) What can you say about the acceleration of the ball?

**35.** A car is stopped at a toll booth on a straight highway. Starting at time $t = 0$ it accelerates at a constant rate of 10 ft/s$^2$ for 10 s. It then travels at a constant speed of 100 ft/s for 90 s. At that time it begins to decelerate at a constant rate of 5 ft/s$^2$ for 20 s, at which point in time it reaches a full stop at a traffic light.
  (a) Sketch the velocity versus time curve.
  (b) Express $v$ as a piecewise function of $t$.

**36.** Make a reasonable sketch of a position versus time curve for a particle that moves in the positive *x*-direction with positive constant acceleration.

**37.** A spring with a natural length of 15 in stretches to a length of 20 in when a 45-lb object is suspended from it.
  (a) Use Hooke's law to find an equation that expresses the length *y* that the spring is stretched (in inches) in terms of the suspended weight *x* (in pounds).
  (b) Graph the equation obtained in part (b).
  (c) Find the length of the spring when a 100-lb object is suspended from it.
  (d) What is the largest weight that can be suspended from the spring if the spring cannot be stretched to more than twice its natural length?

**38.** The spring in a heavy-duty shock absorber has a natural length of 3 ft and is compressed 0.2 ft by a load of 1 ton. An additional load of 5 tons compresses the spring an additional 1 ft.
  (a) Assuming that Hooke's law applies to compression as well as extension, find an equation that expresses the length *y* that the spring is compressed from its natural length (in feet) in terms of the load *x* (in tons).
  (b) Graph the equation obtained in part (a).
  (c) Find the amount that the spring is compressed from its natural length by a load of 3 tons.
  (d) Find the maximum load that can be applied if safety regulations prohibit compressing the spring to less than half its natural length.

In Exercises 39 and 40, confirm that a linear model is appropriate for the relationship between $x$ and $y$. Find a linear equation relating $x$ and $y$, and verify that the data points lie on the graph of your equation.

**39.**

| $x$ | 0 | 1 | 2 | 4 | 6 |
|-----|---|---|---|---|---|
| $y$ | 2 | 3.2 | 4.4 | 6.8 | 9.2 |

**40.**

| $x$ | −1 | 0 | 2 | 5 | 8 |
|-----|-----|-----|-----|---|------|
| $y$ | 12.6 | 10.5 | 6.3 | 0 | −6.3 |

**41.** There are two common systems for measuring temperature, Celsius and Fahrenheit. Water freezes at $0°$ Celsius $(0°C)$ and $32°$ Fahrenheit $(32°F)$; it boils at $100°C$ and $212°F$.
 (a) Assuming that the Celsius temperature $T_C$ and the Fahrenheit temperature $T_F$ are related by a linear equation, find the equation.
 (b) What is the slope of the line relating $T_F$ and $T_C$ if $T_F$ is plotted on the horizontal axis?
 (c) At what temperature is the Fahrenheit reading equal to the Celsius reading?
 (d) Normal body temperature is $98.6°F$. What is it in $°C$?

**42.** Thermometers are calibrated using the so-called "triple point" of water, which is 273.16 K on the Kelvin scale and $0.01°C$ on the Celsius scale. A one-degree difference on the Celsius scale is the same as a one-degree difference on the Kelvin scale, so there is a linear relationship between the temperature $T_C$ in degrees Celsius and the temperature $T_K$ in kelvins.
 (a) Find an equation that relates $T_C$ and $T_K$.
 (b) Absolute zero (0 K on the Kelvin scale) is the temperature below which a body's temperature cannot be lowered. Express absolute zero in $°C$.

**43.** To the extent that water can be assumed to be incompressible, the pressure $p$ in a body of water varies linearly with the distance $h$ below the surface.
 (a) Given that the pressure is 1 atmosphere (1 atm) at the surface and 5.9 atm at a depth of 50 m, find an equation that relates pressure to depth.
 (b) At what depth is the pressure twice that at the surface?

**44.** A resistance thermometer is a device that determines temperature by measuring the resistance of a fine wire whose resistance varies with temperature. Suppose that the resistance $R$ in ohms $(\Omega)$ varies linearly with the temperature $T$ in $°C$ and that $R = 123.4\ \Omega$ when $T = 20°C$ and that $R = 133.9\ \Omega$ when $T = 45°C$.
 (a) Find an equation for $R$ in terms of $T$.
 (b) If $R$ is measured experimentally as $128.6\ \Omega$, what is the temperature?

**45.** Suppose that the mass of a spherical mothball decreases with time, due to evaporation, at a rate that is proportional to its surface area. Assuming that it always retains the shape of a sphere, it can be shown that the radius $r$ of the sphere decreases linearly with the time $t$.
 (a) If, at a certain instant, the radius is 0.80 mm and 4 days later it is 0.75 mm, find an equation for $r$ (in millimeters) in terms of the elapsed time $t$ (in days).
 (b) How long will it take for the mothball to completely evaporate?

**46.** The accompanying figure shows three masses suspended from a spring: a mass of 11 g, a mass of 24 g, and an unknown mass of $W$ g.
 (a) What will the pointer indicate on the scale if no mass is suspended?
 (b) Find $W$.

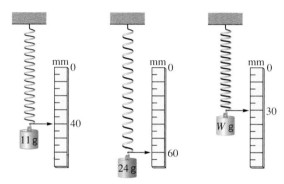

Figure Ex-46

**47.** The price for a round-trip bus ride from a university to center city is $2.00, but it is possible to purchase a monthly commuter pass for $25.00 with which each round-trip ride costs an additional $0.25.
 (a) Find equations for the cost $C$ of making $x$ round-trips per month under both payment plans, and graph the equations for $0 \leq x \leq 30$ (treating $C$ as a continuous function of $x$, even though $x$ assumes only integer values).
 (b) How many round-trips per month would a student have to make for the commuter pass to be worthwhile?

**48.** A student must decide between buying one of two used cars: car $A$ for $4000 or car $B$ for $5500. Car $A$ gets 20 miles per gallon of gas, and car $B$ gets 30 miles per gallon. The student estimates that gas will run $1.25 per gallon. Both cars are in excellent condition, so the student feels that repair costs should be negligible for the foreseeable future. How many miles would the student have to drive before car $B$ becomes the better buy?

**49.** (**The Age of the Universe**) In the early 1900s the astronomer Edwin P. Hubble (1889–1953) noted an unexpected relationship between the radial velocity of a galaxy and its distance $d$ from any reference point (Earth, for example). That relationship, now known as **Hubble's law**, states that the galaxies are receding with a velocity $v$ that is directly proportional to the distance $d$. This is usually expressed as $v = Hd$, where $H$ (the constant of proportionality) is called **Hubble's constant**. When applying this formula it is usual to express $v$ in kilometers per second (km/s) and $d$ in millions of light-years (Mly), in which case $H$ has units of km/s/Mly. The accompanying figure shows an original plot and trend line of the velocity-distance relationship obtained by Hubble and a collaborator Milton L. Humason (1891–1972).

(a)  Use the trend line in the figure to estimate Hubble's constant.

(b)  An estimate of the age of the universe can be obtained by assuming that the galaxies move with constant velocity $v$, in which case $v$ and $d$ are related by $d = vt$. Assuming that the Universe began with a "big bang" that initiated its expansion, show that the Universe is roughly $1.5 \times 10^{10}$ years old. [Take $H = 20$ km/s/Mly,

which is in keeping with current estimates that place $H$ between 15 and 27 km/s/Mly. (Note that the current estimates are significantly less than that resulting from Hubble's data.)]

(c)  In a more realistic model of the Universe, the velocity $v$ would decrease with time. What effect would that have on your estimate in part (b)?

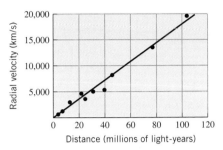

Figure Ex-49

# 1.6 FAMILIES OF FUNCTIONS

*Functions are often grouped into families according to the form of their defining formulas or other common characteristics. In this section we will discuss some of the most basic families of functions.*

This section includes quick reviews of precalculus material on polynomials and trigonometry. Readers who want to review this material in more depth are referred to Appendices E and F. Instructors who want to spend some additional time on precalculus review can divide this section into two parts, covering the trigonometry material in a second lecture.

**FAMILIES OF LINES**

A function $f$ whose values are all the same is called a ***constant function***. For example, the formula $f(x) = c$ defines the constant function whose value is $c$ for all $x$. The graph of the constant function $f(x) = c$ is the horizontal line $y = c$ (Figure 1.6.1a). If we vary $c$, then we obtain a set or ***family*** of horizontal lines (Figure 1.6.1b).

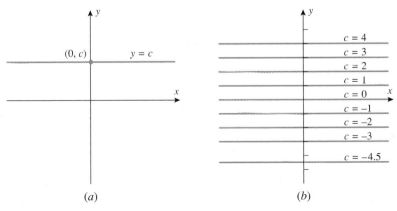

Figure 1.6.1

REMARK. The expression $f(x) = c$ can be confusing because it can be interpreted either as an equation that is satisfied for certain $x$ (as in $x^2 = c$) or as an identity that is satisfied for all $x$; it is the latter interpretation that defines a constant function. Thus, when you see an expression of the form $f(x) = c$, you will have to determine from its context whether it is intended as an equation or a constant function.

The quantities $m$ and $b$ in the equation $y = mx + b$ can be viewed as unspecified constants whose values may change from one application to another; such changeable constants are called **parameters**.

If we keep $b$ fixed and vary the parameter $m$ in the equation $y = mx + b$, then we obtain a family of lines whose members all have $y$-intercept $b$ (Figure 1.6.2$a$); and if we keep $m$ fixed and vary the parameter $b$, then we obtain a family of parallel lines whose members all have slope $m$ (Figure 1.6.2$b$).

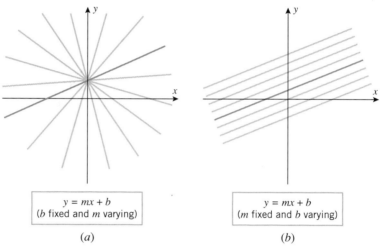

$y = mx + b$
($b$ fixed and $m$ varying)

($a$)

$y = mx + b$
($m$ fixed and $b$ varying)

($b$)

Figure 1.6.2

## Example 1

(a) Find an equation for the family of lines with slope $\frac{1}{2}$.

(b) Find the member of the family in part (a) that passes through the point (4, 1).

(c) Find an equation for the family of lines whose members are perpendicular to the lines in part (a).

*Solution* ($a$). The lines of slope $\frac{1}{2}$ are of the form

$$y = \tfrac{1}{2}x + b \tag{1}$$

where the parameter $b$ can have any real value.

*Solution* ($b$). To find the line in the family that passes through the point (4, 1), we must find the value of $b$ for which the coordinates $x = 4$ and $y = 1$ satisfy (1). Substituting these coordinates into (1) and solving for $b$ yields $b = -1$, and hence the equation of the line is

$$y = \tfrac{1}{2}x - 1 \tag{2}$$

(Figure 1.6.3$a$).

*Solution* ($c$). Since the slopes of perpendicular lines are negative reciprocals, it follows that the lines perpendicular to those in part (a) have slope $-2$ and hence are of the form

$$y = -2x + b$$

Some typical lines in families (1) and (2) are graphed in Figure 1.6.3$b$. ◄

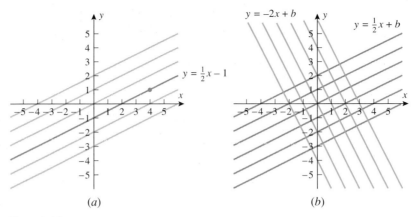

Figure 1.6.3

**THE FAMILY y = xⁿ**

A function of the form $f(x) = x^p$, where $p$ is constant is called a **power function**. If $p$ is a positive integer, say $p = n$, then the power functions have the form $f(x) = x^n$. The graphs of the curves $y = x^n$ for $n = 1, 2, 3, 4,$ and 5 are shown in Figure 1.6.4. The first graph is the line $y = x$ with slope 1 that passes through the origin, and the second is a parabola that opens up and has its vertex at the origin (see Appendix 2).

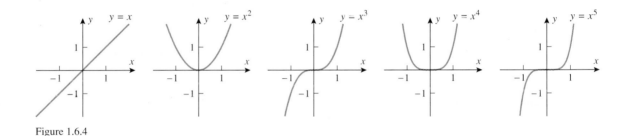

Figure 1.6.4

For $n > 2$ the shape of the graph of $y = x^n$ depends on whether $n$ is even or odd (Figure 1.6.5). For even values of $n$ the graphs have the same general shape as the parabola $y = x^2$ (though they are not actually parabolas if $n > 2$), and for odd values of $n$ greater than 1 they have the same general shape as $y = x^3$. The graphs in the family $y = x^n$ share a number of important characteristics:

- For even values of $n$ the functions $f(x) = x^n$ are even, and their graphs are symmetric about the $y$-axis; for odd values of $n$ the functions $f(x) = x^n$ are odd, and their graphs are symmetric about the origin.

- For all values of $n$ the graphs pass through the origin and the point $(1, 1)$. For even values of $n$ the graphs pass through $(-1, 1)$, and for odd values of $n$ they pass through $(-1, -1)$.

- Increasing $n$ causes the graph to become flatter over the interval $-1 < x < 1$ and steeper over the intervals $x > 1$ and $x < -1$.

REMARK.    The last characteristic can be explained numerically by considering the effect of raising a real number $x$ to successively higher powers. If $x$ is a fraction, that is, $-1 < x < 1$, then the absolute value of $x^n$ *decreases* as $n$ increases (try raising $\frac{1}{2}$ or $-\frac{1}{2}$ to higher and higher powers, for example). This explains why successive graphs in Figure 1.6.5 become flatter over the interval $-1 < x < 1$. On the other hand, if $x > 1$ or $x < -1$, then the

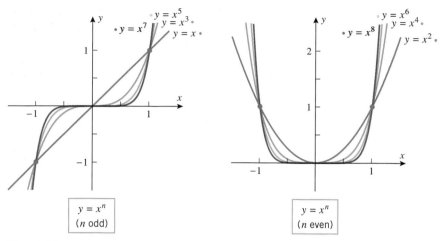

Figure 1.6.5

absolute value of $x^n$ *increases* as $n$ increases (try raising 2 or $-2$ to higher and higher powers). This explains why successive graphs become steeper if $x > 1$ or $x < -1$.

**THE FAMILY $y = x^{-n}$**

If $p$ is a negative integer, say $p = -n$, then the power functions $f(x) = x^p$ have the form $f(x) = x^{-n} = 1/x^n$. Figure 1.6.6*a* shows the graphs of $y = 1/x$ and $y = 1/x^2$, and Figure 1.6.6*b* shows how these graphs relate to other members of the family. The graph of $y = 1/x$ is called an ***equilateral hyperbola*** (for reasons to be discussed later).

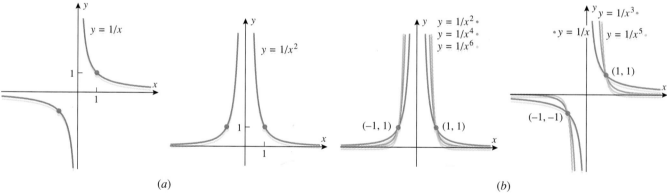

Figure 1.6.6

For odd values of $n$ the graphs have the same general shape as $y = 1/x$, and for even values of $n$ they have the same general shape as $y = 1/x^2$. The graphs in the family $y = 1/x^n$ share a number of important characteristics:

- For even values of $n$ the functions $f(x) = 1/x^n$ are even, and their graphs are symmetric about the $y$-axis; for odd values of $n$ the functions $f(x) = x^n$ are odd, and their graphs are symmetric about the origin.

- For all values of $n$ the graphs pass through the point $(1, 1)$ and have a break (called a ***discontinuity***) at the origin. This is caused by the division by zero that occurs when $x = 0$. For even values of $n$ the graphs pass through $(-1, 1)$, and for odd values of $n$ they pass through $(-1, -1)$.

- Increasing $n$ causes the graph to become steeper over the interval $-1 < x < 1$ and flatter over the intervals $x > 1$ and $x < -1$.

REMARK.   The last characteristic can be explained numerically by considering the effect of raising the reciprocal of a number $x$ to successively higher powers. If $x$ is a nonzero fraction, then it lies in the interval $-1 < x < 1$, and its reciprocal satisfies $1/x > 1$ or $1/x < -1$. Thus, as $n$ increases the absolute value of $1/x^n$ also increases. This explains why successive graphs in Figure 1.6.6 become successively steeper over the interval $-1 < x < 1$. On the other hand, if $x > 1$ or $x < -1$, then $-1 < 1/x < 1$. Thus, as $n$ increases the absolute value of $1/x^n$ *decreases*. This explains why successive graphs in Figure 1.6.6 get successively flatter if $x > 1$ or $x < -1$.

**THE FAMILY $y = x^{1/n}$**

If $p = 1/n$, where $n$ is a positive integer, then the power functions $f(x) = x^p$ have the form $f(x) = x^{1/n} = \sqrt[n]{x}$. In particular, if $n = 2$, then $f(x) = \sqrt{x}$, and if $n = 3$, then $f(x) = \sqrt[3]{x}$. The graphs of these functions are shown in parts (*a*) and (*b*) of Figure 1.6.7. Observe that the graph of $y = \sqrt[3]{x}$ extends over the entire $x$-axis because $f(x) = \sqrt[3]{x}$ is defined for all real values of $x$ (every real number has a cube root); in contrast, the graph of $y = \sqrt{x}$ only extends over the nonnegative $x$-axis (negative numbers have imaginary square roots). Observe also that the graph of $y = \sqrt{x}$ is the upper half of the parabola $x = y^2$ (Figure 1.6.7*c*).

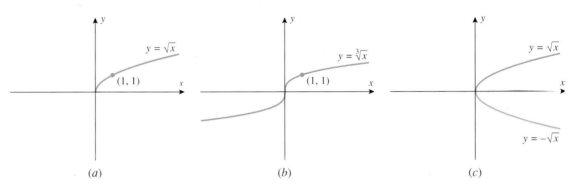

Figure 1.6.7

For even values of $n$ the graphs of $y = \sqrt[n]{x}$ have the same general shape as $y = \sqrt{x}$, and for odd values of $n$ they have the same general shape as $y = \sqrt[3]{x}$.

FOR THE READER.   Sketch the graphs of $y = \sqrt[n]{x}$ for $n = 2, 4, 6$ on one set of axes and for $n = 3, 5, 7$ on another set. Use a graphing device to check your work.

**POWER FUNCTIONS WITH FRACTIONAL AND IRRATIONAL EXPONENTS**

Power functions can also have fractional or irrational exponents. For example,

$$f(x) = x^{2/3}, \quad f(x) = \sqrt[5]{x^3}, \quad f(x) = x^{-7/8}, \quad \text{and} \quad f(x) = x^{\sqrt{2}} \tag{3}$$

are all power functions of this type; we will discuss power functions of these forms in later sections.

FOR THE READER.   Read the note preceding Exercise 29 of Section 1.3, and use a graphing utility to generate complete graphs of the functions in (3).

**MODELS INVOLVING INVERSE PROPORTIONS**

Recall that a variable $y$ is said to be *inversely proportional to a variable $x$* if there is a positive constant $k$, called the *constant of proportionality*, such that

$$y = \frac{k}{x} \tag{4}$$

Since $k$ is assumed to be positive, the graph of this equation has the same basic shape as $y = 1/x$ but is compressed or stretched in the $x$-direction.

Observe that in Formula (4) doubling $x$ decreases $y$ by a factor of $1/2$, tripling $x$ decreases $y$ by a factor of $1/3$, and, more generally, increasing $x$ by a factor of $r$ decreases $y$ by a factor of $1/r$.

Models involving inverse proportion arise in various laws of physics. For example, **Boyle's law** in physics states that at a constant temperature the pressure $P$ exerted by a fixed quantity of an ideal gas is inversely proportional to the volume $V$ occupied by the gas, that is,

$$P = \frac{k}{V}$$

(Figure 1.6.8).

If $y$ is inversely proportional to $x$, then it follows from (4) that the product of $y$ and $x$ is constant, since $yx = k$. This provides a useful way of identifying inverse proportion models in experimental data.

### Example 2

Table 1.6.1 shows some experimental data.

**Table 1.6.1**

EXPERIMENTAL DATA

| $x$ | 0.8 | 1 | 2.5 | 4 | 6.25 | 10 |
|---|---|---|---|---|---|---|
| $y$ | 6.25 | 5 | 2 | 1.25 | 0.8 | 0.5 |

(a)   Explain why the data suggest that $y$ is inversely proportional to $x$.

(b)   Express $y$ as a function of $x$.

(c)   Graph your function and the data together for $x \geq 0$.

*Solution.*   For every data point we have $xy = 5$, so $y$ is inversely proportional to $x$ and $y = 5/x$. The graph of this equation with the data points is shown in Figure 1.6.9.   ◄

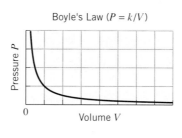

Boyle's Law ($P = k/V$)

As the volume of the gas changes, the temperature control unit adds or removes heat to maintain a constant temperature.

Figure 1.6.8

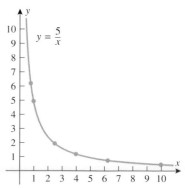

Figure 1.6.9

**A QUICK REVIEW OF POLYNOMIALS**

A detailed review of polynomials is given in Appendix F, but for convenience we will review some of the terminology here.

A *polynomial in x* is a function that is expressible as a sum of finitely many terms of the form $cx^n$, where $c$ is a constant and $n$ is a nonnegative integer. Some examples of polynomials are

$$2x + 1, \quad 3x^2 + 5x - \sqrt{2}, \quad x^3, \quad 4 (= 4x^0), \quad 5x^7 - x^4 + 3$$

The function $(x^2 - 4)^3$ is also a polynomial because it can be expanded by the binomial formula (see the inside front cover) and expressed as a sum of terms of the form $cx^n$:

$$(x^2 - 4)^3 = (x^2)^3 - 3(x^2)^2(4) + 3(x^2)(4^2) - (4^3) = x^6 - 12x^4 + 48x^2 - 64 \qquad (5)$$

A general polynomial can be written in either of the following forms, depending on whether one wants the powers of $x$ in ascending or descending order:

$$c_0 + c_1 x + c_2 x^2 + \cdots + c_n x^n$$
$$c_n x^n + c_{n-1} x^{n-1} + \cdots + c_1 x + c_0$$

The constants $c_0, c_1, \ldots, c_n$ are called the ***coefficients*** of the polynomial. When a polynomial is expressed in one of these forms, the highest power of $x$ that occurs with a nonzero coefficient is called the ***degree*** of the polynomial. Constants are considered to have degree 0, since we can write $c = cx^0$. Polynomials of degree 1, 2, 3, 4, and 5 are described as ***linear***, ***quadratic***, ***cubic***, ***quartic***, and ***quintic***, respectively. For example,

| | |
|---|---|
| $3 + 5x$ | Has degree 1 (linear) |
| $x^2 - 3x + 1$ | Has degree 2 (quadratic) |
| $2x^3 - 7$ | Has degree 3 (cubic) |
| $8x^4 - 9x^3 + 5x - 3$ | Has degree 4 (quartic) |
| $\sqrt{3} + x^3 + x^5$ | Has degree 5 (quintic) |
| $(x^2 - 4)^3$ | Has degree 6 [see (5)] |

The natural domain of a polynomial in $x$ is $(-\infty, +\infty)$, since the only operations involved are multiplication and addition; the range depends on the particular polynomial. We already know that the graphs of polynomials of degree 0 and 1 are lines and that the graphs of polynomials of degree 2 are parabolas. Figure 1.6.10 shows the graphs of some typical polynomials of higher degree. Later, we will discuss polynomial graphs in detail, but for now it suffices to observe that graphs of polynomials are very well behaved in the sense that they have no discontinuities or sharp corners. As illustrated in Figure 1.6.10, the graphs of polynomials wander up and down for awhile in a roller-coaster fashion, but eventually that behavior stops and the graphs steadily rise or fall indefinitely as one travels along the curve in either the positive or negative direction. We will see later that the number of peaks and valleys is determined by the degree of the polynomial.

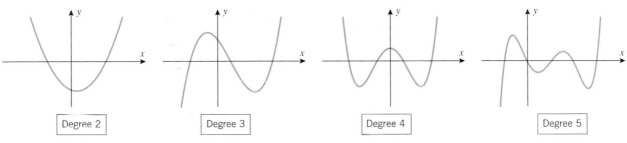

Degree 2    Degree 3    Degree 4    Degree 5

Figure 1.6.10

**RATIONAL FUNCTIONS**

A function that can be expressed as a ratio of two polynomials is called a ***rational function***. If $P(x)$ and $Q(x)$ are polynomials, then the domain of the rational function

$$f(x) = \frac{P(x)}{Q(x)}$$

consists of all values of $x$ such that $Q(x) \neq 0$. For example, the domain of the rational function

$$f(x) = \frac{x^2 + 2x}{x^2 - 1}$$

consists of all values of $x$, except $x = 1$ and $x = -1$. Its graph is shown in Figure 1.6.11 along with the graphs of two other typical rational functions.

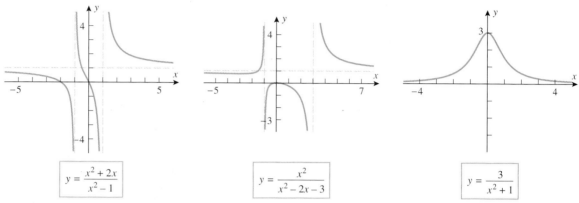

$$y = \frac{x^2 + 2x}{x^2 - 1}$$

$$y = \frac{x^2}{x^2 - 2x - 3}$$

$$y = \frac{3}{x^2 + 1}$$

Figure 1.6.11

The graphs of rational functions with nonconstant denominators differ from the graphs of polynomials in some essential ways:

- Unlike polynomials whose graphs are continuous (unbroken) curves, the graphs of rational functions have discontinuities at the points where the denominator is zero.

- As $x$ gets closer and closer to a point of discontinuity, the graph rises or falls indefinitely, getting closer and closer to a vertical line, called a ***vertical asymptote***; these are represented by the dashed vertical lines in Figure 1.6.11.

- Unlike the graphs of polynomials, which eventually rise or fall indefinitely, the graphs of many (but not all) rational functions eventually get closer and closer to some horizontal line, called a ***horizontal asymptote***, as one travels along the curve in either the positive or negative direction; these are represented by the dashed horizontal lines in the first two parts of Figure 1.6.11. In the third part of the figure the $x$-axis is a horizontal asymptote.

**ALGEBRAIC FUNCTIONS**

Functions that can be constructed from polynomials by applying finitely many algebraic operations (addition, subtraction, division, and root extraction) are called ***algebraic functions***. Some examples are

$$f(x) = \sqrt{x^2 - 4}, \quad f(x) = 3\sqrt[3]{x}(2 + x), \quad f(x) = x^{2/3}(x + 2)^2$$

As illustrated in Figure 1.6.12, the graphs of algebraic functions vary widely, so it is difficult to make general statements about them. Later in this text we will develop general calculus methods for analyzing such functions.

**A QUICK REVIEW OF TRIGONOMETRIC FUNCTIONS**

A detailed review of trigonometric functions is given in Appendix E, but for convenience we will summarize some of the main ideas here.

It is often convenient to think of the trigonometric functions in terms of circles rather than triangles. For this purpose, consider a point that moves either clockwise or counter-

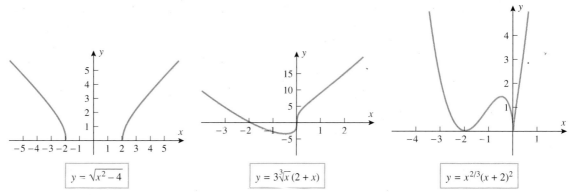

$$y = \sqrt{x^2 - 4}$$

$$y = 3\sqrt[3]{x}\,(2 + x)$$

$$y = x^{2/3}(x + 2)^2$$

Figure 1.6.12

clockwise along the ***unit circle*** $u^2 + v^2 = 1$ in the $uv$-plane, starting at $(1, 0)$ and stopping at a point $P$ (Figure 1.6.13$a$). Let $x$ denote the ***signed*** arc length traveled by the moving point, taking $x$ to be positive for counterclockwise motion and negative for clockwise motion. (We allow for the possibility that the point may traverse the circle more than once.) When convenient, the variable $x$ can also be interpreted as the angle in radians that is swept out by the radial line from the origin to $P$, with the usual convention that angles are positive if generated by counterclockwise rotations and negative if generated by clockwise rotations. We can *define* $\cos x$ to be the $u$-coordinate of $P$ and $\sin x$ to be the $v$-coordinate of $P$ (Figure 1.6.13$b$).

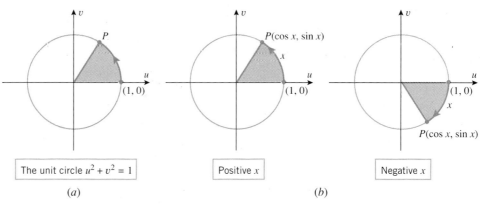

The unit circle $u^2 + v^2 = 1$

Positive $x$

Negative $x$

(a)

(b)

Figure 1.6.13

The remaining trigonometric functions can be defined in terms of the functions $\sin x$ and $\cos x$:

$$\tan x = \frac{\sin x}{\cos x} \qquad \cot x = \frac{\cos x}{\sin x}$$

$$\sec x = \frac{1}{\cos x} \qquad \csc x = \frac{1}{\sin x}$$

The graphs of the six trigonometric functions in Figure 1.6.14 should already be familiar to you, but try generating them using a graphing utility, making sure to use radian measure for $x$.

REMARK.   In this text we will always assume that the independent variable in a trigonometric function is in radians unless specifically stated otherwise.

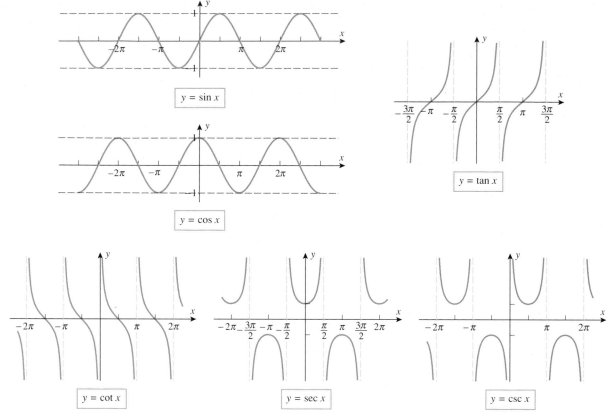

Figure 1.6.14

Many of the basic properties of $\sin x$ and $\cos x$ can be deduced from the circle definitions of these functions. For example:

- As the point $P(\cos x, \sin x)$ moves around the unit circle, its coordinates vary between $-1$ and $1$, and hence

  $$-1 \leq \sin x \leq 1 \quad \text{and} \quad -1 \leq \cos x \leq 1$$

- If $x$ increases or decreases by $2\pi$ radians, then the point $P(\cos x, \sin x)$ makes one complete revolution around the unit circle, and the coordinates return to their starting values. Thus, $\sin x$ and $\cos x$ have period $2\pi$; that is,

  $$\sin(x \pm 2\pi) = \sin x$$
  $$\cos(x \pm 2\pi) = \cos x$$

- As $P(\cos x, \sin x)$ moves around the unit circle, $\sin x$ is zero when $P$ is on the horizontal axis (which occurs when $x$ is an integer multiple of $\pi$), and $\cos x$ is zero when $P$ is on the vertical axis (which occurs when $x$ is an odd multiple of $\pi/2$). Thus,

  $$\sin x = 0 \quad \text{if and only if} \quad x = 0, \pm\pi, \pm2\pi, \pm3\pi, \ldots$$
  $$\cos x = 0 \quad \text{if and only if} \quad x = \pm\pi/2, \pm3\pi/2, \pm5\pi/2, \ldots$$

- As $P(\cos x, \sin x)$ moves around the unit circle $u^2 + v^2 = 1$, its coordinates satisfy this equation for all $x$, which produces the fundamental trigonometric identity

  $$\cos^2 x + \sin^2 x = 1$$

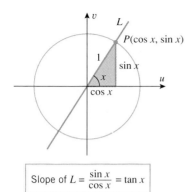

Slope of $L = \dfrac{\sin x}{\cos x} = \tan x$

Figure 1.6.15

Observe that the graph of $y = \tan x$ has vertical asymptotes at the points $x = \pm\pi/2$, $\pm 3\pi/2$, $\pm 5\pi/2$, .... This is to be expected since $\tan x = \sin x/\cos x$, and these are the values of $x$ at which $\cos x$ is zero. What is less obvious, however, is the fact that $\tan x$ repeats every $\pi$ radians (i.e., has period $\pi$), even though $\sin x$ and $\cos x$ have period $2\pi$. This can be explained by interpreting

$$\tan x = \frac{\sin x}{\cos x}$$

as the slope of the line $L$ that passes through the origin and the point $P(\cos x, \sin x)$ on the unit circle in the $uv$-plane (Figure 1.6.15). Each time $x$ increases or decreases by $\pi$ radians, the point $P$ traverses half the circumference, and the line $L$ rotates $\pi$ radians, so its starting and ending slope are the same.

## RADIANS AS A DIMENSIONLESS UNIT

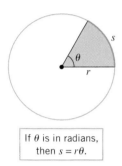

If $\theta$ is in radians, then $s = r\theta$.

Figure 1.6.16

The choice of radian measure as opposed to degree measure depends on the nature of the problem being considered; degree measure is usually chosen in engineering problems involving measurements of angles, and radian measure is usually chosen when the function properties of $\sin x$, $\cos x$, $\tan x$, ... are the primary focus. Radian measure is also usually chosen in problems involving arc lengths on circles because of the basic result in trigonometry which states that the arc length $s$ of a sector with radius $r$ and a central angle of $\theta$ (radians) is given by

$$s = r\theta \tag{6}$$

(Figure 1.6.16).

In applications involving angles, radians require special treatment to ensure that quantities are assigned proper units. To see why this is so, let us rewrite (6) as

$$\theta = \frac{s}{r}$$

The left side of this equation is in radians, and the right side is the ratio of two lengths, say meters/meters or feet/feet. However, because these units of length cancel, the right side of this equation is actually *dimensionless* (has no units). Thus, to ensure consistency between the two sides of the equation, we would have to omit the units of radians on the left side to make it dimensionless as well. In practical terms this means that units of radians can be used in intermediate computations, when convenient, but they need to be omitted in the end result to ensure consistency of units. This is confusing, to say the least, but the following example should clarify the idea.

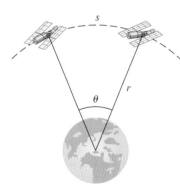

Figure 1.6.17

### Example 3

Suppose that two satellites circle the equator in an orbit of radius $r = 4.23 \times 10^7$ m (Figure 1.6.17). Find the arc length $s$ that separates the satellites if they have an angular separation of $\theta = 2.00°$.

*Solution.* To apply Formula (6), we must convert the angular separation to radians:

$$2.00° = \frac{\pi}{180}(2.00) \approx 0.0349 \text{ rad}$$

Thus, from (6)

$$s = r\theta = (4.23 \times 10^7 \text{ m})(0.0349 \text{ rad}) = 1.48 \times 10^6 \text{ m}$$

In this computation the product $r\theta$ produces units of meters $\times$ radians, but if we treat radians as dimensionless, we have meters $\times$ radians $=$ meters, which correctly produces units of meters (m) for the arc length $s$. ◀

Many important applications lead to trigonometric functions of the form

$$f(x) = A \sin(Bx - C) \quad \text{and} \quad g(x) = A \cos(Bx - C) \tag{7}$$

where $A$, $B$, and $C$ are nonzero constants. The graphs of such functions can be obtained by stretching, compressing, translating, and reflecting the graphs of $y = \sin x$ and $y = \cos x$ appropriately. To see why this is so, let us start with the case where $C = 0$ and consider how the graphs of the equations

$$y = A \sin Bx \quad \text{and} \quad y = A \cos Bx$$

relate to the graphs of $y = \sin x$ and $y = \cos x$. If $A$ and $B$ are positive, then the effect of the constant $A$ is to stretch or compress the graphs of $y = \sin x$ and $y = \cos x$ vertically by a factor of $A$, and the effect of the constant $B$ is to compress or stretch the graphs of $\sin x$ and $\cos x$ horizontally by a factor of $B$. For example, the graph of $y = 2 \sin 4x$ can be obtained by stretching the graph of $y = \sin x$ vertically by a factor of 2 and compressing it horizontally by a factor of 4. (Recall from Section 1.4 that the multiplier of $x$ *stretches* when it is less than 1 and *compresses* when it is greater than 1.) Thus, as shown in Figure 1.6.18, the graph of $y = 2 \sin 4x$ varies between $-2$ and 2, and repeats every $2\pi/4 = \pi/2$ units.

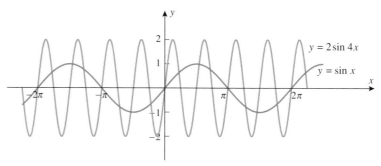

Figure 1.6.18

In general, if $A$ and $B$ are positive numbers, then the graphs of

$$y = A \sin Bx \quad \text{and} \quad y = A \cos Bx$$

oscillate between $-A$ and $A$ and repeat every $2\pi/B$ units, so we say that these functions have **amplitude** $A$ and **period** $2\pi/B$. In addition, we define the **frequency** of these functions to be the reciprocal of the period, that is, the frequency is $B/2\pi$. If $A$ or $B$ is negative, then these constants cause reflections of the graphs about the axes as well as compressing or stretching them; and in this case the amplitude, period, and frequency are given by $|A|$, $2\pi/|B|$, and $|B|/2\pi$, respectively.

### Example 4

Make sketches of the following graphs that show the period and amplitude.

(a)  $y = 3 \sin 2\pi x$      (b)  $y = -3 \cos 0.5x$      (c)  $y = 1 + \sin x$

*Solution (a).* The equation is of the form $y = A \sin Bx$ with $A = 3$ and $B = 2\pi$, so the graph has the shape of a sine function, but with amplitude $A = 3$ and period $2\pi/B = 2\pi/2\pi = 1$ (Figure 1.6.19a).

*Solution (b).* The equation is of the form $y = A \cos Bx$ with $A = -3$ and $B = 0.5$, so the graph has the shape of a cosine function that has been reflected about the $x$-axis (because $A = -3$ is negative), but with amplitude $|A| = 3$ and period $2\pi/B = 2\pi/0.5 = 4\pi$ (Figure 1.6.19b).

*Solution (c).* The graph has the shape of a sine function that has been translated up 1 unit (Figure 1.6.19c).  ◀

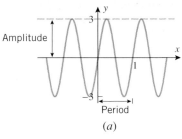
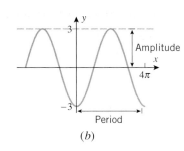
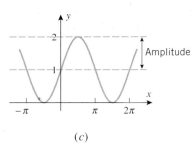

(a)           (b)           (c)

Figure 1.6.19

**THE FAMILIES $y = A \sin(Bx - C)$ AND $y = A \cos(Bx - C)$**

To investigate the graphs of the more general families

$$y = A \sin(Bx - C) \quad \text{and} \quad y = A \cos(Bx - C)$$

it will be helpful to rewrite these equations as

$$y = A \sin\left[ B\left( x - \frac{C}{B} \right) \right] \quad \text{and} \quad y = A \cos\left[ B\left( x - \frac{C}{B} \right) \right]$$

In this form we see that the graphs of these equations can be obtained by translating the graphs of $y = A \sin Bx$ and $y = A \cos Bx$ to the left or right, depending on the sign of $C/B$. For example, if $C/B > 0$, then the graph of

$$y = A \sin[B(x - C/B)] = A \sin(Bx - C)$$

can be obtained by translating the graph of $y = A \sin Bx$ to the right by $C/B$ units (Figure 1.6.20). The quantity $C/B$ is called the **phase shift** of the function; a positive phase shift corresponds to right translation, and a negative phase shift corresponds to a left translation.

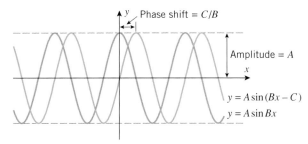

Figure 1.6.20

### Example 5

Find the amplitude, period, and phase shift of

$$y = 3 \cos\left( 2x + \frac{\pi}{2} \right)$$

and confirm your results by graphing the equation on a calculator or computer.

*Solution.* The equation can be rewritten as

$$y = 3 \cos\left[ 2x - \left( -\frac{\pi}{2} \right) \right] = 3 \cos\left[ 2\left( x - \left( -\frac{\pi}{4} \right) \right) \right]$$

which is of the form

$$y = A \cos\left[ B\left( x - \frac{C}{B} \right) \right]$$

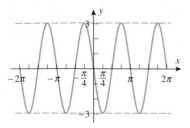

Figure 1.6.21

with $A = 3$, $B = 2$, and $C/B = -\pi/4$; thus, the graph has the shape of a cosine function, but with amplitude $A = 3$, period $2\pi/B = \pi$, and phase shift $C/B = -\pi/4$ (Figure 1.6.21). ◄

### Example 6

Figure 1.6.22a shows a table and scatter plot of temperature data recorded over a 24-hour period in the city of Philadelphia.[*] Find a function that models the data, and graph your function and data together.

PHILADELPHIA TEMPERATURES
FROM 1:00 A.M. TO 12:00 MIDNIGHT ON 27 AUGUST 1993
($t$ = HOURS AFTER MIDNIGHT AND $T$ = DEGREES FAHRENHEIT)

| | A.M. | | P.M. | |
|---|---|---|---|---|
| | $t$ | $T$ | $t$ | $T$ |
| 1:00 | 1 | 78° | 13 | 91° |
| 2:00 | 2 | 77° | 14 | 93° |
| 3:00 | 3 | 77° | 15 | 94° |
| 4:00 | 4 | 76° | 16 | 95° |
| 5:00 | 5 | 76° | 17 | 93° |
| 6:00 | 6 | 75° | 18 | 92° |
| 7:00 | 7 | 75° | 19 | 89° |
| 8:00 | 8 | 77° | 20 | 86° |
| 9:00 | 9 | 79° | 21 | 84° |
| 10:00 | 10 | 83° | 22 | 83° |
| 11:00 | 11 | 87° | 23 | 81° |
| 12:00 | 12 | 90° | 24 | 79° |

Source: Philadelphia Inquirer, 28 August 1993.

Figure 1.6.22

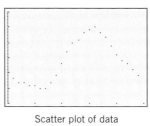

Scatter plot of data
$[0, 25] \times [70, 100]$
$t$    $T$

(a)

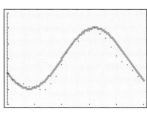

Model for data
$T = 85 + 10 \sin[(\pi/12)(t - 10)]$
$[0, 25] \times [70, 100]$
$t$    $T$

(b)

*Solution.* The pattern of the data suggests that the relationship between the temperature $T$ and the time $t$ can be modeled by a sinusoidal function that has been translated both horizontally and vertically, so we will look for an equation of the form

$$T = D + A \sin[Bt - C] = D + A \sin\left[B\left(t - \frac{C}{B}\right)\right] \tag{8}$$

Since the highest temperature is $95°\text{F}$ and the lowest temperature is $75°\text{F}$, we take $2A = 20$ or $A = 10$. The midpoint between the high and low is $85°\text{F}$, so we have a vertical shift of $D = 85$. The period seems to be about 24, so $2\pi/B = 24$ or $B = \pi/12$. The phase shift appears to be about 10 (verify), so $C/B = 10$. Substituting these values in (8) yields the equation

$$T = 85 + 10 \sin\left[\frac{\pi}{12}(t - 10)\right]$$

(Figure 1.6.22b). ◄

---

[*]This example is based on the article "Everybody Talks About It!—Weather Investigations," by Gloria S. Dion and Iris Brann Fetta, *The Mathematics Teacher*, Vol. 89, No. 2, February 1996, pp. 160–165.

**OTHER FAMILIES**

In addition to the functions mentioned in this section, there are exponential and logarithmic functions, which we will study later, and various special functions that arise in physics and engineering. There are also many kinds of functions that have no names; indeed, one of the important themes of calculus is to provide methods for analyzing new types of functions.

## EXERCISE SET 1.6  ⌢ Graphing Calculator  C CAS

1. (a) Find an equation for the family of lines whose members have slope $m = 3$.
   (b) Find an equation for the member of the family that passes through $(-1, 3)$.
   (c) Sketch some members of the family, and label them with their equations. Include the line in part (b).

2. Find an equation for the family of lines whose members are perpendicular to those in Exercise 1.

3. (a) Find an equation for the family of lines with $y$-intercept $b = 2$.
   (b) Find an equation for the member of the family whose angle of inclination is $135°$.
   (c) Sketch some members of the family, and label them with their equations. Include the line in part (b).

4. Find an equation for
   (a) the family of lines that pass through the origin
   (b) the family of lines with $x$-intercept $a = 1$
   (c) the family of lines that pass through the point $(1, -2)$
   (d) the family of lines parallel to $2x + 4y = 1$.

In Exercises 5 and 6, state a geometric property common to all lines in the family, and sketch five of the lines.

5. (a) The family $y = -x + b$
   (b) The family $y = mx - 1$
   (c) The family $y = m(x + 4) + 2$
   (d) The family $x - ky = 1$

6. (a) The family $y = b$
   (b) The family $Ax + 2y + 1 = 0$
   (c) The family $2x + By + 1 = 0$
   (d) The family $y - 1 = m(x + 1)$

7. Find an equation for the family of lines tangent to the circle with center at the origin and radius 3.

8. Find an equation for the family of lines that pass through the intersection of $5x - 3y + 11 = 0$ and $2x - 9y + 7 = 0$.

9. The U.S. Internal Revenue Service uses a 10-year linear depreciation schedule to determine the value of various business items. This means that an item is assumed to have a

value of zero at the end of the tenth year and that at intermediate times the value is a linear function of the elapsed time. Sketch some typical depreciation lines, and explain the practical significance of the $y$-intercepts.

10. Find all lines through $(6, -1)$ for which the product of the $x$- and $y$-intercepts is 3.

11. In each part, match the equation with one of the accompanying graphs.
    (a) $y = \sqrt[5]{x}$
    (b) $y = 2x^5$
    (c) $y = -1/x^8$
    (d) $y = 8^x$
    (e) $y = \sqrt[4]{x - 2}$
    (f) $y = 1/8^x$

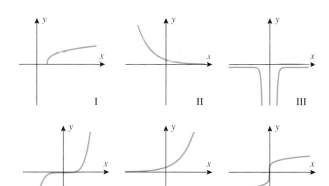

Figure Ex-11

12. The table in the accompanying figure gives approximate values of three functions: one of the form $kx^2$, one of the form $kx^{-3}$, and one of the form $kx^{3/2}$. Identify which is which, and estimate $k$ in each case.

| $x$ | 0.25 | 0.37 | 2.1 | 4.0 | 5.8 | 6.2 | 7.9 | 9.3 |
|---|---|---|---|---|---|---|---|---|
| $f(x)$ | 640 | 197 | 1.08 | 0.156 | 0.0513 | 0.0420 | 0.0203 | 0.0124 |
| $g(x)$ | 0.0312 | 0.0684 | 2.20 | 8.00 | 16.8 | 19.2 | 31.2 | 43.2 |
| $h(x)$ | 0.250 | 0.450 | 6.09 | 16.0 | 27.9 | 30.9 | 44.4 | 56.7 |

Figure Ex-12

In Exercises 13 and 14, sketch the graph of the equation for $n = 1, 3$, and 5 in one coordinate system and for $n = 2, 4$, and 6 in another coordinate system. Check your work with a graphing utility.

13. (a) $y = -x^n$     (b) $y = 2x^{-n}$     (c) $y = (x - 1)^{1/n}$

14. (a) $y = 2x^n$          (b) $y = -x^{-n}$
    (c) $y = -3(x + 2)^{1/n}$

15. (a) Sketch the graph of $y = ax^2$ for $a = \pm 1, \pm 2$, and $\pm 3$ in a single coordinate system.
    (b) Sketch the graph of $y = x^2 + b$ for $b = \pm 1, \pm 2$, and $\pm 3$ in a single coordinate system.
    (c) Sketch some typical members of the family of curves $y = ax^2 + b$.

16. (a) Sketch the graph of $y = a\sqrt{x}$ for $a = \pm 1, \pm 2$, and $\pm 3$ in a single coordinate system.
    (b) Sketch the graph of $y = \sqrt{x} + b$ for $b = \pm 1, \pm 2$, and $\pm 3$ in a single coordinate system.
    (c) Sketch some typical members of the family of curves $y = a\sqrt{x} + b$.

In Exercises 17–20, sketch the graph of the equation by making appropriate transformations to the graph of a basic power function. Check your work with a graphing utility.

17. (a) $y = 2(x + 1)^2$       (b) $y = -3(x - 2)^3$
    (c) $y = \dfrac{-3}{(x + 1)^2}$     (d) $y = \dfrac{1}{(x - 3)^5}$

18. (a) $y = 1 - \sqrt{x + 2}$     (b) $y = 1 - \sqrt[3]{x + 2}$
    (c) $y = \dfrac{5}{(1 - x)^3}$      (d) $y = \dfrac{2}{(4 + x)^4}$

19. (a) $y = \sqrt[3]{x + 1}$       (b) $y = 1 - \sqrt{x - 2}$
    (c) $y = (x - 1)^5 + 2$    (d) $y = \dfrac{x + 1}{x}$

20. (a) $y = 1 + \dfrac{1}{x - 2}$    (b) $y = \dfrac{1}{1 + 2x - x^2}$
    (c) $y = -\dfrac{2}{x^7}$       (d) $y = x^2 + 2x$

21. Sketch the graph of $y = x^2 + 2x$ by completing the square and making appropriate transformations to the graph of $y = x^2$.

22. (a) Use the graph of $y = \sqrt{x}$ to help sketch the graph of $y = \sqrt{|x|}$.
    (b) Use the graph of $y = \sqrt[3]{x}$ to help sketch the graph of $y = \sqrt[3]{|x|}$.

23. The table in the accompanying figure provides data about the relationship between distance $d$ traveled in meters and elapsed time $t$ in seconds for an object dropped near the Earth's surface. Plot time versus distance and make a guess at a "square-root function" that provides a reasonable model for $t$ in terms of $d$. Use a graphing utility to confirm the reasonableness of your guess.

| $d$ (meters) | 0 | 2.5 | 5 | 10 | 15 | 20 | 25 |
|---|---|---|---|---|---|---|---|
| $t$ (seconds) | 0 | 0.7 | 1.0 | 1.4 | 1.7 | 2 | 2.3 |

Figure Ex-23

24. (a) The table below provides data on five moons of the planet Saturn. In this table $r$ is the *orbital radius* (the average distance between the moon and Saturn) and $t$ is the time in days required for the moon to complete one orbit around Saturn. For each data pair calculate $tr^{-3/2}$, and use your results to find a reasonable model for $r$ as a function of $t$.
    (b) Use the model from part (a) to estimate the orbital radius of the moon Enceladus, given that its orbit time is $t \approx 1.370$ days.
    (c) Use the model from part (a) to estimate the orbit time of the moon Tethys, given that its orbital radius is $r \approx 2.9467 \times 10^5$ km.

| Moon | Radius (100,000 km) | Orbit Time (days) |
|---|---|---|
| 1980S28 | 1.3767 | 0.602 |
| 1980S27 | 1.3935 | 0.613 |
| 1980S26 | 1.4170 | 0.629 |
| 1980S3 | 1.5142 | 0.694 |
| 1980S1 | 1.5147 | 0.695 |

25. As discussed in this section, Boyle's law states that at a constant temperature the pressure $P$ exerted by a gas is related to the volume $V$ by the equation $P = k/V$.
    (a) Find the appropriate units for the constant $k$ if pressure (which is force per unit area) is in newtons per square meter ($N/m^2$) and volume is in cubic meters ($m^3$).
    (b) Find $k$ if the gas exerts a pressure of 20,000 $N/m^2$ when the volume is 1 liter (0.001 $m^3$).
    (c) Make a table that shows the pressures for volumes of 0.25, 0.5, 1.0, 1.5, and 2.0 liters.
    (d) Make a graph of $P$ versus $V$.

26. A manufacturer of cardboard drink containers wants to construct a closed rectangular container that has a square base and will hold $\frac{1}{10}$ liter (100 $cm^3$). Estimate the dimension of the container that will require the least amount of material for its manufacture.

A variable $y$ is said to be ***inversely proportional to the square of a variable x*** if $y$ is related to $x$ by an equation of the form $y = k/x^2$, where $k$ is a nonzero constant, called the ***constant of proportionality***. This terminology is used in Exercises 27 and 28.

**27.** According to Coulomb's law, the force $F$ of attraction between positive and negative point charges is inversely proportional to the square of the distance $x$ between them.

(a) Assuming that the force of attraction between two point charges is 0.0005 newton when the distance between them is 0.3 meter, find the constant of proportionality (with proper units).

(b) Find the force of attraction between the point charges when they are 3 meters apart.

(c) Make a graph of force versus distance for the two charges.

(d) What happens to the force as the particles get closer and closer together? What happens as they get farther and farther apart?

**28.** It follows from Newton's Universal Law of Gravitation that the weight $W$ of an object (relative to the Earth) is inversely proportional to the square of the distance $x$ between the object and the center of the Earth, that is, $W = C/x^2$.

(a) Assuming that a weather satellite weighs 2000 pounds on the surface of the Earth and that the Earth is a sphere of radius 4000 miles, find the constant $C$.

(b) Find the weight of the satellite when it is 1000 miles above the surface of the Earth.

(c) Make a graph of the satellite's weight versus its distance from the center of the Earth.

(d) Is there any distance from the center of the Earth at which the weight of the satellite is zero? Explain your reasoning.

**29.** In each part, match the equation with one of the accompanying graphs, and give the equations for the horizontal and vertical asymptotes.

(a) $y = \dfrac{x^2}{x^2 - x - 2}$

(b) $y = \dfrac{x - 1}{x^2 - x - 6}$

(c) $y = \dfrac{2x^4}{x^4 + 1}$

(d) $y = \dfrac{4}{(x + 2)^2}$

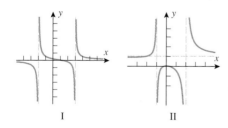

I

II

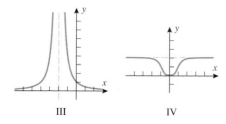

III

IV

Figure Ex-29

**30.** Find an equation of the form $y = k/(x^2 + bx + c)$ whose graph is a reasonable match to that in the accompanying figure. Check your work with a graphing utility.

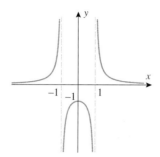

Figure Ex-30

In Exercises 31 and 32, draw a radial line from the origin with the given angle, and determine whether the six trigonometric functions are positive, negative, or undefined for that angle.

**31.** (a) $\dfrac{\pi}{3}$

(b) $-\dfrac{\pi}{2}$

(c) $\dfrac{2\pi}{3}$

(d) $-1$

(e) $\dfrac{5\pi}{4}$

(f) $\dfrac{11\pi}{6}$

**32.** (a) $\dfrac{3\pi}{2}$

(b) $-\dfrac{5\pi}{4}$

(c) $\pi$

(d) $\dfrac{5\pi}{2}$

(e) $4$

(f) $-\dfrac{33\pi}{7}$

In Exercises 33 and 34, use a calculating utility set to the radian mode to confirm the approximations $\sin(\pi/5) \approx 0.588$ and $\cos(\pi/8) \approx 0.924$, and then use these values to approximate the given expressions by hand calculation. Check your answers using the trigonometric function operations of your calculating utility.

**33.** (a) $\sin\dfrac{4\pi}{5}$

(b) $\cos\left(-\dfrac{\pi}{8}\right)$

(c) $\sin\dfrac{11\pi}{5}$

(d) $\cos\dfrac{7\pi}{8}$

(e) $\sin\dfrac{2\pi}{5}$

(f) $\cos^2\dfrac{\pi}{5}$

**34.** (a) $\sin\dfrac{16\pi}{5}$

(b) $\cos\left(-\dfrac{17\pi}{8}\right)$

(c) $\sin\dfrac{41\pi}{5}$

(d) $\sin\left(-\dfrac{\pi}{16}\right)$

(e) $\cos\dfrac{27\pi}{8}$

(f) $\tan^2\dfrac{\pi}{8}$

**35.** Assuming that $\sin\alpha = a$, $\cos\beta = b$, and $\tan\gamma = c$, express the stated quantities in terms of $a$, $b$, and $c$.

(a) $\sin(-\alpha)$

(b) $\cos(-\beta)$

(c) $\tan(-\gamma)$

(d) $\sin\left(\dfrac{\pi}{2} - \alpha\right)$

(e) $\cos(\pi - \beta)$

(f) $\sin(\alpha + \pi)$

(g) $\sin(2\beta)$

(h) $\cos(2\beta)$

(i) $\sec(\beta + 2\pi)$

(j) $\csc(\alpha + \pi)$

(k) $\cot(\gamma + 5\pi)$

(l) $\sin^2\left(\dfrac{\beta}{2}\right)$

**36.** A ship travels from a point near Hawaii at $20°$ N latitude directly north to a point near Alaska at $56°$ N latitude.

    (a) Assuming the Earth to be a sphere of radius 4000 mi, find the actual distance traveled by the ship.

    (b) What fraction of the Earth's circumference did the ship travel?

**37.** The Moon completes one revolution around the Earth in approximately 29.5 days. Assuming that the Moon's orbit is a circle with a radius of $0.38 \times 10^9$ m from the center of the Earth, find the arc length traveled by the Moon in 1 day.

**38.** A spoked wheel with a diameter of 3 ft rolls along a flat road without slipping. How far along the road does the wheel roll if the spokes turn through $225°$?

**39.** As illustrated in the accompanying figure, suppose that you hold one quarter flat against a table while you rotate a second quarter around it without slippage. Through what angle will the second quarter have turned about its own center when it returns to its original location?

Figure Ex-39

**40.** Suppose that you begin cutting wedge-shaped pieces from a pie so that the arc length along the outer crust of each piece is equal to the radius. What fraction of the pie will remain after all pieces that can be cut in this way are eaten?

In Exercises 41 and 42, find an equation for the graph assuming that there is no phase shift.

**41.**

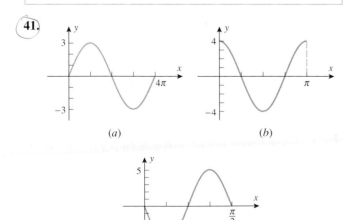

Figure Ex-41

**42.**

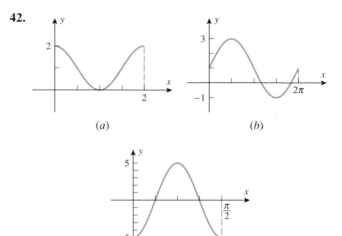

Figure Ex-42

**43.** In each part, find an equation for the graph that has the form $y = y_0 + A \sin(Bx - C)$.

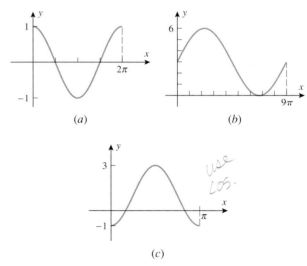

Figure Ex-43

**44.** In the United States, a standard electrical outlet supplies sinusoidal electrical current with a maximum voltage of $V = 120\sqrt{2}$ volts (V) at a frequency of 60 cycles per second. Write an equation that expresses $V$ as a function of the time $t$, assuming that $V = 0$ if $t = 0$.

In Exercises 45 and 46, find the amplitude, period, and phase shift, and sketch at least two periods of the graph by hand. Check your work with a graphing utility.

**45.** (a) $y = 3 \sin 4x$            (b) $y = -2 \cos \pi x$

     (c) $y = 2 + \cos\left(\dfrac{x}{2}\right)$

**46.** (a) $y = -1 - 4\sin 2x$     (b) $y = \frac{1}{2}\cos(3x - \pi)$

(c) $y = -4\sin\left(\dfrac{x}{3} + 2\pi\right)$

**47.** Equations of the form

$$x = A_1 \sin \omega t + A_2 \cos \omega t$$

arise in the study of vibrations and other periodic motion.

(a) Use the trigonometric identity for $\sin(\alpha + \beta)$ to show that this equation can be expressed in the form

$$x = A \sin(\omega t + \theta)$$

(b) State formulas that express $A$ and $\theta$ in terms of the constants $A_1$, $A_2$, and $\omega$.

(c) Express the equation

$$x = 5\sqrt{3}\sin 2\pi t + \tfrac{5}{2}\cos 2\pi t$$

in the form $x = A\sin(\omega t + \theta)$, and use a graphing utility to confirm that both equations have the same graph.

**48.** Determine the number of solutions of $x = 2\sin x$, and use a graphing or calculating utility to estimate them.

# 1.7 PARAMETRIC EQUATIONS

*Thus far, our study of graphs has focused on graphs of functions. However, because such graphs must pass the vertical line test, this limitation precludes curves with self-intersections or even such basic curves as circles. In this section we will study an alternative method for describing curves algebraically that is not subject to the severe restriction of the vertical line test.*

> This material is placed here to provide an early parametric option. However, it can be deferred until Chapter 12, if preferred.

### PARAMETRIC EQUATIONS

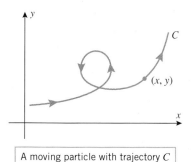

A moving particle with trajectory $C$

Figure 1.7.1

Suppose that a particle moves along a curve $C$ in the $xy$-plane in such a way that its $x$- and $y$-coordinates, as functions of time, are

$$x = f(t), \quad y = g(t)$$

We call these the **parametric equations** of motion for the particle and refer to $C$ as the **trajectory** of the particle or the **graph** of the equations (Figure 1.7.1). The variable $t$ is called the **parameter** for the equations.

### Example 1

Sketch the trajectory over the time interval $0 \leq t \leq 10$ of the particle whose parametric equations of motion are

$$x = t - 3\sin t, \quad y = 4 - 3\cos t \qquad (1)$$

*Solution.* One way to sketch the trajectory is to choose a representative succession of times, plot the $(x, y)$ coordinates of points on the trajectory at those times, and connect the points with a smooth curve. The trajectory in Figure 1.7.2 was obtained in this way from Table 1.7.1 in which the approximate coordinates of the particle are given at time increments of 1 unit. Observe that there is no $t$-axis in the picture; the values of $t$ appear only as labels on the plotted points, and even these are usually omitted unless it is important to emphasize the location of the particle at specific times. ◄

FOR THE READER. Read the documentation for your graphing utility to learn how to graph parametric equations, and then generate the trajectory in Example 1. Explore the behavior of the particle beyond time $t = 10$.

Although parametric equations commonly arise in problems of motion with time as the parameter, they arise in other contexts as well. Thus, unless the problem dictates that the

**Table 1.7.1**

| $t$ | $x$ | $y$ |
|---|---|---|
| 0 | 0.0 | 1.0 |
| 1 | −1.5 | 2.4 |
| 2 | −0.7 | 5.2 |
| 3 | 2.6 | 7.0 |
| 4 | 6.3 | 6.0 |
| 5 | 7.9 | 3.1 |
| 6 | 6.8 | 1.1 |
| 7 | 5.0 | 1.7 |
| 8 | 5.0 | 4.4 |
| 9 | 7.8 | 6.7 |
| 10 | 11.6 | 6.5 |

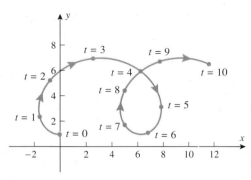

Figure 1.7.2

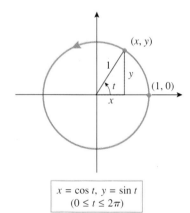

$$x = \cos t, \ y = \sin t$$
$$(0 \le t \le 2\pi)$$

Figure 1.7.3

parameter $t$ in the equations

$$x = f(t), \quad y = g(t)$$

represents time, it should be viewed simply as an independent variable that varies over some interval of real numbers. (In fact, there is no need to use the letter $t$ for the parameter; any letter not reserved for another purpose can be used.) If no restrictions on the parameter are stated explicitly or implied by the equations, then it is understood that it varies from $-\infty$ to $+\infty$. To indicate that a parameter $t$ is restricted to an interval $[a, b]$, we will write

$$x = f(t), \quad y = g(t) \qquad (a \le t \le b)$$

### Example 2

Find the graph of the parametric equations

$$x = \cos t, \quad y = \sin t \qquad (0 \le t \le 2\pi) \tag{2}$$

*Solution.* One way to find the graph is to eliminate the parameter $t$ by noting that

$$x^2 + y^2 = \sin^2 t + \cos^2 t = 1$$

Thus, the graph is the unit circle $x^2 + y^2 = 1$. This result can also be deduced geometrically by interpreting $t$ as the angle swept out by the radial line from the origin to the point $(x, y) = (\cos t, \sin t)$ on the unit circle (Figure 1.7.3). As $t$ increases from 0 to $2\pi$, the point traces the circle counterclockwise, starting at $(1, 0)$ when $t = 0$ and completing one full revolution when $t = 2\pi$. One can obtain different portions of the circle by varying the interval over which the parameter varies. For example,

$$x = \cos t, \quad y = \sin t \qquad (0 \le t \le \pi) \tag{3}$$

represents just the upper semicircle in Figure 1.7.3. ◀

**ORIENTATION**

The direction in which the graph of a pair of parametric equations is traced as the parameter increases is called the ***direction of increasing parameter*** or sometimes the ***orientation*** imposed on the curve by the equations. Thus, we make a distinction between a ***curve***, which is a set of points, and a ***parametric curve***, which is a curve with an orientation imposed on it by a set of parametric equations. For example, we saw in Example 2 that the circle represented parametrically by (2) is traced counterclockwise as $t$ increases and hence has *counterclockwise orientation*. As shown in Figures 1.7.2 and 1.7.3, the orientation of a parametric curve can be indicated by arrowheads.

To obtain parametric equations for the unit circle with *clockwise orientation*, we can replace $t$ by $-t$ in (2), and use the identities $\cos(-t) = \cos t$ and $\sin(-t) = -\sin t$. This yields

$$x = \cos t, \quad y = -\sin t \qquad (0 \le t \le 2\pi)$$

Here, the circle is traced clockwise by a point that starts at $(1, 0)$ when $t = 0$ and completes one full revolution when $t = 2\pi$ (Figure 1.7.4).

FOR THE READER.    When parametric equations are graphed using a calculator, the orientation can often be determined by watching the direction in which the graph is traced on the screen. However, many computers graph so fast that it is often hard to discern the orientation. See if you can use your graphing utility to confirm that (3) has a counterclockwise orientation.

### Example 3

Graph the parametric curve

$$x = 2t - 3, \quad y = 6t - 7$$

by eliminating the parameter, and indicate the orientation on the graph.

*Solution.*    To eliminate the parameter we will solve the first equation for $t$ as a function of $x$, and then substitute this expression for $t$ into the second equation:

$$t = \left(\tfrac{1}{2}\right)(x + 3)$$
$$y = 6\left(\tfrac{1}{2}\right)(x + 3) - 7$$
$$y = 3x + 2$$

Thus, the graph is a line of slope 3 and $y$-intercept 2. To find the orientation we must look to the original equations; the direction of increasing $t$ can be deduced by observing that $x$ increases as $t$ increases *or* by observing that $y$ increases as $t$ increases. Either piece of information tells us that the line is traced left to right as shown in Figure 1.7.5.    ◀

REMARK.    Not all parametric equations produce curves with definite orientations; if the equations are badly behaved, then the point tracing the curve may leap around sporadically or move back and forth, failing to determine a definite direction. For example, if

$$x = \sin t, \quad y = \sin^2 t$$

then the point $(x, y)$ moves along the parabola $y = x^2$. However, the value of $x$ varies periodically between $-1$ and $1$, so the point $(x, y)$ moves periodically back and forth along the parabola between the points $(-1, 1)$ and $(1, 1)$ (as shown in Figure 1.7.6). Later in the text we will discuss restrictions that eliminate such erratic behavior, but for now we will just avoid such complications.

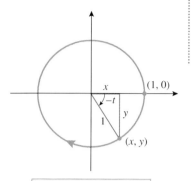

$x = \cos(-t), \ y = \sin(-t)$
$(0 \le t \le 2\pi)$

Figure 1.7.4

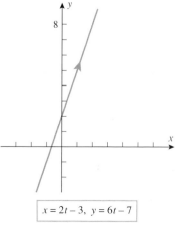

$x = 2t - 3, \ y = 6t - 7$

Figure 1.7.5

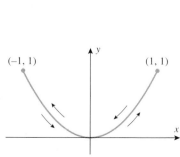

Figure 1.7.6

**EXPRESSING ORDINARY FUNCTIONS PARAMETRICALLY**

An equation $y = f(x)$ can be expressed in parametric form by introducing the parameter $t = x$; this yields the parametric equations $x = t$, $y = f(t)$. For example, the portion of the curve $y = \cos x$ over the interval $[-2\pi, 2\pi]$ can be expressed parametrically as

$$x = t, \quad y = \cos t \quad (-2\pi \leq t \leq 2\pi)$$

(Figure 1.7.7).

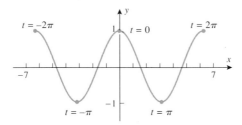

Figure 1.7.7

**GENERATING PARAMETRIC CURVES WITH GRAPHING UTILITIES**

Many graphing utilities allow you to graph equations of the form $y = f(x)$ but not equations of the form $x = g(y)$. Sometimes you will be able to rewrite $x = g(y)$ in the form $y = f(x)$; however, if this is inconvenient or impossible, then you can graph $x = g(y)$ by introducing a parameter $t = y$ and expressing the equation in the parametric form $x = g(t)$, $y = t$. (You may have to experiment with various intervals for $t$ to produce a complete graph.)

### Example 4

Use a graphing utility to graph the equation $x = 3y^5 - 5y^3 + 1$.

*Solution.* If we let $t = y$ be the parameter, then the equation can be written in parametric form as

$$x = 3t^5 - 5t^3 + 1, \quad y = t$$

Figure 1.7.8 shows the graph of these equations for $-1.5 \leq t \leq 1.5$. ◀

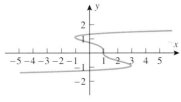

$x = 3t^5 - 5t^3 + 1, \ y = t$
$-1.5 \leq t \leq 1.5$

Figure 1.7.8

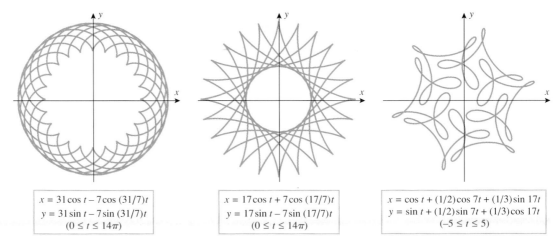

$x = 31\cos t - 7\cos (31/7)t$
$y = 31\sin t - 7\sin (31/7)t$
$(0 \leq t \leq 14\pi)$

$x = 17\cos t + 7\cos (17/7)t$
$y = 17\sin t - 7\sin (17/7)t$
$(0 \leq t \leq 14\pi)$

$x = \cos t + (1/2)\cos 7t + (1/3)\sin 17t$
$y = \sin t + (1/2)\sin 7t + (1/3)\cos 17t$
$(-5 \leq t \leq 5)$

Figure 1.7.9

Some parametric curves are so complex that it is virtually impossible to visualize them without using some kind of graphing utility. Figure 1.7.9 shows three such curves.

FOR THE READER. Without spending too much time, try your hand at generating some parametric curves with a graphing utility that you think are interesting or beautiful.

**TRANSLATION**

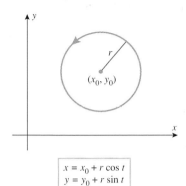

$$x = x_0 + r \cos t$$
$$y = y_0 + r \sin t$$
$$(0 \le t \le 2\pi)$$

Figure 1.7.10

If a parametric curve $C$ is given by the equations $x = f(t)$, $y = g(t)$, then adding a constant to $f(t)$ translates the curve $C$ in the $x$-direction, and adding a constant to $g(t)$ translates it in the $y$-direction. Thus, a circle of radius $r$, centered at $(x_0, y_0)$ can be represented parametrically as

$$x = x_0 + r \cos t, \quad y = y_0 + r \sin t \quad (0 \le t \le 2\pi) \tag{4}$$

(Figure 1.7.10). If desired, we can eliminate the parameter from these equations by noting that

$$(x - x_0)^2 + (y - y_0)^2 = (r \cos t)^2 + (r \sin t)^2 = r^2$$

Thus, we have obtained the familiar equation in rectangular coordinates for a circle of radius $r$, centered at $(x_0, y_0)$:

$$(x - x_0)^2 + (y - y_0)^2 = r^2 \tag{5}$$

FOR THE READER.   Use the parametric capability of your graphing utility to generate a circle of radius 5 that is centered at $(3, -2)$.

**SCALING**

If a parametric curve $C$ is given by the equations $x = f(t)$, $y = g(t)$, then multiplying $f(t)$ by a constant stretches or compresses $C$ in the $x$-direction, and multiplying $g(t)$ by a constant stretches or compresses $C$ in the $y$-direction. For example, we would expect the parametric equations

$$x = 3 \cos t, \quad y = 2 \sin t \quad (0 \le t \le 2\pi)$$

to represent an ellipse, centered at the origin, since the graph of these equations results from stretching the unit circle

$$x = \cos t, \quad y = \sin t \quad (0 \le t \le 2\pi)$$

by a factor of 3 in the $x$-direction and a factor of 2 in the $y$-direction. In general, if $a$ and $b$ are positive constants, then the parametric equations

$$x = a \cos t, \quad y = b \sin t \quad (0 \le t \le 2\pi) \tag{6}$$

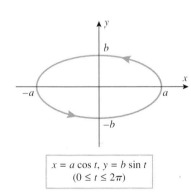

$$x = a \cos t, \ y = b \sin t$$
$$(0 \le t \le 2\pi)$$

Figure 1.7.11

represent an ellipse, centered at the origin, and extending between $-a$ and $a$ on the $x$-axis and between $-b$ and $b$ on the $y$-axis (Figure 1.7.11). The numbers $a$ and $b$ are called the **semiaxes** of the ellipse. If desired, we can eliminate the parameter $t$ in (6) and rewrite the equations in rectangular coordinates as

$$\frac{x^2}{a^2} + \frac{y^2}{b^2} = 1 \tag{7}$$

FOR THE READER.   Use the parametric capability of your graphing utility to generate an ellipse that is centered at the origin and that extends between $-4$ and $4$ in the $x$-direction and between $-3$ and $3$ in the $y$-direction. Generate an ellipse with the same dimensions, but translated so that its center is at $(2, 3)$.

**LISSAJOUS CURVES**

In the mid-1850s the French physicist Jules Antoine Lissajous (1822–1880) became interested in parametric equations of the form

$$x = \sin at, \quad y = \sin bt \tag{8}$$

in the course of studying vibrations that combine two perpendicular sinusoidal motions. The first equation in (8) describes a sinusoidal oscillation in the $x$-direction with frequency $a/2\pi$, and the second describes a sinusoidal oscillation in the $y$-direction with frequency $b/2\pi$. If $a/b$ is a rational number, then the combined effect of the oscillations is a periodic motion along a path called a **Lissajous curve**. Figure 1.7.12 shows some typical Lissajous curves.

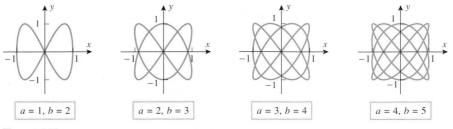

| $a = 1, b = 2$ | $a = 2, b = 3$ | $a = 3, b = 4$ | $a = 4, b = 5$ |

Figure 1.7.12

FOR THE READER.    Generate some Lissajous curves on your graphing utility, and also see if you can figure out when each of the curves in Figure 1.7.12 begins to repeat.

**CYCLOIDS**

If a wheel rolls in a straight line along a flat road, then a point on the rim of the wheel will trace a curve called a *cycloid* (Figure 1.7.13). This curve has a fascinating history, which we will discuss shortly; but first we will show how to obtain parametric equations for it. For this purpose, let us assume that the wheel has radius $a$ and rolls along the positive $x$-axis of a rectangular coordinate system. Let $P(x, y)$ be the point on the rim that traces the cycloid, and assume that $P$ is initially at the origin. We will take as our parameter the angle $\theta$ that is swept out by the radial line to $P$ as the wheel rolls (Figure 1.7.13). It is standard here to regard $\theta$ to be positive, even though it is generated by a clockwise rotation.

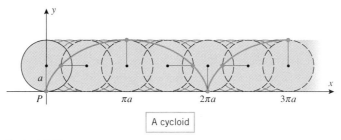

A cycloid

Figure 1.7.13

The motion of $P$ is a combination of the movement of the wheel's center parallel to the $x$-axis and the rotation of $P$ around the center. As the radial line sweeps out an angle $\theta$, the point $P$ traverses an arc of length $a\theta$, and the wheel moves a distance $a\theta$ along the $x$-axis (why?). Thus, as suggested by Figure 1.7.14, the center moves to the point $(a\theta, a)$, and the coordinates of $P(x, y)$ are

$$x = a\theta - a \sin\theta, \quad y = a - a \cos\theta \tag{9}$$

These are the equations of the cycloid in terms of the parameter $\theta$.

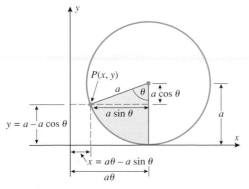

Figure 1.7.14

FOR THE READER.    Use your graphing utility to generate two "arches" of the cycloid produced by a point on the rim of a wheel of radius 1.

**THE ROLE OF THE CYCLOID IN
MATHEMATICS HISTORY**

Figure 1.7.15

The cycloid is of interest because it provides the solution to two famous mathematical problems—the **brachistochrone problem** (from Greek words meaning "shortest time") and the **tautochrone problem** (from Greek words meaning "equal time"). The brachistochrone problem is to determine the shape of a wire along which a bead might slide from a point $P$ to another point $Q$, not directly below, in the *shortest time*. The tautochrone problem is to find the shape of a wire from $P$ to $Q$ such that two beads started at any points on the wire between $P$ and $Q$ reach $Q$ in the same amount of time (Figure 1.7.15). The solution to both problems turns out to be an inverted cycloid.

In June of 1696, Johann Bernoulli[*] posed the brachistochrone problem in the form of a challenge to other mathematicians. At first, one might conjecture that the wire should form a straight line, since that shape results in the shortest distance from $P$ to $Q$. However, the inverted cycloid allows the bead to fall more rapidly at first, building up sufficient initial

---

[*]BERNOULLI. An amazing Swiss family that included several generations of outstanding mathematicians and scientists. Nikolaus Bernoulli (1623–1708), a druggist, fled from Antwerp to escape religious persecution and ultimately settled in Basel, Switzerland. There he had three sons, Jakob I (also called Jacques or James), Nikolaus, and Johann I (also called Jean or John). The Roman numerals are used to distinguish family members with identical names (see the family tree below). Following Newton and Leibniz, the Bernoulli brothers, Jakob I and Johann I, are considered by some to be the two most important founders of calculus. Jakob I was self-taught in mathematics. His father wanted him to study for the ministry, but he turned to mathematics and in 1686 became a professor at the University of Basel. When he started working in mathematics, he knew nothing of Newton's and Leibniz' work. He eventually became familiar with Newton's results, but because so little of Leibniz' work was published, Jakob duplicated many of Leibniz' results.

Jakob's younger brother Johann I was urged to enter into business by his father. Instead, he turned to medicine and studied mathematics under the guidance of his older brother. He eventually became a mathematics professor at Groningen in Holland, and then, when Jakob died in 1705, Johann succeeded him as mathematics professor at Basel. Throughout their lives, Jakob I and Johann I had a mutual passion for criticizing each other's work, which frequently erupted into ugly confrontations. Leibniz tried to mediate the disputes, but Jakob, who resented Leibniz' superior intellect, accused him of siding with Johann, and thus Leibniz became entangled in the arguments. The brothers often worked on common problems that they posed as challenges to one another. Johann, interested in gaining fame, often used unscrupulous means to make himself appear the originator of his brother's results; Jakob occasionally retaliated. Thus, it is often difficult to determine who deserves credit for many results. However, both men made major contributions to the development of calculus. In addition to his work on calculus, Jakob helped establish fundamental principles in probability, including the Law of Large Numbers, which is a cornerstone of modern probability theory.

Among the other members of the Bernoulli family, Daniel, son of Johann I, is the most famous. He was a professor of mathematics at St. Petersburg Academy in Russia and subsequently a professor of anatomy and then physics at Basel. He did work in calculus and probability, but is best known for his work in physics. A basic law of fluid flow, called Bernoulli's principle, is named in his honor. He won the annual prize of the French Academy 10 times for work on vibrating strings, tides of the sea, and kinetic theory of gases.

Johann II succeeded his father as professor of mathematics at Basel. His research was on the theory of heat and sound. Nikolaus I was a mathematician and law scholar who worked on probability and series. On the recommendation of Leibniz, he was appointed professor of mathematics at Padua and then went to Basel as a professor of logic and then law. Nikolaus II was professor of jurisprudence in Switzerland and then professor of mathematics at St. Petersburg Academy. Johann III was a professor of mathematics and astronomy in Berlin and Jakob II succeeded his uncle Daniel as professor of mathematics at St. Petersburg Academy in Russia. Truly an incredible family!

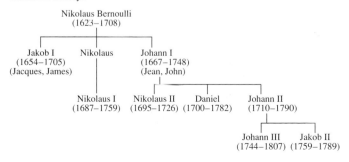

speed to reach $Q$ in the shortest time, even though it travels a longer distance. The problem was solved by Newton and Leibniz as well as by Johann Bernoulli and his older brother Jakob; it was formulated and solved *incorrectly* years earlier by Galileo, who thought the answer was a circular arc.

Newton's solution of the brachistochrone problem in his own handwriting

### EXERCISE SET 1.7  ⌇ Graphing Calculator  [C] CAS

**1.** (a) By eliminating the parameter, sketch the trajectory over the time interval $0 \leq t \leq 5$ of the particle whose parametric equations of motion are

$$x = t - 1, \quad y = t + 1$$

 (b) Indicate the direction of motion on your sketch.

 (c) Make a table of $x$- and $y$-coordinates of the particle at times $t = 0, 1, 2, 3, 4, 5$.

 (d) Mark the position of the particle on the curve at the times in part (c), and label those positions with the values of $t$.

**2.** (a) By eliminating the parameter, sketch the trajectory over the time interval $0 \leq t \leq 1$ of the particle whose parametric equations of motion are

$$x = \cos(\pi t), \quad y = \sin(\pi t)$$

 (b) Indicate the direction of motion on your sketch.

 (c) Make a table of $x$- and $y$-coordinates of the particle at times $t = 0, 0.25, 0.5, 0.75, 1$.

 (d) Mark the position of the particle on the curve at the times in part (c), and label those positions with the values of $t$.

> In Exercises 3–12, sketch the curve by eliminating the parameter, and indicate the direction of increasing $t$.

**3.** $x = 3t - 4, \ y = 6t + 2$

**4.** $x = t - 3, \ y = 3t - 7 \quad (0 \leq t \leq 3)$

**5.** $x = 2 \cos t, \ y = 5 \sin t \quad (0 \leq t \leq 2\pi)$

**6.** $x = \sqrt{t}, \ y = 2t + 4$

**7.** $x = 3 + 2 \cos t, \ y = 2 + 4 \sin t \quad (0 \leq t \leq 2\pi)$

**8.** $x = \sec t, \ y = \tan t \quad (\pi \leq t < 3\pi/2)$

**9.** $x = \cos 2t, \ y = \sin t \quad (-\pi/2 \leq t \leq \pi/2)$

**10.** $x = 4t + 3, \ y = 16t^2 - 9$

**11.** $x = 2 \sin^2 t, \ y = 3 \cos^2 t$

**12.** $x = \sec^2 t, \ y = \tan^2 t$

> In Exercises 13–18, find parametric equations for the curve, and check your work by generating the curve with a graphing utility.

⌇ **13.** A circle of radius 5, centered at the origin, oriented clockwise.

⌇ **14.** The portion of the circle $x^2 + y^2 = 1$ that lies in the third quadrant, oriented counterclockwise.

⌇ **15.** A vertical line intersecting the $x$-axis at $x = 2$, oriented upward.

⌇ **16.** The ellipse $\frac{x^2}{4} + \frac{y^2}{9} = 1$, oriented counterclockwise.

⌇ **17.** The portion of the parabola $x = y^2$ joining $(1, -1)$ and $(1, 1)$, oriented down to up.

⌇ **18.** The circle of radius 4, centered at $(1, -3)$, oriented counterclockwise.

**19.** In each part, match the parametric equation with one of the curves labeled (I)–(VI), and explain your reasoning.

 (a) $x = \sqrt{t}, \ y = \sin 3t$  (b) $x = 2 \cos t, \ y = 3 \sin t$

 (c) $x = t \cos t, \ y = t \sin t$  (d) $x = \dfrac{3t}{1 + t^3}, \ y = \dfrac{3t^2}{1 + t^3}$

 (e) $x = \dfrac{t^3}{1 + t^2}, \ y = \dfrac{2t^2}{1 + t^2}$  (f) $x = 2 \cos t, \ y = \sin 2t$

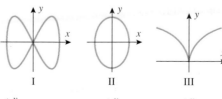

I          II          III

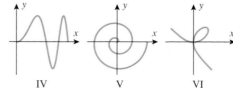

IV          V          VI          Figure Ex-19

**20.** Use a graphing utility to generate the curves in Exercise 19, and in each case identify the orientation.

**21.** (a) Use a graphing utility to generate the trajectory of a particle whose equations of motion over the time interval $0 \le t \le 5$ are

$$x = 6t - \tfrac{1}{2}t^3, \quad y = 1 + \tfrac{1}{2}t^2$$

 (b) Make a table of $x$- and $y$-coordinates of the particle at times $t = 0, 1, 2, 3, 4, 5$.

 (c) At what times is the particle on the $y$-axis?

 (d) During what time interval is $y < 5$?

 (e) At what time is the $x$-coordinate of the particle maximum?

**22.** (a) Use a graphing utility to generate the trajectory of a paper airplane whose equations of motion for $t \ge 0$ are

$$x = t - 2\sin t, \quad y = 3 - 2\cos t$$

 (b) Assuming that the plane flies in a room in which the floor is at $y = 0$, explain why the plane will not crash into the floor. [For simplicity, ignore the physical size of the plane by treating it as a particle.]

 (c) How high must the ceiling be to ensure that the plane does not touch or crash into it?

---

In Exercises 23 and 24, graph the equation using a graphing utility.

---

**23.** (a) $x = y^2 + 2y + 1$

 (b) $x = \sin y$, $-2\pi \le y \le 2\pi$

**24.** (a) $x = y + 2y^3 - y^5$

 (b) $x = \tan y$, $-\pi/2 < y < \pi/2$

**25.** (a) By eliminating the parameter, show that the equations

$$x = x_0 + (x_1 - x_0)t, \quad y = y_0 + (y_1 - y_0)t$$

 represent the line passing through the points $(x_0, y_0)$ and $(x_1, y_1)$.

 (b) Show that if $0 \le t \le 1$, then the equations in part (a) represent the line segment joining $(x_0, y_0)$ and $(x_1, y_1)$, oriented in the direction from $(x_0, y_0)$ to $(x_1, y_1)$.

 (c) Use the result in part (b) to find parametric equations for the line segment joining the points $(1, -2)$ and $(2, 4)$, oriented in the direction from $(1, -2)$ to $(2, 4)$.

 (d) Use the result in part (b) to find parametric equations for the line segment in part (c), but oriented in the direction from $(2, 4)$ to $(1, -2)$.

**26.** Use the result in Exercise 25 to find

 (a) parametric equations for the line segment joining the points $(-3, -4)$ and $(-5, 1)$, oriented from $(-3, -4)$ to $(-5, 1)$

 (b) parametric equations for the line segment traced from $(0, b)$ to $(a, 0)$, oriented from $(0, b)$ to $(a, 0)$.

**27.** (a) Suppose that the line segment from the point $P(x_0, y_0)$ to $Q(x_1, y_1)$ is represented parametrically by

$$x = x_0 + (x_1 - x_0)t,$$
$$y = y_0 + (y_1 - y_0)t \quad (0 \le t \le 1)$$

and that $R(x, y)$ is the point on the line segment corresponding to a specified value of $t$ (see the accompanying figure). Show that $t = r/q$, where $r$ is the distance from $P$ to $R$ and $q$ is the distance from $P$ to $Q$.

 (b) What value of $t$ produces the midpoint between points $P$ and $Q$?

 (c) What value of $t$ produces the point that is three-fourths of the way from $P$ to $Q$?

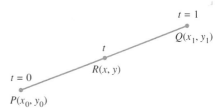

Figure Ex-27

**28.** Find parametric equations for the line segment joining $P(2, -1)$ and $Q(3, 1)$, and use the result in Exercise 27 to find

 (a) the midpoint between $P$ and $Q$

 (b) the point that is one-fourth of the way from $P$ to $Q$

 (c) the point that is three-fourths of the way from $P$ to $Q$.

**29.** Explain why the parametric curve

$$x = t^2, \quad y = t^4 \quad (-1 \le t \le 1)$$

does not have a definite orientation.

**30.** (a) In parts (a) and (b) of Exercise 25 we obtained parametric equations for a line segment in which the parameter varied from $t = 0$ to $t = 1$. Sometimes it is desirable to have parametric equations for a line segment in which the parameter varies over some other interval, say $t_0 \le t \le t_1$. Use the ideas in Exercise 25 to show that the line segment joining the points $(x_0, y_0)$ and $(x_1, y_1)$ can be represented parametrically as

$$x = x_0 + (x_1 - x_0)\frac{t - t_0}{t_1 - t_0},$$
$$ \quad (t_0 \le t \le t_1)$$
$$y = y_0 + (y_1 - y_0)\frac{t - t_0}{t_1 - t_0}$$

 (b) Which way is the line segment oriented?

 (c) Find parametric equations for the line segment traced from $(3, -1)$ to $(1, 4)$ as $t$ varies from 1 to 2, and check your result with a graphing utility.

**31.** (a) By eliminating the parameter, show that if $a$ and $c$ are not both zero, then the graph of the parametric equations

$$x = at + b, \quad y = ct + d \quad (t_0 \le t \le t_1)$$

is a line segment.

 (b) Sketch the parametric curve

$$x = 2t - 1, \quad y = t + 1 \quad (1 \le t \le 2)$$

and indicate its orientation.

**32.** (a) What can you say about the line in Exercise 31 if $a$ or $c$ (but not both) is zero?

(b) What do the equations represent if $a$ and $c$ are both zero?

**33.** Parametric curves can be defined piecewise by using different formulas for different values of the parameter. Sketch the curve that is represented piecewise by the parametric equations

$$x = 2t, \quad y = 4t^2 \qquad \left(0 \le t \le \tfrac{1}{2}\right)$$
$$x = 2 - 2t, \quad y = 2t \qquad \left(\tfrac{1}{2} \le t \le 1\right)$$

**34.** Find parametric equations for the rectangle in the accompanying figure, assuming that the rectangle is traced counterclockwise as $t$ varies from 0 to 1, starting at $\left(\tfrac{1}{2}, \tfrac{1}{2}\right)$ when $t = 0$. [*Hint:* Represent the rectangle piecewise, letting $t$ vary from 0 to $\tfrac{1}{4}$ for the first edge, from $\tfrac{1}{4}$ to $\tfrac{1}{2}$ for the second edge, and so forth.]

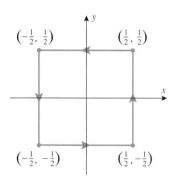

Figure Ex-34

**35.** (a) Find parametric equations for the ellipse that is centered at the origin and has intercepts $(4, 0)$, $(-4, 0)$, $(0, 3)$, and $(0, -3)$.

(b) Find parametric equations for the ellipse that results by translating the ellipse in part (a) so that its center is at $(-1, 2)$.

(c) Confirm your results in parts (a) and (b) using a graphing utility.

**36.** We will show later in the text that if a projectile is fired from ground level with an initial speed of $v_0$ meters per second at an angle $\alpha$ with the horizontal, and if air resistance is neglected, then its position after $t$ seconds, relative to the coordinate system in the accompanying figure is

$$x = (v_0 \cos \alpha)t, \quad y = (v_0 \sin \alpha)t - \tfrac{1}{2}gt^2$$

where $g \approx 9.8 \text{ m/s}^2$.

(a) By eliminating the parameter, show that the trajectory is a parabola.

(b) Sketch the trajectory if $\alpha = 30°$ and $v_0 = 1000 \text{ m/s}$.

Figure Ex-36

**37.** A shell is fired from a cannon at an angle of $\alpha = 45°$ with an initial speed of $v_0 = 800 \text{ m/s}$.

(a) Find parametric equations for the shell's trajectory relative to the coordinate system in Figure Ex-36.

(b) How high does the shell rise?

(c) How far does the shell travel horizontally?

**38.** A robot arm, designed to buff flat surfaces on an automobile, consists of two attached rods, one that moves back and forth horizontally, and a second, with the buffing pad at the end, that moves up and down (see the accompanying figure).

(a) Suppose that the horizontal arm of the robot moves so that the $x$-coordinate of the buffer's center at time $t$ is $x = 25 \sin \pi t$ and the vertical arm moves so that the $y$-coordinate of the buffer's center at time $t$ is $y = 12.5 \sin \pi t$. Graph the trajectory of the center of the buffing pad.

(b) Suppose that the $x$- and $y$-coordinates in part (a) are $x = 25 \sin \pi a t$ and $y = 12.5 \sin \pi b t$, where the constants $a$ and $b$ can be controlled by programming the robot arm. Graph the trajectory of the center of the pad if $a = 4$ and $b = 5$.

(c) Investigate the trajectories that result in part (b) for various choices of $a$ and $b$.

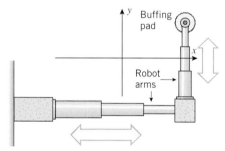

Figure Ex-38

**39.** Describe the family of curves described by the parametric equations

$$x = a \cos t + h, \quad y = b \sin t + k \qquad (0 \le t \le 2\pi)$$

if

(a) $h$ and $k$ are fixed but $a$ and $b$ can vary

(b) $a$ and $b$ are fixed but $h$ and $k$ can vary

(c) $a = 1$ and $b = 1$, but $h$ and $k$ vary so that $h = k + 1$.

**40.** A *hypocycloid* is a curve traced by a point $P$ on the circumference of a circle that rolls inside a larger fixed circle. Suppose that the fixed circle has radius $a$, the rolling circle has radius $b$, and the fixed circle is centered at the origin. Let $\phi$ be the angle shown in the following figure, and assume that the point $P$ is at $(a, 0)$ when $\phi = 0$. Show that the hypocycloid generated is given by the parametric equations

$$x = (a - b) \cos \phi + b \cos \left(\frac{a - b}{b}\phi\right)$$

$$y = (a - b) \sin \phi - b \sin \left(\frac{a - b}{b}\phi\right)$$

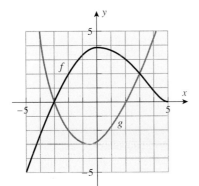

Figure Ex-40

**41.** If $b = \frac{1}{4}a$ in Exercise 40, then the resulting curve is called a four-cusped hypocycloid.

(a) Sketch this curve.
(b) Show that the curve is given by the parametric equations
$$x = a \cos^3 \phi, \quad y = a \sin^3 \phi.$$
(c) Show that the curve is given by the equation
$$x^{2/3} + y^{2/3} = a^{2/3}$$
in rectangular coordinates.

**42.** The parametric equations
$$x - a + \cos t, \quad y = a \tan t + \sin t$$
represent a family of curves called **conchoids of Nicomedes**.
(a) Use a graphing utility to study how the curves in the family change as the parameter $a$ varies from $-2$ to $2$.
(b) Write a brief report on your findings. Your report should include some representative graphs and an explanation of why the curves are called *conchoids*.

## SUPPLEMENTARY EXERCISES

**1.** Referring to the cigarette consumption graph in Figure 1.1.2b, during what 5-year period was the annual cigarette consumption per adult increasing most rapidly on average? Explain your reasoning.

**2.** Use the graphs of the functions $f$ and $g$ in the accompanying figure to solve the following problems.
(a) Find the values of $f(-1)$ and $g(3)$.
(b) For what values of $x$ is $f(x) = g(x)$?
(c) For what values of $x$ is $f(x) < 2$?
(d) What are the domain and range of $f$?
(e) What are the domain and range of $g$?
(f) Find the zeros of $f$ and $g$.

Figure Ex-2

**3.** A glass filled with water that has a temperature of $40°$F is placed in a room in which the temperature is a constant $70°$F. Sketch a rough graph that reasonably describes the temperature of the water in the glass as a function of the elapsed time.

**4.** A student begins driving toward school but 5 minutes into the trip remembers that he forgot his homework. He drives home hurriedly, retrieves his notes, and then drives at great speed toward school, hitting a tree 5 minutes after leaving home. Sketch a rough graph that reasonably describes the student's distance from home as a function of the elapsed time.

**5.** A rectangular storage container with an open top and a square base has a volume of 8 cubic meters. Material for the base costs $5 per square meter, and material for the sides $2 per square meter. Express the total cost of the materials as a function of the length of a side of the base.

**6.** You want to paint the top of a circular table. Find a formula that expresses the amount of paint required as a function of the radius, and discuss all of the assumptions you have made in finding the formula.

**7.** Sketch the graph of the function
$$f(x) = \begin{cases} -1, & x \le -5 \\ \sqrt{25 - x^2}, & -5 < x < 5 \\ x - 5, & x \ge 5 \end{cases}$$

**8.** A ball of radius 3 inches is coated uniformly with plastic. Express the volume of the plastic as a function of its thickness.

**9.** A box with a closed top is to be made from a 6-ft by 10-ft piece of cardboard by cutting out four squares of equal size (see the accompanying figure), folding along the dashed lines, and tucking the two extra flaps inside.
(a) Find a formula that expresses the volume of the box as a function of the length of the sides of the cut-out squares.

(b) Find an inequality that specifies the domain of the function in part (a).

(c) Estimate the dimensions of the box of largest volume.

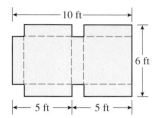

Figure Ex-9

**10.** Let $f(x) = -x^2$ and $g(x) = 1/\sqrt{x}$. Find the natural domains of $f \circ g$ and $g \circ f$.

**11.** Given that $f(x) = 2x - 5$ and $g(x) = 3x - 2$, find a value of $x$ such that $f(g(x)) = g(f(x))$.

**12.** Let $f(x) = (2x - 1)/(x + 1)$ and $g(x) = 1/(x - 1)$.
(a) Find $f(g(x))$.
(b) Is the natural domain of the function $f(g(x))$ obtained in part (a) the same as the domain of $f \circ g$? Explain.

**13.** Find $f(g(h(x)))$, given that

$$f(x) = \frac{x}{x - 1}, \quad g(x) = \frac{1}{x}, \quad h(x) = x^2 - 1$$

**14.** Given that $f(x) = 2x + 1$ and $h(x) = 2x^2 + 4x + 1$, find a function $g$ such that $f(g(x)) = h(x)$.

**15.** Complete the following table.

| $x$ | $-4$ | $-3$ | $-2$ | $-1$ | $0$ | $1$ | $2$ | $3$ | $4$ |
|---|---|---|---|---|---|---|---|---|---|
| $f(x)$ | $0$ | $-1$ | $2$ | $1$ | $3$ | $-2$ | $-3$ | $4$ | $-4$ |
| $g(x)$ | $3$ | $2$ | $1$ | $-3$ | $-1$ | $-4$ | $4$ | $-2$ | $0$ |
| $(f \circ g)(x)$ | | | | | | | | | |
| $(g \circ f)(x)$ | | | | | | | | | |

**16.** (a) Write an equation for the graph that is obtained by reflecting the graph of $y = |x - 1|$ about the $y$-axis, then stretching that graph vertically by a factor of 2, then translating that graph down 3 units, and then reflecting that graph about the $x$-axis.
(b) Sketch the original graph and the final graph.

**17.** In each part, classify the function as even, odd, or neither.
(a) $x^2 \sin x$       (b) $\sin^2 x$
(c) $x + x^2$       (d) $\sin x \tan x$

**18.** (a) Find exact values for all $x$-intercepts of

$$y = \cos x - \sin 2x$$

in the interval $-2\pi \le x \le 2\pi$.

(b) Find the coordinates of all intersections of the graphs of $y = \cos x$ and $y = \sin 2x$ if $-2\pi \le x \le 2\pi$, and use a graphing utility to check your answer.

**19.** (a) A surveyor measures the angle of elevation $\alpha$ of a tower from a point $A$ due south of the tower and also measures the angle of elevation $\beta$ from a point $B$ that is $d$ feet due east of the point $A$ (see the accompanying figure). Show that the height $h$ of the tower in feet is given by

$$h = \frac{d \sin \alpha \sin \beta}{\sqrt{\sin(\alpha + \beta) \sin(\alpha - \beta)}}$$

(b) Use a calculating utility to approximate the height of the tower to the nearest tenth of a foot if $\alpha = 17°$, $\beta = 12°$, and $d = 1000$ ft.

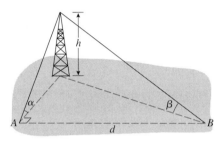

Figure Ex-19

**20.** Suppose that the expected low temperature in Anchorage, Alaska (in $°$F), is modeled by the equation

$$T = 50 \sin \frac{2\pi}{365}(t - 101) + 25$$

where $t$ is in days and $t = 0$ corresponds to January 1.
(a) Sketch the graph of $T$ versus $t$ for $0 \le t \le 365$.
(b) Use the model to predict when the coldest day of the year will occur.
(c) Based on this model, how many days during the year would you expect the temperature to be below $0°$F?

**21.** The accompanying figure shows the graph of the equation $y = \frac{1}{2}x + \sin x$ for $-2\pi \le x \le 2\pi$. Find the coordinates of the points $A, B, C,$ and $D$. Explain your reasoning.

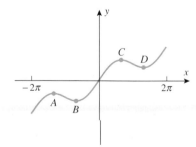

Figure Ex-21

**22.** The accompanying figure shows a model for the tide variation in an inlet to San Francisco Bay during a 24-hour period. Find an equation of the form $y = y_0 + y_1 \sin(at + b)$ for the model, assuming that $t = 0$ corresponds to midnight.

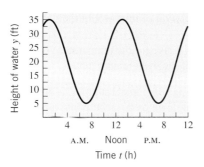

Figure Ex-22

**23.** In each part describe the family of curves.
(a) $(x - a)^2 + (y - a^2)^2 = 1$
(b) $y = a + (x - 2a)^2$

**24.** (a) Suppose that the equations $x = f(t)$, $y = g(t)$ describe a curve $C$ as $t$ increases from 0 to 1. Find parametric equations that describe the same curve $C$ but traced in the opposite direction as $t$ increases from 0 to 1.
(b) Check your work using the parametric graphing feature of a graphing utility by generating the line segment between $(1, 2)$ and $(4, 0)$ in both possible directions as $t$ increases from 0 to 1.

**25.** Sketch the graph of the equation $x^2 - 4y^2 = 0$.

**26.** Find an equation for a parabola that passes through the points $(2, 0)$, $(8, 18)$, and $(-8, 18)$.

**27.** Sketch the curve described by the parametric equations
$$x = t \cos(2\pi t), \quad y = t \sin(2\pi t)$$
and check your result with a graphing utility.

**28.** The electrical resistance $R$ in ohms ($\Omega$) for a pure metal wire is related to its temperature $T$ in $°C$ by the formula
$$R = R_0(1 + kT)$$
in which $R_0$ and $k$ are positive constants.
(a) Make a hand-drawn sketch of the graph of $R$ versus $T$, and explain the geometric significance of $R_0$ and $k$ for your graph.
(b) In theory, the resistance $R$ of a wire drops to zero when the temperature reaches absolute zero ($T = -273°C$). What information does this give you about $k$?
(c) A tungsten bulb filament has a resistance of 1.1 $\Omega$ at a temperature of $20°C$. What information does this give you about $R_0$ for the filament?
(d) At what temperature will a tungsten filament have a resistance of 1.5 $\Omega$?

Most of the following exercises require access to graphing and calculating utilities. When you are asked to *find* an answer or to *solve* an equation, you may choose to find either an exact result or a numerical approximation, depending on the particular technology you are using and your own imagination.

**29.** Find the distance between the point $P(1, 2)$ and an arbitrary point $(x, \sqrt{x})$ on the curve $y = \sqrt{x}$. Graph this distance ver-

sus $x$, and use the graph to find the $x$-coordinate of the point on the curve that is closest to the point $P$.

**30.** Find the distance between the point $P(1, 0)$ and an arbitrary point $(x, 1/x)$ on the curve $y = 1/x$, where $x > 0$. Graph this distance versus $x$, and use the graph to find the $x$-coordinate of the point on the curve that is closest to the point $P$.

In Exercises 31 and 32, use **Archimedes' principle**: *A body wholly or partially immersed in a fluid is buoyed up by a force equal to the weight of the fluid that it displaces.*

**31.** A hollow metal sphere of diameter 5 feet weighs 108 pounds and floats partially submerged in seawater. Assuming that seawater weighs 63.9 pounds per cubic foot, how far below the surface is the bottom of the sphere? [*Hint:* If a sphere of radius $r$ is submerged to a depth $h$, then the volume $V$ of the submerged portion is given by the formula $V = \pi h^2(r - h/3)$.]

**32.** Suppose that a hollow metal sphere of diameter 5 feet and weight $w$ pounds floats in seawater. (See Exercise 31.)
(a) Graph $w$ versus $h$ for $0 \leq h \leq 5$.
(b) Find the weight of the sphere if exactly half of the sphere is submerged.

**33.** A breeding group of 20 bighorn sheep is released in a protected area in Colorado. It is expected that with careful management the number of sheep, $N$, after $t$ years will be given by the formula
$$N = \frac{220}{1 + 10(0.83)^t}$$
and that the sheep population will be able to maintain itself without further supervision once the population reaches a size of 80.
(a) Graph $N$ versus $t$.
(b) How many years must the state of Colorado maintain a program to care for the sheep?
(c) How many bighorn sheep can the environment in the protected area support? [*Hint:* Examine the graph of $N$ versus $t$ for large values of $t$.]

In Exercises 34 and 35, use the following empirical formula for the windchill index (WCI) [see Example 3 of Section 1.2]:
$$\text{WCI} = \begin{cases} T, & 0 \leq v \leq 4 \\ 91.4 + (91.4 - T)(0.0203v - 0.304\sqrt{v} - 0.474), & 4 < v < 45 \\ 1.6T - 55, & v \geq 45 \end{cases}$$
where $T$ is the air temperature in $°F$, $v$ is the wind speed in $mi/h$, and WCI is the equivalent temperature in $°F$.

**34.** (a) Graph $T$ versus $v$ over the interval $4 \leq v \leq 45$ for WCI $= 0$.
(b) Use your graph to estimate the values of $T$ for WCI $= 0$ corresponding to $v = 10, 20, 30$, to the nearest degree.

**35.** (a) Graph WCI versus $v$ over the interval $0 \leq v \leq 50$ for $T = 20$.

(b) Use your graph to estimate the values of the WCI corresponding to $v = 10, 20, 30, 40$, to the nearest degree.

(c) Use your graph to estimate the values of $v$ corresponding to WCI $= -20, -10, 0, 10$, to the nearest mile per hour.

**36.** Find the domain and range of the function
$$f(x) = \frac{\sin x}{x^4 + x^3 + 5}$$

**37.** Find the domain and range of the function
$$f(x) = x^2 - \sqrt{1 + x - x^4}$$

**38.** An oven is preheated and then remains at a constant temperature. A potato is placed in the oven to bake. Suppose that the temperature $T$ (in °F) of the potato $t$ minutes later is given by $T = 400 - 325(0.97)^t$. The potato will be considered done when its temperature is anywhere between 260°F and 280°F.

(a) During what interval of time would the potato be considered done?

(b) How long does it take for the temperature of the potato to reach 95% of the oven temperature?

**39.** Suppose that a package of medical supplies is dropped from a helicopter straight down by parachute into a remote area. The velocity $v$ (in feet per second) of the package $t$ seconds after it is released is given by $v = 24.61(1 - (0.273^t))$.

(a) Graph $v$ versus $t$.

(b) Show that the graph has a horizontal asymptote $v = c$.

(c) The constant $c$ is called the **terminal velocity**. Explain what the terminal velocity means in practical terms.

(d) Can the package actually reach its terminal velocity? Explain.

(e) How long does it take for the package to reach 98% of its terminal velocity?

---

## EXPANDING THE CALCULUS HORIZON

# Iteration and Dynamical Systems

*What do the four figures below have in common? The answer is that all of them are of interest in contemporary research and all involve a mathematical process called **iteration**. In this module we will introduce this concept and touch on some of the fascinating ideas to which it leads.*

Barnsley's fern

The Sierpinski triangle

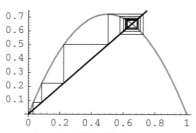
A cobweb diagram

A Julia set

### Iterative Processes

Recall that in the notation $y = f(x)$, the variable $x$ is called an **input** of the function $f$, and the variable $y$ is called the corresponding **output**. Suppose that we start with some input, say $x = c$, and each time we compute an output we feed it back into $f$ as an input. This generates the following sequence of numbers:

$$f(c), \quad f(f(c)), \quad f(f(f(c))), \quad f(f(f(f(c)))), \ldots$$

This is called an **iterated function sequence** for $f$ (from the Latin word *iteratus*, meaning "to repeat"). The number $c$ is called the **seed value** for the sequence, the terms in the sequence are called **iterates**, and each time $f$ is applied we say that we have performed an **iteration**. Iterated function sequences arise in a wide variety of physical processes that are collectively called **dynamical systems**.

..........
*Exercise 1*    Let $f(x) = x^2$.

(a) Calculate the first 10 iterates in the iterated function sequence for $f$, starting with seed values of $c = 0.5, 1$, and 2. In each case make a conjecture about the **long-term behavior** of the iterates, that is, the behavior of the iterates as more and more iterations are performed.

(b) Try your own seed values, and make a conjecture about the effect of a seed value on the long-term behavior of the iterates.

■■■■ **Recursion Formulas**

The proliferation of parentheses in an iterated function sequence can become confusing, so for simplicity let us introduce the following notation for the successive iterates

$$y_0 = c, \quad y_1 = f(c), \quad y_2 = f(f(c)), \quad y_3 = f(f(f(c))), \quad y_4 = f(f(f(f(c)))), \ldots$$

or expressed more simply,

$$y_0 = c, \quad y_1 = f(y_0), \quad y_2 = f(y_1), \quad y_3 = f(y_2), \quad y_4 = f(y_3), \ldots$$

Thus, successive terms in the sequence are related by the formulas

$$y_0 = c, \quad y_{n+1} = f(y_n) \quad (n = 0, 1, 2, 3, \ldots)$$

These two formulas, taken together, comprise what is called a *recursion formula* for the iterated function sequence. In general, a **recursion formula** is any formula or set of formulas that provides a method for generating the terms of a sequence from the preceding terms and a seed value. For example, the recursion formula for the iterated function sequence of $f(x) = x^2$ with seed value $c$ is

$$y_0 = c, \quad y_{n+1} = y_n^2$$

As another example, review the formula in the discussion preceding Exercise 8 in the Introduction. As noted in that discussion, the recursion formula

$$y_0 = 1, \quad y_{n+1} = \frac{1}{2}\left(y_n + \frac{p}{y_n}\right) \tag{1}$$

produces an iterated function sequence whose iterates can be used to approximate $\sqrt{p}$ to any degree of accuracy.

..........
*Exercise 2*    Use (1) to approximate $\sqrt{5}$ by generating successive iterates on a calculator until you encounter two successive iterates that are the same. Compare this approximation of $\sqrt{5}$ to that produced directly by your calculator.

..........
*Exercise 3*

(a) Find iterates $y_1$ up to $y_6$ of the sequence that is generated by the recursion formula

$$y_0 = 1, \quad y_{n+1} = \frac{1}{2}y_n$$

(b) By examining the terms generated in part (a), find a formula that expresses $y_n$ as a function of $n$.

..........
*Exercise 4*    Suppose that you deposit $1000 in a bank at 5% interest per year and allow it to accumulate value without making withdrawals.

(a) If $y_n$ denotes the value of the account at the end of the $n$th year, how could you find the value of $y_{n+1}$ if you knew the value of $y_n$?

(b) Starting with $y_0 = 1000$ (dollars), use the result in part (a) to calculate $y_1, y_2, y_3, y_4$, and $y_5$.

(c) Find a recursion formula for the sequence of yearly account values assuming that $y_0 = 1000$.

(d) Find a formula that expresses $y_n$ as a function of $n$, and use that formula to calculate the value of the account at the end of the 15th year.

### Exploring Iterated Function Sequences

Iterated function sequences for a function $f$ can be explored in various ways. Here are three possibilities:

- Choose a specific seed value, and investigate the long-term behavior of the iterates (as in Exercise 1).

- Let the seed value be a variable $x$ (in which case the iterates become functions of $x$), and investigate what happens to the graphs of the iterates as more and more iterations are performed.

- Choose a specific iterate, say the 10th, and investigate how the value of this iterate varies with different seed values.

*Exercise 5*    Let $f(x) = \sqrt{x}$.

(a) Find formulas for the first five iterates in the iterated function sequence for $f$, taking the seed value to be $x$.

(b) Graph the iterates in part (a) in the same coordinate system, and make a conjecture about the behavior of the graphs as more and more iterations are performed.

### Continued Fractions and Fibonacci Sequences

If $f(x) = 1/x$, and the seed value is $x$, then the iterated function sequence for $f$ flip-flops between $x$ and $1/x$:

$$y_1 = \frac{1}{x}, \quad y_2 = \frac{1}{1/x} = x, \quad y_3 = \frac{1}{x}, \quad y_4 = \frac{1}{1/x} = x, \ldots$$

However, if $f(x) = 1/(x+1)$, then the iterated function sequence becomes a sequence of fractions that, if continued indefinitely, is an example of a **continued fraction**:

$$\frac{1}{1+x}, \quad \frac{1}{1+\dfrac{1}{1+x}}, \quad \frac{1}{1+\dfrac{1}{1+\dfrac{1}{1+x}}}, \quad \frac{1}{1+\dfrac{1}{1+\dfrac{1}{1+\dfrac{1}{1+x}}}}, \ldots$$

*Exercise 6*    Let $f(x) = 1/(x + 1)$ and $c = 1$.

(a) Find *exact values* for the first 10 terms in the iterated function sequence for $f$; that is, express each term as a fraction $p/q$ with no common factors in the numerator and denominator.

(b) Write down the numerators from part (a) in sequence, and see if you can discover how each term after the first two is related to its predecessors. The sequence of numerators is called a **Fibonacci sequence** [in honor of its medieval discoverer Leonardo ("Fibonacci") da Pisa]. Do some research on Fibonacci and his sequence, and write a paper on the subject.

(c) Use the pattern you discovered in part (b) to write down the exact values of the second 10 terms in the iterated function sequence.

(d) Find a recursion formula that will generate all the terms in the Fibonacci sequence after the first two.

(e) It can be proved that the terms in the iterated function sequence for $f$ get closer and closer to one of the two solutions of the equation $q = 1/(1 + q)$. Which solution is it? This solution is a number known as the **golden ratio**. Do some research on the golden ratio, and write a paper on the subject.

### Applications to Ecology

There are numerous models for predicting the growth and decline of populations (flowers, plants, people, animals, etc.). One way to model populations is to give a recursion formula that describes how the number of individuals in each generation relates to the number of individuals in the preceding generation. One of the simplest such models, called the ***exponential model***, assumes that the number of individuals in each generation is a fixed percentage of the number of individuals in the preceding generation. Thus, if there are $c$ individuals initially and if the number of individuals in any generation is $r$ times the number of individuals in the preceding generation, then the growth through successive generations is given by the recursion formula

$$y_0 = c, \quad y_{n+1} = ry_n \quad (n = 0, 1, 2, 3, \ldots)$$

*Exercise 7*    Suppose that a population with an exponential growth model has $c$ individuals initially.

(a) Express the iterates $y_1, y_2, y_3$, and $y_4$ in terms of $c$ and $r$.

(b) Find a formula for $y_{n+1}$ in terms of $c$ and $r$.

(c) Describe the eventual fate of the population if $r = 1, r < 1$, and $r > 1$.

There is a more sophisticated model of population growth, called the ***logistic model***, that takes environmental constraints into account. In this model, it is assumed that there is some maximum population that can be supported by the environment, and the population is expressed as a fraction of the maximum. Thus, in each generation the population is represented as a number in the interval $0 \leq y_n \leq 1$. When $y_n$ is near 0 the population has lots of room to grow, but when $y_n$ is near 1 the population is close to the maximum and the environmental factors tend to inhibit further growth. Models of this type are given by recursion formulas of the form

$$y_0 = c, \quad y_{n+1} = ky_n(1 - y_n) \tag{2}$$

in which $k$ is a positive constant that depends on the ecological conditions.

Figure 1 illustrates a graphical method for tracking the growth of a population described by (2). That figure, which is called a ***cobweb diagram***, shows graphs of the line $y = x$ and the curve $y = kx(1 - x)$.

*Exercise 8*    Explain why the values $y_1, y_2$, and $y_3$ are the populations for the first three generations of the logistic growth model given by (2).

*Exercise 9*    The cobweb diagram in Figure 2 tracks the growth of a population with a logistical growth model given by the recursion formula

$$y_0 = 0.1, \quad y_{n+1} = 2.9y_n(1 - y_n)$$

(a) Find the populations $y_1, y_2, \ldots, y_5$ of the first five generations.

(b) What happens to the population over the long term?

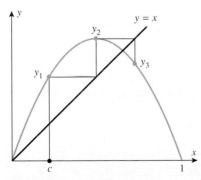

Figure 1

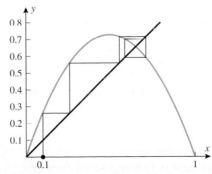

Figure 2

### Chaos and Fractals

Observe that (2) is a recursion formula for the iterated function sequence of $f(x) = kx(1 - x)$. Iterated function sequences of this form are called *iterated quadratic systems*. These are important not only in modeling populations but also in the study of *chaos* and *fractals*—two important fields of contemporary research.

*Module by: C. Lynn Kiaer, Rose-Hulman Institute of Technology*
*Howard Anton, Drexel University*

**Additional material for this module can be found on the World Wide Web at http://www.wiley.com/college/anton**

# 2

# LIMITS AND CONTINUITY

Leonard Euler

$\mathcal{T}$he development of calculus was stimulated by two geometric problems: finding areas of plane regions and finding tangent lines to curves. As discussed in the Introduction, both of these problems require a "limit process" for their general solution. However, limit processes occur in many other applications as well—so many, in fact that the concept of a "limit" is the fundamental building block on which all other calculus concepts are based.

In this chapter we will develop the concept of a limit in stages: In Section 2.1 we will develop the basic ideas informally, relying on our intuition; in Section 2.2 we will discuss methods for calculating limits; and in Section 2.3 we will give the precise mathematical definition of a limit. In Sections 2.4 and 2.5 we will apply limits to the study "continuous" curves. Such curves are important because they model the idea of a smooth flow without breaks or interruptions—the flow of time, the motion of an object in flight, or the gradual warming of a room on a sunny day, for example.

## 2.1 LIMITS (AN INTUITIVE INTRODUCTION)

*As discussed in the introduction to this chapter, the concept of a limit is the fundamental building block on which all other calculus concepts are based. In this section we will study limits informally, with the goal of developing an "intuitive feel" for the basic ideas. In the next two sections we will focus on the computational methods and precise definitions.*

**THE TANGENT LINE, AREA, AND VELOCITY PROBLEMS**

Many of the basic ideas in calculus can be motivated by the following three problems.

> THE TANGENT LINE PROBLEM.    Given a function $f$ and a point $P(x_0, y_0)$ on its graph, find an equation of the line that is tangent to the graph at $P$ (Figure 2.1.1).

> THE AREA PROBLEM.    Given a function $f$, find the area between the graph of $f$ and an interval $[a, b]$ on the $x$-axis (Figure 2.1.2).

> THE INSTANTANEOUS VELOCITY PROBLEM.    Given the position versus time curve for a particle moving along a coordinate line, find the velocity of the particle at a specified instant of time.

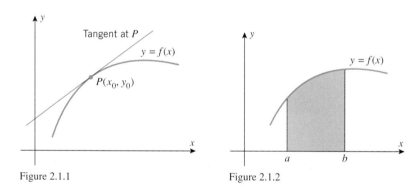

Figure 2.1.1                    Figure 2.1.2

Traditionally, that portion of calculus arising from the tangent line problem is called *differential calculus* and that arising from the area problem is called *integral calculus*. However, we will see later that the tangent line and area problems are so closely related that the distinction between differential and integral calculus is often hard to discern.

In order to solve the three problems posed above, it is necessary to have a more precise understanding of what the terms *tangent line*, *area*, and *velocity at an instant* actually mean. Let us begin with the notion of a tangent line.

**TANGENT LINES AND LIMITS**

In plane geometry, a line is called *tangent* to a circle if it meets the circle at precisely one point (Figure 2.1.3a). However, this definition is not appropriate for more general curves. For example, in Figure 2.1.3b, the line meets the curve exactly once but is obviously not what we would regard to be a tangent line; and in Figure 2.1.3c, the line appears to be tangent to the curve, yet it intersects the curve more than once.

To obtain a definition of a tangent line that applies to curves other than circles, we must view tangent lines another way. For this purpose, suppose that we are interested in the tangent line at a point $P$ on a curve in the $xy$-plane and that $Q$ is any point that lies on the curve and is different from $P$. The line through $P$ and $Q$ is called a *secant line* for the curve at $P$. Intuition suggests that if we move the point $Q$ along the curve toward $P$, then the

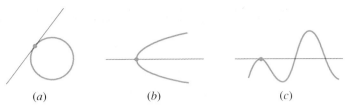

Figure 2.1.3

secant line will rotate toward a *limiting position*. The line in this limiting position is what we will consider to be the **tangent line** at $P$ (Figure 2.1.4a). As suggested by Figure 2.1.4b, this new concept of a tangent line coincides with the traditional concept when applied to circles.

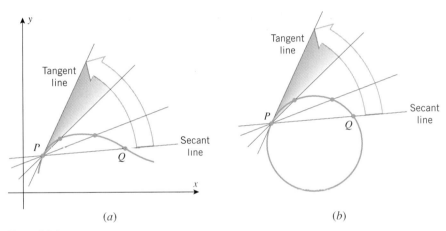

Figure 2.1.4

**AREAS AND LIMITS**

Figure 2.1.5

Just as the general notion of a tangent line leads to the concept of *limit*, so does the general notion of area. For many plane regions with straight-line boundaries, areas can be calculated by subdividing the region into rectangles or triangles and adding the areas of the constituent parts (Figure 2.1.5). However, for regions with curved boundaries, such as that in Figure 2.1.6a, a more general approach is needed. One such approach is to begin by approximating the area of the region by inscribing a number of rectangles of equal width under the curve and adding the areas of these rectangles (Figure 2.1.6b). Intuition suggests that if we repeat that approximation process using more and more rectangles, then the rectangles will tend to fill in the gaps under the curve, and the approximations will get closer and closer to the exact area under the curve (Figure 2.1.6c). This suggests that we can define the area under the curve to be the *limiting value* of these approximations.

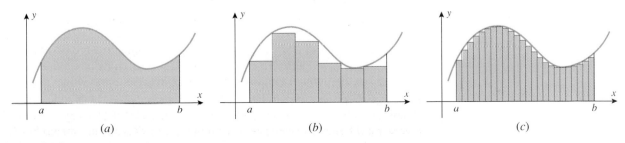

Figure 2.1.6

Recall from Formula (11) of Section 1.5 that if a particle moves along an $s$-axis, then its average velocity $v_{ave}$ over the time interval from $t_0$ to $t_1$ is defined as

$$v_{ave} = \frac{\Delta s}{\Delta t} = \frac{s_1 - s_0}{t_1 - t_0} \tag{1}$$

where $s_0$ and $s_1$ are the coordinates of the particle at times $t_0$ and $t_1$, respectively. Geometrically, $v_{ave}$ is the slope of the secant line joining the points $(t_0, s_0)$ and $(t_1, s_1)$ on the position versus time curve for the particle (Figure 2.1.7).

Suppose, however, that we are not interested in the average velocity over a time interval, but rather the velocity $v_{inst}$ at a specific instant of time. It is not a simple matter of applying Formula (1), since the displacement and the elapsed time in an instant are both 0. However, intuition suggests that over a sufficiently small time interval, the velocity of the particle will not vary much; thus, there should not be much difference between the instantaneous velocity at an instant of time, say $t = t_0$, and the average velocity over a time interval from $t = t_0$ to $t = t_1$, provided that the time interval is small. This suggests that we can approximate $v_{inst}$ as

$$v_{inst} \approx v_{ave} = \frac{s_1 - s_0}{t_1 - t_0} \tag{2}$$

Moreover, the closer $t_1$ is to $t_0$, the better the approximation. However, as $t_1$ gets closer and closer to $t_0$, the slope of the secant line in Figure 2.1.8 will approach the slope of the tangent line to the curve at time $t = t_0$; and this suggests that we can *define* the instantaneous velocity of the particle at time $t = t_0$ to be the slope of the tangent line to the position versus time curve at that point. Thus, once we know how to calculate slopes of tangent lines, we will have a method for calculating instantaneous velocities.

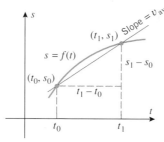
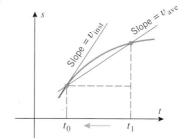

Figure 2.1.7                    Figure 2.1.8

Now that we have seen how the concept of a limit enters into solving the tangent line, area, and instantaneous velocity problems, let us focus on the limit concept itself.

The most basic use of limits is to describe how a function behaves as the independent variable approaches a given value. For example, let us examine the behavior of the function

$$f(x) = x^2 - x + 1$$

as $x$ gets closer and closer to 2. It is evident from the graph and table in Figure 2.1.9 that the values of $f(x)$ get closer and closer to 3 as $x$ gets closer and closer to 2 from either the left side or the right side. Moreover, the graph and table both suggest that we can make the values of $f(x)$ as close as we like to 3 by making $x$ sufficiently close to 2. We describe this by saying that the "limit of $x^2 - x + 1$ is 3 as $x$ approaches 2 from either side," and we write

$$\lim_{x \to 2} (x^2 - x + 1) = 3 \tag{3}$$

Observe that in this limit analysis we are only concerned with the values of $f$ *near* the point $x = 2$ and not the value of $f$ *at* the point $x = 2$.

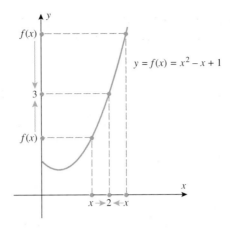

| $x$ | 1.0 | 1.5 | 1.9 | 1.95 | 1.99 | 1.995 | 1.999 | 2 | 2.001 | 2.005 | 2.01 | 2.05 | 2.1 | 2.5 | 3.0 |
|------|------|------|------|------|------|------|------|------|------|------|------|------|------|------|------|
| $f(x)$ | 1.000000 | 1.750000 | 2.710000 | 2.852500 | 2.970100 | 2.985025 | 2.997001 | | 3.003001 | 3.015025 | 3.030100 | 3.152500 | 3.310000 | 4.750000 | 7.000000 |

Left side ⟶ ⟵ Right side

Figure 2.1.9

This leads us to the following general idea.

> **2.1.1** LIMITS (AN INFORMAL VIEW). If the values of $f(x)$ can be made as close as we like to $L$ by making $x$ sufficiently close to $a$ (but not equal to $a$), then we write
> $$\lim_{x \to a} f(x) = L,$$
> (4)
> which is read "the limit of $f(x)$ as $x$ approaches $a$ is $L$."

Expression (4) is also commonly written as

$$f(x) \to L \quad \text{as} \quad x \to a$$

With this notation we can express (3) as

$$x^2 - x + 1 \to 3 \quad \text{as} \quad x \to 2$$

### Example 1

Make a conjecture about the value of the limit

$$\lim_{x \to 0} \frac{x}{\sqrt{x+1}-1}$$
(5)

*Solution.* Observe that this function is undefined at $x = 0$. However, this has no bearing on the limit, since the limit is concerned with the behavior of $f$ for $x$ near, but not equal to, 0. Table 2.1.1 shows successions of $x$-values approaching 0 from the left side and the right side. In both cases the values of $f(x)$, calculated to six decimal places, appear to get closer

**Table 2.1.1**

| $x$ | −0.01 | −0.001 | −0.0001 | −0.00001 | 0 | 0.00001 | 0.0001 | 0.001 | 0.01 |
|------|------|------|------|------|------|------|------|------|------|
| $f(x)$ | 1.994987 | 1.999500 | 1.999950 | 1.999995 | | 2.000005 | 2.000050 | 2.000500 | 2.004988 |

Left side ⟶ ⟵ Right side

and closer to 2, and hence we conjecture that

$$\lim_{x \to 0} \frac{x}{\sqrt{x+1}-1} = 2 \qquad (6)$$

However, it should be kept in mind that this conjecture is based on a limited amount of numerical evidence; we are *guessing* that if we were to extend the table and continue to let $x$ get closer and closer to 0 from either side, then the values of $f(x)$ would continue to get closer and closer to 2. Fortunately, in this example we have other ways of confirming our conjecture. One possibility is to simplify Formula (5) algebraically by rationalizing the denominator. This yields

$$f(x) = \frac{x}{\sqrt{x+1}-1} = \frac{x(\sqrt{x+1}+1)}{(x+1)-1} = \sqrt{x+1}+1 \quad (x \neq 0) \qquad (7)$$

It is evident from this alternative formula for $f$ that as $x$ gets closer and closer to 0, the values of $f(x) = \sqrt{x+1}+1$ get closer and closer to 2, confirming (6). Yet another confirmation of (6) can be obtained from the graph of $f$. It follows from (7) that the graph of $f$ is identical to the graph of $y = \sqrt{x+1}+1$, except for a hole at $x = 0$, where $f$ is undefined (Figure 2.1.10). This figure suggests that as $x$ moves along the $x$-axis toward 0 from either side, the values of $y = f(x)$ get closer and closer to 2, which again agrees with (6).    ◀

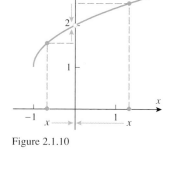

Figure 2.1.10

### Example 2

Make a conjecture about the value of the limit

$$\lim_{x \to 0} \frac{\sin x}{x}$$

*Solution.*    The function $f(x) = (\sin x)/x$ is undefined at $x = 0$, but, as discussed previously, this has no bearing on the limit. With the help of a calculating utility set to radian measure, we obtain Table 2.1.2, which suggests that

$$\lim_{x \to 0} \frac{\sin x}{x} = 1 \qquad (8)$$

This result is consistent with the graph of $f(x) = (\sin x)/x$ shown in Figure 2.1.11; but unlike the preceding example, where we were able to confirm the limit algebraically by simplifying the formula for the function, that is not possible here. However, later in this chapter we will give a geometric argument to prove that our conjecture is correct.    ◀

**Table 2.1.2**

| $x$ (RADIANS) | $y = \dfrac{\sin x}{x}$ |
|---|---|
| ±1.0 | 0.84147 |
| ±0.9 | 0.87036 |
| ±0.8 | 0.89670 |
| ±0.7 | 0.92031 |
| ±0.6 | 0.94107 |
| ±0.5 | 0.95885 |
| ±0.4 | 0.97355 |
| ±0.3 | 0.98507 |
| ±0.2 | 0.99335 |
| ±0.1 | 0.99833 |
| ±0.01 | 0.99998 |

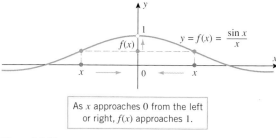

As $x$ approaches 0 from the left or right, $f(x)$ approaches 1.

Figure 2.1.11

FOR THE READER.    Use a calculating utility to confirm limit (8). Does the limit change if $x$ is in degrees?

**NUMERICAL PITFALLS**

Although numerical evidence is helpful for guessing at limits, it can lead to incorrect conclusions. For example, Table 2.1.3 shows values of $f(x) = \sin(\pi/x)$ at selected values of $x$ on both sides of 0. The numerical data in that table suggest that

$$\lim_{x \to 0} \sin\left(\frac{\pi}{x}\right) = 0$$

However, this conclusion is incorrect, as evidenced by the graph of $f$ shown in Figure 2.1.12. This graph shows that as $x \to 0$, the values of $f$ oscillate between $-1$ and $1$ with increasing rapidity, and hence do not approach a limit. The numerical data in Table 2.1.3 deceived us into believing the limit to be zero because we happened to choose values of $x$ that were all $x$-intercepts.

**Table 2.1.3**

| $x$ (RADIANS) | $\dfrac{\pi}{x}$ | $f(x) = \sin\left(\dfrac{\pi}{x}\right)$ |
|---|---|---|
| $x = \pm 1$ | $\pm\pi$ | $\sin(\pm\pi) = 0$ |
| $x = \pm 0.1$ | $\pm 10\pi$ | $\sin(\pm 10\pi) = 0$ |
| $x = \pm 0.01$ | $\pm 100\pi$ | $\sin(\pm 100\pi) = 0$ |
| $x = \pm 0.001$ | $\pm 1000\pi$ | $\sin(\pm 1000\pi) = 0$ |
| $x = \pm 0.0001$ | $\pm 10{,}000\pi$ | $\sin(\pm 10{,}000\pi) = 0$ |
| $\vdots$ | $\vdots$ | $\vdots$ |

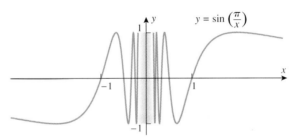

Figure 2.1.12

Numerical evidence can also lead to incorrect conclusions about limits because of round-off error or because the table of values used to find the limit is not extensive enough to reveal the behavior of the function completely. Thus, when a limit is conjectured from numerical data it is important to look for corroborating graphical or algebraic evidence to support the conjecture.

**ONE-SIDED LIMITS**

The limit in (4) is commonly called a **two-sided limit** because it requires the values of $f(x)$ to get closer and closer to $L$ as $x$ approaches $a$ from *either* side. However, some functions exhibit different behaviors on the two sides of a point $a$, in which case it is necessary to distinguish whether $x$ is near $a$ on the left side or the right side for purposes of investigating the limiting behavior. For example, consider the function

$$f(x) = \frac{|x|}{x} = \begin{cases} 1, & x > 0 \\ -1, & x < 0 \end{cases}$$

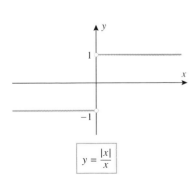

Figure 2.1.13

(Figure 2.1.13). As $x$ approaches 0 from the right side, the values of $f(x)$ approach 1 (in fact, they are exactly 1 for all such $x$), and as $x$ approaches 0 from the left side, the values of $f(x)$ approach $-1$. We describe these two statements by saying that "the limit of $f(x) = |x|/x$ is 1 as $x$ approaches 0 from the right" and that "the limit of $f(x) = |x|/x$ is $-1$ as $x$ approaches 0 from the left"; we denote these limits by writing

$$\lim_{x \to 0^+} \frac{|x|}{x} = 1 \quad \text{and} \quad \lim_{x \to 0^-} \frac{|x|}{x} = -1 \tag{9--10}$$

With this notation, the superscript "$+$" indicates a limit from the right and the superscript "$-$" indicates a limit from the left.

This leads us to the following general idea:

---

**2.1.2** ONE-SIDED LIMITS (AN INFORMAL VIEW). If the values of $f(x)$ can be made as close as we like to $L$ by making $x$ sufficiently close to $a$ (but greater than $a$), then we write

$$\lim_{x \to a^+} f(x) = L \qquad (11)$$

which is read "the limit of $f(x)$ as $x$ approaches $a$ from the right is $L$." Similarly, if the values of $f(x)$ can be made as close as we like to $L$ by making $x$ sufficiently close to $a$ (but less than $a$), then we write

$$\lim_{x \to a^-} f(x) = L \qquad (12)$$

---

Expressions (11) and (12), which are called *one-sided limits*, are also commonly written as

$$f(x) \to L \text{ as } x \to a^+ \quad \text{and} \quad f(x) \to L \text{ as } x \to a^-$$

respectively. With this notation (9) and (10) can be expressed as

$$\frac{|x|}{x} \to 1 \text{ as } x \to 0^+ \quad \text{and} \quad \frac{|x|}{x} \to -1 \text{ as } x \to 0^-$$

**THE RELATIONSHIP BETWEEN ONE-SIDED AND TWO-SIDED LIMITS**

In general, there is no guarantee that a function will have a limit at a specified point, and there is some terminology to describe such situations. If the values of $f(x)$ do not get closer and closer to some *single* number $L$ as $x \to a$, then we say that the limit of $f(x)$ as $x$ approaches $a$ *does not exist* (and similarly for one-sided limits). For example, the two-sided limit of $f(x) = |x|/x$ does not exist as $x \to 0$ because the values of $f(x)$ do not approach a single number—the values approach $-1$ from the left and $1$ from the right.

In general, the following condition must be satisfied for the two-sided limit of a function to exist.

---

**2.1.3** THE RELATIONSHIP BETWEEN ONE-SIDED AND TWO-SIDED LIMITS. The two-sided limit of a function $f$ exists at a point $a$ if and only if the one-sided limits exist at that point and have the same value; that is,

$$\lim_{x \to a} f(x) = L \quad \text{if and only if} \quad \lim_{x \to a^-} f(x) = L = \lim_{x \to a^+} f(x)$$

---

REMARK. Sometimes, one or both of the one-sided limits may fail to exist (which, in turn, implies that the two-sided limit does not exist). For example, we saw earlier that the one-sided limits of $f(x) = \sin(\pi/x)$ do not exist as $x$ approaches 0 because the function keeps oscillating between $-1$ and $1$, failing to settle in on a single value; and this implies that the two-sided limit does not exist as $x$ approaches 0.

**Example 3**

For the functions in Figure 2.1.14, find the one-sided and two-sided limits at $x = a$ if they exist.

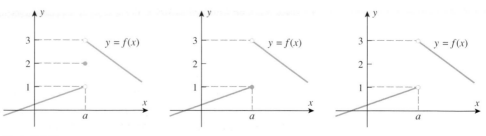

Figure 2.1.14

*Solution.*  The functions in all three figures have the same one-sided limits as $x \to a$, since the functions are identical, except at $x = a$. These limits are

$$\lim_{x \to a^+} f(x) = 3 \quad \text{and} \quad \lim_{x \to a^-} f(x) = 1$$

In all three cases the two-sided limit does not exist as $x \to a$ because the one-sided limits are not equal.  ◄

## Example 4

For the functions in Figure 2.1.15, find the one-sided and two-sided limits at $x = a$ if they exist.

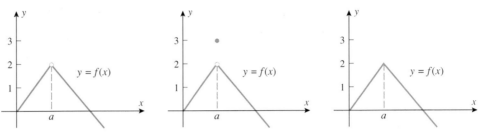

Figure 2.1.15

*Solution.*  As in the preceding example, the value of $f$ at $x = a$ has no bearing on the limits as $x \to a$, so that in all three cases we have

$$\lim_{x \to a^+} f(x) = 2 \quad \text{and} \quad \lim_{x \to a^-} f(x) = 2$$

Since the one-sided limits are equal, the two-sided limit exists and

$$\lim_{x \to a} f(x) = 2 \qquad\qquad ◄$$

**A FIRST LOOK AT CONTINUITY**

Plane curves can be divided into two categories—those that have breaks or holes and those that do not. Breaks or holes in a curve are called *discontinuities*; a curve with no discontinuities is called *continuous* (Figure 2.1.16).

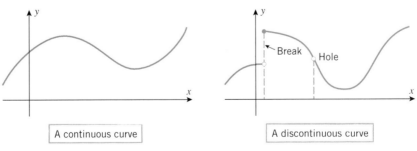

A continuous curve

A discontinuous curve

Figure 2.1.16

Examples 3 and 4 provide some useful insight into what it means for the graph of a function to be continuous. Of the six functions in those examples, only the last one does not have a break or hole in its graph at $x = a$. For the functions in Example 3, the break in the graph at $x = a$ results from the fact that the one-sided limits at that point have different values. A break of this type is called a *jump discontinuity* in the graph. For the first two functions in Example 4, the hole in the graph is caused by a mismatch between the value of the function at $x = a$ and the two-sided limit as $x$ approaches $a$. In the first graph, the function is simply undefined at $x = a$, leaving a hole; and in the second graph, $f(a)$ is defined, but its value is different from the limit, resulting in a point that is displaced from the main part of the graph. A break due to a hole or a displaced point is called a *removable discontinuity* in the graph. The third graph is continuous at $x = a$, since the value of $f$ at $x = a$ is the same as the two-sided limit of $f$ as $x$ approaches $a$, thereby ensuring that there is no jump or hole.

All of this suggests that three conditions must be satisfied to ensure that the graph of a function does not have a discontinuity at a given point:

- The function must be defined at the point.
- The two-sided limit must exist at the point.
- The value of the function and the value of the two-sided limit must be the same.

There will be more on this later.

**INFINITE LIMITS AND VERTICAL ASYMPTOTES**

Sometimes one-sided or two-sided limits will fail to exist because the values of the function increase or decrease indefinitely. For example, consider the behavior of the function $f(x) = 1/x$ as $x$ gets closer and closer to 0. It is evident from the table and graph in Figure 2.1.17 that as $x$ gets closer and closer to 0 from the right, the values of $f(x) = 1/x$ are positive and increase indefinitely; and as $x$ gets closer and closer to 0 from the left, the values of $f(x)$ are negative and decrease indefinitely. We denote these limiting behaviors by writing

$$\lim_{x \to 0^+} \frac{1}{x} = +\infty \quad \text{and} \quad \lim_{x \to 0^-} \frac{1}{x} = -\infty$$

More generally:

---

**2.1.4**    INFINITE LIMITS (AN INFORMAL VIEW).    If the values of $f(x)$ increase indefinitely as $x$ approaches $a$ from the right or left, then we write

$$\lim_{x \to a^+} f(x) = +\infty \quad \text{or} \quad \lim_{x \to a^-} f(x) = +\infty$$

as appropriate, and we say that $f(x)$ *increases without bound* as $x \to a^+$ or $x \to a^-$. Similarly, if the values of $f(x)$ decrease indefinitely as $x$ approaches $a$ from the right or left, then we write

$$\lim_{x \to a^+} f(x) = -\infty \quad \text{or} \quad \lim_{x \to a^-} f(x) = -\infty$$

as appropriate, and say that $f(x)$ *decreases without bound* as $x \to a^+$ or $x \to a^-$. Moreover, if both one-sided limits are $+\infty$, then we write

$$\lim_{x \to a} f(x) = +\infty$$

and if both one-sided limits are $-\infty$, then we write

$$\lim_{x \to a} f(x) = -\infty$$

---

REMARK.    It should be emphasized that the symbols $+\infty$ and $-\infty$, as used here, describe the particular way in which the limits fail to exist; they are not numerical limits and consequently cannot be manipulated using rules of algebra. For example, it is *not* correct to write $(+\infty) - (+\infty) = 0$.

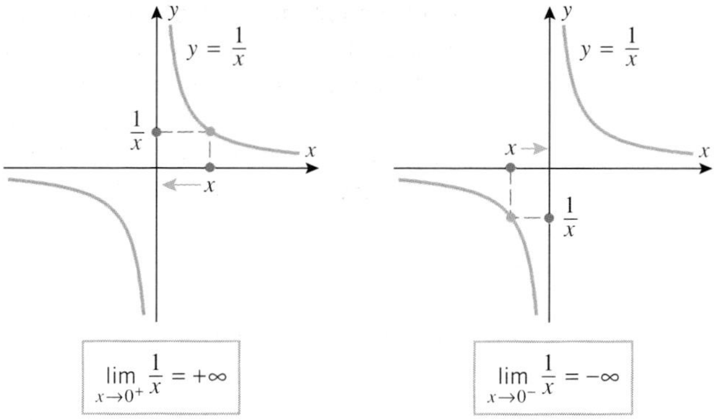

$$y = \frac{1}{x}$$

$$\lim_{x \to 0^+} \frac{1}{x} = +\infty$$

$$y = \frac{1}{x}$$

$$\lim_{x \to 0^-} \frac{1}{x} = -\infty$$

| $x$ | $-1$ | $-0.1$ | $-0.01$ | $-0.001$ | $-0.0001$ | 0 | 0.0001 | 0.001 | 0.01 | 0.1 | 1 |
|---|---|---|---|---|---|---|---|---|---|---|---|
| $\frac{1}{x}$ | $-1$ | $-10$ | $-100$ | $-1000$ | $-10,000$ | | 10,000 | 1000 | 100 | 10 | 1 |

Left side $\longrightarrow$   $\longleftarrow$ Right side

Figure 2.1.17

## Example 5

For the functions in Figure 2.1.18, describe the limits at $x = a$ in appropriate limit notation.

*Solution (a).* In Figure 2.1.18a, the function increases indefinitely as $x$ approaches $a$ from the right and decreases indefinitely as $x$ approaches $a$ from the left. Thus,

$$\lim_{x \to a^+} \frac{1}{x - a} = +\infty \quad \text{and} \quad \lim_{x \to a^-} \frac{1}{x - a} = -\infty$$

*Solution (b).* In Figure 2.1.18b, the function increases indefinitely as $x$ approaches $a$ from both the left and right. Thus,

$$\lim_{x \to a} \frac{1}{(x - a)^2} = \lim_{x \to a^+} \frac{1}{(x - a)^2} = \lim_{x \to a^-} \frac{1}{(x - a)^2} = +\infty$$

*Solution (c).* In Figure 2.1.18c, the function decreases indefinitely as $x$ approaches $a$ from the right and increases indefinitely as $x$ approaches $a$ from the left. Thus,

$$\lim_{x \to a^+} \frac{-1}{x - a} = -\infty \quad \text{and} \quad \lim_{x \to a^-} \frac{-1}{x - a} = +\infty$$

*Solution (d).* In Figure 2.1.18d, the function decreases indefinitely as $x$ approaches $a$ from both the left and right. Thus,

$$\lim_{x \to a} \frac{-1}{(x - a)^2} = \lim_{x \to a^+} \frac{-1}{(x - a)^2} = \lim_{x \to a^-} \frac{-1}{(x - a)^2} = -\infty \quad \blacktriangleleft$$

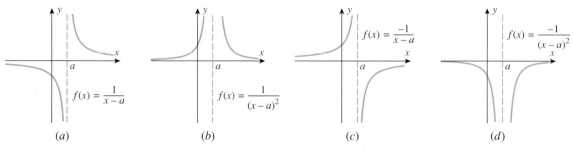

$$f(x) = \frac{1}{x - a}$$

$$(a)$$

$$f(x) = \frac{1}{(x - a)^2}$$

$$(b)$$

$$f(x) = \frac{-1}{x - a}$$

$$(c)$$

$$f(x) = \frac{-1}{(x - a)^2}$$

$$(d)$$

Figure 2.1.18

Geometrically, if $f(x) \rightarrow +\infty$ as $x$ approaches $a$ from the left or right, then the graph of $y = f(x)$ eventually gets closer and closer to the line $x = a$ as the graph is traversed in the positive $y$-direction; and if $f(x) \rightarrow -\infty$ as $x$ approaches $a$ from the left or right, then the graph of $y = f(x)$ eventually gets closer and closer to the line $x = a$ as the graph is traversed in the negative $y$-direction. We call this line a *vertical asymptote* (from the Greek *asymptotos*, meaning "nonintersecting").

---

**2.1.5** DEFINITION. A line $x = a$ is called a ***vertical asymptote*** of the graph of a function $f$ if $f(x)$ approaches $+\infty$ or $-\infty$ as $x$ approaches $a$ from the left or right.

---

### Example 6

The four functions graphed in Figure 2.1.18 all have a vertical asymptote at $x = a$, which is indicated by the dashed vertical lines in the figure. ◄

**LIMITS AT INFINITY AND HORIZONTAL ASYMPTOTES**

Thus far, we have used limits to describe the behavior of $f(x)$ as $x$ approaches a point $x = a$. However, sometimes we will not be concerned with the behavior of $f(x)$ near a specific point, but rather with how the values of $f(x)$ behave as $x$ increases without bound or decreases without bound. This is sometimes called the ***end behavior*** of the function because it describes how the function behaves for values of $x$ that are far from the origin. For example, it is evident from the table and graph in Figure 2.1.19 that as $x$ increases without bound, the values of $f(x) = 1/x$ are positive, but get closer and closer to 0; and similarly, as $x$ decreases without bound, the values of $f(x) = 1/x$ are negative, but also get closer and closer to 0. We denote these limiting behaviors by writing

$$\lim_{x \to +\infty} \frac{1}{x} = 0 \quad \text{and} \quad \lim_{x \to -\infty} \frac{1}{x} = 0$$

More generally:

---

**2.1.6** LIMITS AT INFINITY (AN INFORMAL VIEW). If the values of $f(x)$ eventually get closer and closer to a number $L$ as $x$ increases without bound, then we write

$$\lim_{x \to +\infty} f(x) = L \tag{13}$$

Similarly, if the values of $f(x)$ eventually get closer and closer to a number $L$ as $x$ decreases without bound, then we write

$$\lim_{x \to -\infty} f(x) = L \tag{14}$$

---

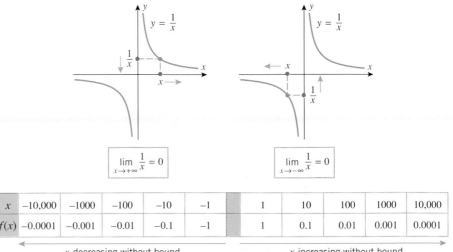

| $x$ | −10,000 | −1000 | −100 | −10 | −1 | 1 | 10 | 100 | 1000 | 10,000 |
|---|---|---|---|---|---|---|---|---|---|---|
| $f(x)$ | −0.0001 | −0.001 | −0.01 | −0.1 | −1 | 1 | 0.1 | 0.01 | 0.001 | 0.0001 |

$x$ decreasing without bound ⟶      ⟶ $x$ increasing without bound

Figure 2.1.19

Geometrically, if $f(x) \to L$ as $x \to +\infty$, then the graph of $y = f(x)$ eventually gets closer and closer to the line $y = L$ as the graph is traversed in the positive direction (Figure 2.1.20a); and if $f(x) \to L$ as $x \to -\infty$, then the graph of $y = f(x)$ eventually gets closer and closer to the line $y = L$ as the graph is traversed in the negative $x$-direction (Figure 2.1.20b). In either case we call the line $y = L$ a *horizontal asymptote* of the graph of $f$. For example, the four functions in Figure 2.1.18 all have $y = 0$ as a horizontal asymptote.

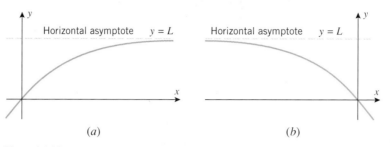

(a)                              (b)

Figure 2.1.20

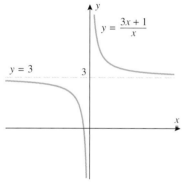

Figure 2.1.21

> **2.1.7**  DEFINITION.  A line $y = L$ is called a ***horizontal asymptote*** of the graph of a function $f$ if $f(x) \to L$ as $x \to +\infty$ or as $x \to -\infty$.

Sometimes the existence of a horizontal asymptote of a function $f$ will be readily apparent from the formula for $f$. For example, it is evident that the function

$$f(x) = \frac{3x + 1}{x} = 3 + \frac{1}{x}$$

has a horizontal asymptote at $y = 3$ (Figure 2.1.21), since the value of $1/x$ approaches 0 as $x \to +\infty$ or $x \to -\infty$. For more complicated functions, algebraic manipulations or special techniques that we will study in the next section may have to be applied to confirm the existence of horizontal asymptotes.

**HOW LIMITS AT INFINITY CAN FAIL TO EXIST**

Limits at infinity can fail to exist for various reasons. One possibility is that the values of $f(x)$ may increase or decrease without bound as $x \to +\infty$ or as $x \to -\infty$. For example, the values of $f(x) = x^3$ increase without bound as $x \to +\infty$ and decrease without bound as $x \to -\infty$; and for $f(x) = -x^3$ the values decrease without bound as $x \to +\infty$ and increase without bound as $x \to -\infty$ (Figure 2.1.22). We denote this by writing

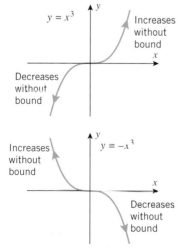

Figure 2.1.22

$$\lim_{x \to +\infty} x^3 = +\infty, \quad \lim_{x \to -\infty} x^3 = -\infty, \quad \lim_{x \to +\infty} (-x^3) = -\infty, \quad \lim_{x \to -\infty} (-x^3) = +\infty$$

More generally:

> **2.1.8**  INFINITE LIMITS AT INFINITY (AN INFORMAL VIEW).  If the values of $f(x)$ increase without bound as $x \to +\infty$ or as $x \to -\infty$, then we write
>
> $$\lim_{x \to +\infty} f(x) = +\infty \quad \text{or} \quad \lim_{x \to -\infty} f(x) = +\infty$$
>
> as appropriate; and if the values of $f(x)$ decrease without bound as $x \to +\infty$ or as $x \to -\infty$, then we write
>
> $$\lim_{x \to +\infty} f(x) = -\infty \quad \text{or} \quad \lim_{x \to -\infty} f(x) = -\infty$$
>
> as appropriate.

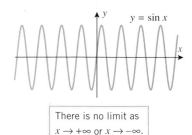

$y = \sin x$

Limits at infinity can also fail to exist because the graph of the function oscillates indef-initely in such a way that the values of the function do not approach a fixed number and do not increase or decrease without bound; the trigonometric functions $\sin x$ and $\cos x$ have this property, for example (Figure 2.1.23). In such cases we say that the limit ***fails to exist because of oscillation***.

There is no limit as $x \to +\infty$ or $x \to -\infty$.

Figure 2.1.23

**EXERCISE SET 2.1**    ⌇ Graphing Calculator    Ⓒ CAS

**1.** For the function $f$ graphed in the accompanying figure, find
   (a) $\lim_{x \to 3^-} f(x)$    (b) $\lim_{x \to 3^+} f(x)$    (c) $\lim_{x \to 3} f(x)$
   (d) $f(3)$    (e) $\lim_{x \to -\infty} f(x)$    (f) $\lim_{x \to +\infty} f(x)$.

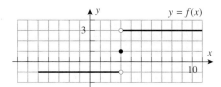

Figure Ex-1

**2.** For the function $f$ graphed in the accompanying figure, find
   (a) $\lim_{x \to 2^-} f(x)$    (b) $\lim_{x \to 2^+} f(x)$    (c) $\lim_{x \to 2} f(x)$
   (d) $f(2)$    (e) $\lim_{x \to -\infty} f(x)$    (f) $\lim_{x \to +\infty} f(x)$.

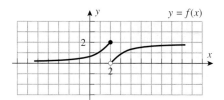

Figure Ex-2

**3.** For the function $g$ graphed in the accompanying figure, find
   (a) $\lim_{x \to 4^-} g(x)$    (b) $\lim_{x \to 4^+} g(x)$    (c) $\lim_{x \to 4} g(x)$
   (d) $g(4)$    (e) $\lim_{x \to -\infty} g(x)$    (f) $\lim_{x \to +\infty} g(x)$.

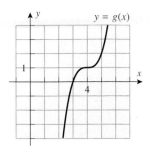

Figure Ex-3

**4.** For the function $g$ graphed in the accompanying figure, find
   (a) $\lim_{x \to 0^-} g(x)$    (b) $\lim_{x \to 0^+} g(x)$    (c) $\lim_{x \to 0} g(x)$
   (d) $g(0)$    (e) $\lim_{x \to -\infty} g(x)$    (f) $\lim_{x \to +\infty} g(x)$.

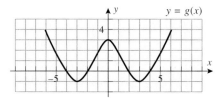

Figure Ex-4

**5.** For the function $F$ graphed in the accompanying figure, find
   (a) $\lim_{x \to -2^-} F(x)$    (b) $\lim_{x \to -2^+} F(x)$    (c) $\lim_{x \to -2} F(x)$
   (d) $F(-2)$    (e) $\lim_{x \to -\infty} F(x)$    (f) $\lim_{x \to +\infty} F(x)$.

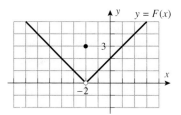

Figure Ex-5

**6.** For the function $F$ graphed in the accompanying figure, find
   (a) $\lim_{x \to 3^-} F(x)$    (b) $\lim_{x \to 3^+} F(x)$    (c) $\lim_{x \to 3} F(x)$
   (d) $F(3)$    (e) $\lim_{x \to -\infty} F(x)$    (f) $\lim_{x \to +\infty} F(x)$.

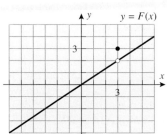

Figure Ex-6

**7.** For the function $\phi$ graphed in the accompanying figure, find

(a) $\lim\limits_{x \to -2^-} \phi(x)$   (b) $\lim\limits_{x \to -2^+} \phi(x)$   (c) $\lim\limits_{x \to -2} \phi(x)$

(d) $\phi(-2)$   (e) $\lim\limits_{x \to -\infty} \phi(x)$   (f) $\lim\limits_{x \to +\infty} \phi(x)$.

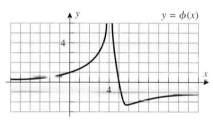

Figure Ex-7

**8.** For the function $\phi$ graphed in the accompanying figure, find

(a) $\lim\limits_{x \to 4^-} \phi(x)$   (b) $\lim\limits_{x \to 4^+} \phi(x)$   (c) $\lim\limits_{x \to 4} \phi(x)$

(d) $\phi(4)$   (e) $\lim\limits_{x \to -\infty} \phi(x)$   (f) $\lim\limits_{x \to +\infty} \phi(x)$.

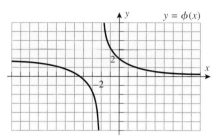

Figure Ex-8

**9.** For the function $f$ graphed in the accompanying figure, find

(a) $\lim\limits_{x \to 3^-} f(x)$   (b) $\lim\limits_{x \to 3^+} f(x)$   (c) $\lim\limits_{x \to 3} f(x)$

(d) $f(3)$   (e) $\lim\limits_{x \to -\infty} f(x)$   (f) $\lim\limits_{x \to +\infty} f(x)$.

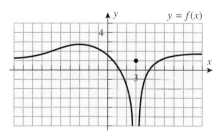

Figure Ex-9

**10.** For the function $f$ graphed in the accompanying figure, find

(a) $\lim\limits_{x \to 0^-} f(x)$   (b) $\lim\limits_{x \to 0^+} f(x)$   (c) $\lim\limits_{x \to 0} f(x)$

(d) $f(0)$   (e) $\lim\limits_{x \to -\infty} f(x)$   (f) $\lim\limits_{x \to +\infty} f(x)$.

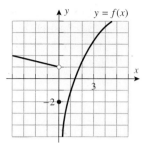

Figure Ex-10

**11.** For the function $G$ graphed in the accompanying figure, find

(a) $\lim\limits_{x \to 0^-} G(x)$   (b) $\lim\limits_{x \to 0^+} G(x)$   (c) $\lim\limits_{x \to 0} G(x)$

(d) $G(0)$   (e) $\lim\limits_{x \to -\infty} G(x)$   (f) $\lim\limits_{x \to +\infty} G(x)$.

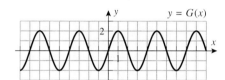

Figure Ex-11

**12.** For the function $G$ graphed in the accompanying figure, find

(a) $\lim\limits_{x \to 0^-} G(x)$   (b) $\lim\limits_{x \to 0^+} G(x)$   (c) $\lim\limits_{x \to 0} G(x)$

(d) $G(0)$   (e) $\lim\limits_{x \to -\infty} G(x)$   (f) $\lim\limits_{x \to +\infty} G(x)$.

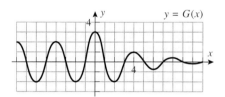

Figure Ex-12

**13.** Consider the function $g$ graphed in the accompanying figure. For what values of $x_0$ does $\lim\limits_{x \to x_0} g(x)$ exist?

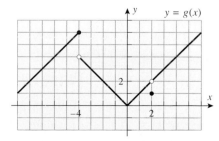

Figure Ex-13

**14.** Consider the function $f$ graphed in the accompanying figure. For what values of $x_0$ does $\lim\limits_{x \to x_0} f(x)$ exist?

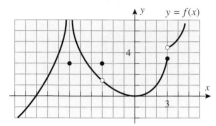

Figure Ex-14

In Exercises 15 and 16, find all points of discontinuity for the function, and for each such point state which of the three condition(s) for continuity fail to hold.

**15.** (a) The function $f$ in Exercise 1
(b) The function $F$ in Exercise 5
(c) The function $f$ in Exercise 9

**16.** (a) The function $f$ in Exercise 2
(b) The function $F$ in Exercise 6
(c) The function $f$ in Exercise 10

In Exercises 17–20: (i) Make a guess at the limit (if it exists) by evaluating the function at the specified points. (ii) Confirm your conclusions about the limit by graphing the function over an appropriate interval. (iii) If you have a CAS, then use it to find the limit. [*Note:* For the trigonometric functions, be sure to set your calculating and graphing utilities to the radian mode.]

c **17.** (a) $\lim\limits_{x \to 1} \dfrac{x-1}{x^3-1}$; $x = 2, 1.5, 1.1, 1.01, 1.001, 0, 0.5, 0.9,$
$0.99, 0.999$

(b) $\lim\limits_{x \to 1^+} \dfrac{x+1}{x^3-1}$; $x = 2, 1.5, 1.1, 1.01, 1.001, 1.0001$

(c) $\lim\limits_{x \to 1^-} \dfrac{x+1}{x^3-1}$; $x = 0, 0.5, 0.9, 0.99, 0.999, 0.9999$

c **18.** (a) $\lim\limits_{x \to 0} \dfrac{\sqrt{x+1}-1}{x}$; $x = \pm 0.25, \pm 0.1, \pm 0.001,$
$\pm 0.0001$

(b) $\lim\limits_{x \to 0^+} \dfrac{\sqrt{x+1}+1}{x}$; $x = 0.25, 0.1, 0.001, 0.0001$

(c) $\lim\limits_{x \to 0^-} \dfrac{\sqrt{x+1}+1}{x}$; $x = -0.25, -0.1, -0.001,$
$-0.0001$

c **19.** (a) $\lim\limits_{x \to 0} \dfrac{\sin 3x}{x}$; $x = \pm 0.25, \pm 0.1, \pm 0.001, \pm 0.0001$

(b) $\lim\limits_{x \to -1} \dfrac{\cos x}{x+1}$; $x = 0, -0.5, -0.9, -0.99, -0.999,$
$-1.5, -1.1, -1.01, -1.001$

c **20.** (a) $\lim\limits_{x \to -1} \dfrac{\tan(x+1)}{x+1}$; $x = 0, -0.5, -0.9, -0.99, -0.999,$
$-1.5, -1.1, -1.01, -1.001$

(b) $\lim\limits_{x \to 0} \dfrac{\sin(5x)}{\sin(2x)}$; $x = \pm 0.25, \pm 0.1, \pm 0.001, \pm 0.0001$

In Exercises 21 and 22: (i) Approximate the $y$-coordinates of all horizontal asymptotes of $y = f(x)$ by evaluating $f$ at the points $\pm 10, \pm 100, \pm 1000, \pm 100,000$, and $\pm 100,000,000$. (ii) Confirm your conclusions by graphing $y = f(x)$ over an appropriate interval. (iii) If you have a CAS, then use it to find the horizontal asymptotes.

c **21.** (a) $f(x) = \dfrac{2x+3}{x+4}$   (b) $f(x) = \left(1 + \dfrac{3}{x}\right)^x$

(c) $f(x) = \dfrac{x^2+1}{x+1}$

c **22.** (a) $f(x) = \dfrac{x^2-1}{5x^2+1}$   (b) $f(x) = \left(2 + \dfrac{1}{x}\right)^x$

(c) $f(x) = \dfrac{\sin x}{x}$

In Exercises 23 and 24, express the limit as an equivalent limit in which $x \to 0^+$ or $x \to 0^-$, as appropriate. [You need not evaluate the limit.]

**23.** (a) $\lim\limits_{x \to +\infty} x \sin\left(\dfrac{1}{x}\right)$   (b) $\lim\limits_{x \to +\infty} \dfrac{1-x}{1+x}$

(c) $\lim\limits_{x \to -\infty} \left(1 + \dfrac{2}{x}\right)^x$

**24.** (a) $\lim\limits_{x \to +\infty} \dfrac{\cos(\pi/x)}{\pi/x}$   (b) $\lim\limits_{x \to +\infty} \dfrac{x}{1+x}$

(c) $\lim\limits_{x \to -\infty} (1+2x)^{1/x}$

**25.** (a) Sketch the graph of a function that has two horizontal asymptotes.

(b) Can the graph of a function intersect its horizontal asymptotes? If not, explain why. If so, sketch such a graph.

**26.** (a) Do any of the trigonometric functions, $\sin x$, $\cos x$, $\tan x$, $\cot x$, $\sec x$, $\csc x$, have horizontal asymptotes?

(b) Do any of them have vertical asymptotes? Where?

c **27.** (a) Let

$$f(x) = x^3 - \frac{3^x}{2000}$$

Make a conjecture about the limit of $f$ as $x \to 0^+$ by evaluating $f$ at the points $x = 1, 0.75, 0.5, 0.25,$ $0.1, 0.05$.

(b) Evaluate $f$ at the points $x = 0.01, 0.001, 0.0001,$ $0.00001, 0.000001$, and make another conjecture.

(c) What flaw does this reveal about using numerical evidence to make conjectures about limits?

(d) If you have a CAS, use it to show that the exact value of the limit is $-1/2000$.

Roundoff error is one source of inaccuracy in calculator and computer computations. Another source of error, called *catastrophic subtraction*, occurs when two nearly equal numbers are subtracted, and the result is used as part of another calculation. For example, by hand calculation we have

$$(0.123456789012345 - 0.123456789012344) \times 10^{15} = 1$$

However, the author's calculator produces a value of 0 for this computation because it can only store 14 decimal digits, and the numbers being subtracted are identical in the first 14 decimal digits. Catastrophic subtraction can sometimes be avoided by rearranging formulas algebraically, but your best defense is to be aware that it can occur. Watch out for it in the next exercise.

[c] **28.** (a) Let

$$f(x) = \frac{x - \sin x}{x^3}$$

Make a conjecture about the limit of $f$ as $x \to 0^+$ by evaluating $f$ at the points $x = 0.1, 0.01, 0.001, 0.0001$.

(b) Evaluate $f$ at the points $x = 0.00001, 0.0000001,$ $0.00000001, 0.000000001, 0.0000000001$, and make another conjecture.

(c) What flaw does this reveal about using numerical evidence to make conjectures about limits?

(d) If you have a CAS, use it to show that the exact value of the limit is $\frac{1}{6}$.

**29.** (a) The accompanying figure shows graphs of the function from Exercise 28 over two different intervals. What is happening?

(b) Use your graphing utility to generate the graphs, and see whether the same problem occurs.

(c) Would you expect a similar problem to occur in the vicinity of $x = 0$ for the function

$$f(x) = \frac{1 - \cos x}{x}?$$

See if it does.

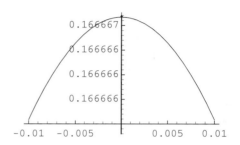

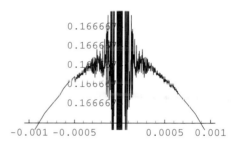

*Erratic graph generated by Mathematica*

Figure Ex-29

## 2.2 LIMITS (COMPUTATIONAL TECHNIQUES)

*In the last section we discussed limits informally, focusing on the basic ideas. In this section, we will discuss algebraic methods for finding limits, reserving the discussion of the underlying theory behind these methods for the next section.*

**SOME BASIC LIMITS**

Our strategy for finding limits algebraically has two parts:

- First we will establish the limits of some simple functions.

- Then we will develop a repertoire of theorems that will enable us to use the limits of those simple functions as building blocks for finding limits of more complicated functions.

The ten limits in the following theorem, all of which should be evident from Figure 2.2.1, will form our building blocks—three involve the constant function $f(x) = k$, three involve the linear function $f(x) = x$, and four involve the rational function $f(x) = 1/x$.

**2.2.1** THEOREM.

$$\lim_{x \to a} k = k \qquad \lim_{x \to +\infty} k = k \qquad \lim_{x \to -\infty} k = k$$

$$\lim_{x \to a} x = a \qquad \lim_{x \to +\infty} x = +\infty \qquad \lim_{x \to -\infty} x = -\infty$$

$$\lim_{x \to 0^+} \frac{1}{x} = +\infty \qquad \lim_{x \to 0^-} \frac{1}{x} = -\infty \qquad \lim_{x \to +\infty} \frac{1}{x} = 0 \qquad \lim_{x \to -\infty} \frac{1}{x} = 0$$

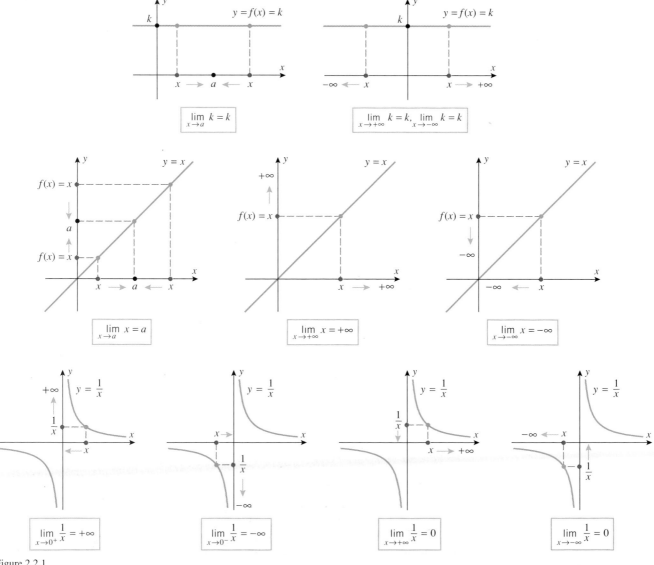

Figure 2.2.1

In the case of the constant function $f(x) = k$, the values of $f(x)$ do not change as $x$ varies, which explains why the limit of $f(x)$ is $k$, regardless of whether the limit is computed at a point $a$ or as $x$ approaches $+\infty$ or $-\infty$. For example,

$$\lim_{x \to 2} 3 = 3, \quad \lim_{x \to -2} 3 = 3, \quad \lim_{x \to +\infty} 3 = 3, \quad \lim_{x \to +\infty} 0 = 0, \quad \lim_{x \to -\infty} 3 = 3, \quad \lim_{x \to -\infty} 0 = 0$$

The limits of the function $f(x) = 1/x$ should make sense to you intuitively, based on your experience with fractions: making the denominator closer to zero increases the numerical size of the fraction (i.e., increases its absolute value), and increasing the numerical size of the denominator makes the numerical size of the fraction closer to zero. This is illustrated in Table 2.2.1.

**Table 2.2.1**

| | VALUES | | | | | | CONCLUSION |
|---|---|---|---|---|---|---|---|
| $x$ | 1 | .1 | .01 | .001 | .0001 | $\cdots$ | As $x \to 0^+$ the value of $1/x$ |
| $1/x$ | 1 | 10 | 100 | 1000 | 10,000 | $\cdots$ | increases without bound. |
| $x$ | $-1$ | $-.1$ | $-.01$ | $-.001$ | $-.0001$ | $\cdots$ | As $x \to 0^-$ the value of $1/x$ |
| $1/x$ | $-1$ | $-10$ | $-100$ | $-1000$ | $-10,000$ | $\cdots$ | decreases without bound. |
| $x$ | 1 | 10 | 100 | 1000 | 10,000 | $\cdots$ | As $x \to +\infty$ the value of $1/x$ |
| $1/x$ | 1 | .1 | .01 | .001 | .0001 | $\cdots$ | decreases toward zero. |
| $x$ | $-1$ | $-10$ | $-100$ | $-1000$ | $-10,000$ | $\cdots$ | As $x \to -\infty$ the value of $1/x$ |
| $1/x$ | $-1$ | $-.1$ | $-.01$ | $-.001$ | $-.0001$ | $\cdots$ | increases toward zero. |

The following theorem, parts of which are proved in Appendix G, will be our basic tool for finding limits algebraically.

---

**2.2.2** THEOREM. *Let* $\lim$ *stand for one of the limits* $\lim\limits_{x \to a}$, $\lim\limits_{x \to a^-}$, $\lim\limits_{x \to a^+}$, $\lim\limits_{x \to +\infty}$, *or* $\lim\limits_{x \to -\infty}$. *If* $L_1 = \lim f(x)$ *and* $L_2 = \lim g(x)$ *both exist, then*

(a) $\lim[f(x) + g(x)] = \lim f(x) + \lim g(x) = L_1 + L_2$

(b) $\lim[f(x) - g(x)] = \lim f(x) - \lim g(x) = L_1 - L_2$

(c) $\lim[f(x)g(x)] = \lim f(x) \lim g(x) = L_1 L_2$

(d) $\lim \dfrac{f(x)}{g(x)} = \dfrac{\lim f(x)}{\lim g(x)} = \dfrac{L_1}{L_2}$ *if* $L_2 \neq 0$

(e) $\lim \sqrt[n]{f(x)} = \sqrt[n]{\lim f(x)} = \sqrt[n]{L_1}$ *provided* $L_1 \geq 0$ *if* $n$ *is even.*

---

In words, this theorem states:

(a) *The limit of a sum is the sum of the limits.*

(b) *The limit of a difference is the difference of the limits.*

(c) *The limit of a product is the product of the limits.*

(d) *The limit of a quotient is the quotient of the limits provided the limit of the denominator is not zero.*

(e) *The limit of an nth root is the nth root of the limits.*

REMARK. Although results $(a)$ and $(c)$ are stated for two functions $f$ and $g$, these results hold as well for any finite number of functions; that is, if the limits $\lim f_1(x)$,

$\lim f_2(x), \dots, \lim f_n(x)$ all exist, then

$$\lim [f_1(x) + f_2(x) + \cdots + f_n(x)] = \lim f_1(x) + \lim f_2(x) + \cdots + \lim f_n(x) \qquad (1)$$

$$\lim [f_1(x) f_2(x) \cdots f_n(x)] = \lim f_1(x) \lim f_2(x) \cdots \lim f_n(x) \qquad (2)$$

In particular, if $f_1, f_2, \dots, f_n$ are all the same function $f$, then (2) reduces to

$$\lim [f(x)]^n = [\lim f(x)]^n \qquad (3)$$

It follows from this result that

$$\lim_{x \to a} x^n = [\lim_{x \to a} x]^n = a^n \qquad (4)$$

and

$$\lim_{x \to +\infty} \frac{1}{x^n} = \left( \lim_{x \to +\infty} \frac{1}{x} \right)^n = 0 \qquad \lim_{x \to -\infty} \frac{1}{x^n} = \left( \lim_{x \to -\infty} \frac{1}{x} \right)^n = 0 \qquad (5)$$

For example,

$$\lim_{x \to 3} x^4 = 3^4 = 81, \qquad \lim_{x \to +\infty} \frac{1}{x^4} = 0, \qquad \lim_{x \to -\infty} \frac{1}{x^4} = 0$$

Another useful result follows from part (c) of Theorem 2.2.2 in the special case where one of the factors is a constant $k$:

$$\lim k f(x) = \lim k \lim f(x) = k \lim f(x) \qquad (6)$$

In words, the first and last expressions in (6) state:

*A constant factor can be moved through a limit sign.*

**LIMITS OF POLYNOMIALS AS $x \to a$**

### Example 1

Find $\lim_{x \to 5} (x^2 - 4x + 3)$ and justify each step.

*Solution.*

$$\lim_{x \to 5} (x^2 - 4x + 3) = \lim_{x \to 5} x^2 - \lim_{x \to 5} 4x + \lim_{x \to 5} 3 \qquad \text{Theorem 2.2.2}(a), (b)$$

$$= \lim_{x \to 5} x^2 - 4 \lim_{x \to 5} x + \lim_{x \to 5} 3 \qquad \text{Equation (6)}$$

$$= 5^2 - 4(5) + 3 \qquad \text{Equation (4)}$$

$$= 8 \qquad \blacktriangleleft$$

Our next result will show that the limit of a polynomial $p(x)$ at a point $x = a$ is the same as the value of the polynomial at that point. This greatly simplifies the computation of limits of polynomials by allowing us to evaluate the polynomial instead. Moreover, as discussed in the last section, this result also establishes that graphs of polynomials are continuous curves (see the discussion in the subsection of Section 2.1 entitled *A First Look at Continuity*).

---

**2.2.3** THEOREM. *For any polynomial*

$$p(x) = c_0 + c_1 x + \cdots + c_n x^n$$

*and any real number $a$,*

$$\lim_{x \to a} p(x) = c_0 + c_1 a + \cdots + c_n a^n = p(a)$$

*Proof.*

$$\lim_{x \to a} p(x) = \lim_{x \to a} \left( c_0 + c_1 x + \cdots + c_n x^n \right)$$

$$= \lim_{x \to a} c_0 + \lim_{x \to a} c_1 x + \cdots + \lim_{x \to a} c_n x^n$$

$$= \lim_{x \to a} c_0 + c_1 \lim_{x \to a} x + \cdots + c_n \lim_{x \to a} x^n$$

$$= c_0 + c_1 a + \cdots + c_n a^n = p(a)$$ ∎

### Example 2

If we apply Theorem 2.2.3 to the problem in Example 1, we can bypass the intermediate steps and write immediately

$$\lim_{x \to 5} (x^2 - 4x + 3) = 5^2 - 4(5) + 3 = 8$$ ◄

**LIMITS OF $x^n$ AS $x \to +\infty$ OR $x \to -\infty$**

In Figure 2.2.2 we have graphed the polynomials of the form $x^n$ for $n = 1, 2, 3$, and 4; and below each figure we have indicated the limits as $x \to +\infty$ and $x \to -\infty$. The results in the figure are special cases of the following general results:

$$\lim_{x \to +\infty} x^n = +\infty, \quad n = 1, 2, 3, \ldots \tag{7}$$

$$\lim_{x \to -\infty} x^n = \begin{cases} +\infty, & n = 2, 4, 6, \ldots \\ -\infty, & n = 1, 3, 5, \ldots \end{cases} \tag{8}$$

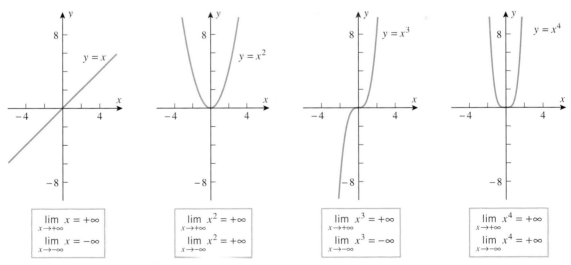

Figure 2.2.2

Multiplying $x^n$ by a positive real number does not affect limits (7) and (8), but multiplying by a negative real number reverses the signs.

### Example 3

$$\lim_{x \to +\infty} 2x^5 = +\infty, \qquad \lim_{x \to -\infty} 2x^5 = -\infty$$

$$\lim_{x \to +\infty} -7x^6 = -\infty, \qquad \lim_{x \to -\infty} -7x^6 = -\infty$$ ◄

There is a useful principle about polynomials which, expressed informally, states that:

*A polynomial behaves like its term of highest degree as $x \to +\infty$ or $x \to -\infty$.*

Stated more precisely, if $c_n \neq 0$, then

$$\lim_{x \to +\infty} \left( c_0 + c_1 x + \cdots + c_n x^n \right) = \lim_{x \to +\infty} c_n x^n \tag{9}$$

$$\lim_{x \to -\infty} \left( c_0 + c_1 x + \cdots + c_n x^n \right) = \lim_{x \to -\infty} c_n x^n \tag{10}$$

We can motivate these results by factoring out the highest power of $x$ from the polynomial and examining the limit of the factored expression. Thus,

$$c_0 + c_1 x + \cdots + c_n x^n = x^n \left( \frac{c_0}{x^n} + \frac{c_1}{x^{n-1}} + \cdots + c_n \right)$$

As $x \to +\infty$ or $x \to -\infty$, it follows from (5) that all of the terms with positive powers of $x$ in the denominator approach 0, so (9) and (10) are certainly plausible.

### Example 4

$$\lim_{x \to -\infty} (7x^5 - 4x^3 + 2x - 9) = \lim_{x \to -\infty} 7x^5 = -\infty$$
$$\lim_{x \to -\infty} (-4x^8 + 17x^3 - 5x + 1) = \lim_{x \to -\infty} -4x^8 = -\infty \qquad \blacktriangleleft$$

Recall that a rational function is the ratio of two polynomials. Theorem 2.2.3 and Theorem 2.2.2($d$) can often be used in combination to compute limits of rational functions.

### Example 5

Find $\lim\limits_{x \to 2} \dfrac{5x^3 + 4}{x - 3}$.

*Solution.*

$$\lim_{x \to 2} \frac{5x^3 + 4}{x - 3} = \frac{\lim\limits_{x \to 2} (5x^3 + 4)}{\lim\limits_{x \to 2} (x - 3)} = \frac{5 \cdot 2^3 + 4}{2 - 3} = -44 \qquad \blacktriangleleft$$

The method of the preceding example will not work if the limit of the denominator is zero, since Theorem 2.2.2($d$) is not applicable in this situation. However, if the numerator and denominator *both* approach zero as $x$ approaches $a$, then the numerator and denominator will have a common factor of $x - a$ and the limit can often be obtained by first canceling the common factors. The following example illustrates this technique.

### Example 6

Find $\lim\limits_{x \to 2} \dfrac{x^2 - 4}{x - 2}$.

*Solution.*   The numerator and denominator both have a limit of zero as $x$ approaches 2, so they share a common factor of $x - 2$. The limit can be obtained as follows:

$$\lim_{x \to 2} \frac{x^2 - 4}{x - 2} = \lim_{x \to 2} \frac{(x - 2)(x + 2)}{x - 2} = \lim_{x \to 2} (x + 2) = 4 \qquad \blacktriangleleft$$

REMARK.   Although correct, the second equality in the preceding computation needs some justification, since canceling the factor $x - 2$ alters the function. However, as discussed in Example 5 of Section 1.2, the two functions are identical, except at $x = 2$ (Figure 1.2.9); and we know from our discussions in the last section that this difference has no effect on the limit as $x$ approaches 2.

## Example 7

Find

$$\text{(a) } \lim_{x \to 3} \frac{x^2 - 6x + 9}{x - 3} \qquad \text{(b) } \lim_{x \to -4} \frac{2x + 8}{x^2 + x - 12}$$

*Solution (a).* The numerator and denominator both have a limit of zero as $x$ approaches 3, so there is a common factor of $x - 3$. We proceed as follows:

$$\lim_{x \to 3} \frac{x^2 - 6x + 9}{x - 3} = \lim_{x \to 3} \frac{(x - 3)^2}{x - 3} = \lim_{x \to 3} (x - 3) = 0$$

*Solution (b).* The numerator and denominator both have a limit of zero as $x$ approaches $-4$, so there is a common factor of $x - (-4) = x + 4$. We proceed as follows:

$$\lim_{x \to -4} \frac{2x + 8}{x^2 + x - 12} = \lim_{x \to -4} \frac{2(x + 4)}{(x + 4)(x - 3)} = \lim_{x \to -4} \frac{2}{x - 3} = -\frac{2}{7} \qquad \blacktriangleleft$$

If the limit of the denominator is zero, but the limit of the numerator is not, then there are three possibilities for the limit of the rational function as $x \to a$:

- The limit may be $+\infty$.
- The limit may be $-\infty$.
- The limit may be $+\infty$ from one side and $-\infty$ from the other.

Figure 2.2.3 illustrates this graphically for functions of the form $1/(x - a)$, $1/(x - a)^2$, and $-1/(x - a)^2$.

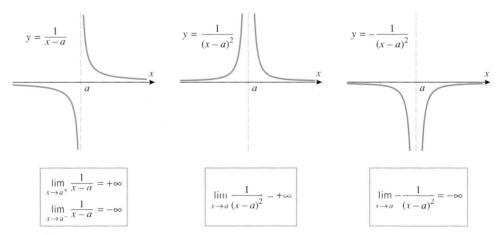

Figure 2.2.3

## Example 8

Find

$$\text{(a) } \lim_{x \to 4^+} \frac{2 - x}{(x - 4)(x + 2)} \qquad \text{(b) } \lim_{x \to 4^-} \frac{2 - x}{(x - 4)(x + 2)} \qquad \text{(c) } \lim_{x \to 4} \frac{2 - x}{(x - 4)(x + 2)}$$

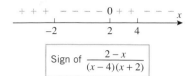

Figure 2.2.4

*Solution.* In all three parts the limit of the numerator is $-2$, and the limit of the denominator is $0$, so the limit of the ratio does not exist. To be more specific than this, we need to analyze the sign of the ratio. The sign of the ratio, which is given in Figure 2.2.4, is determined by the signs of $2 - x$, $x - 4$, and $x + 2$. (The method of test points, discussed in Appendix A, provides a simple way of finding the sign of the ratio here.) It follows from this figure that as $x$ approaches 4 from the right, the ratio is always negative; and as $x$ approaches 4 from the left, the ratio is eventually positive (after $x$ exceeds 2). Thus,

$$\lim_{x \to 4^+} \frac{2-x}{(x-4)(x+2)} - -\infty \quad \text{and} \quad \lim_{x \to 4^-} \frac{2-x}{(x-4)(x+2)} = +\infty$$

Because the one-sided limits have opposite signs, all we can say about the two-sided limit is that it does not exist. ◀

**LIMITS OF RATIONAL FUNCTIONS AS $x \to +\infty$ OR $x \to -\infty$**

If we divide the numerator and denominator of a rational function by the highest power of $x$ that occurs in the denominator, then all the powers of $x$ in the denominator become constants or powers of $1/x$. The following examples show how this observation together with (5), (9), and (10) can be used to find limits of rational functions as $x \to +\infty$ or $x \to -\infty$.

**Example 9**

Find $\lim\limits_{x \to +\infty} \dfrac{3x + 5}{6x - 8}$.

*Solution.* Divide the numerator and denominator by the highest power of $x$ that occurs in the denominator; this is $x^1 = x$. We obtain

$$\lim_{x \to +\infty} \frac{3x+5}{6x-8} = \lim_{x \to +\infty} \frac{3 + 5/x}{6 - 8/x} = \frac{\lim\limits_{x \to +\infty} (3 + 5/x)}{\lim\limits_{x \to +\infty} (6 - 8/x)}$$

$$= \frac{\lim\limits_{x \to +\infty} 3 + \lim\limits_{x \to +\infty} 5/x}{\lim\limits_{x \to +\infty} 6 - \lim\limits_{x \to +\infty} 8/x} = \frac{3 + 5 \lim\limits_{x \to +\infty} 1/x}{6 - 8 \lim\limits_{x \to +\infty} 1/x}$$

$$= \frac{3 + (5 \cdot 0)}{6 - (8 \cdot 0)} = \frac{1}{2} \qquad \blacktriangleleft$$

**Example 10**

Find

(a) $\lim\limits_{x \to -\infty} \dfrac{4x^2 - x}{2x^3 - 5}$     (b) $\lim\limits_{x \to -\infty} \dfrac{5x^3 - 2x^2 + 1}{3x + 5}$

*Solution (a).* Divide the numerator and denominator by the highest power of $x$ that occurs in the denominator, namely $x^3$. We obtain

$$\lim_{x \to -\infty} \frac{4x^2 - x}{2x^3 - 5} = \lim_{x \to -\infty} \frac{4/x - 1/x^2}{2 - 5/x^3} = \frac{\lim\limits_{x \to -\infty} (4/x - 1/x^2)}{\lim\limits_{x \to -\infty} (2 - 5/x^3)}$$

$$= \frac{(4 \cdot 0) - 0}{2 - (5 \cdot 0)} = \frac{0}{2} = 0$$

*Solution (b).* Divide the numerator and denominator by $x$ to obtain

$$\lim_{x \to -\infty} \frac{5x^3 - 2x^2 + 1}{3x + 5} = \lim_{x \to -\infty} \frac{5x^2 - 2x + 1/x}{3 + 5/x} = +\infty$$

where the final step is justified by the fact that

$$5x^2 - 2x \to +\infty, \quad 1/x \to 0, \quad \text{and} \quad 3 + 5/x \to 3$$

as $x \to -\infty$. ◄

**A QUICK METHOD FOR FINDING LIMITS OF RATIONAL FUNCTIONS AS $x \to +\infty$ OR $x \to -\infty$**

Since a polynomial behaves like its term of highest degree as $x \to +\infty$ or $x \to -\infty$, it follows that a rational function behaves like the ratio of the terms of highest degree in the numerator and denominator as $x \to +\infty$ or $x \to -\infty$; that is, if $c_n \neq 0$ and $d_n \neq 0$, then

$$\lim_{x \to +\infty} \frac{c_0 + c_1 x + \cdots + c_n x^n}{d_0 + d_1 x + \cdots + d_m x^m} = \lim_{x \to +\infty} \frac{c_n x^n}{d_m x^m} \tag{11}$$

and

$$\lim_{x \to -\infty} \frac{c_0 + c_1 x + \cdots + c_n x^n}{d_0 + d_1 x + \cdots + d_m x^m} = \lim_{x \to -\infty} \frac{c_n x^n}{d_m x^m} \tag{12}$$

### Example 11

Use Formulas (11) and (12) to find

(a) $\displaystyle \lim_{x \to +\infty} \frac{3x + 5}{6x - 8}$   (b) $\displaystyle \lim_{x \to -\infty} \frac{4x^2 - x}{2x^3 - 5}$   (c) $\displaystyle \lim_{x \to +\infty} \frac{3 - 2x^4}{x + 1}$

*Solution (a).*

$$\lim_{x \to +\infty} \frac{3x + 5}{6x - 8} = \lim_{x \to +\infty} \frac{3x}{6x} = \lim_{x \to +\infty} \frac{1}{2} = \frac{1}{2}$$

which agrees with the result obtained in Example 9.

*Solution (b).*

$$\lim_{x \to -\infty} \frac{4x^2 - x}{2x^3 - 5} = \lim_{x \to -\infty} \frac{4x^2}{2x^3} = \lim_{x \to -\infty} \frac{2}{x} = 0$$

which agrees with the result obtained in Example 10.

*Solution (c).*

$$\lim_{x \to +\infty} \frac{3 - 2x^4}{x + 1} = \lim_{x \to +\infty} \frac{-2x^4}{x} = \lim_{x \to +\infty} -2x^3 = -\infty \qquad ◄$$

REMARK.   We emphasize that Formulas (11) and (12) are only applicable if $x \to +\infty$ or $x \to -\infty$; they do not apply to limits in which $x$ approaches a *finite* number $a$.

**LIMITS INVOLVING RADICALS**

### Example 12

Find $\displaystyle \lim_{x \to +\infty} \sqrt[3]{\frac{3x + 5}{6x - 8}}$.

*Solution.*

$$\lim_{x \to +\infty} \sqrt[3]{\frac{3x + 5}{6x - 8}} = \sqrt[3]{\lim_{x \to +\infty} \frac{3x + 5}{6x - 8}} = \sqrt[3]{\frac{1}{2}}$$

$$\underbrace{\hspace{3cm}}_{\text{Theorem 2.2.2}(e)} \qquad \underbrace{\hspace{2cm}}_{\text{Example 9}} \qquad ◄$$

### Example 13

Find

(a) $\displaystyle \lim_{x \to +\infty} \frac{\sqrt{x^2 + 2}}{3x - 6}$   (b) $\displaystyle \lim_{x \to -\infty} \frac{\sqrt{x^2 + 2}}{3x - 6}$

In both parts it would be helpful to manipulate the function so that the powers of $x$ become powers of $1/x$. This can be achieved in both cases by dividing the numerator and denominator by $|x|$ and using the fact that $\sqrt{x^2} = |x|$.

*Solution (a).* As $x \to +\infty$, the values of $x$ are eventually positive, so we can replace $|x|$ by $x$ where helpful. We obtain

$$\lim_{x \to +\infty} \frac{\sqrt{x^2 + 2}}{3x - 6} = \lim_{x \to +\infty} \frac{\sqrt{x^2 + 2}/|x|}{(3x - 6)/|x|} = \lim_{x \to +\infty} \frac{\sqrt{x^2 + 2}/\sqrt{x^2}}{(3x - 6)/x}$$

$$= \lim_{x \to +\infty} \frac{\sqrt{1 + 2/x^2}}{3 - 6/x} = \frac{\lim\limits_{x \to +\infty} \sqrt{1 + 2/x^2}}{\lim\limits_{x \to +\infty} (3 - 6/x)}$$

$$= \frac{\sqrt{\lim\limits_{x \to +\infty} (1 + 2/x^2)}}{\lim\limits_{x \to +\infty} (3 - 6/x)} = \frac{\sqrt{\lim\limits_{x \to +\infty} 1 + 2 \lim\limits_{x \to +\infty} 1/x^2}}{\lim\limits_{x \to +\infty} 3 - 6 \lim\limits_{x \to +\infty} 1/x}$$

$$= \frac{\sqrt{1 + (2 \cdot 0)}}{3 - (6 \cdot 0)} = \frac{1}{3}$$

*Solution (b).* As $x \to -\infty$, the values of $x$ are eventually negative, so we can replace $|x|$ by $-x$ where helpful. We obtain

$$\lim_{x \to -\infty} \frac{\sqrt{x^2 + 2}}{3x - 6} = \lim_{x \to -\infty} \frac{\sqrt{x^2 + 2}/|x|}{(3x - 6)/|x|} = \lim_{x \to -\infty} \frac{\sqrt{x^2 + 2}/\sqrt{x^2}}{(3x - 6)/(-x)}$$

$$= \lim_{x \to -\infty} \frac{\sqrt{1 + 2/x^2}}{(6/x) - 3} = -\frac{1}{3} \qquad \blacktriangleleft$$

......................................................
**LIMITS OF FUNCTIONS DEFINED PIECEWISE**

For functions that are defined piecewise, a two-sided limit at a point where the formula changes is best obtained by first finding the one-sided limits at the point.

**Example 14**

Find $\lim\limits_{x \to 3} f(x)$ for $f(x) = \begin{cases} x^2 - 5, & x \le 3 \\ \sqrt{x + 13}, & x > 3. \end{cases}$

*Solution.* As $x$ approaches 3 from the left, the formula for $f$ is

$$f(x) = x^2 - 5$$

so that

$$\lim_{x \to 3^-} f(x) = \lim_{x \to 3^-} (x^2 - 5) = 3^2 - 5 = 4$$

As $x$ approaches 3 from the right, the formula for $f$ is

$$f(x) = \sqrt{x + 13}$$

so that

$$\lim_{x \to 3^+} f(x) = \lim_{x \to 3^+} \sqrt{x + 13} = \sqrt{\lim_{x \to 3^+} (x + 13)} = \sqrt{16} = 4$$

Since the one-sided limits are equal, we have

$$\lim_{x \to 3} f(x) = 4 \qquad \blacktriangleleft$$

## EXERCISE SET 2.2

1. Given that

$$\lim_{x \to a} f(x) = 2, \quad \lim_{x \to a} g(x) = -4, \quad \lim_{x \to a} h(x) = 0$$

find the limits that exist. If the limit does not exist, explain why.

(a) $\lim_{x \to a} [f(x) + 2g(x)]$

(b) $\lim_{x \to a} [h(x) - 3g(x) + 1]$

(c) $\lim_{x \to a} [f(x)g(x)]$

(d) $\lim_{x \to a} [g(x)]^2$

(e) $\lim_{x \to a} \sqrt[3]{6 + f(x)}$

(f) $\lim_{x \to a} \dfrac{2}{g(x)}$

(g) $\lim_{x \to a} \dfrac{3f(x) - 8g(x)}{h(x)}$

(h) $\lim_{x \to a} \dfrac{7g(x)}{2f(x) + g(x)}$

2. Use the graphs of $f$ and $g$ in the accompanying figure to find the limits that exist. If the limit does not exist, explain why.

(a) $\lim_{x \to 2} [f(x) + g(x)]$

(b) $\lim_{x \to 0} [f(x) + g(x)]$

(c) $\lim_{x \to 0^+} [f(x) + g(x)]$

(d) $\lim_{x \to 0^-} [f(x) + g(x)]$

(e) $\lim_{x \to 2} \dfrac{f(x)}{1 + g(x)}$

(f) $\lim_{x \to 2} \dfrac{1 + g(x)}{f(x)}$

(g) $\lim_{x \to 0^+} \sqrt{f(x)}$

(h) $\lim_{x \to 0^-} \sqrt{f(x)}$

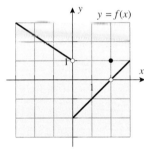 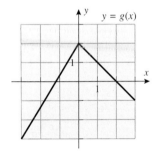

Figure Ex-2

3. In each part, find the limit by inspection.

(a) $\lim_{x \to 8} 7$

(b) $\lim_{x \to -\infty} (-3)$

(c) $\lim_{x \to 0^+} \pi$

(d) $\lim_{x \to -2} 3x$

(e) $\lim_{y \to 3^+} 12y$

(f) $\lim_{h \to +\infty} (-2h)$

4. In each part, find the stated limit of $f(x) = x/|x|$ by inspection.

(a) $\lim_{x \to 5} f(x)$

(b) $\lim_{x \to -5} f(x)$

(c) $\lim_{x \to +\infty} f(x)$

(d) $\lim_{x \to -\infty} f(x)$

(e) $\lim_{x \to 0^+} f(x)$

(f) $\lim_{x \to 0^-} f(x)$

---

Find the limits in Exercises 5–48.

---

5. $\lim_{y \to 2^-} \dfrac{(y - 1)(y - 2)}{y + 1}$

6. $\lim_{x \to 3} \dfrac{x^2 - 2x}{x + 1}$

7. $\lim_{x \to 4} \dfrac{x^2 - 16}{x - 4}$

8. $\lim_{x \to 0} \dfrac{6x - 9}{x^3 - 12x + 3}$

9. $\lim_{x \to 1^+} \dfrac{x^4 - 1}{x - 1}$

10. $\lim_{t \to -2} \dfrac{t^3 + 8}{t + 2}$

11. $\lim_{x \to -1} \dfrac{x^2 + 6x + 5}{x^2 - 3x - 4}$

12. $\lim_{x \to 2} \dfrac{x^2 - 4x + 4}{x^2 + x - 6}$

13. $\lim_{x \to +\infty} \dfrac{3x + 1}{2x - 5}$

14. $\lim_{t \to 1} \dfrac{t^3 + t^2 - 5t + 3}{t^3 - 3t + 2}$

15. $\lim_{y \to -\infty} \dfrac{3}{y + 4}$

16. $\lim_{x \to +\infty} \dfrac{1}{x - 12}$

17. $\lim_{x \to -\infty} \dfrac{x - 2}{x^2 + 2x + 1}$

18. $\lim_{x \to +\infty} \dfrac{5x^2 + 7}{3x^2 - x}$

19. $\lim_{x \to -\infty} \dfrac{\sqrt{5x^2 - 2}}{x + 3}$

20. $\lim_{s \to +\infty} \sqrt[3]{\dfrac{3s^7 - 4s^5}{2s^7 + 1}}$

21. $\lim_{y \to -\infty} \dfrac{2 - y}{\sqrt{7 + 6y^2}}$

22. $\lim_{x \to +\infty} \dfrac{\sqrt{5x^2 - 2}}{x + 3}$

23. $\lim_{x \to -\infty} \dfrac{\sqrt{3x^4 + x}}{x^2 - 8}$

24. $\lim_{y \to +\infty} \dfrac{2 - y}{\sqrt{7 + 6y^2}}$

25. $\lim_{x \to 3^+} \dfrac{x}{x - 3}$

26. $\lim_{x \to +\infty} \dfrac{\sqrt{3x^4 + x}}{x^2 - 8}$

27. $\lim_{x \to 3} \dfrac{x}{x - 3}$

28. $\lim_{x \to 3^-} \dfrac{x}{x - 3}$

29. $\lim_{x \to 2^-} \dfrac{x}{x^2 - 4}$

30. $\lim_{x \to 2^+} \dfrac{x}{x^2 - 4}$

31. $\lim_{y \to 6^+} \dfrac{y + 6}{y^2 - 36}$

32. $\lim_{x \to 2} \dfrac{x}{x^2 - 4}$

33. $\lim_{y \to 6} \dfrac{y + 6}{y^2 - 36}$

34. $\lim_{y \to 6^-} \dfrac{y + 6}{y^2 - 36}$

35. $\lim_{x \to 4^-} \dfrac{3 - x}{x^2 - 2x - 8}$

36. $\lim_{x \to 4^+} \dfrac{3 - x}{x^2 - 2x - 8}$

37. $\lim_{x \to +\infty} \dfrac{7 - 6x^5}{x + 3}$

38. $\lim_{x \to 4} \dfrac{3 - x}{x^2 - 2x - 8}$

39. $\lim_{t \to +\infty} \dfrac{6 - t^3}{7t^3 + 3}$

40. $\lim_{t \to -\infty} \dfrac{5 - 2t^3}{t^2 + 1}$

41. $\lim_{x \to 9} \dfrac{x - 9}{\sqrt{x} - 3}$

42. $\lim_{x \to 3^-} \dfrac{1}{|x - 3|}$

43. $\lim_{x \to +\infty} \sqrt{x}$

44. $\lim_{y \to 4} \dfrac{4 - y}{2 - \sqrt{y}}$

45. $\lim_{x \to -\infty} (3 - x)$

46. $\lim_{x \to -\infty} \sqrt{5 - x}$

47. $\lim_{x \to +\infty} (1 + 2x - 3x^5)$

48. $\lim_{x \to +\infty} (2x^3 - 100x + 5)$

49. Let

$$f(x) = \begin{cases} x - 1, & x \le 3 \\ 3x - 7, & x > 3 \end{cases}$$

Find

(a) $\lim_{x \to 3^-} f(x)$

(b) $\lim_{x \to 3^+} f(x)$

(c) $\lim_{x \to 3} f(x)$.

**50.** Let

$$g(t) = \begin{cases} t^2, & t \geq 0 \\ t - 2, & t < 0 \end{cases}$$

Find

(a) $\lim\limits_{t \to 0^-} g(t)$     (b) $\lim\limits_{t \to 0^+} g(t)$     (c) $\lim\limits_{t \to 0} g(t)$.

**51.** Let $f(x) = \dfrac{x^3 - 1}{x - 1}$.

(a) Find $\lim\limits_{x \to 1} f(x)$.

(b) Sketch the graph of $y = f(x)$.

**52.** Let

$$f(x) = \begin{cases} \dfrac{x^2 - 9}{x + 3}, & x \neq -3 \\ k, & x = -3 \end{cases}$$

(a) Find $k$ so that $F(-3) = \lim\limits_{x \to -3} F(x)$.

(b) With $k$ assigned the value $\lim\limits_{x \to -3} F(x)$, show that $F(x)$ can be expressed as a polynomial.

**53.** (a) Explain why the following calculation is incorrect.

$$\lim_{x \to 0^+} \left( \frac{1}{x} - \frac{1}{x^2} \right) = \lim_{x \to 0^+} \frac{1}{x} - \lim_{x \to 0^+} \frac{1}{x^2}$$
$$= +\infty - (+\infty) = 0$$

(b) Show that $\lim\limits_{x \to 0^+} \left( \dfrac{1}{x} - \dfrac{1}{x^2} \right) = -\infty$.

**54.** Find $\lim\limits_{x \to 0^-} \left( \dfrac{1}{x} + \dfrac{1}{x^2} \right)$.

In Exercises 55 and 56, first rationalize the numerator, then find the limit.

**55.** $\lim\limits_{x \to 0} \dfrac{\sqrt{x + 4} - 2}{x}$     **56.** $\lim\limits_{x \to 0} \dfrac{\sqrt{x^2 + 4} - 2}{x}$

---

Find the limits in Exercises 57–60.

**57.** $\lim\limits_{x \to +\infty} (\sqrt{x^2 + 3} - x)$

**58.** $\lim\limits_{x \to +\infty} (\sqrt{x^2 - 3x} - x)$

**59.** $\lim\limits_{x \to +\infty} (\sqrt{x^2 + ax} - x)$

**60.** $\lim\limits_{x \to +\infty} (\sqrt{x^2 + ax} - \sqrt{x^2 + bx})$

**61.** Discuss the limits of $p(x) = (1 - x)^n$ as $x \to +\infty$ and $x \to -\infty$ for positive integer values of $n$.

**62.** Let $p(x) = (1 - x)^n$ and $q(x) = (1 - x)^m$. Discuss the limits of $p(x)/q(x)$ as $x \to +\infty$ and $x \to -\infty$ for positive integer values of $m$ and $n$.

**63.** Let $p(x)$ be a polynomial of degree $n$. Discuss the limits of $p(x)/x^m$ as $x \to +\infty$ and $x \to -\infty$ for positive integer values of $m$.

**64.** In each part, find examples of polynomials $p(x)$ and $q(x)$ that satisfy the stated condition and such that $p(x) \to +\infty$ and $q(x) \to +\infty$ as $x \to +\infty$.

(a) $\lim\limits_{x \to +\infty} \dfrac{p(x)}{q(x)} = 1$     (b) $\lim\limits_{x \to +\infty} \dfrac{p(x)}{q(x)} = 0$

(c) $\lim\limits_{x \to +\infty} \dfrac{p(x)}{q(x)} = +\infty$     (d) $\lim\limits_{x \to +\infty} [p(x) - q(x)] = 3$

**65.** Let $p(x)$ and $q(x)$ be polynomials, and suppose $q(x_0) = 0$. Discuss the behavior of the graph of $y = p(x)/q(x)$ in the vicinity of the point $x = x_0$. Give examples to support your conclusions.

**66.** Find

$$\lim_{x \to +\infty} \frac{c_0 + c_1 x + \cdots + c_n x^n}{d_0 + d_1 x + \cdots + d_m x^m}$$

where $c_n \neq 0$ and $d_m \neq 0$. [*Hint:* Your answer will depend on whether $m < n$, $m = n$, or $m > n$.]

---

## 2.3 LIMITS (DISCUSSED MORE RIGOROUSLY)

*Thus far, our discussion of limits has been based on our intuitive feeling of what it means for the values of a function to get closer and closer to a limiting value. However, this level of informality can only take us so far, so our goal in this section is to define limits precisely. From a purely mathematical point of view these definitions are needed to establish limits with certainty and to prove theorems about them. However, they will also provide us with a deeper understanding of the limit concept, making it possible for us to visualize some of the more subtle properties of functions.*

---

**DEFINITION OF A LIMIT**

In earlier sections we interpreted the limit

$$\lim_{x \to a} f(x) = L$$

to mean that we can force the values of $f(x)$ closer and closer to $L$ by making $x$ closer and closer (but not equal) to $a$. Our goal here is to try to make the notion of a limit more precise by giving the informal phrase "closer and closer to" a precise mathematical meaning. However, the concept is subtle, so we will build up to it by giving two preliminary definitions that capture the essential ideas, and then giving the final definition as it is commonly stated.

To start, consider the function $f$ graphed in Figure 2.3.1$a$ for which $f(x) \to L$ as $x \to a$. We have intentionally placed a hole in the graph at $x = a$ to emphasize that the function $f$ need not be defined at $x = a$ to have a limit there.

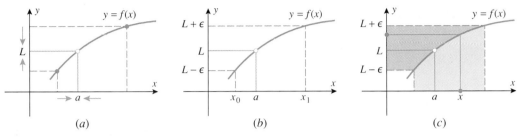

Figure 2.3.1

To motivate an appropriate definition for a two-sided limit, suppose that we choose *any* positive number, say $\epsilon$, and draw horizontal lines from the points $L + \epsilon$ and $L - \epsilon$ on the $y$-axis to the curve $y = f(x)$ and then draw vertical lines from those points on the curve to the $x$-axis. As shown in Figure 2.3.1$b$, let $x_0$ and $x_1$ be points where the vertical lines intersect the $x$-axis.

Next, imagine that $x$ gets closer and closer to $a$ (from either side). Eventually, $x$ will lie inside the interval $(x_0, x_1)$, which is marked by the green band in Figure 2.3.1$c$; and when this happens, the value of $f(x)$ will fall between $L - \epsilon$ and $L + \epsilon$, marked by the red band in the figure. Thus, we conclude:

*If $f(x) \to L$ as $x \to a$, then for any positive number $\epsilon$, we can find an open interval on the x-axis that contains the point $x = a$ and has the property that for each $x$ in that interval (except possibly for $x = a$), the value of $f(x)$ is between $L - \epsilon$ and $L + \epsilon$.*

What is important about this result is that it holds no matter how small we make $\epsilon$. However, making $\epsilon$ smaller and smaller forces $f(x)$ *closer and closer* to $L$—which is precisely the concept we were trying to capture mathematically. This suggests the following definition of a two-sided limit.

---

**2.3.1** LIMIT (FIRST PRELIMINARY DEFINITION). Let $f(x)$ be defined for all $x$ in some open interval containing the number $a$, with the possible exception that $f(x)$ need not be defined at $a$. We will write

$$\lim_{x \to a} f(x) = L$$

if given any number $\epsilon > 0$ we can find an open interval $(x_0, x_1)$ containing the point $a$ such that $f(x)$ satisfies

$$L - \epsilon < f(x) < L + \epsilon$$

for each $x$ in the interval $(x_0, x_1)$, except possibly $x = a$.

---

Observe that in Figure 2.3.1$c$ the interval $(x_0, x_1)$ extends farther on the right side of $a$ than on the left side. However, for many purposes it is preferable to have an interval that extends the same distance on both sides of $a$. For this purpose, let us choose any positive number $\delta$ that is smaller than both $x_1 - a$ and $a - x_0$, and consider the interval $(a - \delta, a + \delta)$. This interval extends the same distance $\delta$ on both sides of $a$ and lies inside of the interval $(x_0, x_1)$ (Figure 2.3.2). Moreover, the condition $L - \epsilon < f(x) < L + \epsilon$ holds for every $x$ in this interval (except possibly $x = a$), since this condition holds on the larger interval $(x_0, x_1)$. This suggests the following reformulation of Definition 2.3.1.

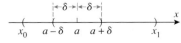

Figure 2.3.2

---

**2.3.2**    LIMIT (SECOND PRELIMINARY DEFINITION).    Let $f(x)$ be defined for all $x$ in some open interval containing the number $a$, with the possible exception that $f(x)$ need not be defined at $a$. We will write

$$\lim_{x \to a} f(x) = L$$

if given any number $\epsilon > 0$ we can find a number $\delta > 0$ such that $f(x)$ satisfies

$$L - \epsilon < f(x) < L + \epsilon$$

for each $x$ in the interval $(a - \delta, a + \delta)$, except possibly $x = a$.

---

For our final version of the limit definition, we note that in Definition 2.3.2 the condition $L - \epsilon < f(x) < L + \epsilon$ can be expressed as

$$|f(x) - L| < \epsilon$$

and the condition that $x$ lies in the interval $(a - \delta, a + \delta)$, but $x \neq a$, can be expressed as

$$0 < |x - a| < \delta$$

Thus, we can rewrite Definition 2.3.2 as follows.

---

**2.3.3**    LIMIT DEFINITION (FINAL FORM).    Let $f(x)$ be defined for all $x$ in some open interval containing the number $a$, with the possible exception that $f(x)$ need not be defined at $a$. We will write

$$\lim_{x \to a} f(x) = L$$

if given any number $\epsilon > 0$ we can find a number $\delta > 0$ such that

$$|f(x) - L| < \epsilon \quad \text{if} \quad 0 < |x - a| < \delta$$

---

REMARK.    This defines a two-sided limit. The definitions for one-sided limits are similar, the difference being that the condition $|f(x) - L| < \epsilon$ is only required to hold on the interval $a < x < a + \delta$ for right-sided limits and on the interval $a - \delta < x < a$ for left-sided limits.

In the preceding sections we illustrated various numerical and graphical methods for *guessing* at limits. Now that we have a precise definition to work with, we can actually confirm the validity of those guesses with mathematical proof. Here is a typical example of such a proof.

### Example 1

Use Definition 2.3.3 to prove that $\lim_{x \to 2} (3x - 5) = 1$.

*Solution.*  We must show that given any positive number $\epsilon$, we can find a positive number $\delta$ such that

$$| \underbrace{(3x - 5)}_{f(x)} - \underbrace{1}_{L} | < \epsilon \quad \text{if} \quad 0 < |x - \underbrace{2}_{a} | < \delta \tag{1}$$

There are two things to do. First, we must *discover* a value of $\delta$ for which this statement holds, and then we must *prove* that the statement holds for that $\delta$. For the discovery part we begin by simplifying (1) and writing it as

$$|3x - 6| < \epsilon \quad \text{if} \quad 0 < |x - 2| < \delta$$

Next, we will rewrite this statement in a form that will facilitate the discovery of an appropriate $\delta$:

$$3|x - 2| < \epsilon \quad \text{if} \quad 0 < |x - 2| < \delta$$
$$|x - 2| < \epsilon/3 \quad \text{if} \quad 0 < |x - 2| < \delta \tag{2}$$

It should be self-evident that this last statement holds if $\delta = \epsilon/3$, which completes the discovery portion of our work. Now we need to prove that (1) holds for this choice of $\delta$. However, statement (1) is equivalent to (2), and (2) holds with $\delta = \epsilon/3$, so (1) also holds with $\delta = \epsilon/3$. This proves that $\lim_{x \to 2} (3x - 5) = 1$. ◀

REMARK.    This example illustrates the general form of a limit proof: We *assume* that we are given a positive number $\epsilon$, and we try to *prove* that we can find a positive number $\delta$ such that

$$|f(x) - L| < \epsilon \quad \text{if} \quad 0 < |x - a| < \delta \tag{3}$$

This is done by first discovering $\delta$, and then proving that the discovered $\delta$ works. Since the argument has to be general enough to work for all positive values of $\epsilon$, the quantity $\delta$ has to be expressed as a function of $\epsilon$. In Example 1 we found the function $\delta = \epsilon/3$ by some simple algebra; however, most limit proofs require a little more algebraic and logical ingenuity. Thus, if you find our ensuing discussion of "$\delta$-$\epsilon$" proofs challenging, do not become discouraged; the concepts and techniques are intrinsically difficult. In fact, a precise understanding of limits evaded the finest mathematical minds for more than 150 years after the basic concepts of calculus were discovered.

## Example 2

Prove that $\lim_{x \to 0^+} \sqrt{x} = 0$.

*Solution.*   We must show that given $\epsilon > 0$, there exists a $\delta > 0$ such that

$$|\sqrt{x} - 0| < \epsilon \quad \text{if} \quad 0 < x < 0 + \delta$$

or more simply,

$$\sqrt{x} < \epsilon \quad \text{if} \quad 0 < x < \delta \tag{4}$$

But, by squaring both sides of the inequality $\sqrt{x} < \epsilon$, we can rewrite (4) as

$$x < \epsilon^2 \quad \text{if} \quad 0 < x < \delta \tag{5}$$

It should be self-evident that (5) is true if $\delta = \epsilon^2$; and since (5) is a reformulation of (4), we have shown that (4) holds with $\delta = \epsilon^2$. This proves that $\lim_{x \to 0^+} \sqrt{x} = 0$. ◀

REMARK.    In this example we were only concerned with the limit from the right because $f(x) = \sqrt{x}$ has imaginary values for $x < 0$. Thus, the limit from the left and the two-sided limit are not applicable at $x = 0$.

**THE VALUE OF $\delta$ IS NOT UNIQUE**

In preparation for our next example, we note that the value of $\delta$ in Definition 2.3.3 is not unique; once we have found a value of $\delta$ that fulfills the requirements of the definition, then any *smaller* positive number $\delta_1$ will also fulfill those requirements. That is, if it is true that

$$|f(x) - L| < \epsilon \quad \text{if} \quad 0 < |x - a| < \delta$$

then it will also be true that

$$|f(x) - L| < \epsilon \quad \text{if} \quad 0 < |x - a| < \delta_1$$

This is because $\{x : 0 < |x - a| < \delta_1\}$ is a subset of $\{x : 0 < |x - a| < \delta\}$ (Figure 2.3.3), and hence if $|f(x) - L| < \epsilon$ is satisfied for all $x$ in the larger set, then it will automatically be satisfied for all $x$ in the subset. Thus, in Example 1, where we used $\delta = \epsilon/3$, we could have used any smaller value of $\delta$ such as $\delta = \epsilon/4$, $\delta = \epsilon/5$, or $\delta = \epsilon/6$.

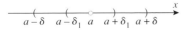

Figure 2.3.3

**Example 3**

Prove that $\lim_{x \to 3} x^2 = 9$.

*Solution.* We must show that given any positive number $\epsilon$, we can find a positive number $\delta$ such that

$$|x^2 - 9| < \epsilon \quad \text{if} \quad 0 < |x - 3| < \delta \tag{6}$$

Because $|x - 3|$ occurs on the right side of this "if statement," it will be helpful to factor the left side to introduce a factor of $|x - 3|$. This yields the following alternative form of (6):

$$|x + 3|\,|x - 3| < \epsilon \quad \text{if} \quad 0 < |x - 3| < \delta \tag{7}$$

To make this statement hold we need to find a $\delta$ that "controls" the size of both factors on the left side. However, the condition on the right side gives us direct control on the size of $|x - 3|$ but not of $|x + 3|$. To circumvent this difficulty, let us *temporarily* replace the factor $|x + 3|$ by a positive constant $k$ and look for a $\delta$ such that

$$k|x - 3| < \epsilon \quad \text{if} \quad 0 < |x - 3| < \delta \tag{8}$$

This statement can be rewritten as

$$|x - 3| < \epsilon/k \quad \text{if} \quad 0 < |x - 3| < \delta$$

which can be satisfied by taking

$$\delta = \epsilon/k \tag{9}$$

Now let us assume that $k$ can be chosen so that

$$|x + 3| < k \tag{10}$$

in which case

$$|x + 3|\,|x - 3| < \epsilon \quad \text{if} \quad k|x - 3| < \epsilon$$

Thus, if we can find $k$ so that (10) holds, then choosing $\delta$ as in (9) will make (8) hold, and this in turn will make (7) hold.

To find $k$, let us *arbitrarily* agree that we will choose $\delta$ so that $\delta \le 1$. This is justified because of our earlier observation that once a value of $\delta$ is found, then any smaller positive value of $\delta$ can be used. Thus, if it so happens that $\delta > 1$ in (9), we can use $\delta = 1$ instead, thereby guaranteeing that $\delta \le 1$. If we impose this restriction on $\delta$, then it will follow from the right side of (8) that

$$|x - 3| < 1 \quad \text{or} \quad 2 < x < 4 \quad \text{or} \quad 5 < x + 3 < 7$$

from which we can conclude that

$$|x + 3| < 7$$

Thus, given $\epsilon > 0$, we can take $k = 7$ in (8), and hence from (9) we can take $\delta = \epsilon/7$ (or smaller), subject to the restriction that $\delta \le 1$. We can achieve this by taking $\delta$ to be the minimum of the numbers $\epsilon/7$ and 1, which is sometimes written as $\delta = \min(\epsilon/7, 1)$. This proves that $\lim_{x \to 3} x^2 = 9$. ◀

REMARK. You may have wondered how we knew to make the restriction $\delta \le 1$ (as opposed to $\delta \le \frac{1}{2}$ or $\delta \le 5$, for example). Actually, it does not matter; any restriction of the form $\delta \le c$ would work equally well.

**LIMITS AS $x \to +\infty$ OR $x \to -\infty$**

In Section 2.1 we discussed the limits

$$\lim_{x \to +\infty} f(x) = L \quad \text{and} \quad \lim_{x \to -\infty} f(x) = L$$

from an intuitive viewpoint. We interpreted the first statement to mean that the values of $f(x)$ eventually get closer and closer to $L$ as $x$ increases indefinitely, and we interpreted the second statement to mean that the values of $f(x)$ eventually get closer and closer to $L$ as $x$ decreases indefinitely. These ideas are captured more precisely in the following definitions and are illustrated in Figure 2.3.4.

---

**2.3.4**  DEFINITION.  Let $f(x)$ be defined for all $x$ in some infinite open interval extending in the positive $x$-direction. We will write

$$\lim_{x \to +\infty} f(x) = L$$

if given any number $\epsilon > 0$, there corresponds a positive number $N$ such that

$$|f(x) - L| < \epsilon \quad \text{if} \quad x > N$$

---

**2.3.5**  DEFINITION.  Let $f(x)$ be defined for all $x$ in some infinite open interval extending in the negative $x$-direction. We will write

$$\lim_{x \to -\infty} f(x) = L$$

if given any number $\epsilon > 0$, there corresponds a negative number $N$ such that

$$|f(x) - L| < \epsilon \quad \text{if} \quad x < N$$

---

To see how these definitions relate to our informal concepts of these limits, suppose that $f(x) \to L$ as $x \to +\infty$, and for a given $\epsilon$ let $N$ be the positive number described in Definition 2.3.4. If $x$ is allowed to increase indefinitely, then eventually $x$ will lie in the interval $(N, +\infty)$, which is marked by the green band in Figure 2.3.4$a$; when this happens, the value of $f(x)$ will fall between $L - \epsilon$ and $L + \epsilon$, marked by the red band in the figure. Since this is true for all positive values of $\epsilon$ (no matter how small), we can force the values of $f(x)$ as close as we like to $L$ by making $N$ sufficiently large. This agrees with our informal concept of this limit. Similarly, Figure 2.3.4$b$ illustrates Definition 2.3.5.

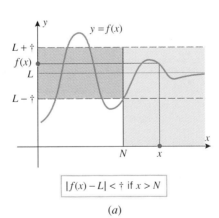

Figure 2.3.4

**Example 4**

Prove that $\displaystyle\lim_{x \to +\infty} \frac{1}{x} = 0.$

*Solution.*  Applying Definition 2.3.4 with $f(x) = 1/x$ and $L = 0$, we must show that given $\epsilon > 0$, we must find a number $N > 0$ such that

$$\left| \frac{1}{x} - 0 \right| < \epsilon \quad \text{if} \quad x > N \tag{11}$$

Because $x \to +\infty$ we can assume that $x > 0$. Thus, we can eliminate the absolute values in this statement and rewrite it as

$$\frac{1}{x} < \epsilon \quad \text{if} \quad x > N$$

or, on taking reciprocals,

$$x > \frac{1}{\epsilon} \quad \text{if} \quad x > N \tag{12}$$

It is self-evident that $N = 1/\epsilon$ satisfies this requirement, and since (12) is equivalent to (11) for $x > 0$, the proof is complete. ◀

**INFINITE LIMITS**

In Section 2.1 we discussed limits of the following type from an intuitive viewpoint:

$$\lim_{x \to a} f(x) = +\infty, \qquad \lim_{x \to a} f(x) = -\infty \tag{13}$$

$$\lim_{x \to a^+} f(x) = +\infty, \qquad \lim_{x \to a^+} f(x) = -\infty \tag{14}$$

$$\lim_{x \to a^-} f(x) = +\infty, \qquad \lim_{x \to a^-} f(x) = -\infty \tag{15}$$

Recall that each of these expressions describes a particular way in which the limit fails to exist. The $+\infty$ indicates that the limit fails to exist because $f(x)$ increases without bound, and the $-\infty$ indicates that the limit fails to exist because $f(x)$ decreases without bound. These ideas are captured more precisely in the following definitions and are illustrated in Figure 2.3.5.

---

**2.3.6** DEFINITION. Let $f(x)$ be defined for all $x$ in some open interval containing $a$, except that $f(x)$ need not be defined at $a$. We will write

$$\lim_{x \to a} f(x) = +\infty$$

if given any positive number $M$, we can find a number $\delta > 0$ such that $f(x)$ satisfies

$$f(x) > M \quad \text{if} \quad 0 < |x - a| < \delta$$

---

**2.3.7** DEFINITION. Let $f(x)$ be defined for all $x$ in some open interval containing $a$, except that $f(x)$ need not be defined at $a$. We will write

$$\lim_{x \to a} f(x) = -\infty$$

if given any negative number $M$, we can find a number $\delta > 0$ such that $f(x)$ satisfies

$$f(x) < M \quad \text{if} \quad 0 < |x - a| < \delta$$

---

To see how these definitions relate to our informal concepts of these limits, suppose that $f(x) \to +\infty$ as $x \to a$, and for a given $M$ let $\delta$ be the corresponding positive number described in Definition 2.3.6. Next, imagine that $x$ gets closer and closer to $a$ (from either side). Eventually, $x$ will lie in the interval $(a - \delta, a + \delta)$, which is marked by the green band in Figure 2.3.5a; when this happens the value of $f(x)$ will be greater than $M$, marked by the red band in the figure. Since this is true for any positive value of $M$ (no matter how large), we can force the values of $f(x)$ to be as large as we like by making $x$ sufficiently close to $a$. This agrees with our informal concept of this limit. Similarly, Figure 2.3.5b illustrates Definition 2.3.7.

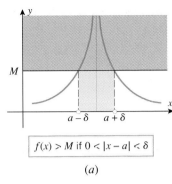 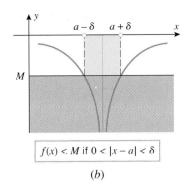

$$f(x) > M \text{ if } 0 < |x - a| < \delta$$

$$f(x) < M \text{ if } 0 < |x - a| < \delta$$

(a)   (b)

Figure 2.3.5

REMARK. The definitions for the one-sided limits in (14) and (15) are similar, the difference being that the conditions $f(x) > M$ and $f(x) < M$ are only required to hold on the interval $a < x < a + \delta$ for the right-sided limits and on the interval $a - \delta < x < a$ for the left-sided limits.

### Example 5

Prove that $\lim\limits_{x \to 0} \dfrac{1}{x^2} = +\infty$.

*Solution.* Applying Definition 2.3.6 with $f(x) = 1/x^2$ and $a = 0$, we must show that given a number $M > 0$, we can find a number $\delta > 0$ such that

$$\frac{1}{x^2} > M \quad \text{if} \quad 0 < |x - 0| < \delta \tag{16}$$

or, on taking reciprocals and simplifying,

$$x^2 < \frac{1}{M} \quad \text{if} \quad 0 < |x| < \delta \tag{17}$$

But $x^2 < 1/M$ if $|x| < 1/\sqrt{M}$, so that $\delta = 1/\sqrt{M}$ satisfies (17). Since (16) is equivalent to (17), the proof is complete. ◀

FOR THE READER. How would you define

$$\lim_{x \to +\infty} f(x) = +\infty, \qquad \lim_{x \to +\infty} f(x) = -\infty$$

$$\lim_{x \to -\infty} f(x) = +\infty, \qquad \lim_{x \to -\infty} f(x) = -\infty? \tag{18}$$

---

**EXERCISE SET 2.3**  ⬭ Graphing Calculator   [c] CAS
. . . . . . . . . . . . . . . . . . . . . . . . . . . . . . . . . . . . . . . . . . . . . . . . . . . . . . . . . . . . . . . . . . . . .

**1.** (a) Find the largest open interval, centered at the origin on the $x$-axis, such that for each point $x$ in the interval the value of the function $f(x) = x + 2$ is within 0.1 unit of the number $f(0) = 2$.

   (b) Find the largest open interval, centered at the point $x = 3$, such that for each point $x$ in the interval the value of the function $f(x) = 4x - 5$ is within 0.01 unit of the number $f(3) = 7$.

   (c) Find the largest open interval, centered at the point $x = 4$, such that for each point $x$ in the interval the value of the function $f(x) = x^2$ is within 0.001 unit of the number $f(4) = 16$.

**2.** In each part, find the largest open interval, centered at the point $x = 0$, such that for each point $x$ in the interval the value of $f(x) = 2x + 3$ is within $\epsilon$ units of the number $f(0) = 3$.

   (a) $\epsilon = 0.1$    (b) $\epsilon = 0.01$

   (c) $\epsilon = 0.0012$

**3.** (a) Find the values of $x_1$ and $x_2$ in the accompanying figure.
(b) Find a positive number $\delta$ such that $|\sqrt{x} - 2| < 0.05$ if $0 < |x - 4| < \delta$.

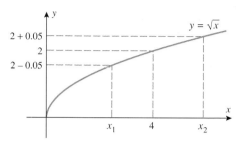

*Not drawn to scale*

Figure Ex-3

**4.** (a) Find the values of $x_1$ and $x_2$ in the accompanying figure.
(b) Find a positive number $\delta$ such that $|(1/x) - 1| < 0.1$ if $0 < |x - 1| < \delta$.

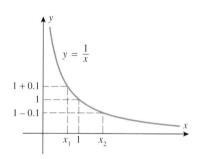

*Not drawn to scale*

Figure Ex-4

---

In Exercises 5–14, a positive number $\epsilon$ and the limit $L$ of a function $f$ at a point $a$ are given. Find a number $\delta$ such that $|f(x) - L| < \epsilon$ if $0 < |x - a| < \delta$.

**5.** $\lim\limits_{x \to 4} 2x = 8; \ \epsilon = 0.1$

**6.** $\lim\limits_{x \to -2} \dfrac{1}{2}x = -1; \ \epsilon = 0.1$

**7.** $\lim\limits_{x \to -1} (7x + 5) = -2; \ \epsilon = 0.01$

**8.** $\lim\limits_{x \to 3} (5x - 2) = 13; \ \epsilon = 0.01$

**9.** $\lim\limits_{x \to 2} \dfrac{x^2 - 4}{x - 2} = 4; \ \epsilon = 0.05$

**10.** $\lim\limits_{x \to -1} \dfrac{x^2 - 1}{x + 1} = -2; \ \epsilon = 0.05$

**11.** $\lim\limits_{x \to 4} x^2 = 16; \ \epsilon = 0.001$

**12.** $\lim\limits_{x \to 9} \sqrt{x} = 3; \ \epsilon = 0.001$

**13.** $\lim\limits_{x \to 5} \dfrac{1}{x} = \dfrac{1}{5}; \ \epsilon = 0.05$

**14.** $\lim\limits_{x \to 0} |x| = 0; \ \epsilon = 0.05$

---

In Exercises 15–28, use Definition 2.3.3 to prove that the stated limit is correct.

**15.** $\lim\limits_{x \to 5} 3x = 15$

**16.** $\lim\limits_{x \to 3} (4x - 5) = 7$

**17.** $\lim\limits_{x \to 2} (2x - 7) = -3$

**18.** $\lim\limits_{x \to -1} (2 - 3x) = 5$

**19.** $\lim\limits_{x \to 0} \dfrac{x^2 + x}{x} = 1$

**20.** $\lim\limits_{x \to -3} \dfrac{x^2 - 9}{x + 3} = -6$

**21.** $\lim\limits_{x \to 1} 2x^2 = 2$

**22.** $\lim\limits_{x \to 3} (x^2 - 5) = 4$

**23.** $\lim\limits_{x \to 1/3} \dfrac{1}{x} = 3$

**24.** $\lim\limits_{x \to -2} \dfrac{1}{x + 1} = -1$

**25.** $\lim\limits_{x \to 4} \sqrt{x} = 2$

**26.** $\lim\limits_{x \to 6} \sqrt{x + 3} = 3$

**27.** $\lim\limits_{x \to 1} f(x) = 3$, where $f(x) = \begin{cases} x + 2, & x \neq 1 \\ 10, & x = 1 \end{cases}$

**28.** $\lim\limits_{x \to 2} (x^2 + 3x - 1) = 9$

**29.** (a) Find the smallest positive number $N$ such that for each point $x$ in the interval $(N, +\infty)$, the value of the function $f(x) = 1/x^2$ is within 0.1 unit of $L = 0$.
(b) Find the smallest positive number $N$ such that for each point $x$ in the interval $(N, +\infty)$, the value of $f(x) = x/(x + 1)$ is within 0.01 unit of $L = 1$.
(c) Find the largest negative number $N$ such that for each point $x$ in the interval $(-\infty, N)$, the value of the function $f(x) = 1/x^3$ is within 0.001 unit of $L = 0$.
(d) Find the largest negative number $N$ such that for each point $x$ in the interval $(-\infty, N)$, the value of the function $f(x) = x/(x + 1)$ is within 0.01 unit of $L = 1$.

**30.** In each part, find the smallest positive value of $N$ such that for each point $x$ in the interval $(N, +\infty)$, the function $f(x) = 1/x^3$ is within $\epsilon$ units of the number $L = 0$.
(a) $\epsilon = 0.1$    (b) $\epsilon = 0.01$    (c) $\epsilon = 0.001$

**31.** (a) Find the values of $x_1$ and $x_2$ in the accompanying figure.
(b) Find a positive number $N$ such that
$$\left| \frac{x^2}{1 + x^2} - 1 \right| < \epsilon$$
for $x > N$.
(c) Find a negative number $N$ such that
$$\left| \frac{x^2}{1 + x^2} - 1 \right| < \epsilon$$
for $x < N$.

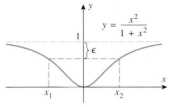

*Not drawn to scale*     Figure Ex-31

**32.** (a) Find the values of $x_1$ and $x_2$ in the accompanying figure.

     (b) Find a positive number $N$ such that

$$\left| \frac{1}{\sqrt[3]{x}} - 0 \right| = \left| \frac{1}{\sqrt[3]{x}} \right| < \epsilon$$

     for $x > N$.

     (c) Find a negative number $N$ such that

$$\left| \frac{1}{\sqrt[3]{x}} - 0 \right| = \left| \frac{1}{\sqrt[3]{x}} \right| < \epsilon$$

     for $x < N$.

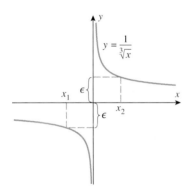

Figure Ex-32

In Exercises 33–36, a positive number $\epsilon$ and the limit $L$ of a function $f$ at $+\infty$ are given. Find a positive number $N$ such that $|f(x) - L| < \epsilon$ if $x > N$.

**33.** $\lim\limits_{x \to +\infty} \dfrac{1}{x^2} = 0$; $\epsilon = 0.01$

**34.** $\lim\limits_{x \to +\infty} \dfrac{1}{x + 2} = 0$; $\epsilon = 0.005$

**35.** $\lim\limits_{x \to +\infty} \dfrac{x}{x + 1} = 1$; $\epsilon = 0.001$

**36.** $\lim\limits_{x \to +\infty} \dfrac{4x - 1}{2x + 5} = 2$; $\epsilon = 0.1$

In Exercises 37–40, a positive number $\epsilon$ and the limit $L$ of a function $f$ at $-\infty$ are given. Find a negative number $N$ such that $|f(x) - L| < \epsilon$ if $x < N$.

**37.** $\lim\limits_{x \to -\infty} \dfrac{1}{x + 2} = 0$; $\epsilon = 0.005$

**38.** $\lim\limits_{x \to -\infty} \dfrac{1}{x^2} = 0$; $\epsilon = 0.01$

**39.** $\lim\limits_{x \to -\infty} \dfrac{4x - 1}{2x + 5} = 2$; $\epsilon = 0.1$

**40.** $\lim\limits_{x \to -\infty} \dfrac{x}{x + 1} = 1$; $\epsilon = 0.001$

In Exercises 41–48, use Definitions 2.3.4 and 2.3.5 to prove that the stated limit is correct.

**41.** $\lim\limits_{x \to +\infty} \dfrac{1}{x^2} = 0$

**42.** $\lim\limits_{x \to -\infty} \dfrac{1}{x} = 0$

**43.** $\lim\limits_{x \to -\infty} \dfrac{1}{x + 2} = 0$

**44.** $\lim\limits_{x \to +\infty} \dfrac{1}{x + 2} = 0$

**45.** $\lim\limits_{x \to +\infty} \dfrac{x}{x + 1} = 1$

**46.** $\lim\limits_{x \to -\infty} \dfrac{x}{x + 1} = 1$

**47.** $\lim\limits_{x \to -\infty} \dfrac{4x - 1}{2x + 5} = 2$

**48.** $\lim\limits_{x \to +\infty} \dfrac{4x - 1}{2x + 5} = 2$

**49.** (a) Find the largest open interval, centered at the origin on the $x$-axis, such that for each point $x$ in the interval, other than the center, the values of $f(x) = 1/x^2$ are greater than 100.

     (b) Find the largest open interval, centered at the point $x = 1$, such that for each point $x$ in the interval, other than the center, the values of the function

$$f(x) = 1/|x - 1|$$

     are greater than 1000.

     (c) Find the largest open interval, centered at the point $x = 3$, such that for each point $x$ in the interval, other than the center, the values of the function

$$f(x) = -1/(x - 3)^2$$

     are less than $-1000$.

     (d) Find the largest open interval, centered at the origin on the $x$-axis, such that for each point $x$ in the interval, other than the center, the values of $f(x) = -1/x^4$ are less than $-10,000$.

**50.** In each part, find the largest open interval, centered at the point $x = 1$, such that for each point $x$ in the interval the value of $f(x) = 1/(x - 1)^2$ is greater than $M$.

     (a) $M = 10$      (b) $M = 1000$      (c) $M = 100,000$

In Exercises 51–56, use Definitions 2.3.6 and 2.3.7 to prove that the stated limit is correct.

**51.** $\lim\limits_{x \to 3} \dfrac{1}{(x - 3)^2} = +\infty$

**52.** $\lim\limits_{x \to 3} \dfrac{-1}{(x - 3)^2} = -\infty$

**53.** $\lim\limits_{x \to 0} \dfrac{1}{|x|} = +\infty$

**54.** $\lim\limits_{x \to 1} \dfrac{1}{|x - 1|} = +\infty$

**55.** $\lim\limits_{x \to 0} \left( -\dfrac{1}{x^4} \right) = -\infty$

**56.** $\lim\limits_{x \to 0} \dfrac{1}{x^4} = +\infty$

In Exercises 57–62, use the remark following Definition 2.3.3 to prove that the stated limit is correct.

**57.** $\lim\limits_{x \to 2^+} (x + 1) = 3$

**58.** $\lim\limits_{x \to 1^-} (3x + 2) = 5$

**59.** $\lim\limits_{x \to 4^+} \sqrt{x - 4} = 0$

**60.** $\lim\limits_{x \to 0^-} \sqrt{-x} = 0$

**61.** $\lim\limits_{x \to 2^+} f(x) = 2$, where $f(x) = \begin{cases} x, & x > 2 \\ 3x, & x \le 2 \end{cases}$

**62.** $\lim\limits_{x \to 2^-} f(x) = 6$, where $f(x) = \begin{cases} x, & x > 2 \\ 3x, & x < 2 \end{cases}$

In Exercises 63 and 64, use the remark following Definitions 2.3.6 and 2.3.7 to prove that the stated limit is correct.

**63.** (a) $\lim\limits_{x \to 1^+} \dfrac{1}{1 - x} = -\infty$    (b) $\lim\limits_{x \to 1^-} \dfrac{1}{1 - x} = +\infty$

**64.** (a) $\lim\limits_{x \to 0^+} \dfrac{1}{x} = +\infty$    (b) $\lim\limits_{x \to 0^-} \dfrac{1}{x} = -\infty$

For Exercises 65 and 66, write out definitions of the four limits in (18), and use your definitions to prove that the stated limit is correct.

**65.** (a) $\lim\limits_{x \to +\infty} (x + 1) = +\infty$    (b) $\lim\limits_{x \to -\infty} (x + 1) = -\infty$

**66.** (a) $\lim\limits_{x \to +\infty} (x^2 - 3) = +\infty$    (b) $\lim\limits_{x \to -\infty} (x^3 + 5) = -\infty$

**67.** Prove the result in Example 3 under the assumption that $\delta \le 2$ rather than $\delta \le 1$.

**68.** (a) In Definition 2.3.3 there is a condition requiring that $f(x)$ be defined for all $x$ in some open interval containing $a$, except possibly at $a$ itself. What is the purpose of this requirement?

(b) Why is $\lim\limits_{x \to 0} \sqrt{x} = 0$ an incorrect statement?

(c) Is $\lim\limits_{x \to 0.01} \sqrt{x} = 0.1$ a correct statement?

**69.** Generate the graph of $f(x) = x^3 - 4x + 5$ with a graphing utility, and use the graph to find a number $\delta$ such that $|f(x) - 2| < 0.05$ if $0 < |x - 1| < \delta$. [*Hint:* Show that the inequality $|f(x) - 2| < 0.05$ can be rewritten as $1.95 < x^3 - 4x + 5 < 2.05$, and estimate the values of $x$ for which $x^3 - 4x + 5 = 1.95$ and $x^3 - 4x + 5 = 2.05$.]

**70.** Use the method of Exercise 69 to find a number $\delta$ such that $|\sqrt{5x + 1} - 4| < 0.5$ if $0 < |x - 3| < \delta$.

## 2.4 CONTINUITY

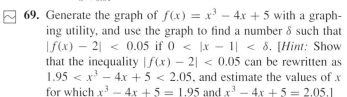

*A moving object cannot vanish at some point and reappear someplace else to continue its motion. Thus, we perceive the path of a moving object as a continuous curve, that is, a curve without gaps, breaks, or holes. Earlier, we discussed continuity from an intuitive viewpoint; in this section we will define this concept precisely and develop some fundamental properties of continuous curves.*

**DEFINITION OF CONTINUITY**

Recall from Section 2.1 that the graph of a function $f$ will have a hole or a break in it at a point $c$ if any of the following situations occur:

- The function $f$ is undefined at $c$ (Figure 2.4.1*a*).
- The limit of $f(x)$ does not exist as $x$ approaches $c$ (Figures 2.4.1*b*, 2.4.1*c*).
- The value of the function and the value of the limit at $c$ are different (Figure 2.4.1*d*).

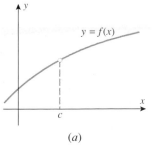
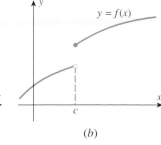
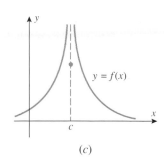
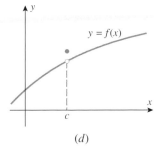

(*a*)    (*b*)    (*c*)    (*d*)

Figure 2.4.1

This suggests the following definition.

---

**2.4.1** DEFINITION. A function $f$ is said to be ***continuous at a point c*** if the following conditions are satisfied:

1. $f(c)$ is defined.
2. $\lim\limits_{x \to c} f(x)$ exists.
3. $\lim\limits_{x \to c} f(x) = f(c)$.

---

If one or more of the conditions in this definition fails to hold, then we will say that $f$ has a ***discontinuity*** at the point $x = c$. If $f$ is continuous at each point of an open interval $(a, b)$, then we will say that $f$ is ***continuous on*** $(a, b)$. This definition also applies to infinite open intervals of the form $(a, +\infty)$, $(-\infty, b)$, and $(-\infty, +\infty)$. In the case where $f$ is continuous on $(-\infty, +\infty)$, we will say that $f$ is ***continuous everywhere***. If $f$ is continuous on an open interval, but the particular interval is not important for the discussion, we will say that $f$ is ***continuous*** (without referencing the interval).

REMARK. The first two conditions in Definition 2.4.1 are actually superfluous, since it is implicit in the third condition that $f(c)$ is defined and the limit exists (otherwise the equality would make no sense). We have included the first two conditions for emphasis and clarity, but, as a practical matter, you need only confirm that the third condition holds when you want to show that a function $f$ is continuous at a point $c$.

**Example 1**

Determine whether the following functions are continuous at the point $x = 2$.

$$f(x) = \frac{x^2 - 4}{x - 2}, \qquad g(x) = \begin{cases} \dfrac{x^2 - 4}{x - 2}, & x \neq 2 \\ 3, & x = 2, \end{cases} \qquad h(x) = \begin{cases} \dfrac{x^2 - 4}{x - 2}, & x \neq 2 \\ 4, & x = 2 \end{cases}$$

*Solution.* In each case we must determine whether the limit of the function as $x \to 2$ is the same as the value of the function at $x = 2$. In all three cases the functions are identical, except at the point $x = 2$, and hence all three have the same limit at $x = 2$, namely

$$\lim_{x \to 2} f(x) = \lim_{x \to 2} g(x) = \lim_{x \to 2} h(x) = \lim_{x \to 2} \frac{x^2 - 4}{x - 2} = \lim_{x \to 2} (x + 2) = 4$$

The function $f$ is undefined at $x = 2$, and hence is not continuous at that point (Figure 2.4.2$a$). The function $g$ is defined at $x = 2$, but its value there is $g(2) = 3$, which is not the same as the limit at that point; hence, $g$ is also not continuous at $x = 2$ (Figure 2.4.2$b$). The value of the function $h$ at $x = 2$ is $h(2) = 4$, which is the same as the limit at that

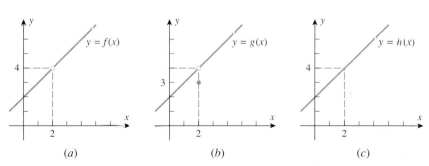

Figure 2.4.2

point; hence, $h$ is continuous at $x = 2$ (Figure 2.4.2$c$). (Note that the function $h$ could have been written more simply as $h(x) = x + 2$, but we wrote it in piecewise form to emphasize its relationship to $f$ and $g$.)    ◀

**CONTINUITY IN APPLICATIONS**

In applications, discontinuities often signal the occurrence of important physical phenomena. For example, Figure 2.4.3$a$ is a graph of voltage versus time for an underground cable that is accidentally cut by a work crew at time $t = t_0$ (the voltage drops to zero when the line is cut). Figure 2.4.3$b$ shows the graph of inventory versus time for a company that restocks its warehouse to $y_1$ units when the inventory falls to $y_0$ units. The discontinuities occur at those times when restocking occurs.

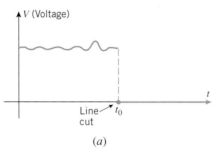
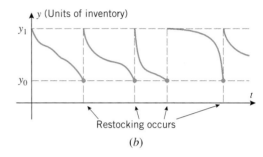

Figure 2.4.3

Given the possible physical significance of discontinuities, it is important to be able to identify points of discontinuity for specific functions, and to be able to make general statements about the continuity properties of entire families of functions. This is our next goal.

**CONTINUITY OF POLYNOMIALS**

The general procedure for showing that a function is continuous everywhere is to show that it is continuous at an *arbitrary* point. For example, we showed in Theorem 2.2.3 that if $p(x)$ is a polynomial and $a$ is any real number, then

$$\lim_{x \to a} p(x) = p(a)$$

Thus, we have the following result.

---

**2.4.2    THEOREM.**    *Polynomials are continuous everywhere.*

---

### Example 2

Show that $|x|$ is continuous everywhere (Figure 1.2.5).

*Solution.*    We can write $|x|$ as

$$|x| = \begin{cases} x & \text{if} \quad x > 0 \\ 0 & \text{if} \quad x = 0 \\ -x & \text{if} \quad x < 0 \end{cases}$$

so $|x|$ is the same as the polynomial $x$ on the interval $(0, +\infty)$ and is the same as the polynomial $-x$ on the interval $(-\infty, 0)$. But polynomials are continuous functions, so $x = 0$ is the only possible point of discontinuity for $|x|$. At this point we have $|0| = 0$, so to prove the continuity at $x = 0$ we must show that

$$\lim_{x \to 0} |x| = 0 \tag{1}$$

Because the formula for $|x|$ changes at 0, it will be helpful to consider the one-sided limits at 0 rather than the two-sided limit. We obtain

$$\lim_{x \to 0^+} |x| = \lim_{x \to 0^+} x = 0 \quad \text{and} \quad \lim_{x \to 0^-} |x| = \lim_{x \to 0^-} (-x) = 0$$

Thus, (1) holds and $|x|$ is continuous at $x = 0$.  ◄

**SOME PROPERTIES OF CONTINUOUS FUNCTIONS**

The following theorem, which is a consequence of Theorem 2.2.2, will enable us to reach conclusions about the continuity of functions that are obtained by adding, subtracting, multiplying, and dividing continuous functions.

---

**2.4.3  THEOREM.**  *If the functions $f$ and $g$ are continuous at $c$, then*

(a)   *$f + g$ is continuous at $c$.*

(b)   *$f - g$ is continuous at $c$.*

(c)   *$fg$ is continuous at $c$.*

(d)   *$f/g$ is continuous at $c$ if $g(c) \neq 0$ and has a discontinuity at $c$ if $g(c) = 0$.*

---

We will prove part ($d$). The remaining proofs are similar and will be omitted.

*Proof.*  First, consider the case where $g(c) = 0$. In this case $f(c)/g(c)$ is undefined, so the function $f/g$ has a discontinuity at $c$.

Next, consider the case where $g(c) \neq 0$. To prove that $f/g$ is continuous at $c$, we must show that

$$\lim_{x \to c} \frac{f(x)}{g(x)} = \frac{f(c)}{g(c)} \tag{2}$$

Since $f$ and $g$ are continuous at $c$,

$$\lim_{x \to c} f(x) = f(c) \quad \text{and} \quad \lim_{x \to c} g(x) = g(c)$$

Thus, by Theorem 2.2.2($d$)

$$\lim_{x \to c} \frac{f(x)}{g(x)} = \frac{\lim\limits_{x \to c} f(x)}{\lim\limits_{x \to c} g(x)} = \frac{f(c)}{g(c)}$$

which proves (2).  ∎

**CONTINUITY OF RATIONAL FUNCTIONS**

Since polynomials are continuous functions, and since rational functions are ratios of polynomials, part ($d$) of Theorem 2.4.3 yields the following result.

---

**2.4.4  THEOREM.**  *A rational function is continuous everywhere except at the points where the denominator is zero.*

---

**Example 3**

For what values of $x$ is there a hole or a gap in the graph of

$$y = \frac{x^2 - 9}{x^2 - 5x + 6}?$$

*Solution.*  The function being graphed is a rational function, and hence is continuous everywhere except at the points where the denominator is zero. Solving the equation

$$x^2 - 5x + 6 = 0$$

yields two points of discontinuity, $x = 2$ and $x = 3$.  ◄

FOR THE READER.    If you use a graphing utility to generate the graph of the equation in this example, then there is a good chance that you will see the discontinuity at $x = 2$ but not at $x = 3$. Try it, and explain what you think is happening.

**CONTINUITY OF COMPOSITIONS**

The following theorem, whose proof is given in Appendix G, will be useful for calculating limits of compositions of functions.

> **2.4.5    THEOREM.**    *Let* lim *stand for one of the limits* $\lim_{x \to c}$, $\lim_{x \to c^-}$, $\lim_{x \to c^+}$, $\lim_{x \to +\infty}$, *or* $\lim_{x \to -\infty}$. *If* $\lim g(x) = L$ *and if the function* $f$ *is continuous at* $L$, *then* $\lim f(g(x)) = f(L)$. *That is,* $\lim f(g(x)) = f(\lim g(x))$.

In words, this theorem states:

> *A limit symbol can be moved through a function sign provided the limit of the expression inside the function sign exists and the function is continuous at this limit.*

**Example 4**

Suppose that $\lim g(x)$ exists, where lim stands for any of the limits in Theorem 2.4.5. We know from Example 2 that the function $|x|$ is continuous everywhere; thus, it follows that

$$\lim |g(x)| = |\lim g(x)| \tag{3}$$

that is, a limit symbol can be moved through an absolute value sign, provided the limit of the expression inside the absolute value signs exists. For example,

$$\lim_{x \to 3} |5 - x^2| = |\lim_{x \to 3} (5 - x^2)| = |-4| = 4 \qquad \blacktriangleleft$$

The following theorem is concerned with the continuity of compositions of functions; the first part deals with continuity at a specific point, and the second part with continuity everywhere.

> **2.4.6    THEOREM.**
> (a)    *If the function* $g$ *is continuous at the point* $c$, *and the function* $f$ *is continuous at the point* $g(c)$, *then the composition* $f \circ g$ *is continuous at* $c$.
> (b)    *If the function* $g$ *is continuous everywhere and the function* $f$ *is continuous everywhere, then the composition* $f \circ g$ *is continuous everywhere.*

*Proof.*    We will prove part (a) only; the proof of part (b) can be obtained by applying part (a) at an arbitrary point $c$. To prove that $f \circ g$ is continuous at $c$, we must show that the value of $f \circ g$ and the value of its limit are the same at $x = c$. But this is so, since we can write

$$\lim_{x \to c} (f \circ g)(x) = \lim_{x \to c} f(g(x)) = f(\lim_{x \to c} g(x)) = f(g(c)) = (f \circ g)(c)$$

| Theorem 2.4.5 | $g$ is continuous at $c$. |

We know from Example 2 that the function $|x|$ is continuous everywhere. Thus, if $g(x)$ is continuous at the point $c$, then by part (a) of Theorem 2.4.6, the function $|g(x)|$ must also be continuous at the point $c$; and, more generally, if $g(x)$ is continuous everywhere, then so is $|g(x)|$. Stated informally:

> *The absolute value of a continuous function is continuous.*

For example, the polynomial $g(x) = 4 - x^2$ is continuous everywhere, so we can conclude that the function $|4 - x^2|$ is also continuous everywhere (Figure 2.4.4).

FOR THE READER.   Can the absolute value of a function that is not continuous be continuous? Justify your answer.

**CONTINUITY FROM THE LEFT AND RIGHT**

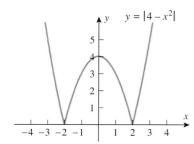

Figure 2.4.4

Because Definition 2.4.1 involves a two-sided limit, that definition does not generally apply at the endpoints of a closed interval $[a, b]$ or at the endpoint of an interval of the form $[a, b)$, $(a, b]$, $(-\infty, b]$, or $[a, +\infty)$. To remedy this problem, we will agree that a function is continuous at an endpoint of an interval if its value at the endpoint is equal to the appropriate one-sided limit at that point. For example, the function graphed in Figure 2.4.5 is continuous at the right endpoint of the interval $[a, b]$ because

$$\lim_{x \to b^-} f(x) = f(b)$$

but it is not continuous at the left endpoint because

$$\lim_{x \to a^+} f(x) \neq f(a)$$

In general, we will say a function $f$ is **continuous from the left** at a point $c$ if

$$\lim_{x \to c^-} f(x) = f(c)$$

and is **continuous from the right** at a point $c$ if

$$\lim_{x \to c^+} f(x) = f(c)$$

Using this terminology we define continuity on a closed interval as follows.

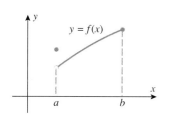

Figure 2.4.5

---

**2.4.7**   DEFINITION.   A function $f$ is said to be **continuous on a closed interval** $[a, b]$ if the following conditions are satisfied:
1.   $f$ is continuous on $(a, b)$.
2.   $f$ is continuous from the right at $a$.
3.   $f$ is continuous from the left at $b$.

---

FOR THE READER.   We leave it for you to modify this definition appropriately so that it applies to intervals of the form $[a, +\infty)$, $(-\infty, b]$, $(a, b]$, and $[a, b)$.

**Example 5**

What can you say about the continuity of the function $f(x) = \sqrt{9 - x^2}$?

*Solution.*   Because the natural domain of this function is the closed interval $[-3, 3]$, we will need to investigate the continuity of $f$ on the open interval $(-3, 3)$ and at the two endpoints. If $c$ is any point in the interval $(-3, 3)$, then it follows from Theorem 2.2.2(e) that

$$\lim_{x \to c} f(x) = \lim_{x \to c} \sqrt{9 - x^2} = \sqrt{\lim_{x \to c} (9 - x^2)} = \sqrt{9 - c^2} = f(c)$$

which proves $f$ is continuous at each point of the interval $(-3, 3)$. The function $f$ is also continuous at the endpoints since

$$\lim_{x \to 3^-} f(x) = \lim_{x \to 3^-} \sqrt{9 - x^2} = \sqrt{\lim_{x \to 3^-} (9 - x^2)} = 0 = f(3)$$

$$\lim_{x \to -3^+} f(x) - \lim_{x \to -3^+} \sqrt{9 - x^2} = \sqrt{\lim_{x \to -3^+} (9 - x^2)} = 0 = f(-3)$$

Thus, $f$ is continuous on the closed interval $[-3, 3]$.   ◀

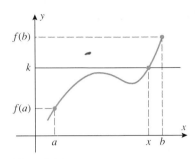

Figure 2.4.6

Figure 2.4.6 shows the graph of a function that is continuous on the closed interval $[a, b]$. The figure suggests that if we draw any horizontal line $y = k$, where $k$ is between $f(a)$ and $f(b)$, then that line will cross the curve $y = f(x)$ line at least once over the $[a, b]$. Stated in numerical terms, if $f$ is continuous on $[a, b]$, then the function $f$ must take on every value $k$ between $f(a)$ and $f(b)$ at least once as $x$ varies from $a$ to $b$. For example, the polynomial $p(x) = x^5 - x + 3$ has a value of 3 at $x = 1$ and a value of 33 at $x = 2$. Thus, it follows from the continuity of $p$ that the equation $x^5 - x + 3 = k$ has at least one solution in the interval $[1, 2]$ for every value of $k$ between 3 and 33. This idea is stated more precisely in the following theorem.

**2.4.8** THEOREM (*Intermediate-Value Theorem*). *If $f$ is continuous on a closed interval $[a, b]$ and $k$ is any number between $f(a)$ and $f(b)$, inclusive, then there is at least one number $x$ in the interval $[a, b]$ such that $f(x) = k$.*

Although this theorem is intuitively obvious, its proof depends on a mathematically precise development of the real number system, which is beyond the scope of this text.

A variety of problems can be reduced to solving an equation $f(x) = 0$ for its roots. Sometimes it is possible to solve for the roots exactly using algebra, but often this is not possible and one must settle for decimal approximations of the roots. One procedure for approximating roots is based on the following consequence of the Intermediate-Value Theorem.

**2.4.9** THEOREM. *If $f$ is continuous on $[a, b]$, and if $f(a)$ and $f(b)$ are nonzero and have opposite signs, then there is at least one solution of the equation $f(x) = 0$ in the interval $(a, b)$.*

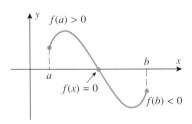

Figure 2.4.7

This result, which is illustrated in Figure 2.4.7 for the case where $f(a) > 0$ and $f(b) < 0$, can be proved as follows.

*Proof.* Since $f(a)$ and $f(b)$ have opposite signs, 0 is between $f(a)$ and $f(b)$. Thus, by the Intermediate-Value Theorem there is at least one number $x$ in the interval $[a, b]$ such that $f(x) = 0$. However, $f(a)$ and $f(b)$ are nonzero, so $x$ must lie in the interval $(a, b)$, which completes the proof. ▮

Before we illustrate how this theorem can be used to approximate roots, it will be helpful to discuss some standard terminology for describing errors in approximations. If $x$ is an approximation to a quantity $x_0$, then we call

$$\epsilon = |x - x_0|$$

the *absolute error* or (less precisely) the *error* in the approximation. The following terminology is used to describe the size of such errors:

| Error | Description |
|---|---|
| $|x - x_0| \leq 0.1$ | $x$ approximates $x_0$ with an error of at most 0.1. |
| $|x - x_0| \leq 0.01$ | $x$ approximates $x_0$ with an error of at most 0.01. |
| $|x - x_0| \leq 0.001$ | $x$ approximates $x_0$ with an error of at most 0.001. |
| $|x - x_0| \leq 0.0001$ | $x$ approximates $x_0$ with an error of at most 0.0001. |
| | |
| $|x - x_0| \leq 0.5$ | $x$ approximates $x_0$ to the nearest integer. |
| $|x - x_0| \leq 0.05$ | $x$ approximates $x_0$ to 1 decimal place (i.e., to the nearest tenth). |
| $|x - x_0| \leq 0.005$ | $x$ approximates $x_0$ to 2 decimal places (i.e., to the nearest hundredth). |
| $|x - x_0| \leq 0.0005$ | $x$ approximates $x_0$ to 3 decimal places (i.e., to the nearest thousandth). |

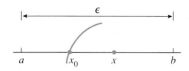

Every number $x$ in the interval $[a, b]$ differs from $x_0$ by at most $\epsilon$, and the midpoint of the interval differs from $x_0$ by at most $\epsilon/2$.

Figure 2.4.8

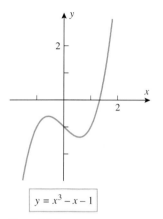

$y = x^3 - x - 1$

Figure 2.4.9

We will also need the following result, which should be evident geometrically from Figure 2.4.8.

**2.4.10** APPROXIMATION PRINCIPLE. Suppose that the equation $f(x) = 0$ has a root $x_0$ in the interval $[a, b]$ and that this interval has length $\epsilon = b - a$. Then any number $x$ in the interval $[a, b]$ approximates $x_0$ with an error of at most $\epsilon$, and the midpoint of the interval approximates $x_0$ with an error of at most $\epsilon/2$.

## Example 6

The equation

$$x^3 - x - 1 = 0$$

cannot be solved algebraically very easily because the left side has no simple factors. However, if we graph $p(x) = x^3 - x - 1$ with a graphing utility (Figure 2.4.9), then we are led to conjecture that there is one real root and that this root lies inside the interval $[1, 2]$. The existence of a root in this interval is also confirmed by Theorem 2.4.8, since $p(1) = -1$ and $p(2) = 5$ have opposite signs. Approximate this root to two decimal-place accuracy.

*Solution.* Our objective is to approximate the unknown root $x_0$ with an error of at most 0.005. It follows from the Approximation Principle (2.4.10) that if we can find an interval of length 0.01 that contains the root, then the midpoint of that interval will approximate the root with an error of at most $0.01/2 = 0.005$, which will achieve the desired accuracy.

We know that the root $x_0$ lies in the interval $[1, 2]$. However, this interval has length 1, which is too large. We can pinpoint the location of the root more precisely by dividing the interval $[1, 2]$ into 10 equal parts and evaluating $p$ at the points of subdivision using a calculating utility (Table 2.4.1). In this table $p(1.3)$ and $p(1.4)$ have opposite signs, so we know that the root lies in the interval $[1.3, 1.4]$. This interval has length 0.1, which is still too large, so we repeat the process by dividing the interval $[1.3, 1.4]$ into 10 parts and evaluating $p$ at the points of subdivision; this yields Table 2.4.2, which tells us that the root is inside the interval $[1.32, 1.33]$. Since this interval has length 0.01, its midpoint 1.325 will approximate the root with an error of at most 0.005. Thus, $x_0 \approx 1.325$ to two decimal-place accuracy. ◄

### Table 2.4.1

| $x$ | 1 | 1.1 | 1.2 | 1.3 | 1.4 | 1.5 | 1.6 | 1.7 | 1.8 | 1.9 | 2 |
|-----|-----|------|------|------|------|------|------|------|------|------|---|
| $f(x)$ | $-1$ | $-0.77$ | $-0.47$ | $-0.10$ | 0.34 | 0.88 | 1.50 | 2.21 | 3.03 | 3.96 | 5 |

### Table 2.4.2

| $x$ | 1.3 | 1.31 | 1.32 | 1.33 | 1.34 | 1.35 | 1.36 | 1.37 | 1.38 | 1.39 | 1.4 |
|-----|-------|--------|--------|-------|-------|-------|-------|-------|-------|-------|-------|
| $f(x)$ | $-0.103$ | $-0.062$ | $-0.020$ | 0.023 | 0.066 | 0.110 | 0.155 | 0.201 | 0.248 | 0.296 | 0.344 |

**APPROXIMATING ROOTS BY ZOOMING WITH A GRAPHING UTILITY**

The method illustrated in Example 6 can also be implemented with a graphing utility as follows.

**Step 1.** Figure 2.4.10*a* shows the graph of $f$ in the window $[-5, 5] \times [-5, 5]$ with $x\mathrm{Scl} = 1$ and $y\mathrm{Scl} = 1$. That graph places the root between $x = 1$ and $x = 2$.

**Step 2.** Since we know that the root lies between $x = 1$ and $x = 2$, we will zoom in by regraphing $f$ over an $x$-interval that extends between these points and in which $x\mathrm{Scl} = .1$. The $y$-interval and $y\mathrm{Scl}$ are not critical, as long as the $y$-interval extends above and below the $x$-axis. Figure 2.4.10*b* shows the graph of $f$ in the window $[1, 2] \times [-1, 1]$ with $x\mathrm{Scl} = .1$ and $y\mathrm{Scl} = .1$. That graph places the root between $x = 1.3$ and $x = 1.4$.

**Step 3.** Since we know that the root lies between $x = 1.3$ and $x = 1.4$, we will zoom in again by regraphing $f$ over an $x$-interval that extends between these points and in which $x\mathrm{Scl} = .01$. Figure 2.4.10*c* shows the graph of $f$ in the window $[1.3, 1.4] \times [-.1, .1]$ with $x\mathrm{Scl} = .01$ and $y\mathrm{Scl} = .01$. That graph places the root between $x = 1.32$ and $x = 1.33$.

**Step 4.** Since the interval in Step 3 has length .01, its midpoint 1.325 approximates the root with an error of at most 0.005, so $x_0 \approx 1.325$ to two decimal-place accuracy.

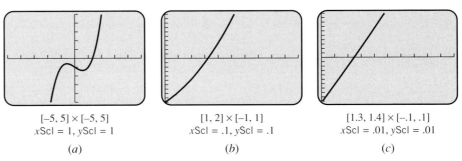

$[-5, 5] \times [-5, 5]$
$x\mathrm{Scl} = 1, y\mathrm{Scl} = 1$

$(a)$

$[1, 2] \times [-1, 1]$
$x\mathrm{Scl} = .1, y\mathrm{Scl} = .1$

$(b)$

$[1.3, 1.4] \times [-.1, .1]$
$x\mathrm{Scl} = .01, y\mathrm{Scl} = .01$

$(c)$

Figure 2.4.10

**FOR THE READER.** Use a graphing or calculating utility to show that the root $x_0$ in Example 6 can be approximated as $x_0 \approx 1.3245$ to three decimal-place accuracy.

## EXERCISE SET 2.4 ⌇ Graphing Calculator   [C] CAS

In Exercises 1–4, let $f$ be the function whose graph is shown. On which of the following intervals, if any, is $f$ continuous?
(a) $[1, 3]$    (b) $(1, 3)$    (c) $[1, 2]$
(d) $(1, 2)$    (e) $[2, 3]$    (f) $(2, 3)$
On those intervals where $f$ is not continuous, state where the discontinuities occur.

**1.**    **2.**

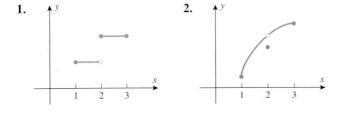

**3.**    **4.**

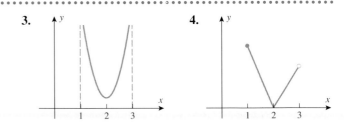

**5.** Suppose that $f$ and $g$ are continuous functions such that $f(2) = 1$ and $\lim\limits_{x \to 2} [f(x) + 4g(x)] = 13$. Find
(a) $g(2)$             (b) $\lim\limits_{x \to 2} g(x)$.

**6.** Suppose that $f$ and $g$ are continuous functions such that $\lim\limits_{x \to 3} g(x) = 5$ and $f(3) = -2$. Find $\lim\limits_{x \to 3} [f(x)/g(x)]$.

**7.** In each part sketch the graph of a function $f$ that satisfies the stated conditions.

(a) $f$ is continuous everywhere except at $x = 3$, at which point it is continuous from the right.

(b) $f$ has a two-sided limit at $x = 3$, but it is not continuous at that point.

(c) $f$ is not continuous at $x = 3$, but if its value at $x = 3$ is changed from $f(3) = 1$ to $f(3) = 0$, it becomes continuous at $x = 3$.

(d) $f$ is continuous on the interval $[0, 3)$ and is defined on the closed interval $[0, 3]$; but $f$ is not continuous on the interval $[0, 3]$.

**8.** Find formulas for some functions that are continuous on the intervals $(-\infty, 0)$ and $(0, +\infty)$, but are not continuous on the interval $(-\infty, +\infty)$.

**9.** A student parking lot at a university charges $2.00 for the first half hour (or any part) and $1.00 for each subsequent half hour (or any part) up to a daily maximum of $10.00.

(a) Sketch a graph of cost as a function of the time parked.

(b) Discuss the significance of the discontinuities in the graph to a student who parks there.

**10.** In each part determine whether the function is continuous or not, and explain your reasoning.

(a) The Earth's population as a function of time

(b) Your exact height as a function of time

(c) The cost of a taxi ride in your city as a function of the distance traveled

(d) The volume of a melting ice cube as a function of time

In Exercises 11–22, find the points of discontinuity, if any.

**11.** $f(x) = x^3 - 2x + 3$

**12.** $f(x) = (x - 5)^{17}$

**13.** $f(x) = \dfrac{x}{x^2 + 1}$

**14.** $f(x) = \dfrac{x}{x^2 - 1}$

**15.** $f(x) = \dfrac{x - 4}{x^2 - 16}$

**16.** $f(x) = \dfrac{3x + 1}{x^2 + 7x - 2}$

**17.** $f(x) = \dfrac{x}{|x| - 3}$

**18.** $f(x) = \dfrac{5}{x} + \dfrac{2x}{x + 4}$

**19.** $f(x) = |x^3 - 2x^2|$

**20.** $f(x) = \dfrac{x + 3}{|x^2 + 3x|}$

**21.** $f(x) = \begin{cases} 2x + 3, & x \le 4 \\ 7 + \dfrac{16}{x}, & x > 4 \end{cases}$

**22.** $f(x) = \begin{cases} \dfrac{3}{x - 1}, & x \ne 1 \\ 3, & x = 1 \end{cases}$

**23.** Find a value for the constant $k$, if possible, that will make the function continuous.

(a) $f(x) = \begin{cases} 7x - 2, & x \le 1 \\ kx^2, & x > 1 \end{cases}$

(b) $f(x) = \begin{cases} kx^2, & x \le 2 \\ 2x + k, & x > 2 \end{cases}$

**24.** On which of the following intervals is

$$f(x) = \frac{1}{\sqrt{x - 2}}$$

continuous?

(a) $[2, +\infty)$      (b) $(-\infty, +\infty)$

(c) $(2, +\infty)$      (d) $[1, 2)$

---

A function $f$ is said to have a ***removable discontinuity*** at $x = c$ if $\lim\limits_{x \to c} f(x)$ exists, but

$$f(c) \ne \lim_{x \to c} f(x)$$

either because $f(c)$ is undefined or the value of $f(c)$ differs from the value of the limit. This terminology will be needed in Exercises 25–28.

---

**25.** (a) Sketch the graph of a function with a removable discontinuity at $x = c$ for which $f(c)$ is undefined.

(b) Sketch the graph of a function with a removable discontinuity at $x = c$ for which $f(c)$ is defined.

**26.** (a) The terminology *removable discontinuity* is appropriate because a removable discontinuity of a function $f$ at a point $x = c$ can be "removed" by redefining the value of $f$ appropriately at $x = c$. What value for $f(c)$ removes the discontinuity?

(b) Show that the following functions have removable discontinuities at $x = 1$, and sketch their graphs.

$$f(x) = \frac{x^2 - 1}{x - 1} \quad \text{and} \quad g(x) = \begin{cases} 1, & x > 1 \\ 0, & x = 1 \\ 1, & x < 1 \end{cases}$$

(c) What values should be assigned to $f(1)$ and $g(1)$ to remove the discontinuities?

---

In Exercises 27 and 28, find the points of discontinuity, and determine whether the discontinuities are removable.

---

**27.** (a) $f(x) = \dfrac{|x|}{x}$      (b) $f(x) = \dfrac{x^2 + 3x}{x + 3}$

(c) $f(x) = \dfrac{x - 2}{|x| - 2}$

**28.** (a) $f(x) = \dfrac{x^2 - 4}{x^3 - 8}$

(b) $f(x) = \begin{cases} 2x - 3, & x \le 2 \\ x^2, & x > 2 \end{cases}$

(c) $f(x) = \begin{cases} 3x^2 + 5, & x \ne 1 \\ 6, & x = 1 \end{cases}$

**29.** (a) Use a graphing utility to generate the graph of the function $f(x) = (x + 3)/(2x^2 + 5x - 3)$, and then use the graph to make a conjecture about the number and location of all discontinuities.

(b) Check your conjecture by factoring the denominator.

⌁ **30.** (a) Use a graphing utility to generate the graph of the function $f(x) = x/(x^3 - x + 2)$, and then use the graph to make a conjecture about the number and location of all discontinuities.

(b) Use the Intermediate-Value Theorem to approximate the location of all points of discontinuity to two decimal places.

**31.** Prove that $f(x) = x^{3/5}$ is continuous everywhere, carefully justifying each step.

**32.** Prove that $f(x) = 1/\sqrt{x^4 + 7x^2 + 1}$ is continuous everywhere, carefully justifying each step.

**33.** Let $f$ and $g$ be discontinuous at $c$. Give examples to show that

(a) $f + g$ can be continuous or discontinuous at $c$

(b) $fg$ can be continuous or discontinuous at $c$.

**34.** Prove Theorem 2.4.4.

**35.** Prove:

(a) part (a) of Theorem 2.4.3

(b) part (b) of Theorem 2.4.3

(c) part (c) of Theorem 2.4.3.

**36.** Prove: If $f$ and $g$ are continuous on $[a, b]$, and $f(a) > g(a)$, $f(b) < g(b)$, then there is at least one solution of the equation $f(x) = g(x)$ in $(a, b)$. [*Hint:* Consider $f(x) - g(x)$.]

**37.** Give an example of a function $f$ that is defined at every point in a closed interval, and whose values at the endpoints have opposite signs, but for which the equation $f(x) = 0$ has no solution in the interval.

**38.** Use the Intermediate-Value Theorem to show that there is a square with a diagonal length that is between $r$ and $2r$ and an area that is half the area of a circle of radius $r$.

**39.** Use the Intermediate-Value Theorem to show that there is a right circular cylinder of height $h$ and radius less than $r$ whose volume is equal to that of a right circular cone of height $h$ and radius $r$.

In Exercises 40 and 41, show that the equation has at least one solution in the given interval.

**40.** $x^3 - 4x + 1 = 0$; $[1, 2]$

**41.** $x^3 + x^2 - 2x = 1$; $[-1, 1]$

**42.** Prove: If $p(x)$ is a polynomial of odd degree, then the equation $p(x) = 0$ has at least one real solution.

**43.** The accompanying figure shows the graph of $y = x^4 + x - 1$. Use the method of Example 6 to approximate the $x$-intercepts with an error of at most 0.05.

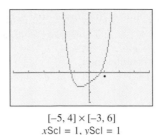

$[-5, 4] \times [-3, 6]$
$x$Scl $= 1$, $y$Scl $= 1$

Figure Ex-43

⌁ **44.** Use a graphing utility to solve the problem in Exercise 43 by zooming.

**45.** The accompanying figure shows the graph of $y = 5 - x - x^4$. Use the method of Example 6 to approximate the roots of the equation $5 - x - x^4 = 0$ to two decimal-place accuracy.

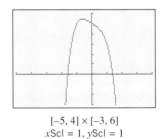

$[-5, 4] \times [-3, 6]$
$x$Scl $= 1$, $y$Scl $= 1$

Figure Ex-45

⌁ **46.** Use a graphing utility to solve the problem in Exercise 45 by zooming.

**47.** Use the fact that $\sqrt{5}$ is a solution of $x^2 - 5 = 0$ to approximate $\sqrt{5}$ with an error of at most 0.005.

**48.** Prove that if $a$ and $b$ are positive, then the equation

$$\frac{a}{x - 1} + \frac{b}{x - 3} = 0$$

has at least one solution in the interval $(1, 3)$.

**49.** A sphere of unknown radius $x$ consists of a spherical core and a coating that is 1 cm thick (see the accompanying figure). Given that the volume of the coating and the volume of the core are the same, approximate the radius of the sphere to three decimal-place accuracy.

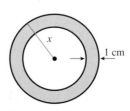

Figure Ex-49

**50.** A monk begins walking up a mountain road at 12:00 noon and reaches the top at 12:00 midnight. He meditates and rests until 12:00 noon the next day, at which time he begins walking down the same road, reaching the bottom at 12:00 midnight. Show that there is at least one point on the road that he reaches at the same time of day on the way up as on the way down.

**51.** Let $f$ be defined at $c$. Prove that $f$ is continuous at $c$ if, given $\epsilon > 0$, there exists a $\delta > 0$ such that $|f(x) - f(c)| < \epsilon$ if $|x - c| < \delta$.

## 2.5 LIMITS AND CONTINUITY OF TRIGONOMETRIC FUNCTIONS

*In this section we will investigate the continuity properties of the trigonometric functions, and we will discuss some important limits involving these functions.*

.......................................
**CONTINUITY OF TRIGONOMETRIC FUNCTIONS**

Before we begin, recall that in the expressions $\sin x$, $\cos x$, $\tan x$, $\cot x$, $\sec x$, and $\csc x$ it is understood that $x$ is in radian measure.

In trigonometry, the graphs of $\sin x$ and $\cos x$ are drawn as continuous curves (Figure 2.5.1). To actually prove that these functions are continuous everywhere, we must show that the following equalities hold for every real number $c$:

$$\lim_{x \to c} \sin x = \sin c \quad \text{and} \quad \lim_{x \to c} \cos x = \cos c \tag{1--2}$$

Although we will not formally prove these results, we can make them plausible by considering the behavior of the point $P(\cos x, \sin x)$ as it moves around the unit circle. For this purpose, view $c$ as a fixed angle in radian measure, and let $Q(\cos c, \sin c)$ be the corresponding point on the unit circle. As $x \to c$ (i.e., as the angle $x$ approaches the angle $c$), the point $P$ moves along the circle toward $Q$, and this implies that the coordinates of $P$ approach the corresponding coordinates of $Q$; that is, $\cos x \to \cos c$, and $\sin x \to \sin c$ (Figure 2.5.2).

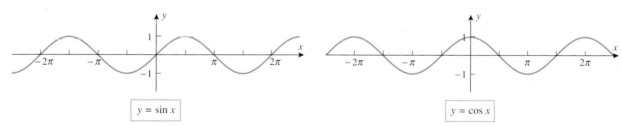

$y = \sin x$

$y = \cos x$

Figure 2.5.1

Formulas (1) and (2) can be used to find limits of the remaining trigonometric functions by expressing them in terms of $\sin x$ and $\cos x$; for example, if $\cos c \neq 0$, then

$$\lim_{x \to c} \tan x = \lim_{x \to c} \frac{\sin x}{\cos x} = \frac{\sin c}{\cos c} = \tan c$$

Thus, we are led to the following theorem.

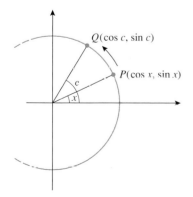

$Q(\cos c, \sin c)$

$P(\cos x, \sin x)$

Figure 2.5.2

---

**2.5.1   THEOREM.**   *If $c$ is any number in the natural domain of the stated trigonometric function, then*

$$\lim_{x \to c} \sin x = \sin c \qquad \lim_{x \to c} \cos x = \cos c \qquad \lim_{x \to c} \tan x = \tan c$$

$$\lim_{x \to c} \csc x = \csc c \qquad \lim_{x \to c} \sec x = \sec c \qquad \lim_{x \to c} \cot x = \cot c$$

---

It follows from this theorem, for example, that $\sin x$ and $\cos x$ are continuous everywhere and that $\tan x$ is continuous, except at the points where it is undefined.

**Example 1**

Find the limit

$$\lim_{x \to 1} \cos\left(\frac{x^2 - 1}{x - 1}\right)$$

**Solution.** Recall from the last section that a limit symbol can be moved through a function sign if the function is continuous and the limit of the expression inside the function sign exists. Thus,

$$\lim_{x \to 1} \cos\left(\frac{x^2 - 1}{x - 1}\right) = \lim_{x \to 1} \cos(x + 1) = \cos(\lim_{x \to 1} (x + 1)) = \cos 2 \quad \blacktriangleleft$$

**OBTAINING LIMITS BY SQUEEZING**

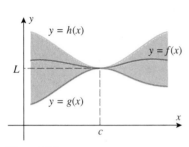

Figure 2.5.3

In Section 2.1 we used the numerical evidence in Table 2.1.2 to *conjecture* that

$$\lim_{x \to 0} \frac{\sin x}{x} = 1 \tag{3}$$

However, it is not a simple matter to establish this limit with certainty. The difficulty is that the numerator and denominator both approach zero as $x \to 0$; such limits are called *indeterminate forms of type* $0/0$. Sometimes indeterminate forms of this type can be established by manipulating the ratio algebraically (as in Example 7 of Section 2.2); but in this case no simple algebraic manipulation will work, so we must look for other methods.

The problem with indeterminate forms of type $0/0$ is that there are two conflicting influences at work: as the numerator approaches 0 it drives the magnitude of the ratio toward 0, and as the denominator approaches 0 it drives the magnitude of the ratio toward $\pm\infty$ (depending on the sign of the expression). The limiting behavior of the ratio is determined by the precise way in which these influences offset each other. Later in this text we will discuss general methods for attacking indeterminate forms, but for the limit in (3) we can use a method called *squeezing*.

In the method of squeezing one proves that a function $f$ has a limit $L$ at a point $c$ by trapping the function between two other functions, $g$ and $h$, whose limits at $c$ are known to be $L$ (Figure 2.5.3). This is the idea behind the following theorem, which we state without proof.

---

**2.5.2** THEOREM (*The Squeezing Theorem*). *Let $f$, $g$, and $h$ be functions satisfying*

$$g(x) \le f(x) \le h(x)$$

*for all $x$ in some open interval containing the point $c$, with the possible exception that the inequalities need not hold at $c$. If $g$ and $h$ have the same limit as $x$ approaches $c$, say*

$$\lim_{x \to c} g(x) = \lim_{x \to c} h(x) = L$$

*then $f$ also has this limit as $x$ approaches $c$, that is,*

$$\lim_{x \to c} f(x) = L$$

---

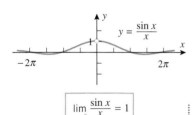

$$\lim_{x \to 0} \frac{\sin x}{x} = 1$$

FOR THE READER. The Squeezing Theorem also holds for one-sided limits and limits at $+\infty$ and $-\infty$. How do you think the hypotheses of the theorem would change in those cases?

The usefulness of the Squeezing Theorem will be evident in our proof of the following theorem (Figure 2.5.4).

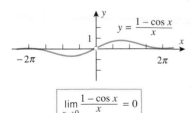

$$\lim_{x \to 0} \frac{1 - \cos x}{x} = 0$$

Figure 2.5.4

---

**2.5.3** THEOREM.

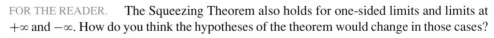

(*a*) $\displaystyle\lim_{x \to 0} \frac{\sin x}{x} = 1$  (*b*) $\displaystyle\lim_{x \to 0} \frac{1 - \cos x}{x} = 0$

---

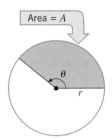

Figure 2.5.5

However, before giving the proof, it will be helpful to review the formula for the area $A$ of a sector with radius $r$ and a central angle of $\theta$ radians (Figure 2.5.5). The area of the sector can be derived by setting up the following proportion to the area of the entire circle:

$$\frac{A}{\pi r^2} = \frac{\theta}{2\pi} \qquad \left[\frac{\text{area of the sector}}{\text{area of the circle}} = \frac{\text{central angle of the sector}}{\text{central angle of the circle}}\right]$$

From this we obtain the formula

$$A = \tfrac{1}{2}r^2\theta \tag{4}$$

Now we are ready for the proof of Theorem 2.5.3.

*Proof (a).* In this proof we will interpret $x$ as an angle in radian measure, and we will assume to start that $0 < x < \pi/2$. It follows from Formula (4) that the area of a sector of radius 1 and central angle $x$ is $x/2$. Moreover, it is suggested by Figure 2.5.6 that the area of this sector lies between the areas of two triangles, one with area $(\tan x)/2$ and one with area $(\sin x)/2$. Thus,

$$\frac{\tan x}{2} \geq \frac{x}{2} \geq \frac{\sin x}{2}$$

Multiplying through by $2/(\sin x)$ yields

$$\frac{1}{\cos x} \geq \frac{x}{\sin x} \geq 1$$

and then taking reciprocals and reversing the inequalities yields

$$\cos x \leq \frac{\sin x}{x} \leq 1 \tag{5}$$

Moreover, these inequalities also hold for $-\pi/2 < x < 0$, since replacing $x$ by $-x$ in (5) and using the identities $\sin(-x) = -\sin x$ and $\cos(-x) = \cos x$ leaves the inequalities unchanged (verify). Finally, since the functions $\cos x$ and 1 both have limits of 1 as $x \to 0$, it follows from the Squeezing Theorem that $(\sin x)/x$ also has a limit of 1 as $x \to 0$.

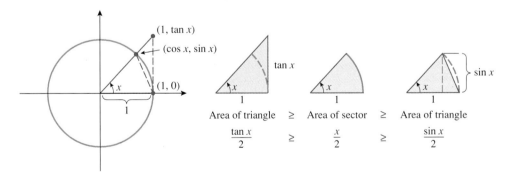

Figure 2.5.6

*Proof (b).* For this proof we will use the limit in part $(a)$, the continuity of the sine function, and the trigonometric identity $\sin^2 x = 1 - \cos^2 x$. We obtain

$$\lim_{x \to 0} \frac{1 - \cos x}{x} = \lim_{x \to 0}\left[\frac{1 - \cos x}{x} \cdot \frac{1 + \cos x}{1 + \cos x}\right] = \lim_{x \to 0} \frac{\sin^2 x}{(1 + \cos x)x}$$

$$= \left(\lim_{x \to 0} \frac{\sin x}{x}\right)\left(\lim_{x \to 0} \frac{\sin x}{1 + \cos x}\right) = (1)\left(\frac{0}{1 + 1}\right) = 0 \qquad\blacksquare$$

**Example 2**

Find

(a) $\displaystyle\lim_{x\to 0}\frac{\tan x}{x}$  (b) $\displaystyle\lim_{\theta\to 0}\frac{\sin 2\theta}{\theta}$  (c) $\displaystyle\lim_{x\to 0}\frac{\sin 3x}{\sin 5x}$

*Solution (a).*

$$\lim_{x\to 0}\frac{\tan x}{x}=\lim_{x\to 0}\left(\frac{\sin x}{x}\cdot\frac{1}{\cos x}\right)=(1)(1)=1$$

*Solution (b).* The trick is to multiply and divide by 2, which will make the denominator the same as the argument of the sine function [ just as in Theorem 2.5.3(*a*)]:

$$\lim_{\theta\to 0}\frac{\sin 2\theta}{\theta}=\lim_{\theta\to 0}2\cdot\frac{\sin 2\theta}{2\theta}=2\lim_{\theta\to 0}\frac{\sin 2\theta}{2\theta}$$

Now make the substitution $x=2\theta$, and use the fact that $x\to 0$ as $\theta\to 0$. This yields

$$\lim_{\theta\to 0}\frac{\sin 2\theta}{\theta}=2\lim_{\theta\to 0}\frac{\sin 2\theta}{2\theta}=2\lim_{x\to 0}\frac{\sin x}{x}=2(1)=2$$

*Solution (c).*

$$\lim_{x\to 0}\frac{\sin 3x}{\sin 5x}=\lim_{x\to 0}\frac{\dfrac{\sin 3x}{x}}{\dfrac{\sin 5x}{x}}=\lim_{x\to 0}\frac{3\cdot\dfrac{\sin 3x}{3x}}{5\cdot\dfrac{\sin 5x}{5x}}=\frac{3\cdot 1}{5\cdot 1}=\frac{3}{5}$$ ◀

FOR THE READER.  Use a graphing utility to confirm the limits in the last example graphically, and if you have a CAS, then use it to obtain the limits.

**Example 3**

Make conjectures about the limits

(a) $\displaystyle\lim_{x\to 0}\sin\left(\frac{1}{x}\right)$  (b) $\displaystyle\lim_{x\to 0}x\sin\left(\frac{1}{x}\right)$

and confirm your conclusions by generating the graphs of the functions near $x=0$ using a graphing utility.

*Solution (a).* Since $1/x\to+\infty$ as $x\to 0^+$, we can view $\sin(1/x)$ as the sine of an angle that increases indefinitely as $x\to 0^+$. As this angle increases, the function $\sin(1/x)$ keeps oscillating between $-1$ and $1$ without approaching a limit. Similarly, there is no limit from the left since $1/x\to-\infty$ as $x\to 0^-$. These conclusions are consistent with the graph of $y=\sin(1/x)$ shown in Figure 2.5.7*a*. Observe that the oscillations become more and more rapid as $x$ approaches 0 because $1/x$ increases (or decreases) more and more rapidly as $x$ approaches 0.

*Solution (b).* The values of $x\sin(1/x)$ oscillate between $x$ and $-x$, both of which approach 0 as $x$ approaches 0. Thus, the Squeezing Theorem suggests that $x\sin(1/x)\to 0$ as $x\to 0$. This is consistent with Figure 2.5.7*b*. ◀

REMARK.  It follows from part (b) of this example that the function

$$f(x)=\begin{cases}x\sin(1/x), & x\neq 0\\ 0, & x=0\end{cases}$$

is continuous at $x=0$, since the value of the function and the value of the limit are the same at that point. This shows that the behavior of a function can be very complex at a point of continuity.

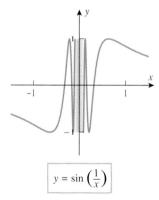

$y=\sin\left(\frac{1}{x}\right)$

(*a*)

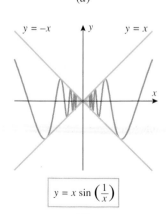

$y=-x$  $y=x$

$y=x\sin\left(\frac{1}{x}\right)$

(*b*)

Figure 2.5.7

**EXERCISE SET 2.5**  ⬚ Graphing Calculator  ⬚c CAS

In Exercises 1–10, find the points of discontinuity, if any.

**1.** $f(x) = \sin(x^2 - 2)$

**2.** $f(x) = \cos\left(\dfrac{x}{x - \pi}\right)$

**3.** $f(x) = \cot x$

**4.** $f(x) = \sec x$

**5.** $f(x) = \csc x$

**6.** $f(x) = \dfrac{1}{1 + \sin^2 x}$

**7.** $f(x) = |\cos x|$

**8.** $f(x) = \sqrt{2 + \tan^2 x}$

**9.** $f(x) = \dfrac{1}{1 - 2\sin x}$

**10.** $f(x) = \dfrac{3}{5 + 2\cos x}$

**11.** Use Theorem 2.4.6 to show that the following functions are continuous everywhere by expressing them as compositions of simpler functions that are known to be continuous.
(a) $\sin(x^3 + 7x + 1)$  (b) $|\sin x|$
(c) $\cos^3(x + 1)$  (d) $\sqrt{3 + \sin 2x}$
(e) $\sin(\sin x)$  (f) $\cos^5 x - 2\cos^3 x + 1$

**12.** (a) Prove that if $g(x)$ is continuous everywhere, then so are $\sin(g(x))$, $\cos(g(x))$, $g(\sin(x))$, and $g(\cos(x))$.
(b) Illustrate the result in part (a) with some of your own choices for $g$.

Find the limits in Exercises 13–35.

**13.** $\lim\limits_{x \to +\infty} \cos\left(\dfrac{1}{x}\right)$

**14.** $\lim\limits_{x \to +\infty} \sin\left(\dfrac{2}{x}\right)$

**15.** $\lim\limits_{x \to +\infty} \sin\left(\dfrac{\pi x}{2 - 3x}\right)$

**16.** $\lim\limits_{h \to 0} \dfrac{\sin h}{2h}$

**17.** $\lim\limits_{\theta \to 0} \dfrac{\sin 3\theta}{\theta}$

**18.** $\lim\limits_{\theta \to 0^+} \dfrac{\sin \theta}{\theta^2}$

**19.** $\lim\limits_{x \to 0^-} \dfrac{\sin x}{|x|}$

**20.** $\lim\limits_{x \to 0} \dfrac{\sin^2 x}{3x^2}$

**21.** $\lim\limits_{x \to 0^+} \dfrac{\sin x}{5\sqrt{x}}$

**22.** $\lim\limits_{x \to 0} \dfrac{\sin 6x}{\sin 8x}$

**23.** $\lim\limits_{x \to 0} \dfrac{\tan 7x}{\sin 3x}$

**24.** $\lim\limits_{\theta \to 0} \dfrac{\sin^2 \theta}{\theta}$

**25.** $\lim\limits_{h \to 0} \dfrac{h}{\tan h}$

**26.** $\lim\limits_{h \to 0} \dfrac{\sin h}{1 - \cos h}$

**27.** $\lim\limits_{\theta \to 0} \dfrac{\theta^2}{1 - \cos \theta}$

**28.** $\lim\limits_{x \to 0} \dfrac{x}{\cos\left(\frac{1}{2}\pi - x\right)}$

**29.** $\lim\limits_{\theta \to 0} \dfrac{\theta}{\cos \theta}$

**30.** $\lim\limits_{t \to 0} \dfrac{t^2}{1 - \cos^2 t}$

**31.** $\lim\limits_{h \to 0} \dfrac{1 - \cos 5h}{\cos 7h - 1}$

**32.** $\lim\limits_{x \to 0^+} \sin\left(\dfrac{1}{x}\right)$

**33.** $\lim\limits_{x \to 0^+} \cos\left(\dfrac{1}{x}\right)$

**34.** $\lim\limits_{x \to 0} \dfrac{x^2 - 3\sin x}{x}$

**35.** $\lim\limits_{x \to 0} \dfrac{2x + \sin x}{x}$

**36.** Find a value for the constant $k$ that makes
$$f(x) = \begin{cases} \dfrac{\sin 3x}{x}, & x \neq 0 \\ k, & x = 0 \end{cases}$$
continuous at $x = 0$.

**37.** Find a nonzero value for the constant $k$ that makes
$$f(x) = \begin{cases} \dfrac{\tan kx}{x}, & x < 0 \\ 3x + 2k^2, & x \geq 0 \end{cases}$$
continuous at $x = 0$.

**38.** Is
$$f(x) = \begin{cases} \dfrac{\sin x}{|x|}, & x \neq 0 \\ 1, & x = 0 \end{cases}$$
continuous at $x = 0$?

**39.** In each part, find the limit by making the indicated substitution.
(a) $\lim\limits_{x \to +\infty} x \sin \dfrac{1}{x}$.  $\left[\text{Hint: Let } t = \dfrac{1}{x}.\right]$
(b) $\lim\limits_{x \to -\infty} x\left(1 - \cos \dfrac{1}{x}\right)$.  $\left[\text{Hint: Let } t = \dfrac{1}{x}.\right]$
(c) $\lim\limits_{x \to \pi} \dfrac{\pi - x}{\sin x}$.  [Hint: Let $t = \pi - x$.]

**40.** Find $\lim\limits_{x \to 2} \dfrac{\cos(\pi/x)}{x - 2}$.  $\left[\text{Hint: Let } t = \dfrac{\pi}{2} - \dfrac{\pi}{x}.\right]$

**41.** Find $\lim\limits_{x \to 1} \dfrac{\sin(\pi x)}{x - 1}$.

**42.** Find $\lim\limits_{x \to \pi/4} \dfrac{\tan x - 1}{x - \pi/4}$.

**43.** Use the Squeezing Theorem to show that
$$\lim\limits_{x \to 0} x \cos \dfrac{50\pi}{x} = 0$$
and illustrate the principle involved by using a graphing utility to graph $y = x$, $y = -x$, and $y = x\cos(50\pi/x)$ on the same screen over the $x$-interval from $-1$ to $1$.

**44.** Use the Squeezing Theorem to show that
$$\lim\limits_{x \to 0} x^2 \sin\left(\dfrac{50\pi}{\sqrt[3]{x}}\right) = 0$$
and illustrate the principle involved by using a graphing utility to graph $y = x^2$, $y = -x^2$, and $y - x^2 \sin(50\pi/\sqrt[3]{x})$ on the same screen over the $x$-interval from $-0.5$ to $0.5$.

**45.** Sketch the graphs of $y = 1 - x^2$, $y = \cos x$, and $y = f(x)$, where $f$ is any continuous function that satisfies the inequalities
$$1 - x^2 \leq f(x) \leq \cos x$$
for all $x$ in the interval $(-\pi/2, \pi/2)$. What can you say about the limit of $f(x)$ as $x \to 0$? Explain your reasoning.

**46.** Sketch the graphs of $y = 1/x$, $y = -1/x$, and $y = f(x)$ in one coordinate system, where $f$ is any continuous function that satisfies the inequalities

$$-\frac{1}{x} \le f(x) \le \frac{1}{x}$$

for all $x$ in the interval $[1, +\infty)$. What can you say about the limit of $f(x)$ as $x \to +\infty$? Explain your reasoning.

**47.** Find formulas for functions $g$ and $h$ such that $g(x) \to 0$ and $h(x) \to 0$ as $x \to +\infty$ and such that

$$g(x) \le \frac{\sin x}{x} \le h(x)$$

for positive values of $x$. What can you say about the limit

$$\lim_{x \to +\infty} \frac{\sin x}{x}?$$

Explain your reasoning.

**48.** Draw pictures analogous to Figure 2.5.3 that illustrate the Squeezing Theorem for limits of the form $\lim_{x \to +\infty} f(x)$ and $\lim_{x \to -\infty} f(x)$.

---

Recall that unless stated otherwise the variable $x$ in trigonometric functions such as $\sin x$ and $\cos x$ are assumed to be in radian measure. The limits in Theorem 2.5.3 are based on that assumption. Exercises 49 and 50 explore what happens to those limits if degree measure is used for $x$.

---

**49. (a)** Show that if $x$ is in degrees, then

$$\lim_{x \to 0} \frac{\sin x}{x} = \frac{\pi}{180}$$

**(b)** Confirm that the limit in part (a) is consistent with the results produced by your calculating utility by setting the utility to degree measure and calculating $(\sin x)/x$ for some values of $x$ that get closer and closer to 0.

**50.** What is the limit of $(1 - \cos x)/x$ as $x \to 0$ if $x$ is in degrees?

**51.** It follows from part $(a)$ of Theorem 2.5.3 that if $\theta$ is small (near zero) and measured in radians, then one should expect the approximation

$$\sin \theta \approx \theta$$

to be good.

**(a)** Find $\sin 10°$ using a calculating utility.

**(b)** Find $\sin 10°$ using the approximation above.

**52. (a)** Use the approximation of $\sin \theta$ that is given in Exercise 51 together with the identity $\cos 2\alpha = 1 - 2\sin^2 \alpha$ with $\alpha = \theta/2$ to show that if $\theta$ is small (near zero) and measured in radians, then one should expect the approximation

$$\cos \theta \approx 1 - \tfrac{1}{2}\theta^2$$

to be good.

**(b)** Find $\cos 10°$ using a calculating utility.

**(c)** Find $\cos 10°$ using the approximation above.

**53.** It follows from part (a) of Example 2 that if $\theta$ is small (near zero) and measured in radians, then one should expect the

approximation

$$\tan \theta \approx \theta$$

to be good.

**(a)** Find $\tan 5°$ using a calculating utility.

**(b)** Find $\tan 5°$ using the approximation above.

**54.** Referring to the accompanying figure, suppose that the angle of elevation of the top of a building, as measured from a point $L$ feet from its base, is found to be $\alpha$ degrees.

**(a)** Use the relationship $h = L \tan \alpha$ to calculate the height of a building for which $L = 500$ ft and $\alpha = 6°$.

**(b)** Show that if $L$ is large compared to the building height $h$, then one should expect good results in approximating $h$ by $h \approx \pi L\alpha/180$.

**(c)** Use the result in part (b) to approximate the building height $h$ in part (a).

Figure Ex-54

**55. (a)** Use the Intermediate-Value Theorem to show that the equation $x = \cos x$ has at least one solution in the interval $[0, \pi/2]$.

**(b)** Show graphically that there is exactly one solution in the interval.

**(c)** Approximate the solution to three decimal places.

**56. (a)** Use the Intermediate-Value Theorem to show that the equation $x + \sin x = 1$ has at least one solution in the interval $[0, \pi/6]$.

**(b)** Show graphically that there is exactly one solution in the interval.

**(c)** Approximate the solution to three decimal places.

**57.** In the study of falling objects near the surface of the Earth, the *acceleration g due to gravity* is commonly taken to be $9.8 \text{ m/s}^2$ or $32 \text{ ft/s}^2$. However, the elliptical shape of the Earth and other factors cause variations in this constant that are latitude dependent. The following formula, known as the Geodetic Reference Formula of 1967, is commonly used to predict the value of $g$ at a latitude of $\phi$ degrees (either north or south of the equator):

$$g = 9.7803185(1.0 + 0.005278895 \sin^2 \phi$$
$$- 0.000023462 \sin^4 \phi) \text{ m/s}^2$$

**(a)** Observe that $g$ is an even function of $\phi$. What does this suggest about the shape of the Earth, as modeled by the Geodetic Reference Formula?

**(b)** Show that $g = 9.8 \text{ m/s}^2$ somewhere between latitudes of $38°$ and $39°$.

**58.** Let

$$f(x) = \begin{cases} 1 & \text{if } x \text{ is a rational number} \\ 0 & \text{if } x \text{ is an irrational number} \end{cases}$$

**(a)** Make a conjecture about the limit of $f(x)$ as $x \to 0$.

**(b)** Make a conjecture about the limit of $xf(x)$ as $x \to 0$.

**(c)** Prove your conjectures.

## SUPPLEMENTARY EXERCISES

**1.** For the function $f$ graphed in the accompanying figure, find the limit if it exists.

(a) $\lim\limits_{x \to 1} f(x)$    (b) $\lim\limits_{x \to 2} f(x)$    (c) $\lim\limits_{x \to 3} f(x)$

(d) $\lim\limits_{x \to 4} f(x)$    (e) $\lim\limits_{x \to +\infty} f(x)$    (f) $\lim\limits_{x \to -\infty} f(x)$

(g) $\lim\limits_{x \to 3^+} f(x)$    (h) $\lim\limits_{x \to 3^-} f(x)$    (i) $\lim\limits_{x \to 0} f(x)$

Figure Ex-1

**2.** (a) Find a formula for a rational function that has a vertical asymptote at $x = 1$ and a horizontal asymptote at $y = 2$.

  (b) Check your work by using a graphing utility to graph the function.

**3.** (a) Write a paragraph or two that describes how the limit of a function can fail to exist at a point $x = a$. Accompany your description with some specific examples.

  (b) Write a paragraph or two that describes how the limit of a function can fail to exist as $x \to +\infty$ or $x \to -\infty$. Also, accompany your description with some specific examples.

  (c) Write a paragraph or two that describes how a function can fail to be continuous at a point $x = a$. Accompany your description with some specific examples.

**4.** Show that the Intermediate-Value Theorem is false if $f$ is not continuous on the interval $[a, b]$.

**5.** In each part, evaluate the function for the stated values of $x$, and make a conjecture about the value of the limit. Confirm your conjecture by finding the limit algebraically.

  (a) $f(x) = \dfrac{x - 2}{x^2 - 4}$;    $\lim\limits_{x \to 2^+} f(x)$;  $x = 2.5, 2.1, 2.01,$
  $$2.001, 2.0001, 2.00001$$

  (b) $f(x) = \dfrac{\tan 4x}{x}$;    $\lim\limits_{x \to 0} f(x)$;  $x = \pm 1.0, \pm 0.1, \pm 0.01,$
  $$\pm 0.001, \pm 0.0001, \pm 0.00001$$

**6.** In each part, find the horizontal asymptotes, if any.

  (a) $y = \dfrac{2x - 7}{x^2 - 4x}$    (b) $y = \dfrac{x^3 - x^2 + 10}{3x^2 - 4x}$

  (c) $y = \dfrac{2x^2 - 6}{x^2 + 5x}$

**7.** (a) Approximate the value for the limit
  $$\lim\limits_{x \to 0} \frac{3^x - 2^x}{x}$$
  to three decimal places by constructing an appropriate table of values.

  (b) Confirm your approximation using graphical evidence.

**8.** According to Ohm's law, when a voltage of $V$ volts is applied across a resistor with a resistance of $R$ ohms, a current of $I = V/R$ amperes flows through the resistor.

  (a) How much current flows if a voltage of 3.0 volts is applied across a resistance of 7.5 ohms?

  (b) If the resistance varies by $\pm 0.1$ ohm, and the voltage remains constant at 3.0 volts, what is the resulting range of values for the current?

  (c) If temperature variations cause the resistance to vary by $\pm \delta$ from its value of 7.5 ohms, and the voltage remains constant at 3.0 volts, what is the resulting range of values for the current?

  (d) If the current is not allowed to vary by more than $\epsilon = \pm 0.001$ ampere at a voltage of 3.0 volts, what variation of $\pm \delta$ from the value of 7.5 ohms is allowable?

  (e) Certain alloys become **superconductors** as their temperature approaches absolute zero $(-273°\,\mathrm{C})$, meaning that their resistance approaches zero. If the voltage remains constant, what happens to the current in a superconductor as $R \to 0^+$?

**9.** Suppose that $f$ is continuous on the interval $[0, 1]$ and that $0 \le f(x) \le 1$ for all $x$ in this interval.

  (a) Sketch the graph of $y = x$ together with a possible graph for $f$ over the interval $[0, 1]$.

  (b) Use the Intermediate-Value Theorem to help prove that there is at least one number $c$ in the interval $[0, 1]$ such that $f(c) = c$.

**10.** Use algebraic methods to find

  (a) $\lim\limits_{\theta \to 0} \tan \left( \dfrac{1 - \cos \theta}{\theta} \right)$    (b) $\lim\limits_{t \to 1} \dfrac{t - 1}{\sqrt{t} - 1}$

  (c) $\lim\limits_{x \to +\infty} \dfrac{(2x - 1)^5}{(3x^2 + 2x - 7)(x^3 - 9x)}$

  (d) $\lim\limits_{\theta \to 0} \cos \left( \dfrac{\sin(\theta + \pi)}{2\theta} \right)$.

**11.** Suppose that $f$ is continuous on the interval $[0, 1]$, that $f(0) = 2$, and that $f$ has no zeros in the interval. Prove that $f(x) > 0$ for all $x$ in $[0, 1]$.

**12.** Suppose that
$$f(x) = \begin{cases} -x^4 + 3, & x \le 2 \\ x^2 + 9, & x > 2 \end{cases}$$

Is $f$ continuous everywhere? Justify your conclusion.

**13.** Show that the equation $x^4 + 5x^3 + 5x - 1 = 0$ has at least two real solutions in the interval $[-6, 2]$.

**14.** Use the Intermediate-Value Theorem to approximate $\sqrt{11}$ to three decimal places, and check your answer by finding the root directly with a calculating utility.

**15.** Suppose that $f$ is continuous and that $f(x_0) > 0$. Give either a $\delta$-$\epsilon$ proof or a convincing verbal argument to show that there must be an open interval containing $x_0$ on which $f(x) > 0$.

**16.** Sketch the graph of $f(x) = |x^2 - 4|/(x^2 - 4)$.

**17.** In each part, approximate the points of discontinuity of $f$ to three decimal places.

(a)  $f(x) = \dfrac{\sqrt{x+1}}{x^2 + 2x - 5}$

(b)  $f(x) = \dfrac{x+3}{|2\sin x - x|}$

**18.** In Example 3 of Section 2.5 we used the Squeezing Theorem to prove that

$$\lim_{x \to 0} x \sin\left(\frac{1}{x}\right) = 0$$

Why couldn't we have obtained the same result by writing

$$\lim_{x \to 0} x \sin\left(\frac{1}{x}\right) = \lim_{x \to 0} x \cdot \lim_{x \to 0} \sin\left(\frac{1}{x}\right)$$

$$= 0 \cdot \lim_{x \to 0} \sin\left(\frac{1}{x}\right) = 0?$$

In Exercises 19 and 20, find $\lim\limits_{x \to a} f(x)$, if it exists, for

$a = 0,\ 5^+,\ -5^-,\ -5,\ 5,\ -\infty,\ +\infty$

**19.** (a)  $f(x) = \sqrt{5-x}$        (b)  $f(x) = (x^2 - 25)/(x - 5)$

**20.** (a)  $f(x) = (x+5)/(x^2 - 25)$

(b)  $f(x) = \begin{cases} (x-5)/|x-5|, & x \neq 5 \\ 0, & x = 5 \end{cases}$

In Exercises 21–28, find the indicated limit, if it exists.

**21.** $\lim\limits_{x \to 0} \dfrac{\tan ax}{\sin bx}$   $(a \neq 0, b \neq 0)$

**22.** $\lim\limits_{x \to 0} \dfrac{\sin 3x}{\tan 3x}$

**23.** $\lim\limits_{\theta \to 0} \dfrac{\sin 2\theta}{\theta^2}$         **24.** $\lim\limits_{x \to 0} \dfrac{x \sin x}{1 - \cos x}$

**25.** $\lim\limits_{x \to 0^+} \dfrac{\sin x}{\sqrt{x}}$          **26.** $\lim\limits_{x \to 0} \dfrac{\sin^2(kx)}{x^2},\quad k \neq 0$

**27.** $\lim\limits_{x \to 0} \dfrac{3x - \sin(kx)}{x},\quad k \neq 0$

**28.** $\lim\limits_{x \to +\infty} \dfrac{2x + x \sin 3x}{5x^2 - 2x + 1}$

**29.** The author's dictionary describes a continuous function as "one whose value at each point is closely approached by its values at neighboring points."

(a)  How would you explain the meaning of the terms "neighboring points" and "closely approached" to a nonmathematician?

(b)  Write a paragraph that explains why the dictionary definition is consistent with the definition given in the text.

**30.** (a)  Show by rationalizing the numerator that

$$\lim_{x \to 0} \frac{\sqrt{x^2 + 4} - 2}{x^2} = \frac{1}{4}$$

(b)  Evaluate $f(x)$ for

$x = \pm 1.0,\ \pm 0.1,\ \pm 0.01,\ \pm 0.001,\ \pm 0.0001,\ \pm 0.00001$

and explain why the values are not getting closer and closer to the limit.

(c)  The accompanying figure shows the graph of $f$ generated with a graphing utility and zooming in on the origin. Explain what is happening.

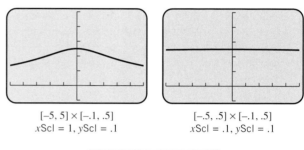

$[-5, 5] \times [-.1, .5]$
$x\text{Scl} = 1, y\text{Scl} = .1$

$[-.5, .5] \times [-.1, .5]$
$x\text{Scl} = .1, y\text{Scl} = .1$

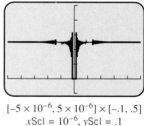

$[-5 \times 10^{-6}, 5 \times 10^{-6}] \times [-.1, .5]$
$x\text{Scl} = 10^{-6}, y\text{Scl} = .1$

Figure Ex-30

In Exercises 31–36, approximate the limit of the function by looking at its graph and calculating values for some appropriate choices of $x$. Compare your answer with the value produced by a CAS.

**31.** $\boxed{c}$ $\lim\limits_{x \to 0} (1 + x)^{1/x}$        **32.** $\boxed{c}$ $\lim\limits_{x \to 3} \dfrac{2^x - 8}{x - 3}$

**33.** $\boxed{c}$ $\lim\limits_{x \to 1} \dfrac{\sin x - \sin 1}{x - 1}$     **34.** $\boxed{c}$ $\lim\limits_{x \to 0^+} x^{-2}(1.001)^{-1/x}$

**35.** $\boxed{c}$ $\lim\limits_{x \to +\infty} \left( \sqrt{x + \sqrt{x}} - \sqrt{x} \right)$

**36.** $\boxed{c}$ $\lim\limits_{x \to +\infty} \left( 3^x + 5^x \right)^{1/x}$

**37.** The limit

$$\lim_{x \to 0} \frac{\sin x}{x} = 1$$

ensures that there is a number $\delta$ such that

$$\left| \frac{\sin x}{x} - 1 \right| < 0.001$$

if $0 < |x| < \delta$. Estimate the largest such $\delta$.

**38.** If $1000 is invested in an account that pays 7% interest compounded $n$ times each year, then in 10 years there will be $1000(1 + 0.07/n)^{10n}$ dollars in the account. How much money will be in the account in 10 years if the interest is compounded quarterly ($n = 4$)? Monthly ($n = 12$)? Daily ($n = 365$)? How much money will be in the account in 10 years if the interest is compounded *continuously*, that is, as $n \to +\infty$?

**39.** There are various numerical methods other than the method discussed in Section 2.4 to obtain approximate solutions of equations of the form $f(x) = 0$. One such method requires that the equation be expressed in the form $x = g(x)$, so that a solution $x = c$ can be interpreted as the value of $x$ where the line $y = x$ intersects the curve $y = g(x)$, as shown in the accompanying figure. If $x_1$ is an initial estimate of $c$ and the graph of $y = g(x)$ is not too steep in the vicinity of $c$, then a better approximation can be obtained from $x_2 = g(x_1)$ (see the figure). An even better approximation is obtained from $x_3 = g(x_2)$, and so forth. The formula $x_{n+1} = g(x_n)$ for $n = 1, 2, 3, \ldots$ generates successive approximations $x_2, x_3, x_4, \ldots$ that get closer and closer to $c$.

(a) The equation $x^3 - x - 1 = 0$ has only one real solution. Show that this equation can be written as

$$x = g(x) = \sqrt[3]{x + 1}$$

(b) Graph $y = x$ and $y = g(x)$ in the same coordinate system for $-1 \le x \le 3$.

(c) Starting with an arbitrary point $x_1$, make a sketch that shows the location of the successive iterates

$$x_2 = g(x_1), \quad x_3 = g(x_2), \ldots$$

(d) Use $x_1 = 1$ and calculate $x_2, x_3, \ldots$, continuing until you obtain two consecutive values that differ by less than $10^{-4}$. Experiment with other starting values such as $x_1 = 2$ or $x_1 = 1.5$.

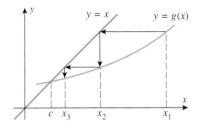

Figure Ex-39

**40.** The method described in Exercise 39 will not always work.

(a) The equation $x^3 - x - 1 = 0$ can be expressed as $x = g(x) = x^3 - 1$. Graph $y = x$ and $y = g(x)$ in the same coordinate system. Starting with an arbitrary point $x_1$, make a sketch illustrating the location of the successive iterates $x_2 = g(x_1)$, $x_3 = g(x_2)$, ....

(b) Use $x_1 = 1$ and calculate the successive iterates $x_n$ for $n = 2, 3, 4, 5, 6$.

> In Exercises 41 and 42, use the method of Exercise 39 to approximate the roots of the equation.

**41.** $x^5 - x - 2 = 0$ **42.** $x - \cos x = 0$

---

### EXPANDING THE CALCULUS HORIZON

**For additional material relating to this chapter, visit the Anton Website at http://www.wiley.com/college/anton**

# 3

Sir Isaac Newton

# THE DERIVATIVE

$\mathcal{M}$any physical phenomena involve changing quantities—the speed of a rocket, the inflation of currency, the number of bacteria in a culture, the shock intensity of an earthquake, the voltage of an electrical signal, and so forth. In this chapter we will develop the concept of a *derivative*, which is the mathematical tool that is used to study rates at which physical quantities change. In Section 3.1 we will show that there is a close relationship between rates of change and tangent lines to graphs, and we will show how the familiar idea of velocity can be viewed as a rate of change. In Sections 3.2 to 3.5 we will define the concept of a derivative precisely and develop the mathematical tools for calculating them.

One of the important themes in applied science is developing methods for approximating quantities that are difficult to calculate exactly. In Section 3.6 we will show how derivatives can be applied to certain kinds of approximation problems.

## 3.1 TANGENT LINES AND RATES OF CHANGE

---

*In this section we will establish a basic relationship between tangent lines and rates of change. Our work here is intended to be informal and introductory, and all of the ideas that we develop will be revisited in more detail in later sections.*

---

**SLOPE OF A TANGENT LINE**

In Section 2.1 we observed informally that if a secant line is drawn between two distinct points $P$ and $Q$ on a curve $y = f(x)$, and if $Q$ is allowed to move along the curve toward $P$, then we can expect the secant line to rotate toward a *limiting position*, which can be regarded as the tangent line to the curve at the point $P$ (Figure 3.1.1). In the next section we will give a precise mathematical definition of a tangent line, but for now this intuitive idea will suffice.

In many problems we will be more concerned with the *slope* of the tangent line than with the tangent line itself, so it will be helpful to understand the relationship between the slope $m_{\text{tan}}$ of the tangent line at $P$ and the slope $m_{\text{sec}}$ of the secant line between $P$ and $Q$ as the point $Q$ moves along the curve $y = f(x)$ toward $P$. For this purpose, suppose that the secant line passes through the distinct points $P(x_0, f(x_0))$ and $Q(x_1, f(x_1))$, in which case its slope is

$$m_{\text{sec}} = \frac{f(x_1) - f(x_0)}{x_1 - x_0} \tag{1}$$

(Figure 3.1.2). As this figure suggests, the point $Q$ moves along the curve toward $P$ if and only if $x_1$ approaches $x_0$. Thus, from (1) the slope of the tangent line at $P$ is

$$m_{\text{tan}} = \lim_{x_1 \to x_0} \frac{f(x_1) - f(x_0)}{x_1 - x_0}$$

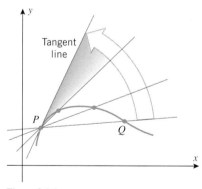

Figure 3.1.1

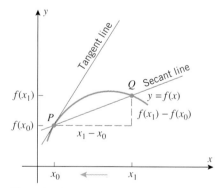

Figure 3.1.2

**AVERAGE VERSUS INSTANTANEOUS VELOCITY**

Although tangent lines are of interest as a matter of pure geometry, much of the impetus for studying them arose in the seventeenth century when scientists recognized their importance in studying the motion of objects that move with nonconstant velocity. Some of the relevant ideas were discussed in Section 1.5, but it will be helpful to review them here.

Recall that a particle moving along a line, say an $s$-axis, is said to have ***rectilinear motion***. In the most general kind of rectilinear motion the particle may move back and forth on the line; however, here, as in Section 1.5, we will assume that the particle moves in one direction only—the positive direction of the $s$-axis. As discussed in Section 1.5, this allows us to use the terms *speed* and *velocity* interchangeably, since there is only one possible direction of motion. General rectilinear motion will be discussed later.

We showed in Section 1.5 that if a particle has uniform rectilinear motion, that is, it moves with constant velocity $v$ along a line, then its position versus time curve is a line of slope $v$; conversely, if the position versus time curve for a particle in rectilinear motion is a line of slope $v$, then the particle has constant velocity $v$. Here, we will consider the more general case of a particle moving in the positive $s$-direction with *variable* velocity, in which case the position versus time curve need not be linear. For this purpose we will need to examine the meaning of the term *velocity* more critically.

If a car travels 75 miles over a straight road in a 3-hour period, then its average velocity during the trip is $75/3 = 25$ mi/h. However, this does not mean that the car travels at 25 mi/h for the entire trip; it may speed up and slow down at various times. Thus, the average velocity provides information about the velocity of the car over the entire trip but no information about its velocity at specific times during the trip.

Although average velocity is useful for many purposes, there are many situations in which it is of no help. For example, if a car strikes a tree during a trip, the damage sustained is not determined by the average velocity up to the time of impact, but rather by the *instantaneous velocity* at the precise moment of impact. However, the concept of instantaneous velocity is subtle, and a clear understanding of its meaning evaded scientists until the advent of calculus in the seventeenth century.

A nice explanation of the difficulty in defining and calculating instantaneous velocity was given by Morris Kline[*] who wrote:

> *In contrasting average velocity with instantaneous velocity we implicitly utilize a distinction between interval and instant. . . . An average velocity is one that concerns what happens over an interval of time—3 hours, 5 seconds, one-half second, and so forth. The interval may be small or large, but it does represent the passage of a definite amount of time. We use the word instant, however, to state the fact that something happens so fast that no time elapses. The event is momentary. When we say, for example, that it is 3 o'clock, we refer to an instant, a precise moment. If the lapse of time is pictured by length along a line, then an interval (of time) is represented by a line segment, whereas an instant corresponds to a point. The notion of an instant, although it is used in everyday life, is strictly a mathematical idealization.*
>
> *Our ways of thinking about real events cause us to speak in terms of instants and velocity at an instant, but closer examination shows that the concept of velocity at an instant presents difficulties. Average velocity, which is simply the distance traveled during some interval of time divided by that amount of time, is easily calculated. Suppose, however, that we try to carry over this process to instantaneous velocity. The distance an automobile travels in one instant is 0 and the time that elapses during one instant is also 0. Hence the distance divided by the time is 0/0, which is meaningless. Thus, although instantaneous velocity is a physical reality, there seems to be a difficulty in calculating it, and unless we can calculate it, we cannot work with it mathematically.*

Our goal, then, is to define the concept of instantaneous velocity in a way that it can be calculated and worked with mathematically. For this purpose, consider a car that moves in a single direction along a straight road, and assume that an $s$-axis has been introduced with its positive direction in the direction of motion. As shown in Figure 3.1.3, suppose that a clock tracks the elapsed time $t$, starting at $t = 0$, and that the coordinate of the car as a function of $t$ is $s = f(t)$. The function $f$ is called the **position function** of the car, and the graph of $s = f(t)$ is what we have been calling the position versus time curve. The third part of Figure 3.1.3 shows a typical position versus time curve for a car whose coordinate

[*]MORRIS KLINE (1908–1992). American mathematician, scholar, and educator. Kline made numerous contributions to mathematical thought, wrote extensively on education, especially mathematics education, and taught, lectured, and served as a consultant throughout his very active career. He was the author of many popular books including *Mathematical Thought from Ancient to Modern Times* and *Why Johnny Can't Add: The Failure of the New Mathematics.*

at time $t = 0$ is $s_0$. Observe that we have drawn the curve so that $s$ increases with $t$. This is because we have assumed the car to be traveling in the positive direction, and decreasing values of $s$ would imply a motion in the negative direction.

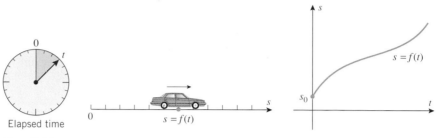

Figure 3.1.3

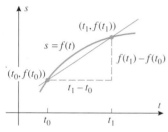

Figure 3.1.4

The position versus time curve provides a simple geometric interpretation of the average velocity of the car over a time interval, say from $t_0$ to $t_1$. If the car has a coordinate $s_0 = f(t_0)$ at time $t_0$ and coordinate $s_1 = f(t_1)$ at time $t_1$, where $t_1 > t_0$, then the distance traveled during the time interval is $s_1 - s_0$ and the time elapsed is $t_1 - t_0$. Thus, the average velocity during the time interval, denoted by $v_{ave}$, is

$$v_{ave} = \frac{s_1 - s_0}{t_1 - t_0} = \frac{f(t_1) - f(t_0)}{t_1 - t_0} \tag{2}$$

which is just the slope of the secant line connecting the points $(t_0, s_0)$ and $(t_1, s_1)$ on the position versus time curve (Figure 3.1.4).

Now suppose that we are interested in the instantaneous velocity of the car at time $t_0$. Intuition suggests that over a small time interval the velocity of the car cannot vary much, so if $t_1$ is close to $t_0$, then the average velocity of the car over the time interval from $t_0$ to $t_1$ should closely approximate the instantaneous velocity of the car at time $t_0$. Moreover, the smaller the time interval between $t_0$ and $t_1$, the better the approximation. This suggests that if we let $t_1$ get closer and closer to $t_0$, then the average velocity of the car over the time interval from $t_0$ to $t_1$ should get closer and closer to the instantaneous velocity at time $t_0$. Thus, if we denote the instantaneous velocity of the car at time $t_0$ by $v_{inst}$, we have

$$v_{inst} = \lim_{t_1 \to t_0} v_{ave} = \lim_{t_1 \to t_0} \frac{f(t_1) - f(t_0)}{t_1 - t_0} \tag{3}$$

Since $v_{ave}$ is the slope of the secant line joining the points $(t_0, f(t_0))$ and $(t_1, f(t_1))$ on the position versus time curve $s = f(t)$, and since the point $(t_1, f(t_1))$ moves along this curve toward $(t_0, f(t_0))$ as $t_1 \to t_0$ (Figure 3.1.5), it follows from (3) that $v_{inst}$ can be interpreted as the slope of the tangent line to the position versus time curve at the point $(t_0, f(t_0))$.

These ideas are illustrated numerically in Table 3.1.1. The first part of the table shows the coordinates of a particle moving along an $s$-axis over the time interval from $t = 4.00$ to

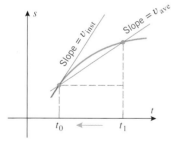

Figure 3.1.5

**Table 3.1.1**

| $t$ (s) | 4.00 | 4.50 | 5.00 | 5.50 | 5.80 | 5.90 | 5.95 | 5.98 | 6.00 |
|---|---|---|---|---|---|---|---|---|---|
| $s$ (ft) | 1.00 | 1.25 | 2.00 | 3.25 | 4.24 | 4.61 | 4.80 | 4.92 | 5.00 |

| TIME INTERVAL | [4.00, 6.00] | [4.50, 6.00] | [5.00, 6.00] | [5.50, 6.00] | [5.80, 6.00] | [5.90, 6.00] | [5.95, 6.00] | [5.98, 6.00] |
|---|---|---|---|---|---|---|---|---|
| AVERAGE VELOCITY (ft/s) | 1.00 | 2.50 | 3.00 | 3.50 | 3.80 | 3.90 | 4.00 | 4.00 |

$t = 6.00$. From these values we can calculate the average velocity of the particle over a succession of shrinking time intervals ending at time $t = 6.00$ s. For example, the calculations for the average velocity over the time interval [4.50, 6.00] are

$$v_{ave} = \frac{5.00 - 1.25}{6.00 - 4.50} = \frac{3.75}{1.50} = 2.50 \text{ ft/s}$$

The resulting average velocities in the second part of the table suggest that to two decimal places the instantaneous velocity at time $t = 6.00$ s is 4.00 ft/s.

The main ideas in the preceding discussion can be summarized as follows.

> **3.1.1**  GEOMETRIC INTERPRETATION OF AVERAGE VELOCITY.  *If a particle moves in the positive direction along an s-axis, and if the position versus time curve is $s = f(t)$, then the average velocity of the particle between times $t_0$ and $t_1$ is represented geometrically by the slope of the secant line joining the points $(t_0, f(t_0))$ and $(t_1, f(t_1))$.*

> **3.1.2**  GEOMETRIC INTERPRETATION OF INSTANTANEOUS VELOCITY.  *If a particle moves in the positive direction along an s-axis, and if the position versus time curve is $s = f(t)$, then the instantaneous velocity of the particle at time $t_0$ is represented geometrically by the slope of the tangent line to the curve at the point $(t_0, f(t_0))$.*

**AVERAGE AND INSTANTANEOUS RATES OF CHANGE**

Velocity can be viewed as a *rate of change*—the rate of change of position with time, or in algebraic terms, the rate of change of $s$ with $t$. Rates of change occur in many applications. For example:

- A microbiologist might be interested in the rate at which the number of bacteria in a colony changes with time.

- An engineer might be interested in the rate at which the length of a metal rod changes with temperature.

- An economist might be interested in the rate at which production cost changes with the quantity of a product that is manufactured.

- A medical researcher might be interested in the rate at which the radius of an artery changes with the concentration of alcohol in the bloodstream.

In general, if $x$ and $y$ are any quantities related by an equation $y = f(x)$, we can consider the rate at which $y$ changes with $x$. As with velocity, we distinguish between an average rate of change represented by the slope of a secant line and an instantaneous rate of change represented by the slope of the tangent line. More precisely, we make the following definitions.

> **3.1.3**  DEFINITION.  If $y = f(x)$, then the ***average rate of change of y with respect to x over the interval*** $[x_0, x_1]$ is the slope $m_{sec}$ of the secant line joining the points $(x_0, f(x_0))$ and $(x_1, f(x_1))$ on the graph of $f$ (Figure 3.1.6a); that is,
>
> $$m_{sec} = \frac{f(x_1) - f(x_0)}{x_1 - x_0} \tag{4}$$

**3.1.4** DEFINITION. If $y = f(x)$, then the ***instantaneous rate of change of y with respect to x at the point*** $x_0$ is the slope $m_{tan}$ of the tangent line to the graph of $f$ at the point $x_0$ (Figure 3.1.6$b$); that is,

$$m_{tan} = \lim_{x_1 \to x_0} \frac{f(x_1) - f(x_0)}{x_1 - x_0} \tag{5}$$

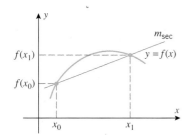

$m_{sec}$ is the average rate of change of $y$ with respect to $x$ over the interval $[x_0, x_1]$.

($a$)

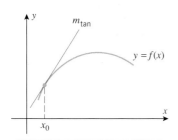

$m_{tan}$ is the instantaneous rate of change of $y$ with respect to $x$ at the point $x_0$.

($b$)

Figure 3.1.6

## Example 1

Let $y = x^2 + 1$.

(a) Find the average rate of change of $y$ with respect to $x$ over the interval $[3, 5]$.

(b) Find the instantaneous rate of change of $y$ with respect to $x$ at the point $x = -4$.

(c) Find the instantaneous rate of change of $y$ with respect to $x$ at a general point $x = x_0$.

*Solution* ($a$). We will apply Formula (4) with $f(x) = x^2 + 1$, $x_0 = 3$, and $x_1 = 5$. This yields

$$m_{sec} = \frac{f(x_1) - f(x_0)}{x_1 - x_0} = \frac{f(5) - f(3)}{5 - 3} = \frac{26 - 10}{5 - 3} = 8$$

Thus, on the average, $y$ increases 8 units per unit increase in $x$ over the interval $[3, 5]$.

*Solution* ($b$). We will apply Formula (5) with $f(x) = x^2 + 1$ and $x_0 = -4$. This yields

$$m_{tan} = \lim_{x_1 \to x_0} \frac{f(x_1) - f(x_0)}{x_1 - x_0} = \lim_{x_1 \to -4} \frac{(x_1^2 + 1) - 17}{x_1 + 4}$$

$$= \lim_{x_1 \to -4} \frac{x_1^2 - 16}{x_1 + 4} = \lim_{x_1 \to -4} (x_1 - 4) = -8$$

Because the instantaneous rate of change is negative, $y$ is *decreasing* at the point $x = -4$; it is decreasing at a rate of 8 units per unit increase in $x$.

*Solution* ($c$). We proceed as in part ($b$).

$$m_{tan} = \lim_{x_1 \to x_0} \frac{f(x_1) - f(x_0)}{x_1 - x_0} = \lim_{x_1 \to x_0} \frac{(x_1^2 + 1) - (x_0^2 + 1)}{x_1 - x_0}$$

$$= \lim_{x_1 \to x_0} \frac{x_1^2 - x_0^2}{x_1 - x_0} = \lim_{x_1 \to x_0} (x_1 + x_0) = 2x_0$$

Thus, the instantaneous rate of change of $y$ with respect to $x$ at $x = x_0$ is $2x_0$. Observe that the result in part ($b$) can be obtained from this more general result by letting $x_0 = -4$. ◀

**RATES OF CHANGE IN APPLICATIONS**

In applied problems, average and instantaneous rates of change must be accompanied by appropriate units. In general, the units for a rate of change of $y$ with respect to $x$ are obtained by "dividing" the units of $y$ by the units of $x$ and then simplifying according to the standard rules of algebra. Here are some examples:

- If $y$ is in degrees Fahrenheit ($°F$) and $x$ is in inches (in), then a rate of change of $y$ with respect to $x$ has units of degrees Fahrenheit per inch ($°F/in$).

- If $y$ is in feet per second (ft/s) and $x$ is in seconds (s), then a rate of change of $y$ with respect to $x$ has units of feet per second per second (ft/s/s), which would usually be written as ft/s².

- If $y$ is in newton-meters (N·m) and $x$ is in meters (m), then a rate of change of $y$ with respect to $x$ has units of newtons (N), since N·m/m = N.

- If $y$ is in foot-pounds (ft·lb) and $x$ is in hours (h), then a rate of change of $y$ with respect to $x$ has units of foot-pounds per hour (ft·lb/h).

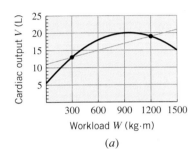

Figure 3.1.7

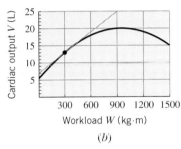

Figure 3.1.8

## Example 2

The limiting factor in athletic endurance is cardiac output, that is, the volume of blood that the heart can pump per unit of time during an athletic competition. Figure 3.1.7 shows a stress-test graph of cardiac output $V$ in liters (L) of blood versus workload $W$ in kilogram-meters (kg·m) for 1 minute of weight lifting. This graph illustrates the known medical fact that cardiac output increases with the workload, but after reaching a peak value begins to decrease.

(a) Use the secant line shown in Figure 3.1.8a to estimate the average rate of change of cardiac output with respect to workload as the workload increases from 300 to 1200 kg·m.

(b) Use the tangent line shown in Figure 3.1.8b to estimate the instantaneous rate of change of cardiac output with respect to workload at the point where the workload is 300 kg·m.

*Solution (a).* Using the estimated points (300, 13) and (1200, 19), the slope of the secant line indicated in Figure 3.1.8a is

$$m_{sec} \approx \frac{19 - 13}{1200 - 300} \approx 0.0067 \frac{L}{kg \cdot m}$$

Thus, the average rate of change of cardiac output with respect to workload over the interval is approximately 0.0067 L/kg·m. This means that on the average a 1-unit increase in workload produced a 0.0067-L increase in cardiac output over the interval.

*Solution (b).* Using the estimated tangent line in Figure 3.1.8b and the estimated points (0, 7) and (900, 25) on this tangent line, we obtain

$$m_{tan} \approx \frac{25 - 7}{900 - 0} \approx 0.02 \frac{L}{kg \cdot m}$$

Thus, the instantaneous rate of change of cardiac output with respect to workload is approximately 0.02 L/kg·m. ◄

## EXERCISE SET 3.1

In Exercises 1–4, a function $y = f(x)$ and values of $x_0$ and $x_1$ are given.
(a) Find the average rate of change of $y$ with respect to $x$ over the interval $[x_0, x_1]$.
(b) Find the instantaneous rate of change of $y$ with respect to $x$ at the given value of $x_0$.
(c) Find the instantaneous rate of change of $y$ with respect to $x$ at a general point $x_0$.
(d) Sketch the graph of $y = f(x)$ together with the secant and tangent lines whose slopes are given by the results in parts (a) and (b).

**1.** $y = \frac{1}{2}x^2$; $x_0 - 3$, $x_1 = 4$

**2.** $y = x^3$; $x_0 = 1$, $x_1 = 2$

**3.** $y = 1/x$; $x_0 = 2$, $x_1 = 3$

**4.** $y = 1/x^2$; $x_0 = 1$, $x_1 = 2$

In Exercises 5–8, a function $f$ and a value of $x_0$ are given.
(a) Find the slope of the tangent to the graph of $f$ at a general point $x_0$.
(b) Use the result in part (a) to find the slope of the tangent line at the given value of $x_0$.

**5.** $f(x) = x^2 + 1$; $x_0 = 2$

**6.** $f(x) = x^2 + 3x + 2$; $x_0 = 2$

**7.** $f(x) = \sqrt{x}$; $x_0 = 1$

**8.** $f(x) = 1/\sqrt{x}$; $x_0 = 4$

**9.** The accompanying figure shows the position versus time curve for an elevator that moves upward a distance of 60 m and then discharges its passengers.

(a) Estimate the instantaneous velocity of the elevator at $t = 10$ s.

(b) Sketch a velocity versus time curve for the motion of the elevator for $0 \le t \le 20$.

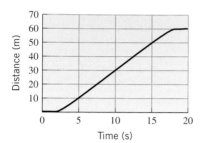

Figure Ex-9

**10.** The accompanying figure shows the position versus time curve for a certain particle moving along a straight line. Estimate each of the following from the graph:

(a) the average velocity over the interval $0 \le t \le 3$

(b) the values of $t$ at which the instantaneous velocity is zero

(c) the values of $t$ at which the instantaneous velocity is either a maximum or a minimum

(d) the instantaneous velocity when $t = 3$ s.

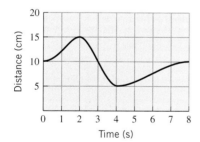

Figure Ex-10

**11.** The accompanying figure shows the position versus time curve for a certain particle moving on a straight line.

(a) Is the particle moving faster at time $t_0$ or time $t_2$? Explain.

(b) At the origin, the tangent is horizontal. What does this tell us about the initial velocity of the particle?

(c) Is the particle speeding up or slowing down in the interval $[t_0, t_1]$? Explain.

(d) Is the particle speeding up or slowing down in the interval $[t_1, t_2]$? Explain.

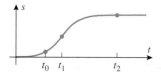

Figure Ex-11

**12.** An automobile, initially at rest, begins to move along a straight track. The velocity increases steadily until suddenly the driver sees a concrete barrier in the road and applies the brakes sharply at time $t_0$. The car decelerates rapidly, but it is too late—the car crashes into the barrier at time $t_1$ and instantaneously comes to rest. Sketch a position versus time curve that might represent the motion of the car.

**13.** If a particle moves at constant velocity, what can you say about its position versus time curve?

**14.** The accompanying figure shows the position versus time curves of four different particles moving on a straight line. For each particle, determine whether its instantaneous velocity is increasing or decreasing with time.

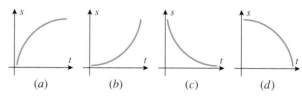

Figure Ex-14

**15.** Suppose that the outside temperature versus time curve over a 24-hour period is as shown in the accompanying figure.

(a) Estimate the maximum temperature and the time at which it occurs.

(b) The temperature rise is fairly linear from 8 A.M. to 2 P.M. Estimate the rate at which the temperature is increasing during this time period.

(c) Estimate the time at which the temperature is decreasing most rapidly. Estimate the instantaneous rate of change of temperature with respect to time at this instant.

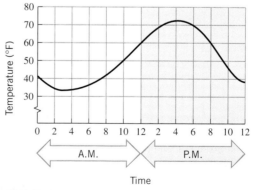

Figure Ex-15

**16.** The accompanying figure shows the graph of the pressure $p$ in atmospheres (atm) versus the volume $V$ in liters (L) of 1 mole of an ideal gas at a constant temperature of 300 K (kelvins). Use the tangent lines shown in the figure to estimate the rate of change of pressure with respect to volume at the points where $V = 10$ L and $V = 25$ L.

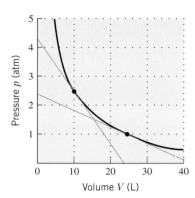

Figure Ex-16

In Exercises 18–21, use Formulas (2) and (3) to find the average and instantaneous velocity.

**18.** A rock is dropped from a height of 576 ft and falls toward Earth in a straight line. In $t$ seconds the rock drops a distance of $s = 16t^2$ ft.
   (a) How many seconds after release does the rock hit the ground?
   (b) What is the average velocity of the rock during the time it is falling?
   (c) What is the average velocity of the rock for the first 3 s?
   (d) What is the instantaneous velocity of the rock when it hits the ground?

**19.** During the first 40 s of a rocket flight, the rocket is propelled straight up so that in $t$ seconds it reaches a height of $s = 5t^3$ ft.
   (a) How high does the rocket travel in 40 s?
   (b) What is the average velocity of the rocket during the first 40 s?
   (c) What is the average velocity of the rocket during the first 135 ft of its flight?
   (d) What is the instantaneous velocity of the rocket at the end of 40 s?

**17.** The accompanying figure shows the graph of the height $h$ in centimeters versus the age $t$ in years of an individual from birth to age 20.
   (a) When is the growth rate greatest?
   (b) Estimate the growth rate at age 5.
   (c) At approximately what age between 10 and 20 is the growth rate greatest? Estimate the growth rate at this age.
   (d) Draw a rough graph of the growth rate versus age.

**20.** A particle moves on a line away from its initial position so that after $t$ hours it is $s = 3t^2 + t$ miles from its initial position.
   (a) Find the average velocity of the particle over the interval $[1, 3]$.
   (b) Find the instantaneous velocity at $t = 1$.

**21.** A particle moves in the positive direction along a straight line so that after $t$ minutes its distance is $s = 6t^4$ feet from the origin.
   (a) Find the average velocity of the particle over the interval $[2, 4]$.
   (b) Find the instantaneous velocity at $t = 2$.

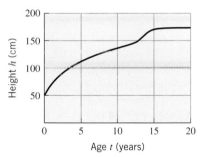

Figure Ex-17

## 3.2 THE DERIVATIVE

*In this section we will introduce the concept of a* derivative, *which is the primary mathematical tool that is used to calculate rates of change and slopes of tangent lines.*

**TANGENT LINES DEFINED PRECISELY**

In the preceding section we showed informally that the slope of the tangent line to the graph of $y = f(x)$ at the point $x_0$ is given by

$$m_{\tan} = \lim_{x_1 \to x_0} \frac{f(x_1) - f(x_0)}{x_1 - x_0} \tag{1}$$

However, for computational purposes it will be more convenient to express this formula in a different form by introducing a new variable $h = x_1 - x_0$. It follows that $x_1 = x_0 + h$, and consequently $x_1 \to x_0$ as $h \to 0$. Thus, (1) can be expressed as

$$m_{\tan} = \lim_{h \to 0} \frac{f(x_0 + h) - f(x_0)}{h}$$

(Figure 3.2.1). This suggests the following formal definition of a tangent line.

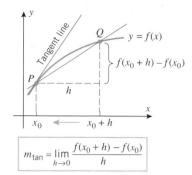

Figure 3.2.1

**3.2.1 DEFINITION.** If $P(x_0, y_0)$ is a point on the graph of a function $f$, then the *tangent line to the graph of $f$ at $P$*, also called the *tangent line to the graph of $f$ at $x_0$*, is defined to be the line through $P$ with slope

$$m_{\tan} = \lim_{h \to 0} \frac{f(x_0 + h) - f(x_0)}{h} \tag{2}$$

provided this limit exists. If the limit does not exist, then by agreement the graph has no tangent line at $P$.

It follows from this definition that the point-slope form of the equation of the tangent line at $x_0$ is

$$y - y_0 = m_{\tan}(x - x_0) \tag{3}$$

### Example 1

Find the equation of the tangent line to the graph of $y = x^2 + 1$ at the point $(2, 5)$ (Figure 3.2.2).

*Solution.* First, we will find the slope of the tangent line using (2) with $f(x) = x^2 + 1$ and $x_0 = 2$, and then we will find the equation by using (3). We obtain

$$m_{\tan} = \lim_{h \to 0} \frac{f(2 + h) - f(2)}{h} = \lim_{h \to 0} \frac{[(2 + h)^2 + 1] - 5}{h}$$

$$= \lim_{h \to 0} \frac{(5 + 4h + h^2) - 5}{h} = \lim_{h \to 0} \frac{4h + h^2}{h}$$

$$= \lim_{h \to 0} (4 + h) = 4$$

Thus, from (3) with $x_0 = 2$, $y_0 = 5$, and $m_{\tan} = 4$, the point-slope form of the equation of the tangent line is

$$y - 5 = 4(x - 2)$$

which we can write in slope-intercept form as $y = 4x - 3$. ◀

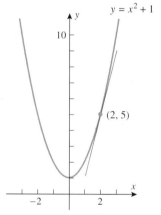

Figure 3.2.2

### SLOPES OF TANGENT LINES BY ZOOMING

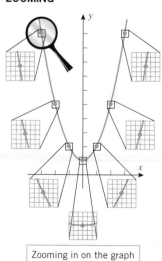

Zooming in on the graph of $y = x^2 + 1$

Figure 3.2.3

Slopes of tangent lines can be estimated by zooming with graphing utilities. The idea is to zoom in on the point of tangency until the surrounding curve segment appears to be a straight line that nearly coincides with the tangent line (Figure 3.2.3). The utility's trace operation can then be used to estimate the slope. Figure 3.2.4 illustrates this procedure for the tangent line in Example 1. The first part of the figure shows the graph of $y = x^2 + 1$ in the window[*] $[-6.3, 6.3] \times [0, 6.2]$, and the second part shows the graph after we have zoomed in on the point $(2, 5)$ by a factor of 10. The trace operation produces the points $(2.05, 5.2025)$ and $(1.95, 4.8025)$ on the line, so the slope of the tangent line can be approximated as

$$m \approx \frac{5.2025 - 4.8025}{2.05 - 1.95} = \frac{0.4}{0.1} = 4.0$$

which happens to agree exactly with the result in Example 1. It is important to understand, however, that the exact agreement in this case is accidental; in general, this method will not produce exact results because of roundoff errors in the computations, and also because the magnified curve segment may have a slight curvature, even though it appears to be a straight line.

---

[*]The window $[-6.3, 6.3] \times [0, 6.2]$ was chosen because it contains the point of tangency $(2, 5)$ and produces convenient steps on the author's calculator when the trace operation is applied. Books on graphing calculators sometimes call these "friendly windows."

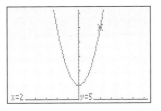   

Figure 3.2.4

**THE DERIVATIVE**

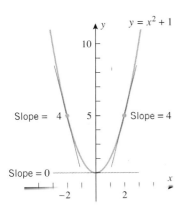

Figure 3.2.5

In general, the slope of a tangent line to a curve $y = f(x)$ will depend on the point $x$ at which the slope is being computed; thus, the slope is itself a function of $x$. To illustrate this, let us use (2) to compute $m_{tan}$ at a general point $x$ for the curve $y = x^2 + 1$. The computations are similar to those in Example 1, except that now we let $x_0$ have an arbitrary value $x_0 = x$, whereas in Example 1 we had $x_0 = 2$. We obtain

$$m_{tan} = \lim_{h \to 0} \frac{f(x + h) - f(x)}{h} = \lim_{h \to 0} \frac{[(x + h)^2 + 1] - [x^2 + 1]}{h}$$

$$= \lim_{h \to 0} \frac{x^2 + 2xh + h^2 + 1 - x^2 - 1}{h} = \lim_{h \to 0} \frac{2xh + h^2}{h}$$

$$= \lim_{h \to 0} (2x + h) = 2x \tag{4}$$

Now we can use the general formula $m_{tan} = 2x$ to compute the slope of the tangent line at any point along the curve $y = x^2 + 1$ simply by substituting the appropriate value for $x$. For example, if $x = 2$, then we obtain $m_{tan} = 2x = 4$, which agrees with the result in Example 1. Similarly, if $x = 0$, then $m_{tan} = 0$; and if $x = -2$, then $m_{tan} = -4$ (Figure 3.2.5).

To generalize this idea, the slope of the tangent line to the graph of $y = f(x)$ at a general point $x$ can be obtained by setting $x_0 = x$ in (2), which yields the formula

$$m_{tan} = \lim_{h \to 0} \frac{f(x + h) - f(x)}{h}$$

This "slope-producing function" is so important that it has some notation and terminology associated with it.

---

**3.2.2 DEFINITION.** The function $f'$ defined by the formula

$$f'(x) = \lim_{h \to 0} \frac{f(x + h) - f(x)}{h} \tag{5}$$

is called the ***derivative of f with respect to x***. The domain of $f'$ consists of all $x$ for which the limit exists.

---

Recalling from the last section that the slope of a tangent line to the graph of $y = f(x)$ can be interpreted as the instantaneous rate of change of $y$ with respect to $x$, it follows that the derivative of a function $f$ can be interpreted in two ways:

***Two interpretations of the Derivative.*** The derivative $f'$ of a function $f$ can be interpreted either as a function whose value at $x$ is the slope of the tangent line to the graph of $y = f(x)$ at $x$, or, alternatively, it can be interpreted as a function whose value at $x$ is the instantaneous rate of change of $y$ with respect to $x$ at the point $x$.

**Example 2**

(a) Find the derivative with respect to $x$ of $f(x) = x^3 - x$.

(b) Graph $f$ and $f'$ together, and discuss the relationship between the two graphs.

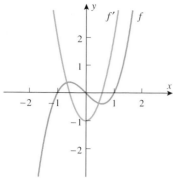

Figure 3.2.6

*Solution (a).* Later in this chapter we will develop efficient methods for finding derivatives, but for now we will find the derivative directly from Formula (5) in the definition of $f'$. The computations are as follows:

$$f'(x) = \lim_{h \to 0} \frac{f(x+h) - f(x)}{h} = \lim_{h \to 0} \frac{[(x+h)^3 - (x+h)] - (x^3 - x)}{h}$$

$$= \lim_{h \to 0} \frac{x^3 + 3x^2h + 3xh^2 + h^3 - x - h - x^3 + x}{h}$$

$$= \lim_{h \to 0} \frac{3x^2h + 3xh^2 + h^3 - h}{h}$$

$$= \lim_{h \to 0} (3x^2 + 3xh + h^2 - 1) = 3x^2 - 1$$

*Solution (b).* Since $f'(x)$ can be interpreted as the slope of the tangent line to the graph of $y = f(x)$ at the point $x$, the derivative $f'(x)$ is positive where the tangent line $y = f(x)$ has positive slope, it is negative where the tangent line has negative slope, and it is zero where the tangent line is horizontal. We leave it for the reader to verify that this is consistent with the graphs of $f(x) = x^3 - x$ and $f'(x) = 3x^2 - 1$ shown in Figure 3.2.6. ◀

## Example 3

At each point $x$, the tangent line to a line $y = mx + b$ coincides with the line itself (Figure 3.2.7), and hence all tangent lines have slope $m$. This suggests geometrically that if $f(x) = mx + b$, then $f'(x) = m$ for all $x$. This is confirmed by the following computations:

$$f'(x) = \lim_{h \to 0} \frac{f(x+h) - f(x)}{h} = \lim_{h \to 0} \frac{[m(x+h) + b] - (mx + b)}{h}$$

$$= \lim_{h \to 0} \frac{mx + mh + b - mx - b}{h} = \lim_{h \to 0} \frac{mh}{h} = \lim_{h \to 0} m = m \qquad ◀$$

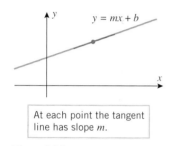

At each point the tangent line has slope $m$.

Figure 3.2.7

## Example 4

(a)  Find the derivative with respect to $x$ of $f(x) = \sqrt{x}$.

(b)  Find the slope of the tangent line to $y = \sqrt{x}$ at $x = 9$.

(c)  Find the limits of $f'(x)$ as $x \to 0^+$ and as $x \to +\infty$, and explain what those limits say about the graph of $f$.

*Solution (a).* From Definition 3.2.2,

$$f'(x) = \lim_{h \to 0} \frac{f(x+h) - f(x)}{h} = \lim_{h \to 0} \frac{\sqrt{x+h} - \sqrt{x}}{h}$$

$$= \lim_{h \to 0} \frac{(\sqrt{x+h} - \sqrt{x})(\sqrt{x+h} + \sqrt{x})}{h(\sqrt{x+h} + \sqrt{x})} = \lim_{h \to 0} \frac{(x+h) - x}{h(\sqrt{x+h} + \sqrt{x})}$$

$$= \lim_{h \to 0} \frac{h}{h(\sqrt{x+h} + \sqrt{x})} = \lim_{h \to 0} \frac{1}{\sqrt{x+h} + \sqrt{x}}$$

$$= \frac{1}{\sqrt{x} + \sqrt{x}} = \frac{1}{2\sqrt{x}}$$

*Solution (b).* The slope of the tangent line at $x = 9$ is $f'(9)$, and thus from part (a) this slope is $f'(9) = 1/(2\sqrt{9}) = \frac{1}{6}$.

*Solution (c).* The graphs of $f(x) = \sqrt{x}$ and $f'(x) = 1/(2\sqrt{x})$ are shown in Figure 3.2.8. Observe that $f'(x) > 0$ if $x > 0$, which means that all tangent lines to the graph of $y = \sqrt{x}$

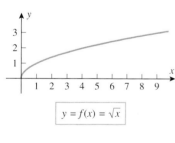

$y = f(x) = \sqrt{x}$

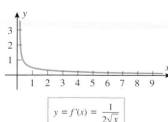

$y = f'(x) = \dfrac{1}{2\sqrt{x}}$

Figure 3.2.8

have positive slope over this interval. Since

$$\lim_{x \to 0^+} \frac{1}{2\sqrt{x}} = +\infty \quad \text{and} \quad \lim_{x \to +\infty} \frac{1}{2\sqrt{x}} = 0$$

the tangent lines become more and more vertical as $x \to 0^+$, and they become more and more horizontal as $x \to +\infty$. ◀

FOR THE READER.  Use a graphing utility to estimate the slope of the tangent line to $y = \sqrt{x}$ at $x = 9$ by zooming, and compare your result to the exact value obtained in the last example. If you have a CAS, read the documentation to determine how it can be used to find derivatives, and then use it to confirm the derivatives obtained in Examples 2, 3, and 4.

**DIFFERENTIABILITY**

Recall from Definition 3.2.2 that the derivative of a function $f$ is defined at those points where the limit (5) exists. Points where this limit exists are called ***points of differentiability*** for $f$, and points where this limit does not exist are called ***points of nondifferentiability*** for $f$.

If $x_0$ is a point of differentiability for $f$, then we say that $f$ is ***differentiable at $x_0$*** or that ***the derivative of f exists at $x_0$***; and if $x_0$ is a point of nondifferentiability for $f$, then we say that ***the derivative of f does not exist at $x_0$***. If $f$ is differentiable at every point in an open interval $(a, b)$, then we will say that $f$ is ***differentiable on $(a, b)$***. This definition also applies to infinite open intervals of the form $(a, +\infty)$, $(-\infty, b)$, and $(-\infty, +\infty)$. In the case where $f$ is differentiable on $(-\infty, +\infty)$ we will say that $f$ is ***differentiable everywhere***. If $f$ is differentiable on an open interval but the particular interval is not important for the discussion, then we will say that $f$ is ***differentiable*** (without referencing the interval).

Geometrically, the points of differentiability of $f$ are the points where the curve $y = f(x)$ has a tangent line, and the points of nondifferentiability are the points where the curve does not have a tangent line. Informally stated, the most commonly encountered points of nondifferentiability can be classified as

- Corners
- Points of vertical tangency
- Points of discontinuity

Figure 3.2.9 illustrates each of these situations.

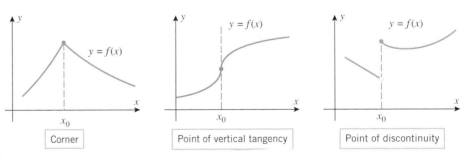

Figure 3.2.9

It makes sense intuitively that corners are points of nondifferentiability, since there is no reasonable way to draw a unique tangent line at such points. For example, Figure 3.2.10a shows a typical corner point $P(x_0, f(x_0))$ on the graph of a function $f$. At this point the secant lines joining $P$ and $Q$ have different limiting positions, depending on whether $Q$ approaches $P$ from the left or right; hence the slopes of the secant lines do not have a two-sided limit.

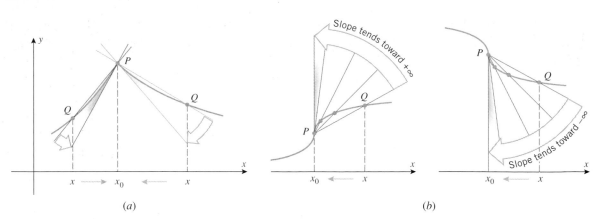

(a)                                              (b)

Figure 3.2.10

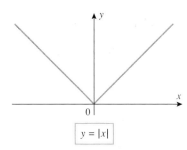

$y = |x|$

Figure 3.2.11

By a point of *vertical tangency* we mean a place on the curve where the secant lines approach a vertical limiting position. At such points, the only reasonable candidate for the tangent line is the vertical line at the point. But vertical lines have infinite slope, so the derivative (were it to exist) would not have a finite real value there, which explains intuitively why the derivative does not exist at points of vertical tangency (Figure 3.2.10b).

### Example 5

The graph of $y = |x|$ in Figure 3.2.11 suggests that there is a corner at $x = 0$, and this implies that $f(x) = |x|$ is not differentiable at that point.

(a)   Prove that $f(x) = |x|$ is not differentiable at $x = 0$ by showing that the limit in Definition 3.2.2 does not exist at that point.

(b)   Find a formula for $f'(x)$.

*Solution* (*a*).  From Formula (5) with $x = 0$, the value of $f'(0)$, if it were to exist, would be given by

$$f'(0) = \lim_{h \to 0} \frac{f(0+h) - f(0)}{h} = \lim_{h \to 0} \frac{f(h) - f(0)}{h} = \lim_{h \to 0} \frac{|h| - |0|}{h} = \lim_{h \to 0} \frac{|h|}{h}$$

But

$$\frac{|h|}{h} = \begin{cases} 1, & h > 0 \\ -1, & h < 0 \end{cases}$$

so that

$$\lim_{h \to 0^-} \frac{|h|}{h} = -1 \quad \text{and} \quad \lim_{h \to 0^+} \frac{|h|}{h} = 1$$

Thus,

$$f'(0) = \lim_{h \to 0} \frac{|h|}{h}$$

does not exist because the one-sided limits are not equal. Consequently, $f(x) = |x|$ is not differentiable at $x = 0$.

*Solution* (*b*).  A formula for the derivative of $f(x) = |x|$ can be obtained by writing $|x|$ in piecewise form and treating the cases $x > 0$ and $x < 0$ separately. If $x > 0$, then $f(x) = x$ and $f'(x) = 1$; and if $x < 0$, then $f(x) = -x$ and $f'(x) = -1$. Thus,

$$f'(x) = \begin{cases} 1, & x > 0 \\ -1, & x < 0 \end{cases}$$

The graph of $f'$ is shown in Figure 3.2.12. Observe that $f'$ is not a continuous function, so this example shows that the derivative of a continuous function need not be continuous. ◀

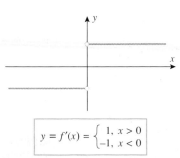

$y = f'(x) = \begin{cases} 1, & x > 0 \\ -1, & x < 0 \end{cases}$

Figure 3.2.12

**RELATIONSHIP BETWEEN DIFFERENTIABILITY AND CONTINUITY**

It makes sense intuitively that a function $f$ cannot be differentiable at a point of discontinuity, since there is no reasonable way to draw a unique tangent line at such points. The following theorem shows that a function $f$ must be continuous at each point where it is differentiable (or stated another way, a function $f$ cannot be differentiable at a point of discontinuity).

**3.2.3** THEOREM. *If $f$ is differentiable at a point $x_0$, then $f$ is also continuous at $x_0$.*

*Proof.* We are given that $f$ is differentiable at $x_0$, so it follows from (5) that $f'(x_0)$ exists and is given by

$$f'(x_0) = \lim_{h \to 0} \left[ \frac{f(x_0 + h) - f(x_0)}{h} \right] \tag{6}$$

To show that $f$ is continuous at $x_0$, we must show that $\lim_{x \to x_0} f(x) = f(x_0)$, or equivalently,

$$\lim_{x \to x_0} [f(x) - f(x_0)] = 0$$

Expressing this in terms of the variable $h = x - x_0$, we must prove that

$$\lim_{h \to 0} [f(x_0 + h) - f(x_0)] = 0$$

However, this can be proved using (6) as follows:

$$\lim_{h \to 0} [f(x_0 + h) - f(x_0)] = \lim_{h \to 0} \left[ \frac{f(x_0 + h) - f(x_0)}{h} \cdot h \right]$$
$$= \lim_{h \to 0} \left[ \frac{f(x_0 + h) - f(x_0)}{h} \right] \cdot \lim_{h \to 0} h$$
$$= f'(x_0) \cdot 0 = 0 \qquad \blacksquare$$

REMARK. Theorem 3.2.3 shows that differentiability at a point implies continuity at that point. However, the converse is false; that is, *a function may be continuous at a point but not differentiable there.* In fact, this occurs at any point where the function is continuous and has a corner. For example, we saw in Example 5 that the function $f(x) = |x|$ is continuous at $x = 0$, yet not differentiable there.

The relationship between continuity and differentiability was of great historical significance in the development of calculus. In the early nineteenth century mathematicians believed that the graph of a continuous function could not have too many points of nondifferentiability bunched up. They felt that if a continuous function had many points of nondifferentiability, these points, like the tips of a sawblade, would have to be separated from each other and joined by smooth curve segments (Figure 3.2.13). This misconception was shattered by a series of discoveries beginning in 1834. In that year a Bohemian priest, philosopher, and mathematician named Bernhard Bolzano[*] discovered a procedure for constructing a continuous function that is not differentiable at any point. Later, in 1860, the great

[*]BERNHARD BOLZANO (1781–1848). Bolzano, the son of an art dealer, was born in Prague, Bohemia (Czech Republic). He was educated at the University of Prague, and eventually won enough mathematical fame to be recommended for a mathematics chair there. However, Bolzano became an ordained Roman Catholic priest, and in 1805 he was appointed to a chair of Philosophy at the University of Prague. Bolzano was a man of great human compassion; he spoke out for educational reform, he voiced the right of individual conscience over government demands, and he lectured on the absurdity of war and militarism. His views so disenchanted Emperor Franz I of Austria that the emperor pressed the Archbishop of Prague to have Bolzano recant his statements. Bolzano refused and was then forced to retire in 1824 on a small pension. Bolzano's main contribution to mathematics was philosophical. His work helped convince mathematicians that sound mathematics must ultimately rest on rigorous proof rather than intuition. In addition to his work in mathematics, Bolzano investigated problems concerning space, force, and wave propagation.

German mathematician, Karl Weierstrass[*] produced the first formula for such a function. The graphs of such functions are impossible to draw; it is as if the corners are so numerous that any segment of the curve, when suitably enlarged, reveals more corners. The discovery of these pathological functions was important in that it made mathematicians distrustful of their geometric intuition and more reliant on precise mathematical proof. However, they remained only mathematical curiosities until the early 1980s, when applications of them began to emerge. During the past 10 years they have started to play a fundamental role in the study of geometric objects called **fractals**. Fractals have revealed an order to natural phenomena that were previously dismissed as random and chaotic.

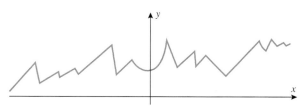

Figure 3.2.13

**DERIVATIVE NOTATION**

The process of finding a derivative is called **differentiation**. You can think of differentiation as an operation on functions that associates a function $f'$ with a function $f$. When the independent variable is $x$, the differentiation operation is often denoted by

$$\frac{d}{dx}[f(x)]$$

which is read "**the derivative of $f(x)$ with respect to $x$.**" Thus,

$$\frac{d}{dx}[f(x)] = f'(x) \tag{7}$$

For example, with this notation the derivatives obtained in Examples 2, 3, and 4 can be expressed as

$$\frac{d}{dx}[x^3 - x] = 3x^2 - 1, \quad \frac{d}{dx}[mx + b] = m, \quad \frac{d}{dx}[\sqrt{x}] = \frac{1}{2\sqrt{x}} \tag{8}$$

To denote the value of the derivative at a specific point $x_0$ with the notation in (7), we would

---

[*] KARL WEIERSTRASS (1815–1897). Weierstrass, the son of a customs officer, was born in Ostenfelde, Germany. As a youth Weierstrass showed outstanding skills in languages and mathematics. However, at the urging of his dominant father, Weierstrass entered the law and commerce program at the University of Bonn. To the chagrin of his family, the rugged and congenial young man concentrated instead on fencing and beer drinking. Four years later he returned home without a degree. In 1839 Weierstrass entered the Academy of Münster to study for a career in secondary education, and he met and studied under an excellent mathematician named Christof Gudermann. Gudermann's ideas greatly influenced the work of Weierstrass. After receiving his teaching certificate, Weierstrass spent the next 15 years in secondary education teaching German, geography, and mathematics. In addition, he taught handwriting to small children. During this period much of Weierstrass's mathematical work was ignored because he was a secondary schoolteacher and not a college professor. Then, in 1854, he published a paper of major importance that created a sensation in the mathematics world and catapulted him to international fame overnight. He was immediately given an honorary Doctorate at the University of Königsberg and began a new career in college teaching at the University of Berlin in 1856. In 1859 the strain of his mathematical research caused a temporary nervous breakdown and led to spells of dizziness that plagued him for the rest of his life. Weierstrass was a brilliant teacher and his classes overflowed with multitudes of auditors. In spite of his fame, he never lost his early beer-drinking congeniality and was always in the company of students, both ordinary and brilliant. Weierstrass was acknowledged as the leading mathematical analyst in the world. He and his students opened the door to the modern school of mathematical analysis.

write

$$\frac{d}{dx}[f(x)]\bigg|_{x=x_0} = f'(x_0) \tag{9}$$

For example, from (8)

$$\frac{d}{dx}[x^3 - x]\bigg|_{x=1} = 3(1^2) - 1 = 2, \quad \frac{d}{dx}[mx + b]\bigg|_{x=5} = m, \quad \frac{d}{dx}[\sqrt{x}]\bigg|_{x=9} = \frac{1}{2\sqrt{9}} = \frac{1}{6}$$

Notations (7) and (9) are convenient when no dependent variable is involved. However, if there is a dependent variable, say $y = f(x)$, then (7) and (9) can be written as

$$\frac{d}{dx}[y] = f'(x) \quad \text{and} \quad \frac{d}{dx}[y]\bigg|_{x=x_0} = f'(x_0)$$

It is common to omit the brackets on the left side and write these expressions as

$$\frac{dy}{dx} = f'(x) \qquad \text{and} \qquad \frac{dy}{dx}\bigg|_{x=x_0} = f'(x_0)$$

where $dy/dx$ is read as "the derivative of $y$ with respect to $x$." For example, if $y = \sqrt{x}$, then

$$\frac{dy}{dx} = \frac{1}{2\sqrt{x}}, \quad \frac{dy}{dx}\bigg|_{x=x_0} = \frac{1}{2\sqrt{x_0}}, \quad \frac{dy}{dx}\bigg|_{x=9} = \frac{1}{2\sqrt{9}} = \frac{1}{6}$$

REMARK. Later, the symbols $dy$ and $dx$ will be defined separately. However, for the time being, $dy/dx$ should not be regarded as a ratio; rather, it should be considered as a single symbol denoting the derivative.

When letters other than $x$ and $y$ are used for the independent and dependent variables, then the various notations for the derivative must be adjusted accordingly. For example, if $y = f(u)$, then the derivative with respect to $u$ would be written as

$$\frac{d}{du}[f(u)] = f'(u) \quad \text{and} \quad \frac{dy}{du} = f'(u)$$

In particular, if $y = \sqrt{u}$, then

$$\frac{dy}{du} = \frac{1}{2\sqrt{u}}, \quad \frac{dy}{du}\bigg|_{u=u_0} = \frac{1}{2\sqrt{u_0}}, \quad \frac{dy}{du}\bigg|_{u=9} = \frac{1}{2\sqrt{9}} = \frac{1}{6}$$

**OTHER NOTATIONS**

Some writers denote the derivative as $D_x[f(x)] = f'(x)$, but we will not use this notation in this text. In problems where the name of the independent variable is clear from the context, there are some other possible notations for the derivative. For example, if $y = f(x)$, but it is clear from the problem that the independent variable is $x$, then the derivative with respect to $x$ might be denoted by $y'$ or $f'$.

Often, you will see Definition 3.2.2 expressed using $\Delta x$ (delta $x$) rather than $h$ for the varying quantity, in which case (5) has the form

$$f'(x) = \lim_{\Delta x \to 0} \frac{f(x + \Delta x) - f(x)}{\Delta x} \tag{10}$$

If $y = f(x)$, then it is also common to let

$$\Delta y = f(x + \Delta x) - f(x)$$

in which case

$$\frac{dy}{dx} = \lim_{\Delta x \to 0} \frac{\Delta y}{\Delta x} = \lim_{\Delta x \to 0} \frac{f(x + \Delta x) - f(x)}{\Delta x} \qquad (11)$$

The geometric interpretations of $\Delta x$ and $\Delta y$ are shown in Figure 3.2.14.

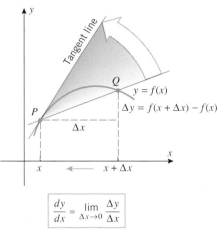

$$\boxed{\frac{dy}{dx} = \lim_{\Delta x \to 0} \frac{\Delta y}{\Delta x}}$$

Figure 3.2.14

**DERIVATIVES AT THE ENDPOINTS OF AN INTERVAL**

If a function $f$ is defined on a closed interval $[a, b]$ and is not defined outside of that interval, then the derivative $f'(x)$ is not defined at the endpoints $a$ and $b$ because

$$f'(x) = \lim_{h \to 0} \frac{f(x + h) - f(x)}{h}$$

is a two-sided limit and only one-sided limits make sense at the endpoints. To deal with this situation, we define **derivatives from the left and right**. These are denoted by $f'_-$ and $f'_+$, respectively, and are defined by

$$f'_-(x) = \lim_{h \to 0^-} \frac{f(x + h) - f(x)}{h} \quad \text{and} \quad f'_+(x) = \lim_{h \to 0^+} \frac{f(x + h) - f(x)}{h}$$

At points where $f'_+(x)$ exists we say that the function $f$ is **differentiable from the right**, and at points where $f'_-(x)$ exists we say that $f$ is **differentiable from the left**. Geometrically, $f'_+(x)$ is the limit of the slopes of the secant lines approaching $x$ from the right, and $f'_-(x)$ is the limit of the slopes of the secant lines approaching $x$ from the left (Figure 3.2.15).

It can be proved that a function $f$ is continuous from the left at those points where it is differentiable from the left and is continuous from the right at those points where it is differentiable from the right.

We will call a function $f$ **differentiable** on an interval of the form $[a, b], [a, +\infty), (-\infty, b]$, $[a, b)$, or $(a, b]$ if it is differentiable at all points inside the interval, and it is differentiable at the endpoint(s) from the left or right, as appropriate.

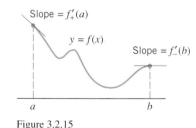

Figure 3.2.15

---

**EXERCISE SET 3.2**   ⌁ Graphing Calculator   ☐c CAS

1. Use the graph of $y = f(x)$ in the accompanying figure to estimate the value of $f'(1)$, $f'(3)$, $f'(5)$, and $f'(6)$.

2. For the function graphed in the accompanying figure, arrange the numbers $0$, $f'(-3)$, $f'(0)$, $f'(2)$, and $f'(4)$ in increasing order.

3. (a) If you are given an equation for the tangent line at the point $(a, f(a))$ on a curve $y = f(x)$, how would you go about finding $f'(a)$?

   (b) Given that the tangent line to the graph of $y = f(x)$ at the point $(2, 5)$ has the equation $y = 3x + 1$, find $f'(2)$.

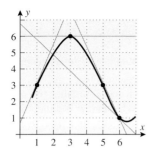

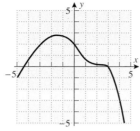

Figure Ex-1  Figure Ex-2

(c) For the equation in part (b), what is the instantaneous rate of change of $y$ with respect to $x$ at $x = 2$?

4. Given that the tangent line to $y = f(x)$ at the point $(-1, 3)$ passes through the point $(0, 4)$, find $f'(-1)$.

5. Sketch the graph of a function $f$ for which $f(0) = 1$, $f'(0) = 0$, $f'(x) > 0$ if $x < 0$, and $f'(x) < 0$ if $x > 0$.

6. Sketch the graph of a function $f$ for which $f(0) = 0$, $f'(0) = 0$, and $f'(x) > 0$ if $x < 0$ or $x > 0$.

7. Given that $f(3) = -1$ and $f'(3) = 5$, find an equation for the tangent line to the graph of $y = f(x)$ at the point where $x = 3$.

8. Given that $f(-2) = 3$ and $f'(-2) = -4$, find an equation for the tangent line to the graph of $y = f(x)$ at the point where $x = -2$.

> In Exercises 9–14, use Definition 3.2.2 to find $f'(x)$, and then find the equation of the tangent line to $y = f(x)$ at the point $x = a$.

9. $f(x) = 3x^2$; $a = 3$

10. $f(x) = x^2 - x$; $a = 2$

11. $f(x) = x^3$; $a = 0$

12. $f(x) = 2x^3 + 1$; $a = -1$

13. $f(x) = \sqrt{x + 1}$; $a = 8$    14. $f(x) = x^4$; $a = -2$

> In Exercises 15–20, use Formula (11) to find $dy/dx$.

15. $y = \dfrac{1}{x}$    16. $y = \dfrac{1}{x^2}$

17. $y = ax^2 + b$ ($a, b$ constants)    18. $y = \dfrac{1}{x + 1}$

19. $y = \dfrac{1}{\sqrt{x}}$    20. $y = x^{1/3}$

> In Exercises 21 and 22, use Definition 3.2.2 (with the appropriate change in notation) to obtain the derivative requested.

21. Find $f'(t)$ if $f(t) = 4t^2 + t$.

22. Find $dV/dr$ if $V = \frac{4}{3}\pi r^3$.

23. Match the graphs of the functions shown in (a)–(f) with the graphs of their derivatives in (A)–(F).

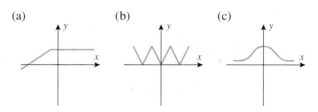

(a)    (b)    (c)

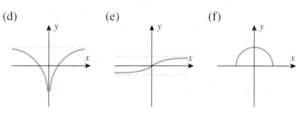

(d)    (e)    (f)

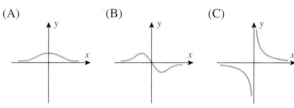

(A)    (B)    (C)

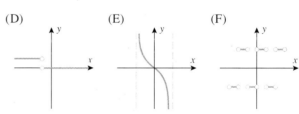

(D)    (E)    (F)

24. Find a function $f$ such that $f'(x) = 1$ for all $x$, and give an informal argument to justify your answer.

> In Exercises 25 and 26, sketch the graph of the derivative of the function whose graph is shown.

25. (a)    (b)    (c)

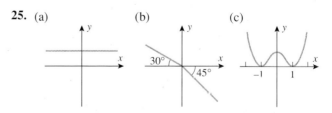

26. (a)    (b)    (c)

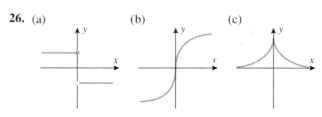

In Exercises 27 and 28, the limit represents $f'(a)$ for some function $f$ and some number $a$. Find $f(x)$ and $a$ in each case.

**27.** (a) $\displaystyle\lim_{h \to 0} \frac{(3+h)^2 - 9}{h}$    (b) $\displaystyle\lim_{\Delta x \to 0} \frac{\sqrt{1 + \Delta x} - 1}{\Delta x}$

**28.** (a) $\displaystyle\lim_{h \to 0} \frac{\cos(\pi + h) + 1}{h}$    (b) $\displaystyle\lim_{x \to 1} \frac{x^7 - 1}{x - 1}$

**29.** Find $dy/dx|_{x=1}$, given that $y = 4x^2 + 1$.

**30.** Find $dy/dx|_{x=-2}$, given that $y = (5/x) + 1$.

**31.** Find an equation for the line that is tangent to the curve $y = x^3 - 2x + 1$ at the point $(0, 1)$, and use a graphing utility to graph the curve and its tangent line on the same screen.

**32.** Use a graphing utility to graph the following on the same screen: the curve $y = x^2/4$, the tangent line to this curve at $x = 1$, and the secant line joining the points $(0, 0)$ and $(2, 1)$ on this curve.

**33.** Let $f(x) = 2^x$. Estimate $f'(1)$ by
(a) using a graphing utility to zoom in at an appropriate point until the graph looks like a straight line, and then estimating the slope
(b) using a calculating utility to estimate the limit in Definition 3.2.2 by making a table of values for a succession of smaller and smaller values of $h$.

**34.** Let $f(x) = \sin x$. Estimate $f'(\pi/4)$ by
(a) using a graphing utility to zoom in at an appropriate point until the graph looks like a straight line, and then estimating the slope
(b) using a calculating utility to estimate the limit in Definition 3.2.2 by making a table of values for a succession of smaller and smaller values of $h$.

**35.** Suppose that the cost of drilling $x$ feet for an oil well is $C = f(x)$ dollars.
(a) What are the units of $f'(x)$?
(b) In practical terms, what does $f'(x)$ mean in this case?
(c) What can you say about the sign of $f'(x)$?
(d) Estimate the cost of drilling an additional foot, starting at a depth of 300 ft, given that $f'(300) = 1000$.

**36.** A paint manufacturing company estimates that it can sell $g = f(p)$ gallons of paint at a price of $p$ dollars.
(a) What are the units of $dg/dp$?
(b) In practical terms, what does $dg/dp$ mean in this case?
(c) What can you say about the sign of $dg/dp$?
(d) Given that $dg/dp|_{p=10} = -100$, what can you say about the effect of increasing the price from $10 per gallon to $11 per gallon?

**37.** It is a fact that when a flexible rope is wrapped around a rough cylinder, a small force of magnitude $F_0$ at one end can resist a large force of magnitude $F$ at the other end. The size of $F$ depends on the angle $\theta$ through which the rope is wrapped around the cylinder (see the accompanying figure). That figure shows the graph of $F$ (in pounds) versus $\theta$ (in

radians), where $F$ is the magnitude of the force that can be resisted by a force with magnitude $F_0 = 10$ lb for a certain rope and cylinder.
(a) Estimate the values of $F$ and $dF/d\theta$ when the angle $\theta = 10$ radians.
(b) It can be shown that the force $F$ satisfies the equation $dF/d\theta = \mu F$, where the constant $\mu$ is called the **coefficient of friction**. Use the results in part (a) to estimate the value of $\mu$.

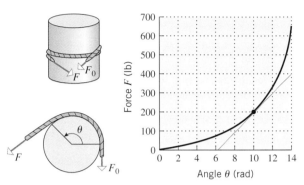

Figure Ex-37

**38.** According to *The World Almanac and the Book of Facts* (1987), the estimated world population, $N$, in millions for the years 1850, 1900, 1950, and 1985 was 1175, 1600, 2490, and 4843, respectively. Although the increase in population is not a continuous function of the time $t$, we can apply the ideas in this section if we are willing to approximate the graph of $N$ versus $t$ by a continuous curve, as shown in the accompanying figure.
(a) Use the estimated tangent line shown in the figure at the point where $t = 1950$ to approximate the value of $dN/dt$ there. Describe your result as a rate of change.
(b) At any instant, the **growth rate** is defined as

$$\frac{dN/dt}{N}$$

Use your answer to part (a) to approximate the growth rate in 1950. Express the result as a percentage and include the proper units.

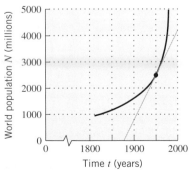

Figure Ex-38

**39.** According to ***Newton's Law of Cooling***, the rate of change of an object's temperature is proportional to the difference between the temperature of the object and that of the surrounding medium. The accompanying figure shows the graph of the temperature $T$ (in degrees Fahrenheit) versus time $t$ (in minutes) for a cup of coffee, initially with a temperature of $200°$F, that is allowed to cool in a room with a constant temperature of $75°$F.

(a) Estimate $T$ and $dT/dt$ when $t = 10$ min.

(b) Newton's Law of Cooling can be expressed as

$$\frac{dT}{dt} = k(T - T_0)$$

where $k$ is the constant of proportionality and $T_0$ is the temperature (assumed constant) of the surrounding medium. Use the results in part (a) to estimate the value of $k$.

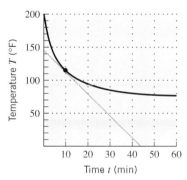

Figure Ex-39

**40.** Write a paragraph that explains what it means for a function to be differentiable. Include some examples of functions that are not differentiable, and explain the relationship between differentiability and continuity.

**41.** Show that $f(x) = \sqrt[3]{x}$ is continuous at $x = 0$ but not differentiable at $x = 0$. Sketch the graph of $f$.

**42.** Show that $f(x) = \sqrt[3]{(x-2)^2}$ is continuous at $x = 2$ but not differentiable at $x = 2$. Sketch the graph of $f$.

**43.** Show that

$$f(x) = \begin{cases} x^2 + 1, & x \le 1 \\ 2x, & x > 1 \end{cases}$$

is continuous and differentiable at $x = 1$. Sketch the graph of $f$.

**44.** Show that

$$f(x) = \begin{cases} x^2 + 2, & x \le 1 \\ x + 2, & x > 1 \end{cases}$$

is continuous but not differentiable at $x = 1$. Sketch the graph of $f$.

**45.** Suppose that a function $f$ is differentiable at $x = 1$ and $\lim_{h \to 0} \dfrac{f(1 + h)}{h} = 5$. Find $f(1)$ and $f'(1)$.

**46.** Suppose that $f$ is a differentiable function with the property that $f(x + y) = f(x) + f(y) + 5xy$ and $\lim_{h \to 0} \dfrac{f(h)}{h} = 3$. Find $f(0)$ and $f'(x)$.

**47.** Suppose that $f$ has the property $f(x + y) = f(x)f(y)$ for all values of $x$ and $y$ and that $f(0) = f'(0) = 1$. Show that $f$ is differentiable and $f'(x) = f(x)$. [*Hint:* Start by expressing $f'(x)$ as a limit.]

## 3.3 TECHNIQUES OF DIFFERENTIATION

*In the last section we defined the derivative of a function $f$ as a limit, and we used that limit to calculate a few simple derivatives. In this section we will develop some important theorems that will enable us to calculate derivatives more efficiently.*

**DERIVATIVE OF A CONSTANT**

The graph of a constant function $f(x) = c$ is the horizontal line $y = c$, and hence the tangent line to this graph has slope 0 at every point $x$ (Figure 3.3.1). Thus, we should expect the derivative of a constant function to be 0 for all $x$.

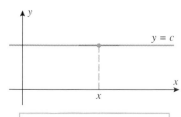

The tangent line to the graph of $f(x) = c$ has slope 0 for all $x$.

Figure 3.3.1

**3.3.1 THEOREM** *The derivative of a constant function is 0; that is, if $c$ is any real number, then*

$$\frac{d}{dx}[c] = 0$$

**Proof.** Let $f(x) = c$. Then from the definition of a derivative,

$$\frac{d}{dx}[c] = f'(x) = \lim_{h \to 0} \frac{f(x+h) - f(x)}{h} = \lim_{h \to 0} \frac{c - c}{h} = \lim_{h \to 0} 0 = 0 \qquad \blacksquare$$

### Example 1

If $f(x) = 5$ for all $x$, then $f'(x) = 0$ for all $x$; that is,

$$\frac{d}{dx}[5] = 0$$

◀

**DERIVATIVE OF $x$ TO A POWER**

3.3.2 THEOREM (*The Power Rule*). *If $n$ is a positive integer, then*

$$\frac{d}{dx}[x^n] = nx^{n-1}$$

*Proof.* Let $f(x) = x^n$. Thus, from the definition of a derivative and the binomial theorem for expanding the expression $(x + h)^n$, we obtain

$$\frac{d}{dx}[x^n] = f'(x) = \lim_{h \to 0} \frac{f(x+h) - f(x)}{h} = \lim_{h \to 0} \frac{(x+h)^n - x^n}{h}$$

$$= \lim_{h \to 0} \frac{\left[ x^n + nx^{n-1}h + \frac{n(n-1)}{2!}x^{n-2}h^2 + \cdots + nxh^{n-1} + h^n \right] - x^n}{h}$$

$$= \lim_{h \to 0} \frac{nx^{n-1}h + \frac{n(n-1)}{2!}x^{n-2}h^2 + \cdots + nxh^{n-1} + h^n}{h}$$

$$= \lim_{h \to 0} \left[ nx^{n-1} + \frac{n(n-1)}{2!}x^{n-2}h + \cdots + nxh^{n-2} + h^{n-1} \right]$$

$$= nx^{n-1} + 0 + \cdots + 0 + 0$$

$$= nx^{n-1}$$

REMARK. In words, *to differentiate $x$ to a positive integer power, multiply that power by $x$ raised to the next lower integer power.*

### Example 2

$$\frac{d}{dx}[x^5] = 5x^4, \quad \frac{d}{dx}[x] = 1 \cdot x^0 = 1, \quad \frac{d}{dx}[x^{12}] = 12x^{11}$$

◀

**DERIVATIVE OF A CONSTANT TIMES A FUNCTION**

3.3.3 THEOREM. *If $f$ is differentiable at $x$ and $c$ is any real number, then $cf$ is also differentiable at $x$ and*

$$\frac{d}{dx}[cf(x)] = c\frac{d}{dx}[f(x)]$$

*Proof.*

$$\frac{d}{dx}[cf(x)] = \lim_{h \to 0} \frac{cf(x+h) - cf(x)}{h} = \lim_{h \to 0} c\left[ \frac{f(x+h) - f(x)}{h} \right]$$

$$= c \lim_{h \to 0} \frac{f(x+h) - f(x)}{h} = c\frac{d}{dx}[f(x)]$$

A constant factor can be moved through a limit sign.

In function notation, Theorem 3.3.3 states

$$(cf)' = cf'$$

REMARK.    In words, *a constant factor can be moved through a derivative sign.*

**Example 3**

$$\frac{d}{dx}[4x^8] = 4\frac{d}{dx}[x^8] = 4[8x^7] = 32x^7$$

$$\frac{d}{dx}[-x^{12}] = (-1)\frac{d}{dx}[x^{12}] = -12x^{11}$$

$$\frac{d}{dx}\left[\frac{x}{\pi}\right] = \frac{1}{\pi}\frac{d}{dx}[x] = \frac{1}{\pi}$$    ◄

**DERIVATIVES OF SUMS AND DIFFERENCES**

**3.3.4**    THEOREM.    *If f and g are differentiable at x, then so are f + g and f − g and*

$$\frac{d}{dx}[f(x) + g(x)] = \frac{d}{dx}[f(x)] + \frac{d}{dx}[g(x)]$$

$$\frac{d}{dx}[f(x) - g(x)] = \frac{d}{dx}[f(x)] - \frac{d}{dx}[g(x)]$$

*Proof.*

$$\frac{d}{dx}[f(x) + g(x)] = \lim_{h \to 0} \frac{[f(x+h) + g(x+h)] - [f(x) + g(x)]}{h}$$

$$= \lim_{h \to 0} \frac{[f(x+h) - f(x)] + [g(x+h) - g(x)]}{h}$$

$$= \lim_{h \to 0} \frac{f(x+h) - f(x)}{h} + \lim_{h \to 0} \frac{g(x+h) - g(x)}{h}$$    The limit of a sum is the sum of the limits.

$$= \frac{d}{dx}[f(x)] + \frac{d}{dx}[g(x)]$$

The proof for $f - g$ is similar.    ■

In function notation, Theorem 3.3.4 states

$$(f + g)' = f' + g' \qquad (f - g)' = f' - g'$$

REMARK.    In words, *the derivative of a sum equals the sum of the derivatives,* and *the derivative of a difference equals the difference of the derivatives.*

**Example 4**

$$\frac{d}{dx}[x^4 + x^2] = \frac{d}{dx}[x^4] + \frac{d}{dx}[x^2] = 4x^3 + 2x$$

$$\frac{d}{dx}[6x^{11} - 9] = \frac{d}{dx}[6x^{11}] - \frac{d}{dx}[9] = 66x^{10} - 0 = 66x^{10}$$

Although Theorem 3.3.4 was stated for sums and differences of two terms, it can be extended to any mixture of finitely many sums and differences of differentiable functions. For example,

$$\frac{d}{dx}[3x^8 - 2x^5 + 6x + 1] = \frac{d}{dx}[3x^8] + \frac{d}{dx}[-2x^5] + \frac{d}{dx}[6x] + \frac{d}{dx}[1]$$

$$= 24x^7 - 10x^4 + 6 \qquad \blacktriangleleft$$

**DERIVATIVE OF A PRODUCT**

**3.3.5** THEOREM (*The Product Rule*). *If f and g are differentiable at x, then so is the product f · g, and*

$$\frac{d}{dx}[f(x)g(x)] = f(x)\frac{d}{dx}[g(x)] + g(x)\frac{d}{dx}[f(x)]$$

*Proof.* The earlier proofs in this section were straightforward applications of the definition of the derivative. However, this proof requires a trick—adding and subtracting the quantity $f(x + h)g(x)$ to the numerator in the derivative definition as follows:

$$\frac{d}{dx}[f(x)g(x)] = \lim_{h \to 0} \frac{f(x + h) \cdot g(x + h) - f(x) \cdot g(x)}{h}$$

$$= \lim_{h \to 0} \frac{f(x + h)g(x + h) - f(x + h)g(x) + f(x + h)g(x) - f(x)g(x)}{h}$$

$$= \lim_{h \to 0} \left[ f(x + h) \cdot \frac{g(x + h) - g(x)}{h} + g(x) \cdot \frac{f(x + h) - f(x)}{h} \right]$$

$$= \lim_{h \to 0} f(x + h) \cdot \lim_{h \to 0} \frac{g(x + h) - g(x)}{h} + \lim_{h \to 0} g(x) \cdot \lim_{h \to 0} \frac{f(x + h) - f(x)}{h}$$

$$= [\lim_{h \to 0} f(x + h)]\frac{d}{dx}[g(x)] + [\lim_{h \to 0} g(x)]\frac{d}{dx}[f(x)]$$

$$= f(x)\frac{d}{dx}[g(x)] + g(x)\frac{d}{dx}[f(x)]$$

[*Note:* In the last step $f(x + h) \to f(x)$ as $h \to 0$ because $f$ is continuous at $x$ by Theorem 3.2.3, and $g(x) \to g(x)$ as $h \to 0$ because $g(x)$ does not involve $h$ and hence remains constant.] ∎

The product rule can be written in function notation as

$$(f \cdot g)' = f \cdot g' + g \cdot f'$$

REMARK. In words, *the derivative of a product of two functions is the first function times the derivative of the second plus the second function times the derivative of the first.*

WARNING. Note that it is *not* true in general that $(f \cdot g)' = f' \cdot g'$; that is, the derivative of a product is *not* generally the product of the derivatives!

### Example 5

Find $dy/dx$ if $y = (4x^2 - 1)(7x^3 + x)$.

*Solution.* There are two methods that can be used to find $dy/dx$. We can either use the product rule or we can multiply out the factors in $y$ and then differentiate. We will give both methods.

**Method I.** (*Using the Product Rule*)

$$\frac{dy}{dx} = \frac{d}{dx}[(4x^2 - 1)(7x^3 + x)]$$

$$= (4x^2 - 1)\frac{d}{dx}[7x^3 + x] + (7x^3 + x)\frac{d}{dx}[4x^2 - 1]$$

$$= (4x^2 - 1)(21x^2 + 1) + (7x^3 + x)(8x) = 140x^4 - 9x^2 - 1$$

**Method II.** (*Multiplying First*)

$$y = (4x^2 - 1)(7x^3 + x) = 28x^5 - 3x^3 - x$$

Thus,

$$\frac{dy}{dx} = \frac{d}{dx}[28x^5 - 3x^3 - x] = 140x^4 - 9x^2 - 1$$

which agrees with the result obtained using the product rule. ◄

**DERIVATIVE OF A QUOTIENT**

**3.3.6** THEOREM (*The Quotient Rule*). *If $f$ and $g$ are differentiable at $x$ and $g(x) \neq 0$, then $f/g$ is differentiable at $x$ and*

$$\frac{d}{dx}\left[\frac{f(x)}{g(x)}\right] = \frac{g(x)\dfrac{d}{dx}[f(x)] - f(x)\dfrac{d}{dx}[g(x)]}{[g(x)]^2}$$

*memorized*

*Proof.*

$$\frac{d}{dx}\left[\frac{f(x)}{g(x)}\right] = \lim_{h \to 0} \frac{\dfrac{f(x+h)}{g(x+h)} - \dfrac{f(x)}{g(x)}}{h} = \lim_{h \to 0} \frac{f(x+h) \cdot g(x) - f(x) \cdot g(x+h)}{h \cdot g(x) \cdot g(x+h)}$$

Adding and subtracting $f(x) \cdot g(x)$ in the numerator yields

$$\frac{d}{dx}\left[\frac{f(x)}{g(x)}\right] = \lim_{h \to 0} \frac{f(x+h) \cdot g(x) - f(x) \cdot g(x) - f(x) \cdot g(x+h) + f(x) \cdot g(x)}{h \cdot g(x) \cdot g(x+h)}$$

$$= \lim_{h \to 0} \frac{\left[g(x) \cdot \dfrac{f(x+h) - f(x)}{h}\right] - \left[f(x) \cdot \dfrac{g(x+h) - g(x)}{h}\right]}{g(x) \cdot g(x+h)}$$

$$= \frac{\displaystyle\lim_{h \to 0} g(x) \cdot \lim_{h \to 0} \frac{f(x+h) - f(x)}{h} - \lim_{h \to 0} f(x) \cdot \lim_{h \to 0} \frac{g(x+h) - g(x)}{h}}{\displaystyle\lim_{h \to 0} g(x) \cdot \lim_{h \to 0} g(x+h)}$$

$$= \frac{[\displaystyle\lim_{h \to 0} g(x)] \cdot \dfrac{d}{dx}[f(x)] - [\displaystyle\lim_{h \to 0} f(x)] \cdot \dfrac{d}{dx}[g(x)]}{\displaystyle\lim_{h \to 0} g(x) \cdot \lim_{h \to 0} g(x+h)}$$

$$= \frac{g(x)\dfrac{d}{dx}[f(x)] - f(x)\dfrac{d}{dx}[g(x)]}{[g(x)]^2}$$

[See the note at the end of the proof of Theorem 3.3.5 for an explanation of the last step.] ▮

The quotient rule can be written in function notation as

*memorized*

$$\left(\frac{f}{g}\right)' = \frac{g \cdot f' - f \cdot g'}{g^2}$$

REMARK.   In words, *the derivative of a quotient of two functions is the denominator times the derivative of the numerator minus the numerator times the derivative of the denominator, all divided by the denominator squared.*

WARNING.   Note that it is *not* generally true that $(f/g)' = f'/g'$; that is, the derivative of a quotient is *not* generally the quotient of the derivatives.

### Example 6

Let $f(x) = \dfrac{x^2 - 1}{x^4 + 1}$.

(a)   Graph $y = f(x)$, and use your graph to make rough estimates of the locations of all horizontal tangent lines.

(b)   By differentiating, find the exact locations of the horizontal tangent lines.

*Solution (a).* Figure 3.3.2 shows the graph of $y = f(x)$ in the window $[-2, 2] \times [-1, 1]$. This graph suggests that horizontal tangent lines occur at $x = 0$, $x \approx 1.5$, and $x \approx -1.5$.

*Solution (b).* To find the exact location of the horizontal tangent lines, we must find the points where $dy/dx = 0$. We start by finding $dy/dx$:

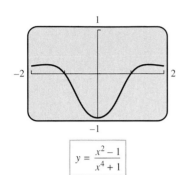

$$y = \frac{x^2 - 1}{x^4 + 1}$$

Figure 3.3.2

$$\frac{dy}{dx} = \frac{d}{dx}\left[\frac{x^2 - 1}{x^4 + 1}\right] = \frac{(x^4 + 1)\dfrac{d}{dx}[x^2 - 1] - (x^2 - 1)\dfrac{d}{dx}[x^4 + 1]}{\left(x^4 + 1\right)^2}$$

$$= \frac{(x^4 + 1)(2x) - (x^2 - 1)(4x^3)}{\left(x^4 + 1\right)^2}$$

> The differentiation is complete. The rest is simplification.

$$= \frac{-2x^5 + 4x^3 + 2x}{\left(x^4 + 1\right)^2} = -\frac{2x(x^4 - 2x^2 - 1)}{\left(x^4 + 1\right)^2}$$

Now we will set $dy/dx = 0$ and solve for $x$. We obtain

$$-\frac{2x(x^4 - 2x^2 - 1)}{\left(x^4 + 1\right)^2} = 0$$

The solutions of this equation are the values of $x$ for which the numerator is 0:

$$2x(x^4 - 2x^2 - 1) = 0$$

The first factor yields the solution $x = 0$. Other solutions can be found by solving the equation

$$x^4 - 2x^2 - 1 = 0$$

This can be treated as a quadratic equation in $x^2$ and solved by the quadratic formula. This yields

$$x^2 = \frac{2 \pm \sqrt{8}}{2} = 1 \pm \sqrt{2}$$

The minus sign yields imaginary values of $x$, which we ignore since they are not relevant to the problem. The plus sign yields the solutions

$$x = \pm\sqrt{1 + \sqrt{2}}$$

In summary, horizontal tangent lines occur at

$$x = 0, \quad x = \sqrt{1 + \sqrt{2}} \approx 1.55, \quad \text{and} \quad x = -\sqrt{1 + \sqrt{2}} \approx -1.55$$

which is consistent with the rough estimates that we obtained graphically in part (a).   ◀

**DERIVATIVE OF A RECIPROCAL**

The special case of Theorem 3.3.6 in which $f$ is the constant function 1 is of interest in its own right. We leave it for the reader to deduce the following result from Theorem 3.3.6.

---

**3.3.7**   THEOREM (*The Reciprocal Rule*).   *If $g$ is differentiable at $x$ and $g(x) \neq 0$, then $1/g$ is differentiable at $x$ and*

$$\frac{d}{dx}\left[\frac{1}{g(x)}\right] = -\frac{\dfrac{d}{dx}[g(x)]}{[g(x)]^2}$$

---

The reciprocal rule can be written in function notation as

$$\left(\frac{1}{g}\right)' = -\frac{g'}{g^2}$$

REMARK.   In words, *the derivative of the reciprocal of a function is the negative of the derivative of the function divided by the function squared.*

**Example 7**

$$\frac{d}{dx}\left[\frac{1}{x}\right] = -\frac{\dfrac{d}{dx}[x]}{x^2} = -\frac{1}{x^2}$$

$$\frac{d}{dx}\left[\frac{1}{x^3 + 2x - 3}\right] = -\frac{\dfrac{d}{dx}[x^3 + 2x - 3]}{\left(x^3 + 2x - 3\right)^2} = -\frac{3x^2 + 2}{\left(x^3 + 2x - 3\right)^2} \qquad \blacktriangleleft$$

REMARK.   The computations in the preceding example could have been done using the quotient rule, but this would have been more work. Where it applies, the reciprocal rule is preferable to the quotient rule.

**THE POWER RULE FOR INTEGER EXPONENTS**

In Theorem 3.3.2 we established the formula

$$\frac{d}{dx}[x^n] = nx^{n-1}$$

for *positive* integer values of $n$. Eventually, we will show that this formula applies if $n$ is any real number. As our first step in this direction we will show that it applies for *all integer* values of $n$.

---

**3.3.8**   THEOREM.   *If $n$ is any integer, then*

$$\frac{d}{dx}[x^n] = nx^{n-1} \qquad\qquad\qquad (1)$$

---

*Proof.*   The result has already been established in the case where $n > 0$. If $n < 0$, then let $m = -n$ so that

$$f(x) = x^{-m} = \frac{1}{x^m}$$

From Theorem 3.3.7,

$$f'(x) = \frac{d}{dx}\left[\frac{1}{x^m}\right] = -\frac{\frac{d}{dx}[x^m]}{(x^m)^2}$$

Since $n < 0$, it follows that $m > 0$, so $x^m$ can be differentiated using Theorem 3.3.2. Thus,

$$f'(x) = -\frac{mx^{m-1}}{x^{2m}} = -mx^{m-1-2m} = -mx^{-m-1} = nx^{n-1}$$

which proves (1). In the case $n = 0$ Formula (1) reduces to

$$\frac{d}{dx}[1] = 0 \cdot x^{-1} = 0$$

which is correct by Theorem 3.3.1.  ∎

## Example 8

$$\frac{d}{dx}[x^{-9}] = -9x^{-9-1} = -9x^{-10}$$

$$\frac{d}{dx}\left[\frac{1}{x}\right] = \frac{d}{dx}[x^{-1}] = (-1)x^{-1-1} = -x^{-2} = -\frac{1}{x^2}$$

Note that the last result agrees with that obtained in Example 7.  ◄

In Example 4 of Section 3.2 we showed that

$$\frac{d}{dx}[\sqrt{x}] = \frac{1}{2\sqrt{x}} \tag{2}$$

which shows that Formula (1) also works with $n = \frac{1}{2}$, since

$$\frac{d}{dx}[x^{1/2}] = \frac{1}{2x^{1/2}} = \frac{1}{2}x^{-1/2}$$

**HIGHER DERIVATIVES**

If the derivative $f'$ of a function $f$ is itself differentiable, then the derivative of $f'$ is denoted by $f''$ and is called the *second derivative* of $f$. As long as we have differentiability, we can continue the process of differentiating derivatives to obtain third, fourth, fifth, and even higher derivatives of $f$. The successive derivatives of $f$ are denoted by

$$f', \quad f'' = (f')', \quad f''' = (f'')', \quad f^{(4)} = (f''')', \quad f^{(5)} = (f^{(4)})', \dots$$

These are called the first derivative, the second derivative, the third derivative, and so forth. Beyond the third derivative, it is too clumsy to continue using primes, so we switch from primes to integers in parentheses to denote the *order* of the derivative. In this notation it is easy to denote a derivative of arbitrary order by writing

$$f^{(n)} \qquad \boxed{\text{The } n\text{th derivative of } f}$$

The significance of the derivatives of order 2 and higher will be discussed later.

## Example 9

If $f(x) = 3x^4 - 2x^3 + x^2 - 4x + 2$, then

$$\begin{aligned}
f'(x) &= 12x^3 - 6x^2 + 2x - 4 \\
f''(x) &= 36x^2 - 12x + 2 \\
f'''(x) &= 72x - 12 \\
f^{(4)}(x) &= 72 \\
f^{(5)}(x) &= 0 \\
&\;\vdots \\
f^{(n)}(x) &= 0 \quad (n \geq 5)
\end{aligned}$$

◄

Successive derivatives can also be denoted as follows:

$$f'(x) = \frac{d}{dx}[f(x)]$$

$$f''(x) = \frac{d}{dx}\left[\frac{d}{dx}[f(x)]\right] = \frac{d^2}{dx^2}[f(x)]$$

$$f'''(x) = \frac{d}{dx}\left[\frac{d^2}{dx^2}[f(x)]\right] = \frac{d^3}{dx^3}[f(x)]$$

$$\vdots \qquad\qquad \vdots$$

In general, we write

$$f^{(n)}(x) = \frac{d^n}{dx^n}[f(x)]$$

which is read "the $n$th derivative of $f$ with respect to $x$."

When a dependent variable is involved, say $y = f(x)$, then successive derivatives can be denoted by writing

$$\frac{dy}{dx}, \quad \frac{d^2y}{dx^2}, \quad \frac{d^3y}{dx^3}, \quad \frac{d^4y}{dx^4}, \quad \ldots, \quad \frac{d^ny}{dx^n}, \quad \ldots$$

or more briefly,

$$y', \quad y'', \quad y''', \quad y^{(4)}, \ldots, y^{(n)}, \ldots$$

**EXERCISE SET 3.3**  $\boxdot$ Graphing Calculator   $\boxed{c}$ CAS

*0 — Mr. Light' class HW.*

In Exercises 1–12, find $dy/dx$.

**1.** $y = 4x^7$      **2.** $y = -3x^{12}$

**3.** $y = 3x^8 + 2x + 1$      **4.** $y = \frac{1}{2}(x^4 + 7)$

**5.** $y = \pi^3$      **6.** $y = \sqrt{2}x + (1/\sqrt{2})$

**7.** $y = -\frac{1}{3}(x^7 + 2x - 9)$      **8.** $y = \dfrac{x^2 + 1}{5}$

**9.** $y = ax^3 + bx^2 + cx + d$    ($a, b, c, d$ constant)

**10.** $y = \dfrac{1}{a}\left(x^2 + \dfrac{1}{b}x + c\right)$    ($a, b, c$ constant)

**11.** $y = -3x^{-8} + 2\sqrt{x}$      **12.** $y = 7x^{-6} - 5\sqrt{x}$

In Exercises 13–20, find $f'(x)$.

**13.** $f(x) = x^{-3} + \dfrac{1}{x^7}$      **14.** $f(x) = \sqrt{x} + \dfrac{1}{x}$

**15.** $f(x) = (3x^2 + 6)\left(2x - \frac{1}{4}\right)$

**16.** $f(x) = (2 - x - 3x^3)(7 + x^5)$

**17.** $f(x) = (x^3 + 7x^2 - 8)(2x^{-3} + x^{-4})$

**18.** $f(x) = \left(\dfrac{1}{x} + \dfrac{1}{x^2}\right)(3x^3 + 27)$

**19.** $f(x) = (3x^2 + 1)^2$      **20.** $f(x) = (x^5 + 2x)^2$

In Exercises 21 and 22, find $y'(1)$.

**21.** $y = \dfrac{1}{5x - 3}$      **22.** $y = \dfrac{3}{\sqrt{x} + 2}$

In Exercises 23 and 24, find $dx/dt$.

**23.** $x = \dfrac{3t}{2t + 1}$      **24.** $x = \dfrac{t^2 + 1}{3t}$

In Exercises 25–28, find $dy/dx|_{x=1}$.

**25.** $y = \dfrac{2x - 1}{x + 3}$      **26.** $y = \dfrac{4x + 1}{x^2 - 5}$

**27.** $y = \left(\dfrac{3x + 2}{x}\right)(x^{-5} + 1)$

**28.** $y = (2x^7 - x^2)\left(\dfrac{x - 1}{x + 1}\right)$

In Exercises 29–32, find the indicated derivative.

**29.** $\dfrac{d}{dt}[16t^2]$      **30.** $\dfrac{dC}{dr}$, where $C = 2\pi r$

**31.** $V'(r)$, where $V = \pi r^3$      **32.** $\dfrac{d}{d\alpha}[2\alpha^{-1} + \alpha]$

**33.** A spherical balloon is being inflated.
(a) Find a general formula for the instantaneous rate of change of the volume $V$ with respect to the radius $r$.
(b) Find the rate of change of $V$ with respect to $r$ at the instant when the radius is $r = 5$.

**c** **34.** Use a CAS to check the answers to the problems you solved in Exercises 1–32.

**35.** Find $g'(4)$ given that $f(4) = 3$ and $f'(4) = -5$.
(a) $g(x) = \sqrt{x} f(x)$      (b) $g(x) = \dfrac{f(x)}{x}$

**36.** Find $g'(3)$ given that $f(3) = -2$ and $f'(3) = 4$.
(a) $g(x) = 3x^2 - 5f(x)$     (b) $g(x) = \dfrac{2x + 1}{f(x)}$

**37.** Find $F'(2)$ given that $f(2) = -1$, $f'(2) = 4$, $g(2) = 1$, and $g'(2) = -5$.
(a) $F(x) = 5f(x) + 2g(x)$   (b) $F(x) = f(x) - 3g(x)$
(c) $F(x) = f(x)g(x)$         (d) $F(x) = f(x)/g(x)$

**38.** Find an equation for the line that is tangent to the curve $y = (1 - x)/(1 + x)$ at the point where $x = 2$.

**39.** Find an equation of the tangent line to the graph of $y = f(x)$ at the point where $x = -3$ if $f(-3) = 2$ and $f'(-3) = 5$.

**40.** Find $\dfrac{d}{d\lambda} \left[ \dfrac{\lambda \lambda_0 + \lambda^6}{2 - \lambda_0} \right]$    ($\lambda_0$ is constant).

In Exercises 41 and 42, find $d^2 y/dx^2$.

**41.** (a) $y = 7x^3 - 5x^2 + x$      (b) $y = 12x^2 - 2x + 3$
(c) $y = \dfrac{x + 1}{x}$            (d) $y = (5x^2 - 3)(7x^3 + x)$

**42.** (a) $y = 4x^7 - 5x^3 + 2x$    (b) $y = 3x + 2$
(c) $y = \dfrac{3x - 2}{5x}$         (d) $y = (x^3 - 5)(2x + 3)$

In Exercises 43 and 44, find $y'''$.

**43.** (a) $y = x^{-5} + x^5$        (b) $y = 1/x$
(c) $y = ax^3 + bx + c$   ($a, b, c$ constant)

**44.** (a) $y = 5x^2 - 4x + 7$     (b) $y = 3x^{-2} + 4x^{-1} + x$
(c) $y = ax^4 + bx^2 + c$   ($a, b, c$ constant)

**45.** Find
(a) $f'''(2)$, where $f(x) = 3x^2 - 2$
(b) $\dfrac{d^2 y}{dx^2} \bigg|_{x=1}$, where $y = 6x^5 - 4x^2$
(c) $\dfrac{d^4}{dx^4} [x^{-3}] \bigg|_{x=1}$

**46.** Find
(a) $y'''(0)$, where $y = 4x^4 + 2x^3 + 3$
(b) $\dfrac{d^4 y}{dx^4} \bigg|_{x=1}$, where $y = \dfrac{6}{x^4}$.

**47.** Show that $y = x^3 + 3x + 1$ satisfies $y''' + xy'' - 2y' = 0$.

**48.** Show that if $x \neq 0$, then $y = 1/x$ satisfies the equation $x^3 y'' + x^2 y' - xy = 0$.

**49.** Find a general formula for $F''(x)$ if $F(x) = xf(x)$ and $f$ and $f'$ are differentiable at $x$.

**c** **50.** Use a CAS to check the answers to the problems you solved in Exercises 41–46.

In Exercises 51 and 52, use a graphing utility to make rough estimates of the locations of all horizontal tangent lines, and then find their exact locations by differentiating.

**51.** $y = \frac{1}{3} x^3 - \frac{3}{2} x^2 + 2x$    **52.** $y = \dfrac{x}{x^2 + 9}$

In Exercises 53 and 54, use Definition 3.2.2 to approximate $f'(1)$ by choosing a small value of $h$ to approximate the limit, and then find the exact value of $f'(1)$ by differentiating.

**53.** $f(x) = x^3 - 3x + 1$      **54.** $f(x) = x\sqrt{x}$

In Exercises 55 and 56, estimate the value of $f'(1)$ by zooming in on the graph of $f$, and then compare your estimate to the exact value obtained by differentiating.

**55.** $f(x) = \dfrac{x}{x^2 + 1}$      **56.** $f(x) = \dfrac{x^2 - 1}{x^2 + 1}$

**57.** Find a function $y = ax^2 + bx + c$ whose graph has an $x$-intercept of 1, a $y$-intercept of $-2$, and a tangent line with a slope of $-1$ at the $y$-intercept.

**58.** Find $k$ if the curve $y = x^2 + k$ is tangent to the line $y = 2x$.

**59.** Find the $x$-coordinate of the point on the graph of $y = x^2$ where the tangent line is parallel to the secant line that cuts the curve at $x = -1$ and $x = 2$.

**60.** Find the $x$-coordinate of the point on the graph of $y = \sqrt{x}$ where the tangent line is parallel to the secant line that cuts the curve at $x = 1$ and $x = 4$.

**61.** Find the coordinate of all points on the graph of $y = 1 - x^2$ at which the tangent line passes through the point $(2, 0)$.

**62.** Show that any two tangent lines to the parabola $y = ax^2$, $a \neq 0$, intersect at a point that is on the vertical line halfway between the points of tangency.

**63.** Suppose that $L$ is the tangent line at $x = x_0$ to the graph of the cubic equation $y = ax^3 + bx$. Find the $x$-coordinate of the point where $L$ intersects the graph a second time.

**64.** Show that the segment of the tangent line to the graph of $y = 1/x$ that is cut off by the coordinate axes is bisected by the point of tangency.

**65.** Show that the triangle that is formed by any tangent line to the graph of $y = 1/x$, $x > 0$, and the coordinate axes has an area of 2 square units.

**66.** Find conditions on $a$, $b$, $c$, and $d$ so that the graph of the polynomial $f(x) = ax^3 + bx^2 + cx + d$ has
(a) exactly two horizontal tangents
(b) exactly one horizontal tangent
(c) no horizontal tangents.

**67.** Newton's Law of Gravitation states that the magnitude $F$ of the force exerted by a point with mass $M$ on a point with mass $m$ is

$$F = \frac{GmM}{r^2}$$

where $G$ is a constant and $r$ is the distance between the bodies. Assuming that the points are moving, find a formula for the instantaneous rate of change of $F$ with respect to $r$.

**68.** In the temperature range between $0°C$ and $700°C$ the resistance $R$ [in ohms $(\Omega)$] of a certain platinum resistance thermometer is given by

$$R = 10 + 0.04124T - 1.779 \times 10^{-5}T^2$$

where $T$ is the temperature in degrees Celsius. Where in the interval from $0°C$ to $700°C$ is the resistance of the thermometer most sensitive and least sensitive to temperature changes? [*Hint:* Consider the size of $dR/dT$ in the interval $0 \le T \le 700$.]

In Exercises 69 and 70, use a graphing utility to make rough estimates of the intervals on which $f'(x) > 0$, and then find those intervals exactly by differentiating.

**69.** $f(x) = x - \dfrac{1}{x}$    **70.** $f(x) = \dfrac{5x}{x^2 + 4}$

**71.** Apply the product rule (3.3.5) twice to show that if $f$, $g$, and $h$ are differentiable functions, then $f \cdot g \cdot h$ is differentiable, and

$$(f \cdot g \cdot h)' = f' \cdot g \cdot h + f \cdot g' \cdot h + f \cdot g \cdot h'$$

**72.** Based on the result in Exercise 71, make a conjecture about a formula for differentiating a product of $n$ functions.

**73.** Use the formula in Exercise 71 to find

(a) $\dfrac{d}{dx}\left[(2x+1)\left(1 + \dfrac{1}{x}\right)(x^{-3} + 7)\right]$

(b) $\dfrac{d}{dx}\left[(x^7 + 2x - 3)^3\right].$

**74.** Use the formula you obtained in Exercise 72 to find

(a) $\dfrac{d}{dx}\left[x^{-5}(x^2 + 2x)(4 - 3x)(2x^9 + 1)\right]$

(b) $\dfrac{d}{dx}\left[(x^2 + 1)^{50}\right].$

In Exercises 75–79, you will have to determine whether a function $f$ is differentiable at a point $x_0$ where the formula for $f$ changes. Use the following result:

***Theorem.*** *Let* $f$ *be continuous at* $x_0$ *and suppose that*

$$\lim_{x \to x_0^+} f'(x) \quad and \quad \lim_{x \to x_0^-} f'(x)$$

*exist. Then* $f$ *is differentiable at* $x_0$ *if and only if these limits are equal. Moreover, in the case of equality*

$$f'(x_0) = \lim_{x \to x_0^+} f'(x) = \lim_{x \to x_0^-} f'(x)$$

**75.** Let

$$f(x) = \begin{cases} x^2, & x \le 1 \\ \sqrt{x}, & x > 1 \end{cases}$$

Determine whether $f$ is differentiable at $x = 1$. If so, find the value of the derivative there.

**76.** Let

$$f(x) = \begin{cases} x^3 + \frac{1}{16}, & x < \frac{1}{2} \\ \frac{3}{4}x^2, & x \ge \frac{1}{2} \end{cases}$$

Determine whether $f$ is differentiable at $x = \frac{1}{2}$. If so, find the value of the derivative there.

**77.** Let

$$f(x) = \begin{cases} 3x^2, & x \le 1 \\ ax + b, & x > 1 \end{cases}$$

Find the values of $a$ and $b$ so that $f$ will be differentiable at $x = 1$.

**78.** (a) Let

$$f(x) = \begin{cases} x^2, & x \le 0 \\ x^2 + 1, & x > 0 \end{cases}$$

Show that

$$\lim_{x \to 0^-} f'(x) = \lim_{x \to 0^+} f'(x)$$

but that $f'(0)$ does not exist.

(b) Let

$$f(x) = \begin{cases} x^2, & x \le 0 \\ x^3, & x > 0 \end{cases}$$

Show that $f'(0)$ exists but $f''(0)$ does not.

**79.** Find all points where $f$ fails to be differentiable. Justify your answer.

(a) $f(x) = |3x - 2|$    (b) $f(x) = |x^2 - 4|$

**80.** In each part compute $f'$, $f''$, $f'''$ and then state the formula for $f^{(n)}$.

(a) $f(x) = 1/x$    (b) $f(x) = 1/x^2$

[*Hint:* The expression $(-1)^n$ has a value of 1 if $n$ is even and $-1$ if $n$ is odd. Use this expression in your answer.]

**81.** (a) Prove:

$$\frac{d^2}{dx^2}[cf(x)] = c\frac{d^2}{dx^2}[f(x)]$$

$$\frac{d^2}{dx^2}[f(x) + g(x)] = \frac{d^2}{dx^2}[f(x)] + \frac{d^2}{dx^2}[g(x)]$$

(b) Do the results in part (a) generalize to $n$th derivatives? Justify your answer.

**82.** Prove:

$$(f \cdot g)'' = f'' \cdot g + 2f' \cdot g' + f \cdot g''$$

**83.** (a) Find $f^{(n)}(x)$ if $f(x) = x^n$.

(b) Find $f^{(n)}(x)$ if $f(x) = x^k$ and $n > k$, where $k$ is a positive integer.

(c) Find $f^{(n)}(x)$ if

$$f(x) = a_0 + a_1 x + a_2 x^2 + \cdots + a_n x^n$$

**84.** Let $f(x) = x^8 - 2x + 3$; find

$$\lim_{h \to 0} \frac{f'(2+h) - f'(2)}{h}$$

**85.** (a) Prove: If $f''(x)$ exists for each $x$ in $(a, b)$, then both $f$ and $f'$ are continuous on $(a, b)$.

(b) What can be said about the continuity of $f$ and its derivatives if $f^{(n)}(x)$ exists for each $x$ in $(a, b)$?

## 3.4 DERIVATIVES OF TRIGONOMETRIC FUNCTIONS

*The main objective of this section is to obtain formulas for the derivatives of trigonometric functions.*

**DERIVATIVES OF THE TRIGONOMETRIC FUNCTIONS**

For the purpose of finding derivatives of the trigonometric functions $\sin x$, $\cos x$, $\tan x$, $\cot x$, $\sec x$, and $\csc x$, we will assume that $x$ is measured in radians. We will also need the following limits, which were stated in Theorem 2.5.3 (with $x$ rather than $h$ as the variable):

$$\lim_{h \to 0} \frac{\sin h}{h} = 1 \quad \text{and} \quad \lim_{h \to 0} \frac{1 - \cos h}{h} = 0$$

We begin with the problem of differentiating $\sin x$. From the definition of a derivative we have

$$\frac{d}{dx}[\sin x] = \lim_{h \to 0} \frac{\sin(x + h) - \sin x}{h}$$

$$= \lim_{h \to 0} \frac{\sin x \cos h + \cos x \sin h - \sin x}{h}$$

$$= \lim_{h \to 0} \left[ \sin x \left( \frac{\cos h - 1}{h} \right) + \cos x \left( \frac{\sin h}{h} \right) \right]$$

$$= \lim_{h \to 0} \left[ \cos x \left( \frac{\sin h}{h} \right) - \sin x \left( \frac{1 - \cos h}{h} \right) \right]$$

Since $\sin x$ and $\cos x$ do not involve $h$, they remain constant as $h \to 0$; thus,

$$\lim_{h \to 0} (\sin x) = \sin x \quad \text{and} \quad \lim_{h \to 0} (\cos x) = \cos x$$

Consequently,

$$\frac{d}{dx}[\sin x] = \cos x \cdot \lim_{h \to 0} \left( \frac{\sin h}{h} \right) - \sin x \cdot \lim_{h \to 0} \left( \frac{1 - \cos h}{h} \right)$$

$$= \cos x \cdot (1) - \sin x \cdot (0) = \cos x$$

Thus, we have shown that

$$\frac{d}{dx}[\sin x] = \cos x \tag{1}$$

The derivative of $\cos x$ can be obtained similarly, resulting in the formula

$$\frac{d}{dx}[\cos x] = -\sin x \tag{2}$$

The derivatives of the remaining trigonometric functions are

$$\frac{d}{dx}[\tan x] = \sec^2 x \qquad \frac{d}{dx}[\sec x] = \sec x \tan x \tag{3-4}$$

$$\frac{d}{dx}[\cot x] = -\csc^2 x \qquad \frac{d}{dx}[\csc x] = -\csc x \cot x \qquad (5\text{–}6)$$

These can all be obtained from (1) and (2) using the relationships

$$\tan x = \frac{\sin x}{\cos x}, \quad \cot x = \frac{\cos x}{\sin x}, \quad \sec x = \frac{1}{\cos x}, \quad \csc x = \frac{1}{\sin x}.$$

For example,

$$\frac{d}{dx}[\tan x] = \frac{d}{dx}\left[\frac{\sin x}{\cos x}\right] = \frac{\cos x \cdot \frac{d}{dx}[\sin x] - \sin x \cdot \frac{d}{dx}[\cos x]}{\cos^2 x}$$

$$= \frac{\cos x \cdot \cos x - \sin x \cdot (-\sin x)}{\cos^2 x} = \frac{\cos^2 x + \sin^2 x}{\cos^2 x} = \frac{1}{\cos^2 x} = \sec^2 x$$

REMARK. The derivative formulas for the trigonometric functions should be memorized. An easy way of doing this is discussed in Exercise 42. Moreover, we emphasize again that in all of the derivative formulas for the trigonometric functions, $x$ is measured in radians.

**Example 1**

Find $f'(x)$ if $f(x) = x^2 \tan x$.

*Solution.* Using the product rule and Formula (3), we obtain

$$f'(x) = x^2 \cdot \frac{d}{dx}[\tan x] + \tan x \cdot \frac{d}{dx}[x^2] = x^2 \sec^2 x + 2x \tan x \qquad \blacktriangleleft$$

**Example 2**

Find $dy/dx$ if $y = \dfrac{\sin x}{1 + \cos x}$.

*Solution.* Using the quotient rule together with Formulas (1) and (2) we obtain

$$\frac{dy}{dx} = \frac{(1 + \cos x) \cdot \frac{d}{dx}[\sin x] - \sin x \cdot \frac{d}{dx}[1 + \cos x]}{(1 + \cos x)^2}$$

$$= \frac{(1 + \cos x)(\cos x) - (\sin x)(-\sin x)}{(1 + \cos x)^2}$$

$$= \frac{\cos x + \cos^2 x + \sin^2 x}{(1 + \cos x)^2} = \frac{\cos x + 1}{(1 + \cos x)^2} = \frac{1}{1 + \cos x} \qquad \blacktriangleleft$$

**Example 3**

Find $y''(\pi/4)$ if $y(x) = \sec x$.

*Solution.*

$$y'(x) = \sec x \tan x$$

$$y''(x) = \sec x \cdot \frac{d}{dx}[\tan x] + \tan x \cdot \frac{d}{dx}[\sec x]$$

$$= \sec x \cdot \sec^2 x + \tan x \cdot \sec x \tan x$$

$$= \sec^3 x + \sec x \tan^2 x$$

Thus,

$$y''(\pi/4) = \sec^3(\pi/4) + \sec(\pi/4) \tan^2(\pi/4)$$

$$= (\sqrt{2})^3 + (\sqrt{2})(1)^2 = 3\sqrt{2} \qquad \blacktriangleleft$$

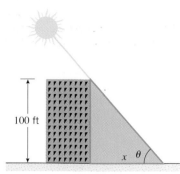

100 ft

$x$ $\theta$

Figure 3.4.1

## Example 4

Suppose that the rising Sun passes directly over a building that is 100 feet high, and let $\theta$ be the Sun's angle of elevation (Figure 3.4.1). Find the rate at which the length $x$ of the building's shadow is changing with respect to $\theta$ when $\theta = 45°$. Express the answer in units of feet/degree.

*Solution.*  The variables $x$ and $\theta$ are related by $\tan\theta = 100/x$, or equivalently,

$$x = 100\cot\theta \tag{7}$$

If $\theta$ is measured in radians, then Formula (5) is applicable, which yields

$$\frac{dx}{d\theta} = -100\csc^2\theta$$

which is the rate of change of shadow length with respect to the elevation angle $\theta$ in units of feet/radian. When $\theta = 45°$ (or equivalently, $\theta = \pi/4$ radians), we obtain

$$\left.\frac{dx}{d\theta}\right|_{\theta=\pi/4} = -100\csc^2(\pi/4) = -200 \text{ feet/radian}$$

Converting radians (rad) to degrees (deg) yields

$$-200\frac{\text{ft}}{\text{rad}} \cdot \frac{\pi}{180}\frac{\text{rad}}{\text{deg}} = -\frac{10}{9}\pi \approx -3.49 \text{ ft/deg}$$

Thus, when $\theta = 45°$, the shadow length is decreasing (because of the minus sign) at an approximate rate of 3.49 ft/deg increase in the angle of elevation.  ◄

## EXERCISE SET 3.4  ⊠ Graphing Calculator  © CAS

In Exercises 1–18, find $f'(x)$.

**1.** $f(x) = 2\cos x - 3\sin x$   **2.** $f(x) = \sin x\cos x$

**3.** $f(x) = \dfrac{\sin x}{x}$   **4.** $f(x) = x^2\cos x$

**5.** $f(x) = x^3\sin x - 5\cos x$   **6.** $f(x) = \dfrac{\cos x}{x\sin x}$

**7.** $f(x) = \sec x - \sqrt{2}\tan x$   **8.** $f(x) = (x^2 + 1)\sec x$

**9.** $f(x) = \sec x\tan x$   **10.** $f(x) = \dfrac{\sec x}{1 + \tan x}$

**11.** $f(x) = \csc x\cot x$

**12.** $f(x) = x - 4\csc x + 2\cot x$

**13.** $f(x) = \dfrac{\cot x}{1 + \csc x}$   **14.** $f(x) = \dfrac{\csc x}{\tan x}$

**15.** $f(x) = \sin^2 x + \cos^2 x$   **16.** $f(x) = \dfrac{1}{\cot x}$

**17.** $f(x) = \dfrac{\sin x\sec x}{1 + x\tan x}$

**18.** $f(x) = \dfrac{(x^2 + 1)\cot x}{3 - \cos x\csc x}$

In Exercises 19–24, find $d^2y/dx^2$.

**19.** $y = x\cos x$   **20.** $y = \csc x$

**21.** $y = x\sin x - 3\cos x$   **22.** $y = x^2\cos x + 4\sin x$

**23.** $y = \sin x\cos x$   **24.** $y = \tan x$

© **25.** Use a CAS to check the answers to the problems you solved in Exercises 1–24.

**26.** Find the equation of the line tangent to the graph of $\sin x$ at the point where
(a) $x = 0$      (b) $x = \pi$      (c) $x = \pi/4$.

**27.** Find the equation of the line tangent to the graph of $\tan x$ at the point where
(a) $x = 0$      (b) $x = \pi/4$      (c) $x = -\pi/4$.

**28.** (a) Show that $y = \cos x$ and $y = \sin x$ are solutions of the equation $y'' + y = 0$.
(b) Show that $y = A\sin x + B\cos x$ is a solution for all constants $A$ and $B$.

**29.** Find all points in the interval $[-2\pi, 2\pi]$ at which the graph of $f$ has a horizontal tangent line.
(a) $f(x) = \sin x$      (b) $f(x) = x + \cos x$
(c) $f(x) = \tan x$      (d) $f(x) = \sec x$

⊠ **30.** (a) Use a graphing utility to make rough estimates of the points in the interval $[0, 2\pi]$ at which the graph of $y = \sin x\cos x$ has a horizontal tangent line.

(b) Find the exact locations of the points where the graph has a horizontal tangent line.

**31.** A 10-ft ladder leans against a wall at an angle $\theta$ with the horizontal, as shown in the accompanying figure. The top of the ladder is $x$ feet above the ground. If the bottom of the ladder is pushed toward the wall, find the rate at which $x$ changes with respect to $\theta$ when $\theta = 60°$. Express the answer in units of feet/degree.

**32.** An airplane is flying on a horizontal path at a height of 3800 ft, as shown in the accompanying figure. At what rate is the distance $s$ between the airplane and the fixed point $P$ changing with respect to $\theta$ when $\theta = 30°$? Express the answer in units of feet/degree.

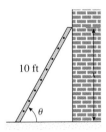

10 ft

Figure Ex-31

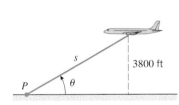

$s$

3800 ft

$P$

Figure Ex-32

**33.** A searchlight is located 50 m from a straight wall, as shown in the accompanying figure. Find the rate at which the distance $D$ is changing with $\theta$ when $\theta = 45°$. Express the answer in units of meters/degree.

**34.** An Earth-observing satellite can see only a portion of the Earth's surface. The satellite has horizon sensors that can detect the angle $\theta$ shown in the accompanying figure. Let $r$ be the radius of the Earth (assumed spherical) and $h$ the distance of the satellite from the Earth's surface.

(a) Show that $h = r(\csc\theta - 1)$.

(b) Using $r = 6378$ km, and assuming that the satellite is getting closer to the Earth, find the rate at which $h$ is changing with respect to $\theta$ when $\theta = 30°$. Express the answer in units of kilometers/degree. [Adapted from *Space Mathematics*, NASA, 1985.]

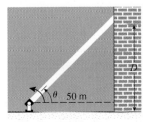

$D$

$\theta$  50 m

Figure Ex-33

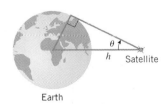

$\theta$

$h$  Satellite

Earth

Figure Ex-34

In Exercises 35 and 36, make a conjecture about the derivative by calculating the first few derivatives and observing the resulting pattern.

**35.** (a) $\dfrac{d^{87}}{dx^{87}}[\sin x]$   (b) $\dfrac{d^{100}}{dx^{100}}[\cos x]$

**36.** $\dfrac{d^{17}}{dx^{17}}[x\sin x]$

**37.** In each part, determine where $f$ is differentiable.

(a) $f(x) = \sin x$   (b) $f(x) = \cos x$

(c) $f(x) = \tan x$   (d) $f(x) = \cot x$

(e) $f(x) = \sec x$   (f) $f(x) = \csc x$

(g) $f(x) = \dfrac{1}{1 + \cos x}$   (h) $f(x) = \dfrac{1}{\sin x \cos x}$

(i) $f(x) = \dfrac{\cos x}{2 - \sin x}$

**38.** (a) Derive Formula (2) using the definition of a derivative.

(b) Use Formulas (1) and (2) to obtain (5).

(c) Use Formula (2) to obtain (4).

(d) Use Formula (1) to obtain (6).

**39.** Let $f(x) = \cos x$. Find all positive integers $n$ for which $f^{(n)}(x) = \sin x$.

**40.** (a) Show that $\lim\limits_{h\to 0} \dfrac{\tan h}{h} = 1$.

(b) Use the result in part (a) to help derive the formula for the derivative of $\tan x$ directly from the definition of a derivative.

**41.** Without using any trigonometric identities, find

$$\lim_{x\to 0} \frac{\tan(x+y) - \tan y}{x}$$

[*Hint:* Relate the given limit to the definition of the derivative of an appropriate function of $y$.]

**42.** Let us agree to call the functions $\cos x$, $\cot x$, and $\csc x$ the *cofunctions* of $\sin x$, $\tan x$, and $\sec x$, respectively. Convince yourself that the derivative of any cofunction can be obtained from the derivative of the corresponding function by introducing a minus sign and replacing each function in the derivative by its cofunction. Memorize the derivatives of $\sin x$, $\tan x$, and $\sec x$ and then use the above observation to deduce the derivatives of the cofunctions.

**43.** The derivative formulas for $\sin x$, $\cos x$, $\tan x$, $\cot x$, $\sec x$, and $\csc x$ were obtained under the assumption that $x$ is measured in radians. This exercise shows that different (more complicated) formulas result if $x$ is measured in degrees. Prove that if $h$ and $x$ are degree measures, then

(a) $\lim\limits_{h\to 0} \dfrac{\cos h - 1}{h} = 0$   (b) $\lim\limits_{h\to 0} \dfrac{\sin h}{h} = \dfrac{\pi}{180}$

(c) $\dfrac{d}{dx}[\sin x] = \dfrac{\pi}{180}\cos x$.

## 3.5 THE CHAIN RULE

*In this section we will derive a formula that expresses the derivative of a composition $f \circ g$ in terms of the derivatives of $f$ and $g$. This formula will enable us to differentiate complicated functions using known derivatives of simpler functions.*

**DERIVATIVES OF COMPOSITIONS**

**3.5.1**   PROBLEM.   *If we know the derivatives of $f$ and $g$, how can we use this information to find the derivative of the composition $f \circ g$?*

The key to solving this problem is to introduce dependent variables

$$y = (f \circ g)(x) = f(g(x)) \quad \text{and} \quad u = g(x)$$

so that $y = f(u)$. We are interested in using the known derivatives

$$\frac{dy}{du} = f'(u) \quad \text{and} \quad \frac{du}{dx} = g'(x)$$

to find the unknown derivative

$$\frac{dy}{dx} = \frac{d}{dx}[f(g(x))]$$

Stated another way, we are interested in using the known rates of change $dy/du$ and $du/dx$ to find the unknown rate of change $dy/dx$. But intuition suggests that rates of change multiply. For example, if $y$ changes at 4 times the rate of change of $u$ and $u$ changes at 2 times the rate of change of $x$, then $y$ changes at $4 \times 2 = 8$ times the rate of change of $x$. Thus, Figure 3.5.1 suggests that

$$\frac{dy}{dx} = \frac{dy}{du} \cdot \frac{du}{dx}$$

These ideas are formalized in the following theorem.

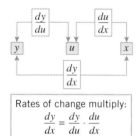

Rates of change multiply:
$$\frac{dy}{dx} = \frac{dy}{du} \cdot \frac{du}{dx}$$

Figure 3.5.1

**3.5.2**   THEOREM (*The Chain Rule*).   *If $g$ is differentiable at the point $x$ and $f$ is differentiable at the point $g(x)$, then the composition $f \circ g$ is differentiable at the point $x$. Moreover, if*

$$y = f(g(x)) \quad \text{and} \quad u = g(x)$$

*then $y = f(u)$ and*

$$\frac{dy}{dx} = \frac{dy}{du} \cdot \frac{du}{dx} \tag{1}$$

The proof of this result is given in Appendix G.

### Example 1

Find $dy/dx$ if $y = 4\cos(x^3)$.

*Solution.*   Let $u = x^3$ so that

$$y = 4\cos u$$

By the chain rule,

$$\frac{dy}{dx} = \frac{dy}{du} \cdot \frac{du}{dx} = \frac{d}{du}[4\cos u] \cdot \frac{d}{dx}[x^3]$$
$$= (-4\sin u) \cdot (3x^2) = (-4\sin(x^3)) \cdot (3x^2) = -12x^2 \sin(x^3) \quad \blacktriangleleft$$

REMARK.  Formula (1) is easy to remember because the left side is exactly what results if we "cancel" the $du$'s on the right side. This "canceling" device provides a good way to remember the chain rule when variables other than $x$, $y$, and $u$ are used.

### Example 2

Find $dw/dt$ if $w = \tan x$ and $x = 4t^3 + t$.

*Solution.*  In this case the chain rule takes the form

$$\frac{dw}{dt} = \frac{dw}{dx} \cdot \frac{dx}{dt} = \frac{d}{dx}[\tan x] \cdot \frac{d}{dt}[4t^3 + t]$$

$$= (\sec^2 x)(12t^2 + 1) = (12t^2 + 1)\sec^2(4t^3 + t) \qquad \blacktriangleleft$$

**GENERALIZED DERIVATIVE FORMULAS**

Although Formula (1) is useful, it is sometimes unwieldy because it involves so many dependent variables. A simpler version of the chain rule can be obtained by noting that $y = f(u)$ in (1), so

$$\frac{dy}{dx} = \frac{d}{dx}[f(u)] \quad \text{and} \quad \frac{dy}{du} = f'(u)$$

Substituting these expressions in (1) yields the following alternative form of the chain rule:

$$\frac{d}{dx}[f(u)] = f'(u)\frac{du}{dx} \qquad (2)$$

This very powerful formula vastly extends our differentiation capabilities. For example, to differentiate the function

$$f(x) = \left(x^2 - x + 1\right)^{23} \qquad (3)$$

we can let $u = x^2 - x + 1$, so (3) becomes $f(u) = u^{23}$, then apply (2) to obtain

$$\frac{d}{dx}\left[(x^2 - x + 1)^{23}\right] = \frac{d}{dx}[u^{23}] = \underbrace{23u^{22}}_{f'(u)}\frac{du}{dx}$$

$$= 23\left(x^2 - x + 1\right)^{22}\frac{d}{dx}[x^2 - x + 1]$$

$$= 23\left(x^2 - x + 1\right)^{22} \cdot (2x - 1)$$

More generally, if $u$ were any other differentiable function of $x$, the pattern of computations would be virtually the same. For example, if $u = \cos x$, then

$$\frac{d}{dx}[\cos^{23} x] = \frac{d}{dx}[u^{23}] = 23u^{22}\frac{du}{dx} = 23\cos^{22} x \frac{d}{dx}[\cos x]$$

$$= 23\cos^{22} x \cdot (-\sin x) = -23\sin x \cos^{22} x$$

In both of the preceding computations, the chain rule took the form

$$\frac{d}{dx}[u^{23}] = 23u^{22}\frac{du}{dx} \qquad (4)$$

This formula is a generalization of the more basic formula

$$\frac{d}{dx}[x^{23}] = 23x^{22} \qquad (5)$$

In fact, in the special case where $u = x$, Formula (4) reduces to (5) since

$$\frac{d}{dx}[u^{23}] = 23u^{22}\frac{du}{dx} = 23x^{22}\frac{d[x]}{dx} = 23x^{22}$$

Table 3.5.1 contains a list of *generalized derivative formulas* that are consequences of (2).

**Table 3.5.1**

GENERALIZED DERIVATIVE FORMULAS

$$\frac{d}{dx}[u^n] = nu^{n-1}\frac{du}{dx} \quad (n \text{ an integer}) \qquad \frac{d}{dx}[\sqrt{u}] = \frac{1}{2\sqrt{u}}\frac{du}{dx}$$

$$\frac{d}{dx}[\sin u] = \cos u\frac{du}{dx} \qquad \frac{d}{dx}[\cos u] = -\sin u\frac{du}{dx}$$

$$\frac{d}{dx}[\tan u] = \sec^2 u\frac{du}{dx} \qquad \frac{d}{dx}[\cot u] = -\csc^2 u\frac{du}{dx}$$

$$\frac{d}{dx}[\sec u] = \sec u \tan u\frac{du}{dx} \qquad \frac{d}{dx}[\csc u] = -\csc u \cot u\frac{du}{dx}$$

## Example 3

Find

(a) $\dfrac{d}{dx}[\sin(2x)]$  (b) $\dfrac{d}{dx}[\tan(x^2 + 1)]$

(c) $\dfrac{d}{dx}[\sqrt{x^3 + \csc x}]$  (d) $\dfrac{d}{dx}\left[(1 + x^5 \cot x)^{-8}\right]$

*Solution (a).* Taking $u = 2x$ in the generalized derivative formula for $\sin u$ yields

$$\frac{d}{dx}[\sin(2x)] = \frac{d}{dx}[\sin u] = \cos u\frac{du}{dx} = \cos 2x \cdot \frac{d}{dx}[2x] = \cos 2x \cdot 2 = 2\cos 2x$$

*Solution (b).* Taking $u = x^2 + 1$ in the generalized derivative formula for $\tan u$ yields

$$\frac{d}{dx}[\tan(x^2 + 1)] = \frac{d}{dx}[\tan u] = \sec^2 u\frac{du}{dx}$$

$$= \sec^2(x^2 + 1) \cdot \frac{d}{dx}[x^2 + 1] = \sec^2(x^2 + 1) \cdot 2x$$

$$= 2x \sec^2(x^2 + 1)$$

*Solution (c).* Taking $u = x^3 + \csc x$ in the generalized derivative formula for $\sqrt{u}$ yields

$$\frac{d}{dx}[\sqrt{x^3 + \csc x}] = \frac{d}{dx}[\sqrt{u}] = \frac{1}{2\sqrt{u}}\frac{du}{dx} = \frac{1}{2\sqrt{x^3 + \csc x}} \cdot \frac{d}{dx}[x^3 + \csc x]$$

$$= \frac{1}{2\sqrt{x^3 + \csc x}} \cdot (3x^2 - \csc x \cot x) = \frac{3x^2 - \csc x \cot x}{2\sqrt{x^3 + \csc x}}$$

*Solution (d).* Taking $u = 1 + x^5 \cot x$ in the generalized derivative formula for $u^{-8}$ yields

$$\frac{d}{dx}\left[(1 + x^5 \cot x)^{-8}\right] = \frac{d}{dx}[u^{-8}] = -8u^{-9}\frac{du}{dx}$$

$$= -8\left(1 + x^5 \cot x\right)^{-9} \cdot \frac{d}{dx}[1 + x^5 \cot x]$$

$$= -8\left(1 + x^5 \cot x\right)^{-9} \cdot (x^5(-\csc^2 x) + 5x^4 \cot x)$$

$$= (8x^5 \csc^2 x - 40x^4 \cot x)\left(1 + x^5 \cot x\right)^{-9} \qquad \blacktriangleleft$$

Sometimes you will have to make adjustments in notation or apply the chain rule more than once to calculate a derivative.

## Example 4

Find

(a) $\dfrac{d}{dx}[\sin(\sqrt{1 + \cos x})]$  (b) $\dfrac{d\mu}{dt}$ if $u = \sec\sqrt{\omega t}$  ($\omega$ constant)

*Solution* (*a*). Taking $u = \sqrt{1 + \cos x}$ in the generalized derivative formula for $\sin u$ yields

$$\frac{d}{dx}[\sin(\sqrt{1 + \cos x})] = \frac{d}{dx}[\sin u] = \cos u \frac{du}{dx}$$

$$= \cos(\sqrt{1 + \cos x}) \cdot \frac{d}{dx}[\sqrt{1 + \cos x}]$$

> We use the generalized derivative formula for $\sqrt{u}$ with $u = 1 + \cos x$.

$$= \cos(\sqrt{1 + \cos x}) \cdot \frac{-\sin x}{2\sqrt{1 + \cos x}}$$

$$= -\frac{\sin x \cos(\sqrt{1 + \cos x})}{2\sqrt{1 + \cos x}}$$

*Solution* (*b*).

$$\frac{d\mu}{dt} = \frac{d}{dt}[\sec \sqrt{\omega t}] = \sec \sqrt{\omega t} \tan \sqrt{\omega t} \frac{d}{dt}[\sqrt{\omega t}]$$

> We used the generalized derivative formula for $\sec u$ with $u = \sqrt{\omega t}$.

$$= \sec \sqrt{\omega t} \tan \sqrt{\omega t} \frac{\omega}{2\sqrt{\omega t}}$$

> We used the generalized derivative formula for $\sqrt{u}$ with $u = \omega t$.

◀

**AN ALTERNATIVE APPROACH TO USING THE CHAIN RULE**

As you become more comfortable with the chain rule, you may want to dispense with actually writing out the expression for $u$ in your computations. To accomplish this, it is helpful to express Formula (2) in words. If we call $u$ the "inside function" and $f$ the "outside function" in the composition $f(u)$, then (2) states:

> *The derivative of $f(u)$ is the derivative of the outside function evaluated at the inside function times the derivative of the inside function.*

For example,

$$\frac{d}{dx}[\cos(x^2 + 9)] = \underbrace{-\sin(x^2 + 9)}_{\substack{\text{Derivative of the} \\ \text{outside evaluated} \\ \text{at the inside}}} \cdot \underbrace{2x}_{\substack{\text{Derivative} \\ \text{of the inside}}}$$

$$\frac{d}{dx}[\tan^2 x] = \frac{d}{dx}\left[(\tan x)^2\right] = \underbrace{(2\tan x)}_{\substack{\text{Derivative of} \\ \text{the outside} \\ \text{evaluated at} \\ \text{the inside}}} \cdot \underbrace{(\sec^2 x)}_{\substack{\text{Derivative} \\ \text{of the inside}}} = 2\tan x \sec^2 x$$

In general, if $f(g(x))$ is a composition of functions in which the inside function $g$ and the outside function $f$ are differentiable, then

$$\frac{d}{dx}[f(g(x))] = \underbrace{f'(g(x))}_{\substack{\text{Derivative of} \\ \text{the outside} \\ \text{evaluated at} \\ \text{the inside}}} \cdot \underbrace{g'(x)}_{\substack{\text{Derivative} \\ \text{of the inside}}}$$

$$(6)$$

**DIFFERENTIATING USING COMPUTER ALGEBRA SYSTEMS**

Although the chain rule makes it possible to differentiate extremely complicated functions, the computations can be time-consuming to execute by hand. For complicated derivatives engineers and scientists often use computer algebra systems such as *Mathematica*, *Maple*, and *Derive*. For example, although we have all of the mathematical tools to perform the differentiation

$$\frac{d}{dx}\left[\frac{(x^2+1)^{10}\sin^3(\sqrt{x})}{\sqrt{1+\csc x}}\right] \tag{7}$$

by hand, the computations are sufficiently tedious that it would be more efficient to use a computer algebra system.

FOR THE READER.    If you have a CAS, use it to obtain the derivatives in Examples 2, 3, and 4, and also to perform the differentiation in (7).

## EXERCISE SET 3.5 ☑ Graphing Calculator  [c] CAS

In Exercises 1–24, find $f'(x)$.

**1.** $f(x) = (x^3 + 2x)^{37}$

**2.** $f(x) = (3x^2 + 2x - 1)^6$

**3.** $f(x) = \left(x^3 - \dfrac{7}{x}\right)^{-2}$

**4.** $f(x) = \dfrac{1}{(x^5 - x + 1)^9}$

**5.** $f(x) = \dfrac{4}{(3x^2 - 2x + 1)^3}$

**6.** $f(x) = \sqrt{x^3 - 2x + 5}$

**7.** $f(x) = \sqrt{4 + \sqrt{3x}}$

**8.** $f(x) = \sin^3 x$

**9.** $f(x) = \sin(x^3)$

**10.** $f(x) = \cos^2(3\sqrt{x})$

**11.** $f(x) = \tan(4x^2)$

**12.** $f(x) = 3\cot^4 x$

**13.** $f(x) = 4\cos^5 x$

**14.** $f(x) = \csc(x^3)$

**15.** $f(x) = \sin\left(\dfrac{1}{x^2}\right)$

**16.** $f(x) = \tan^4(x^3)$

**17.** $f(x) = 2\sec^2(x^7)$

**18.** $f(x) = \cos^3\left(\dfrac{x}{x+1}\right)$

**19.** $f(x) = \sqrt{\cos(5x)}$

**20.** $f(x) = \sqrt{3x - \sin^2(4x)}$

**21.** $f(x) = \left[x + \csc(x^3 + 3)\right]^{-3}$

**22.** $f(x) = \left[x^4 - \sec(4x^2 - 2)\right]^{-4}$

**23.** $f(x) = x^2\sqrt{5 - x^2}$

**24.** $f(x) = \dfrac{x}{\sqrt{1 - x^2}}$

In Exercises 25–39, find $dy/dx$.

**25.** $y = x^3 \sin^2(5x)$

**26.** $y = \sqrt{x}\tan^3(\sqrt{x})$

**27.** $y = x^5 \sec(1/x)$

**28.** $y = \dfrac{\sin x}{\sec(3x + 1)}$

**29.** $y = \cos(\cos x)$

**30.** $y = \sin(\tan 3x)$

**31.** $y = \cos^3(\sin 2x)$

**32.** $y = \dfrac{1 + \csc(x^2)}{1 - \cot(x^2)}$

**33.** $y = (5x + 8)^{13}\left(x^3 + 7x\right)^{12}$

**34.** $y = (2x - 5)^2\left(x^2 + 4\right)^3$

**35.** $y = \left(\dfrac{x - 5}{2x + 1}\right)^3$

**36.** $y = \left(\dfrac{1 + x^2}{1 - x^2}\right)^{17}$

**37.** $y = \dfrac{(2x + 3)^3}{(4x^2 - 1)^8}$

**38.** $y = \left[1 + \sin^3(x^5)\right]^{12}$

**39.** $y = \left[x\sin 2x + \tan^4(x^7)\right]^5$

In Exercises 40–43, find $d^2y/dx^2$.

**40.** $y = \sin(3x^2)$

**41.** $y = x\cos(5x) - \sin^2 x$

**42.** $y = x\tan\left(\dfrac{1}{x}\right)$

**43.** $y = \dfrac{1 + x}{1 - x}$

[c] **44.** Use a CAS to check the answers to the problems you solved in Exercises 1–43.

In Exercises 45–48, find an equation for the tangent line to the graph at the specified point.

**45.** $y = x\cos 3x,\ x = \pi$

**46.** $y = \sin(1 + x^3),\ x = -3$

**47.** $y = \sec^3\left(\dfrac{\pi}{2} - x\right),\ x = -\dfrac{\pi}{2}$

**48.** $y = \left(x - \dfrac{1}{x}\right)^3,\ x = 2$

In Exercises 49–52, find the indicated derivative.

**49.** $y = \cot^3(\pi - \theta)$; find $\dfrac{dy}{d\theta}$.

**50.** $\lambda = \left(\dfrac{au + b}{cu + d}\right)^6$; find $\dfrac{d\lambda}{du}$   $(a, b, c, d$ constants$)$.

**51.** $\dfrac{d}{d\omega}[a\cos^2 \pi\omega + b\sin^2 \pi\omega]$   $(a, b$ constants$)$.

**52.** $x = \csc^2 \left( \dfrac{\pi}{3} - y \right)$; find $\dfrac{dx}{dy}$.

**53.** (a) Use a graphing utility to obtain the graph of the function $f(x) = x\sqrt{4 - x^2}$.
   (b) Use the graph in part (a) to make a rough sketch of the graph of $f'$.
   (c) Find $f'(x)$, and then check your work in part (b) by using the graphing utility to obtain the graph of $f'$.
   (d) Find the equation of the tangent line to the graph of $f$ at $x = 1$, and graph $f$ and the tangent line together.

**54.** (a) Use a graphing utility to obtain the graph of the function $f(x) = \sin x^2 \cos x$ over the interval $[-\pi/2, \pi/2]$.
   (b) Use the graph in part (a) to make a rough sketch of the graph of $f'$ over the interval.
   (c) Find $f'(x)$, and then check your work in part (b) by using the graphing utility to obtain the graph of $f'$ over the interval.
   (d) Find the equation of the tangent line to the graph of $f$ at $x = 1$, and graph $f$ and the tangent line together over the interval.

**55.** If an object suspended from a spring is displaced vertically from its equilibrium position by a small amount and released, and if the air resistance and the mass of the spring are ignored, then the resulting oscillation of the object is called **simple harmonic motion**. Under appropriate conditions the displacement $y$ from equilibrium in terms of time $t$ is given by

$$y = A \cos \omega t$$

where $A$ is the initial displacement at time $t = 0$, and $\omega$ is a constant that depends on the mass of the object and the stiffness of the spring (see the accompanying figure). The constant $|A|$ is called the **amplitude** of the motion and $\omega$ the **angular frequency**.
   (a) Show that
$$\frac{d^2 y}{dt^2} = -\omega^2 y$$
   (b) The **period** $T$ is the time required to make one complete oscillation. Show that $T = 2\pi/\omega$.
   (c) The **frequency** $f$ of the vibration is the number of oscillations per unit time. Find $f$ in terms of the period $T$.
   (d) Find the amplitude, period, and frequency of an object that is executing simple harmonic motion given by $y = 0.6 \cos 15t$, where $t$ is in seconds and $y$ is in centimeters.

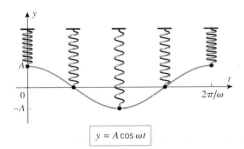

$y = A \cos \omega t$

Figure Ex-55

**56.** Find the value of the constant $A$ so that $y = A \sin 3t$ satisfies the equation

$$\frac{d^2 y}{dt^2} + 2y = 4 \sin 3t$$

**57.** The accompanying figure shows the graph of atmospheric pressure $p$ (lb/in$^2$) versus the altitude $h$ (mi) above sea level.
   (a) From the graph and the tangent line at $h = 2$ shown on the graph, estimate the values of $p$ and $dp/dh$ at an altitude of 2 mi.
   (b) If the altitude of a space vehicle is increasing at the rate of 0.3 mi/s at the instant when it is 2 mi above sea level, how fast is the pressure changing with time at this instant?

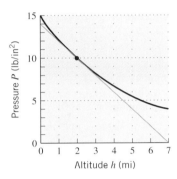

Figure Ex-57

**58.** The force $F$ (in pounds) acting at an angle $\theta$ with the horizontal that is needed to drag a crate weighing $W$ pounds along a horizontal surface at a constant velocity is given by

$$F = \frac{\mu W}{\cos \theta + \mu \sin \theta}$$

where $\mu$ is a constant called the **coefficient of sliding friction** between the crate and the surface (see the accompanying figure). Suppose that the crate weighs 150 lb and that $\mu = 0.3$.
   (a) Find $dF/d\theta$ when $\theta = 30°$. Express the answer in units of pounds/degree.
   (b) Find $dF/dt$ when $\theta = 30°$ if $\theta$ is decreasing at the rate of 0.5°/s at this instant.

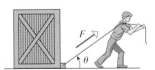

Figure Ex-58

**59.** Recall that

$$\frac{d}{dx}(|x|) = \begin{cases} 1, & x > 0 \\ -1, & x < 0 \end{cases}$$

Use this result and the chain rule to find

$$\frac{d}{dx}(|\sin x|)$$

for nonzero $x$ in the interval $(-\pi, \pi)$.

**60.** Use the derivative formula for $\sin x$ and the identity

$$\cos x = \sin \left( \frac{\pi}{2} - x \right)$$

to obtain the derivative formula for $\cos x$.

**61.** Let

$$f(x) = \begin{cases} x \sin \dfrac{1}{x}, & x \neq 0 \\ 0, & x = 0 \end{cases}$$

(a) Find $f'(x)$ for $x \neq 0$.
(b) Show that $f$ is continuous at $x = 0$.
(c) Use Definition 3.2.2 to show that $f'(0)$ does not exist.

**62.** Let

$$f(x) = \begin{cases} x^2 \sin \dfrac{1}{x}, & x \neq 0 \\ 0, & x = 0 \end{cases}$$

(a) Find $f'(x)$ for $x \neq 0$.
(b) Show that $f$ is continuous at $x = 0$.
(c) Use Definition 3.2.2 to find $f'(0)$.
(d) Show that $f'$ is not continuous at $x = 0$.

**63.** Given the following table of values, find the indicated derivatives in parts (a) and (b).

| $x$ | $f(x)$ | $f'(x)$ |
|---|---|---|
| 2 | 1 | 7 |
| 8 | 5 | -3 |

(a) $g'(2)$, where $g(x) = [f(x)]^3$
(b) $h'(2)$, where $h(x) = f(x^3)$

**64.** Given the following table of values, find the indicated derivatives in parts (a) and (b).

| $x$ | $f(x)$ | $f'(x)$ | $g(x)$ | $g'(x)$ |
|---|---|---|---|---|
| -1 | 2 | 3 | 2 | -3 |
| 2 | 0 | 4 | 1 | -5 |

(a) $F'(-1)$, where $F(x) = f(g(x))$
(b) $G'(-1)$, where $G(x) = g(f(x))$

**65.** Given that $f'(0) = 2$, $g(0) = 0$, and $g'(0) = 3$, find $(f \circ g)'(0)$.

**66.** Given that $f'(x) = \sqrt{3x + 4}$ and $g(x) = x^2 - 1$, find $F'(x)$ if $F(x) = f(g(x))$.

**67.** Given that $f'(x) = \dfrac{x}{x^2 + 1}$ and $g(x) = \sqrt{3x - 1}$, find $F'(x)$ if $F(x) = f(g(x))$.

**68.** Find $f'(x^2)$ if $\dfrac{d}{dx}[f(x^2)] = x^2$.

**69.** Find $\dfrac{d}{dx}[f(x)]$ if $\dfrac{d}{dx}[f(3x)] = 6x$.

**70.** Recall that a function $f$ is *even* if $f(-x) = f(x)$ and *odd* if $f(-x) = -f(x)$, for all $x$ in the domain of $f$. Assuming that $f$ is differentiable, prove:
(a) $f'$ is odd if $f$ is even
(b) $f'$ is even if $f$ is odd.

**71.** Draw some pictures to illustrate the results in Exercise 70, and write a paragraph that gives an informal explanation of why the results are true.

**72.** Let $y = f_1(u)$, $u = f_2(v)$, $v = f_3(w)$, and $w = f_4(x)$. Express $dy/dx$ in terms of $dy/du$, $dw/dx$, $du/dv$, and $dv/dw$.

**73.** Find a formula for

$$\frac{d}{dx}[f(g(h(x)))]$$

## 3.6 LOCAL LINEAR APPROXIMATION; DIFFERENTIALS

*Up to now we have been interpreting $dy/dx$ as a single entity representing the derivative of $y$ with respect to $x$. In this section we will give the quantities $dy$ and $dx$ separate meanings that will allow us to treat $dy/dx$ as a ratio. We will also show how derivatives can be used to approximate functions by simpler linear functions.*

**INCREMENTS**

If the value of a variable changes from one number to another, then the final value minus the initial value is called an ***increment*** in the variable. It is traditional in calculus to denote an increment in a variable $x$ by $\Delta x$ (read "delta $x$"). Thus, if the initial value of $x$ is $x_0$ and the final value is $x_1$, then

$$\Delta x = x_1 - x_0$$

In this notation the expression $\Delta x$ is not the product of $\Delta$ and $x$; rather, it is a single entity representing the *change* in the value of $x$. This notation can be used with any variable; for example, increments in $y$, $t$, and $\theta$ would be denoted as $\Delta y$, $\Delta t$, and $\Delta \theta$.

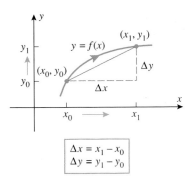

$$\boxed{\begin{array}{l} \Delta x = x_1 - x_0 \\ \Delta y = y_1 - y_0 \end{array}}$$

Figure 3.6.1

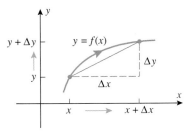

Figure 3.6.2

If $y = f(x)$, and if $x$ changes from an initial value $x_0$ to a final value $x_1$, then there is a corresponding change in the value of $y$ from $y_0 = f(x_0)$ to $y_1 = f(x_1)$. Stated another way, the increment $\Delta x = x_1 - x_0$ in $x$ produces a corresponding increment $\Delta y = y_1 - y_0$ in $y$, where

$$\Delta y = y_1 - y_0 = f(x_1) - f(x_0) \tag{1}$$

(Figure 3.6.1).

Increments can be positive, negative, or zero, depending on the relative positions of the initial and final points—an increment is positive if the final point is to the right of the initial point, negative if the final point is to the left of the initial point, and zero if the initial and final points coincide. In Figure 3.6.1, both $\Delta x$ and $\Delta y$ are positive.

Observe that the expressions $\Delta x = x_1 - x_0$ and $\Delta y = y_1 - y_0$ can be rewritten as

$$x_1 = x_0 + \Delta x \quad \text{and} \quad y_1 = y_0 + \Delta y$$

which simply states that the *final value of a variable is equal to its initial value plus its increment*. With this notation we can express (1) as

$$\Delta y = f(x_0 + \Delta x) - f(x_0) \tag{2}$$

Sometimes, it is convenient to dispense with subscripts on the initial and final values of a variable, in which case the initial and final values of $x$ would be denoted as $x$ and $x + \Delta x$, and the initial and final values of the variable $y$ would be denoted as $y$ and $y + \Delta y$ (Figure 3.6.2). With this notation the symbols $x$ and $y$ play dual roles—they serve as the names as well as the initial values of the variables. However, this rarely causes any confusion.

With the subscripts omitted, Formula (2) becomes

$$\Delta y = f(x + \Delta x) - f(x) \tag{3}$$

The ratio $\Delta y / \Delta x$ can be interpreted as the slope of the secant line joining the points $(x, y)$ and $(x + \Delta x, y + \Delta y)$, and hence the derivative of $y$ with respect to $x$ can be expressed as

$$\frac{dy}{dx} = \lim_{\Delta x \to 0} \frac{\Delta y}{\Delta x} = \lim_{\Delta x \to 0} \frac{f(x + \Delta x) - f(x)}{\Delta x} \tag{4}$$

(Figure 3.6.3). This is consistent with (11) in Section 3.2.

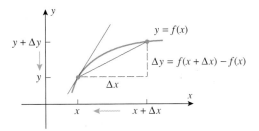

Figure 3.6.3

**DIFFERENTIALS**

When Newton and Leibniz independently published their discoveries of calculus, they each used different notations for the derivative, and battles raged for more than 50 years over which notation was better. In the end the ***Leibniz notation*** $dy/dx$ won out because it produced correct formulas in a natural way; the chain rule

$$\frac{dy}{dx} = \frac{dy}{du} \cdot \frac{du}{dx} ,$$

is a good example.

The symbols "$dy$" and "$dx$" that appear in the derivative $dy/dx$ are called ***differentials***, and our next objective is to define these symbols so that $dy/dx$ can actually be treated as a

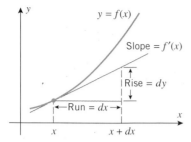

Figure 3.6.4

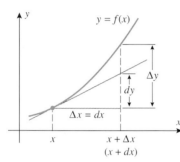

Figure 3.6.5

ratio. For this purpose, regard $x$ as fixed and *define* $dx$ to be an independent variable that can be assigned an arbitrary value. If $f$ is differentiable at $x$, then we *define* $dy$ by the formula

$$dy = f'(x)\, dx \qquad (5)$$

If $dx \neq 0$, then we can divide both sides of (5) by $dx$ to obtain

$$\frac{dy}{dx} = f'(x)$$

Thus, we have achieved our goal of defining $dy$ and $dx$ so that their ratio is $f'(x)$.
Because

$$\frac{dy}{dx} = f'(x) = m_{\text{tan}}$$

where $m_{\text{tan}}$ is the slope of the tangent to $y = f(x)$ at $x$, the differentials $dy$ and $dx$ can be viewed as a corresponding rise and run of this tangent line (Figure 3.6.4).

It is important to understand the distinction between the increment $\Delta y$ and the differential $dy$. To see the difference, let us assign the independent variables $dx$ and $\Delta x$ the same value, so $dx = \Delta x$. Then $\Delta y$ represents the change in $y$ that occurs when we start at $x$ and travel *along the curve* $y = f(x)$ until we have moved $\Delta x \, (= dx)$ units in the $x$-direction, while $dy$ represents the change in $y$ that occurs if we start at $x$ and travel *along the tangent line* until we have moved $dx \, (= \Delta x)$ units in the $x$-direction (Figure 3.6.5).

### Example 1

If $y = x^2$, then the relation $dy/dx = 2x$ can be written in the *differential form*

$$dy = 2x \, dx$$

When $x = 3$, this becomes

$$dy = 6 \, dx$$

This tells us that if we travel along the tangent to the curve $y = x^2$ at $x = 3$, then a change of $dx$ units in $x$ produces a change of $6 \, dx$ units in $y$. For example, if the change in $x$ is $dx = 4$, then the change in $y$ along the tangent is

$$dy = 6(4) = 24 \quad \text{units} \qquad \blacktriangleleft$$

### Example 2

Let $y = \sqrt{x}$. Find $dy$ and $\Delta y$ at $x = 4$ with $dx = \Delta x = 3$. Then make a sketch of $y = \sqrt{x}$, showing $dy$ and $\Delta y$ in the picture.

*Solution.* From (3) with $f(x) = \sqrt{x}$ we obtain

$$\Delta y = \sqrt{x + \Delta x} - \sqrt{x} = \sqrt{7} - \sqrt{4} \approx 0.65$$

If $y = \sqrt{x}$, then

$$\frac{dy}{dx} = \frac{1}{2\sqrt{x}}, \quad \text{so} \quad dy = \frac{1}{2\sqrt{x}} \, dx = \frac{1}{2\sqrt{4}}(3) = \frac{3}{4} = 0.75$$

Figure 3.6.6 shows the curve $y = \sqrt{x}$ together with $dy$ and $\Delta y$. $\qquad \blacktriangleleft$

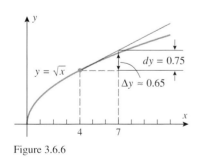

Figure 3.6.6

#### LOCAL LINEAR APPROXIMATION

Points of differentiability for a function $f$ can be described informally in terms of the behavior of the graph of $f$ under magnification: If $P$ is a point of differentiability for a function $f$, then stronger and stronger magnifications at $P$ eventually make the curve segment containing $P$ look more and more like a nonvertical line, the line being the tangent line at $P$. For this reason, a function that is differentiable at a point $P$ is said to be ***locally linear*** at $P$ (Figure 3.6.7).

It follows from the preceding observations that if $f$ is differentiable at $x_0$, then the tangent line through $(x_0, f(x_0))$ closely approximates the graph of $f$ for values of $x$ near

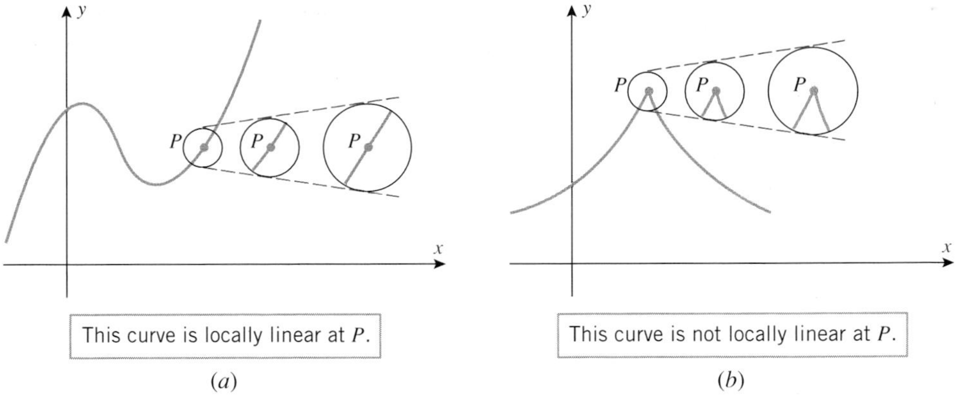

This curve is locally linear at $P$.

(a)

This curve is not locally linear at $P$.

(b)

Figure 3.6.7

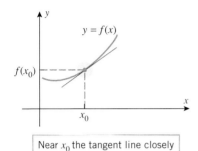

Near $x_0$ the tangent line closely approximates the curve.

Figure 3.6.8

$x_0$ (Figure 3.6.8). To capture this intuitive idea analytically, observe that the tangent line through the point $(x_0, f(x_0))$ has slope $f'(x_0)$, so the point-slope form of its equation is

$$y - f(x_0) = f'(x_0)(x - x_0)$$

which we can rewrite as

$$y = f(x_0) + f'(x_0)(x - x_0)$$

To say that this line closely approximates the curve $y = f(x)$ for values of $x$ near $x_0$, we mean that the approximation

$$f(x) \approx f(x_0) + f'(x_0)(x - x_0) \tag{6}$$

gets better and better as $x \to x_0$. We call (6) the ***local linear approximation of $f$ at $x_0$***. An alternative version of this formula can be obtained by letting $\Delta x = x - x_0$, in which case (6) can be expressed as

$$f(x_0 + \Delta x) \approx f(x_0) + f'(x_0)\,\Delta x \tag{7}$$

### Example 3

(a)   Find the local linear approximation of $f(x) = \sin x$ at $x_0 = 0$.

(b)   Use the local linear approximation obtained in part (a) to approximate $\sin 2°$, and compare your approximation to the result produced directly by your calculating device.

*Solution (a).* Since $f'(x) = \cos x$, it follows from (6) that the local linear approximation of $\sin x$ at a point $x_0$ is

$$\sin x \approx \sin x_0 + (\cos x_0)(x - x_0)$$

Thus, the local linear approximation at $x_0 = 0$ is

$$\sin x \approx \sin 0 + (\cos 0)(x - 0)$$

which simplifies to

$$\sin x \approx x \tag{8}$$

*Solution (b).* In (8), the variable $x$ is in radian measure, so we must first convert $2°$ to radians before we can apply this formula. Since

$$2° = 2(\pi/180) = \pi/90 \text{ radians}$$

it follows from (8) with $x = \pi/90$ that

$$\sin 2° = \sin(\pi/90) \approx \pi/90 \approx 0.0349066$$

This is quite close to the value

$$\sin 2° \approx 0.0348995$$

produced directly on the author's calculator.  ◀

### Example 4

(a) Find the local linear approximation of $f(x) = \sqrt{x}$ at $x_0 = 1$.
(b) Use the local linear approximation obtained in part (a) to approximate $\sqrt{1.1}$, and compare your approximation to the result produced directly by your calculating device.

*Solution* (*a*). Since $f'(x) = 1/(2\sqrt{x})$, it follows from (6) that the local linear approximation of $\sqrt{x}$ at a point $x_0$ is

$$\sqrt{x} \approx \sqrt{x_0} + \frac{1}{2\sqrt{x_0}}(x - x_0)$$

Thus, the local linear approximation at $x_0 = 1$ is

$$\sqrt{x} \approx 1 + \tfrac{1}{2}(x - 1) \tag{9}$$

*Solution* (*b*). Applying Formula (9) with $x = 1.1$ yields

$$\sqrt{1.1} \approx 1 + \tfrac{1}{2}(0.1) = 1.05$$

which compares favorably with the approximation

$$\sqrt{1.1} \approx 1.04881$$

produced directly on the author's calculator.  ◀

REMARK.    In the last two examples we used Formula (6) for the local linear approximation. We could just as well have used Formula (7). For example, with this formula the local linear approximation of $f(x) = \sqrt{x}$ at $x_0$ is

$$\sqrt{x_0 + \Delta x} \approx \sqrt{x_0} + \frac{1}{2\sqrt{x_0}}\Delta x$$

Thus, to approximate $\sqrt{1.1}$ with this formula, we take $x_0 = 1$ and $\Delta x = 0.1$, which yields

$$\sqrt{1.1} \approx 1 + \tfrac{1}{2}(0.1) = 1.05$$

This agrees with the result in Example 4.

**ERROR IN LOCAL LINEAR APPROXIMATIONS**

As a general rule, the accuracy of the local linear approximation to $f(x)$ at a point $x_0$ will deteriorate as $x$ gets progressively farther from $x_0$. To illustrate this for approximation (8) in Example 3, let us graph the function

$$E(x) = |\sin x - x|$$

which is the absolute value of the error in the approximation (Figure 3.6.9).

In Figure 3.6.9, the graph shows how the absolute error in the local linear approximation of $\sin x$ at 0 increases as $x$ moves progressively farther from 0 in either the positive or negative direction. The graph also tells us that for values of $x$ between the two vertical lines the absolute error does not exceed 0.01. Thus, for example, we could use the local linear approximation $\sin x \approx x$ for all values of $x$ in the interval $-0.35 < x < 0.35$ (radians) with confidence that the approximation is within $\pm 0.01$ of the exact value.

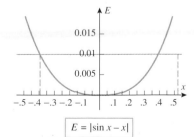

$$E = |\sin x - x|$$

Figure 3.6.9

In applications, small errors invariably occur in measured quantities. When these quantities are used in computations, those errors are propagated in turn to the computed quantities; this is called ***error propagation***. We will now show how to use a local linear approximation to estimate the error in a computed quantity from estimates of the error in the measured quantity. For this purpose, suppose that

$x$ is the quantity being measured

$y = f(x)$ is the quantity being computed

$x_0$ is the true value of $x$

$y_0$ is the true value of $y$

$\Delta x$ is the measurement error in $x$

$\Delta y$ is the propagated error in $y$

Thus, the measured value of $x$ is $x_0 + \Delta x$, and the computed value of $y$ is $y_0 + \Delta y$; and we are interested in using an estimate of $\Delta x$ to find an estimate of $\Delta y$. To do this, we will start with version (7) of the local linear approximation of $f$ at $x_0$:

$$f(x_0 + \Delta x) \approx f(x_0) + f'(x_0)\,\Delta x \tag{10}$$

In this formula, $f(x_0) = y_0$ is the true value of $y$, and $f(x_0 + \Delta x) = y_0 + \Delta y$ is the computed value of $y$, so we can rewrite (10) as

$$y_0 + \Delta y \approx y_0 + f'(x_0)\,\Delta x$$

or

$$\Delta y \approx f'(x_0)\,\Delta x$$

Moreover, if we agree to let $dx = \Delta x$, then we can rewrite this as

$$\Delta y \approx f'(x_0)\,dx = dy \tag{11}$$

which tells us that *the propagated error in $y$ can be estimated by the differential of $y$ at $x_0$ with $dx$ interpreted as the measurement error in $x$.*

Although Formula (11) looks nice on the surface, it is useless in applied problems because the true value $x_0$ is unknown! (Keep in mind that the only value of $x$ that is available to the researcher is the measured value $x_0 + dx = x_0 + \Delta x$.) To get around this roadblock researchers use the observed value $x_0 + dx$ rather than the true value $x_0$ in computing the differential. This is usually satisfactory if $dx$ is small, since $x_0$ and $x_0 + dx$ are close in value. We will illustrate how this works in the next example, but it will simplify our computations if we drop the subscript in (11) and write the formula as

$$\Delta y \approx dy = f'(x)\,dx \tag{12}$$

### Example 5

The radius of a sphere is measured to be 50 cm with a measurement error of $\pm 0.02$ cm. Estimate the error in the computed volume of the sphere.

*Solution.* The volume of a sphere is

$$V = \tfrac{4}{3}\pi r^3$$

We are given that the error in the radius is $\Delta r = \pm 0.02$, and we want to find the error $\Delta V$ in $V$. If we consider $\Delta r$ to be small and if we let $dr = \Delta r$, then $\Delta V$ can be approximated by $dV$. Thus, from (12),

$$\Delta V \approx dV = 4\pi r^2\,dr \tag{13}$$

Substituting $r = 50$ and $dr = \pm 0.02$ in (13), we obtain

$$\Delta V \approx 4\pi(2500)(\pm 0.02) \approx \pm 628.32$$

Therefore, the error in the calculated volume is approximately $\pm 628.32$ cubic centimeters ($cm^3$). ◄

If the true value of a quantity is $q$ and a measurement or calculation produces an error $\Delta q$, then $\Delta q/q$ is called the ***relative error*** in the measurement or calculation; when expressed as a percentage, $\Delta q/q$ is called the ***percentage error***. As a practical matter, the true value $q$ is usually unknown, so that the measured or calculated value of $q$ is used instead; and the relative error is approximated by $dq/q$.

### Example 6

The side of a square is measured with a percentage error of $\pm 5\%$. Estimate the percentage error in the calculated area of the square.

*Solution.* The area $A$ of a square with side $x$ is

$$A = x^2$$

so

$$dA = 2x\, dx$$

We are given that $dx/x = \pm 0.05$, and we want to find $dA/A$. But it follows from the two preceding formulas that

$$\frac{dA}{A} = \frac{2x\, dx}{A} = \frac{2x\, dx}{x^2} = 2\frac{dx}{x} = 2(\pm 0.05) = \pm 0.1 \tag{14}$$

Thus, the percentage error in the calculated area of the square is $\pm 10\%$. ◄

FOR THE READER. In (14) we saw that $dA/A = 2(dx/x)$, which tells us that as a rule of thumb the percentage error in the calculated area of a square is twice the percentage error in the measured side. What rule of thumb relates the percentage error in the computed volume of a cube to the percentage error in the measured side? Why?

**DIFFERENTIAL FORMULAS**

Now that we have defined differentials, every derivative formula has a corresponding differential formula. For example, if $y = \sin x$, then the derivative formula $dy/dx = \cos x$ can also be expressed as

$$dy = \cos x\, dx$$

Moreover, all of the general rules of differentiation have corresponding differential versions:

| DERIVATIVE FORMULA | DIFFERENTIAL FORMULA |
|---|---|
| $\dfrac{d}{dx}[c] = 0$ | $d[c] = 0$ |
| $\dfrac{d}{dx}[cf] = c\dfrac{df}{dx}$ | $d[cf] = c\, df$ |
| $\dfrac{d}{dx}[f + g] = \dfrac{df}{dx} + \dfrac{dg}{dx}$ | $d[f + g] = df + dg$ |
| $\dfrac{d}{dx}[fg] = f\dfrac{dg}{dx} + g\dfrac{df}{dx}$ | $d[fg] = f\, dg + g\, df$ |
| $\dfrac{d}{dx}\left[\dfrac{f}{g}\right] = \dfrac{g\dfrac{df}{dx} - f\dfrac{dg}{dx}}{g^2}$ | $d\left[\dfrac{f}{g}\right] = \dfrac{g\, df - f\, dg}{g^2}$ |

**EXERCISE SET 3.6** ⊠ Graphing Calculator [c] CAS

**1.** (a) Let $y = x^2$. Find $dy$ and $\Delta y$ at $x = 2$ with $dx = \Delta x = 1$.
   (b) Sketch the graph of $y = x^2$, showing $dy$ and $\Delta y$ in the picture.

**2.** (a) Let $y = x^3$. Find $dy$ and $\Delta y$ at $x = 1$ with $dx = \Delta x = 1$.
   (b) Sketch the graph of $y = x^3$, showing $dy$ and $\Delta y$ in the picture.

**3.** (a) Let $y = 1/x$. Find $dy$ and $\Delta y$ at $x = 1$ with $dx = \Delta x = -0.5$.
   (b) Sketch the graph of $y = 1/x$, showing $dy$ and $\Delta y$ in the picture.

**4.** (a) Let $y = \sqrt{x}$. Find $dy$ and $\Delta y$ at $x = 9$ with $dx = \Delta x = -1$.
   (b) Sketch the graph of $y = \sqrt{x}$, showing $dy$ and $\Delta y$ in the picture.

In Exercises 5–8, find formulas for $dy$ and $\Delta y$ at a general point $x$.

**5.** $y = x^3$                           **6.** $y = 8x - 4$

**7.** $y = x^2 - 2x + 1$                  **8.** $y = \sin x$

In Exercises 9–12, find the differential $dy$.

**9.** (a) $y = 4x^3 - 7x^2$               (b) $y = x \cos x$

**10.** (a) $y = 1/x$                      (b) $y = 5 \tan x$

**11.** (a) $y = x\sqrt{1 - x}$            (b) $y = (1 + x)^{-17}$

**12.** (a) $y = \dfrac{1}{x^3 - 1}$       (b) $y = \dfrac{1 - x^3}{2 - x}$

**13.** (a) Use Formula (6) to obtain the local linear approximation of $x^3$ at $x_0 = 1$.
   (b) Use Formula (7) to rewrite the approximation obtained in part (a) in terms of $\Delta x$.
   (c) Use the result obtained in part (a) to approximate $(1.02)^3$, and confirm that the formula obtained in part (b) produces the same result.

**14.** (a) Use Formula (6) to obtain the local linear approximation of $1/x$ at $x_0 = 2$.
   (b) Use Formula (7) to rewrite the approximation obtained in part (a) in terms of $\Delta x$.
   (c) Use the result obtained in part (a) to approximate $1/2.05$, and confirm that the formula obtained in part (b) produces the same result.

**15.** (a) Find the local linear approximation of $f(x) = \sqrt{1 + x}$ at $x_0 = 0$, and use it to approximate $\sqrt{0.9}$ and $\sqrt{1.1}$.
   (b) Graph $f$ and its tangent line at $x_0$ together, and use the graphs to illustrate the relationship between the exact values and the approximations of $\sqrt{0.9}$ and $\sqrt{1.1}$.

**16.** (a) Find the local linear approximation of $f(x) = 1/\sqrt{x}$ at $x_0 = 4$, and use it to approximate $1/\sqrt{3.9}$ and $1/\sqrt{4.1}$.

   (b) Graph $f$ and its tangent line at $x_0$ together, and use the graphs to illustrate the relationship between the exact values and the approximations of $1/\sqrt{3.9}$ and $1/\sqrt{4.1}$.

In Exercises 17–20, confirm that the stated formula is the local linear approximation at $x_0 = 0$.

**17.** $(1 + x)^{15} \approx 1 + 15x$     **18.** $\dfrac{1}{\sqrt{1 - x}} \approx 1 + \tfrac{1}{2}x$

**19.** $\tan x \approx x$                 **20.** $\dfrac{1}{1 + x} \approx 1 - x$

In Exercises 21–24, confirm that the stated formula is the local linear approximation of $f$ at $x_0 = 1$, where $\Delta x = x - 1$.

**21.** $f(x) = x^4$; $(1 + \Delta x)^4 \approx 1 + 4x^3 \Delta x$

**22.** $f(x) = \sqrt{x}$; $\sqrt{1 + \Delta x} \approx 1 + \tfrac{1}{2}\Delta x$

**23.** $f(x) = \dfrac{1}{2 + x}$; $\dfrac{1}{3 + \Delta x} \approx \dfrac{1}{3} - \dfrac{1}{9}\Delta x$

**24.** $f(x) = (4 + x)^3$; $(5 + \Delta x)^3 \approx 125 + 75\Delta x$

**25.** (a) Use the local linear approximation of $\sin x$ at $x_0 = 0$ obtained in Example 3 to approximate $\sin 1°$, and compare the approximation to the result produced directly by your calculating device.
   (b) How would you choose $x_0$ to approximate $\sin 44°$?
   (c) Approximate $\sin 44°$; compare the approximation to the result produced directly by your calculating device.

**26.** (a) Use the local linear approximation of $\tan x$ at $x_0 = 0$ to approximate $\tan 2°$, and compare the approximation to the result produced directly by your calculating device.
   (b) How would you choose $x_0$ to approximate $\tan 61°$?
   (c) Approximate $\tan 61°$; compare the approximation to the result produced directly by your calculating device.

In Exercises 27–35, use an appropriate local linear approximation to estimate the value of the given quantity.

**27.** $(3.02)^4$        **28.** $(1.97)^3$        **29.** $\sqrt{65}$

**30.** $\sqrt{24}$       **31.** $\sqrt{80.9}$      **32.** $\sqrt{36.03}$

**33.** $\sin 0.1$        **34.** $\tan 0.2$         **35.** $\cos 31°$

**36.** The approximation $(1 + x)^k \approx 1 + kx$ is commonly used by engineers for quick calculations.
   (a) Derive this result, and use it to make a rough estimate of $(1.001)^{37}$.
   (b) Compare your estimate to that produced directly by your calculating device.
   (c) Show that this formula produces a very bad estimate of $(1.1)^{37}$, and explain why.

In Exercises 37–40, confirm that the formula is a local linear approximation at $x_0 = 0$, and use a graphing utility to estimate an interval of $x$-values on which the error in the approximation is at most $\pm 0.1$.

**37.** $\sqrt{x+3} \approx \sqrt{3} + \dfrac{1}{2\sqrt{3}}x$

**38.** $\dfrac{1}{\sqrt{9-x}} \approx \dfrac{1}{3} + \dfrac{1}{54}x$

**39.** $\tan x \approx x$

**40.** $\dfrac{1}{(1+2x)^5} \approx 1 - 10x$

In Exercises 41–44, use $dy$ to approximate $\Delta y$ when $x$ changes as indicated.

**41.** $y = \sqrt{3x - 2}$; from $x = 2$ to $x = 2.03$

**42.** $y = \sqrt{x^2 + 8}$; from $x = 1$ to $x = 0.97$

**43.** $y = \dfrac{x}{x^2 + 1}$; from $x = 2$ to $x = 1.96$

**44.** $y = x\sqrt{8x + 1}$; from $x = 3$ to $x = 3.05$

**45.** The side of a square is measured to be 10 ft, with a possible error of $\pm 0.1$ ft.
  (a) Use differentials to estimate the error in the calculated area.
  (b) Estimate the percentage errors in the side and the area.

**46.** The side of a cube is measured to be 25 cm, with a possible error of $\pm 1$ cm.
  (a) Use differentials to estimate the error in the calculated volume.
  (b) Estimate the percentage errors in the side and volume.

**47.** The hypotenuse of a right triangle is known to be 10 in exactly, and one of the acute angles is measured to be $30°$, with a possible error of $\pm 1°$.
  (a) Use differentials to estimate the errors in the sides opposite and adjacent to the measured angle.
  (b) Estimate the percentage errors in the sides.

**48.** One side of a right triangle is known to be 25 cm exactly. The angle opposite to this side is measured to be $60°$, with a possible error of $\pm 0.5°$.
  (a) Use differentials to estimate the errors in the adjacent side and the hypotenuse.
  (b) Estimate the percentage errors in the adjacent side and hypotenuse.

**49.** The electrical resistance $R$ of a certain wire is given by $R = k/r^2$, where $k$ is a constant and $r$ is the radius of the wire. Assuming that the radius $r$ has a possible error of $\pm 5\%$, use differentials to estimate the percentage error in $R$. (Assume $k$ is exact.)

**50.** A 12-foot ladder leaning against a wall makes an angle $\theta$ with the floor. If the top of the ladder is $h$ feet up the wall, express $h$ in terms of $\theta$ and then use $dh$ to estimate the change in $h$ if $\theta$ changes from $60°$ to $59°$.

**51.** The area of a right triangle with a hypotenuse of $H$ is calculated using the formula $A = \frac{1}{4}H^2 \sin 2\theta$, where $\theta$ is one of the acute angles. Use differentials to approximate the error in calculating $A$ if $H = 4$ cm (exactly) and $\theta = 30° \pm 15'$.

**52.** The side of a square is measured with a possible percentage error of $\pm 1\%$. Use differentials to estimate the percentage error in the area.

**53.** The side of a cube is measured with a possible percentage error of $\pm 2\%$. Use differentials to estimate the percentage error in the volume.

**54.** The volume of a sphere is to be computed from a measured value of its radius. Estimate the maximum permissible percentage error in the measurement if the percentage error in the volume must be kept within $\pm 3\%$. ($V = \frac{4}{3}\pi r^3$ is the volume of a sphere of radius $r$.)

**55.** The area of a circle is to be computed from a measured value of its diameter. Estimate the maximum permissible percentage error in the measurement if the percentage error in the area must be kept within $\pm 1\%$.

**56.** A steel cube with 1-in sides is coated with 0.01 in of copper. Use differentials to estimate the volume of copper in the coating. [*Hint:* Let $\Delta V$ be the change in the volume of the cube.]

**57.** A metal rod 15 cm long and 5 cm in diameter is to be covered (except for the ends) with insulation that is 0.001 cm thick. Use differentials to estimate the volume of insulation. [*Hint:* Let $\Delta V$ be the change in volume of the rod.]

**58.** The time required for one complete oscillation of a pendulum is called its ***period***. If the length $L$ of the pendulum is measured in feet and the period $P$ in seconds, then the period is given by $P = 2\pi\sqrt{L/g}$, where $g$ is a constant called *the acceleration due to gravity*. Use differentials to show that the percentage error in $P$ is approximately half the percentage error in $L$.

**59.** If the temperature $T$ of a metal rod of length $L$ is changed by an amount $\Delta T$, then the length will change by the amount $\Delta L = \alpha L \, \Delta T$, where $\alpha$ is called the ***coefficient of linear expansion***. For moderate changes in temperature $\alpha$ is taken as constant.
  (a) Suppose that a rod 40 cm long at $20°$C is found to be 40.006 cm long when the temperature is raised to $30°$C. Find $\alpha$.
  (b) If an aluminum pole is 180 cm long at $15°$C, how long is the pole if the temperature is raised to $40°$C? [Take $\alpha = 2.3 \times 10^{-5}/°$C.]

**60.** If the temperature $T$ of a solid or liquid of volume $V$ is changed by an amount $\Delta T$, then the volume will change by the amount $\Delta V = \beta V \, \Delta T$, where $\beta$ is called the ***coefficient of volume expansion***. For moderate changes in temperature $\beta$ is taken as constant. Suppose that a tank truck loads 4000 gallons of ethyl alcohol at a temperature of $35°$C and delivers its load sometime later at a temperature of $15°$C. Using $\beta = 7.5 \times 10^{-4}/°$C for ethyl alcohol, find the number of gallons delivered.

# SUPPLEMENTARY EXERCISES

1. State the definition of a derivative, and give two interpretations of it.

2. Explain the difference between average and instantaneous rate of change, and discuss how they are calculated.

3. Given that $y = f(x)$, explain the difference between $dy$ and $\Delta y$. Draw a picture that illustrates the relationship between these quantities.

4. Use the definition of a derivative to find $dy/dx$, and check your answer by calculating the derivative using appropriate derivative formulas.
   (a) $y = \sqrt{9 - 4x}$
   (b) $y = \dfrac{x}{x + 1}$

In Exercises 5–8, find the values of $x$ at which the curve $y = f(x)$ has a horizontal tangent line.

5. $f(x) = (2x + 7)^6(x - 2)^5$

6. $f(x) = \dfrac{(x - 3)^4}{x^2 + 2x}$

7. $f(x) = \sqrt{3x + 1}(x - 1)^2$

8. $f(x) = \left(\dfrac{3x + 1}{x^2}\right)^3$

9. The accompanying figure shows the graph of $y = f'(x)$ for an unspecified function $f$.
   (a) For what values of $x$ does the curve $y = f(x)$ have a horizontal tangent line?
   (b) Over what intervals does the curve $y = f(x)$ have tangent lines with positive slope?
   (c) Over what intervals does the curve $y = f(x)$ have tangent lines with negative slope?
   (d) Given that $g(x) = f(x) \sin x$, and $f(0) = -1$, find $g''(0)$.

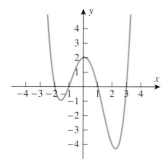

Figure Ex-9

10. In each part, evaluate the expression given that $f(1) = 1$, $g(1) = -2$, $f'(1) = 3$, and $g'(1) = -1$.
    (a) $\dfrac{d}{dx}[f(x)g(x)]\Big|_{x=1}$
    (b) $\dfrac{d}{dx}\left[\dfrac{f(x)}{g(x)}\right]\Big|_{x=1}$
    (c) $\dfrac{d}{dx}[\sqrt{f(x)}]\Big|_{x=1}$
    (d) $\dfrac{d}{dx}[f(1)g'(1)]$

11. Find the equations of all lines through the origin that are tangent to the curve $y = x^3 - 9x^2 - 16x$.

12. Find all values of $x$ for which the tangent line to $y = 2x^3 - x^2$ is perpendicular to the line $x + 4y = 10$.

13. Find all values of $x$ for which the line that is tangent to $y = 3x - \tan x$ is parallel to the line $y - x = 2$.

14. Suppose that $f(x) = \begin{cases} x^2 - 1, & x \le 1 \\ k(x - 1), & x > 1. \end{cases}$
    For what values of $k$ is $f$
    (a) continuous
    (b) differentiable?

15. Let $f(x) = x^2$. Show that for any distinct values of $a$ and $b$, the slope of the tangent line to $y = f(x)$ at $x = \frac{1}{2}(a + b)$ is equal to the slope of the secant line through the points $(a, a^2)$ and $(b, b^2)$. Draw a picture to illustrate this result.

16. A car is traveling on a straight road that is 120 mi long. For the first 100 mi the car travels at an average velocity of 50 mi/h. Show that no matter how fast the car travels for the final 20 mi it cannot bring the average velocity up to 60 mi/h for the entire trip.

17. In each part, use the given information to find $\Delta x$, $\Delta y$, and $dy$.
    (a) $y = 1/(x - 1)$; $x$ decreases from 2 to 1.5.
    (b) $y = \tan x$; $x$ increases from $-\pi/4$ to 0.
    (c) $y = \sqrt{25 - x^2}$; $x$ increases from 0 to 3.

18. Use the formula $V = l^3$ for the volume of a cube of side $l$ to find
    (a) the average rate at which the volume of a cube changes with $l$ as $l$ increases from $l = 2$ to $l = 4$
    (b) the instantaneous rate at which the volume of a cube changes with $l$ when $l = 5$.

19. The amount of water in a tank $t$ minutes after it has started to drain is given by $W = 100(t - 15)^2$ gal.
    (a) At what rate is the water running out at the end of 5 min?
    (b) What is the average rate at which the water flows out during the first 5 min?

20. Use an appropriate local linear approximation to estimate the value of $\cot 46°$, and compare your answer to the value obtained with a calculating device.

21. The base of the Great Pyramid at Giza is a square that is 230 m on each side.
    (a) As illustrated in the accompanying figure, suppose that an archaeologist standing at the center of a side measures the angle of elevation of the apex to be $\phi = 51°$ with an error of $\pm 0.5°$. What can the archaeologist reasonably say about the height of the pyramid?
    (b) Use differentials to estimate the allowable error in the elevation angle that will ensure an error in the height is at most $\pm 5$ m.

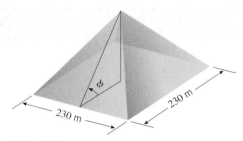

Figure Ex-21

**22.** The period $T$ of a clock pendulum (i.e., the time required for one back-and-forth movement) is given in terms of its length $L$ by $T = 2\pi\sqrt{L/g}$, where $g$ is the gravitational constant.
(a) Assuming that the length of a clock pendulum can vary (say, due to temperature changes), find the rate of change of the period $T$ with respect to the length $L$.
(b) If $L$ is in meters (m) and $T$ is in seconds (s), what are the units for the rate of change in part (a)?
(c) If a pendulum clock is running slow, should the length of the pendulum be increased or decreased to correct the problem?
(d) The constant $g$ generally decreases with altitude. If you move a pendulum clock from sea level to a higher elevation, will it run faster or slower?
(e) Assuming the length of the pendulum to be constant, find the rate of change of the period $T$ with respect to $g$.
(f) Assuming that $T$ is in seconds (s) and $g$ is in meters per second per second (m/s²), find the units for the rate of change in part (e).

In Exercises 23 and 24, zoom in on the graph of $f$ on an interval containing $x = x_0$ until the graph looks like a straight line. Estimate the slope of this line and then check your answer by finding the exact value of $f'(x_0)$.

**23.** (a) $f(x) = x^2 - 1$, $x_0 = 1.8$
(b) $f(x) = \dfrac{x^2}{x-2}$, $x_0 = 3.5$

**24.** (a) $f(x) = x^3 - x^2 + 1$, $x_0 = 2.3$
(b) $f(x) = \dfrac{x}{x^2+1}$, $x_0 = -0.5$

In Exercises 25 and 26, approximate $f'(2)$ by using the limit in Definition 3.2.2 with small values of $h$. If you have a CAS, see if it can find the exact value of the limit.

**25.** $f(x) = 2^x$

**26.** $f(x) = x^{\sin x}$

**27.** At time $t = 0$ a car moves into the passing lane to pass a slow-moving truck. The average velocity of the car from $t = 1$ to $t = 1+h$ is

$$v_{\text{ave}} = \frac{3(h+1)^{2.5} + 580h - 3}{10h}$$

Estimate the instantaneous velocity of the car at $t = 1$, where time is in seconds and distance is in feet.

**28.** A sky diver jumps from an airplane. Suppose that the distance she falls during the period before her parachute opens is $s(t) = 986((0.835)^t - 1) + 176t$, where $s$ is in feet, $t$ is in seconds, and $t \geq 1$. Graph $s$ versus $t$ for $1 \leq t \leq 20$, and use your graph to estimate the instantaneous velocity at $t = 15$.

**29.** Approximate the values of $x$ at which the tangent line to the graph of $y = x^3 - \sin x$ is horizontal.

**30.** Use a graphing utility to graph the function

$$f(x) = |x^4 - x - 1| - x$$

and find the values of $x$ where the derivative of this function does not exist.

**31.** Use a CAS to find the derivative of $f$ from the definition

$$f'(x) = \lim_{h \to 0} \frac{f(x+h) - f(x)}{h}$$

and check the result by finding the derivative by hand.
(a) $f(x) = x^5$
(b) $f(x) = 1/x$
(c) $f(x) = 1/\sqrt{x}$
(d) $f(x) = \dfrac{2x+1}{x-1}$
(e) $f(x) = \sqrt{3x^2+5}$
(f) $f(x) = \sin 3x$

In Exercises 32–37: (a) use a CAS to find $f'(x)$, and check the result by hand; (b) use the CAS to find $f''(x)$.

**32.** $f(x) = x^2 \sin x$

**33.** $f(x) = \sqrt{x} + \cos^2 x$

**34.** $f(x) = \dfrac{2x^2 - x + 5}{3x+2}$

**35.** $f(x) = \dfrac{\tan x}{1+x^2}$

**36.** $f(x) = \dfrac{1}{x} \sin \sqrt{x}$

**37.** $f(x) = \dfrac{\sqrt{x^4 - 3x + 2}}{x(2 - \cos x)}$

## EXPANDING THE CALCULUS HORIZON

# Robotics

*Robin designs and sells room dividers to defray college expenses. She is soon overwhelmed with orders and decides to build a robot to spray paint her dividers. As in most engineering projects, Robin begins with a simplified model that she will eventually refine to be more realistic. However, Robin quickly discovers that robotics (the design and control of robots) involves a considerable amount of mathematics, some of which we will discuss in this module.*

### The Design Plan

Robin's plan is to develop a two-dimensional version of the robot arm in Figure 1. As shown in Figure 2, Robin's robot arm will consist of two links of fixed length, each of which will rotate independently about a pivot point. A paint sprayer will be attached to the end of the second link, and a computer will vary the angles $\theta_1$ and $\theta_2$, thereby allowing the robot to paint a region of the $xy$-plane.

### The Mathematical Analysis

To analyze the motion of the robot arm, Robin denotes the coordinates of the paint sprayer by $(x, y)$, as in Figure 3, and she derives the following equations that express $x$ and $y$ in terms of the angles $\theta_1$ and $\theta_2$ and the lengths $l_1$ and $l_2$ of the links:

$$x = l_1 \cos \theta_1 + l_2 \cos(\theta_1 + \theta_2)$$
$$y = l_1 \sin \theta_1 + l_2 \sin(\theta_1 + \theta_2)$$

(1)

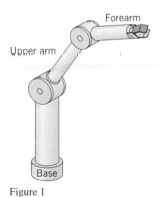

Figure 1

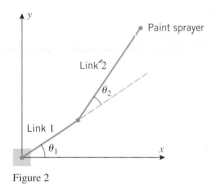

Figure 2

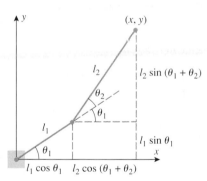

Figure 3

**Exercise 1**   Use Figure 3 to confirm the equations in (1).

In the language of robotics, $\theta_1$ and $\theta_2$ are called the ***control angles***, the point $(x, y)$ is called the ***end effector***, and the equations in (1) are called the ***forward kinematic equations*** (from the Greek word *kinema*, meaning "motion").

**Exercise 2**   What is the region of the plane that can be reached by the end effector if:
(a) $l_1 = l_2$, (b) $l_1 > l_2$, and (c) $l_1 < l_2$?

**Exercise 3**   What are the coordinates of the end effector if $l_1 = 2$, $l_2 = 3$, $\theta_1 = \pi/4$, and $\theta_2 = \pi/6$?

### Simulating Paint Patterns

Robin recognizes that if $\theta_1$ and $\theta_2$ are regarded as functions of time, then the forward kinematic equations can be expressed as

$$x = l_1 \cos\theta_1(t) + l_2 \cos(\theta_1(t) + \theta_2(t))$$
$$y = l_1 \sin\theta_1(t) + l_2 \sin(\theta_1(t) + \theta_2(t))$$

which are parametric equations for the curve traced by the end effector. For example, if the arms extend horizontally along the positive $x$-axis at time $t = 0$, and if links 1 and 2 rotate at the constant rates of $\omega_1$ and $\omega_2$ radians per second (rad/s), respectively, then

$$\theta_1(t) = \omega_1 t \quad \text{and} \quad \theta_2(t) = \omega_2 t$$

and the parametric equations of motion for the end effector become

$$x = l_1 \cos\omega_1 t + l_2 \cos(\omega_1 t + \omega_2 t)$$
$$y = l_1 \sin\omega_1 t + l_2 \sin(\omega_1 t + \omega_2 t)$$

*Exercise 4*   Show that if $l_1 = l_2 = 1$, and if $\omega_1 = 2$ rad/s and $\omega_2 = 3$ rad/s, then the parametric equations of motion are

$$x = \cos 2t + \cos 5t$$
$$y = \sin 2t + \sin 5t$$

Use a graphing utility to show that the curve traced by the end effector over the time interval $0 \le t \le 12$ is as shown in Figure 4. This would be the painting pattern of Robin's paint sprayer.

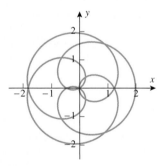

Figure 4

*Exercise 5*   Use a graphing utility to explore how the rotation rates of the links affect the spray patterns of a robot arm for which $l_1 = l_2 = 1$.

*Exercise 6*   Suppose that $l_1 = l_2 = 1$, and a malfunction in the robot arm causes the second link to lock at $\theta_2 = 0$, while the first link rotates at a constant rate of 1 rad/s. Make a conjecture about the path of the end effector, and confirm your conjecture by finding parametric equations for its motion.

### Controlling the Position of the End Effector

Robin's plan is to make the robot paint the dividers in vertical strips, sweeping from the bottom up. After a strip is painted, she will have the arm return to the bottom of the divider and then move horizontally to position itself for the next upward sweep. Since the sections of her dividers will be 3 ft wide by 5 ft high, Robin decides on a robot with two 3-ft links whose base is positioned near the lower left corner of a divider section, as in Figure 5a. Since the fully extended links span a radius of 6 ft, she feels that this arrangement will work.

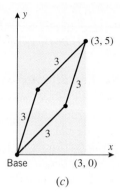

Figure 5

Robin starts with the problem of painting the far right edge from $(3, 0)$ to $(3, 5)$. With the help of some basic geometry (Figure 5$b$), she determines that the end effector can be placed at the point $(3, 0)$ by taking the control angles to be $\theta_1 = \pi/3 \ (= 60°)$ and $\theta_2 = -2\pi/3 \ (= -120°)$ (verify). However, the problem of finding the control angles that correspond to the point $(3, 5)$ is more complicated, so she starts by substituting the link lengths $l_1 = l_2 = 3$ into the forward kinematic equations in (1) to obtain

$$x = 3\cos\theta_1 + 3\cos(\theta_1 + \theta_2)$$
$$y = 3\sin\theta_1 + 3\sin(\theta_1 + \theta_2)$$

(2)

Thus, to put the end effector at the point $(3, 5)$, the control angles must satisfy the equations

$$\cos\theta_1 + \cos(\theta_1 + \theta_2) = 1$$
$$3\sin\theta_1 + 3\sin(\theta_1 + \theta_2) = 5$$

(3)

Solving these equations for $\theta_1$ and $\theta_2$ challenges Robin's algebra and trigonometry skills, but she manages to do it using the procedure in the following exercise.

· · · · · · · · · · · ·
*Exercise 7*

(a) Use the equations in (3) and the identity

$$\sin^2(\theta_1 + \theta_2) + \cos^2(\theta_1 + \theta_2) = 1$$

to show that

$$15\sin\theta_1 + 9\cos\theta_1 = 17$$

(b) Solve the last equation for $\sin\theta_1$ in terms of $\cos\theta_1$ and substitute in the identity

$$\sin^2\theta_1 + \cos^2\theta_1 = 1$$

to obtain

$$153\cos^2\theta_1 - 153\cos\theta_1 + 32 = 0$$

(c) Treat this as a quadratic equation in $\cos\theta_1$, and use the quadratic formula to obtain

$$\cos\theta_1 = \frac{1}{2} \pm \frac{5\sqrt{17}}{102}$$

(d) Use the arccosine (inverse cosine) operation of a calculating utility to solve the equations in part (c) to obtain

$$\theta_1 \approx 0.792436 \text{ rad} \approx 45.4032° \quad \text{and} \quad \theta_1 \approx 1.26832 \text{ rad} \approx 72.6694°$$

(e) Substitute each of these angles into the first equation in (3), and solve for the corresponding values of $\theta_2$.

At first, Robin was surprised that the solutions for $\theta_1$ and $\theta_2$ were not unique, but her sketch in Figure 5c quickly made it clear that there will always be two ways of positioning the links to put the end effector at a specified point.

## Controlling the Motion of the End Effector

Now that Robin has figured out how to place the end effector at the points $(3, 0)$ and $(3, 5)$, she turns to the problem of making the robot paint the vertical line segment between those points. She recognizes that not only must she make the end effector move on a vertical line, but she must control its velocity—if the end effector moves too quickly, the paint will be too thin, and if it moves too slowly, the paint will be too thick.

After some experimentation, she decides that the end effector should have a constant velocity of 1 ft/s. Thus, Robin's mathematical problem is to determine the rotation rates $d\theta_1/dt$ and $d\theta_2/dt$ (in rad/s) that will make $dx/dt = 0$ and $dy/dt = 1$. The first condition will ensure that the end effector moves vertically (no horizontal velocity), and the second condition will ensure that it moves upward at 1 ft/s.

To find formulas for $dx/dt$ and $dy/dt$, Robin uses the chain rule to differentiate the forward kinematic equations in (2) and obtains

$$\frac{dx}{dt} = -3 \sin\theta_1 \frac{d\theta_1}{dt} - [3 \sin(\theta_1 + \theta_2)]\left(\frac{d\theta_1}{dt} + \frac{d\theta_2}{dt}\right)$$

$$\frac{dy}{dt} = 3 \cos\theta_1 \frac{d\theta_1}{dt} + [3 \cos(\theta_1 + \theta_2)]\left(\frac{d\theta_1}{dt} + \frac{d\theta_2}{dt}\right)$$

She uses the forward kinematic equations again to simplify these formulas and she then substitutes $dx/dt = 0$ and $dy/dt = 1$ to obtain

$$-y\frac{d\theta_1}{dt} - 3 \sin(\theta_1 + \theta_2)\frac{d\theta_2}{dt} = 0$$

$$x\frac{d\theta_1}{dt} + 3 \cos(\theta_1 + \theta_2)\frac{d\theta_2}{dt} = 1$$

(4)

· · · · · · · · · ·
*Exercise 8*    Confirm Robin's computations.

The equations in (4) will be used in the following way: At a given time $t$, the robot will report the control angles $\theta_1$ and $\theta_2$ of its links to the computer, the computer will use the forward kinematic equations in (2) to calculate the $x$- and $y$-coordinates of the end effector, and then the values of $\theta_1, \theta_2, x$, and $y$ will be substituted into (4) to produce two equations in the two unknowns $d\theta_1/dt$ and $d\theta_2/dt$. The computer will solve these equations to determine the required rotation rates for the links.

· · · · · · · · · ·
*Exercise 9*    In each part, use the given information to sketch the position of the links, and then calculate the rotation rates for the links in rad/s that will make the end effector of Robin's robot move upward with a velocity of 1 ft/s from that position.

  (a) $\theta_1 = \pi/3, \; \theta_2 = -2\pi/3$      (b) $\theta_1 = \pi/2, \; \theta_2 = -\pi/2$

*Module by Mary Ann Connors, USMA, West Point, and Howard Anton, Drexel University, and based on the article "Moving a Planar Robot Arm" by Walter Meyer, MAA Notes Number 29, The Mathematical Association of America, 1993.*

Additional material for this module can be found on the World Wide Web at http://www.wiley.com/college/anton

# 4

Pierre de Fermat

# LOGARITHMIC AND EXPONENTIAL FUNCTIONS

*I*n this chapter we will study logarithms and exponents from the function point of view. These functions have applications in the study of population growth, sound, heating and cooling, earthquakes, and carbon dating, to name a few. We will review the algebraic aspects of logarithms and exponents, but we will focus mainly on those aspects of logarithmic and exponential functions that relate to calculus. The heart of this chapter is Section 4.1 on inverse functions, in which we develop fundamental ideas that link logarithmic and exponential functions together numerically, algebraically, and graphically. We also apply inverse functions to the study of inverse trigonometric functions (Section 4.5) and to the problem of differentiating functions whose formulas cannot be expressed in the form $y = f(x)$ (Section 4.3). We show how these methods of differentiation can be applied to problems involving rates of change (Section 4.6); and finally, we develop a powerful tool for evaluating limits, especially limits involving logarithmic and exponential functions.

## 4.1 INVERSE FUNCTIONS

*In everyday language the term "inversion" conveys the idea of a reversal. For example, in meteorology a temperature inversion is a reversal in the usual temperature properties of air layers; in music an inversion is a recurring theme that uses the same notes in reverse order; and in grammar an inversion is a reversal of the normal order of words. In mathematics the term **inverse** is used to describe functions that are reverses of one another in the sense that each undoes the effect of the other. The purpose of this section is to discuss this fundamental mathematical idea.*

**INVERSE FUNCTIONS**

The idea of solving an equation $y = f(x)$ for $x$ as a function of $y$, say $x = g(y)$, is one of the most important ideas in mathematics. Sometimes, solving an equation is a simple process; for example, using basic algebra the equation

$$y = x^3 + 1 \qquad \boxed{y = f(x)}$$

can be solved for $x$ as a function of $y$:

$$x = \sqrt[3]{y - 1} \qquad \boxed{x = g(y)}$$

The first equation is better for computing $y$ if $x$ is known, and the second is better for computing $x$ if $y$ is known (Figure 4.1.1).

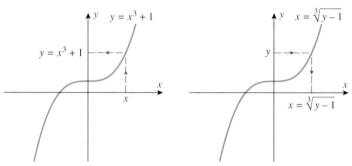

Figure 4.1.1

Our primary interest in this section is to identify relationships that may exist between the functions $f$ and $g$ when an equation $y = f(x)$ is expressed as $x = g(y)$, or conversely. For example, consider the functions $f(x) = x^3 + 1$ and $g(y) = \sqrt[3]{y - 1}$ discussed above. When these functions are composed in either order they cancel out the effect of one another in the sense that

$$g(f(x)) = \sqrt[3]{f(x) - 1} = \sqrt[3]{(x^3 + 1) - 1} = x$$
$$f(g(y)) = [g(y)]^3 + 1 = (\sqrt[3]{y - 1})^3 + 1 = y$$

(1)

The first of these equations states that each output of the composition $g(f(x))$ is the same as the input, and the second states that each output of the composition $f(g(y))$ is the same as the input. Pairs of functions with these two properties are so important that there is some terminology for them.

**4.1.1**  DEFINITION.  If the functions $f$ and $g$ satisfy the two conditions

$$g(f(x)) = x \text{ for every } x \text{ in the domain of } f$$
$$f(g(y)) = y \text{ for every } y \text{ in the domain of } g$$

then we say that $f$ and $g$ are **inverse functions**. Moreover, we call **$f$ an inverse of $g$** and **$g$ an inverse of $f$**.

### Example 1

It follows from (1) that $f(x) = x^3 + 1$ and $g(y) = \sqrt[3]{y-1}$ are inverse functions. ◀

It can be shown that a function cannot have two different inverses. Thus, if a function $f$ has an inverse, then the inverse is unique, and we are entitled to talk about *the* inverse of $f$. The inverse of a function $f$ is commonly denoted by $f^{-1}$ (read "$f$ inverse"). Thus, instead of using $g$ in Example 1, the inverse of $f(x) = x^3$ could have been expressed as $f^{-1}(y) = \sqrt[3]{y-1}$.

WARNING. The symbol $f^{-1}$ should always be interpreted as the inverse of $f$ and *never* as the reciprocal $1/f$.

It is important to understand that a function is determined by the relationship that it establishes between its inputs and outputs and not by the letter used for the independent variable. Thus, even though the formulas $f(x) = 3x$ and $f(y) = 3y$ use different independent variables, they define the *same* function $f$, since the two formulas have the same "form" and hence assign the same value to each input; for example, in either notation $f(2) = 6$. As we progress through this text, there will be certain occasions on which we will want the independent variables for $f$ and $f^{-1}$ to be the same, and other occasions on which we will want them to be different. Thus, in Example 1 we could have expressed the inverse of $f(x) = x^3 + 1$ as $f^{-1}(x) = \sqrt[3]{x-1}$ had we wanted $f$ and $f^{-1}$ to have the same independent variable.

If we use the notation $f^{-1}$ (rather than $g$) in Definition 4.1.1, and if we use $x$ as the independent variable in the formulas for both $f$ and $f^{-1}$, then the defining equations relating these functions are

$$f^{-1}(f(x)) = x \quad \text{for every } x \text{ in the domain of } f$$
$$f(f^{-1}(x)) = x \quad \text{for every } x \text{ in the domain of } f^{-1} \tag{2}$$

### Example 2

Confirm each of the following.

(a)  The inverse of $f(x) = 2x$ is $f^{-1}(x) = \frac{1}{2}x$.

(b)  The inverse of $f(x) = x^3$ is $f^{-1}(x) = x^{1/3}$.

*Solution (a).*

$$f^{-1}(f(x)) = f^{-1}(2x) = \tfrac{1}{2}(2x) = x$$
$$f(f^{-1}(x)) = f\left(\tfrac{1}{2}x\right) = 2\left(\tfrac{1}{2}x\right) = x$$

*Solution (b).*

$$f^{-1}(f(x)) = f^{-1}(x^3) = \left(x^3\right)^{1/3} = x$$
$$f(f^{-1}(x)) = f(x^{1/3}) = \left(x^{1/3}\right)^3 = x$$ ◀

REMARK. The results in Example 2 should make sense to you intuitively, since the operations of multiplying by 2 and multiplying by $\frac{1}{2}$ in either order cancel the effect of one another, as do the operations of cubing and taking a cube root.

**DOMAIN AND RANGE OF INVERSE FUNCTIONS**

The equations in (2) imply certain relationships between the domains and ranges of $f$ and $f^{-1}$. For example, in the first equation the quantity $f(x)$ is an input of $f^{-1}$, so points in the range of $f$ lie in the domain of $f^{-1}$; and in the second equation the quantity $f^{-1}(x)$ is an input of $f$, so points in the range of $f^{-1}$ lie in the domain of $f$. All of this suggests the following relationships, which we state without formal proof:

$$\text{domain of } f^{-1} = \text{range of } f$$
$$\text{range of } f^{-1} = \text{domain of } f \tag{3}$$

At the beginning of this section we solved the equation $y = f(x) = x^3 + 1$ for $x$ as a function of $y$ to obtain $x = g(y) = \sqrt[3]{y-1}$, and we observed in Example 1 that $g$ is the inverse of $f$. This was not accidental—whenever an equation $y = f(x)$ is solved for $x$ as a function of $y$, say $x = g(y)$, then $f$ and $g$ will be inverses. We can see why this is so by making two substitutions:

- Substitute $y = f(x)$ into $x = g(y)$. This yields $x = g(f(x))$, which is the first equation in Definition 4.1.1.

- Substitute $x = g(y)$ into $y = f(x)$. This yields $y = f(g(y))$, which is the second equation in Definition 4.1.1.

Since $f$ and $g$ satisfy the two conditions in Definition 4.1.1, we conclude that they are inverses. Thus, we have the following result.

> **4.1.2  THEOREM.**  *If an equation $y = f(x)$ can be solved for $x$ as a function of $y$, then $f$ has an inverse and the resulting equation is $x = f^{-1}(y)$.*

**A METHOD FOR FINDING INVERSES**

**Example 3**

Find the inverse of $f(x) = \sqrt{3x-2}$.

*Solution.*  From Theorem 4.1.2 we can find a formula for $f^{-1}(y)$ by solving the equation

$$y = \sqrt{3x-2}$$

for $x$ as a function of $y$. The computations are

$$y^2 = 3x - 2$$
$$x = \tfrac{1}{3}(y^2 + 2)$$

from which it follows that

$$f^{-1}(y) = \tfrac{1}{3}(y^2 + 2)$$

At this point we have successfully produced a formula for $f^{-1}$; however, we are not quite done, since there is no guarantee that the natural domain associated with this formula is the correct domain for $f^{-1}$. To determine whether this is so, we will examine the range of $y = f(x) = \sqrt{3x-2}$. The range consists of all $y$ in the interval $[0, +\infty)$, so from (3) this interval is also the domain of $f^{-1}(y)$; thus, the inverse of $f$ is given by the formula

$$f^{-1}(y) = \tfrac{1}{3}(y^2 + 2), \quad y \geq 0 \qquad \blacktriangleleft$$

REMARK.   When a formula for $f^{-1}$ is obtained by solving the equation $y = f(x)$ for $x$ as a function of $y$, the resulting formula has $y$ as the independent variable. If it is preferable to have $x$ as the independent variable for $f^{-1}$, then there are two ways to proceed: you can solve $y = f(x)$ for $x$ as a function of $y$, and then replace $y$ by $x$ in the *final* formula for $f^{-1}$, or you can interchange $x$ and $y$ in the *original* equation and solve the equation $x = f(y)$ for $y$ in terms of $x$, in which case the final equation will be $y = f^{-1}(x)$. In Example 3, either of these procedures will produce $f^{-1}(x) = \tfrac{1}{3}(x^2 + 2), x \geq 0$.

Theorem 4.1.2 not only provides a method for finding the inverse of a function $f$, but it also provides an interpretation of what the values of $f^{-1}$ represent. The theorem tells us that for a given $y$, the quantity $f^{-1}(y)$ is that number $x$ with the property that $f(x) = y$. For example, if $f^{-1}(1) = 4$, then you know that $f(4) = 1$; and similarly, if $f(3) = 7$, then you know that $f^{-1}(7) = 3$.

**EXISTENCE OF INVERSE FUNCTIONS**

Not every function has an inverse. In general, in order for a function $f$ to have an inverse it must assign distinct outputs to distinct inputs. To see why this is so, consider the function $f(x) = x^2$. Since $f(2) = f(-2) = 4$, the function $f$ assigns the same output to two distinct inputs. If $f$ were to have an inverse, then the equation $f(2) = 4$ would imply that $f^{-1}(4) = 2$, and the equation $f(-2) = 4$ would imply that $f^{-1}(4) = -2$. This is obviously impossible, since we cannot have two different values for $f^{-1}(4)$. Thus, $f(x) = x^2$ has no inverse. Another way to see that $f(x) = x^2$ has no inverse is to attempt to find the inverse by solving the equation $y = x^2$ for $x$ in terms of $y$. We run into trouble immediately because the resulting equation, $x = \pm\sqrt{y}$, does not express $x$ as a *single* function of $y$.

Functions that assign distinct outputs to distinct inputs are sufficiently important that there is a name for them—they are said to be ***one-to-one*** or ***invertible***. Stated algebraically, a function $f$ is one-to-one if $f(x_1) \neq f(x_2)$ whenever $x_1 \neq x_2$; and stated geometrically, a function $f$ is one-to-one if the graph of $y = f(x)$ is cut at most once by any horizontal line (Figure 4.1.2).

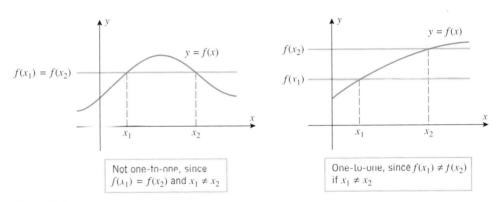

Figure 4.1.2

One can prove that a function $f$ has an inverse if and only if it is one-to-one, and this provides us with the following geometric test for determining whether a function has an inverse.

> **4.1.3** THEOREM (***The Horizontal Line Test***). *A function $f$ has an inverse if and only if its graph is cut at most once by any horizontal line.*

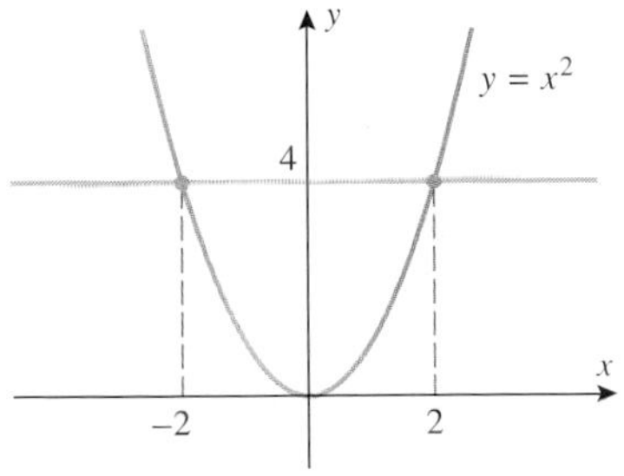

### Example 4

We observed above that the function $f(x) = x^2$ does not have an inverse. This is confirmed by the horizontal line test, since the graph of $y = x^2$ is cut more than once by certain horizontal lines (Figure 4.1.3). ◀

figure 4.1.3

### Example 5

We saw in Example 2(b) that the function $f(x) = x^3$ has an inverse [namely, $f^{-1}(x) = x^{1/3}$]. The existence of an inverse is confirmed by the horizontal line test, since the graph of $y = x^3$ is cut at most once by any horizontal line (Figure 4.1.4). ◀

Figure 4.1.3

### Example 6

Explain why the function $f$ that is graphed in Figure 4.1.5 has an inverse, and find $f^{-1}(3)$.

*Solution.*  The function $f$ has an inverse since its graph passes the horizontal line test. To evaluate $f^{-1}(3)$, we view $f^{-1}(3)$ as that number $x$ for which $f(x) = 3$. From the graph we see that $f(2) = 3$, so $f^{-1}(3) = 2$.  ◀

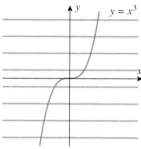

Figure 4.1.4

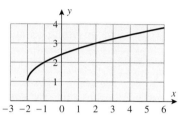

Figure 4.1.5

**GRAPHS OF INVERSE FUNCTIONS**

Our next objective is to explore the relationship between the graphs of $f$ and $f^{-1}$. For this purpose, it will be desirable to use $x$ as the independent variable for both functions, which means that we will be comparing the graphs of $y = f(x)$ and $y = f^{-1}(x)$.

If $(a, b)$ is a point on the graph $y = f(x)$, then $b = f(a)$. This is equivalent to the statement that $a = f^{-1}(b)$, which means that $(b, a)$ is a point on the graph of $y = f^{-1}(x)$. In short, reversing the coordinates of a point on the graph of $f$ produces a point on the graph of $f^{-1}$. Similarly, reversing the coordinates of a point on the graph of $f^{-1}$ produces a point on the graph of $f$ (verify). However, the geometric effect of reversing the coordinates of a point is to reflect that point about the line $y = x$ (Figure 4.1.6), and hence the graphs of $y = f(x)$ and $y = f^{-1}(x)$ are reflections of one another about this line (Figure 4.1.7). In summary, we have the following result.

> **4.1.4**   THEOREM.   *If $f$ has an inverse, then the graphs of $y = f(x)$ and $y = f^{-1}(x)$ are reflections of one another about the line $y = x$; that is, each is the mirror image of the other with respect to that line.*

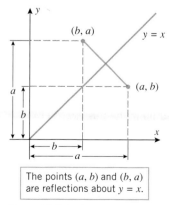

The points $(a, b)$ and $(b, a)$ are reflections about $y = x$.

Figure 4.1.6

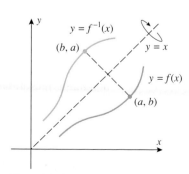

Figure 4.1.7

### Example 7

Figure 4.1.8 shows the graphs of the inverse functions discussed in Examples 2 and 3. ◀

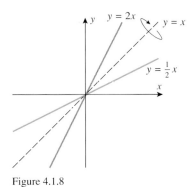 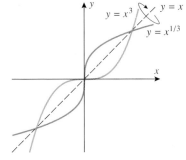 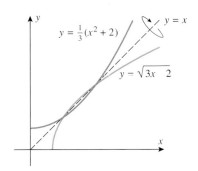

Figure 4.1.8

**INCREASING OR DECREASING FUNCTIONS HAVE INVERSES**

If the graph of a function $f$ is always increasing or always decreasing over the domain of $f$, then the graph of $f$ can be cut at most once by any horizontal line and consequently the function $f$ must have an inverse. One way to tell whether the graph of a function is increasing or decreasing over an interval is by examining the slopes of its tangent lines. We will prove in the next chapter that the graph of $f$ must be increasing on any interval where $f'(x) > 0$ (since the tangent lines have positive slope) and must be decreasing on any interval where $f'(x) < 0$ (since the tangent lines have negative slope) (Figure 4.1.9). These intuitive observations suggest the following theorem, which we state without formal proof.

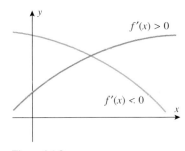

> **4.1.5** THEOREM. *If the domain of $f$ is an interval on which $f'(x) > 0$ or on which $f'(x) < 0$, then the function $f$ has an inverse.*

Figure 4.1.9

### Example 8

The graph of $f(x) = x^5 + x + 1$ is always increasing on $(-\infty, +\infty)$, since

$$f'(x) = 5x^4 + 1 > 0$$

for all $x$. However, there is no easy way to solve the equation $y = x^5 + x + 1$ for $x$ in terms of $y$ (try it), so even though we know that $f$ has an inverse, we cannot produce a formula for it. ◀

REMARK. What is important to understand here is that our inability to find a formula for the inverse does not negate the existence of the inverse; indeed, one of our goals in later sections will be to develop ways of finding properties of functions in which there are no explicit formulas for the functions to work with.

**RESTRICTING DOMAINS TO MAKE FUNCTIONS INVERTIBLE**

Sometimes a function that is not one-to-one can be made one-to-one by restricting its domain. For example, although the function $f(x) = x^2$ is not one-to-one, the functions

$$g(x) = x^2, \quad x \geq 0$$
$$h(x) = x^2, \quad x \leq 0$$

which result from restricting the domain of $f$, are one-to-one since their graphs pass the horizontal line test [the graph of $g$ is the right half of the parabola $y = x^2$ and the graph of $h$ is the left half (Figure 4.1.10)]. The inverses of $g$ and $h$ can be found by solving each of the equations $y = g(x)$ and $y = h(x)$ for $x$ as a function of $y$. For example, to find the inverse of $g$ we solve

$$y = x^2, \quad x \geq 0$$

for $x$, which yields $x = \sqrt{y}$; hence, $g^{-1}(y) = \sqrt{y}$. Similarly, $h^{-1}(y) = -\sqrt{y}$. Geometrically, the graphs of $g(x) = x^2$, $x \geq 0$ and $g^{-1}(x) = \sqrt{x}$ are reflections of one another about the line $y = x$ (Figure 4.1.11), which reveals that the graph of $y = \sqrt{x}$ is a portion of a reflected parabola.

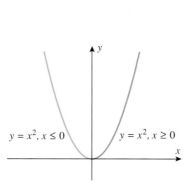

Figure 4.1.10                    Figure 4.1.11

**CONTINUITY OF INVERSE FUNCTIONS**

Because the graphs of $f$ and $f^{-1}$ are reflections of one another about the line $y = x$, it is intuitively obvious that if the graph of $f$ has no breaks, then neither will the graph of $f^{-1}$. This suggests the following result, which we state without proof.

> **4.1.6**   THEOREM.   *If a function $f$ is continuous and has an inverse, then $f^{-1}$ is also continuous.*

For example, even though we cannot find a formula for $f^{-1}$ in Example 8, the continuity of the polynomial $f$ guarantees that $f^{-1}$ is a continuous function.

**DIFFERENTIABILITY OF INVERSE FUNCTIONS**

Suppose that $f$ is a continuous one-to-one function. Speaking informally, the points of nondifferentiability of $f^{-1}$ occur most commonly at corners or points of vertical tangency in the graph of $y = f^{-1}(x)$. However, the graph of $y = f^{-1}(x)$ is the reflection about $y = x$ of the graph of $y = f(x)$; hence, corners in the graph of $f^{-1}$ are reflections of corners in the graph of $f$, and points of vertical tangency in the graph of $f^{-1}$ are reflections of points of horizontal tangency in the graph of $f$. This suggests that if $f$ is a differentiable function whose derivative is nonzero, then $f^{-1}$ will be a differentiable function. The following theorem, which we state without proof, makes this idea precise.

> **4.1.7**   THEOREM (***Differentiability of Inverse Functions***).    *Suppose that the function $f$ is invertible and differentiable on an interval I. Then $f^{-1}$ is differentiable at any point $x$ where $f'(f^{-1}(x)) \neq 0$.*

**Example 9**

We showed in Example 8 that the function $f(x) = x^5 + x + 1$ has an inverse. Use Theorem 4.1.7 to show that $f^{-1}$ is differentiable on the interval $(-\infty, +\infty)$.

*Solution.* Let $I$ denote the interval $(-\infty, +\infty)$. We must show that for each $x$ in $I$, the function $f$ has a nonzero derivative at the point $f^{-1}(x)$. But this is so because the derivative of $f$ is

$$f'(x) = 5x^4 + 1$$

which is nonzero for all $x$.   ◀

**GRAPHING INVERSE FUNCTIONS WITH GRAPHING UTILITIES**

Most graphing utilities cannot graph inverse functions directly. However, there is a way of graphing inverse functions by expressing the graph parametrically. To see how this can be done, suppose that we are interested in graphing the inverse of a one-to-one function $f$. We observed in Section 1.7 that the equation $y = f(x)$ can be expressed parametrically as

$$x = t, \quad y = f(t) \tag{4}$$

Moreover, we know that the graph of $f^{-1}$ can be obtained by interchanging $x$ and $y$, since this reflects the graph of $f$ about the line $y = x$. Thus, from (4) the graph of $f^{-1}$ can be represented parametrically as

$$x = f(t), \quad y = t \tag{5}$$

For example, Figure 4.1.12 shows the graph of $f(x) = x^5 + x + 1$ and its inverse generated with a graphing utility. The graph of $f$ was generated from the parametric equations

$$x = t, \quad y = t^5 + t + 1$$

and the graph of $f^{-1}$ was generated from the parametric equations

$$x = t^5 + t + 1, \quad y = t$$

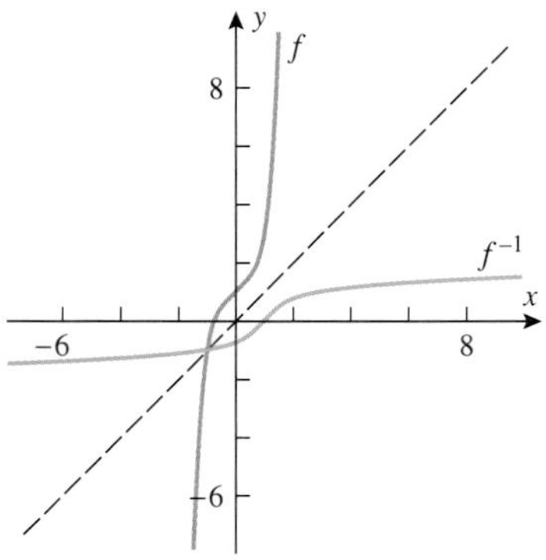

Figure 4.1.12

---

**EXERCISE SET 4.1**   ⌇ Graphing Calculator   © CAS

1. In (a)–(d), determine whether $f$ and $g$ are inverse functions.
   (a) $f(x) = 4x$, $g(x) = \frac{1}{4}x$
   (b) $f(x) = 3x + 1$, $g(x) = 3x - 1$
   (c) $f(x) = \sqrt[3]{x - 2}$, $g(x) = x^3 + 2$
   (d) $f(x) = x^4$, $g(x) = \sqrt[4]{x}$

⌇ 2. Check your answers to Exercise 1 with a graphing utility by determining whether the graphs of $f$ and $g$ are reflections of one another about the line $y = x$.

3. In each part, determine whether the function $f$ defined by the table is one-to-one.

   (a)

   | $x$ | 1 | 2 | 3 | 4 | 5 | 6 |
   |---|---|---|---|---|---|---|
   | $f(x)$ | −2 | −1 | 0 | 1 | 2 | 3 |

   (b)

   | $x$ | 1 | 2 | 3 | 4 | 5 | 6 |
   |---|---|---|---|---|---|---|
   | $f(x)$ | 4 | −7 | 6 | −3 | 1 | 4 |

4. In each part, determine whether the function $f$ is one-to-one, and justify your answer.
   (a) $f(t)$ is the number of people in line at a movie theater at time $t$.
   (b) $f(x)$ is your weight on your $x$th birthday.
   (c) $f(v)$ is the weight of $v$ cubic inches of lead.

5. In each part, use the horizontal line test to determine whether the function $f$ is one-to-one.
   (a) $f(x) = 3x + 2$           (b) $f(x) = \sqrt{x - 1}$
   (c) $f(x) = |x|$              (d) $f(x) = x^3$
   (e) $f(x) = x^2 - 2x + 2$     (f) $f(x) = \sin x$

⌇ 6. In each part, generate the graph of the function $f$ with a graphing utility, and determine whether $f$ is one-to-one.
   (a) $f(x) = x^3 - 3x + 2$     (b) $f(x) = x^3 - 3x^2 + 3x - 1$

7. In each part, determine whether $f$ is one-to-one.
   (a) $f(x) = \tan x$
   (b) $f(x) = \tan x$, $-\pi < x < \pi$
   (c) $f(x) = \tan x$, $-\pi/2 < x < \pi/2$

**8.** In each part, determine whether $f$ is one-to-one.
(a) $f(x) = \cos x$
(b) $f(x) = \cos x$, $-\pi/2 \le x \le \pi/2$
(c) $f(x) = \cos x$, $0 \le x \le \pi$

**9.** (a) The accompanying figure shows the graph of a function $f$ over its domain $-8 \le x \le 8$. Explain why $f$ has an inverse, and use the graph to find $f^{-1}(2)$, $f^{-1}(-1)$, and $f^{-1}(0)$.
(b) Find the domain and range of $f^{-1}$.
(c) Sketch the graph of $f^{-1}$.

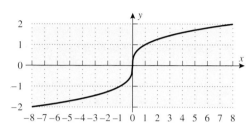

Figure Ex-9

**10.** (a) Explain why the function $f$ graphed in the accompanying figure has no inverse on its domain $-3 \le x \le 4$.
(b) Subdivide the domain into three adjacent intervals on each of which the function $f$ has an inverse.

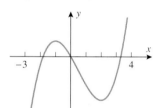

Figure Ex-10

In Exercises 11 and 12, determine whether the function $f$ is one-to-one by examining the sign of $f'(x)$.

**11.** (a) $f(x) = x^2 + 8x + 1$
(b) $f(x) = 2x^5 + x^3 + 3x + 2$
(c) $f(x) = 2x + \sin x$

**12.** (a) $f(x) = x^3 + 3x^2 - 8$
(b) $f(x) = x^5 + 8x^3 + 2x - 1$
(c) $f(x) = \dfrac{x}{x+1}$

In Exercises 13–23, find a formula for $f^{-1}(x)$.

**13.** $f(x) = x^5$

**14.** $f(x) = 6x$

**15.** $f(x) = 7x - 6$

**16.** $f(x) = \dfrac{x+1}{x-1}$

**17.** $f(x) = 3x^3 - 5$

**18.** $f(x) = \sqrt[5]{4x + 2}$

**19.** $f(x) = \sqrt[3]{2x - 1}$

**20.** $f(x) = 5/(x^2 + 1)$, $x \ge 0$

**21.** $f(x) = 3/x^2$, $x < 0$

**22.** $f(x) = \begin{cases} 2x, & x \le 0 \\ x^2, & x > 0 \end{cases}$

**23.** $f(x) = \begin{cases} 5/2 - x, & x < 2 \\ 1/x, & x \ge 2 \end{cases}$

**24.** Find a formula for $p^{-1}(x)$, given that
$$p(x) = x^3 - 3x^2 + 3x - 1$$

In Exercises 25–29, find a formula for $f^{-1}(x)$, and state the domain of $f^{-1}$.

**25.** $f(x) = (x + 2)^4$, $x \ge 0$

**26.** $f(x) = \sqrt{x + 3}$

**27.** $f(x) = -\sqrt{3 - 2x}$

**28.** $f(x) = 3x^2 + 5x - 2$, $x \ge 0$

**29.** $f(x) = x - 5x^2$, $x \ge 1$

**30.** The formula $F = \frac{9}{5}C + 32$, where $C \ge -273.15°C$ expresses the Fahrenheit temperature $F$ as a function of the Celsius temperature $C$.
(a) Find a formula for the inverse function.
(b) In words, what does the inverse function tell you?
(c) Find the domain and range of the inverse function.

**31.** (a) One meter is about $6.214 \times 10^{-4}$ miles. Find a formula $y = f(x)$ that expresses a length $x$ in meters as a function of the same length $y$ in miles.
(b) Find a formula for the inverse of $f$.
(c) In practical terms, what does the formula $x = f^{-1}(y)$ tell you?

**32.** Suppose that $f$ is a one-to-one, continuous function such that $\lim_{x \to 3} f(x) = 7$. Find $\lim_{x \to 7} f^{-1}(x)$, and justify your reasoning.

**33.** Let $f(x) = x^2$, $x > 1$, and $g(x) = \sqrt{x}$.
(a) Show that $f(g(x)) = x$, $x > 1$, and $g(f(x)) = x$, $x > 1$.
(b) Show that $f$ and $g$ are *not* inverses of one another by showing that the graphs of $y = f(x)$ and $y = g(x)$ are not reflections of one another about $y = x$.
(c) Do parts (a) and (b) contradict one another? Explain.

**34.** Let $f(x) = ax^2 + bx + c$, $a > 0$. Find $f^{-1}$ if the domain of $f$ is restricted to
(a) $x \ge -b/(2a)$
(b) $x \le -b/(2a)$.

**35.** (a) Show that $f(x) = (3 - x)/(1 - x)$ is its own inverse.
(b) What does the result in part (a) tell you about the graph of $f$?

**36.** Suppose that a line of nonzero slope $m$ intersects the $x$-axis at $(x_0, 0)$. Find an equation for the reflection of this line about $y = x$.

**37.** (a) Show that $f(x) = x^3 - 3x^2 + 2x$ is not one-to-one on $(-\infty, +\infty)$.
(b) Find the largest value of $k$ such that $f$ is one-to-one on the interval $(-k, k)$.

**38.** (a) Show that the function $f(x) = x^4 - 2x^3$ is not one-to-one on $(-\infty, +\infty)$.
(b) Find the smallest value of $k$ such that $f$ is one-to-one on the interval $[k, +\infty)$.

**39.** Let $f(x) = 2x^3 + 5x + 3$. Find $x$ if $f^{-1}(x) = 1$.

**40.** Let $f(x) = \dfrac{x^3}{x^2 + 1}$. Find $x$ if $f^{-1}(x) = 2$.

> In Exercises 41–44, use a graphing utility and parametric equations to display the graphs of $f$ and $f^{-1}$ on the same screen.

$\sim$ **41.** $f(x) = x^3 + 0.2x - 1, \quad -1 \le x \le 2$

$\sim$ **42.** $f(x) = \sqrt{x^2 + 2} + x, \quad -5 \le x \le 5$

$\sim$ **43.** $f(x) = \cos(\cos 0.5x), \quad 0 \le x \le 3$

$\sim$ **44.** $f(x) = x + \sin x, \quad 0 \le x \le 6$

**45.** Prove that if $a^2 + bc \ne 0$, then the graph of
$$f(x) = \frac{ax + b}{cx - a}$$
is symmetric about the line $y = x$.

**46.** (a) Prove: If $f$ and $g$ are one-to-one, then so is the composition $f \circ g$.
(b) Prove: If $f$ and $g$ are one-to-one, then
$$(f \circ g)^{-1} = g^{-1} \circ f^{-1}$$

**47.** Sketch the graph of a function that is one-to-one on $(-\infty, +\infty)$, yet not increasing on $(-\infty, +\infty)$ and not decreasing on $(-\infty, +\infty)$.

**48.** Prove: A one-to-one function $f$ cannot have two different inverses.

**49.** Let $F(x) = f(2g(x))$ where $f(x) = x^4 + x^3 + 1$ for $0 \le x \le 2$, and $g(x) = f^{-1}(x)$. Find $F(3)$.

## 4.2 LOGARITHMIC AND EXPONENTIAL FUNCTIONS

---

*When logarithms were introduced in the seventeenth century as a computational tool, they provided scientists of that period computing power that was previously unimaginable. Although computers and calculators have largely replaced logarithms for numerical calculations, the logarithmic functions and their relatives have wide-ranging applications in mathematics and science. Some of these will be introduced in this section.*

---

**IRRATIONAL EXPONENTS**

In algebra, integer and rational powers of a number $b$ are defined by
$$b^n = b \times b \times \cdots \times b \quad (n \text{ factors}), \qquad b^{-n} = \frac{1}{b^n}, \qquad b^0 = 1,$$
$$b^{p/q} = \sqrt[q]{b^p} = (\sqrt[q]{b})^p, \qquad b^{-p/q} = \frac{1}{b^{p/q}}$$

If $b$ is negative, then some of the fractional powers of $b$ will have imaginary values; for example, $(-2)^{1/2} = \sqrt{-2}$. To avoid this complication we will assume throughout this section that $b \ge 0$, even if it is not stated explicitly.

Observe that the preceding definitions do not include *irrational* powers of $b$ such as
$$2^\pi, \quad 3^{\sqrt{2}}, \quad \text{and} \quad \pi^{-\sqrt{7}}$$

There are various methods for defining irrational powers. One approach is to define irrational powers of $b$ as limits of rational powers of $b$. For example, to define $2^\pi$ we can start with the decimal representation of $\pi$, namely,
$$3.1415926\ldots$$

From this decimal we can form a sequence of rational numbers that gets closer and closer to $\pi$, namely,
$$3.1, \quad 3.14, \quad 3.141, \quad 3.1415, \quad 3.14159$$

and from these we can form a sequence of *rational* powers of 2:
$$2^{3.1}, \quad 2^{3.14}, \quad 2^{3.141}, \quad 2^{3.1415}, \quad 2^{3.14159}$$

**Table 4.2.1**

| $x$ | $2^x$ |
|---|---|
| 3 | 8.000000 |
| 3.1 | 8.574188 |
| 3.14 | 8.815241 |
| 3.141 | 8.821353 |
| 3.1415 | 8.824411 |
| 3.14159 | 8.824962 |
| 3.141592 | 8.824974 |

Since the exponents of the terms in this sequence approach a limit of $\pi$, it seems plausible that the terms themselves approach a limit, and it would seem reasonable to *define* $2^\pi$ to be this limit. Table 4.2.1 provides numerical evidence that the sequence does, in fact, have a limit and that to four decimal places the value of this limit is $2^\pi \approx 8.8250$. More generally, for any irrational exponent $p$ and positive number $b$, we can define $b^p$ as the limit of the rational powers of $b$ created from the decimal expansion of $p$.

FOR THE READER. Confirm the approximation $2^\pi \approx 8.8250$ by computing $2^\pi$ directly using your calculating utility.

Although our definition of $b^p$ for irrational $p$ certainly seems reasonable, there is a lot of tedious mathematical detail required to make the definition precise. We will not be concerned with such matters here and will accept without proof that the following familiar laws hold for all real exponents:

$$b^p b^q = b^{p+q}, \quad \frac{b^p}{b^q} = b^{p-q}, \quad \left(b^p\right)^q = b^{pq}$$

**THE FAMILY OF EXPONENTIAL FUNCTIONS**

A function of the form $f(x) = b^x$, where $b > 0$ and $b \neq 1$, is called an *exponential function with base b*. Some examples are

$$f(x) = 2^x, \quad f(x) = \left(\tfrac{1}{2}\right)^x, \quad f(x) = \pi^x$$

Note that an exponential function has a constant base and variable exponent. Thus, functions such as $f(x) = x^2$ and $f(x) = x^\pi$ would not be classified as exponential functions, since they have a variable base and a constant exponent. Functions of this type, which are called *power functions*, will be studied later.

It can be shown that exponential functions are continuous and have one of the basic two shapes shown in Figure 4.2.1a, depending on whether $0 < b < 1$ or $b > 1$. Figure 4.2.1b shows the graphs of some specific exponential functions.

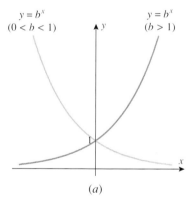

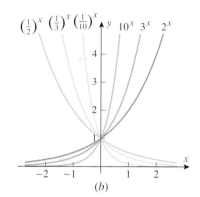

Figure 4.2.1

REMARK. If $b = 1$, then the function $b^x$ is constant, since $b^x = 1^x = 1$. This case is of no interest to us here, so we have excluded it from the family of exponential functions.

FOR THE READER. Use your graphing utility to confirm that the graphs $y = \left(\tfrac{1}{2}\right)^x$ and $y = 2^x$ agree with Figure 4.2.1b, and explain why the two graphs are reflections of one another about the $y$-axis.

Since it is not our objective in this section to develop the properties of exponential functions in rigorous mathematical detail, we will simply observe without proof that the following properties of exponential functions are consistent with the graphs shown in Figure 4.2.1.

> **4.2.1**   THEOREM.   *If* $b > 0$ *and* $b \neq 1$, *then*:
>
> *(a)*   *The function* $f(x) = b^x$ *is defined for all real values of* $x$, *so its natural domain is* $(-\infty, +\infty)$.
>
> *(b)*   *The function* $f(x) = b^x$ *is continuous on the interval* $(-\infty, +\infty)$, *and its range is* $(0, +\infty)$.

**LOGARITHMS**

Recall from algebra that a logarithm is an exponent. More precisely, if $b > 0$ and $b \neq 1$, then for positive values of $x$ the **logarithm to the base b of x** is denoted by

$$\log_b x$$

and is defined to be that exponent to which $b$ must be raised to produce $x$. For example,

$$\log_{10} 100 = 2, \quad \log_{10}(1/1000) = -3, \quad \log_2 16 = 4, \quad \log_b 1 = 0, \quad \log_b b = 1$$

| $10^2 = 100$ | $10^{-3} = 1/1000$ | $2^4 = 16$ | $b^0 = 1$ | $b^1 = b$ |

Historically, the first logarithms ever studied were the logarithms with base 10, called **common logarithms**. For such logarithms it is usual to suppress explicit reference to the base and write $\log x$ rather than $\log_{10} x$. More recently, logarithms with base 2 have played a role in computer science, since they arise naturally in the binary number system. However, the most widely used logarithms in applications are the **natural logarithms**, which have an irrational base denoted by the letter $e$ in honor of the Swiss mathematician Leonard Euler (p. 19), who first suggested its application to logarithms in an unpublished paper written in 1728. This constant, whose value to six decimal places is

$$e \approx 2.718282 \tag{1}$$

arises as the horizontal asymptote of the graph of the equation

$$y = \left(1 + \frac{1}{x}\right)^x \tag{2}$$

(Figure 4.2.2).

THE VALUES OF $(1 + 1/x)^x$ APPROACH $e$

| $x$ | $1 + \frac{1}{x}$ | $\left(1 + \frac{1}{x}\right)^x$ |
| --- | --- | --- |
| 1 | 2 | $\approx 2.000000$ |
| 10 | 1.1 | 2.593742 |
| 100 | 1.01 | 2.704814 |
| 1000 | 1.001 | 2.716924 |
| 10,000 | 1.0001 | 2.718146 |
| 100,000 | 1.00001 | 2.718268 |
| 1,000,000 | 1.000001 | 2.718280 |

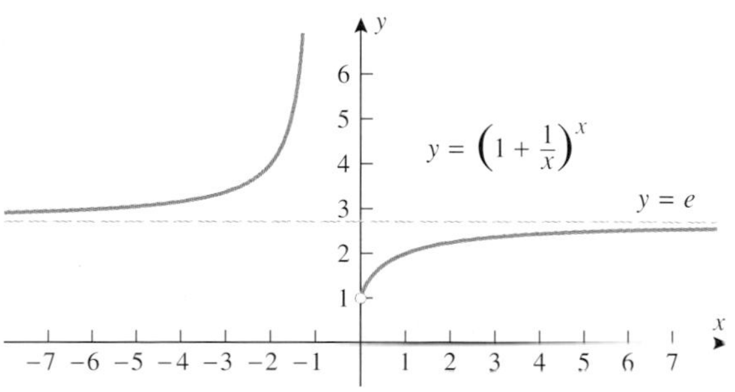

Figure 4.2.2

The fact that $y = e$ is a horizontal asymptote of (2) as $x \to +\infty$ and as $x \to -\infty$ is expressed by the limits

$$e = \lim_{x \to +\infty} \left(1 + \frac{1}{x}\right)^x \qquad \text{and} \qquad e = \lim_{x \to -\infty} \left(1 + \frac{1}{x}\right)^x \tag{3–4}$$

Later, we will show that these limits can be derived from the limit

$$e = \lim_{x \to 0} (1 + x)^{1/x} \tag{5}$$

which is sometimes taken as the definition of the number $e$.

It is standard to denote the natural logarithm of $x$ by $\ln x$ (read "ell en of $x$"), rather than $\log_e x$. Thus, $\ln x$ can be viewed as that power to which $e$ must be raised to produce $x$. For example,

$$\ln 1 = 0, \qquad \ln e = 1, \qquad \ln 1/e = -1, \qquad \ln(e^2) = 2$$

| Since $e^0 = 1$ | Since $e^1 = e$ | Since $e^{-1} = 1/e$ | Since $e^2 = e^2$ |

In general, the statements

$$y = \ln x \quad \text{and} \quad x = e^y$$

are equivalent.

The exponential function $f(x) = e^x$ is called the ***natural exponential function***. To simplify typography, this function is sometimes written as $\exp x$. Thus, for example, you might see the relationship $e^{x_1 + x_2} = e^{x_1} e^{x_2}$ expressed as

$$\exp(x_1 + x_2) = \exp(x_1) \exp(x_2)$$

This notation is also used by graphing and calculating utilities, and it is typical to access the function $e^x$ with some variation of the command EXP.

FOR THE READER.   Most scientific calculating utilities provide some way of evaluating common logarithms, natural logarithms, and powers of $e$. Check your documentation to see how this is done, and then confirm the approximation $e \approx 2.718282$ and the values that appear in the table in Figure 4.2.2.

......................................
**LOGARITHMIC FUNCTIONS**

Figure 4.2.1$a$ suggests that if $b > 0$ and $b \neq 1$, then the graph of $y = b^x$ passes the horizontal line test, and this implies that the function $f(x) = b^x$ has an inverse. To find a formula for this inverse (with $x$ as the independent variable), we can solve the equation $x = b^y$ for $y$ as a function of $x$. This can be done by taking the logarithm to the base $b$ of both sides of this equation. This yields

$$\log_b x = \log_b(b^y) \tag{6}$$

However, if we think of $\log_b(b^y)$ as that exponent to which $b$ must be raised to produce $b^y$, then it becomes evident that $\log_b(b^y) = y$. Thus, (6) can be rewritten as

$$y = \log_b x$$

from which we conclude that the inverse of $f(x) = b^x$ is $f^{-1}(x) = \log_b x$. This implies that the graphs of $y = b^x$ and $y = \log_b x$ are reflections of one another about the line $y = x$ (Figure 4.2.3). We call $\log_b x$ the ***logarithmic function with base b***.

Recall from Section 4.1 that a one-to-one function $f$ and its inverse satisfy the equations

$$f^{-1}(f(x)) = x \quad \text{for every } x \text{ in the domain of } f$$
$$f(f^{-1}(x)) = x \quad \text{for every } x \text{ in the domain of } f^{-1}$$

In particular, if we take $f(x) = b^x$ and $f^{-1}(x) = \log_b x$, and if we keep in mind that the domain of $f^{-1}$ is the same as the range of $f$, then we obtain

$$\begin{aligned} \log_b(b^x) &= x \quad \text{for all real values of } x \\ b^{\log x} &= x \quad \text{for } x > 0 \end{aligned} \tag{7}$$

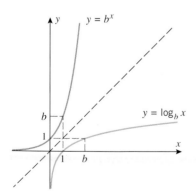

Figure 4.2.3

In the special case where $b = e$, these equations become

$$\ln(e^x) = x \quad \text{for all real values of } x$$
$$e^{\ln x} = x \quad \text{for } x > 0 \tag{8}$$

In words, the equations in (7) tell us that the functions $b^x$ and $\log_b x$ cancel out the effect of one another when composed in either order; for example,

$$\log 10^x = x, \quad 10^{\log x} = x, \quad \ln e^x = x, \quad e^{\ln x} = x, \quad \ln e^5 = 5, \quad e^{\ln \pi} = \pi$$

FOR THE READER.    Figure 4.2.4 shows computer-generated tables and graphs of $y = e^x$ and $y = \ln x$. Use your calculating and graphing utilities to generate the graphs and table values.

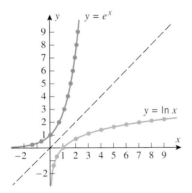

| $x$ | $y = \ln x$ | $x$ | $y = e^x$ |
|---|---|---|---|
| 0.25 | −1.39 | −1.39 | 0.25 |
| 0.50 | −0.69 | −0.69 | 0.50 |
| 1 | 0 | 0 | 1 |
| 2 | 0.69 | 0.69 | 2 |
| 3 | 1.10 | 1.10 | 3 |
| 4 | 1.39 | 1.39 | 4 |
| 5 | 1.61 | 1.61 | 5 |
| 6 | 1.79 | 1.79 | 6 |
| 7 | 1.95 | 1.95 | 7 |
| 8 | 2.08 | 2.08 | 8 |
| 9 | 2.20 | 2.20 | 9 |

Figure 4.2.4

The inverse relationship between $b^x$ and $\log_b x$ allows us to translate properties of exponential functions into properties of logarithmic functions, and vice versa.

---

**4.2.2**    THEOREM (*Comparison of Exponential and Logarithmic Functions for $b > 1$*).

| | |
|---|---|
| $b^0 = 1$ | $\log_b 1 = 0$ |
| $b^1 = b$ | $\log_b b = 1$ |
| range $b^x = (0, +\infty)$ | domain $\log_b x = (0, +\infty)$ |
| domain $b^x = (-\infty, +\infty)$ | range $\log_b x = (-\infty, +\infty)$ |
| $0 < b^x < 1 \quad$ if $x < 0$ | $\log_b x < 0 \quad$ if $0 < x < 1$ |

---

In addition, you should recall the following algebraic properties of logarithms from your earlier studies.

---

**4.2.3**    THEOREM (*Algebraic Properties of Logarithms*).

| | |
|---|---|
| $\log_b(ac) = \log_b a + \log_b c$ | Product property |
| $\log_b(a/c) = \log_b a - \log_b c$ | Quotient property |
| $\log_b(a^r) = r \log_b a$ | Power property |
| $\log_b(1/c) = -\log_b c$ | Reciprocal property |

---

These properties are often used to expand a single logarithm into sums, differences, and multiples of other logarithms and, conversely, to condense sums, differences, and multiples of logarithms into a single logarithm. For example,

$$\log \frac{xy^5}{\sqrt{z}} = \log xy^5 - \log \sqrt{z} = \log x + \log y^5 - \log z^{1/2} = \log x + 5 \log y - \tfrac{1}{2} \log z$$

$$5 \log 2 + \log 3 - \log 8 = \log 32 + \log 3 - \log 8 = \log \frac{32 \cdot 3}{8} = \log 12$$

$$\tfrac{1}{3} \ln x - \ln(x^2 - 1) + 2 \ln(x+3) = \ln x^{1/3} - \ln(x^2 - 1) + \ln(x+3)^2 = \ln \frac{\sqrt[3]{x}(x+3)^2}{x^2 - 1}$$

REMARK. Expressions of the form $\log_b(u+v)$ and $\log_b(u-v)$ have no useful simplifications in terms of $\log_b u$ and $\log_b v$. In particular,

$$\log_b(u+v) \neq \log_b u + \log_b v$$
$$\log_b(u-v) \neq \log_b u - \log_b v$$

**SOLVING EQUATIONS INVOLVING EXPONENTIALS AND LOGARITHMS**

Equations of the form $\log_b x = k$ can be solved by converting them to the exponential form $x = b^k$, and equations of the form $b^x = k$ can be solved by taking a logarithm of both sides (usually log or ln).

### Example 1

Find $x$ such that

(a) $\log x = \sqrt{2}$    (b) $\ln(x+1) = 5$    (c) $5^x = 7$

*Solution (a).* Converting the equation to exponential form yields

$$x = 10^{\sqrt{2}} \approx 25.95$$

*Solution (b).* Converting the equation to exponential form yields

$$x + 1 = e^5 \quad \text{or} \quad x = e^5 - 1 \approx 147.41$$

*Solution (c).* Taking the natural logarithm of both sides and using the power property of logarithms yields

$$x \ln 5 = \ln 7 \quad \text{or} \quad x = \frac{\ln 7}{\ln 5} \approx 1.21 \qquad \blacktriangleleft$$

### Example 2

A satellite that requires 7 watts of power to operate at full capacity is equipped with a radioisotope power supply whose power output in watts is given by the equation

$$P = 75e^{-t/125}$$

where $t$ is the time in days that the supply is used. How long can the satellite operate at full capacity?

*Solution.* The power $P$ will fall to 7 watts when

$$7 = 75e^{-t/125}$$

The solution for $t$ is as follows:

$$7/75 = e^{-t/125}$$
$$\ln(7/75) = \ln(e^{-t/125})$$
$$\ln(7/75) = -t/125$$
$$t = -125 \ln(7/75) \approx 296.4$$

so the satellite can operate at full capacity for about 296 days.  $\blacktriangleleft$

Here is a more complicated example.

**Example 3**

Solve $\dfrac{e^x - e^{-x}}{2} = 1$ for $x$.

*Solution.* Multiplying both sides of the given equation by 2 yields

$$e^x - e^{-x} = 2$$

or equivalently,

$$e^x - \frac{1}{e^x} = 2$$

Multiplying through by $e^x$ yields

$$e^{2x} - 1 = 2e^x \quad \text{or} \quad e^{2x} - 2e^x - 1 = 0$$

This is really a quadratic equation in disguise, as can be seen by rewriting it in the form

$$\left(e^x\right)^2 - 2e^x - 1 = 0$$

and letting $u = e^x$ to obtain

$$u^2 - 2u - 1 = 0$$

Solving for $u$ by the quadratic formula yields

$$u = \frac{2 \pm \sqrt{4+4}}{2} = \frac{2 \pm \sqrt{8}}{2} = 1 \pm \sqrt{2}$$

or, since $u = e^x$,

$$e^x = 1 \pm \sqrt{2}$$

But $e^x$ cannot be negative, so we discard the negative value $1 - \sqrt{2}$; thus,

$$e^x = 1 + \sqrt{2}$$

$$\ln e^x = \ln(1 + \sqrt{2})$$

$$x = \ln(1 + \sqrt{2}) \approx 0.881 \qquad \blacktriangleleft$$

**CHANGE OF BASE FORMULA FOR LOGARITHMS**

Scientific calculators generally provide keys for evaluating common logarithms and natural logarithms but have no keys for evaluating logarithms with other bases. However, this is not a serious deficiency because it is possible to express a logarithm with any base in terms of logarithms with any other base (see Exercise 40). For example, the following formula expresses a logarithm with base $b$ in terms of natural logarithms:

$$\log_b x = \frac{\ln x}{\ln b} \tag{9}$$

We can derive this result by letting $y = \log_b x$, from which it follows that $b^y = x$. Taking the natural logarithm of both sides of this equation we obtain $y \ln b = \ln x$, from which (9) follows.

**Example 4**

Use a calculating utility to evaluate $\log_2 5$ by expressing this logarithm in terms of natural logarithms.

*Solution.* From (9) we obtain

$$\log_2 5 = \frac{\ln 5}{\ln 2} \approx 2.321928 \qquad \blacktriangleleft$$

## LOGARITHMIC SCALES IN SCIENCE AND ENGINEERING

Logarithms are used in science and engineering to deal with quantities whose units vary over an excessively wide range of values. For example, the "loudness" of a sound can be measured by its **intensity** $I$ (in watts per square meter), which is related to the energy transmitted by the sound wave—the greater the intensity, the greater the transmitted energy, and the louder the sound is perceived by the human ear. However, intensity units are unwieldy because they vary over an enormous range. For example, a sound at the threshold of human hearing has an intensity of about $10^{-12}$ W/m$^2$, a close whisper has an intensity that is about 100 times the hearing threshold, and a jet engine at 50 meters has an intensity that is about $1,000,000,000,000 = 10^{12}$ times the hearing threshold. To see how logarithms can be used to reduce this wide spread, observe that if

$$y = \log x$$

then increasing $x$ by a *factor* of 10 *adds* 1 unit to $y$ since

$$\log 10x = \log 10 + \log x = 1 + y$$

Physicists and engineers take advantage of this property by measuring loudness in terms of the **sound level** $\beta$, which is defined by

$$\beta = 10 \log(I/I_0)$$

where $I_0 = 10^{-12}$ W/m$^2$ is a reference intensity close to the threshold of human hearing. The units of $\beta$ are **decibels** (dB), named in honor of the telephone inventor Alexander Graham Bell. With this scale of measurement, *multiplying* the intensity $I$ by a factor of 10 *adds* 10 dB to the sound level $\beta$ (verify). This results in a more tractable scale than intensity for measuring sound loudness (Table 4.2.2). Some other familiar logarithmic scales are the **Richter scale** used to measure earthquake intensity and the **pH** *scale* used to measure acidity in chemistry, both of which are discussed in the exercises.

**Table 4.2.2**

| $\beta$ (dB) | $I/I_0$ |
|---|---|
| 0 | $10^0 = 1$ |
| 10 | $10^1 = 10$ |
| 20 | $10^2 = 100$ |
| 30 | $10^3 = 1,000$ |
| 40 | $10^4 = 10,000$ |
| 50 | $10^5 = 100,000$ |
| $\vdots$ | $\vdots$ |
| 120 | $10^{12} = 1,000,000,000,000$ |

### Example 5

In 1976 the rock group The Who set the record for the loudest concert: 120 dB. By comparison, a jackhammer positioned at the same spot as The Who would have produced a sound level of 92 dB. What is the ratio of the sound intensity of The Who to the sound intensity of a jackhammer?

*Solution.* Let $I_1$ and $\beta_1 (= 120$ dB) denote the intensity and sound level of The Who, and let $I_2$ and $\beta_2 (= 92$ dB) denote the intensity and sound level of the jackhammer. Then

$$I_1/I_2 = (I_1/I_0)/(I_2/I_0)$$
$$\log(I_1/I_2) = \log(I_1/I_0) - \log(I_2/I_0)$$
$$10 \log(I_1/I_2) = 10 \log(I_1/I_0) - 10 \log(I_2/I_0)$$
$$10 \log(I_1/I_2) = \beta_1 - \beta_2 = 120 - 92 = 28$$
$$\log(I_1/I_2) = 2.8$$

Thus, $I_1/I_2 = 10^{2.8} \approx 631$, which tells us that the sound intensity of The Who was 631 times greater than a jackhammer!  ◄

Peter Townsend of the Who sustained permanent hearing reduction due to the high decibel level of his band's music.

## EXPONENTIAL AND LOGARITHMIC GROWTH

The growth patterns of $e^x$ and $\ln x$ illustrated by Table 4.2.3 are worth noting. Both functions increase as $x$ increases, but they increase in dramatically different ways—$e^x$ increases extremely rapidly and $\ln x$ increases extremely slowly. For example, at $x = 10$ the value of $e^x$ is over 22,000, but at $x = 1000$ the value of $\ln x$ has not even reached 7.

The table strongly suggests that $e^x \to +\infty$ as $x \to +\infty$. However, the growth of $\ln x$ is so slow that its limiting behavior as $x \to +\infty$ is not clear from the table. However, in spite of its slow growth, it is still true that $\ln x \to +\infty$ as $x \to +\infty$. To see that this is so, choose any positive number $M$ (as large as you like). The value of $\ln x$ will reach $M$ when $x = e^M$, since

$$\ln x = \ln(e^M) = M$$

**Table 4.2.3**

| $x$ | $e^x$ | $\ln x$ |
|---|---|---|
| 1 | 2.72 | 0.00 |
| 2 | 7.39 | 0.69 |
| 3 | 20.09 | 1.10 |
| 4 | 54.60 | 1.39 |
| 5 | 148.41 | 1.61 |
| 6 | 403.43 | 1.79 |
| 7 | 1096.63 | 1.95 |
| 8 | 2980.96 | 2.08 |
| 9 | 8103.08 | 2.20 |
| 10 | 22026.47 | 2.30 |
| 100 | $2.69 \times 10^{43}$ | 4.61 |
| 1000 | $1.97 \times 10^{434}$ | 6.91 |

Since $\ln x$ increases as $x$ increases, we can conclude that $\ln x > M$ for $x > e^M$; hence, $\ln x \to +\infty$ as $x \to +\infty$ since the values of $\ln x$ eventually exceed any positive number $M$ (Figure 4.2.5).

In summary,

$$\lim_{x \to +\infty} e^x = +\infty \qquad \lim_{x \to +\infty} \ln x = +\infty \qquad (10\text{--}11)$$

The following limits, which are consistent with Figure 4.2.5, can be deduced numerically by constructing appropriate tables of values (verify):

$$\lim_{x \to -\infty} e^x = 0 \qquad \lim_{x \to 0^+} \ln x = -\infty \qquad (12\text{--}13)$$

The following limits can be deduced numerically, but they can be seen more readily by noting that the graph of $y = e^{-x}$ is the reflection about the $y$-axis of the graph of $y = e^x$ (Figure 4.2.6):

$$\lim_{x \to +\infty} e^{-x} = 0 \qquad \lim_{x \to -\infty} e^{-x} = +\infty \qquad (14\text{--}15)$$

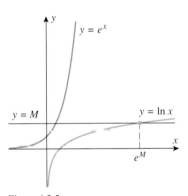

Figure 4.2.5

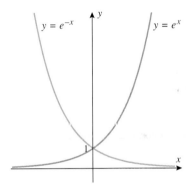

Figure 4.2.6

**EXERCISE SET 4.2** ⬚ Graphing Calculator ⬚ CAS

In Exercises 1 and 2, simplify the expression without using a calculating utility.

**1.** (a) $-8^{2/3}$      (b) $(-8)^{2/3}$      (c) $8^{-2/3}$

**2.** (a) $2^{-4}$      (b) $4^{1.5}$      (c) $9^{-0.5}$

In Exercises 3 and 4, use a calculating utility to approximate the expression. Round your answer to four decimal places.

**3.** (a) $2^{1.57}$      (b) $5^{-2.1}$

**4.** (a) $\sqrt[5]{24}$      (b) $\sqrt[8]{0.6}$

In Exercises 5 and 6, find the exact value of the expression without using a calculating utility.

**5.** (a) $\log_2 16$      (b) $\log_2 \left(\frac{1}{32}\right)$
     (c) $\log_4 4$      (d) $\log_9 3$

**6.** (a) $\log_{10}(0.001)$      (b) $\log_{10}(10^4)$
     (c) $\ln(e^3)$      (d) $\ln(\sqrt{e})$

In Exercises 7 and 8, use a calculating utility to approximate the expression. Round your answer to four decimal places.

**7.** (a) $\log 23.2$      (b) $\ln 0.74$

**8.** (a) $\log 0.3$  (b) $\ln \pi$

In Exercises 9 and 10 use the logarithm properties in Theorem 4.2.3 to rewrite the expression in terms of $r$, $s$, and $t$, where $r = \ln a$, $s = \ln b$, and $t = \ln c$.

**9.** (a) $\ln a^2 \sqrt{bc}$  (b) $\ln \dfrac{b}{a^3 c}$

**10.** (a) $\ln \dfrac{\sqrt[3]{c}}{ab}$  (b) $\ln \sqrt{\dfrac{ab^3}{c^2}}$

In Exercises 11 and 12, expand the logarithm in terms of sums, differences, and multiples of simpler logarithms.

**11.** (a) $\log(10x\sqrt{x-3})$  (b) $\ln \dfrac{x^2 \sin^3 x}{\sqrt{x^2+1}}$

**12.** (a) $\log \dfrac{\sqrt[3]{x+2}}{\cos 5x}$  (b) $\ln \sqrt{\dfrac{x^2+1}{x^3+5}}$

In Exercises 13–15, rewrite the expression as a single logarithm.

**13.** $4\log 2 - \log 3 + \log 16$

**14.** $\frac{1}{2}\log x - 3\log(\sin 2x) + 2$

**15.** $2\ln(x+1) + \frac{1}{3}\ln x - \ln(\cos x)$

In Exercises 16–25, solve for $x$ without using a calculating utility.

**16.** $\log_{10}(1+x) = 3$  **17.** $\log_{10}(\sqrt{x}) = -1$

**18.** $\ln(x^2) = 4$  **19.** $\ln(1/x) = -2$

**20.** $\log_3(3^x) = 7$  **21.** $\log_5(5^{2x}) = 8$

**22.** $\log_{10} x^2 + \log_{10} x = 30$

**23.** $\log_{10} x^{3/2} - \log_{10} \sqrt{x} = 5$

**24.** $\ln 4x - 3\ln(x^2) = \ln 2$

**25.** $\ln(1/x) + \ln(2x^3) = \ln 3$

In Exercises 26–31, solve for $x$ without using a calculating utility. Use the natural logarithm anywhere that logarithms are needed.

**26.** $3^x = 2$  **27.** $5^{-2x} = 3$

**28.** $3e^{-2x} = 5$  **29.** $2e^{3x} = 7$

**30.** $e^x - 2xe^x = 0$  **31.** $xe^{-x} + 2e^{-x} = 0$

In Exercises 32 and 33, rewrite the given equation as a quadratic equation in $u$, where $u = e^x$; then solve for $x$.

**32.** $e^{2x} - e^x = 6$  **33.** $e^{-2x} - 3e^{-x} = -2$

In Exercises 34–36, sketch the graph of the equation without using a graphing utility.

**34.** (a) $y = 1 + \ln(x-2)$  (b) $y = 3 + e^{x-2}$

**35.** (a) $y = \left(\frac{1}{2}\right)^{x-1} - 1$  (b) $y = \ln|x|$

**36.** (a) $y = 1 - e^{-x+1}$  (b) $y = 3\ln\sqrt[3]{x-1}$

**37.** Use a calculating utility and the change of base formula (9) to find the values of $\log_2 7.35$ and $\log_5 0.6$, rounded to four decimal places.

In Exercises 38 and 39, graph the functions on the same screen of a graphing utility. [Use the change of base formula (9), where needed].

**38.** $y = \ln x$, $y = e^x$, $\log x$, $10^x$

**39.** $y = \log_2 x$, $\ln x$, $\log_5 x$, $\log x$

**40.** (a) Derive the general change of base formula

$$\log_b x = \frac{\log_a x}{\log_a b}$$

(b) Use the result in part (a) to find the exact value of $(\log_2 81)(\log_3 32)$ without using a calculating utility. [*Hint:* Take $x = a$.]

**41.** Use a graphing utility to estimate where the graphs of $y = x^{0.2}$ and $y = \ln x$ intersect.

**42.** The United States public debt $D$, in billions of dollars, has been modeled as $D = 0.051517(1.1306727)^x$, where $x$ is the number of years since 1900. Based on this model, when did the debt first reach one trillion dollars?

**43.** (a) Is the curve in the accompanying figure the graph of an exponential function? Explain your reasoning.
(b) Find the equation of an exponential function that passes through the point $(4, 2)$.
(c) Find the equation of an exponential function that passes through the point $\left(2, \frac{1}{4}\right)$.
(d) Use a graphing utility to generate the graph of an exponential function that passes through the point $(2, 5)$.

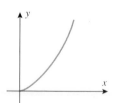

Figure Ex-43

**44.** (a) Make a conjecture about the general shape of the graph of $y = \log(\log x)$, and sketch the graph of this equation and $y = \log x$ in the same coordinate system.
(b) Check your work in part (a) with a graphing utility.

**45.** Find the fallacy in the following "proof" that $\frac{1}{8} > \frac{1}{4}$. Multiply both sides of the inequality $3 > 2$ by $\log \frac{1}{2}$ to get

$$3 \log \tfrac{1}{2} > 2 \log \tfrac{1}{2}$$
$$\log \left(\tfrac{1}{2}\right)^3 > \log \left(\tfrac{1}{2}\right)^2$$
$$\log \tfrac{1}{8} > \log \tfrac{1}{4}$$
$$\tfrac{1}{8} > \tfrac{1}{4}$$

**46.** Prove the four algebraic properties of logarithms in Theorem 4.2.3.

**47.** If equipment in the satellite of Example 2 requires 15 watts to operate correctly, what is the operational lifetime of the power supply?

**48.** The equation $Q = 12e^{-0.055t}$ gives the mass $Q$ in grams of radioactive potassium-42 that will remain from some initial quantity after $t$ hours of radioactive decay.
(a) How many grams were there initially?
(b) How many grams remain after 4 hours?
(c) How long will it take to reduce the amount of radioactive potassium-42 to half of the initial amount?

**49.** The acidity of a substance is measured by its pH value, which is defined by the formula

$$\text{pH} = -\log[H^+]$$

where the symbol $[H^+]$ denotes the concentration of hydrogen ions measured in moles per liter. Distilled water has a pH of 7; a substance is called *acidic* if it has pH $< 7$ and *basic* if it has pH $> 7$. Find the pH of each of the following substances and state whether it is acidic or basic.

| | SUBSTANCE | $[H^+]$ |
|---|---|---|
| (a) | Arterial blood | $3.9 \times 10^{-8}$ mol/L |
| (b) | Tomatoes | $6.3 \times 10^{-5}$ mol/L |
| (c) | Milk | $4.0 \times 10^{-7}$ mol/L |
| (d) | Coffee | $1.2 \times 10^{-6}$ mol/L |

**50.** Use the definition of pH in Exercise 49 to find $[H^+]$ in a solution having a pH equal to
(a) 2.44  (b) 8.06

**51.** The perceived loudness $\beta$ of a sound in decibels (dB) is related to its intensity $I$ in watts/square meter (W/m$^2$) by the equation

$$\beta = 10 \log(I/I_0)$$

where $I_0 = 10^{-12}$ W/m$^2$. Damage to the average ear occurs at 90 dB or greater. Find the decibel level of each of the following sounds and state whether it will cause ear damage.

| | SOUND | $I$ |
|---|---|---|
| (a) | Jet aircraft (from 500 ft) | $1.0 \times 10^2$ W/m$^2$ |
| (b) | Amplified rock music | $1.0$ W/m$^2$ |
| (c) | Garbage disposal | $1.0 \times 10^{-4}$ W/m$^2$ |
| (d) | TV (mid volume from 10 ft) | $3.2 \times 10^{-5}$ W/m$^2$ |

In Exercises 52–54, use the definition of the decibel level of a sound (see Exercise 51).

**52.** If one sound is three times as intense as another, how much greater is its decibel level?

**53.** According to one source, the noise inside a moving automobile is about 70 dB, while an electric blender generates 93 dB. Find the ratio of the intensity of the noise of the blender to that of the automobile.

**54.** Suppose that the decibel level of an echo is $\frac{2}{3}$ the decibel level of the original sound. If each echo results in another echo, how many echoes will be heard from a 120-dB sound given that the average human ear can hear a sound as low as 10 dB?

**55.** On the **Richter scale**, the magnitude $M$ of an earthquake is related to the released energy $E$ in joules (J) by the equation

$$\log E = 4.4 + 1.5M$$

(a) Find the energy $E$ of the 1906 San Francisco earthquake that registered $M = 8.2$ on the Richter scale.
(b) If the released energy of one earthquake is 10 times that of another, how much greater is its magnitude on the Richter scale?

**56.** Suppose that the magnitudes of two earthquakes differ by 1 on the Richter scale. Find the ratio of the released energy of the larger earthquake to that of the smaller earthquake. [*Note:* See Exercise 55 for terminology.]

In Exercises 57 and 58, use Formula (3) or (5), as appropriate, to find the limit.

**57.** Find $\lim\limits_{x \to 0} (1 - 2x)^{1/x}$. [*Hint:* Let $t = -2x$.]

**58.** Find $\lim\limits_{x \to +\infty} (1 + 3/x)^x$. [*Hint:* Let $t = 3/x$.]

## 4.3 IMPLICIT DIFFERENTIATION

*In earlier sections we were concerned with differentiating functions that were given by equations of the form $y = f(x)$. In this section we will consider methods for differentiating functions for which it is inconvenient or impossible to express them in this form.*

**FUNCTIONS DEFINED EXPLICITLY AND IMPLICITLY**

Up to now, we have been concerned with differentiating functions that are expressed in the form $y = f(x)$. An equation of this form is said to define $y$ **explicitly** as a function of $x$, because the variable $y$ appears alone on one side of the equation. However, sometimes functions are defined by equations in which $y$ is not alone on one side; for example, the equation

$$yx + y + 1 = x \tag{1}$$

is not of the form $y = f(x)$. However, this equation still defines $y$ as a function of $x$ since it can be rewritten as

$$y = \frac{x - 1}{x + 1}$$

Thus, we say that (1) defines $y$ **implicitly** as a function of $x$, the function being

$$f(x) = \frac{x - 1}{x + 1}$$

An equation in $x$ and $y$ can implicitly define more than one function of $x$; for example, if we solve the equation

$$x^2 + y^2 = 1 \tag{2}$$

for $y$ in terms of $x$, we obtain $y = \pm\sqrt{1 - x^2}$, so we have found two functions that are defined implicitly by (2), namely

$$f_1(x) = \sqrt{1 - x^2} \quad \text{and} \quad f_2(x) = -\sqrt{1 - x^2} \tag{3}$$

The graphs of these functions are the upper and lower semicircles of the circle $x^2 + y^2 = 1$ (Figure 4.3.1).

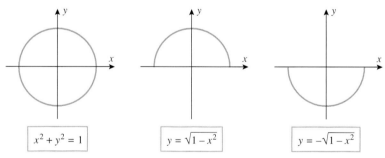

$$x^2 + y^2 = 1 \qquad y = \sqrt{1 - x^2} \qquad y = -\sqrt{1 - x^2}$$

Figure 4.3.1

Observe that the complete circle $x^2 + y^2 = 1$ does not pass the vertical line test, and hence is not itself the graph of a function of $x$. However, the upper and lower semicircles (which are only portions of the entire circle) do pass the vertical line test, and hence are graphs of functions. In general, if we have an equation in $x$ and $y$, then any segment of its graph that passes the vertical line test can be viewed as the graph of a function defined by the equation. Thus, we make the following definition.

**4.3.1** DEFINITION. We will say that a given equation in $x$ and $y$ defines the function $f$ **implicitly** if the graph of $y = f(x)$ coincides with some segment of the graph of the equation.

Thus, for example, the equation $x^2 + y^2 = 1$ defines the functions $f_1(x) = \sqrt{1 - x^2}$ and $f_2(x) = -\sqrt{1 - x^2}$ implicitly, since the graphs of these functions are segments of the circle $x^2 + y^2 = 1$.

Sometimes it may be difficult or impossible to solve an equation in $x$ and $y$ for $y$ in terms of $x$. For example, with persistence the equation

$$x^3 + y^3 = 3xy \tag{4}$$

can be solved for $y$ in terms of $x$, but the algebra is tedious and the resulting formulas are complicated. On the other hand, the equation

$$\sin(xy) = y$$

cannot be solved for $y$ in terms of $x$ by any elementary method. Thus, even though an equation in $x$ and $y$ may define one or more functions of $x$, it may not be practical or possible to find explicit formulas for those functions.

**GRAPHS OF EQUATIONS IN
x AND y**

When an equation in $x$ and $y$ cannot be solved for $y$ in terms of $x$ (or $x$ in terms of $y$), it may be difficult or time-consuming to obtain even a rough sketch of the graph, so the graphing of such equations is usually best left for graphing utilities. In particular, the CAS programs *Mathematica* and *Maple* both have "implicit plot" capabilities for graphing such equations. For example, Figure 4.3.2 shows the graph of Equation (4), which is called the **Folium of Descartes**.

FOR THE READER.    Figure 4.3.3 shows the graphs of two functions (in solid color) that are defined implicitly by (4). Sketch some more.

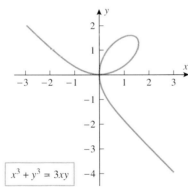

$x^3 + y^3 = 3xy$

Figure 4.3.2

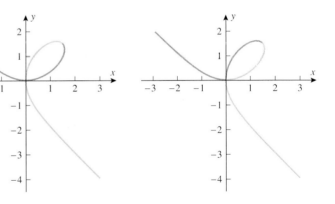

Figure 4.3.3

**IMPLICIT DIFFERENTIATION**

In general, it is not necessary to solve an equation for $y$ in terms of $x$ in order to differentiate the functions defined implicitly by the equation. To illustrate this, let us consider the simple equation

$$xy = 1 \tag{5}$$

One way to find $dy/dx$ is to rewrite this equation as

$$y = \frac{1}{x} \tag{6}$$

from which it follows that

$$\frac{dy}{dx} = -\frac{1}{x^2} \tag{7}$$

However, there is another way to obtain this derivative. We can differentiate both sides of

(5) *before* solving for $y$ in terms of $x$, treating $y$ as a (temporarily unspecified) differentiable function of $x$. With this approach we obtain

$$\frac{d}{dx}[xy] = \frac{d}{dx}[1]$$

$$x\frac{d}{dx}[y] + y\frac{d}{dx}[x] = 0$$

$$x\frac{dy}{dx} + y = 0$$

$$\frac{dy}{dx} = -\frac{y}{x}$$

If we now substitute (6) into the last expression, we obtain

$$\frac{dy}{dx} = -\frac{1}{x^2}$$

which agrees with (7). This method of obtaining derivatives is called ***implicit differentiation***.

### Example 1

Use implicit differentiation to find $dy/dx$ if $5y^2 + \sin y = x^2$.

$$\frac{d}{dx}[5y^2 + \sin y] = \frac{d}{dx}[x^2]$$

$$5\frac{d}{dx}[y^2] + \frac{d}{dx}[\sin y] = 2x$$

$$5\left(2y\frac{dy}{dx}\right) + (\cos y)\frac{dy}{dx} = 2x \qquad \boxed{\begin{array}{l}\text{The chain rule was}\\ \text{used here because}\\ y \text{ is a function of } x.\end{array}}$$

$$10y\frac{dy}{dx} + (\cos y)\frac{dy}{dx} = 2x$$

Solving for $dy/dx$ we obtain

$$\frac{dy}{dx} = \frac{2x}{10y + \cos y} \qquad\qquad (8)$$

Note that this formula involves both $x$ and $y$. In order to obtain a formula for $dy/dx$ that involves $x$ alone, we would have to solve the original equation for $y$ in terms of $x$ and then substitute in (8). However, it is impossible to do this, so we are forced to leave the formula for $dy/dx$ in terms of $x$ and $y$. ◄

### Example 2

Use implicit differentiation to find $d^2y/dx^2$ if $4x^2 - 2y^2 = 9$.

*Solution.*   Differentiating both sides of $4x^2 - 2y^2 = 9$ implicitly yields

$$8x - 4y\frac{dy}{dx} = 0$$

from which we obtain

$$\frac{dy}{dx} = \frac{2x}{y} \qquad\qquad (9)$$

Differentiating both sides of (9) implicitly yields

$$\frac{d^2y}{dx^2} = \frac{(y)(2) - (2x)(dy/dx)}{y^2} \qquad\qquad (10)$$

Substituting (9) into (10) and simplifying using the original equation, we obtain

$$\frac{d^2y}{dx^2} = \frac{2y - 2x(2x/y)}{y^2} = \frac{2y^2 - 4x^2}{y^3} = -\frac{9}{y^3} \qquad\qquad ◄$$

In Examples 1 and 2, the resulting formulas for $dy/dx$ involved both $x$ and $y$. Although it is usually more desirable to have the formula for $dy/dx$ expressed in terms of $x$ alone, having the formula in terms of $x$ and $y$ is not an impediment to finding slopes and equations of tangent lines provided the $x$- and $y$-coordinates of the point of tangency are known. This is illustrated in the following example.

### Example 3

Find the slopes of the tangent lines at $(2, -1)$ and $(2, 1)$ to $y^2 - x + 1 = 0$.

*Solution.* We could proceed by solving the equation for $y$ in terms of $x$, and then evaluating the derivative of $y = \sqrt{x - 1}$ at $(2, 1)$ and the derivative of $y = -\sqrt{x - 1}$ at $(2, -1)$ (Figure 4.3.4). However, implicit differentiation is more efficient since it gives the slopes of *both* functions. Differentiating implicitly yields

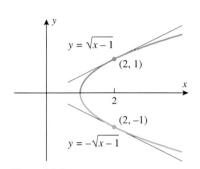

$y = \sqrt{x - 1}$

$(2, 1)$

$(2, -1)$

$y = -\sqrt{x - 1}$

Figure 4.3.4

$$\frac{d}{dx}[y^2 - x + 1] = \frac{d}{dx}[0]$$

$$\frac{d}{dx}[y^2] - \frac{d}{dx}[x] + \frac{d}{dx}[1] = \frac{d}{dx}[0]$$

$$2y\frac{dy}{dx} - 1 = 0$$

$$\frac{dy}{dx} = \frac{1}{2y}$$

At $(2, -1)$ we have $y = -1$, and at $(2, 1)$ we have $y = 1$, so the slopes of the tangent lines at those points are

$$m_{\text{tan}} = \frac{dy}{dx}\bigg|_{\substack{x=2 \\ y=-1}} = \frac{1}{2} \quad \text{and} \quad m_{\text{tan}} = \frac{dy}{dx}\bigg|_{\substack{x=2 \\ y=1}} -\frac{1}{2}$$  ◀

### Example 4

(a)  Use implicit differentiation to find $dy/dx$ for the Folium of Descartes $x^3 + y^3 = 3xy$.

(b)  Find an equation for the tangent line to the Folium of Descartes at the point $\left(\frac{3}{2}, \frac{3}{2}\right)$.

(c)  At what points is the tangent line to the Folium of Descartes horizontal?

*Solution (a).* Differentiating both sides of the given equation implicitly yields

$$\frac{d}{dx}[x^3 + y^3] = \frac{d}{dx}[3xy]$$

$$3x^2 + 3y^2\frac{dy}{dx} = 3x\frac{dy}{dx} + 3y$$

$$x^2 + y^2\frac{dy}{dx} = x\frac{dy}{dx} + y$$

$$(y^2 - x)\frac{dy}{dx} = y - x^2$$

$$\frac{dy}{dx} = \frac{y - x^2}{y^2 - x} \tag{11}$$

*Solution (b).* At the point $\left(\frac{3}{2}, \frac{3}{2}\right)$, we have $x = \frac{3}{2}$ and $y = \frac{3}{2}$, so from (11) the slope $m_{\text{tan}}$ of the tangent line at this point is

$$m_{\text{tan}} = \frac{dy}{dx}\bigg|_{\substack{x=3/2 \\ y=3/2}} = \frac{(3/2) - (3/2)^2}{(3/2)^2 - (3/2)} = -1$$

Thus, the equation of the tangent line at the point $\left(\frac{3}{2}, \frac{3}{2}\right)$ is

$$y - \tfrac{3}{2} = -1\left(x - \tfrac{3}{2}\right) \quad \text{or} \quad x + y = 3$$

which is consistent with Figure 4.3.5.

*Solution (c).* The tangent line is horizontal at the points where $dy/dx = 0$, and from (11) this occurs where $y - x^2 = 0$ or

$$y = x^2 \tag{12}$$

Substituting this expression for $y$ in the equation $x^3 + y^3 = 3xy$ for the curve yields

$$x^3 + \left(x^2\right)^3 = 3x^3$$
$$x^6 - 2x^3 = 0$$
$$x^3(x^3 - 2) = 0$$

whose solutions are $x = 0$ and $x = 2^{1/3}$. Thus, from (12), the tangent line is horizontal at the points $(0, 0)$ and $(2^{1/3}, 2^{2/3}) \approx (1.26, 1.59)$, which is consistent with Figure 4.3.6. ◀

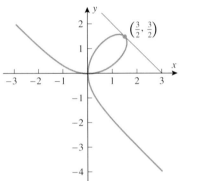

Figure 4.3.5                Figure 4.3.6

**DIFFERENTIABILITY OF FUNCTIONS DEFINED IMPLICITLY**

When differentiating implicitly, it is assumed that $y$ represents a differentiable function of $x$. If this is not so, then the resulting calculations may be nonsense. For example, if we differentiate the equation

$$x^2 + y^2 + 1 = 0 \tag{13}$$

we obtain

$$2x + 2y\frac{dy}{dx} = 0 \quad \text{or} \quad \frac{dy}{dx} = -\frac{x}{y}$$

However, this derivative is meaningless because (13) does not define a function of $x$. (The left side of the equation is greater than zero.)

Sometimes it is possible to identify points of nondifferentiability graphically. For example, the first function in Figure 4.3.3 is differentiable at each point of its domain because there are no corners, discontinuities, or points of vertical tangency; however, the second function is not differentiable at the origin.

In general, it can be difficult to determine analytically whether functions defined implicitly are differentiable, so we will leave such matters for more advanced courses.

**DERIVATIVES OF RATIONAL POWERS OF x**

In Theorem 3.3.8 and the discussion immediately following it, we showed that the formula

$$\frac{d}{dx}[x^n] = nx^{n-1} \tag{14}$$

holds for integer values of $n$ and for $n = \frac{1}{2}$. We will now use implicit differentiation to show that this formula holds for any rational exponent. More precisely, we will show that if $r$ is a rational number, then

$$\frac{d}{dx}[x^r] = rx^{r-1} \tag{15}$$

wherever $x^r$ and $x^{r-1}$ are defined. For now, we will assume without proof that $x^r$ is differentiable; the justification for this will be considered later.

Let $y = x^r$. Since $r$ is a rational number, it can be expressed as a ratio of integers $r = m/n$. Thus, $y = x^r = x^{m/n}$ can be written as

$$y^n = x^m \quad \text{so that} \quad \frac{d}{dx}[y^n] = \frac{d}{dx}[x^m]$$

By differentiating implicitly with respect to $x$ and using (14), we obtain

$$ny^{n-1}\frac{dy}{dx} = mx^{m-1} \tag{16}$$

But

$$y^{n-1} = \left[x^{m/n}\right]^{n-1} = x^{m-(m/n)}$$

Thus, (16) can be written as

$$nx^{m-(m/n)}\frac{dy}{dx} = mx^{m-1}$$

so that

$$\frac{dy}{dx} = \frac{m}{n}x^{(m/n)-1} = rx^{r-1}$$

which establishes (15).

## Example 5

From (15)

$$\frac{d}{dx}[x^{4/5}] = \frac{4}{5}x^{(4/5)-1} = \frac{4}{5}x^{-1/5}$$

$$\frac{d}{dx}[x^{-7/8}] = -\frac{7}{8}x^{(-7/8)-1} = -\frac{7}{8}x^{-15/8}$$

$$\frac{d}{dx}[\sqrt[3]{x}] = \frac{d}{dx}[x^{1/3}] = \frac{1}{3}x^{-2/3} = \frac{1}{3\sqrt[3]{x^2}} \qquad \blacktriangleleft$$

If $u$ is a differentiable function of $x$, and $r$ is a rational number, then the chain rule yields the following generalization of (15):

$$\frac{d}{dx}[u^r] = ru^{r-1} \cdot \frac{du}{dx} \tag{17}$$

## Example 6

$$\frac{d}{dx}\left[x^2 - x + 2\right]^{3/4} = \frac{3}{4}\left(x^2 - x + 2\right)^{-1/4} \cdot \frac{d}{dx}[x^2 - x + 2]$$

$$= \frac{3}{4}\left(x^2 - x + 2\right)^{-1/4}(2x - 1)$$

$$\frac{d}{dx}[(\sec \pi x)^{-4/5}] = -\frac{4}{5}(\sec \pi x)^{-9/5} \cdot \frac{d}{dx}[\sec \pi x]$$

$$= -\frac{4}{5}(\sec \pi x)^{-9/5} \cdot \sec \pi x \tan \pi x \cdot \pi$$

$$= -\frac{4\pi}{5}(\sec \pi x)^{-4/5} \tan \pi x \qquad \blacktriangleleft$$

We conclude this section with a brief discussion of the general relationship between the derivatives of $f$ and $f^{-1}$. For this purpose, suppose that both functions are differentiable, and let

$$y = f^{-1}(x) \tag{18}$$

Rewriting this equation as

$$x = f(y) \tag{19}$$

and differentiating implicitly with respect to $x$ yields

$$\frac{d}{dx}[x] = \frac{d}{dx}[f(y)]$$

$$1 = f'(y) \cdot \frac{dy}{dx}$$

$$\frac{dy}{dx} = \frac{1}{f'(y)} \tag{20}$$

Thus, from (18) we obtain the following formula that relates the derivative of $f^{-1}$ to the derivative of $f$.

$$\frac{d}{dx}[f^{-1}(x)] = \frac{1}{f'(f^{-1}(x))} \tag{21}$$

For example, if $f^{-1}(x) = \sqrt{x}$, then $f(x) = x^2$, so $f'(x) = 2x$; this formula implies that

$$\frac{d}{dx}[\sqrt{x}] = \frac{1}{2(f^{-1}(x))} = \frac{1}{2\sqrt{x}}$$

which is consistent with the known derivative formula for $\sqrt{x}$.

An alternative version of Formula (21) that uses dependent variables can be obtained by using (19) to rewrite $f'(y)$ as $dx/dy$, in which case (21) becomes

$$\frac{dy}{dx} = \frac{1}{dx/dy} \tag{22}$$

For example, if $y = \sqrt{x}$, then $x = y^2$. Thus, $dx/dy = 2y$, and (22) implies that

$$\frac{dy}{dx} = \frac{1}{2y} = \frac{1}{2\sqrt{x}}$$

which again is consistent with the known derivative formula for $y = \sqrt{x}$.

If an explicit formula can be obtained for the inverse of a function, then the differentiability and the derivative of the inverse can usually be deduced from that formula. However, if no explicit formula for the inverse can be obtained, then Theorem 4.1.7 is the primary mathematical tool for establishing differentiability of the inverse. Once differentiability has been established, a formula for the derivative of the inverse can be obtained either by differentiating implicitly or by using Formulas (21) or (22). The following example illustrates this.

### Example 7

We showed in Example 9 of Section 4.1 that the inverse of the function $f(x) = x^5 + x + 1$ is differentiable on the interval $(-\infty, +\infty)$. However, there is no way to obtain an explicit formula for $f^{-1}$, so we must resort to indirect methods to differentiate this function.

(a)  Find the derivative of $f^{-1}$ by using Formula (22).

(b)  Find the derivative of $f^{-1}$ by differentiating implicitly.

*Solution* (*a*).  If we let $y = f^{-1}(x)$, then

$$x = f(y) = y^5 + y + 1 \tag{23}$$

from which it follows that

$$\frac{dx}{dy} = 5y^4 + 1$$

$$\frac{dy}{dx} = \frac{1}{dx/dy} = \frac{1}{5y^4 + 1} \tag{24}$$

Although it would be preferable to have $dy/dx$ expressed as a function of $x$, we are forced to leave it in terms of $y$, since we cannot solve (23) for $y$ in terms of $x$.

*Solution (b).* Differentiating (23) implicitly with respect to $x$ yields

$$\frac{d}{dx}[x] = \frac{d}{dx}[y^5 + y + 1]$$

$$1 = 5y^4 \frac{dy}{dx} + \frac{dy}{dx}$$

$$1 = (5y^4 + 1)\frac{dy}{dx}$$

$$\frac{dy}{dx} = \frac{1}{5y^4 + 1}$$

which agrees with (24). ◀

## EXERCISE SET 4.3  ◻ Graphing Calculator   [C] CAS

In Exercises 1–8, find $dy/dx$.

**1.** $y = \sqrt[3]{2x - 5}$

**2.** $y = \sqrt[3]{2 + \tan(x^2)}$

**3.** $y = \left(\dfrac{x-1}{x+2}\right)^{3/2}$

**4.** $y = \sqrt{\dfrac{x^2 + 1}{x^2 - 5}}$

**5.** $y = x^3 \left(5x^2 + 1\right)^{-2/3}$

**6.** $y = \dfrac{(3 - 2x)^{4/3}}{x^2}$

**7.** $y = [\sin(3/x)]^{5/2}$

**8.** $y = \left[\cos(x^3)\right]^{-1/2}$

In Exercises 9 and 10: (a) Find $dy/dx$ by differentiating implicitly. (b) Solve the equation for $y$ as a function of $x$, and find $dy/dx$ from that equation. (c) Confirm that the two results are consistent by expressing the derivative in part (a) as a function of $x$ alone.

**9.** $x^3 + xy - 2x = 1$

**10.** $\sqrt{y} - e^x = 2$

In Exercises 11–20, find $dy/dx$ by implicit differentiation.

**11.** $x^2 + y^2 = 100$

**12.** $x^3 - y^3 = 6xy$

**13.** $x^2 y + 3xy^3 - x = 3$

**14.** $x^3 y^2 - 5x^2 y + x = 1$

**15.** $\dfrac{1}{y} + \dfrac{1}{x} = 1$

**16.** $x^2 = \dfrac{x + y}{x - y}$

**17.** $\sin(x^2 y^2) = x$

**18.** $x^2 = \dfrac{\cot y}{1 + \csc y}$

**19.** $\tan^3(xy^2 + y) = x$

**20.** $\dfrac{xy^3}{1 + \sec y} = 1 + y^4$

In Exercises 21–26, find $d^2 y/dx^2$ by implicit differentiation.

**21.** $3x^2 - 4y^2 = 7$

**22.** $x^3 + y^3 = 1$

**23.** $x^3 y^3 - 4 = 0$

**24.** $2xy - y^2 = 3$

**25.** $y + \sin y = x$

**26.** $x \cos y = y$

In Exercises 27 and 28, find the slope of the tangent line to the curve at the given points in two ways: first by solving for $y$ in terms of $x$ and differentiating and then by implicit differentiation.

**27.** $x^2 + y^2 = 1$; $(1/\sqrt{2}, 1/\sqrt{2})$, $(1/\sqrt{2}, -1/\sqrt{2})$

**28.** $y^2 - x + 1 = 0$; $(10, 3)$, $(10, -3)$

In Exercises 29–32, use implicit differentiation to find the slope of the tangent line to the curve at the specified point, and check that your answer is consistent with the accompanying graph.

**29.** $x^4 + y^4 = 16$; $(1, \sqrt[4]{15})$   [**Lamé's special quartic**]

**30.** $y^3 + yx^2 + x^2 - 3y^2 = 0$; $(0, 3)$   [**trisectrix**]

**31.** $2(x^2 + y^2)^2 = 25(x^2 - y^2)$; $(3, 1)$   [**lemniscate**]

**32.** $x^{2/3} + y^{2/3} = 4$; $(-1, 3\sqrt{3})$   [**four-cusped hypocycloid**]

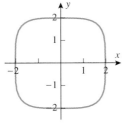

Figure Ex-29

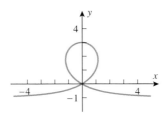

Figure Ex-30

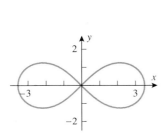

Figure Ex-31

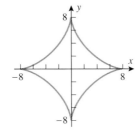

Figure Ex-32

C **33.** If you have a CAS, read the documentation on "implicit plotting," and then generate the four curves in Exercises 29–32.

C **34.** Curves with equations of the form $y^2 = x(x - a)(x - b)$, where $a < b$ are called **bipartite cubics**.
   (a) Use the implicit plotting capability of a CAS to graph the bipartite cubic $y^2 = x(x - 1)(x - 2)$.
   (b) At what points does the curve in part (a) have a horizontal tangent line?
   (c) Solve the equation in part (a) for $y$ in terms of $x$, and use the result to explain why the graph consists of two separate parts (i.e., is *bipartite*).
   (d) Graph the equation in part (a) without using the implicit plotting capability of the CAS.

C **35.** (a) Use the implicit plotting capability of a CAS to graph the rotated ellipse $x^2 - xy + y^2 = 4$.
   (b) Use the graph to estimate the $x$-coordinates of all horizontal tangent lines.
   (c) Find the exact values for the $x$-coordinates in part (b).

In Exercises 36–39, use implicit differentiation to find the specified derivative.

**36.** $\sqrt{u} + \sqrt{v} = 5$; $du/dv$      **37.** $a^4 - t^4 = 6a^2r$; $da/dt$

**38.** $y = \sin x$; $dx/dy$.

**39.** $a^2\omega^2 + b^2\lambda^2 = 1$   ($a, b$ constants); $d\omega/d\lambda$

**40.** At what point(s) is the tangent line to the curve $y^2 = 2x^3$ perpendicular to the line $4x - 3y + 1 = 0$?

**41.** Find the values of $a$ and $b$ for the curve $x^2y + ay^2 = b$ if the point $(1, 1)$ is on its graph and the tangent line at $(1, 1)$ has the equation $4x + 3y = 7$.

**42.** Find the coordinates of the point in the first quadrant at which the tangent line to the curve $x^3 - xy + y^3 = 0$ is parallel to the $x$-axis.

**43.** Find equations for two lines through the origin that are tangent to the curve $x^2 - 4x + y^2 + 3 = 0$.

**44.** Use implicit differentiation to show that the equation of the tangent line to the curve $y^2 = kx$ at $(x_0, y_0)$ is

$$y_0 y = \tfrac{1}{2}k(x + x_0)$$

**45.** Find $dy/dx$ if

$$2y^3t + t^3y = 1 \quad \text{and} \quad \frac{dt}{dx} = \frac{1}{\cos t}$$

In Exercises 46 and 47, find $dy/dt$ in terms of $x$, $y$, and $dx/dt$, assuming that $x$ and $y$ are differentiable functions of the variable $t$. [*Hint:* Differentiate both sides of the given equation with respect to $t$.]

**46.** $x^3y^2 + y = 3$      **47.** $xy^2 = \sin 3x$

**48.** (a) Show that $f(x) = x^{4/3}$ is differentiable at 0, but not twice differentiable at 0.
   (b) Show that $f(x) = x^{7/3}$ is twice differentiable at 0, but not three times differentiable at 0.
   (c) Find an exponent $k$ such that $f(x) = x^k$ is $(n - 1)$ times differentiable at 0, but not $n$ times differentiable at 0.

In Exercises 49 and 50, find all rational values of $r$ such that $y = x^r$ satisfies the given equation.

**49.** $3x^2y'' + 4xy' - 2y = 0$      **50.** $16x^2y'' + 24xy' + y = 0$

Two curves are said to be **orthogonal** if their tangent lines are perpendicular at each point of intersection, and two families of curves are said to be **orthogonal trajectories** of one another if each member of one family is orthogonal to each member of the other family. This terminology is used in Exercises 51 and 52.

**51.** The accompanying figure shows some typical members of the families of circles $x^2 + (y - c)^2 = c^2$ (black curves) and $(x - k)^2 + y^2 = k^2$ (gray curves). Show that these families are orthogonal trajectories of one another. [*Hint:* For the tangent lines to be perpendicular at a point of intersection, the slopes of those tangent lines must be negative reciprocals of one another.]

**52.** The accompanying figure shows some typical members of the families of hyperbolas $xy = c$ (black curves) and $x^2 - y^2 = k$ (gray curves), where $c \neq 0$ and $k \neq 0$. Use the hint in Exercise 51 to show that these families are orthogonal trajectories of one another.

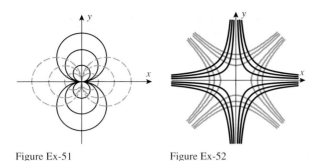

Figure Ex-51          Figure Ex-52

In Exercises 53–56, find the derivative of $f^{-1}$ by using Formula (22), and check your result by differentiating implicitly.

**53.** $f(x) = 5x^3 + x - 7$          **54.** $f(x) = 1/x^2, \ x > 0$

**55.** $f(x) = 2x^5 + x^3 + 1$          **56.** $f(x) = 5x - \sin 2x$

## 4.4 DERIVATIVES OF LOGARITHMIC AND EXPONENTIAL FUNCTIONS

*In this section we will obtain derivative formulas for logarithmic and exponential functions, and we will discuss the general relationship between the derivative of a one-to-one function and its inverse.*

**DERIVATIVES OF LOGARITHMIC FUNCTIONS**

The natural logarithm plays a special role in calculus that can be motivated by differentiating $\log_b x$, where $b$ is an arbitrary base. For this purpose, we will *assume* that $\log_b x$ is differentiable, and hence continuous, for $x > 0$. We will also need the limit

$$\lim_{v \to 0} (1 + v)^{1/v} = e$$

that was given in Formula (5) of Section 4.2 (with $x$ rather than $v$ as the variable).

Using the definition of a derivative, we obtain

$$\frac{d}{dx}[\log_b x] = \lim_{h \to 0} \frac{\log_b(x + h) - \log_b x}{h}$$

$$= \lim_{h \to 0} \frac{1}{h} \log_b\left(\frac{x + h}{x}\right) \qquad \begin{array}{l}\text{The quotient property of} \\ \text{logarithms in Theorem 4.2.3}\end{array}$$

$$= \lim_{h \to 0} \frac{1}{h} \log_b\left(1 + \frac{h}{x}\right)$$

$$= \lim_{v \to 0} \frac{1}{vx} \log_b(1 + v) \qquad \begin{array}{l}\text{Let } v = h/x \text{ and note} \\ \text{that } v \to 0 \text{ as } h \to 0.\end{array}$$

$$= \frac{1}{x} \lim_{v \to 0} \frac{1}{v} \log_b(1 + v) \qquad \begin{array}{l}1/x \text{ does not vary with } v, \text{ so it can} \\ \text{be moved through the limit sign.}\end{array}$$

$$= \frac{1}{x} \lim_{v \to 0} \log_b(1 + v)^{1/v} \qquad \begin{array}{l}\text{The power property of} \\ \text{logarithms in Theorem 4.2.3}\end{array}$$

$$= \frac{1}{x} \log_b\left[\lim_{v \to 0} (1 + v)^{1/v}\right] \qquad \begin{array}{l}\log_b x \text{ is continuous, so we can move} \\ \text{the limit through the function symbol.}\end{array}$$

$$= \frac{1}{x} \log_b e$$

Thus,

$$\frac{d}{dx}[\log_b x] = \frac{1}{x} \log_b e, \quad x > 0$$

But from Formula (9) of Section 4.2 we have $\log_b e = 1/\ln b$, so we can rewrite this derivative formula as

$$\frac{d}{dx}[\log_b x] = \frac{1}{x \ln b}, \quad x > 0 \tag{1}$$

In the special case where $b = e$, we have $\log_b e = \ln e = 1$, so this formula becomes

$$\frac{d}{dx}[\ln x] = \frac{1}{x}, \quad x > 0 \tag{2}$$

Thus, among all possible bases, the base $b = e$ produces the simplest derivative formula for $\log_b x$. This is one of the reasons why the natural logarithm function is preferred over other logarithms in calculus.

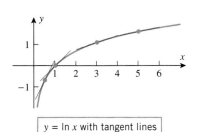

$y = \ln x$ with tangent lines

Figure 4.4.1

### Example 1

(a)   Figure 4.4.1 shows the graph of $y = \ln x$ and its tangent lines at the points $x = \frac{1}{2}, 1,$ 3, and 5. Find the slopes of those tangent lines.

(b)   Do you think that the graph of $y = \ln x$ has any horizontal tangent lines? Use the derivative of $\ln x$ to justify your answer.

*Solution (a).* From (2), the slopes of the tangent lines at the points $x = \frac{1}{2}, 1, 3,$ and 5 are $1/x = 2, 1, \frac{1}{3},$ and $\frac{1}{5}$, which is consistent with Figure 4.4.1.

*Solution (b).* From the graph of $y = \ln x$, it does not appear that there are any horizontal tangent lines. This is confirmed by the fact that $dy/dx = 1/x$ is not equal to zero for any real value of $x$.   ◀

If $u$ is a differentiable function of $x$, and if $u(x) > 0$, then applying the chain rule to (1) and (2) produces the following generalized derivative formulas:

$$\frac{d}{dx}[\log_b u] = \frac{1}{u \ln b} \cdot \frac{du}{dx} \quad \text{and} \quad \frac{d}{dx}[\ln u] = \frac{1}{u} \cdot \frac{du}{dx} \tag{3-4}$$

### Example 2

Find $\dfrac{d}{dx}[\ln(x^2 + 1)]$.

*Solution.* From (4) with $u = x^2 + 1$,

$$\frac{d}{dx}[\ln(x^2 + 1)] = \frac{1}{x^2 + 1} \cdot \frac{d}{dx}[x^2 + 1] = \frac{1}{x^2 + 1} \cdot 2x = \frac{2x}{x^2 + 1} \quad \blacktriangleleft$$

When possible, the properties of logarithms in Theorem 4.2.3 should be used to convert products, quotients, and exponents into sums, differences, and constant multiples *before* differentiating a function involving logarithms.

### Example 3

$$\frac{d}{dx}\left[\ln\left(\frac{x^2 \sin x}{\sqrt{1 + x}}\right)\right] = \frac{d}{dx}\left[2 \ln x + \ln(\sin x) - \frac{1}{2}\ln(1 + x)\right]$$

$$= \frac{2}{x} + \frac{\cos x}{\sin x} - \frac{1}{2(1 + x)}$$

$$= \frac{2}{x} + \cot x - \frac{1}{2 + 2x} \quad \blacktriangleleft$$

### Example 4

Find $\dfrac{d}{dx}[\ln |x|]$.

*Solution.* The function $\ln |x|$ is defined for all $x$, except $x = 0$; we will consider the cases $x > 0$ and $x < 0$ separately.

If $x > 0$, then $|x| = x$, so

$$\frac{d}{dx}[\ln |x|] = \frac{d}{dx}[\ln x] = \frac{1}{x}$$

If $x < 0$, then $|x| = -x$, so from (4) we have

$$\frac{d}{dx}[\ln |x|] = \frac{d}{dx}[\ln(-x)] = \frac{1}{(-x)} \cdot \frac{d}{dx}[-x] = \frac{1}{x}$$

Since the same formula results in both cases, we have shown that

$$\frac{d}{dx}[\ln |x|] = \frac{1}{x} \quad \text{if } x \neq 0 \tag{5}$$

### Example 5

From (5) and the chain rule,

$$\frac{d}{dx}[\ln |\sin x|] = \frac{1}{\sin x} \cdot \frac{d}{dx}[\sin x] = \frac{\cos x}{\sin x} = \cot x \qquad \blacktriangleleft$$

**LOGARITHMIC DIFFERENTIATION**

We now consider a technique called *logarithmic differentiation* that is useful for differentiating functions that are composed of products, quotients, and powers.

### Example 6

The derivative of

$$y = \frac{x^2 \sqrt[3]{7x - 14}}{\left(1 + x^2\right)^4} \tag{6}$$

is messy to calculate directly. However, if we first take the natural logarithm of both sides and then use its properties, we can write

$$\ln y = 2 \ln x + \tfrac{1}{3} \ln(7x - 14) - 4 \ln(1 + x^2)$$

Differentiating both sides with respect to $x$ yields

$$\frac{1}{y}\frac{dy}{dx} = \frac{2}{x} + \frac{7/3}{7x - 14} - \frac{8x}{1 + x^2} \tag{7}$$

Thus, on solving for $dy/dx$ and using (6) we obtain

$$\frac{dy}{dx} = \frac{x^2 \sqrt[3]{7x - 14}}{\left(1 + x^2\right)^4}\left[\frac{2}{x} + \frac{1}{3x - 6} - \frac{8x}{1 + x^2}\right] \tag{8}$$

REMARK.  Since $\ln y$ is defined only for $y > 0$, logarithmic differentiation of $y = f(x)$ is valid only on intervals where $f(x)$ is positive. Thus, the derivative obtained in the preceding example is valid on the interval $(2, +\infty)$, since the given function is positive for $x > 2$. However, the formula is actually valid on the interval $(-\infty, 2)$ as well. This can be seen by taking absolute values before proceeding with the logarithmic differentiation and noting that $\ln |y|$ is defined for all $y$ except $y = 0$. If we do this and simplify using properties of logarithms and absolute values, we obtain

$$\ln |y| = 2 \ln |x| + \tfrac{1}{3} \ln |7x - 14| - 4 \ln |1 + x^2|$$

Differentiating both sides with respect to $x$ yields (7), and hence results in (8).

In general, if the derivative of $y = f(x)$ is to be obtained by logarithmic differentiation, then the same formula for $dy/dx$ will result regardless of whether one first takes absolute values or not. Thus, a derivative formula obtained by logarithmic differentiation will be valid except perhaps at points where $f(x)$ is zero. The formula may, in fact, be valid at those points as well, but it is not guaranteed.

**DERIVATIVES OF IRRATIONAL POWERS OF $x$**

We know from Formula (15) of Section 4.3 that the differentiation formula

$$\frac{d}{dx}[x^r] = rx^{r-1} \tag{9}$$

holds for rational values of $r$. We will now use logarithmic differentiation to show that this formula holds if $r$ is *any* real number (rational or irrational). In our computations we will assume that $x^r$ is a differentiable function and that the familiar laws of exponents hold for real exponents.

Let $y = x^r$, where $r$ is a real number. The derivative $dy/dx$ can be obtained by logarithmic differentiation as follows:

$$\ln y = \ln x^r = r \ln x$$

$$\frac{d}{dx}[\ln y] = \frac{d}{dx}[r \ln x]$$

$$\frac{1}{y}\frac{dy}{dx} = \frac{r}{x}$$

$$\frac{dy}{dx} = \frac{r}{x}y = \frac{r}{x}x^r = rx^{r-1}$$

which establishes (9) for real values of $r$. Thus, for example,

$$\frac{d}{dx}[x^\pi] = \pi x^{\pi-1} \quad \text{and} \quad \frac{d}{dx}\left[x^{\sqrt{2}}\right] = \sqrt{2}x^{\sqrt{2}-1} \tag{10}$$

**DERIVATIVES OF EXPONENTIAL FUNCTIONS**

To obtain a derivative formula for the exponential function

$$y = b^x \tag{11}$$

we rewrite this equation as

$$x = \log_b y$$

and differentiate implicitly using (3) to obtain

$$1 = \frac{1}{y \ln b} \cdot \frac{dy}{dx}$$

which we can rewrite using (11) as

$$\frac{dy}{dx} = y \ln b = b^x \ln b$$

Thus, we have shown that if $b^x$ is a differentiable function, then its derivative with respect to $x$ is

$$\frac{d}{dx}[b^x] = b^x \ln b \tag{12}$$

In the special case where $b = e$ we have $\ln e = 1$, so that (12) becomes

$$\frac{d}{dx}[e^x] = e^x \tag{13}$$

Moreover, if $u$ is a differentiable function of $x$, then it follows from (12) and (13) that

$$\frac{d}{dx}[b^u] = b^u \ln b \cdot \frac{du}{dx} \qquad \text{and} \qquad \frac{d}{dx}[e^u] = e^u \cdot \frac{du}{dx} \tag{14–15}$$

REMARK.  It is important to distinguish between differentiating $b^x$ (variable exponent and constant base) and $x^b$ (variable base and constant exponent). For example, compare the derivative of $x^\pi$ in (10) to the following derivative of $\pi^x$, which is obtained from (12):

$$\frac{d}{dx}[\pi^x] = \pi^x \ln \pi$$

## Example 7

The following computations use (14) and (15).

$$\frac{d}{dx}[2^{\sin x}] = (2^{\sin x})(\ln 2) \cdot \frac{d}{dx}[\sin x] = (2^{\sin x})(\ln 2)(\cos x)$$

$$\frac{d}{dx}[e^{-2x}] = e^{-2x} \cdot \frac{d}{dx}[-2x] = -2e^{-2x}$$

$$\frac{d}{dx}\left[e^{x^3}\right] = e^{x^3} \cdot \frac{d}{dx}[x^3] = 3x^2 e^{x^3}$$

$$\frac{d}{dx}[e^{\cos x}] = e^{\cos x} \cdot \frac{d}{dx}[\cos x] = -(\sin x)e^{\cos x} \qquad \blacktriangleleft$$

## Example 8

A glass of lemonade with a temperature of $40°$F sits in a room whose temperature is a constant $70°$F. Using a principle of physics, called **Newton's Law of Cooling**, one can show that if the temperature of the lemonade reaches $52°$F in 1 hour, then the temperature $T$ of the lemonade as a function of the elapsed time $t$ is modeled approximately by the equation

$$T = 70 - 30e^{-0.5t}$$

where $T$ is in $°$F and $t$ is in hours. The graph of this equation, shown in Figure 4.4.2, confirms our everyday experience that the temperature of the lemonade gradually approaches the temperature of the room.

(a)  In words, what happens to the *rate* of temperature rise over time?

(b)  Use a derivative to confirm your conclusion.

*Solution (a).*  The rate of change of temperature with respect to time is the slope of the tangent line to the graph of $T$ versus $t$. As $t$ increases, these slopes decrease, so the temperature rises at an ever-decreasing rate.

*Solution (b).*  The rate of change of temperature with respect to time is

$$\frac{dT}{dt} = \frac{d}{dt}[70 - 30e^{-0.5t}] = -30(-0.5)e^{-0.5t} = 15e^{-0.5t}$$

As $t$ increases, this derivative decreases, which confirms the conclusion in part (a). $\qquad \blacktriangleleft$

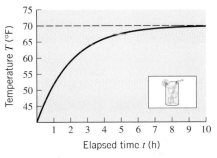

Figure 4.4.2

## EXERCISE SET 4.4  ~ Graphing Calculator  [c] CAS

In Exercises 1–30, find $dy/dx$.

**1.** $y = \ln 2x$

**2.** $y = \ln(x^3)$

**3.** $y = (\ln x)^2$

**4.** $y = \ln(\sin x)$

**5.** $y = \ln|\tan x|$

**6.** $y = \ln(2 + \sqrt{x})$

**7.** $y = \ln\left(\dfrac{x}{1 + x^2}\right)$

**8.** $y = \ln(\ln x)$

**9.** $y = \ln|x^3 - 7x^2 - 3|$

**10.** $y = x^3 \ln x$

**11.** $y = \sqrt{\ln x}$

**12.** $y = \sqrt{1 + \ln^2 x}$

**13.** $y = \cos(\ln x)$

**14.** $y = \sin^2(\ln x)$

**15.** $y = x^3 \log_2(3 - 2x)$

**16.** $y = x\left[\log_2(x^2 - 2x)\right]^3$

**17.** $y = \dfrac{x^2}{1 + \log x}$

**18.** $y = \dfrac{\log x}{1 + \log x}$

**19.** $y = e^{7x}$

**20.** $y = e^{-5x^2}$

**21.** $y = x^3 e^x$

**22.** $y = e^{1/x}$

**23.** $y = \dfrac{e^x - e^{-x}}{e^x + e^{-x}}$

**24.** $y = \sin(e^x)$

**25.** $y = e^{x \tan x}$

**26.** $y = \dfrac{e^x}{\ln x}$

**27.** $y = e^{(x - e^{3x})}$

**28.** $y = \exp(\sqrt{1 + 5x^3})$

**29.** $y = \ln(1 - xe^{-x})$

**30.** $y = \ln(\cos e^x)$

In Exercises 31 and 32, find $dy/dx$ by implicit differentiation.

**31.** $y + \ln xy = 1$

**32.** $y = \ln(x \tan y)$

In Exercises 33 and 34, use the method of Example 3 to help perform the indicated differentiation.

**33.** $\dfrac{d}{dx}\left[\ln\dfrac{\cos x}{\sqrt{4 - 3x^2}}\right]$

**34.** $\dfrac{d}{dx}\left[\ln\sqrt{\dfrac{x - 1}{x + 1}}\right]$

In Exercises 35–38, find $dy/dx$ using the method of logarithmic differentiation.

**35.** $y = x\sqrt[3]{1 + x^2}$

**36.** $y = \sqrt[5]{\dfrac{x - 1}{x + 1}}$

**37.** $y = \dfrac{(x^2 - 8)^{1/3}\sqrt{x^3 + 1}}{x^6 - 7x + 5}$

**38.** $y = \dfrac{\sin x \cos x \tan^3 x}{\sqrt{x}}$

In Exercises 39–42, find $f'(x)$ by Formula (14) and then by logarithmic differentiation.

**39.** $f(x) = 2^x$

**40.** $f(x) = 3^{-x}$

**41.** $f(x) = \pi^{\sin x}$

**42.** $f(x) = \pi^{x \tan x}$

In Exercises 43–46, find $dy/dx$ using the method of logarithmic differentiation.

**43.** $y = (x^3 - 2x)^{\ln x}$

**44.** $y = x^{\sin x}$

**45.** $y = (\ln x)^{\tan x}$

**46.** $y = (x^2 + 3)^{\ln x}$

**47.** Show that for any constants $A$ and $B$, the function

$$y = Ae^{2x} + Be^{-4x}$$

satisfies the equation

$$y'' + 2y' - 8y = 0$$

**48.** Show that for any constants $A$ and $k$, the function $y = Ae^{kt}$ satisfies the equation $dy/dt = ky$.

**49.** Let $f(x) = e^{kx}$ and $g(x) = e^{-kx}$. Find
(a) $f^{(n)}(x)$ 
(b) $g^{(n)}(x)$.

**50.** Find $dy/dt$ if $y = e^{-\lambda t}(A \sin \omega t + B \cos \omega t)$, where $A$, $B$, $\lambda$, and $\omega$ are constants.

**51.** Find $f'(x)$ if

$$f(x) = \dfrac{1}{\sqrt{2\pi}\sigma}\exp\left[-\dfrac{1}{2}\left(\dfrac{x - \mu}{\sigma}\right)^2\right]$$

where $\mu$ and $\sigma$ are constants and $\sigma \neq 0$.

**52.** Show that
(a) $y = xe^{-x}$ satisfies the equation $xy' = (1 - x)y$
(b) $y = xe^{-x^2/2}$ satisfies the equation $xy' = (1 - x^2)y$.

**53.** Find
(a) $\dfrac{d}{dx}[\log_x e]$
(b) $\dfrac{d}{dx}[\log_x 2]$.

**54.** Recall from Section 4.2 that the loudness $\beta$ of a sound in decibels (db) is given by $\beta = 10 \log(I/I_0)$, where $I$ is the intensity of the sound in watts per square meter (W/m²) and $I_0$ is a constant that is approximately the intensity of a sound at the threshold of human hearing. Find the rate of change of $\beta$ with respect to $I$ at the point where
(a) $I/I_0 = 10$ 
(b) $I/I_0 = 100$ 
(c) $I/I_0 = 1000$

**55.** The equilibrium constant $k$ of a balanced chemical reaction changes with the absolute temperature $T$ according to the law

$$k = k_0 \exp\left(-\dfrac{q(T - T_0)}{2T_0 T}\right)$$

where $k_0$, $q$, and $T_0$ are constants. Find the rate of change of $k$ with respect to $T$.

**56.** (a) Explain why Formula (12) cannot be used to find $(d/dx)[x^x]$.
(b) Find this derivative by logarithmic differentiation.

**57.** Find $f'(x)$ if $f(x) = x^e$.

**58.** Find a point on the graph of $y = e^{3x}$ at which the tangent line passes through the origin.

In Exercises 59 and 60, find the limit by interpreting the expression as an appropriate derivative.

**59.** (a) $\displaystyle \lim_{h \to 0} \frac{\ln(1 + h)}{h}$  (b) $\displaystyle \lim_{h \to 0} \frac{10^h - 1}{h}$

**60.** (a) $\displaystyle \lim_{h \to 0} \frac{\ln(e^2 + h) - 2}{h}$  (b) $\displaystyle \lim_{x \to 1} \frac{2^x - 2}{x - 1}$

**61.** (a) Make a conjecture about the shape of the graph of $y = \frac{1}{2}x - \ln x$, and draw a rough sketch.
 (b) Check your conjecture by graphing the equation over the interval $0 < x < 5$ with a graphing utility.
 (c) Show that the slopes of the tangent lines to the curve at $x = 1$ and $x = e$ have opposite signs.
 (d) What does part (c) imply about the existence of a horizontal tangent line to the curve? Explain your reasoning.
 (e) Find the exact $x$-coordinates of all horizontal tangent lines to the curve.

**62.** (a) Use a graphing utility to graph the function
$$f(x) = 2x^3 + x^2 - 20x + 4$$
over the interval $-5 < x < 5$.
 (b) Working with the graph in part (a), make a rough sketch of the graph of $f'(x)$ over the interval $-5 < x < 5$.
 (c) Check your work in part (b) by generating the graph of $f'(x)$ with a graphing utility.
 (d) Find the exact locations of the horizontal tangent lines to the graph of $f$ over the interval $-5 < x < 5$.
 (e) Confirm that the result in part (d) is consistent with the graph of $f'(x)$ in part (c).

**63.** (a) Sketch the curves $y = e^x$ and $y = -e^x$ in the same coordinate system; then make a conjecture about the general shape of the equation $y = e^x \cos \pi x$ for $x \geq 0$, and sketch its graph in the same coordinate system as the two exponential functions.
 (b) Check your conjecture in part (a) by using a graphing utility to generate the graphs of $y = e^x$, $y = -e^x$, and $y = e^x \cos \pi x$ in the same window for $0 \leq x \leq 3$.

**64.** Suppose that the population of oxygen-dependent bacteria in a pond is modeled by the equation
$$P(t) = \frac{60}{5 + 7e^{-t}}$$
where $P(t)$ is the population (in billions) $t$ days after an initial observation at time $t = 0$.
 (a) Use a graphing utility to graph the function $P(t)$.
 (b) In words, explain what happens to the population over time? Check your conclusion by finding $\lim_{t \to +\infty} P(t)$.
 (c) In words, what happens to the *rate* of population growth over time? Check your conclusion by graphing $P'(t)$.

**65.** Suppose that the population of deer on an island is modeled by the equation
$$P(t) = \frac{95}{5 - 4e^{-t/4}}$$
where $P(t)$ is the number of deer $t$ weeks after an initial observation at time $t = 0$.
 (a) Use a graphing utility to graph the function $P(t)$.
 (b) In words, explain what happens to the population over time. Check your conclusion by finding $\lim_{t \to +\infty} P(t)$.
 (c) In words, what happens to the *rate* of population growth over time? Check your conclusion by graphing $P'(t)$.

## 4.5 DERIVATIVES OF INVERSE TRIGONOMETRIC FUNCTIONS

*A common problem in trigonometry is to find an angle whose trigonometric functions are known. As you may recall, problems of this type involve the computation of "arc functions" such as* arcsin $x$, arccos $x$, arctan $x$, *and so forth. In this section we will consider this idea from the viewpoint of inverse functions, with the goal of developing derivative formulas for the inverse trigonometric functions.*

**INVERSE TRIGONOMETRIC FUNCTIONS**

None of the six basic trigonometric functions is one-to-one because they all repeat periodically and hence do not pass the horizontal line test. Thus, to define inverse trigonometric functions we must first restrict the domains of the trigonometric functions to make them one-to-one. The top part of Figure 4.5.1 shows how these restrictions are made for $\sin x$, $\cos x$, $\tan x$, and $\sec x$. (Inverses of $\cot x$ and $\csc x$ are of lesser importance and will be left for the exercises.) The inverses of these restricted functions are denoted by

$$\sin^{-1} x, \quad \cos^{-1} x, \quad \tan^{-1} x, \quad \sec^{-1} x$$

(or alternatively by arcsin $x$, arccos $x$, arctan $x$, arcsec $x$) and are defined as follows:

**4.5.1**   DEFINITION.   $\sin^{-1} x$ is the inverse of the restricted sine function

$$\sin x, \quad -\pi/2 \leq x \leq x/2$$

**4.5.2**   DEFINITION.   $\cos^{-1} x$ is the inverse of the restricted cosine function

$$\cos x, \quad 0 \leq x \leq \pi$$

**4.5.3**   DEFINITION.   $\tan^{-1} x$ is the inverse of the restricted tangent function

$$\tan x, \quad -\pi/2 < x < \pi/2$$

**4.5.4**   DEFINITION.*   $\sec^{-1} x$ is the inverse of the restricted secant function

$$\sec x, \quad 0 \leq x \leq \pi \text{ with } x \neq \pi/2$$

REMARK.   The notations $\sin^{-1} x$, $\cos^{-1} x, \ldots$ are reserved exclusively for the inverse trigonometric functions and are not used for reciprocals of the trigonometric functions. For example, to denote the reciprocal $1/\sin x$ in exponent form, we would write $(\sin x)^{-1}$ and *never* $\sin^{-1} x$.

The graphs of the inverse trigonometric functions, which are shown in the bottom part of Figure 4.5.1, are obtained by reflecting the graphs in the top part of the figure about the line $y = x$. If you have trouble visualizing these relationships, then look at Figure 4.5.2

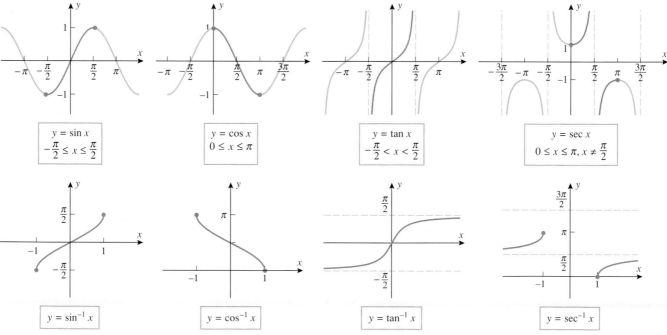

Figure 4.5.1

---

*There is no universal agreement on the definition of $\sec^{-1} x$, and some mathematicians prefer to restrict the domain of $\sec x$ so that $0 \leq x < \pi/2$ or $\pi \leq x < 3\pi/2$, which was the definition used in earlier editions of this text. Each definition has advantages and disadvantages, but we have changed to the current definition to conform with the conventions used by the CAS programs *Mathematica*, *Maple*, and *Derive*.

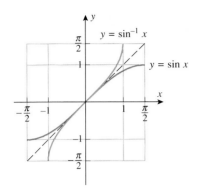

Figure 4.5.2

for a more detailed illustration for the inverse sine. It may also help to keep in mind that reflection about $y = x$ converts vertical lines to horizontal lines, and vice versa, and that $x$-intercepts reflect into $y$-intercepts, and vice versa.

Table 4.5.1 summarizes the basic properties of the inverse sine, cosine, tangent, and secant functions. You should confirm that the domains and ranges listed in this table are consistent with the graphs in the bottom part of Figure 4.5.1.

**Table 4.5.1**

| FUNCTION | DOMAIN | RANGE | BASIC RELATIONSHIPS |
|---|---|---|---|
| $\sin^{-1}$ | $[-1, 1]$ | $[-\pi/2, \pi/2]$ | $\sin^{-1}(\sin x) = x$ if $-\pi/2 \le x \le \pi/2$ <br> $\sin(\sin^{-1} x) = x$ if $-1 \le x \le 1$ |
| $\cos^{-1}$ | $[-1, 1]$ | $[0, \pi]$ | $\cos^{-1}(\cos x) = x$ if $0 \le x \le \pi$ <br> $\cos(\cos^{-1} x) = x$ if $-1 \le x \le 1$ |
| $\tan^{-1}$ | $(-\infty, +\infty)$ | $(-\pi/2, \pi/2)$ | $\tan^{-1}(\tan x) = x$ if $-\pi/2 < x < \pi/2$ <br> $\tan(\tan^{-1} x) = x$ if $-\infty < x < +\infty$ |
| $\sec^{-1}$ | $(-\infty, -1] \cup [1, +\infty)$ | $[0, \pi/2) \cup (\pi/2, \pi]$ | $\sec^{-1}(\sec x) = x$ if $0 \le x \le \pi, x \ne \pi/2$ <br> $\sec(\sec^{-1} x) = x$ if $|x| \ge 1$ |

**EVALUATING INVERSE TRIGONOMETRIC FUNCTIONS**

A common problem in trigonometry is to find an angle whose sine is known. For example, you might want to find an angle $x$ in radian measure such that

$$\sin x = \tfrac{1}{2} \tag{1}$$

and, more generally, for a given value of $y$ in the interval $-1 \le y \le 1$ you might want to solve the equation

$$\sin x = y \tag{2}$$

Because $\sin x$ repeats periodically, such equations have infinitely many solutions for $x$; however, if we solve this equation as

$$x = \sin^{-1} y$$

then we isolate the specific solution that lies in the interval $[-\pi/2, \pi/2]$, since this is the range of the inverse sine. For example, Figure 4.5.3 shows four solutions of Equation (1), namely, $-11\pi/6$, $-7\pi/6$, $\pi/6$, and $5\pi/6$. Of these, $\pi/6$ is the solution in the interval $[-\pi/2, \pi/2]$, so

$$\sin^{-1}\left(\tfrac{1}{2}\right) = \pi/6 \tag{3}$$

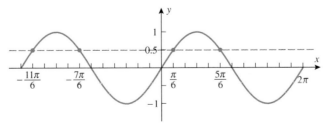

Figure 4.5.3

FOR THE READER. Refer to the documentation for your calculating utility to determine how to calculate inverse sines, inverse cosines, and inverse tangents; and then confirm Equation (3) numerically by showing that

$$\sin^{-1}(0.5) \approx 0.523598775598\ldots \approx \pi/6$$

In general, if we view $x = \sin^{-1} y$ as an angle in radian measure whose sine is $y$, then the restriction $-\pi/2 \le x \le \pi/2$ imposes the geometric requirement that the angle $x$ terminate in either the first or fourth quadrant or on an axis adjacent to those quadrants.

### Example 1

Find exact values of

(a)  $\sin^{-1}(1/\sqrt{2})$      (b)  $\sin^{-1}(-1)$

by inspection, and confirm your results numerically using a calculating utility.

*Solution (a).*  Because $\sin^{-1}(1/\sqrt{2}) > 0$, we can view $x = \sin^{-1}(1/\sqrt{2})$ as that angle in the first quadrant such that $\sin x = 1/\sqrt{2}$. Thus, $\sin^{-1}(1/\sqrt{2}) = \pi/4$. You can confirm this with your calculating utility by showing that $\sin^{-1}(1/\sqrt{2}) \approx 0.785 \approx \pi/4$.

*Solution (b).*  Because $\sin^{-1}(-1) < 0$, we can view $x = \sin^{-1}(-1)$ as an angle in the fourth quadrant (or an adjacent axis) such that $\sin x = -1$. Thus, $\sin^{-1}(-1) = -\pi/2$. You can confirm this with your calculating utility by showing that $\sin^{-1}(-1) \approx -1.57 \approx -\pi/2$.  ◀

FOR THE READER.    If $x = \cos^{-1} y$ is viewed as an angle in radian measure whose cosine is $y$, in what possible quadrants can $x$ lie? Answer the same question for $x = \tan^{-1} y$ and $x = \sec^{-1} y$.

FOR THE READER.    Most calculators do not provide a direct method for calculating inverse secants. In such situations the identity

$$\sec^{-1} x = \cos^{-1}(1/x) \tag{4}$$

is useful (Exercise 16). Use this formula to show that

$$\sec^{-1}(2.25) \approx 1.11 \quad \text{and} \quad \sec^{-1}(-2.25) \approx 2.03$$

If you have a calculating utility (such as a CAS) that can find $\sec^{-1} x$ directly, use it to check these values.

**IDENTITIES FOR INVERSE TRIGONOMETRIC FUNCTIONS**

If we interpret $\sin^{-1} x$ as an angle in radian measure whose sine is $x$, and if that angle is *nonnegative*, then we can represent $\sin^{-1} x$ geometrically as an angle in a right triangle in which the hypotenuse has length 1 and the side opposite to the angle $\sin^{-1} x$ has length $x$ (Figure 4.5.4a). By the Theorem of Pythagoras the side adjacent to the angle $\sin^{-1} x$ has length $\sqrt{1 - x^2}$. Moreover, the angle opposite to $\sin^{-1} x$ is $\cos^{-1} x$, since the cosine of that angle is $x$ (Figure 4.5.4b). This triangle motivates a number of useful identities involving inverse trigonometric functions that are valid for $-1 \le x \le 1$; for example,

$$\sin^{-1} x + \cos^{-1} x = \frac{\pi}{2} \tag{5}$$

$$\cos(\sin^{-1} x) = \sqrt{1 - x^2} \tag{6}$$

$$\sin(\cos^{-1} x) = \sqrt{1 - x^2} \tag{7}$$

$$\tan(\sin^{-1} x) = \frac{x}{\sqrt{1 - x^2}} \tag{8}$$

In a similar manner, $\tan^{-1} x$ and $\sec^{-1} x$ can be represented as angles in the right triangles shown in Figures 4.5.4c and 4.5.4d (verify). Those triangles reveal more useful identities; for example,

$$\sec(\tan^{-1} x) = \sqrt{1 + x^2} \tag{9}$$

$$\sin(\sec^{-1} x) = \frac{\sqrt{x^2 - 1}}{x} \qquad (x \ge 1) \tag{10a}$$

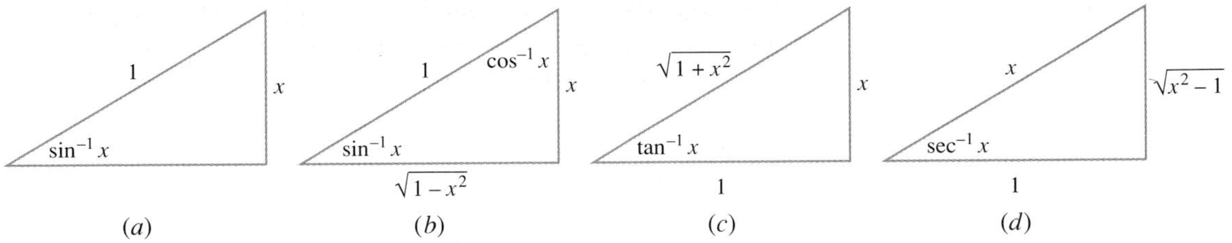

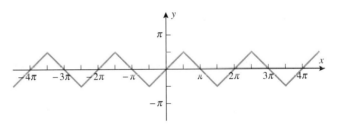

Figure 4.5.4

REMARK.    We leave it as an exercise to use (4) and (7) to obtain the following identity that is valid for $x \geq 1$ and $x \leq 1$ (Exercise 48):

$$\sin(\sec^{-1} x) = \frac{\sqrt{x^2 - 1}}{|x|} \qquad (|x| \geq 1) \tag{10b}$$

REMARK.    There is nothing to be gained by memorizing these identities; what is important to understand is the *method* that was used to obtain them.

Referring to Figure 4.5.1, observe that the inverse sine and inverse tangent are odd functions; that is,

$$\sin^{-1}(-x) = -\sin^{-1}(x) \quad \text{and} \quad \tan^{-1}(-x) = -\tan^{-1}(x) \tag{11–12}$$

### Example 2

Figure 4.5.5 shows a computer-generated graph of $y = \sin^{-1}(\sin x)$. One might think that this graph should be the line $y = x$, since $\sin^{-1}(\sin x) = x$. Why isn't it?

Figure 4.5.5

*Solution.*  The relationship $\sin^{-1}(\sin x) = x$ is valid on the interval $-\pi/2 \leq x \leq \pi/2$, so we can say with certainty that the graphs of $y = \sin^{-1}(\sin x)$ and $y = x$ coincide on this interval (which is confirmed by Figure 4.5.5). However, outside of this interval the relationship $\sin^{-1}(\sin x) = x$ need not hold. For example, if $x$ lies in the interval $\pi/2 \leq x \leq 3\pi/2$, then the quantity $x - \pi$ lies in the interval $-\pi/2 \leq x \leq \pi/2$, so

$$\sin^{-1}[\sin(x - \pi)] = x - \pi$$

Thus, by using the identity $\sin(x - \pi) = -\sin x$ and the fact that $\sin^{-1}$ is an odd function, we can express $\sin^{-1}(\sin x)$ as

$$\sin^{-1}(\sin x) = \sin^{-1}[-\sin(x - \pi)] = -\sin^{-1}[\sin(x - \pi)] = -(x - \pi)$$

This shows that on the interval $\pi/2 \leq x \leq 3\pi/2$ the graph of $y = \sin^{-1}(\sin x)$ coincides with the line $y = -(x - \pi)$, which has slope $-1$ and an $x$-intercept at $x = \pi$. This agrees with Figure 4.5.5.    ◀

**DERIVATIVES OF THE INVERSE TRIGONOMETRIC FUNCTIONS**

Recall that if $f$ is a one-to-one function whose derivative is known, then there are two basic ways to obtain a derivative formula for $f^{-1}(x)$—we can rewrite the equation $y = f^{-1}(x)$ as $x = f(y)$, and differentiate implicitly, or we can apply Formula (21) or (22) of Section 4.3. Here we will use implicit differentiation to obtain the derivative formula for $y = \sin^{-1} x$. Rewriting this equation as $x = \sin y$ and differentiating implicitly, we obtain

$$\frac{d}{dx}[x] = \frac{d}{dx}[\sin y]$$

$$1 = \cos y \cdot \frac{dy}{dx}$$

$$\frac{dy}{dx} = \frac{1}{\cos y} = \frac{1}{\cos(\sin^{-1} x)}$$

At this point we have succeeded in obtaining the derivative; however, this derivative formula can be simplified by applying Formula (6), which is derived from the triangle in Figure 4.5.6. This yields

$$\frac{dy}{dx} = \frac{1}{\sqrt{1 - x^2}}$$

Thus, we have shown that

$$\frac{d}{dx}[\sin^{-1} x] = \frac{1}{\sqrt{1 - x^2}} \tag{13}$$

If $u$ is a differentiable function of $x$, then (13) and the chain rule produce the following generalized derivative formula:

$$\frac{d}{dx}[\sin^{-1} u] = \frac{1}{\sqrt{1 - u^2}} \frac{du}{dx} \tag{14}$$

The method used to obtain this formula can also be used to obtain generalized derivative formulas for the other inverse trigonometric functions. These formulas are

$$\frac{d}{dx}[\sin^{-1} u] = \frac{1}{\sqrt{1 - u^2}} \frac{du}{dx}, \qquad \frac{d}{dx}[\cos^{-1} u] = -\frac{1}{\sqrt{1 - u^2}} \frac{du}{dx} \tag{15--16}$$

$$\frac{d}{dx}[\tan^{-1} u] = \frac{1}{1 + u^2} \frac{du}{dx}, \qquad \frac{d}{dx}[\cot^{-1} u] = -\frac{1}{1 + u^2} \frac{du}{dx} \tag{17--18}$$

$$\frac{d}{dx}[\sec^{-1} u] = \frac{1}{|u|\sqrt{u^2 - 1}} \frac{du}{dx}, \qquad \frac{d}{dx}[\csc^{-1} u] = -\frac{1}{|u|\sqrt{u^2 - 1}} \frac{du}{dx} \tag{19--20}$$

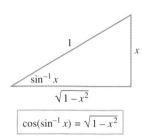

$$\cos(\sin^{-1} x) = \sqrt{1 - x^2}$$

Figure 4.5.6

**DIFFERENTIABILITY OF THE INVERSE TRIGONOMETRIC FUNCTIONS**

In the derivation of (13) we *assumed* that $\sin^{-1} x$ is differentiable. However, we can establish the differentiability with the help of Theorem 4.1.7. Since $f(x) = \sin x$ and $f'(x) = \cos x$, it follows from that theorem that the function $f^{-1}(x) = \sin^{-1} x$ will be differentiable at any point $x$ where $\cos(\sin^{-1} x) \neq 0$ or from (6) where $\sqrt{1 - x^2} \neq 0$. Thus, $\sin^{-1} x$ is differentiable on the interval $(-1, 1)$. The differentiability of the remaining inverse trigonometric functions can be deduced similarly.

REMARK. Observe that $\sin^{-1} x$ is only differentiable on the interval $(-1, 1)$, even though its domain is $[-1, 1]$. However, it can be seen geometrically that $\sin^{-1}$ cannot be differentiable at $x = \pm 1$. Just observe that the graph of $y = \sin x$ has horizontal tangent lines at $(\pi/2, 1)$ and $(-\pi/2, -1)$ and that these become points of vertical tangency for $y = \sin^{-1} x$ when reflected around the line $y = x$.

### Example 3

Find $dy/dx$ if

$$\text{(a)} \quad y = \sin^{-1}(x^3) \qquad \text{(b)} \quad y = \sec^{-1}(e^x)$$

*Solution* (*a*). From (14)

$$\frac{dy}{dx} = \frac{1}{\sqrt{1 - (x^3)^2}}(3x^2) = \frac{3x^2}{\sqrt{1 - x^6}}$$

*Solution* (*b*). From (19)

$$\frac{dy}{dx} = \frac{1}{e^x\sqrt{(e^x)^2 - 1}}(e^x) = \frac{1}{\sqrt{e^{2x} - 1}}$$

◀

## EXERCISE SET 4.5   ⊠ Graphing Calculator   [C] CAS

1. Find the exact value of
   (a) $\sin^{-1}(-1)$
   (b) $\cos^{-1}(-1)$
   (c) $\tan^{-1}(-1)$
   (d) $\sec^{-1}(1)$.

2. Find the exact value of
   (a) $\sin^{-1}\left(\frac{1}{2}\sqrt{3}\right)$
   (b) $\cos^{-1}\left(\frac{1}{2}\right)$
   (c) $\tan^{-1}(1)$
   (d) $\sec^{-1}(-2)$.

3. Given that $\theta = \sin^{-1}\left(-\frac{1}{2}\sqrt{3}\right)$, find the exact values of $\cos\theta$, $\tan\theta$, $\cot\theta$, $\sec\theta$, and $\csc\theta$.

4. Given that $\theta = \cos^{-1}\left(\frac{1}{2}\right)$, find the exact values of $\sin\theta$, $\tan\theta$, $\cot\theta$, $\sec\theta$, and $\csc\theta$.

5. Given that $\theta = \tan^{-1}\left(\frac{4}{3}\right)$, find the exact values of $\sin\theta$, $\cos\theta$, $\cot\theta$, $\sec\theta$, and $\csc\theta$.

6. Make a table that lists the six inverse trigonometric functions together with their domains and ranges.

7. Find the exact value of
   (a) $\sin^{-1}(\sin\pi/7)$
   (b) $\sin^{-1}(\sin\pi)$
   (c) $\sin^{-1}(\sin 5\pi/7)$
   (d) $\sin^{-1}(\sin 630)$.

8. Find the exact value of
   (a) $\cos^{-1}(\cos\pi/7)$
   (b) $\cos^{-1}(\cos\pi)$
   (c) $\cos^{-1}(\cos 12\pi/7)$
   (d) $\cos^{-1}(\cos 200)$.

9. For which values of $x$ is it true that
   (a) $\cos^{-1}(\cos x) = x$
   (b) $\cos(\cos^{-1}x) = x$
   (c) $\tan^{-1}(\tan x) = x$
   (d) $\tan(\tan^{-1}x) = x$

In Exercises 10 and 11, find the exact value of the given quantity.

10. $\sec\left[\sin^{-1}\left(-\frac{3}{4}\right)\right]$

11. $\sin\left[2\cos^{-1}\left(\frac{3}{5}\right)\right]$

In Exercises 12 and 13, complete the identities using the triangle method (Figure 4.5.4).

12. (a) $\sin(\cos^{-1}x) = ?$
    (b) $\tan(\cos^{-1}x) = ?$
    (c) $\csc(\tan^{-1}x) = ?$
    (d) $\sin(\tan^{-1}x) = ?$

13. (a) $\cos(\tan^{-1}x) = ?$
    (b) $\tan(\cot^{-1}x) = ?$
    (c) $\sin(\sec^{-1}x) = ?$
    (d) $\cot(\csc^{-1}x) = ?$

⊠ 14. (a) Use a calculating utility set to radian measure to make tables of values of $y = \sin^{-1}x$ and $y = \cos^{-1}x$ for $x = -1, -0.8, -0.6, \ldots, 0, 0.2, \ldots, 1$. Round your answers to two decimal places.
    (b) Plot the points obtained in part (a), and use the points to sketch the graphs of $y = \sin^{-1}x$ and $y = \cos^{-1}x$. Confirm that your sketches agree with those in Figure 4.5.1.
    (c) Use your graphing utility to graph $y = \sin^{-1}x$ and $y = \cos^{-1}x$; confirm that the graphs agree with those in Figure 4.5.1.

The function $\cot^{-1}x$ is defined to be the inverse of the restricted cotangent function

$$\cot x, \quad 0 < x < \pi$$

and the function $\csc^{-1}x$ is defined to be the inverse of the restricted cosecant function

$$\csc x, \quad -\pi/2 < x < \pi/2, \quad x \neq 0$$

Use these definitions in Exercises 15 and 16 and in all subsequent exercises that involve these functions.

15. (a) Sketch the graphs of $\cot^{-1}x$ and $\csc^{-1}x$.
    (b) Find the domain and range of $\cot^{-1}x$ and $\csc^{-1}x$.

**16.** Show that

(a) $\cot^{-1} x = \tan^{-1} \dfrac{1}{x}$, if $x > 0$

(b) $\sec^{-1} x = \cos^{-1} \dfrac{1}{x}$, if $|x| \geq 1$

(c) $\csc^{-1} x = \sin^{-1} \dfrac{1}{x}$, if $|x| \geq 1$.

**17.** Most scientific calculators have keys for the values of only $\sin^{-1} x$, $\cos^{-1} x$, and $\tan^{-1} x$. The formulas in Exercise 16 show how a calculator can be used to obtain values of $\cot^{-1} x$, $\sec^{-1} x$, and $\csc^{-1} x$ for positive values of $x$. Use these formulas and a calculator to find numerical values for each of the following inverse trigonometric functions. Express your answers in degrees, rounded to the nearest tenth of a degree.

(a) $\cot^{-1} 0.7$      (b) $\sec^{-1} 1.2$      (c) $\csc^{-1} 2.3$

---

In Exercises 18–20, use a calculating utility to approximate the solution of the equation. Where radians are used, express your answer to four decimal places, and where degrees are used, express it to the nearest tenth of a degree. [*Note:* In each part, the solution is not in the range of the relevant inverse trigonometric function.]

---

**18.** (a) $\sin x = 0.37$, $\pi/2 < x < \pi$

(b) $\sin \theta = -0.61$, $180° < \theta < 270°$

**19.** (a) $\cos x = -0.85$, $\pi < x < 3\pi/2$

(b) $\cos \theta = 0.23$, $-90° < \theta < 0°$

**20.** (a) $\tan x = 3.16$, $-\pi < x < -\pi/2$

(b) $\tan \theta = -0.45$, $90° < \theta < 180°$

---

In Exercises 21–28, find $dy/dx$.

---

**21.** (a) $y = \sin^{-1}\left(\frac{1}{3}x\right)$      (b) $y = \cos^{-1}(2x + 1)$

**22.** (a) $y = \tan^{-1}(x^2)$      (b) $y = \cot^{-1}(\sqrt{x})$

**23.** (a) $y = \sec^{-1}(x^7)$      (b) $y = \csc^{-1}(e^x)$

**24.** (a) $y = (\tan x)^{-1}$      (b) $y = \dfrac{1}{\tan^{-1} x}$

**25.** (a) $y = \sin^{-1}(1/x)$      (b) $y = \cos^{-1}(\cos x)$

**26.** (a) $y = \ln(\cos^{-1} x)$      (b) $y = \sqrt{\cot^{-1} x}$

**27.** (a) $y = e^x \sec^{-1} x$      (b) $y = x^2 \left(\sin^{-1} x\right)^3$

**28.** (a) $y = \sin^{-1} x + \cos^{-1} x$      (b) $y = \sec^{-1} x + \csc^{-1} x$

---

In Exercises 29 and 30, find $dy/dx$ by implicit differentiation.

---

**29.** $x^3 + x \tan^{-1} y = e^y$

**30.** $\sin^{-1}(xy) = \cos^{-1}(x - y)$

**31.** (a) Referring to the graph of $y = \sin^{-1} x$ in Figure 4.5.1, make a rough sketch of the graph of $dy/dx$.

(b) Check your work in part (a) using a graphing utility to generate the graph of $dy/dx$.

**32.** (a) Referring to the graph of $y = \tan^{-1} x$ in Figure 4.5.1, make a rough sketch of the graph of $dy/dx$.

(b) Check your work in part (a) using a graphing utility to generate the graph of $dy/dx$.

**33.** (a) Make a conjecture about the shape of the graph of

$$y = \cos^{-1}(\cos x)$$

and sketch the graph for $-4\pi \leq x \leq 4\pi$.

(b) Check your work in part (a) using a graphing utility to generate the graph.

**34.** (a) Use a calculating utility to evaluate $\sin^{-1}(\sin^{-1} 0.25)$ and $\sin^{-1}(\sin^{-1} 0.9)$, and explain what you think is happening in the second calculation.

(b) For what values of $x$ in the interval $-1 \leq x \leq 1$ will your calculating utility produce a real value for the function $\sin^{-1}(\sin^{-1} x)$?

**35.** In each part, sketch the graph and check your work with a graphing utility.

(a) $y = \sin^{-1} 2x$      (b) $y = \tan^{-1} \frac{1}{2}x$

**36.** In each part, express $x$ in terms of $k$ and an appropriate inverse trigonometric function. [*Note:* $x$ may not be in the range of the inverse trigonometric function.]

(a) $\cos x = k$,   if $0 < k < 1$ and $3\pi/2 < x < 2\pi$

(b) $\tan x = k$,   if $k < 0$ and $\pi/2 < x < \pi$

(c) $\sin 2x = k$,   if $0 < k < 1$ and $0 < x < \pi/2$.

[*Hint:* Consider the following cases: $0 < 2x < \pi/2$ and $\pi/2 < 2x < \pi$.]

**37.** An Earth-observing satellite has horizon sensors that can measure the angle $\theta$ shown in the accompanying figure. Let $R$ be the radius of the Earth (assumed spherical) and $h$ the distance between the satellite and the Earth's surface.

(a) Show that $\sin \theta = \dfrac{R}{R + h}$.

(b) Find $\theta$, to the nearest degree, for a satellite that is 10,000 km from the Earth's surface (use $R = 6378$ km).

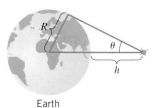

Earth

Figure Ex-37

**38.** The number of hours of daylight on a given day at a given point on the Earth's surface depends on the latitude $\lambda$ of the point, the angle $\gamma$ through which the Earth has moved in its orbital plane during the time period from the vernal equinox (March 21), and the angle of inclination $\phi$ of the Earth's axis of rotation measured from ecliptic north ($\phi \approx 23.55°$). The number of hours of daylight $h$ can be approximated by the formula

$$h = \begin{cases} 24, & D \geq 1 \\ 12 + \frac{2}{15} \sin^{-1} D, & |D| < 1 \\ 0, & D \leq -1 \end{cases}$$

where

$$D = \frac{\sin \phi \sin \gamma \tan \lambda}{\sqrt{1 - \sin^2 \phi \sin^2 \gamma}}$$

and $\sin^{-1} D$ is in degree measure. Given that Fairbanks, Alaska, is located at a latitude of $\lambda = 65°$ N and also that $\gamma = 90°$ on June 20 and $\gamma = 270°$ on December 20, approximate

(a) the maximum number of daylight hours at Fairbanks to one decimal place

(b) the minimum number of daylight hours at Fairbanks to one decimal place.

[*Note:* This problem was adapted from *TEAM, A Path to Applied Mathematics*, The Mathematical Association of America, Washington, D.C., 1985.]

**39.** A soccer player kicks a ball with an initial speed of 14 m/s at an angle $\theta$ with the horizontal (see the accompanying figure). The ball lands 18 m down the field. If air resistance is neglected, then the ball will have a parabolic trajectory and the horizontal range $R$ will be given by

$$R = \frac{v^2}{g} \sin 2\theta$$

where $v$ is the initial speed of the ball and $g$ is the acceleration due to gravity. Using $g = 9.8$ m/s$^2$, approximate two values of $\theta$, to the nearest degree, at which the ball could have been kicked. Which angle results in the shorter time of flight? Why?

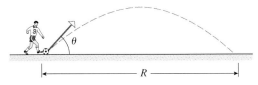

Figure Ex-39

**40.** The **law of cosines** states that

$$c^2 = a^2 + b^2 - 2ab \cos \theta$$

where $a, b$, and $c$ are the lengths of the sides of a triangle and $\theta$ is the angle formed by sides $a$ and $b$. Find $\theta$, to the nearest degree, for the triangle with $a = 2, b = 3$, and $c = 4$.

**41.** An airplane is flying at a constant height of 3000 ft above water at a speed of 400 ft/s. The pilot is to release a survival package so that it lands in the water at a sighted point $P$. If air resistance is neglected, then the package will follow a parabolic trajectory whose equation relative to the coordinate system in the accompanying figure is

$$y = 3000 - \frac{g}{2v^2} x^2$$

where $g$ is the acceleration due to gravity and $v$ is the speed

of the airplane. Using $g = 32$ ft/s$^2$, find the "line of sight" angle $\theta$, to the nearest degree, that will result in the package hitting the target point.

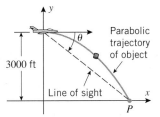

Figure Ex-41

**42.** A camera is positioned $x$ feet from the base of a missile launching pad (see the accompanying figure). If a missile of length $a$ feet is launched vertically, show that when the base of the missile is $b$ feet above the camera lens, the angle $\theta$ subtended at the lens by the missile is

$$\theta = \cot^{-1} \frac{x}{a+b} - \cot^{-1} \frac{x}{b}$$

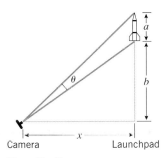

Figure Ex-42

**43.** Prove:

(a) $\sin^{-1}(-x) = -\sin^{-1} x$

(b) $\tan^{-1}(-x) = -\tan^{-1} x$.

**44.** Prove:

(a) $\cos^{-1}(-x) = \pi - \cos^{-1} x$

(b) $\sec^{-1}(-x) = \pi - \sec^{-1} x$, if $|x| \geq 1$.

**45.** Prove:

(a) $\sin^{-1} x = \tan^{-1} \dfrac{x}{\sqrt{1 - x^2}}$

(b) $\cos^{-1} x = \dfrac{\pi}{2} - \tan^{-1} \dfrac{x}{\sqrt{1 - x^2}}$.

**46.** Prove:

$$\tan^{-1} x + \tan^{-1} y = \tan^{-1} \left( \frac{x + y}{1 - xy} \right)$$

provided $-\pi/2 < \tan^{-1} x + \tan^{-1} y < \pi/2$. [*Hint:* Use an identity for $\tan(\alpha + \beta)$.]

**47.** Use the result in Exercise 46 to show that

(a) $\tan^{-1} \frac{1}{2} + \tan^{-1} \frac{1}{3} = \pi/4$

(b) $2 \tan^{-1} \frac{1}{3} + \tan^{-1} \frac{1}{7} = \pi/4$.

**48.** Use identities (4) and (7) to obtain identity (10b).

## 4.6 RELATED RATES

---

*In this section we will study related rates problems. In such problems one tries to find the rate at which some quantity is changing by relating it to other quantities whose rates of change are known.*

---

**RATES OF CHANGE USING THE CHAIN RULE**

Figure 4.6.1 shows a liquid draining through a conical filter. As the liquid drains, its volume $V$, height $h$, and radius $r$ are functions of the elapsed time $t$, and at each instant these variables are related by the equation

$$V = \frac{\pi}{3}r^2h$$

If we differentiate both sides of this equation implicitly with respect to $t$, then we obtain

$$\frac{dV}{dt} = \frac{\pi}{3}\left[r^2\frac{dh}{dt} + h\left(2r\frac{dr}{dt}\right)\right] = \frac{\pi}{3}\left(r^2\frac{dh}{dt} + 2rh\frac{dr}{dt}\right)$$

Thus, if the values of $r, h, dh/dt$, and $dr/dt$ are known, then this equation can be used to find $dV/dt$. Here are some specific examples that use this basic idea.

Figure 4.6.1

### Example 1

Assume that oil spilled from a ruptured tanker spreads in a circular pattern whose radius increases at a constant rate of 2 ft/s. How fast is the area of the spill increasing when the radius of the spill is 60 ft?

*Solution.* Let

$t$ = number of seconds elapsed from the time of the spill
$r$ = radius of the spill in feet after $t$ seconds
$A$ = area of the spill in square feet after $t$ seconds

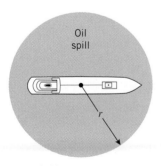

Figure 4.6.2

(Figure 4.6.2). We know the rate at which the radius is increasing, and we want to find the rate at which the area is increasing at the instant when $r = 60$; that is, we want to find

$$\left.\frac{dA}{dt}\right|_{r=60} \quad \text{given that} \quad \frac{dr}{dt} = 2 \text{ ft/s}$$

From the formula for the area of a circle we obtain

$$A = \pi r^2 \tag{1}$$

Because $A$ and $r$ are functions of $t$, we can differentiate both sides of (1) implicitly with respect to $t$ to obtain

$$\frac{dA}{dt} = 2\pi r\frac{dr}{dt}$$

Thus, when $r = 60$ the area of the spill is increasing at the rate of

$$\left.\frac{dA}{dt}\right|_{r=60} = 2\pi(60)(2) = 240\pi \text{ ft}^2/\text{s}$$

or approximately 754 ft$^2$/s.   ◀

With only minor variations, the method used in Example 1 can be used to solve a variety of related rates problems. The method consists of five steps:

> ### A Strategy for Solving Related Rates Problems
>
> **Step 1.**   Draw a figure and label the quantities that vary.
>
> **Step 2.**   Identify the rates of change that are known and the rate of change that is to be found.
>
> **Step 3.**   Find an equation that relates the quantity whose rate of change is to be found to the quantities whose rates of change are known.
>
> **Step 4.**   Differentiate both sides of this equation with respect to time and solve for the derivative that will give the unknown rate of change.
>
> **Step 5.**   Evaluate this derivative at the appropriate point.

### Example 2

A baseball diamond is a square whose sides are 90 ft long (Figure 4.6.3). Suppose that a player running from second base to third base has a speed of 30 ft/s at the instant when he is 20 ft from third base. At what rate is the player's distance from home plate changing at that instant?

*Solution.* Let

$t = $ number of seconds after the player leaves second base
$x = $ distance in feet from third base
$y = $ distance in feet from home plate

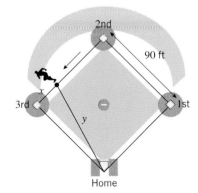

Figure 4.6.3

(Figure 4.6.3). The rate at which the distance from third base is changing is $dx/dt$, and the rate at which the distance from home plate is changing is $dy/dt$. We want to find

$$\left.\frac{dy}{dt}\right|_{x=20} \quad \text{given that} \quad \left.\frac{dx}{dt}\right|_{x=20} = -30 \text{ ft/s}$$

(Note that $dx/dt$ is negative because $x$ is decreasing with respect to $t$.) From the Theorem of Pythagoras we have

$$x^2 + 90^2 = y^2 \tag{2}$$

Differentiating both sides of this equation with respect to $t$ using the chain rule yields

$$2x\frac{dx}{dt} = 2y\frac{dy}{dt} \quad \text{or} \quad \frac{dy}{dt} = \frac{x}{y}\frac{dx}{dt} \tag{3}$$

When $x = 20$, it follows from (2) that

$$y = \sqrt{20^2 + 90^2} = \sqrt{8500} = 10\sqrt{85}$$

so that (3) yields

$$\left.\frac{dy}{dt}\right|_{x=20} = \frac{20}{10\sqrt{85}}(-30) = -\frac{60}{\sqrt{85}} \approx -6.51 \text{ ft/s}$$

The negative sign in the answer tells us that $y$ is decreasing, which makes sense physically from Figure 4.6.3.   ◀

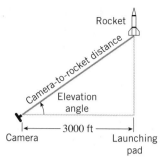

Figure 4.6.4

In Figure 4.6.4 we have shown a camera mounted at a point 3000 ft from the base of a rocket launching pad. Let us assume that the rocket rises vertically and the camera is to take a series of photographs of the rocket. Because the rocket will be rising, the elevation angle of the camera will have to vary at just the right rate to keep the rocket in sight. Moreover, because the camera-to-rocket distance will be changing constantly, the camera focusing mechanism will also have to vary at just the right rate to keep the picture sharp. The focusing problem is considered in the exercises, and the elevation problem is addressed in the following example:

### Example 3

If the rocket shown in Figure 4.6.4 is rising vertically at 880 ft/s when it is 4000 ft up, how fast must the camera elevation angle change at that instant to keep the rocket in sight?

*Solution.*   Let

$t$ = number of seconds elapsed from the time of launch
$\phi$ = camera elevation angle in radians after $t$ seconds
$x$ = height of the rocket in feet after $t$ seconds

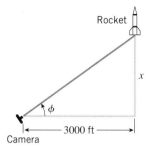

Figure 4.6.5

(Figure 4.6.5). At each instant the rate at which the camera elevation angle must change is $d\phi/dt$, and the rate at which the rocket is rising is $dx/dt$. We want to find

$$\left.\frac{d\phi}{dt}\right|_{x=4000} \qquad \text{given that} \qquad \left.\frac{dx}{dt}\right|_{x=4000} = 880 \text{ ft/s}$$

From Figure 4.6.5 we see that

$$\tan\phi = \frac{x}{3000} \tag{4}$$

Because $\phi$ and $x$ are functions of $t$, we can differentiate both sides of (4) with respect to $t$ to obtain

$$(\sec^2\phi)\frac{d\phi}{dt} = \frac{1}{3000}\frac{dx}{dt} \qquad \text{or} \qquad \frac{d\phi}{dt} = \frac{1}{3000\sec^2\phi}\frac{dx}{dt} \tag{5}$$

When $x = 4000$, it follows that

$$\sec\phi = \frac{5000}{3000} = \frac{5}{3}$$

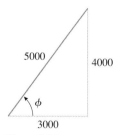

Figure 4.6.6

(Figure 4.6.6), so that from (5)

$$\left.\frac{d\phi}{dt}\right|_{x=4000} = \frac{1}{3000\left(\frac{5}{3}\right)^2}\cdot 880 = \frac{66}{625} \approx 0.11 \text{ radian/s} \approx 6.05 \text{ degrees/s}$$

*Alternative Solution.*   Instead of differentiating both sides of (4), we could have first solved the equation for $\phi$ and then differentiated:

$$\phi = \tan^{-1}\left(\frac{x}{3000}\right)$$

so

$$\frac{d\phi}{dt} = \frac{1}{1+\left(\frac{x}{3000}\right)^2}\cdot\frac{1}{3000}\frac{dx}{dt}$$

Thus,

$$\left.\frac{d\phi}{dt}\right|_{x=4000} = \frac{1}{1+\left(\frac{4000}{3000}\right)^2}\cdot\frac{880}{3000} = \frac{66}{625} \approx 0.11 \text{ radian/s} \approx 6.05 \text{ degrees/s}$$

which agrees with our previous result.   ◄

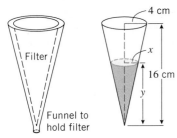

Figure 4.6.7

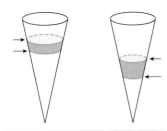

The same volume has drained, but the change in height is greater near the bottom than near the top.

Figure 4.6.8

### Example 4

Suppose that liquid is to be cleared of sediment by pouring it through a conical filter that is 16 cm high and has a radius of 4 cm at the top (Figure 4.6.7). Suppose also that the liquid flows out of the cone at a constant rate of 2 cm³/min.

(a) Do you think that the depth of the liquid will decrease at a constant rate? Give a verbal argument that justifies your conclusion.

(b) Find a formula that expresses the rate of change to the depth of the liquid in terms of the depth, and use that formula to determine whether your conclusion in part (a) is correct.

(c) At what rate is the depth of the liquid changing at the instant when the level is 8 cm deep?

*Solution (a).* For the volume of liquid to decrease by a *fixed amount*, it requires a greater decrease in depth when the cone is near empty than when it is near full (Figure 4.6.8). This suggests that for the volume to decrease at a constant rate, the depth must decrease at an increasing rate.

*Solution (b).* Let

$$t = \text{time elapsed from the initial observation (min)}$$
$$V = \text{volume of liquid in the cone at time } t \text{ (cm}^3)$$
$$y = \text{depth of the liquid in the cone at time } t \text{ (cm)}$$
$$x = \text{radius of the liquid surface at time } t \text{ (cm)}$$

(Figure 4.6.7). At each instant the rate at which the volume of liquid is changing is $dV/dt$, and the rate at which the depth is changing is $dy/dt$. We want to express $dy/dt$ in terms of $y$ given that $dV/dt$ has a constant value of $dV/dt = -2$. (We must use a minus sign here because $V$ *decreases* as $t$ increases.)

From the formula for the volume of a cone, the volume $V$, the radius $x$, and the depth $y$ are related by

$$V = \tfrac{1}{3}\pi x^2 y \tag{6}$$

If we differentiate both sides of (6) with respect to $t$, the right side will involve the quantity $dx/dt$. Since we have no direct information about $dx/dt$, it is desirable to eliminate $x$ from (6) before differentiating. This can be done using similar triangles. From Figure 4.6.7 we see that

$$\frac{x}{y} = \frac{4}{16} \quad \text{or} \quad x = \frac{1}{4}y$$

Substituting this expression in (6) gives

$$V = \frac{\pi}{48} y^3 \tag{7}$$

Differentiating both sides of (7) with respect to $t$ we obtain

$$\frac{dV}{dt} = \frac{\pi}{48}\left(3y^2 \frac{dy}{dt}\right)$$

or

$$\frac{dy}{dt} = \frac{16}{\pi y^2}\frac{dV}{dt} = \frac{16}{\pi y^2}(-2) = -\frac{32}{\pi y^2} \tag{8}$$

which expresses $dy/dt$ in terms of $y$. The minus sign tells us that $y$ is decreasing with time, and

$$\left|\frac{dy}{dt}\right| = \frac{32}{\pi y^2}$$

tells us how fast $y$ is decreasing. From this formula we see that $|dy/dt|$ increases as $y$ decreases, which confirms our conjecture in part (a) that the depth of the liquid decreases at an increasing rate as the liquid drains through the filter.

*Solution (c).* The rate at which the depth is changing when the depth is 8 cm can be obtained from (8) with $y = 8$:

$$\left.\frac{dy}{dt}\right|_{y=8} = -\frac{32}{\pi(8^2)} = -\frac{1}{2\pi} \approx -0.16 \text{ cm/min} \qquad \blacktriangleleft$$

## EXERCISE SET 4.6

1. Let $A$ be the area of a square whose sides have length $x$, and assume that $x$ varies with the time $t$.
   (a) Draw a picture of the square with the labels $A$ and $x$ placed appropriately.
   (b) Write an equation that relates $A$ and $x$.
   (c) Use the equation in part (b) to find an equation that relates $dA/dt$ and $dx/dt$.
   (d) At a certain instant the sides are 3 ft long and increasing at a rate of 2 ft/min. How fast is the area increasing at that instant?

2. Let $A$ be the area of a circle of radius $r$, and assume that $r$ increases with the time $t$.
   (a) Draw a picture of the circle with the labels $A$ and $r$ placed appropriately.
   (b) Write an equation that relates $A$ and $r$.
   (c) Use the equation in part (b) to find an equation that relates $dA/dt$ and $dr/dt$.
   (d) At a certain instant the radius is 5 cm and increasing at the rate of 2 cm/s. How fast is the area increasing at that instant?

3. Let $V$ be the volume of a cylinder having height $h$ and radius $r$, and assume that $h$ and $r$ vary with time.
   (a) How are $dV/dt$, $dh/dt$, and $dr/dt$ related?
   (b) At a certain instant, the height is 6 in and increasing at 1 in/s, while the radius is 10 in and decreasing at 1 in/s. How fast is the volume changing at that instant? Is the volume increasing or decreasing at that instant?

4. Let $l$ be the length of a diagonal of a rectangle whose sides have lengths $x$ and $y$, and assume that $x$ and $y$ vary with time.
   (a) How are $dl/dt$, $dx/dt$, and $dy/dt$ related?
   (b) If $x$ increases at a constant rate of $\frac{1}{2}$ ft/s and $y$ decreases at a constant rate of $\frac{1}{4}$ ft/s, how fast is the size of the diagonal changing when $x = 3$ ft and $y = 4$ ft? Is the diagonal increasing or decreasing at that instant?

5. Let $\theta$ (in radians) be an acute angle in a right triangle, and let $x$ and $y$, respectively, be the lengths of the sides adjacent and opposite $\theta$. Suppose also that $x$ and $y$ vary with time.
   (a) How are $d\theta/dt$, $dx/dt$, and $dy/dt$ related?

(b) At a certain instant, $x = 2$ units and is increasing at 1 unit/s, while $y = 2$ units and is decreasing at $\frac{1}{4}$ unit/s. How fast is $\theta$ changing at that instant? Is $\theta$ increasing or decreasing at that instant?

6. Suppose that $z = x^3 y^2$, where both $x$ and $y$ are changing with time. At a certain instant when $x = 1$ and $y = 2$, $x$ is decreasing at the rate of 2 units/s, and $y$ is increasing at the rate of 3 units/s. How fast is $z$ changing at this instant? Is $z$ increasing or decreasing?

7. The minute hand of a certain clock is 4 in long. Starting from the moment when the hand is pointing straight up, how fast is the area of the sector that is swept out by the hand increasing at any instant during the next revolution of the hand?

8. A stone dropped into a still pond sends out a circular ripple whose radius increases at a constant rate of 3 ft/s. How rapidly is the area enclosed by the ripple increasing at the end of 10 s?

9. Oil spilled from a ruptured tanker spreads in a circle whose area increases at a constant rate of 6 mi²/h. How fast is the radius of the spill increasing when the area is 9 mi²?

10. A spherical balloon is inflated so that its volume is increasing at the rate of 3 ft³/min. How fast is the diameter of the balloon increasing when the radius is 1 ft?

11. A spherical balloon is to be deflated so that its radius decreases at a constant rate of 15 cm/min. At what rate must air be removed when the radius is 9 cm?

12. A 17-ft ladder is leaning against a wall. If the bottom of the ladder is pulled along the ground away from the wall at a constant rate of 5 ft/s, how fast will the top of the ladder be moving down the wall when it is 8 ft above the ground?

13. A 13-ft ladder is leaning against a wall. If the top of the ladder slips down the wall at a rate of 2 ft/s, how fast will the foot be moving away from the wall when the top is 5 ft above the ground?

14. A 10-ft plank is leaning against a wall. If at a certain instant the bottom of the plank is 2 ft from the wall and is being

pushed toward the wall at the rate of 6 in/s, how fast is the acute angle that the plank makes with the ground increasing?

**15.** A softball diamond is a square whose sides are 60 ft long. Suppose that a player running from first to second base has a speed of 25 ft/s at the instant when she is 10 ft from second base. At what rate is the player's distance from home plate changing at that instant?

**16.** A rocket, rising vertically, is tracked by a radar station that is on the ground 5 mi from the launchpad. How fast is the rocket rising when it is 4 mi high and its distance from the radar station is increasing at a rate of 2000 mi/h?

**17.** For the camera and rocket shown in Figure 4.6.4, at what rate is the camera-to-rocket distance changing when the rocket is 4000 ft up and rising vertically at 880 ft/s?

**18.** For the camera and rocket shown in Figure 4.6.4, at what rate is the rocket rising when the elevation angle is $\pi/4$ radians and increasing at a rate of 0.2 radian/s?

**19.** A satellite is in an elliptical orbit around the Earth. Its distance $r$ (in miles) from the center of the Earth is given by

$$r = \frac{4995}{1 + 0.12 \cos \theta}$$

where $\theta$ is the angle measured from the point on the orbit nearest the Earth's surface (see the accompanying figure).
(a) Find the altitude of the satellite at *perigee* (the point nearest the surface of the Earth) and at *apogee* (the point farthest from the surface of the Earth). Use 3960 mi as the radius of the Earth.
(b) At the instant when $\theta$ is $120°$, the angle $\theta$ is increasing at the rate of $2.7°/\text{min}$. Find the altitude of the satellite and the rate at which the altitude is changing at this instant. Express the rate in units of mi/min.

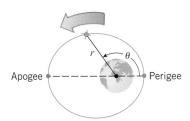

Figure Ex-19

**20.** An aircraft is flying horizontally at a constant height of 4000 ft above a fixed observation point (see the accompanying figure). At a certain instant the angle of elevation $\theta$ is $30°$ and decreasing, and the speed of the aircraft is 300 mi/h.
(a) How fast is $\theta$ decreasing at this instant? Express the result in units of degrees/s.
(b) How fast is the distance between the aircraft and the observation point changing at this instant? Express the result in units of ft/s. Use 1 mi = 5280 ft.

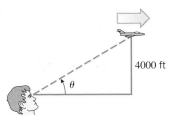

Figure Ex-20

**21.** A conical water tank with vertex down has a radius of 10 ft at the top and is 24 ft high. If water flows into the tank at a rate of 20 ft³/min, how fast is the depth of the water increasing when the water is 16 ft deep?

**22.** Grain pouring from a chute at the rate of 8 ft³/min forms a conical pile whose altitude is always twice its radius. How fast is the altitude of the pile increasing at the instant when the pile is 6 ft high?

**23.** Sand pouring from a chute forms a conical pile whose height is always equal to the diameter. If the height increases at a constant rate of 5 ft/min, at what rate is sand pouring from the chute when the pile is 10 ft high?

**24.** Wheat is poured through a chute at the rate of 10 ft³/min, and falls in a conical pile whose bottom radius is always half the altitude. How fast will the circumference of the base be increasing when the pile is 8 ft high?

**25.** An aircraft is climbing at a $30°$ angle to the horizontal. How fast is the aircraft gaining altitude if its speed is 500 mi/h?

**26.** A boat is pulled into a dock by means of a rope attached to a pulley on the dock (see the accompanying figure). The rope is attached to the bow of the boat at a point 10 ft below the pulley. If the rope is pulled through the pulley at a rate of 20 ft/min, at what rate will the boat be approaching the dock when 125 ft of rope is out?

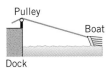

Figure Ex-26

**27.** For the boat in Exercise 26, how fast must the rope be pulled if we want the boat to approach the dock at a rate of 12 ft/min at the instant when 125 ft of rope is out?

**28.** A man 6 ft tall is walking at the rate of 3 ft/s toward a streetlight 18 ft high (see the accompanying figure).
(a) At what rate is his shadow length changing?
(b) How fast is the tip of his shadow moving?

Figure Ex-28

**29.** A beacon that makes one revolution every 10 s is located on a ship anchored 4 kilometers from a straight shoreline. How fast is the beam moving along the shoreline when it makes an angle of 45° with the shore?

**30.** An aircraft is flying at a constant altitude with a constant speed of 600 mi/h. An antiaircraft missile is fired on a straight line perpendicular to the flight path of the aircraft so that it will hit the aircraft at a point $P$ (see the accompanying figure). At the instant the aircraft is 2 mi from the impact point $P$ the missile is 4 mi from $P$ and flying at 1200 mi/h. At that instant, how rapidly is the distance between missile and aircraft decreasing?

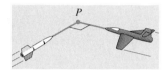

Figure Ex-30

**31.** Solve Exercise 30 under the assumption that the angle between the flight paths is 120° instead of the assumption that the paths are perpendicular. [*Hint:* Use the law of cosines.]

**32.** A police helicopter is flying due north at 100 mi/h and at a constant altitude of $\frac{1}{2}$ mi. Below, a car is traveling west on a highway at 75 mi/h. At the moment the helicopter crosses over the highway the car is 2 mi east of the helicopter.
(a) How fast is the distance between the car and helicopter changing at the moment the helicopter crosses the highway?
(b) Is the distance between the car and helicopter increasing or decreasing at that moment?

**33.** A particle is moving along the curve whose equation is

$$\frac{xy^3}{1+y^2} = \frac{8}{5}$$

Assume that the $x$-coordinate is increasing at the rate of 6 units/s when the particle is at the point $(1, 2)$.
(a) At what rate is the $y$-coordinate of the point changing at that instant?
(b) Is the particle rising or falling at that instant?

**34.** A point $P$ is moving along the curve whose equation is $y = \sqrt{x^3 + 17}$. When $P$ is at $(2, 5)$, $y$ is increasing at the rate of 2 units/s. How fast is $x$ changing?

**35.** A point $P$ is moving along the line whose equation is $y = 2x$. How fast is the distance between $P$ and the point $(3, 0)$ changing at the instant when $P$ is at $(3, 6)$ if $x$ is decreasing at the rate of 2 units/s at that instant?

**36.** A point $P$ is moving along the curve whose equation is $y = \sqrt{x}$. Suppose that $x$ is increasing at the rate of 4 units/s when $x = 3$.
(a) How fast is the distance between $P$ and the point $(2, 0)$ changing at this instant?
(b) How fast is the angle of inclination of the line segment from $P$ to $(2, 0)$ changing at this instant?

**37.** A particle is moving along the curve $y = x \ln x$. Find all values of $x$ at which the rate of change of $y$ with respect to time is three times that of $x$. [Assume that $dx/dt$ is never zero.]

**38.** A particle is moving along the curve $16x^2 + 9y^2 = 144$. Find all points $(x, y)$ at which the rates of change of $x$ and $y$ with respect to time are equal. [Assume that $dx/dt$ and $dy/dt$ are never both zero at the same point.]

**39.** The **thin lens equation** in physics is

$$\frac{1}{s} + \frac{1}{S} = \frac{1}{f}$$

where $s$ is the object distance from the lens, $S$ is the image distance from the lens, and $f$ is the focal length of the lens. Suppose that a certain lens has a focal length of 6 cm and that an object is moving toward the lens at the rate of 2 cm/s. How fast is the image distance changing at the instant when the object is 10 cm from the lens? Is the image moving away from the lens or toward the lens?

**40.** Water is stored in a cone-shaped reservoir (vertex down). Assuming the water evaporates at a rate proportional to the surface area exposed to the air, show that the depth of the water will decrease at a constant rate that does not depend on the dimensions of the reservoir.

**41.** A meteorite enters the Earth's atmosphere and burns up at a rate that, at each instant, is proportional to its surface area. Assuming that the meteorite is always spherical, show that the radius decreases at a constant rate.

**42.** On a certain clock the minute hand is 4 in long and the hour hand is 3 in long. How fast is the distance between the tips of the hands changing at 9 o'clock?

**43.** Coffee is poured at a uniform rate of 20 cm³/s into a cup whose inside is shaped like a truncated cone (see the accompanying figure). If the upper and lower radii of the cup are 4 cm and 2 cm and the height of the cup is 6 cm, how fast will the coffee level be rising when the coffee is halfway up? [*Hint:* Extend the cup downward to form a cone.]

Figure Ex-43

## 4.7 L'HÔPITAL'S RULE; INDETERMINATE FORMS

*In this section we will discuss a general method for using derivatives to find limits. This method will enable us to establish limits with certainty that earlier in the text we were only able to conjecture using numerical or graphical evidence. The method that we will discuss in this section is an extremely powerful tool that is used internally by many computer programs to calculate limits of various types.*

**INDETERMINATE FORMS OF TYPE 0/0**

In earlier sections we discussed limits that can be determined by inspection or by some appropriate algebraic manipulation. In this section we will be concerned with limits that cannot be obtained by such methods. For example, in Theorem 2.5.3 we were able to show that

$$\lim_{x \to 0} \frac{\sin x}{x} = 1 \tag{1}$$

but it required the Squeezing Theorem (2.5.2) and some tricky manipulation of inequalities. Our goal here is to develop a more straightforward method.

What makes the limit in (1) bothersome is the fact that the numerator and denominator both approach 0 as $x \to 0$. Such limits are called **indeterminate forms of type 0/0**. In limits of this type there are two tendencies working against each other: as the numerator approaches 0 it tends to drive the ratio toward 0, and as the denominator approaches 0 it tends to drive the ratio toward $+\infty$ or $-\infty$. What happens in (1) is that these conflicting tendencies offset each other in such a way that the limit is 1.

Although the limit in (1) is not self-evident, it can be conjectured from numerical evidence, as in Table 2.1.2. However, it can also be conjectured from the local linear approximation of $\sin x$ at 0. To see this, recall from Formula (5) of Section 3.6 that if a function $f$ is differentiable at a point $x_0$, then for values of $x$ near $x_0$, the values of $f(x)$ can be approximated as

$$f(x) \approx f(x_0) + f'(x_0)(x - x_0)$$

where the approximation tends to get better and better as $x \to x_0$. In particular, we showed in Example 3 of Section 3.6 that the local linear approximation of $\sin x$ at $x_0 = 0$ is

$$\sin x \approx x$$

This suggests that the value of $(\sin x)/x$ gets closer and closer to 1 as $x \to 0$, and hence we can reasonably conclude that

$$\lim_{x \to 0} \frac{\sin x}{x} = \lim_{x \to 0} \frac{x}{x} = 1$$

**L'HÔPITAL'S RULE**

The idea of using local linear approximations to evaluate indeterminate forms of type 0/0 can be used to motivate a more general procedure for finding such limits. For this purpose, suppose that

$$\lim_{x \to x_0} \frac{f(x)}{g(x)}$$

is an indeterminate form of type 0/0, that is,

$$\lim_{x \to x_0} f(x) = 0 \quad \text{and} \quad \lim_{x \to x_0} g(x) = 0 \tag{2}$$

For simplicity, let us also assume that $f$ and $g$ are differentiable at $x = x_0$ and that $f'$ and $g'$ are continuous at $x = x_0$. The differentiability of $f$ and $g$ at $x = x_0$ implies that $f$ and $g$ are continuous at $x = x_0$, and hence from (2)

$$f(x_0) = \lim_{x \to x_0} f(x) = 0 \quad \text{and} \quad g(x_0) = \lim_{x \to x_0} g(x) = 0 \tag{3}$$

Moreover, the continuity of $f'$ and $g'$ at $x = x_0$ implies that

$$\lim_{x \to x_0} f'(x) = f'(x_0) \quad \text{and} \quad \lim_{x \to x_0} g'(x) = g'(x_0) \tag{4}$$

Thus, from (3) and (4) and the local linear approximations of $f$ and $g$ at $x = x_0$, we have

$$\lim_{x \to x_0} \frac{f(x)}{g(x)} = \lim_{x \to x_0} \frac{f(x_0) + f'(x_0)(x - x_0)}{g(x_0) + g'(x_0)(x - x_0)}$$

$$= \lim_{x \to x_0} \frac{f'(x_0)(x - x_0)}{g'(x_0)(x - x_0)} = \frac{f'(x_0)}{g'(x_0)}$$

which from (4) can be expressed as

$$\lim_{x \to x_0} \frac{f(x)}{g(x)} = \lim_{x \to x_0} \frac{f'(x)}{g'(x)} \tag{5}$$

This result, called **L'Hôpital's** * **rule**, converts an indeterminate form of type $0/0$ into a new limit involving derivatives that in many situations can be evaluated by inspection or by algebraic methods. For example,

$$\lim_{x \to 0} \frac{1 - \cos x}{x} = \lim_{x \to 0} \frac{\dfrac{d}{dx}[1 - \cos x]}{\dfrac{d}{dx}[x]} = \lim_{x \to 0} \frac{\sin x}{1} = \sin 0 = 0$$

which agrees with the result in Theorem 2.5.3.

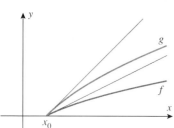

The graphs of $f$ and $g$ together with their local linear approximations at the point $x_0$

Figure 4.7.1

REMARK. Figure 4.7.1 provides a geometric explanation of (5). That figure shows the graphs $f$ and $g$ and the graphs of their local linear approximations at $x_0$. Note that $f(x) \to 0$ and $g(x) \to 0$ as $x \to x_0$ in the figure because the limit of $f/g$ is an indeterminate form of type $0/0$. The figure strongly suggests that for values of $x$ near $x_0$ there is little difference between the ratio of $f(x)$ and $g(x)$, and the ratio of the corresponding values in the local linear approximations, which is what we showed algebraically.

Although we motivated Formula (5) by assuming that $f$ and $g$ have continuous derivatives at $x = x_0$, the result is true without this assumption. Moreover, the result is also valid for one-sided limits and limits at $+\infty$ and $-\infty$. We omit the formal proof.

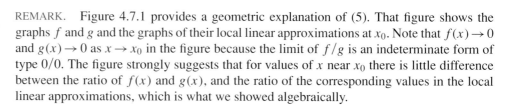

**4.7.1** THEOREM (*L'Hôpital's Rule for Form* **0/0**). *Let* lim *stand for one of the limits* $\lim\limits_{x \to a}$, $\lim\limits_{x \to a^+}$, $\lim\limits_{x \to a^-}$, $\lim\limits_{x \to +\infty}$, *or* $\lim\limits_{x \to -\infty}$, *and suppose that* $\lim f(x) = 0$ *and* $\lim g(x) = 0$. *If* $\lim [f'(x)/g'(x)]$ *has a finite value* $L$, *or if this limit is* $+\infty$ *or* $-\infty$, *then*

$$\lim \frac{f(x)}{g(x)} = \lim \frac{f'(x)}{g'(x)}$$

REMARK. Note that in L'Hôpital's rule the numerator and denominator are differentiated separately, which is not the same as differentiating $f(x)/g(x)$.

---

*GUILLAUME FRANCOIS ANTOINE DE L'HÔPITAL (1661–1704). French mathematician. L'Hôpital, born to parents of the French high nobility, held the title of Marquis de Sainte-Mesme Comte d'Autrement. He showed mathematical talent quite early and at age 15 solved a difficult problem about cycloids posed by Pascal. As a young man he served briefly as a cavalry officer, but resigned because of nearsightedness. In his own time he gained fame as the author of the first textbook ever published on differential calculus, *L'Analyse des Infiniment Petits pour l'Intelligence des Lignes Courbes* (1696). L'Hôpital's rule appeared for the first time in that book. Actually, L'Hôpital's rule and most of the material in the calculus text were due to John Bernoulli, who was L'Hôpital's teacher. L'Hôpital dropped his plans for a book on integral calculus when Leibniz informed him that he intended to write such a text. L'Hôpital was apparently generous and personable, and his many contacts with major mathematicians provided the vehicle for disseminating major discoveries in calculus throughout Europe.

In the following examples we will apply L'Hôpital's rule using the following three-step process:

**Step 1.** Check that $\lim f(x)/g(x)$ is an indeterminate form. If it is not, then L'Hôpital's rule cannot be used.

**Step 2.** Differentiate $f$ and $g$ separately.

**Step 3.** Find $\lim f'(x)/g'(x)$. If this limit is finite, $+\infty$, or $-\infty$, then it is equal to $\lim f(x)/g(x)$.

## Example 1

In each part confirm that the limit is an indeterminate form of type $0/0$, and evaluate it using L'Hôpital's rule.

(a) $\displaystyle\lim_{x \to 2} \frac{x^2 - 4}{x - 2}$    (b) $\displaystyle\lim_{x \to 0} \frac{\sin 2x}{x}$    (c) $\displaystyle\lim_{x \to \pi/2} \frac{1 - \sin x}{\cos x}$    (d) $\displaystyle\lim_{x \to 0} \frac{e^x - 1}{x^3}$

(e) $\displaystyle\lim_{x \to 0^-} \frac{\tan x}{x^2}$    (f) $\displaystyle\lim_{x \to 0} \frac{1 - \cos x}{x^2}$    (g) $\displaystyle\lim_{x \to +\infty} \frac{x^{-4/3}}{\sin(1/x)}$

*Solution* (*a*). The numerator and denominator have a limit of 0, so L'Hôpital's rule applies and yields

$$\lim_{x \to 2} \frac{x^2 - 4}{x - 2} = \lim_{x \to 2} \frac{\dfrac{d}{dx}[x^2 - 4]}{\dfrac{d}{dx}[x - 2]} = \lim_{x \to 2} \frac{2x}{1} = 4$$

Observe that this particular limit could also have been obtained by factoring

$$\lim_{x \to 2} \frac{x^2 - 4}{x - 2} = \lim_{x \to 2} \frac{(x - 2)(x + 2)}{x - 2} = \lim_{x \to 2} (x + 2) = 4$$

*Solution* (*b*). The numerator and denominator have a limit of 0, so L'Hôpital's rule applies and yields

$$\lim_{x \to 0} \frac{\sin 2x}{x} = \lim_{x \to 0} \frac{\dfrac{d}{dx}[\sin 2x]}{\dfrac{d}{dx}[x]} = \lim_{x \to 0} \frac{2 \cos 2x}{1} = 2$$

Observe that this result agrees with that obtained by substitution in Example 2(b) of Section 2.5.

*Solution* (*c*). The numerator and denominator have a limit of 0, so L'Hôpital's rule applies and yields

$$\lim_{x \to \pi/2} \frac{1 - \sin x}{\cos x} = \lim_{x \to \pi/2} \frac{\dfrac{d}{dx}[1 - \sin x]}{\dfrac{d}{dx}[\cos x]} = \lim_{x \to \pi/2} \frac{-\cos x}{-\sin x} = \frac{0}{-1} = 0$$

*Solution* (*d*). The numerator and denominator have a limit of 0, so L'Hôpital's rule applies and yields

$$\lim_{x \to 0} \frac{e^x - 1}{x^3} = \lim_{x \to 0} \frac{\dfrac{d}{dx}[e^x - 1]}{\dfrac{d}{dx}[x^3]} = \lim_{x \to 0} \frac{e^x}{3x^2} = +\infty$$

*Solution (e).* The numerator and denominator have a limit of 0, so L'Hôpital's rule applies and yields

$$\lim_{x \to 0^-} \frac{\tan x}{x^2} = \lim_{x \to 0^-} \frac{\sec^2 x}{2x} = -\infty$$

*Solution (f).* The numerator and denominator have a limit of 0, so L'Hôpital's rule applies and yields

$$\lim_{x \to 0} \frac{1 - \cos x}{x^2} = \lim_{x \to 0} \frac{\sin x}{2x}$$

Since the new limit is another indeterminate form of type 0/0, we apply L'Hôpital's rule again:

$$\lim_{x \to 0} \frac{1 - \cos x}{x^2} = \lim_{x \to 0} \frac{\sin x}{2x} = \lim_{x \to 0} \frac{\cos x}{2} = \frac{1}{2}$$

*Solution (g).* The numerator and denominator have a limit of 0, so L'Hôpital's rule applies and yields

$$\lim_{x \to +\infty} \frac{x^{-4/3}}{\sin(1/x)} = \lim_{x \to +\infty} \frac{-\frac{4}{3}x^{-7/3}}{(-1/x^2)\cos(1/x)} = \lim_{x \to +\infty} \frac{\frac{4}{3}x^{-1/3}}{\cos(1/x)} = \frac{0}{1} = 0 \quad \blacktriangleleft$$

WARNING. Applying L'Hôpital's rule to limits that are not indeterminate forms can lead to incorrect results. For example, in the limit

$$\lim_{x \to 0} \frac{x + 6}{x + 2} = \frac{6}{2} = 3$$

the numerator approaches 6 and the denominator approaches 2, so the limit is not an indeterminate form of type 0/0. However, if we ignore this and blindly apply L'Hôpital's rule, we reach the following *erroneous* conclusion:

$$\lim_{x \to 0} \frac{\frac{d}{dx}[x + 6]}{\frac{d}{dx}[x + 2]} = \lim_{x \to 0} \frac{1}{1} = 1$$

**INDETERMINATE FORMS OF TYPE $\infty/\infty$**

When we want to indicate that the limit (or the one-sided limits) of a function are $+\infty$ or $-\infty$ without being specific about the sign, we will say that the limit is $\infty$. For example,

$$\lim_{x \to a^+} f(x) = \infty \quad \text{means} \quad \lim_{x \to a^+} f(x) = +\infty \quad \text{or} \quad \lim_{x \to a^+} f(x) = -\infty$$

$$\lim_{x \to +\infty} f(x) = \infty \quad \text{means} \quad \lim_{x \to +\infty} f(x) = +\infty \quad \text{or} \quad \lim_{x \to +\infty} f(x) = -\infty$$

$$\lim_{x \to a} f(x) = \infty \quad \text{means} \quad \lim_{x \to a^+} f(x) = \pm\infty \quad \text{and} \quad \lim_{x \to a^-} f(x) = \pm\infty$$

The limit of a ratio, $f(x)/g(x)$, in which the numerator has limit $\infty$ and the denominator has limit $\infty$ is called an **indeterminate form of type $\infty/\infty$**. The following version of L'Hôpital's rule, which we state without proof, can often be used to evaluate limits of this type.

**4.7.2 THEOREM** (*L'Hôpital's Rule for Form $\infty/\infty$*). *Let* lim *stand for one of the limits* $\lim_{x \to a}$, $\lim_{x \to a^+}$, $\lim_{x \to a^-}$, $\lim_{x \to +\infty}$, *or* $\lim_{x \to -\infty}$, *and suppose that* $\lim f(x) = \infty$ *and* $\lim g(x) = \infty$. *If* $\lim [f'(x)/g'(x)]$ *has a finite value L, or if this limit is* $+\infty$ *or* $-\infty$, *then*

$$\lim \frac{f(x)}{g(x)} = \lim \frac{f'(x)}{g'(x)}$$

### Example 2

In each part confirm that the limit is an indeterminate form of type $\infty/\infty$ and apply L'Hôpital's rule.

$$(a)\ \lim_{x \to +\infty} \frac{x}{e^x} \qquad (b)\ \lim_{x \to 0^+} \frac{\ln x}{\csc x}$$

*Solution (a).* The numerator and denominator both have a limit of $+\infty$, so we have an indeterminate form of type $\infty/\infty$. Applying L'Hôpital's rule yields

$$\lim_{x \to +\infty} \frac{x}{e^x} = \lim_{x \to +\infty} \frac{1}{e^x} = 0$$

*Solution (b).* The numerator has a limit of $-\infty$ and the denominator has a limit of $+\infty$, so we have an indeterminate form of type $\infty/\infty$. Applying L'Hôpital's rule yields

$$\lim_{x \to 0^+} \frac{\ln x}{\csc x} = \lim_{x \to 0^+} \frac{1/x}{-\csc x \cot x} \tag{6}$$

This last limit is again an indeterminate form of type $\infty/\infty$. Moreover, any additional applications of L'Hôpital's rule will yield powers of $1/x$ in the numerator and expressions involving $\csc x$ and $\cot x$ in the denominator; thus, repeated application of L'Hôpital's rule simply produces new indeterminate forms. We must try something else. The last limit in (6) can be rewritten as

$$\lim_{x \to 0^+} \left( -\frac{\sin x}{x} \tan x \right) = -\lim_{x \to 0^+} \frac{\sin x}{x} \cdot \lim_{x \to 0^+} \tan x = -(1)(0) = 0$$

Thus,

$$\lim_{x \to 0^+} \frac{\ln x}{\csc x} = 0 \qquad \blacktriangleleft$$

**ANALYZING THE GROWTH OF EXPONENTIAL FUNCTIONS USING L'HÔPITAL'S RULE**

If $n$ is any positive integer, then $x^n \to +\infty$ as $x \to +\infty$. Such integer powers of $x$ are sometimes used as "measuring sticks" to describe how rapidly other functions grow. For example, we know that $e^x \to +\infty$ as $x \to +\infty$ and that the growth of $e^x$ is very rapid (Table 4.2.3); however, the growth of $x^n$ is also rapid when $n$ is a high power, so it is reasonable to ask whether high powers of $x$ grow more or less rapidly than $e^x$. One way to investigate this is to examine the behavior of the ratio $x^n/e^x$ as $x \to +\infty$. For example, Figure 4.7.2a shows the graph of $y = x^5/e^x$. This graph suggests that $x^5/e^x \to 0$ as $x \to +\infty$, and this implies that the growth of the function $e^x$ is sufficiently rapid that its values eventually overtake those of $x^5$ and force the ratio toward zero. Stated informally, "$e^x$ eventually grows more rapidly than $x^5$." The same conclusion could have been reached by putting $e^x$ on top and examining the behavior of $e^x/x^5$ as $x \to +\infty$ (Figure 4.7.2b). In this case the values of $e^x$ eventually overtake those of $x^5$ and force the ratio toward $+\infty$. More generally, we can use L'Hôpital's rule to show that $e^x$ *eventually grows more rapidly than any positive integer power of $x$*, that is,

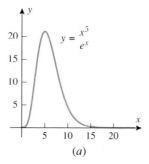

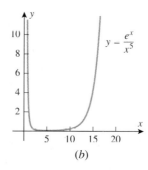

Figure 4.7.2

$$\lim_{x \to +\infty} \frac{x^n}{e^x} = 0 \qquad \text{and} \qquad \lim_{x \to +\infty} \frac{e^x}{x^n} = +\infty \tag{7-8}$$

Both limits are indeterminate forms of type $\infty/\infty$ that can be evaluated using L'Hôpital's rule. For example, to establish (7), we will need to apply L'Hôpital's rule $n$ times. For this purpose, observe that successive differentiations of $x^n$ reduce the exponent by 1 each time, thus producing a constant for the $n$th derivative. For example, the successive derivatives of $x^3$ are $3x^2, 6x$, and 6. In general, the $n$th derivative of $x^n$ is the constant $n(n-1)(n-2) \cdots 1 = n!$ (verify).[*] Thus, applying L'Hôpital's rule $n$ times to (7) yields

$$\lim_{x \to +\infty} \frac{x^n}{e^x} = \lim_{x \to +\infty} \frac{n!}{e^x} = 0$$

Limit (8) can be established similarly.

**INDETERMINATE FORMS OF TYPE $0 \cdot \infty$**

Thus far we have discussed indeterminate forms of type $0/0$ and $\infty/\infty$. However, these are not the only possibilities; in general, the limit of an expression that has one of the forms

$$\frac{f(x)}{g(x)}, \quad f(x) \cdot g(x), \quad f(x)^{g(x)}, \quad f(x) - g(x), \quad f(x) + g(x)$$

is called an *indeterminate form* if the limits of $f(x)$ and $g(x)$ individually exert conflicting influences on the limit of the entire expression. For example, the limit

$$\lim_{x \to 0^+} x \ln x$$

is an **indeterminate form of type $0 \cdot \infty$** because the limit of the first factor is 0, the limit of the second factor is $-\infty$, and these two limits exert conflicting influences on the product. On the other hand, the limit

$$\lim_{x \to +\infty} [\sqrt{x}(1 - x^2)]$$

is not an indeterminate form because the first factor has a limit of $+\infty$, the second factor has a limit of $-\infty$, and these influences work together to produce a limit of $-\infty$ for the product.

WARNING.   It is tempting to argue that an indeterminate form of type $0 \cdot \infty$ has value 0 since "zero times anything is zero." However, this is fallacious since $0 \cdot \infty$ is not a product of numbers, but rather a statement about limits. For example, the following limits are of the form $0 \cdot \infty$:

$$\lim_{x \to 0^+} x \cdot \frac{1}{x} = 1, \quad \lim_{x \to 0^+} x^2 \cdot \frac{1}{x} = 0, \quad \lim_{x \to 0^+} \sqrt{x} \cdot \frac{1}{x} = +\infty$$

Indeterminate forms of type $0 \cdot \infty$ can sometimes be evaluated by rewriting the product as a ratio, and then applying L'Hôpital's rule for indeterminate forms of type $0/0$ or $\infty/\infty$.

### Example 3

Evaluate

(a) $\lim\limits_{x \to 0^+} x \ln x$    (b) $\lim\limits_{x \to \pi/4} (1 - \tan x) \sec 2x$

*Solution (a).* The factor $x$ has a limit of 0 and the factor $\ln x$ has a limit of $-\infty$, so the stated problem is an indeterminate form of type $0 \cdot \infty$. There are two possible approaches: we can rewrite the limit as

$$\lim_{x \to 0^+} \frac{\ln x}{1/x} \quad \text{or} \quad \lim_{x \to 0^+} \frac{x}{1/\ln x}$$

the first being an indeterminate form of type $\infty/\infty$ and the second an indeterminate form of

---

[*] Recall that for $n \geq 1$ the expression $n!$ is read **n-factorial** and denotes the product of the first $n$ integers.

type $0/0$. However, the first form is the preferred initial choice because the derivative of $1/x$ is less complicated than the derivative of $1/\ln x$. That choice yields

$$\lim_{x \to 0^+} x \ln x = \lim_{x \to 0^+} \frac{\ln x}{1/x} = \lim_{x \to 0^+} \frac{1/x}{-1/x^2} = \lim_{x \to 0^+} (-x) = 0$$

***Solution*** (***b***)***.*** The stated problem is an indeterminate form of type $0 \cdot \infty$. We will convert it to an indeterminate form of type $\infty/\infty$:

$$\lim_{x \to \pi/4} (1 - \tan x) \sec 2x = \lim_{x \to \pi/4} \frac{1 - \tan x}{1/\sec 2x} = \lim_{x \to \pi/4} \frac{1 - \tan x}{\cos 2x}$$

$$= \lim_{x \to \pi/4} \frac{-\sec^2 x}{-2 \sin 2x} = \frac{-2}{-2} = 1 \qquad \blacktriangleleft$$

**INDETERMINATE FORMS OF TYPE $\infty - \infty$**

A limit problem that leads to one of the expressions

$$(+\infty) - (+\infty), \quad (-\infty) - (-\infty),$$

$$(+\infty) + (-\infty), \quad (-\infty) + (+\infty)$$

is called an ***indeterminate form of type*** $\infty - \infty$. Such limits are indeterminate because the two terms exert conflicting influences on the expression: one pushes it in the positive direction and the other pushes it in the negative direction. However, limit problems that lead to one of the expressions

$$(+\infty) + (+\infty), \quad (+\infty) - (-\infty),$$

$$(-\infty) + (-\infty), \quad (-\infty) - (+\infty)$$

are not indeterminate, since the two terms work together (those on the top produce a limit of $+\infty$ and those on the bottom produce a limit of $-\infty$).

Indeterminate forms of type $\infty - \infty$ can sometimes be evaluated by combining the terms and manipulating the result to produce an indeterminate form of type $0/0$ or $\infty/\infty$.

## Example 4

Evaluate $\displaystyle\lim_{x \to 0^+} \left( \frac{1}{x} - \frac{1}{\sin x} \right)$.

***Solution.*** Both terms have a limit of $+\infty$, so the stated problem is an indeterminate form of type $\infty - \infty$. Combining the two terms yields

$$\lim_{x \to 0^+} \left( \frac{1}{x} - \frac{1}{\sin x} \right) = \lim_{x \to 0^+} \left( \frac{\sin x - x}{x \sin x} \right)$$

which is an indeterminate form of type $0/0$. Applying L'Hôpital's rule twice yields

$$\lim_{x \to 0^+} \left( \frac{\sin x - x}{x \sin x} \right) = \lim_{x \to 0^+} \frac{\cos x - 1}{\sin x + x \cos x}$$

$$= \lim_{x \to 0^+} \frac{-\sin x}{\cos x + \cos x - x \sin x} = \frac{0}{2} = 0 \qquad \blacktriangleleft$$

**INDETERMINATE FORMS OF TYPE $0^0$, $\infty^0$, $1^\infty$**

Limits of the form

$$\lim f(x)^{g(x)}$$

give rise to ***indeterminate forms of the types*** $0^0$, $\infty^0$, ***and*** $1^\infty$. (The meaning of these symbols should be clear.) For example, the limit

$$\lim_{x \to 0^+} (1 + x)^{1/x}$$

whose value we know to be $e$ [see Formula (5) of Section 4.2] is an indeterminate form of type $1^\infty$. It is indeterminate because the expressions $1 + x$ and $1/x$ exert two conflicting influences: the first approaches 1, which drives the expression toward 1, and the second approaches $+\infty$, which drives the expression toward $+\infty$.

Indeterminate forms of types $0^0$, $\infty^0$, and $1^\infty$ can sometimes be evaluated by first introducing a dependent variable

$$y = f(x)^{g(x)}$$

and then calculating the limit of $\ln y$ by expressing it as

$$\lim \ln y = \lim \left[ \ln (f(x)^{g(x)}) \right] = \lim \left[ g(x) \ln f(x) \right]$$

Once the limit of $\ln y$ is known, the limit of $y = f(x)^{g(x)}$ itself can generally be obtained by a method that we will illustrate in the next example.

### Example 5

Show that $\lim\limits_{x \to 0} (1 + x)^{1/x} = e$.

*Solution.* As discussed above, we begin by introducing a dependent variable

$$y = (1 + x)^{1/x}$$

and taking the natural logarithm of both sides:

$$\ln y = \ln (1 + x)^{1/x} = \frac{1}{x} \ln (1 + x) = \frac{\ln (1 + x)}{x}$$

Thus,

$$\lim_{x \to 0} \ln y = \lim_{x \to 0} \frac{\ln (1 + x)}{x}$$

which is an indeterminate form of type $0/0$, so by L'Hôpital's rule

$$\lim_{x \to 0} \ln y = \lim_{x \to 0} \frac{\ln (1 + x)}{x} = \lim_{x \to 0} \frac{1/(1 + x)}{1} = 1$$

Since we have shown that $\ln y \to 1$ as $x \to 0$, the continuity of the exponential function implies that $e^{\ln y} \to e^1$ as $x \to 0$, and this implies that $y \to e$ as $x \to 0$. Thus,

$$\lim_{x \to 0} (1 + x)^{1/x} = e$$  ◀

### EXERCISE SET 4.7  ⌁ Graphing Calculator   [c] CAS
· · · · · · · · · · · · · · · · · · · · · · · · · · · · · · · · · · · · · · · · · · · · · · · · · · · · · · · · · · · ·

In Exercises 1 and 2, evaluate the given limit without using L'Hôpital's rule, and then check that your answer is correct using L'Hôpital's rule.

**1.** (a) $\lim\limits_{x \to 2} \dfrac{x^2 - 4}{x^2 + 2x - 8}$   (b) $\lim\limits_{x \to +\infty} \dfrac{2x - 5}{3x + 7}$

**2.** (a) $\lim\limits_{x \to 0} \dfrac{\sin x}{\tan x}$   (b) $\lim\limits_{x \to 1} \dfrac{x^2 - 1}{x^3 - 1}$

In Exercises 3–36, find the limit.

**3.** $\lim\limits_{x \to 1} \dfrac{\ln x}{x - 1}$

**4.** $\lim\limits_{x \to 0} \dfrac{\sin 2x}{\sin 5x}$

**5.** $\lim\limits_{x \to 0} \dfrac{e^x - 1}{\sin x}$

**6.** $\lim\limits_{x \to 3} \dfrac{x - 3}{3x^2 - 13x + 12}$

**7.** $\lim\limits_{\theta \to 0} \dfrac{\tan \theta}{\theta}$

**8.** $\lim\limits_{t \to 0} \dfrac{te^t}{1 - e^t}$

**9.** $\lim\limits_{x \to \pi^+} \dfrac{\sin x}{x - \pi}$

**10.** $\lim\limits_{x \to 0^+} \dfrac{\sin x}{x^2}$

**11.** $\lim\limits_{x \to +\infty} \dfrac{\ln x}{x}$

**12.** $\lim\limits_{x \to +\infty} \dfrac{e^{3x}}{x^2}$

**13.** $\lim\limits_{x \to 0^+} \dfrac{\cot x}{\ln x}$

**14.** $\lim\limits_{x \to 0^+} \dfrac{1 - \ln x}{e^{1/x}}$

**15.** $\lim\limits_{x \to +\infty} \dfrac{x^{100}}{e^x}$

**16.** $\lim\limits_{x \to 0^+} \dfrac{\ln (\sin x)}{\ln (\tan x)}$

**17.** $\lim\limits_{x \to 0} \dfrac{\sin^{-1} 2x}{x}$

**18.** $\lim\limits_{x \to 0} \dfrac{x - \tan^{-1} x}{x^3}$

**19.** $\lim\limits_{x \to +\infty} x e^{-x}$

**20.** $\lim\limits_{x \to \pi^-} (x - \pi) \tan \tfrac{1}{2} x$

**21.** $\lim\limits_{x \to +\infty} x \sin \dfrac{\pi}{x}$

**22.** $\lim\limits_{x \to 0^+} \tan x \ln x$

**23.** $\lim\limits_{x \to \pi/2^-} \sec 3x \cos 5x$

**24.** $\lim\limits_{x \to \pi} (x - \pi) \cot x$

**25.** $\lim\limits_{x \to +\infty} (1 - 3/x)^x$

**26.** $\lim\limits_{x \to 0} (1 + 2x)^{-3/x}$

**27.** $\lim\limits_{x \to 0} (e^x + x)^{1/x}$

**28.** $\lim\limits_{x \to +\infty} (1 + a/x)^{bx}$

**29.** $\lim\limits_{x \to 1} (2 - x)^{\tan(\pi/2)x}$

**30.** $\lim\limits_{x \to +\infty} [\cos(2/x)]^{x^2}$

**31.** $\lim\limits_{x \to 0} (\csc x - 1/x)$

**32.** $\lim\limits_{x \to 0} \left( \dfrac{1}{x^2} - \dfrac{\cos 3x}{x^2} \right)$

**33.** $\lim\limits_{x \to +\infty} (\sqrt{x^2 + x} - x)$

**34.** $\lim\limits_{x \to 0} \left( \dfrac{1}{x} - \dfrac{1}{e^x - 1} \right)$

**35.** $\lim\limits_{x \to +\infty} [x - \ln(x^2 + 1)]$

**36.** $\lim\limits_{x \to +\infty} [\ln x - \ln(1 + x)]$

**c** **37.** Use a CAS to check the answers you obtained in Exercises 31–36.

**38.** Show that for any positive integer $n$

(a) $\lim\limits_{x \to +\infty} \dfrac{\ln x}{x^n} = 0$

(b) $\lim\limits_{x \to +\infty} \dfrac{x^n}{\ln x} = +\infty$

**39.** (a) Find the error in the following calculation:

$$\lim\limits_{x \to 1} \dfrac{x^3 - x^2 + x - 1}{x^3 - x^2} = \lim\limits_{x \to 1} \dfrac{3x^2 - 2x + 1}{3x^2 - 2x}$$
$$= \lim\limits_{x \to 1} \dfrac{6x - 2}{6x - 2} = 1$$

(b) Find the correct answer.

**40.** Find $\lim\limits_{x \to 1} \dfrac{x^4 - 4x^3 + 6x^2 - 4x + 1}{x^4 - 3x^3 + 3x^2 - x}$.

In Exercises 41–44, make a conjecture about the limit by graphing the function involved with a graphing utility; then check your conjecture using L'Hôpital's rule.

**41.** $\lim\limits_{x \to +\infty} \dfrac{\ln(\ln x)}{\sqrt{x}}$

**42.** $\lim\limits_{x \to 0^+} x^x$

**43.** $\lim\limits_{x \to 0^+} (\sin x)^{3/\ln x}$

**44.** $\lim\limits_{x \to (1/2)\pi^-} \dfrac{4 \tan x}{1 + \sec x}$

In Exercises 45–48, make a conjecture about the equations of horizontal asymptotes, if any, by graphing the equation with a graphing utility; then check your answer using L'Hôpital's rule.

**45.** $y = \ln x - e^x$

**46.** $y = x - \ln(1 + 2e^x)$

**47.** $y = (\ln x)^{1/x}$

**48.** $y = \left( \dfrac{x + 1}{x + 2} \right)^x$

**49.** Limits of the type

$$0/\infty, \quad \infty/0, \quad 0^\infty, \quad \infty \cdot \infty, \quad +\infty + (+\infty),$$
$$+\infty - (-\infty), \quad -\infty + (-\infty), \quad -\infty - (+\infty)$$

are *not* indeterminate forms. Find the following limits by inspection.

(a) $\lim\limits_{x \to 0^+} \dfrac{x}{\ln x}$

(b) $\lim\limits_{x \to +\infty} \dfrac{x^3}{e^{-x}}$

(c) $\lim\limits_{x \to (1/2)\pi^-} (\cos x)^{\tan x}$

(d) $\lim\limits_{x \to 0^+} (\ln x) \cot x$

(e) $\lim\limits_{x \to 0^+} \left( \dfrac{1}{x} - \ln x \right)$

(f) $\lim\limits_{x \to -\infty} (x + x^3)$

**50.** There is a myth that circulates among beginning calculus students which states that all indeterminate forms of types $0^0$, $\infty^0$, and $1^\infty$ have value 1 because "anything to the zero power is 1" and "1 to any power is 1." The fallacy is that $0^0$, $\infty^0$, and $1^\infty$ are not powers of numbers, but rather descriptions of limits. The following examples, which were transmitted to me by Prof. Jack Staib of Drexel University, show that such indeterminate forms can have any positive real value:

(a) $\lim\limits_{x \to 0^+} \left[ x^{(\ln a)/(1 + \ln x)} \right] = 0^0 = a$

(b) $\lim\limits_{x \to +\infty} \left[ x^{(\ln a)/(1 + \ln x)} \right] = \infty^0 = a$

(c) $\lim\limits_{x \to 0} \left[ (x + 1)^{(\ln a)/x} \right] = 1^\infty = a$.

Prove these results.

In Exercises 51–54, verify that L'Hôpital's rule is of no help in finding the limit, then find the limit, if it exists, by some other method.

**51.** $\lim\limits_{x \to +\infty} \dfrac{x + \sin 2x}{x}$

**52.** $\lim\limits_{x \to +\infty} \dfrac{2x - \sin x}{3x + \sin x}$

**53.** $\lim\limits_{x \to +\infty} \dfrac{x(2 + \sin 2x)}{x + 1}$

**54.** $\lim\limits_{x \to +\infty} \dfrac{x(2 + \sin x)}{x^2 + 1}$

**55.** The accompanying schematic diagram represents an electrical circuit consisting of an electromotive force that produces a voltage $V$, a resistor with resistance $R$, and an inductor with inductance $L$. It is shown in electrical circuit theory that if the voltage is first applied at time $t = 0$, then the current $I$ flowing through the circuit at time $t$ is given by

$$I = \dfrac{V}{R}(1 - e^{-Rt/L})$$

What is the effect on the current at a fixed time $t$ if the resistance approaches 0 (i.e., $R \to 0^+$)?

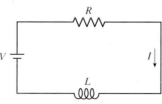

Figure Ex-55

**56.** (a) Show that $\lim\limits_{x \to \pi/2} (\pi/2 - x) \tan x = 1$.

(b) Show that

$$\lim_{x \to \pi/2} \left( \frac{1}{\pi/2 - x} - \tan x \right) = 0$$

(c) It follows from part (b) that the approximation

$$\tan x \approx \frac{1}{\pi/2 - x}$$

should be good for values of $x$ near $\pi/2$. Use a calculator to find $\tan x$ and $1/(\pi/2 - x)$ for $x = 1.57$; compare the results.

**57.** (a) Use a CAS to show that if $k$ is a positive constant, then

$$\lim_{x \to +\infty} x(k^{1/x} - 1) = \ln k$$

(b) Confirm this result using L'Hôpital's rule. [*Hint:* Express the limit in terms of $t = 1/x$.]

(c) If $n$ is a positive integer, then it follows from part (a) with $x = n$ that the approximation

$$n(\sqrt[n]{k} - 1) \approx \ln k$$

should be good when $n$ is large. Use this result and the square root key on a calculator to approximate the values of $\ln 0.3$ and $\ln 2$ with $n = 1024$, then compare

the values obtained with values of the logarithms generated directly from the calculator. [*Hint:* The $n$th roots for which $n$ is a power of 2 can be obtained as successive square roots.]

**58.** Let $f(x) = x^2 \sin(1/x)$.

(a) Are the limits $\lim_{x \to 0^+} f(x)$ and $\lim_{x \to 0^-} f(x)$ indeterminate forms?

(b) Use a graphing utility to generate the graph of $f$, and use the graph to make conjectures about the limits in part (a).

(c) Use the Squeezing Theorem (2.5.2) to confirm that your conjectures in part (b) are correct.

**59.** Find all values of $k$ and $l$ such that

$$\lim_{x \to 0} \frac{k + \cos lx}{x^2} = -4$$

**60.** (a) Explain why L'Hôpital's rule does not apply to the problem

$$\lim_{x \to 0} \frac{x^2 \sin(1/x)}{\sin x}$$

(b) Find the limit.

**61.** Find $\lim_{x \to 0^+} \frac{x \sin(1/x)}{\sin x}$ if it exists.

---

## SUPPLEMENTARY EXERCISES

**1.** (a) State conditions under which two functions, $f$ and $g$, will be inverses, and give several examples of such functions.

(b) In words, what is the relationship between the graphs of $y = f(x)$ and $y = g(x)$ when $f$ and $g$ are inverse functions?

(c) What is the relationship between the domains and ranges of inverse functions $f$ and $g$?

(d) What condition must be satisfied for a function $f$ to have an inverse? Give some examples of functions that do not have inverses.

(e) If $f$ and $g$ are inverse functions and $f$ is continuous, must $g$ be continuous? Give a reasonable informal argument to support your answer.

(f) If $f$ and $g$ are inverse functions and $f$ is differentiable, must $g$ be differentiable? Give a reasonable informal argument to support your answer.

**2.** (a) State the restrictions on the domains of $\sin x$, $\cos x$, $\tan x$, and $\sec x$ that are imposed to make those functions one-to-one in the definitions of $\sin^{-1} x$, $\cos^{-1} x$, $\tan^{-1} x$, and $\sec^{-1} x$.

(b) Sketch the graphs of the restricted trigonometric functions in part (a) and their inverses.

**3.** (a) Under what conditions will a limit of the form

$$\lim [f(x)/g(x)]$$

be an indeterminate form?

(b) If $\lim g(x) = 0$, must $\lim [f(x)/g(x)]$ be an indeterminate form? Give some examples to support your answer.

**4.** Suppose that $\lim f(x) = \pm\infty$ and $\lim g(x) = \pm\infty$. In each of the four possible cases, state whether $\lim [f(x) - g(x)]$ is an indeterminate form, and give a reasonable informal argument to support your answer.

**5.** In each part, find $f^{-1}(x)$ if the inverse exists.

(a) $f(x) = 8x^3 - 1$      (b) $f(x) = x^2 - 2x + 1$

(c) $f(x) = (e^x)^2 + 1$      (d) $f(x) = (x + 2)/(x - 1)$

**6.** Let $f(x) = (ax + b)/(cx + d)$. What conditions on $a, b, c, d$ guarantee that $f^{-1}$ exists? Find $f^{-1}(x)$.

**7.** In each part, find the equation of the tangent line at the specified point.

(a) $x^{2/3} - y^{2/3} - y = 1$; $(1, -1)$

(b) $\sin xy = y$; $(\pi/2, 1)$

8. In each part, find the exact numerical value of the given expression.
   (a) $\cos[\cos^{-1}(4/5) + \sin^{-1}(5/13)]$
   (b) $\sin[\sin^{-1}(4/5) + \cos^{-1}(5/13)]$

9. Express the following function as a rational function of $x$:

$$3\ln\left(e^{2x}(e^x)^3\right) + 2\exp(\ln 1)$$

10. Suppose that $y = Ce^{kt}$, where $C$ and $k$ are constants, and let $Y = \ln y$. Show that the graph of $Y$ versus $t$ is a line, and state its slope and $Y$-intercept.

11. In each part, find the limit.

   (a) $\displaystyle\lim_{x \to +\infty} (e^x - x^2)$   (b) $\displaystyle\lim_{x \to 1} \sqrt{\frac{\ln x}{x^4 - 1}}$

   (c) $\displaystyle\lim_{x \to 0} \frac{a^x - 1}{x}, \quad a > 0$

12. Show that the function $y = e^{ax} \sin bx$ satisfies

$$y'' - 2ay' + (a^2 + b^2)y = 0$$

   for any real constants $a$ and $b$.

13. Show that the function $y = \tan^{-1} x$ satisfies

$$y'' = -2\sin y \cos^3 y$$

14. Show that the rate of change of $y = 3^{2x} 5^{7x}$ is proportional to $y$.

15. The hypotenuse of a right triangle is growing at a rate of $a$ cm/s and one leg is decreasing at a rate of $b$ cm/s. How fast is the acute angle between the hypotenuse and the other leg changing at the instant when both legs are 1 cm?

16. In each part, find $(f^{-1})'(x)$ using Formula (21) of Section 4.3, and check your answer by differentiating $f^{-1}$ directly.
   (a) $f(x) = 3/(x + 1)$   (b) $f(x) = \sqrt{e^x}$

17. (a) Sketch the curves $y = \pm e^{-x/2}$ and $y = e^{-x/2} \sin 2x$ for $-\pi/2 \le x \le 3\pi/2$ in the same coordinate system, and check your work using a graphing utility.
   (b) Find all $x$-intercepts of the curve $y = e^{-x/2} \sin 2x$ in the stated interval, and find the $x$-coordinates of all points where this curve intersects the curves $y = \pm e^{-x/2}$.

18. In each part, sketch the graph, and check your work with a graphing utility.
   (a) $f(x) = 3\sin^{-1}(x/2)$

   (b) $f(x) = \cos^{-1} x - \pi/2$
   (c) $f(x) = 2\tan^{-1}(-3x)$
   (d) $f(x) = \cos^{-1} x + \sin^{-1} x$

19. In each part, use any appropriate method to find $dy/dx$.
   (a) $y = (1 + x)^{1/x}$   (b) $y = x^{(e^x)}$
   (c) $y = e^{\ln(x^3 + 1)}$   (d) $y = \dfrac{a}{1 + be^{-x}}$
   (e) $xy^{2/3} + yx^{2/3} = x^2$   (f) $y = \ln\left(\dfrac{\sqrt{x}\sqrt[3]{x+1}}{\sin x \sec x}\right)$

20. (a) Suppose that the graph of $y = \log x$ is drawn with equal scales of 1 inch per unit in both the $x$- and $y$-directions. If a bug wants to walk along the graph until it reaches a height of 5 ft above the $x$-axis, how many miles to the right of the origin will it have to travel?
   (b) Suppose that the graph of $y = 10^x$ is drawn with equal scales of 1 inch per unit in both the $x$- and $y$-directions. If a bug wants to walk along the graph until it reaches a height of 100 mi above the $x$-axis, how many feet to the right of the origin will it have to travel?

21. (a) Show that the graphs of $y = \ln x$ and $y = x^{0.2}$ intersect.
   (b) Approximate the solution(s) of the equation $\ln x = x^{0.2}$ to three decimal places.

22. (a) Show that for $x > 0$ and $k \ne 0$ the equations

$$x^k = e^x \quad\text{and}\quad \frac{\ln x}{x} = \frac{1}{k}$$

   have the same solutions.
   (b) Use the graph of $y = (\ln x)/x$ to determine the values of $k$ for which the equation $x^k = e^x$ has two distinct positive solutions.
   (c) Find the positive solution(s) of $x^8 = e^x$.

23. Find the value of $b$ so that the line $y = x$ is tangent to the graph of $y = \log_b x$. Confirm your result by graphing both $y = x$ and $y = \log_b x$ in the same coordinate system.

24. In each part, find the value of $k$ for which the graphs of $y = f(x)$ and $y = \ln x$ share a common tangent line at their point of intersection. Confirm your result by graphing $y = f(x)$ and $y = \ln x$ in the same coordinate system.
   (a) $f(x) = \sqrt{x} + k$   (b) $f(x) = k\sqrt{x}$

## EXPANDING THE CALCULUS HORIZON

For additional material relating to this chapter, visit the Anton Website at http://www.wiley.com/college/anton

Maria Agnesi

# 5

# ANALYSIS OF FUNCTIONS AND THEIR GRAPHS

*I*n this chapter we will use methods of calculus to analyze functions and their graphs. We will be concerned here with such matters as identifying where the graph of a function is increasing or decreasing, where its high and low points occur, which way it bends, and what its limiting behavior is at important points.

One of the major goals of this chapter is to show how calculus and graphing utilities, working together, can provide most of the important information about the behavior of functions. Although graphing utilities can give us general information about the shape of a graph, such graphs lack perfect precision, since they are based on numerical approximations that can be affected by compression, distortion, and sampling error—it requires calculus to pin down the *exact* location of the key features and to reveal the nature of the fine detail. On the other hand, graphs produced by graphing utilities often provide information that is useful in pointing the calculus analysis in the right direction.

## 5.1 ANALYSIS OF FUNCTIONS I: INCREASE, DECREASE, AND CONCAVITY

*Although graphing utilities are useful for determining the general shape of a graph, many problems require more precision than graphing utilities are capable of producing. The purpose of this section is to develop mathematical tools that can be used to determine the exact shape of a graph and the precise location of its key features.*

**INCREASING AND DECREASING FUNCTIONS**

The terms *increasing*, *decreasing*, and *constant* are used to describe the behavior of a function over an interval as we travel left to right along its graph. For example, the function graphed in Figure 5.1.1 can be described as increasing on the interval $(-\infty, 0]$, decreasing on the interval $[0, 2]$, increasing again on the interval $[2, 4]$, and constant on the interval $[4, +\infty)$.

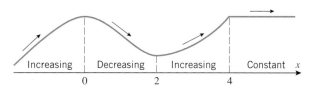

Figure 5.1.1

The following definition, which is illustrated in Figure 5.1.2, expresses these intuitive ideas precisely.

---

**5.1.1**    DEFINITION.    Let $f$ be defined on an interval, and let $x_1$ and $x_2$ denote points in that interval.

(a)    $f$ is ***increasing*** on the interval if $f(x_1) < f(x_2)$ whenever $x_1 < x_2$.

(b)    $f$ is ***decreasing*** on the interval if $f(x_1) > f(x_2)$ whenever $x_1 < x_2$.

(c)    $f$ is ***constant*** on the interval if $f(x_1) = f(x_2)$ for all points $x_1$ and $x_2$.

---

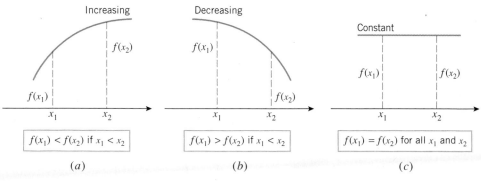

Figure 5.1.2

Figure 5.1.3 suggests that a differentiable function $f$ is increasing on any interval where its graph has tangent lines with positive slope, is decreasing on any interval where its graph has tangent lines with negative slope, and is constant on any interval where its graph has tangent lines with zero slope. This intuitive observation suggests the following important theorem that will be proved in Section 6.5.

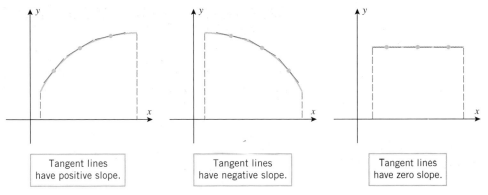

Tangent lines
have positive slope.

Tangent lines
have negative slope.

Tangent lines
have zero slope.

Figure 5.1.3

**5.1.2**  THEOREM.  *Let $f$ be a function that is continuous on a closed interval $[a, b]$ and differentiable on the open interval $(a, b)$.*
(a)  *If $f'(x) > 0$ for every value of $x$ in $(a, b)$, then $f$ is increasing on $[a, b]$.*
(b)  *If $f'(x) < 0$ for every value of $x$ in $(a, b)$, then $f$ is decreasing on $[a, b]$.*
(c)  *If $f'(x) = 0$ for every value of $x$ in $(a, b)$, then $f$ is constant on $[a, b]$.*

REMARK.    Observe that in Theorem 5.1.2 it is only necessary to examine the derivative of $f$ on the open interval $(a, b)$ to determine whether $f$ is increasing, decreasing, or constant on the closed interval $[a, b]$. Moreover, although this theorem was stated for a closed interval $[a, b]$, it is applicable to any interval $I$ on which $f$ is continuous and inside of which $f$ is differentiable. For example, if $f$ is continuous on $(a, +\infty)$ and $f'(x) > 0$ for each $x$ in the interval $(a, +\infty)$, then $f$ is increasing on $[a, +\infty)$; and if $f'(x) < 0$ on $(-\infty, +\infty)$, then $f$ is decreasing on $(-\infty, +\infty)$ [the continuity on $(-\infty, +\infty)$ follows from the differentiability].

### Example 1

Find the intervals on which the following functions are increasing and the intervals on which they are decreasing.

(a)  $f(x) = x^2 - 4x + 3$     (b)  $f(x) = x^3$

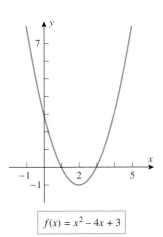

$f(x) = x^2 - 4x + 3$

Figure 5.1.4

*Solution (a).*  The graph of $f$ in Figure 5.1.4 suggests that $f$ is decreasing for $x \leq 2$ and increasing for $x \geq 2$. To confirm this, we differentiate $f$ to obtain

$$f'(x) = 2x - 4 = 2(x - 2)$$

It follows that

$$f'(x) < 0 \quad \text{if} \quad -\infty < x < 2$$
$$f'(x) > 0 \quad \text{if} \quad 2 < x < +\infty$$

Since $f$ is continuous at $x = 2$, it follows from Theorem 5.1.2 and the subsequent remark that

$f$ is decreasing on $(-\infty, 2]$

$f$ is increasing on $[2, +\infty)$

These conclusions are consistent with the graph of $f$ in Figure 5.1.4.

*Solution (b).*  The graph of $f$ in Figure 5.1.5 suggests that $f$ is increasing over the entire $x$-axis. To confirm this, we differentiate $f$ to obtain $f'(x) = 3x^2$. Thus,

$$f'(x) > 0 \quad \text{if} \quad -\infty < x < 0$$
$$f'(x) > 0 \quad \text{if} \quad 0 < x < +\infty$$

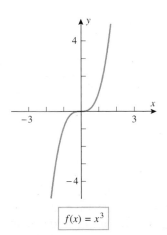

$f(x) = x^3$

Figure 5.1.5

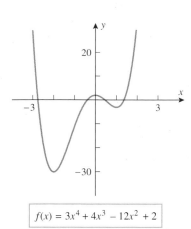

$$f(x) = 3x^4 + 4x^3 - 12x^2 + 2$$

Figure 5.1.6

Since $f$ is continuous at $x = 0$,

> $f$ is increasing on $(-\infty, 0]$
>
> $f$ is increasing on $[0, +\infty)$

Hence $f$ is increasing over the entire interval $(-\infty, +\infty)$, which is consistent with the graph in Figure 5.1.5 (see Exercise 51). ◀

## Example 2

(a) Use the graph of $f(x) = 3x^4 + 4x^3 - 12x^2 + 2$ in Figure 5.1.6 to make a conjecture about the intervals on which $f$ is increasing or decreasing.

(b) Use Theorem 5.1.2 to determine whether your conjecture is correct.

*Solution* (*a*). The graph suggests that $f$ is decreasing if $x \leq -2$, increasing if $-2 \leq x \leq 0$, decreasing if $0 \leq x \leq 1$, and increasing if $x \geq 1$.

*Solution* (*b*). Differentiating $f$ we obtain

$$f'(x) = 12x^3 + 12x^2 - 24x = 12x(x^2 + x - 2) = 12x(x + 2)(x - 1)$$

The sign analysis of $f'$ in Table 5.1.1 can be obtained using the method of test points discussed in Appendix A. The conclusions in that table confirm the conjecture in part (a). ◀

**Table 5.1.1**

| INTERVAL | $12x$ | $x + 2$ | $x - 1$ | $f'$ | CONCLUSION |
|---|---|---|---|---|---|
| $x < -2$ | – | – | – | – | $f$ is decreasing on $(-\infty, -2]$ |
| $-2 < x < 0$ | – | + | – | + | $f$ is increasing on $[-2, 0]$ |
| $0 < x < 1$ | + | + | – | – | $f$ is decreasing on $[0, 1]$ |
| $1 < x$ | + | + | + | + | $f$ is increasing on $[1, +\infty)$ |

## CONCAVITY

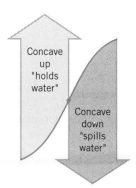

Figure 5.1.7

Although the sign of the derivative of $f$ reveals where the graph of $f$ is increasing or decreasing, it does not reveal the direction of *curvature*. For example, on both sides of the point in Figure 5.1.7 the graph is increasing, but on the left side it has an upward curvature ("holds water") and on the right side it has a downward curvature ("spills water"). On intervals where the graph of $f$ has upward curvature we say that $f$ is *concave up*, and on intervals where the graph has downward curvature we say that $f$ is *concave down*.

For differentiable functions, the direction of curvature can be characterized in terms of the tangent lines in two ways: As suggested by Figure 5.1.8, the graph of a function $f$ has upward curvature on intervals where the graph lies above its tangent lines, and it has downward curvature on intervals where it lies below its tangent lines. Alternatively, the graph has upward curvature on intervals where the tangent lines have increasing slopes and downward curvature on intervals where they have decreasing slopes. We will use this latter characterization as our formal definition.

> **5.1.3** DEFINITION. If $f$ is differentiable on an open interval $I$, then $f$ is said to be *concave up* on $I$ if $f'$ is increasing on $I$, and $f$ is said to be *concave down* on $I$ if $f'$ is decreasing on $I$.

To apply this definition we need some way to determine the intervals on which $f'$ is increasing or decreasing. One way to do this is to apply Theorem 5.1.2 (and the remark that follows it) to the function $f'$. It follows from that theorem and remark that $f'$ will be

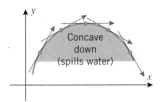

Figure 5.1.8

increasing where its derivative $f''$ is positive and will be decreasing where its derivative $f''$ is negative. This is the idea behind the following theorem.

---

**5.1.4** THEOREM. *Let $f$ be twice differentiable on an open interval $I$.*
*(a) If $f''(x) > 0$ on $I$, then $f$ is concave up on $I$.*
*(b) If $f''(x) < 0$ on $I$, then $f$ is concave down on $I$.*

---

## Example 3

Find open intervals on which the following functions are concave up and open intervals on which they are concave down.

$$\text{(a) } f(x) = x^2 - 4x + 3 \qquad \text{(b) } f(x) = x^3 \qquad \text{(c) } f(x) = x^3 - 3x^2 + 1$$

*Solution (a).* Calculating the first two derivatives we obtain

$$f'(x) = 2x - 4 \quad \text{and} \quad f''(x) = 2$$

Since $f''(x) > 0$ for all $x$, the function $f$ is concave up on $(-\infty, +\infty)$. This is consistent with Figure 5.1.4.

*Solution (b).* Calculating the first two derivatives we obtain

$$f'(x) = 3x^2 \quad \text{and} \quad f''(x) = 6x$$

Since $f''(x) < 0$ if $x < 0$ and $f''(x) > 0$ if $x > 0$, the function $f$ is concave down on $(-\infty, 0)$ and concave up on $(0, +\infty)$. This is consistent with Figure 5.1.5.

*Solution (c).* Calculating the first two derivatives we obtain

$$f'(x) = 3x^2 - 6x \quad \text{and} \quad f''(x) = 6x - 6 = 6(x - 1)$$

Since $f''(x) > 0$ if $x > 1$ and $f''(x) < 0$ if $x < 1$, we conclude that

$f$ is concave up on $(1, +\infty)$

$f$ is concave down on $(-\infty, 1)$

which is consistent with the graph in Figure 5.1.9. ◀

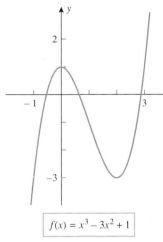

$$f(x) = x^3 - 3x^2 + 1$$

Figure 5.1.9

## INFLECTION POINTS

Points where a graph changes from concave up to concave down, or vice versa, are of special interest, so there is some terminology associated with them.

---

**5.1.5** DEFINITION. If $f$ is continuous on an open interval containing the point $x_0$, and if $f$ changes the direction of its concavity at that point, then we say that $f$ has an ***inflection point at $x_0$***, and we call the point $(x_0, f(x_0))$ on the graph of $f$ an ***inflection point*** of $f$ (Figure 5.1.10).

---

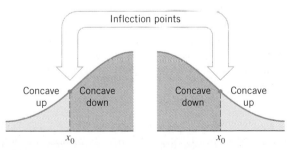

Figure 5.1.10

For example, the function $f(x) = x^3$ has an inflection point at $x = 0$ (Figure 5.1.5), the function $f(x) = x^3 - 3x^2 + 1$ has an inflection point at $x = 1$ (Figure 5.1.9), and the function $f(x) = x^2 - 4x + 3$ has no inflection points (Figure 5.1.4).

### Example 4

Use the graph in Figure 5.1.6 to make rough estimates of the locations of the inflection points of $f(x) = 3x^4 + 4x^3 - 12x^2 + 2$, and check your estimates by finding the exact location of the inflection points.

*Solution.* The graph changes from concave up to concave down somewhere between $-2$ and $-1$, say roughly at $x = -1.25$; and the graph changes from concave down to concave up somewhere between 0 and 1, say roughly at $x = 0.5$. To find the exact location of the inflection points, we start by calculating the second derivative of $f$:

$$f'(x) = 12x^3 + 12x^2 - 24x$$
$$f''(x) = 36x^2 + 24x - 24 = 12(3x^2 + 2x - 2)$$

We could analyze the sign of $f''$ by factoring this function and applying the method of test points (as in Table 5.1.1). However, here is another approach. The graph of $f''$ is a parabola that opens up, and the quadratic formula shows that the equation $f'' = 0$ has the roots

$$x = \frac{-1 - \sqrt{7}}{3} \approx -1.22 \quad \text{and} \quad x = \frac{-1 + \sqrt{7}}{3} \approx 0.55 \tag{1}$$

(verify). Thus, from the rough graph of $f''$ in Figure 5.1.11 we obtain the sign analysis of $f''$ in Table 5.1.2; this implies that $f$ has inflection points at the points in (1). ◀

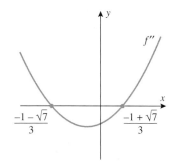

Figure 5.1.11

**Table 5.1.2**

| INTERVAL | SIGN OF $f''$ | CONCLUSION |
|---|---|---|
| $x < \dfrac{-1 - \sqrt{7}}{3}$ | $+$ | $f$ is concave up |
| $\dfrac{-1 - \sqrt{7}}{3} < x < \dfrac{-1 + \sqrt{7}}{3}$ | $-$ | $f$ is concave down |
| $x > \dfrac{-1 + \sqrt{7}}{3}$ | $+$ | $f$ is concave up |

In the preceding example the inflection points of $f$ occurred at points where $f''(x) = 0$. However, inflection points do not always occur at points where $f''(x) = 0$. For example, if the graph of $f''$ happens to touch the $x$-axis at a point without crossing over it, then $f''$ will not change sign at that point, and hence no change in the concavity of $f$ will occur at that point. Here is a specific example.

### Example 5

Find the inflection points of $f(x) = x^4$.

*Solution.* Calculating the first two derivatives of $f$ we obtain

$$f'(x) = 4x^3, \quad f''(x) = 12x^2$$

Here $f''(x) > 0$ for $x < 0$ and for $x > 0$, which implies that $f$ is concave up for $x < 0$ and for $x > 0$. Thus, there are no inflection points; and in particular, there is no inflection point at $x = 0$, even though $f''(0) = 0$ (Figure 5.1.12). ◀

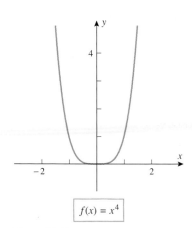

Figure 5.1.12

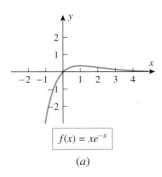

$f(x) = xe^{-x}$

(a)

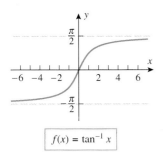

$f(x) = \sin x, \ 0 \le x \le 2\pi$

(b)

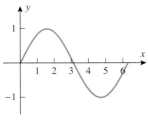

$f(x) = \tan^{-1} x$

(c)

Figure 5.1.13

### Example 6

Find the inflection points of the following functions, and confirm that your results are consistent with the graphs of the functions.

(a) $f(x) = xe^{-x}$    (b) $f(x) = \sin x, \quad 0 \le x \le 2\pi$    (c) $f(x) = \tan^{-1} x$

*Solution* (*a*). Calculating the first two derivatives of $f$ we obtain

$$f'(x) = (1 - x)e^{-x}, \quad f''(x) = (x - 2)e^{-x}$$

(verify). Keeping in mind that $e^{-x}$ is always positive, it follows that the sign of $f''$ is determined by the factor $x - 2$. Thus, $f''(x) < 0$ if $x < 2$, and $f''(x) > 0$ if $x > 2$, which implies that the graph is concave down for $x < 2$ and concave up for $x > 2$. Thus, there is an inflection point at $x = 2$ (Figure 5.1.13*a*).

*Solution* (*b*). Calculating the first two derivatives of $f$ we obtain

$$f'(x) = \cos x, \quad f''(x) = -\sin x$$

Thus, $f''(x) < 0$ if $0 < x < \pi$, and $f''(x) > 0$ if $\pi < x < 2\pi$, which implies that the graph is concave down for $0 < x < \pi$ and concave up for $\pi < x < 2\pi$. Thus, there is an inflection point at $x = \pi \approx 3.14$ (Figure 5.1.13*b*).

*Solution* (*c*). Calculating the first two derivatives of $f$ we obtain

$$f'(x) = \frac{1}{1 + x^2}, \quad f''(x) = -\frac{2x}{\left(1 + x^2\right)^2}$$

(verify). Thus, $f''(x) > 0$ if $x < 0$, and $f''(x) < 0$ if $x > 0$, which implies that the graph is concave up for $x < 0$ and concave down for $x > 0$. Thus, there is an inflection point at $x = 0$ (Figure 5.1.13*c*). ◄

FOR THE READER.    If you have a CAS, devise a method for using it to find exact values for the inflection points of a function $f$, and use your method to find the inflection points of $f(x) = x/(x^2 + 1)$. Verify that your results are consistent with the graph of $f$.

**INFLECTION POINTS IN APPLICATIONS**

Up to now we have viewed the inflection points of a curve $y = f(x)$ as those points where the curve changes the direction of its concavity. However, inflection points also mark the points on the curve where the slopes of the tangent lines change from increasing to decreasing, or vice versa (Figure 5.1.14); stated another way:

*Inflection points mark the places on the curve $y = f(x)$ where the rate of change of $y$ with respect to $x$ changes from increasing to decreasing, or vice versa.*

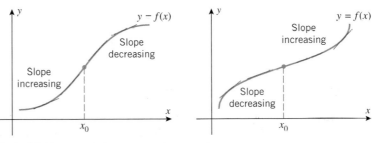

Figure 5.1.14

Note that we are dealing with a rather subtle concept here—a change of a rate of change. However, the following physical example should help to clarify the idea: Suppose that water is added to the flask in Figure 5.1.15 in such a way that the volume increases at a constant rate, and let us examine the rate at which the water level $y$ rises with the time $t$. Initially, the level $y$ will rise at a slow rate because of the wide base. However, as the diameter of the flask narrows, the rate at which the level $y$ rises will increase until the level is at the narrow point in the neck. From that point on the rate at which the level rises will decrease as the diameter gets wider and wider. Thus, the narrow point in the neck is the point at which the rate of change of $y$ with respect to $t$ changes from increasing to decreasing.

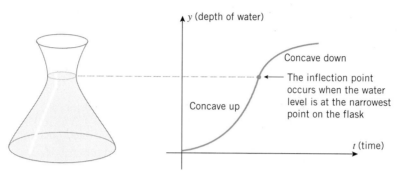

Figure 5.1.15

---

# EXERCISE SET 5.1 ⌁ Graphing Calculator  [C] CAS

1. In each part, sketch the graph of a function $f$ with the stated properties, and discuss the signs of $f'$ and $f''$.
   (a) The function $f$ is concave up and increasing on the interval $(-\infty, +\infty)$.
   (b) The function $f$ is concave down and increasing on the interval $(-\infty, +\infty)$.
   (c) The function $f$ is concave up and decreasing on the interval $(-\infty, +\infty)$.
   (d) The function $f$ is concave down and decreasing on the interval $(-\infty, +\infty)$.

2. In each part, sketch the graph of a function $f$ with the stated properties.
   (a) $f$ is increasing on $(-\infty, +\infty)$, has an inflection point at the origin, and is concave up on $(0, +\infty)$.
   (b) $f$ is increasing on $(-\infty, +\infty)$, has an inflection point at the origin, and is concave down on $(0, +\infty)$.
   (c) $f$ is decreasing on $(-\infty, +\infty)$, has an inflection point at the origin, and is concave up on $(0, +\infty)$.
   (d) $f$ is decreasing on $(-\infty, +\infty)$, has an inflection point at the origin, and is concave down on $(0, +\infty)$.

3. Use the graph of the equation $y = f(x)$ in the accompanying figure to find the signs of $dy/dx$ and $d^2y/dx^2$ at the points $A$, $B$, and $C$.

4. Use the graph of the equation $y = f'(x)$ in the accompanying figure to find the signs of $dy/dx$ and $d^2y/dx^2$ at the points $A$, $B$, and $C$.

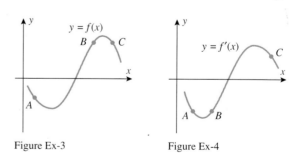

Figure Ex-3          Figure Ex-4

5. Use the graph of $y = f''(x)$ in the accompanying figure to determine the $x$-coordinates of all inflection points of $f$. Explain your reasoning.

6. Use the graph of $y = f'(x)$ in the accompanying figure to replace the question mark with $<$, $=$, or $>$, as appropriate. Explain your reasoning.
   (a) $f(0)$ ? $f(1)$   (b) $f(1)$ ? $f(2)$   (c) $f'(0)$ ? 0
   (d) $f'(1)$ ? 0      (e) $f''(0)$ ? 0     (f) $f''(2)$ ? 0

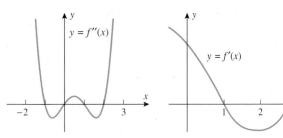

Figure Ex-5                Figure Ex-6

**7.** In each part, use the graph of $y = f(x)$ in the accompanying figure to find the requested information.
 (a) Find the intervals on which $f$ is increasing.
 (b) Find the intervals on which $f$ is decreasing.
 (c) Find the open intervals on which $f$ is concave up.
 (d) Find the open intervals on which $f$ is concave down.
 (e) Find all values of $x$ at which $f$ has an inflection point.

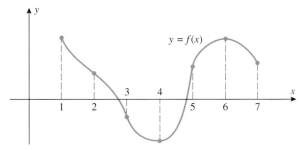

Figure Ex-7

**8.** Use the graph in Exercise 7 to make a table that shows the signs of $f'$ and $f''$ over the intervals $(1, 2)$, $(2, 3)$, $(3, 4)$, $(4, 5)$, $(5, 6)$, and $(6, 7)$.

In Exercises 9–24, find: (a) the intervals on which $f$ is increasing, (b) the intervals on which $f$ is decreasing, (c) the open intervals on which $f$ is concave up, (d) the open intervals on which $f$ is concave down, and (e) the $x$-coordinates of all inflection points.

**9.** $f(x) = x^2 - 5x + 6$    **10.** $f(x) = 4 - 3x - x^2$

**11.** $f(x) = (x + 2)^3$    **12.** $f(x) = 5 + 12x - x^3$

**13.** $f(x) = 3x^4 - 4x^3$    **14.** $f(x) = x^4 - 8x^2 + 16$

**15.** $f(x) = \dfrac{x^2}{x^2 + 2}$    **16.** $f(x) = \dfrac{x}{x^2 + 2}$

**17.** $f(x) = \sqrt[3]{x + 2}$    **18.** $f(x) = x^{2/3}$

**19.** $f(x) = x^{1/3}(x + 4)$    **20.** $f(x) = x^{4/3} - x^{1/3}$

**21.** $f(x) = e^{-x^2/2}$    **22.** $f(x) = xe^{x^2}$

**23.** $f(x) = \ln(1 + x^2)$    **24.** $f(x) = x^2 \ln x$

In Exercises 25–30, analyze the trigonometric function $f$ over the specified interval, stating where $f$ is increasing, decreasing, concave up, and concave down, and stating the $x$-coordinates of all inflection points. Confirm that your results are consistent with the graph of $f$ generated with a graphing utility.

**25.** $f(x) = \cos x$; $[0, 2\pi]$

**26.** $f(x) = \sin^2 2x$; $[0, \pi]$

**27.** $f(x) = \tan x$; $(-\pi/2, \pi/2)$

**28.** $f(x) = 2x + \cot x$; $(0, \pi)$

**29.** $f(x) = \sin x \cos x$; $[0, \pi]$

**30.** $f(x) = \cos^2 x - 2 \sin x$; $[0, 2\pi]$

**31.** In each part sketch a continuous curve $y = f(x)$ with the stated properties.
 (a) $f(2) = 4$, $f'(2) = 0$, $f''(x) > 0$ for all $x$
 (b) $f(2) = 4$, $f'(2) = 0$, $f''(x) < 0$ for $x < 2$, $f''(x) > 0$ for $x > 2$
 (c) $f(2) = 4$, $f''(x) < 0$ for $x \neq 2$ and $\lim\limits_{x \to 2^+} f'(x) = +\infty$, $\lim\limits_{x \to 2^-} f'(x) = -\infty$

**32.** In each part sketch a continuous curve $y = f(x)$ with the stated properties.
 (a) $f(2) = 4$, $f'(2) = 0$, $f''(x) < 0$ for all $x$
 (b) $f(2) = 4$, $f'(2) = 0$, $f''(x) > 0$ for $x < 2$, $f''(x) < 0$ for $x > 2$
 (c) $f(2) = 4$, $f''(x) > 0$ for $x \neq 2$ and $\lim\limits_{x \to 2^+} f'(x) = -\infty$, $\lim\limits_{x \to 2^-} f'(x) = +\infty$

**33.** In each part, assume that $a$ is a constant and find the inflection points, if any.
 (a) $f(x) = (x - a)^3$    (b) $f(x) = (x - a)^4$

**34.** Given that $a$ is a constant and $n$ is a positive integer, what can you say about the existence of inflection points of the function $f(x) = (x - a)^n$? Justify your answer.

If $f$ is increasing on an interval $[0, b)$, then it follows from Definition 5.1.1 that $f(0) < f(x)$ for each $x$ in the interval. Use this result in Exercises 35–38.

**35.** Show that $\sqrt[3]{1 + x} < 1 + \frac{1}{3}x$ if $x > 0$, and confirm the inequality with a graphing utility. [*Hint:* Show that the function $f(x) = 1 + \frac{1}{3}x - \sqrt[3]{1 + x}$ is increasing on $[0, +\infty)$.]

**36.** Show that $x < \tan x$ if $0 < x < \pi/2$, and confirm the inequality with a graphing utility. [*Hint:* Show that the function $f(x) = \tan x - x$ is increasing on $[0, \pi/2)$.]

**37.** Use a graphing utility to make a conjecture about the relative sizes of $x$ and $\sin x$ for $x \geq 0$, and prove your conjecture.

**38.** (a) Show that $e^x \geq 1 + x$ if $x \geq 0$.
 (b) Show that $e^x \geq 1 + x + \frac{1}{2}x^2$ if $x \geq 0$.
 (c) Confirm the inequalities in parts (a) and (b) with a graphing utility.

In Exercises 39 and 40, use a graphing utility to generate the graphs of $f'$ and $f''$ over the stated interval; then use those graphs to estimate the $x$-coordinates of the inflection points of $f$, the intervals on which $f$ is concave up or down, and the intervals on which $f$ is increasing or decreasing. Check your estimates by graphing $f$.

**39.** $f(x) = x^4 - 24x^2 + 12x$, $-5 \leq x \leq 5$

**40.** $f(x) = \dfrac{1}{1 + x^2}$, $-5 \leq x \leq 5$

**41.** For the function $f(x) = e^x/(1 + x^2)$, use the method of Example 6 in Section 2.4 to approximate the $x$-coordinates of the inflection points to two decimal places.

**42.** For the function $f$ in Exercise 40, use the method of Example 6 in Section 2.4 to approximate the $x$-coordinates of the inflection points to two decimal places.

In Exercises 43 and 44, use a CAS to find $f''$, and then use the method of Example 6 in Section 2.4 to approximate the $x$-coordinates of the inflection points to one decimal place. Confirm that your answer is consistent with the graph of $f$.

**43.** $f(x) = \dfrac{10x - 3}{3x^2 - 5x + 8}$    **44.** $f(x) = \dfrac{x^3 - 8x + 7}{\sqrt{x^2 + 1}}$

**45.** Use Definition 5.1.1 to prove that $f(x) = x^2$ is increasing on $[0, +\infty)$.

**46.** Use Definition 5.1.1 to prove that $f(x) = 1/x$ is decreasing on $(0, +\infty)$.

**47.** In each part, determine whether the statement is true or false. If it is false, find functions for which the statement fails to hold.
(a) If $f$ and $g$ are increasing on an interval, then so is $f + g$.
(b) If $f$ and $g$ are increasing on an interval, then so is $f \cdot g$.

**48.** In each part, find functions $f$ and $g$ that are increasing on $(-\infty, +\infty)$ and for which $f - g$ has the stated property.
(a) $f - g$ is decreasing on $(-\infty, +\infty)$.
(b) $f - g$ is constant on $(-\infty, +\infty)$.
(c) $f - g$ is increasing on $(-\infty, +\infty)$.

**49.** (a) Prove that a general cubic polynomial
$$f(x) = ax^3 + bx^2 + cx + d \quad (a \neq 0)$$
has exactly one inflection point.
(b) Prove that if a cubic polynomial has three $x$-intercepts, then the inflection point occurs at the average value of the intercepts.

(c) Use the result in part (b) to find the inflection point of the cubic polynomial $f(x) = x^3 - 3x^2 + 2x$, and check your result by using $f''$ to determine where $f$ is concave up and concave down.

**50.** From Exercise 49, the polynomial $f(x) = x^3 + bx^2 + 1$ has one inflection point. Use a graphing utility to reach a conclusion about the effect of the constant $b$ on the location of the inflection point. Use $f''$ to explain what you have observed graphically.

**51.** Use Definition 5.1.1 to prove:
(a) If $f$ is increasing on the intervals $(a, c]$ and $[c, b)$, then $f$ is increasing on $(a, b)$.
(b) If $f$ is decreasing on the intervals $(a, c]$ and $[c, b)$, then $f$ is decreasing on $(a, b)$.

**52.** Use part (a) of Exercise 51 to show that $f(x) = x + \sin x$ is increasing on the interval $(-\infty, +\infty)$.

**53.** Suppose that the spread of a flu virus on a college campus is modeled by the function
$$y(t) = \frac{1000}{1 + 999e^{-0.9t}}$$
where $y(t)$ is the number of infected students at time $t$ (in days, starting with $t = 0$). Use a graphing utility to estimate the day on which the virus is spreading most rapidly.

**54.** Let $y = 1/(1 + x^2)$. Find the values of $x$ for which $y$ is increasing and decreasing most rapidly.

In Exercises 55 and 56, suppose that water is flowing at a constant rate into the container shown. Make a rough sketch of the graph of the water level $y$ versus the time $t$. Make sure that your sketch conveys where the graph is concave up and concave down, and label the $y$-coordinates of the inflection points.

**55.**

**56.**

## 5.2  ANALYSIS OF FUNCTIONS II: RELATIVE EXTREMA; FIRST AND SECOND DERIVATIVE TESTS

*In this section we will discuss methods for finding the high and low points on the graph of a function. The ideas we develop here will have important applications.*

**RELATIVE MAXIMA AND MINIMA**

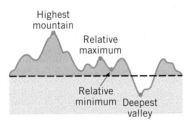

Figure 5.2.1

If we imagine the graph of a function $f$ to be a two-dimensional mountain range with hills and valleys, then the tops of the hills are called *relative maxima*, and the bottoms of the valleys are called *relative minima* (Figure 5.2.1).

The relative maxima are the high points in their *immediate vicinity*, and the relative minima are the low points. Note that a relative maximum need not be the highest point in the entire mountain range, and a relative minimum need not be the lowest point—they are just high and low points *relative* to the nearby terrain. These ideas are captured in the following definition.

**5.2.1**  DEFINITION.  A function $f$ is said to have a ***relative maximum*** at $x_0$ if there is an open interval containing $x_0$ on which $f(x_0)$ is the largest value, that is, $f(x_0) \geq f(x)$ for all $x$ in the interval. Similarly, $f$ is said to have a ***relative minimum*** at $x_0$ if there is an open interval containing $x_0$ on which $f(x_0)$ is the smallest value, that is, $f(x_0) \leq f(x)$ for all $x$ in the interval. If $f$ has either a relative maximum or a relative minimum at $x_0$, then $f$ is said to have a ***relative extremum*** at $x_0$.

**Example 1**

Locate the relative extrema of the four functions graphed in Figure 5.2.2.

*Solution.*

(a)  The function $f(x) = x^2$ has a relative minimum at $x = 0$ but no relative maxima.

(b)  The function $f(x) = x^3$ has no relative extrema.

(c)  The function $f(x) = x^3 - 3x + 3$ has a relative maximum at $x = -1$ and a relative minimum at $x = 1$.

(d)  The function $f(x) = \cos x$ has relative maxima at all even multiples of $\pi$ and relative minima at all odd multiples of $\pi$.  ◀

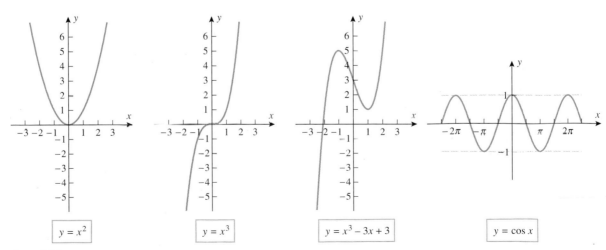

Figure 5.2.2

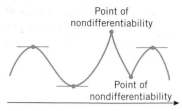

Figure 5.2.3

Relative extrema can be viewed as the transition points that separate the regions where a graph is increasing from those where it is decreasing. As suggested by Figure 5.2.3, the relative extrema of a continuous function $f$ occur either at corners or at points where the graph of $f$ has a horizontal tangent line. This is the content of the following theorem, whose proof is given in Appendix G.

**5.2.2** THEOREM. *If a function $f$ has any relative extrema, then they occur either at points where $f'(x) = 0$ or at points where $f$ is not differentiable.*

The points at which either $f'(x) = 0$ or $f$ is not differentiable are called the **critical points** of $f$, so that Theorem 5.2.2 can be rephrased as follows:

*The relative extrema of a function, if any, occur at critical points.*

**CRITICAL POINTS**

Sometimes we will want to distinguish the critical points at which $f'(x) = 0$ from those points where $f$ is not differentiable, in which case we will call the critical points at which $f'(x) = 0$ the **stationary points** of $f$.

It is important not to read too much into Theorem 5.2.2—the theorem asserts that the relative extrema must occur at critical points, but it does not say that a relative extremum occurs at *every* critical point; that is, there may be critical points at which a relative extremum does not occur. For example, for the eight critical points shown in Figure 5.2.4, relative extrema occur at all of the points in the top row, but not at any of the points in the bottom row.

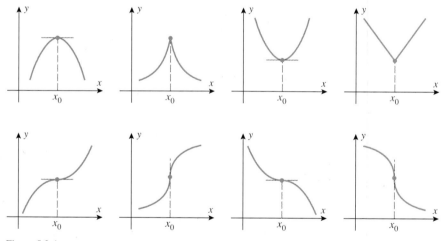

Figure 5.2.4

**FIRST DERIVATIVE TEST**

To develop an effective method for finding critical points of a function $f$, we need some criteria that will enable us to distinguish between the critical points where relative extrema occur and those where they do not. One such criterion can be motivated by examining the sign of the first derivative of $f$ on each side of the eight critical points in Figure 5.2.4:

- At the two relative maxima in the top row, $f'$ is positive to the left of $x_0$ and negative to the right.
- At the two relative minima in the top row, $f'$ is negative to the left of $x_0$ and positive to the right.
- At the first two critical points in the bottom row, $f'$ is positive on both sides of $x_0$.
- At the last two critical points in the bottom row, $f'$ is negative on both sides of $x_0$.

These observations suggest that relative extrema of a function $f$ occur at those critical points, and only those critical points, where $f'$ changes sign. Moreover, if the sign changes from positive to negative, then a relative maximum occurs; and if the sign changes from negative to positive, then a relative minimum occurs. This is the content of the following theorem, whose proof is given at the end of this section.

---

**5.2.3**    THEOREM (*First Derivative Test*).    *Suppose $f$ is continuous at a critical point $x_0$.*

(a)    *If $f'(x) > 0$ on an open interval extending left from $x_0$ and $f'(x) < 0$ on an open interval extending right from $x_0$, then $f$ has a relative maximum at $x_0$.*

(b)    *If $f'(x) < 0$ on an open interval extending left from $x_0$ and $f'(x) > 0$ on an open interval extending right from $x_0$, then $f$ has a relative minimum at $x_0$.*

(c)    *If $f'(x)$ has the same sign [either $f'(x) > 0$ or $f'(x) < 0$] on an open interval extending left from $x_0$ and on an open interval extending right from $x_0$, then $f$ does not have a relative extremum at $x_0$.*

---

## Example 2

(a)    Locate the relative maxima and minima of $f(x) = 3x^{5/3} - 15x^{2/3}$.

(b)    Confirm that the results in part (a) agree with the graph of $f$.

*Solution (a).* The function $f$ is defined and continuous for all real values of $x$, and its derivative is

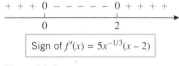

$$f'(x) = 5x^{2/3} - 10x^{-1/3} = 5x^{-1/3}(x - 2) = \frac{5(x - 2)}{x^{1/3}}$$

Sign of $f'(x) = 5x^{-1/3}(x - 2)$

Figure 5.2.5

Since $f'(x)$ does not exist if $x = 0$, and since $f'(x) = 0$ if $x = 2$, there are critical points at $x = 0$ and $x = 2$. To apply the first derivative test, we examine the sign of $f'(x)$ on intervals extending to the left and right of the critical points (Figure 5.2.5). Since the sign of the derivative changes from positive to negative at $x = 0$, there is a relative maximum there, and since it changes from negative to positive at $x = 2$, there is a relative minimum there.

*Solution (b).* The result in part (a) agrees with the graph of $f$ shown in Figure 5.2.6.    ◀

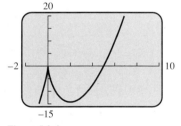

Figure 5.2.6

FOR THE READER.    As discussed in the subsection of Section 1.3 entitled Errors of Omission, many graphing utilities omit portions of the graphs of functions with fractional exponents and must be "tricked" into producing complete graphs; and indeed, for the function in the last example the author's calculator and CAS both failed to produce the portion of the graph over the negative $x$-axis. To generate the graph in Figure 5.2.6, the author had to apply the techniques discussed in Exercise 29 of Section 1.3 to each term in the formula for $f$. Use a graphing utility to generate this graph.

## Example 3

Locate the relative extrema of $f(x) = x^3 - 3x^2 + 3x - 1$, if any.

*Solution.* Since $f$ is differentiable everywhere, the only possible critical points are stationary points. Differentiating $f$ yields

$$f'(x) = 3x^2 - 6x + 3 = 3(x - 1)^2$$

Solving $f'(x) = 0$ yields $x = 1$ as the only stationary point. However, $3(x - 1)^2 \geq 0$ for all $x$, so $f'(x)$ does not change sign at $x = 1$; consequently, $f$ does not have a relative extremum at $x = 1$. Thus, $f$ has no relative extrema (Figure 5.2.7).    ◀

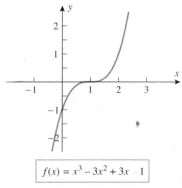

$f(x) = x^3 - 3x^2 + 3x - 1$

Figure 5.2.7

FOR THE READER.    How many relative extrema can a polynomial of degree $n$ have? Explain your reasoning.

**SECOND DERIVATIVE TEST**

There is another test for relative extrema that is often easier to apply than the first derivative test. It is based on the geometric observation that a function $f$ has a relative maximum at a stationary point if the graph of $f$ is concave down on an open interval containing the point, and it has a relative minimum if it is concave up (Figure 5.2.8).

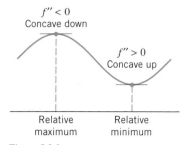

$f'' < 0$
Concave down

$f'' > 0$
Concave up

Relative maximum   Relative minimum

Figure 5.2.8

**5.2.4  THEOREM (*Second Derivative Test*).**  *Suppose that $f$ is twice differentiable at the point $x_0$.*

(*a*)  *If $f'(x_0) = 0$ and $f''(x_0) > 0$, then $f$ has a relative minimum at $x_0$.*

(*b*)  *If $f'(x_0) = 0$ and $f''(x_0) < 0$, then $f$ has a relative maximum at $x_0$.*

(*c*)  *If $f'(x_0) = 0$ and $f''(x_0) = 0$, then the test is inconclusive; that is, $f$ may have a relative maximum, a relative minimum, or neither at $x_0$.*

REMARK.  The proof of parts (*a*) and (*b*) is given at the end of this section. For part (*c*), consider the functions $f(x) = x^3$, $f(x) = x^4$, and $f(x) = -x^4$. In all three cases we have $f'(0) = 0$ and $f''(0) = 0$ (verify); but from Figure 1.6.4, $f(x) = x^4$ has a relative minimum at $x = 0$, $f(x) = -x^4$ has a relative maximum at $x = 0$ (why?), and $f(x) = x^3$ has neither a relative maximum nor a relative minimum at $x = 0$.

## Example 4

Locate the relative maxima and minima of $f(x) = x^4 - 2x^2$, and confirm that your results are consistent with the graph of $f$.

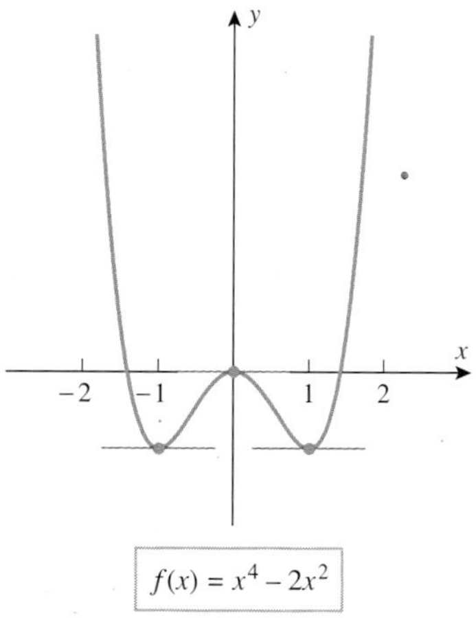

*Solution.*

$$f'(x) = 4x^3 - 4x = 4x(x - 1)(x + 1)$$
$$f''(x) = 12x^2 - 4$$

Solving $f'(x) = 0$ yields the stationary points $x = 0$, $x = 1$, and $x = -1$. Evaluating $f''$ at these points yields

$$f''(0) = -4 < 0$$
$$f''(1) = 8 > 0$$
$$f''(-1) = 8 > 0$$

so there is a relative maximum at $x = 0$ and relative minima at $x = 1$ and $x = -1$ (Figure 5.2.9).  ◄

$f(x) = x^4 - 2x^2$

Figure 5.2.9

**MORE ON THE SIGNIFICANCE OF INFLECTION POINTS**

In Section 5.1 we observed that the inflection points of a curve $y = f(x)$ mark the points where the slopes of the tangent lines change from increasing to decreasing, or vice versa. Thus, in the case where $f$ is twice differentiable, the inflection points mark the places on the curve $y = f(x)$ where $f'(x)$ has a relative maximum or minimum (Figure 5.2.10); stated another way:

*Inflection points mark the places on the curve $y = f(x)$ at which the rate of change of $y$ with respect to $x$ has a relative maximum or minimum; that is, they are the places where $y$ is increasing or decreasing most rapidly in the immediate vicinity.*

As an illustration of this principle, consider the flask shown in Figure 5.1.15. We observed in Section 5.1 that if water is poured into the flask so that the volume increases at a constant rate, then the graph of $y$ versus $t$ has an inflection point when $y$ is at the narrow point in the neck. However, this is also the place where the water level is rising most rapidly.

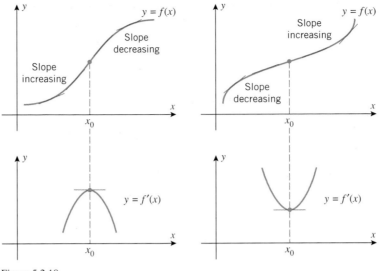

Figure 5.2.10

### PROOF OF THE FIRST DERIVATIVE TEST (*Theorem 5.2.3*)

*Proof.*  We will prove part (*a*) and leave parts (*b*) and (*c*) as exercises. We are assuming that $f'(x) > 0$ on the interval $(a, x_0)$ and that $f'(x) < 0$ on the interval $(x_0, b)$, and we want to show that

$$f(x_0) \geq f(x)$$

for all $x$ in the interval $(a, b)$. However, the two hypotheses, together with Theorem 5.1.2 (and its following remark) imply that $f$ is increasing on the interval $(a, x_0]$ and decreasing on the interval $[x_0, b)$. Thus, $f(x_0) \geq f(x)$ for all $x$ in $(a, b)$ with equality only at $x_0$.  ∎

### PROOF OF THE SECOND DERIVATIVE TEST (*Theorem 5.2.4*)

*Proof.*  We will prove part (*a*) and leave part (*b*) as an exercise. We want to show that if $f'(x_0) = 0$ and $f''(x_0) > 0$, then $f$ has a relative minimum at $x_0$; that is, there is an open interval $(a, b)$ containing $x_0$ on which

$$f(x) \geq f(x_0)$$

For simplicity, we will assume that $f''$ is continuous at $x_0$. The proof for the case where $f$ is twice differentiable at $x_0$ is left for more advanced courses. Observe first that the tangent line at $x_0$ is horizontal [since $f'(x_0) = 0$], and hence its equation is $y = f(x_0)$. Moreover, since $f''(x_0) > 0$, and since $f''$ is continuous at $x_0$, there is an open interval $(a, b)$ containing $x_0$ on which $f''(x) > 0$. This implies that $f$ is concave up on $(a, b)$, and hence its graph lies above the tangent line $y = f(x_0)$ over the interval $(a, b)$. This shows that $f(x) \geq f(x_0)$ on the interval $(a, b)$.  ∎

**EXERCISE SET 5.2** ⌇ Graphing Calculator    [c] CAS

1. In each part, sketch the graph of a continuous function $f$ with the stated properties.
   (a) $f$ is concave up on the interval $(-\infty, +\infty)$ and has exactly one relative extremum.
   (b) $f$ is concave up on the interval $(-\infty, +\infty)$ and has no relative extrema.
   (c) The function $f$ has exactly two relative extrema on the interval $(-\infty, +\infty)$, and $f(x) \to +\infty$ as $x \to +\infty$.
   (d) The function $f$ has exactly two relative extrema on the interval $(-\infty, +\infty)$, and $f(x) \to -\infty$ as $x \to +\infty$.

2. In each part, sketch the graph of a continuous function $f$ with the stated properties.
   (a) $f$ has exactly one relative extremum on $(-\infty, +\infty)$, and $f(x) \to 0$ as $x \to +\infty$ and as $x \to -\infty$.
   (b) $f$ has exactly two relative extrema on $(-\infty, +\infty)$, and $f(x) \to 0$ as $x \to +\infty$ and as $x \to -\infty$.
   (c) $f$ has exactly one inflection point and one relative extremum on $(-\infty, +\infty)$.
   (d) $f$ has infinitely many relative extrema, and $f(x) \to 0$ as $x \to +\infty$ and as $x \to -\infty$.

3. (a) Use both the first and second derivative tests to show that $f(x) = 3x^2 - 6x + 1$ has a relative minimum at $x = 1$.
   (b) Use both the first and second derivative tests to show that $f(x) = x^3 - 3x + 3$ has a relative minimum at $x = 1$ and a relative maximum at $x = -1$.

4. (a) Use both the first and second derivative tests to show that $f(x) = \sin^2 x$ has a relative minimum at $x = 0$.
   (b) Use both the first and second derivative tests to show that $g(x) = \tan^2 x$ has a relative minimum at $x = 0$.
   (c) Give an informal verbal argument to explain without calculus why the functions in parts (a) and (b) have relative minima at $x = 0$.

5. (a) Show that both of the functions $f(x) = (x - 1)^4$ and $g(x) = x^3 - 3x^2 + 3x - 2$ have stationary points at $x = 1$.
   (b) What does the second derivative test tell you about the nature of these stationary points?
   (c) What does the first derivative test tell you about the nature of these stationary points?

6. (a) Show that $f(x) = 1 - x^5$ and $g(x) = 3x^4 - 8x^3$ both have stationary points at $x = 0$.
   (b) What does the second derivative test tell you about the nature of these stationary points?
   (c) What does the first derivative test tell you about the nature of these stationary points?

In Exercises 7–12, locate the critical points, and classify them as stationary points or points of nondifferentiability.

7. (a) $f(x) = x^3 + 3x^2 - 9x + 1$
   (b) $f(x) = x^4 - 6x^2 - 3$

8. (a) $f(x) = 2x^3 - 6x + 7$   (b) $f(x) = 3x^4 - 4x^3$

9. (a) $f(x) = \dfrac{x}{x^2 + 2}$   (b) $f(x) = x^{2/3}$

10. (a) $f(x) = \dfrac{x^2 - 3}{x^2 + 1}$   (b) $f(x) = \sqrt[3]{x + 2}$

11. (a) $f(x) = x^{1/3}(x + 4)$   (b) $f(x) = \cos 3x$

12. (a) $f(x) = x^{4/3} - 6x^{1/3}$   (b) $f(x) = |\sin x|$

In Exercises 13 and 14, use the graph of $f'$ shown in the figure to estimate all values of $x$ at which $f$ has (a) relative minima, (b) relative maxima, and (c) inflection points.

13.
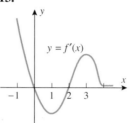

14.
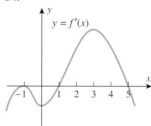

In Exercises 15 and 16, use the given derivative to find the $x$-coordinates of all critical points of $f$, and determine whether a relative maximum, relative minimum, or neither occurs there.

15. (a) $f'(x) = x^3(x^2 - 5)$   (b) $f'(x) = xe^{-x}$

16. (a) $f'(x) = x^2(2x + 1)(x - 1)$
    (b) $f'(x) = \dfrac{9 - 4x^2}{\sqrt[3]{x + 1}}$

In Exercises 17–20, find the relative extrema using both the first and second derivative tests.

17. $f(x) = 1 - 4x - x^2$   18. $f(x) = 2x^3 - 9x^2 + 12x$

19. $f(x) = \sin^2 x, \quad 0 < x < 2\pi$

20. $f(x) = \tfrac{1}{2}x - \sin x, \quad 0 < x < 2\pi$

In Exercises 21–34, use any method to find the relative extrema of the function $f$.

21. $f(x) = x^3 + 5x - 2$   22. $f(x) = x^4 - 2x^2 + 7$

23. $f(x) = x(x - 1)^2$   24. $f(x) = x^4 + 2x^3$

25. $f(x) = 2x^2 - x^4$   26. $f(x) = (2x - 1)^5$

27. $f(x) = x^{4/5}$   28. $f(x) = 2x + x^{2/3}$

29. $f(x) = \dfrac{x^2}{x^2 + 1}$   30. $f(x) = \dfrac{x}{x + 2}$

**31.** $f(x) = \ln(1 + x^2)$

**32.** $f(x) = x^2 e^x$

**33.** $f(x) = |x^2 - 4|$

**34.** $f(x) = \begin{cases} 9 - x, & x \leq 3 \\ x^2 - 3, & x > 3 \end{cases}$

In Exercises 35–38, find the relative extrema in the interval $0 < x < 2\pi$, and confirm that your results are consistent with the graph of $f$ generated with a graphing utility.

**35.** $f(x) = |\sin 2x|$

**36.** $f(x) = \sqrt{3}x + 2\sin x$

**37.** $f(x) = \cos^2 x$

**38.** $f(x) = \dfrac{\sin x}{2 - \cos x}$

In Exercises 39–42, use a graphing utility to make a conjecture about the relative extrema of $f$, and then check your conjecture using either the first or second derivative test.

**39.** $f(x) = x \ln x$

**40.** $f(x) = \dfrac{2}{e^x + e^{-x}}$

**41.** $f(x) = x^2 e^{-2x}$

**42.** $f(x) = 10 \ln x - x$

In Exercises 43 and 44, use a graphing utility to generate the graphs of $f'$ and $f''$ over the stated interval, and then use those graphs to estimate the $x$-coordinates of the relative extrema of $f$. Check that your estimates are consistent with the graph of $f$.

**43.** $f(x) = x^4 - 24x^2 + 12x$, $\quad -5 \leq x \leq 5$

**44.** $f(x) = \sin \frac{1}{2}x \cos x$, $\quad -\pi/2 \leq x \leq \pi/2$

**45.** For the function $f$ in Exercise 43, use the method of Example 6 in Section 2.4 to approximate the $x$-coordinates of the relative maxima to two decimal places.

**46.** For the function $f$ in Exercise 44, use the method of Example 6 in Section 2.4 to approximate the $x$-coordinates of the relative maxima to two decimal places.

In Exercises 47 and 48, use a CAS to graph $f'$ and $f''$ over the stated interval. Use those graphs to make a conjecture about the locations and nature of the relative extrema of $f$, and check your conjecture by graphing $f$.

**47.** $f(x) = \dfrac{10x - 3}{3x^2 - 5x + 8}$

**48.** $f(x) = \dfrac{x^3 - 8x + 7}{\sqrt{x^2 + 1}}$

**49.** In each part, find $k$ so that $f$ has a relative extremum at the point $x = 3$.

(a) $f(x) = x^2 + \dfrac{k}{x}$

(b) $f(x) = \dfrac{x}{x^2 + k}$

**50.** Functions of the form

$$f(x) = cx^n e^{-x}, \quad x > 0$$

where $n$ is a positive integer and $c = 1/n!$, arise in the statistical study of traffic flow.

(a) Use a graphing utility to generate the graph of $f$ for $n = 2, 3, 4$, and 5, and make a conjecture about the number and locations of the relative extrema of $f$.

(b) Confirm your conjecture using the first derivative test.

**51.** Functions of the form

$$f(x) = \dfrac{1}{\sqrt{2\pi}} e^{-x^2/2}$$

arise in a wide variety of statistical problems.

(a) Use the first derivative test to show that $f$ has a relative maximum at $x = 0$, and confirm this by using a graphing utility to graph $f$.

(b) Sketch the graph of

$$f(x) = \dfrac{1}{\sqrt{2\pi}} e^{-(x-\mu)^2/2}$$

where $\mu$ is a constant, and label the coordinates of the relative extrema.

**52.** (a) Use a CAS to graph the function

$$f(x) = \dfrac{x^4 + 1}{x^2 + 1}$$

and use the graph to estimate the $x$-coordinates of the relative extrema.

(b) Find the exact $x$-coordinates by using the CAS to solve the equation $f'(x) = 0$.

**53.** Find values of $a, b, c$, and $d$ so that the function

$$f(x) = ax^3 + bx^2 + cx + d$$

has a relative minimum at $(0, 0)$ and a relative maximum at $(1, 1)$.

**54.** Let $h$ and $g$ have relative maxima at $x_0$. Prove or disprove:

(a) $h + g$ has a relative maximum at $x_0$

(b) $h - g$ has a relative maximum at $x_0$.

**55.** Sketch some curves that show that the three parts of the first derivative test (Theorem 5.2.3) can be false without the assumption that $f$ is continuous at $x_0$.

## 5.3 ANALYSIS OF FUNCTIONS III: APPLYING TECHNOLOGY AND THE TOOLS OF CALCULUS

*In this section we will discuss how to use technology and the tools of calculus that we developed in the last two sections to analyze various types of graphs that occur in applications.*

This section contains a brief review of material on polynomials. Readers who want to review this material in more depth are referred to Appendix F. Instructors who want to spend more time on this section can divide the section into two parts, treating the analysis of polynomials and rational functions in one lecture and the remaining topics in a second lecture.

**PROPERTIES OF GRAPHS**

In many problems, the properties of interest in the graph of a function are:

- symmetries
- $x$-intercepts
- relative extrema
- intervals of increase and decrease
- asymptotes

- periodicity
- $y$-intercepts
- inflection points
- concavity
- behavior as $x \to +\infty$ or $x \to -\infty$

Some of these properties may not be relevant in certain cases; for example, asymptotes are characteristic of rational functions but not of polynomials, and periodicity is characteristic of trigonometric functions but not of logarithmic or exponential functions. Thus, when analyzing the graph of a function $f$, it helps to know something about the general properties of the family to which it belongs.

In a given problem you will usually have a definite objective for your analysis. For example, you may be interested in finding a ***complete graph*** of $y = f(x)$, that is, a graph that shows all of the important characteristics of $f$; or you may be interested in something specific, say the exact location of the relative extrema or the behavior of the graph as $x \to +\infty$ or $x \to -\infty$. However, regardless of your objectives, you will usually find it helpful to begin your analysis by generating the graph with a graphing utility. As discussed in Section 1.3, this graph may or may not be complete, and some of the important characteristics may be obscured by compression or resolution problems. However, with this graph as a starting point, you can often use calculus to complete the analysis and resolve any ambiguities.

**A PROCEDURE FOR ANALYZING GRAPHS**

There are no hard and fast rules that are guaranteed to produce all of the information you may need about the graph of a function $f$, but here is one possible way of organizing the analysis of a function (the order of the steps can be varied).

**Step 1.** Use a graphing utility to generate the graph of $f$ in some reasonable window, taking advantage of any general knowledge you have about the function to help in choosing the window.

**Step 2.** See if the graph suggests the existence of symmetries, periodicity, or domain restrictions. If so, try to confirm those properties analytically.

**Step 3.** Find the intercepts, if needed.

**Step 4.** Investigate the behavior of the graph as $x \to +\infty$ and as $x \to -\infty$, and identify all horizontal and vertical asymptotes, if any.

**Step 5.** Calculate $f'(x)$ and $f''(x)$, and use these derivatives to determine the critical points, the intervals on which $f$ is increasing or decreasing, the intervals on which $f$ is concave up and concave down, and the inflection points.

**Step 6.** If you have discovered that some of the significant features did not fall within the graphing window in Step 1, then try adjusting the window to include them. However, it is possible that compression or resolution problems may prevent you from showing all of the features of interest in a single window, in which case you may need to use different windows to focus on different features. In some cases you may even find that a hand-drawn sketch labeled with the location of the significant features is clearer or more informative than a graph generated with a graphing utility.

**A BRIEF REVIEW OF POLYNOMIALS**

Recall that if $n$ is a nonnegative integer, then a ***polynomial of degree n*** is a function that can be written in the following forms, depending on whether you want the powers of $x$ in ascending or descending order:

$$c_0 + c_1 x + c_2 x^2 + \cdots + c_n x^n \quad (c_n \neq 0)$$
$$c_n x^n + c_{n-1} x^{n-1} + \cdots + c_1 x + c_0 \quad (c_n \neq 0)$$

The numbers $c_0, c_1, \ldots, c_n$ are called the ***coefficients*** of the polynomial. The coefficient $c_n$ (which multiplies the highest power of $x$) is called the ***leading coefficient***, the term $c_n x^n$ is called the ***leading term***, and the coefficient $c_0$ is called the ***constant term***. Polynomials of degree 1, 2, 3, 4, and 5 are called ***linear***, ***quadratic***, ***cubic***, ***quartic***, and ***quintic***, respectively. For simplicity, general polynomials of low degree are often written without subscripts on the coefficients:

$$p(x) = a \qquad \text{Constant polynomial}$$
$$p(x) = ax + b \quad (a \neq 0) \qquad \text{Linear polynomial}$$
$$p(x) = ax^2 + bx + c \quad (a \neq 0) \qquad \text{Quadratic polynomial}$$
$$p(x) = ax^3 + bx^2 + cx + d \quad (a \neq 0) \qquad \text{Cubic polynomial}$$

When you attempt to factor a polynomial completely, one of three things can happen:

- You may be able to decompose the polynomial into distinct linear factors using only real numbers; for example,

$$x^3 + x^2 - 2x = x(x^2 + x - 2) = x(x - 1)(x + 2)$$

- You may be able to decompose the polynomial into linear factors using only real numbers, but some of the factors may be repeated; for example,

$$x^6 - 3x^4 + 2x^3 = x^3(x^3 - 3x + 2) = x^3(x - 1)^2(x + 2) \tag{1}$$

- You may be able to decompose the polynomial into linear and quadratic factors using only real numbers, but you may not be able to decompose the quadratic factors into linear factors without using imaginary numbers (such quadratic factors are said to be ***irreducible*** over the real numbers); for example,

$$x^4 - 1 = (x^2 - 1)(x^2 + 1) = (x - 1)(x + 1)(x^2 + 1)$$
$$= (x - 1)(x + 1)(x - i)(x + i)$$

Here, the factor $x^2 + 1$ is irreducible over the real numbers.

In general, if $p(x)$ is a polynomial of degree $n$ with leading coefficient $a$, and if imaginary numbers are allowed, then $p(x)$ can be factored as

$$p(x) = a(x - r_1)(x - r_2) \cdots (x - r_n) \tag{2}$$

where $r_1, r_2, \ldots, r_n$ are called the **zeros** of $p(x)$ or the **roots** of the equation $p(x) = 0$, and (2) is called the **complete linear factorization** of $p(x)$. If some of the factors in (2) are repeated, then they can be combined; for example, if the first $k$ factors are distinct and the rest are repetitions of the first $k$, then (2) can be expressed in the form

$$p(x) = a(x - r_1)^{m_1}(x - r_2)^{m_2} \cdots (x - r_k)^{m_k} \tag{3}$$

where $r_1, r_2, \ldots, r_k$ are the *distinct* roots of $p(x) = 0$. The exponents $m_1, m_2, \ldots, m_k$ tell us how many times the various factors occur in the complete linear factorization; for example, in (3) the factor $(x - r_1)$ occurs $m_1$ times, the factor $(x - r_2)$ occurs $m_2$ times, and so forth. Some techniques for factoring polynomials are discussed in Appendix F. In general, if a factor $(x - r)$ occurs $m$ times in the complete linear factorization of a polynomial, then we say that $r$ is a root or zero of **multiplicity m**, and if $(x - r)$ has no repetitions (i.e., $r$ has multiplicity 1), then we say that $r$ is a **simple** root or zero. For example, it follows from (1) that the equation $x^6 - 3x^4 + 2x^3 = 0$ can be expressed as

$$x^3(x - 1)^2(x + 2) = 0 \tag{4}$$

so this equation has three distinct roots—a root $x = 0$ of multiplicity 3, a root $x = 1$ of multiplicity 2, and a simple root $x = -2$.

Note that in (3) the multiplicities of the roots must add up to $n$, since $p(x)$ has degree $n$; that is,

$$m_1 + m_2 + \cdots + m_k = n$$

For example, in (4) the multiplicities add up to 6, which is the same as the degree of the polynomial.

It follows from (2) that a polynomial of degree $n$ can have at most $n$ distinct roots; if all of the roots are simple, then there will be *exactly* $n$, but if some are repeated, then there will be fewer than $n$. However, when counting the roots of a polynomial, it is standard practice to count multiplicities, since that convention allows us to say that a polynomial of degree $n$ has $n$ roots. For example, from (1) the six roots of the polynomial $p(x) = x^6 - 3x^4 + 2x^3$ are

$$r = 0, \ 0, \ 0, \ 1, \ 1, \ -2$$

In summary, we have the following important theorem.

**5.3.1**    THEOREM.    *If imaginary roots are allowed, and if roots are counted according to their multiplicity, then a polynomial of degree n has exactly n roots.*

**ANALYSIS OF POLYNOMIALS**

Polynomials are among the simplest functions to graph and analyze, since their only significant features are intercepts, relative extrema, inflection points, and the behavior as $x \to +\infty$ and $x \to -\infty$. Figure 5.3.1 shows the graphs of four typical polynomials in $x$.

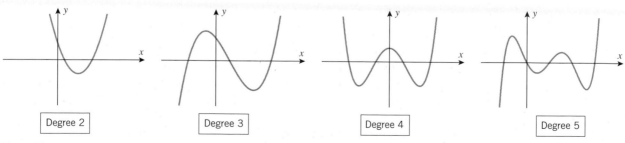

Figure 5.3.1

The graphs in Figure 5.3.1 have properties that are common to all polynomials:

- The natural domain of a polynomial in $x$ is the entire $x$-axis, since the only operations involved in its formula are additions, subtractions, and multiplications; the range depends on the particular polynomial.
- Graphs of polynomials are continuous since polynomials are continuous functions.
- Graphs of polynomials have no sharp corners or points of vertical tangency, since polynomials are differentiable functions.
- The graph of a polynomial eventually increases or decreases without bound as $x \rightarrow +\infty$ or $x \rightarrow -\infty$, since the limit of a polynomial as $x \rightarrow +\infty$ or $x \rightarrow -\infty$ is $\pm\infty$ (see the subsection in Section 2.2 entitled Limits of Polynomials as $x \rightarrow +\infty$ or $x \rightarrow -\infty$).
- The graph of a polynomial of degree $n$ has at most $n$ $x$-intercepts, at most $n - 1$ relative extrema, and at most $n - 2$ inflection points.

The last property is a consequence of the fact that the $x$-intercepts, relative extrema, and inflection points occur at real roots of $p(x) = 0$, $p'(x) = 0$, and $p''(x) = 0$, respectively, so if $p(x)$ has degree $n$ greater than 1, then $p'(x)$ has degree $n - 1$ and $p''(x)$ has degree $n - 2$. Thus, for example, the graph of a quadratic polynomial has at most two $x$-intercepts, one relative extremum, and no inflection points; and the graph of a cubic polynomial has at most three $x$-intercepts, two relative extrema, and one inflection point.

FOR THE READER. For each of the graphs in Figure 5.3.1, count the number of $x$-intercepts, relative extrema, and inflection points, and confirm that your count is consistent with the degree of the polynomial.

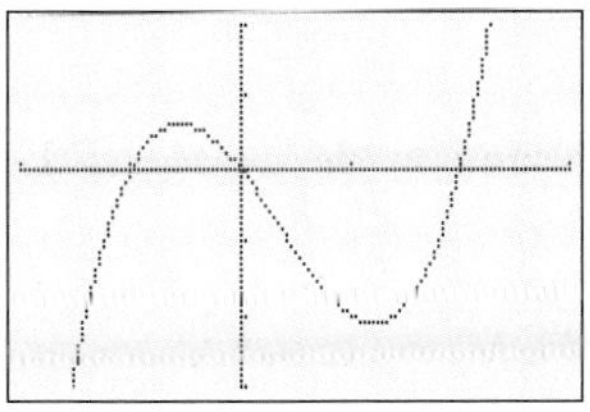

### Example 1

Figure 5.3.2 shows the graph of

$$y = x^3 - x^2 - 2x$$

produced on a graphing calculator. Confirm that the graph is complete, that is, it is not missing any significant features.

*Solution.* We can be confident that the graph is complete because the polynomial has degree 3, and three roots, two relative extrema, and one inflection point are accounted for. Moreover, the graph exhibits the correct behavior as $x \rightarrow +\infty$ and $x \rightarrow -\infty$, since

$$\lim_{x \rightarrow +\infty} (x^3 - x^2 - 2x) = \lim_{x \rightarrow +\infty} x^3 = +\infty$$

$$\lim_{x \rightarrow -\infty} (x^3 - x^2 - 2x) = \lim_{x \rightarrow -\infty} x^3 = -\infty$$

◄

$[-2, 3] \times [-3, 2]$
$x\text{Scl} = 1, y\text{Scl} = 1$

Figure 5.3.2

**GEOMETRIC IMPLICATIONS OF MULTIPLICITY**

For polynomials, there is a close relationship between the multiplicity of a root and the behavior of the graph in the vicinity of the root. For example, observe that the polynomial $p(x) = x^n$ has a root of multiplicity $n$ at $x = 0$, and observe that the graphs in Figure 1.6.4 have the following geometric properties:

- When $n$ is even, the graph of $y = p(x)$ is tangent to the $x$-axis at the origin but does not cross the $x$-axis there.
- When $n$ is odd and greater than 1, the graph is tangent to the $x$-axis at the origin, has an inflection point at the origin, and crosses the $x$-axis there.
- When $n = 1$, the graph crosses the $x$-axis at the origin but is not tangent to the $x$-axis there.

These properties of $p(x) = x^n$ at $x = 0$ are special cases of the following more general result, which we state without formal proof (Figure 5.3.3).

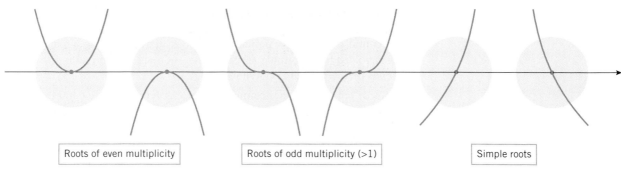

Roots of even multiplicity | Roots of odd multiplicity (>1) | Simple roots

Figure 5.3.3

---

**5.3.2** THE GEOMETRIC IMPLICATIONS OF MULTIPLICITY. *Suppose that $p(x)$ is a polynomial with a root of multiplicity $m$ at $x = r$.*

(a)  *If $m$ is even, then the graph of $y = p(x)$ is tangent to the $x$-axis at $x = r$ and does not cross the $x$-axis there.*

(b)  *If $m$ is odd and greater than 1, then the graph is tangent to the $x$-axis at $x = r$, has an inflection point there, and also crosses the $x$-axis there.*

(c)  *If $m = 1$ (so that the root is simple), then the graph crosses the $x$-axis at $x = r$ but is not tangent to the $x$-axis there.*

---

### Example 2

Make a conjecture about the behavior of the graph of

$$y = x^3(3x - 4)(x + 2)^2$$

in the vicinity of its $x$-intercepts, and test your conjecture by generating the graph.

*Solution.*   The $x$-intercepts occur at $x = 0$, $x = \frac{4}{3}$, and $x = -2$. The root $x = 0$ has multiplicity 3, which is odd, so at that point the graph should be tangent to the $x$-axis, cross the $x$-axis, and have an inflection point. The root $x = -2$ has multiplicity 2, which is even, so the graph should be tangent to but not cross the $x$-axis there. The root $x = \frac{4}{3}$ is simple, so at that point the curve should cross the $x$-axis without being tangent to it. All of this is consistent with the graph in Figure 5.3.4.   ◀

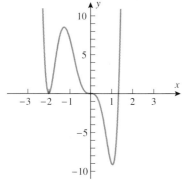

Figure 5.3.4

### Example 3

Generate or sketch a complete graph of the equation

$$y = x^3 - 3x + 2$$

and identify the exact location of the intercepts, relative extrema, and inflection points.

*Solution.*   Figure 5.3.5 shows a graph of the given equation produced with a graphing utility. We can be reasonably confident that the graph is complete since the polynomial has degree 3, and all roots, relative extrema, and inflection points are accounted for in the graph: There are three roots (a simple negative root and a positive root of multiplicity 2), and there are two relative extrema and one inflection point. The following analysis will confirm that the graph is complete and identify the exact location of the intercepts, relative extrema, and inflection points.

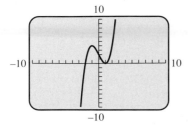

Figure 5.3.5

• *x-intercepts:* Setting $y = 0$ yields

$$x^3 - 3x + 2 = (x + 2)(x^2 - 2x + 1) = (x + 2)(x - 1)^2 = 0$$

so there is a simple root at $x = -2$ and a root of multiplicity 2 at $x = 1$.

- *y-intercept:* Setting $x = 0$ yields $y = 2$.
- *Behavior as $x \to +\infty$ and $x \to -\infty$:* The graph in Figure 5.3.5 suggests that the graph increases without bound as $x \to +\infty$ and decreases without bound as $x \to -\infty$. This is confirmed by the limits

$$\lim_{x \to +\infty} (x^3 - 3x + 2) = \lim_{x \to +\infty} x^3 = +\infty$$

$$\lim_{x \to -\infty} (x^3 - 3x + 2) = \lim_{x \to -\infty} x^3 = -\infty$$

- *Derivatives:*

$$\frac{dy}{dx} = 3x^2 - 3 = 3(x - 1)(x + 1)$$

$$\frac{d^2 y}{dx^2} = 6x$$

- *Intervals of increase and decrease; relative extrema:* Figure 5.3.6 shows the sign pattern of the first and second derivatives and what they imply about the graph shape. In the first part of the figure the upward arrows indicate intervals where the graph is increasing, the downward arrows indicate intervals where the graph is decreasing, and the horizontal arrows indicate the stationary points. The second part of the figure shows what the sign pattern of the second derivative implies about the concavity. The third part of the figure shows what the first and second derivatives together imply about the graph shape.

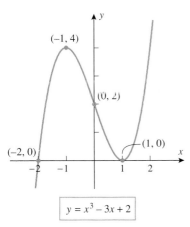

(−1, 4)

(0, 2)

(−2, 0)    (1, 0)

$$y = x^3 - 3x + 2$$

Figure 5.3.7

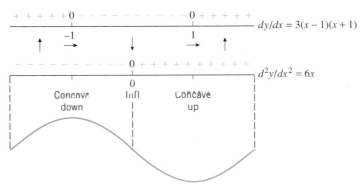

$dy/dx = 3(x - 1)(x + 1)$

$d^2y/dx^2 = 6x$

Concave down    Infl    Concave up

Figure 5.3.6

Figure 5.3.7 shows the complete graph labeled with the coordinates of the intercepts, relative extrema, and inflection point.    ◀

**GRAPHING RATIONAL FUNCTIONS**

Rational functions (ratios of polynomials) are more complicated to graph than polynomials because they have discontinuities and asymptotes.

### Example 4

Generate or sketch a complete graph of the equation

$$y = \frac{2x^2 - 8}{x^2 - 16}$$

and identify the exact location of the intercepts, relative extrema, inflection points, and asymptotes.

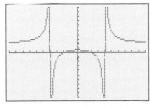

[−10, 10] × [−10, 10]
xScl = 1, yScl = 1

Figure 5.3.8

*Solution.* Figure 5.3.8 shows a calculator-generated graph of the equation in the window $[-10, 10] \times [-10, 10]$. The figure suggests that the graph is symmetric about the $y$-axis and has two vertical asymptotes and a horizontal asymptote. The figure also suggests that there is a relative maximum at $x = 0$ and two $x$-intercepts. There do not seem to be any inflection points. The following analysis will identify the exact location of the key features and help us determine whether the graph in Figure 5.3.8 is complete.

- *Symmetries:* Replacing $x$ by $-x$ does not change the equation, so the graph is symmetric about the $y$-axis.
- *x-intercepts:* Setting $y = 0$ yields the $x$-intercepts $x = -2$ and $x = 2$.
- *y-intercept:* Setting $x = 0$ yields the $y$-intercept $y = 1/2$.
- *Vertical asymptotes:* Setting $x^2 - 16 = 0$ yields the vertical asymptotes $x = -4$ and $x = 4$.
- *Horizontal asymptotes:* The limits

$$\lim_{x \to +\infty} \frac{2x^2 - 8}{x^2 - 16} = \lim_{x \to +\infty} \frac{2x^2}{x^2} = 2$$

$$\lim_{x \to -\infty} \frac{2x^2 - 8}{x^2 - 16} = \lim_{x \to -\infty} \frac{2x^2}{x^2} = 2$$

yield the horizontal asymptote $y = 2$.

The set of points where $x$-intercepts or vertical asymptotes occur is $\{-4, -2, 2, 4\}$. These points divide the $x$-axis into the open intervals

$$(-\infty, -4), \quad (-4, -2), \quad (-2, 2), \quad (2, 4), \quad (4, +\infty)$$

Over each of these intervals, $y$ cannot change sign (why?). We can find the sign of $y$ on each interval by choosing an arbitrary test point in the interval and evaluating $y = f(x)$ at the test points (Table 5.3.1).

**Table 5.3.1**

| INTERVAL | TEST POINT | $y = \dfrac{2x^2 - 8}{x^2 - 16}$ | SIGN OF $y$ |
|---|---|---|---|
| $(-\infty, -4)$ | $x = -5$ | $y = 14/3$ | $+$ |
| $(-4, -2)$ | $x = -3$ | $y = -10/7$ | $-$ |
| $(-2, 2)$ | $x = 0$ | $y = 1/2$ | $+$ |
| $(2, 4)$ | $x = 3$ | $y = -10/7$ | $-$ |
| $(4, +\infty)$ | $x = 5$ | $y = 14/3$ | $+$ |

The information in Table 5.3.1 is consistent with Figure 5.3.8, so we can be certain that the calculator graph has not missed any sign changes. The next step is to use the first and second derivatives to determine whether the calculator graph has missed any relative extrema or changes in concavity.

- *Derivatives:*

$$\frac{dy}{dx} = \frac{(x^2 - 16)(4x) - (2x^2 - 8)(2x)}{(x^2 - 16)^2} = -\frac{48x}{(x^2 - 16)^2}$$

$$\frac{d^2y}{dx^2} = \frac{48(16 + 3x^2)}{(x^2 - 16)^3} \quad \text{(verify)}$$

- *Intervals of increase and decrease; relative extrema:* A sign analysis of $dy/dx$ yields

Thus, the graph is increasing on the intervals $(-\infty, -4)$ and $(-4, 0]$; and it is decreasing on the intervals $[0, 4)$ and $(4, +\infty)$. There is a relative maximum at $x = 0$.

- *Concavity:* A sign analysis of $d^2y/dx^2$ yields

There are changes in concavity at the vertical asymptotes, $x = -4$ and $x = 4$, but there are no inflection points.

This analysis confirms that our calculator-generated graph was, in fact, complete. Figure 5.3.9 shows a complete graph of the equation with the asymptotes, intercepts, and relative maximum identified. ◀

### Example 5

Generate or sketch a complete graph of

$$y = \frac{x^2 - 1}{x^3}$$

and identify the exact location of all asymptotes, intercepts, relative extrema, and inflection points.

*Solution.* Figure 5.3.10*a* shows a calculator-generated graph of the given equation in the window $[-10, 10] \times [-10, 10]$, and Figure 5.3.10*b* shows a second version of the graph that gives more detail in the vicinity of the $x$-axis. These figures suggest that the graph is symmetric about the origin. They also suggest that there are two relative extrema, two inflection points, two $x$-intercepts, a vertical asymptote at $x = 0$, and a horizontal asymptote at $y = 0$. The following analysis will identify the exact location of all the key features and will determine whether the calculator-generated graphs in Figure 5.3.10 have missed any of these features.

- *Symmetries:* Replacing $x$ by $-x$ and $y$ by $-y$ yields an equation that simplifies back to the original equation, so the graph is symmetric about the origin.
- *x-intercepts:* Setting $y = 0$ yields the $x$-intercepts $x = -1$ and $x = 1$.
- *y-intercept:* Setting $x = 0$ leads to a division by zero, so that there is no $y$-intercept.
- *Vertical asymptotes:* Setting $x^3 = 0$ yields the vertical asymptote $x = 0$.
- *Horizontal asymptotes:* The limits

$$\lim_{x \to +\infty} \frac{x^2 - 1}{x^3} = \lim_{x \to +\infty} \frac{x^2}{x^3} = \lim_{x \to +\infty} \frac{1}{x} = 0$$

$$\lim_{x \to -\infty} \frac{x^2 - 1}{x^3} = \lim_{x \to -\infty} \frac{x^2}{x^3} = \lim_{x \to -\infty} \frac{1}{x} = 0$$

yield the horizontal asymptote $y = 0$.

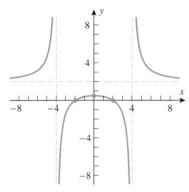

Figure 5.3.9

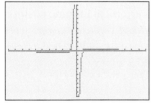

$[-10, 10] \times [-10, 10]$
$x$Scl $= 1$, $y$Scl $= 1$

(*a*)

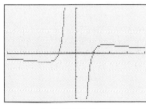

$[-4, 4] \times [-2, 2]$
$x$Scl $= 1$, $y$Scl $= 1$

(*b*)

Figure 5.3.10

- *Derivatives:*

$$\frac{dy}{dx} = \frac{x^3(2x) - (x^2 - 1)(3x^2)}{\left(x^3\right)^2} = \frac{3 - x^2}{x^4}$$

$$\frac{d^2y}{dx^2} = \frac{x^4(-2x) - (3 - x^2)(4x^3)}{\left(x^4\right)^2} = \frac{2(x^2 - 6)}{x^5}$$

- *Intervals of increase and decrease; relative extrema:*

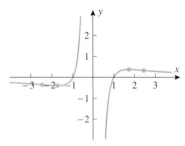

| $x$ | $y = \dfrac{x^2 - 1}{x^3}$ |
|---|---|
| $-\sqrt{6} \approx -2.45$ | $-\dfrac{5\sqrt{6}}{36} \approx -0.34$ |
| $-\sqrt{3} \approx -1.73$ | $-\dfrac{2\sqrt{3}}{9} \approx -0.38$ |
| $\sqrt{3} \approx 1.73$ | $\dfrac{2\sqrt{3}}{9} \approx 0.38$ |
| $\sqrt{6} \approx 2.45$ | $\dfrac{5\sqrt{6}}{36} \approx 0.34$ |

This analysis reveals a relative minimum at $x = -\sqrt{3}$ and a relative maximum at $x = \sqrt{3}$.

- *Concavity:*

This analysis reveals that changes in concavity occur at the vertical asymptote $x = 0$ and at the inflection points $x = -\sqrt{6}$ and $x = \sqrt{6}$.

Figure 5.3.11 shows a table of coordinate values at the relative extrema and inflection points together with a complete graph of the equation on which we have emphasized these points. ◀

Figure 5.3.11

**GRAPHS WITH VERTICAL TANGENTS AND CUSPS**

Figure 5.3.12 shows four curve elements that are commonly found in the graphs of functions that involve radicals or fractional exponents. In all four cases $x_0$ is a point of nondifferentiability, and in all four cases the tangent line at a point $x$ approaches a vertical limiting position as $x$ approaches $x_0$ from either side. Thus, we will call $x_0$ a *point of vertical tangency* for the function. In parts $(c)$ and $(d)$ of the figure the curve segments form what is called a *cusp*.

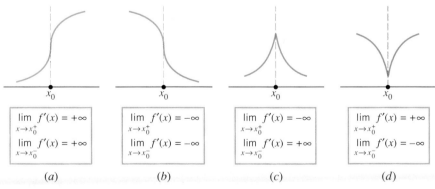

Figure 5.3.12

The following definition makes these ideas precise.

**5.3.3** DEFINITION. The graph of a function $f$ is said to have a *vertical tangent line* at $x_0$, and $x_0$ is called a *point of vertical tangency* for $f$ if $f$ is continuous at $x_0$ and $f'(x)$ approaches either $+\infty$ or $-\infty$ as $x \to x_0^+$ and as $x \to x_0^-$. In the case where $f'(x)$ approaches $+\infty$ from one side and $-\infty$ from the other side, the function $f$ is said to have a *cusp* at $x_0$.

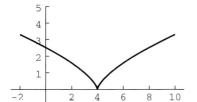

*Generated by Mathematica*

Figure 5.3.13

REMARK.   It is important to observe that vertical tangent lines occur at points of nondifferentiability, whereas nonvertical tangent lines occur at points of differentiability.

## Example 6

Generate or sketch a complete graph of $y = (x - 4)^{2/3}$.

*Solution.*   Figure 5.3.13 shows a computer-generated graph of the equation $y = (x-4)^{2/3}$. (As suggested in the discussion preceding Exercise 29 of Section 1.3, we had to trick the computer into producing the left branch by graphing the equation $y = |x - 4|^{2/3}$.) To determine whether this graph is complete, we let $f(x) = (x-4)^{2/3}$ and proceed as follows.

- *Symmetries:* There are no symmetries about the coordinate axes or the origin (verify). However, the graph of $y = (x - 4)^{2/3}$ is symmetric about the line $x = 4$, since it is a translation (four units to the right) of the graph of $y = x^{2/3}$, which is symmetric about the $y$-axis.

- *x-intercepts:* Setting $y = 0$ yields the $x$-intercept $x = 4$.

- *y-intercepts:* Setting $x = 0$ yields the $y$-intercept $y = \sqrt[3]{16}$.

- *Vertical asymptotes:* None, since $f(x) = (x - 4)^{2/3}$ is a continuous function.

- *Horizontal asymptotes:* None, since

$$\lim_{x \to +\infty} (x - 4)^{2/3} = +\infty \quad \text{and} \quad \lim_{x \to -\infty} (x - 4)^{2/3} = +\infty$$

- *Derivatives:*

$$\frac{dy}{dx} = f'(x) = \frac{2}{3}(x - 4)^{-1/3} = \frac{2}{3(x - 4)^{1/3}}$$

$$\frac{d^2 y}{dx^2} = f''(x) = -\frac{2}{9}(x - 4)^{-4/3} = -\frac{2}{9(x - 4)^{4/3}}$$

- *Relative extrema; concavity:* There is a critical point at $x = 4$, since $f$ is not differentiable there; and by the first derivative test there is a relative minimum at that critical point, since $f'(x) < 0$ if $x < 4$ and $f'(x) > 0$ if $x > 4$. Since $f''(x) < 0$ if $x \neq 4$, the graph is concave down for $x < 4$ and for $x > 4$.

- *Vertical tangent lines:* There is a vertical tangent line and cusp at $x = 4$ of the type in Figure 5.3.12d since $f(x) = (x - 4)^{2/3}$ is continuous at $x = 4$ and

$$\lim_{x \to 4^+} f'(x) = \lim_{x \to 4^+} \frac{2}{3(x - 4)^{1/3}} = +\infty$$

$$\lim_{x \to 4^-} f'(x) = \lim_{x \to 4^-} \frac{2}{3(x - 4)^{1/3}} = -\infty$$

Combining the preceding information with a sign analysis of the first and second derivatives yields Figure 5.3.14. This confirms that the computer-generated graph in Figure 5.3.13 is complete.   ◀

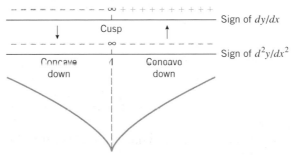

Figure 5.3.14

### Example 7

Generate or sketch a complete graph of $y = 6x^{1/3} + 3x^{4/3}$.

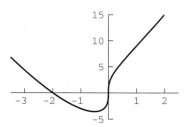

*Generated by Mathematica*

Figure 5.3.15

**Solution.** Figure 5.3.15 shows a computer-generated graph of the equation. Once again, we had to call on the discussion preceding Exercise 29 of Section 1.3 to trick the computer into graphing a portion of the graph over the negative $x$-axis. (See if you can figure out how to do this.) To determine whether the graph in Figure 5.3.15 is complete, we let

$$f(x) = 6x^{1/3} + 3x^{4/3} = 3x^{1/3}(2 + x)$$

and proceed as follows.

- *Symmetries:* There are no symmetries about the coordinate axes or the origin (verify).
- *x-intercepts:* Setting $y = 3x^{1/3}(2 + x) = 0$ yields the $x$-intercepts $x = 0$ and $x = -2$.
- *y-intercept:* Setting $x = 0$ yields the $y$-intercept $y = 0$.
- *Vertical asymptotes:* None, since $f(x) = 6x^{1/3} + 3x^{4/3}$ is continuous.
- *Horizontal asymptotes:* None, since

$$\lim_{x \to +\infty} (6x^{1/3} + 3x^{4/3}) = \lim_{x \to +\infty} 3x^{1/3}(2 + x) = +\infty$$

$$\lim_{x \to -\infty} (6x^{1/3} + 3x^{4/3}) = \lim_{x \to -\infty} 3x^{1/3}(2 + x) = +\infty$$

- *Derivatives:*

$$\frac{dy}{dx} = f'(x) = 2x^{-2/3} + 4x^{1/3} = 2x^{-2/3}(1 + 2x) = \frac{2(2x + 1)}{x^{2/3}}$$

$$\frac{d^2y}{dx^2} = f''(x) = -\frac{4}{3}x^{-5/3} + \frac{4}{3}x^{-2/3} = \frac{4}{3}x^{-5/3}(-1 + x) = \frac{4(x - 1)}{3x^{5/3}}$$

There are critical points at $x = 0$ and $x = -\frac{1}{2}$. From the first derivative test and the sign analysis of $dy/dx$ in Figure 5.3.16, there is a relative minimum at $x = -\frac{1}{2}$. There is a point of vertical tangency at $x = 0$, since

$$\lim_{x \to 0^+} f'(x) = \lim_{x \to 0^+} \frac{2(2x + 1)}{x^{2/3}} = +\infty$$

$$\lim_{x \to 0^-} f'(x) = \lim_{x \to 0^-} \frac{2(2x + 1)}{x^{2/3}} = +\infty$$

From the sign analysis of $d^2y/dx^2$ in Figure 5.3.16, the graph is concave up for $x < 0$, concave down for $0 < x < 1$, and concave up again for $x > 1$.

- *Intervals of increase and decrease; concavity:* Combining the preceding information with a sign analysis of the first and second derivatives yields the graph shape shown in Figure 5.3.16.

This confirms that the computer-generated graph in Figure 5.3.15 is complete, except for the fact that it did not reveal the very subtle inflection point at $x = 1$. In this case the artistic rendering of the curve in Figure 5.3.16 describes the subtleties of the graph shape more effectively than the computer-generated graph.  ◀

### Example 8

Generate or sketch a complete graph of $y = e^{-x^2/2}$ and identify the exact location of all relative extrema and inflection points.

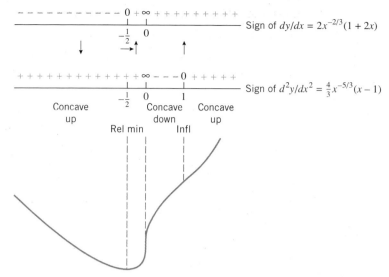

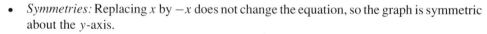

Figure 5.3.16

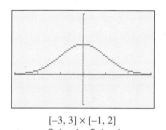

[−3, 3] × [−1, 2]
$x$Scl = 1, $y$Scl = 1

Figure 5.3.17

*Solution.*  Figure 5.3.17 shows a calculator-generated graph of the equation in the window $[−3, 3] \times [−1, 2]$. This figure suggests that the graph is symmetric about the $y$-axis and has a relative maximum at $x = 0$, a horizontal asymptote at $y = 0$, and two inflection points. The following analysis will identify the exact location of these features and determine whether the graph is complete.

- *Symmetries:* Replacing $x$ by $−x$ does not change the equation, so the graph is symmetric about the $y$-axis.
- *x-intercepts:* Setting $y = 0$ yields the equation $e^{-x^2/2} = 0$, which has no solutions since all powers of $e$ have positive values; thus, there are no $x$-intercepts.
- *y-intercepts:* Setting $x = 0$ yields the $y$-intercept $y = 1$.
- *Vertical asymptotes:* None, since $e^{-x^2/2}$ is a continuous function.
- *Horizontal asymptotes:* Since $x^2/2 \to +\infty$ as $x \to +\infty$ or $x \to −\infty$, it follows from Formula (14) of Section 4.2 that
$$\lim_{x \to +\infty} e^{-x^2/2} = \lim_{x \to -\infty} e^{-x^2/2} = 0$$
Thus, $y = 0$ is a horizontal asymptote.
- *Derivatives:*
$$\frac{dy}{dx} = e^{-x^2/2} \frac{d}{dx}\left[ -\frac{x^2}{2} \right] = -xe^{-x^2/2}$$
$$\frac{d^2y}{dx^2} = -x \frac{d}{dx}\left[ e^{-x^2/2} \right] + e^{-x^2/2} \frac{d}{dx}[-x]$$
$$= x^2 e^{-x^2/2} - e^{-x^2/2}$$
$$= (x^2 - 1)e^{-x^2/2}$$
- *Intervals of increase and decrease:* Since $e^{-x^2/2} > 0$ for all $x$, the sign of $dy/dx$ is the same as that of $-x$.

```
+ + + + + 0 − − − − −
          |
          0
  ↑    →     ↓
```
Sign of $-x$ and $dy/dx$

This analysis reveals a relative maximum at $x = 0$.

• *Concavity:* Since $e^{-x^2/2} > 0$ for all $x$, the sign of $d^2y/dx^2$ is the same as that of $x^2 - 1$.

$$
\begin{array}{ccccc}
+\ +\ +\ + & 0 & -\ -\ -\ - & 0 & +\ +\ +\ + \\
\hline
 & -1 & & 1 & \\
\text{Concave} & \text{Infl} & \text{Concave} & \text{Infl} & \text{Concave} \\
\text{up} & & \text{down} & & \text{up}
\end{array}
\qquad \text{Sign of } x^2 - 1 \text{ and } d^2y/dx^2
$$

Thus, the inflection points occur at $x = -1$ and $x = 1$. At these points the corresponding $y$-values are $y = e^{-1/2} \approx 0.61$, which seems consistent with Figure 5.3.17.  ◀

······················
**LOGISTIC CURVES**

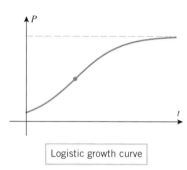

Logistic growth curve

Figure 5.3.18

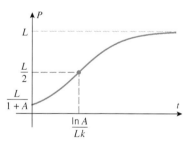

Figure 5.3.19

When a population grows in an environment in which space or food is limited, the graph of population versus time is typically an S-shaped curve of the form shown in Figure 5.3.18. The scenario described by this curve is a population that grows slowly at first and then more and more rapidly as the number of individuals producing offspring increases. However, at a certain point in time (where the inflection point occurs) the environmental factors begin to show their effect, and the growth rate begins a steady decline. Over an extended period of time the population approaches a limiting value that represents the upper limit on the number of individuals that the available space or food can sustain. Population growth curves of this type are called *logistic growth curves*.

## Example 9

We will show in a later chapter that logistic growth curves arise from equations of the form

$$
P = \frac{L}{1 + Ae^{-kLt}} \tag{5}
$$

where $P$ is the population at time $t$ ($t \geq 0$) and $A, k$, and $L$ are positive constants. Show that Figure 5.3.19 correctly describes the graph of this equation.

*Solution.*  We leave it for you to confirm that at time $t = 0$ the value of $P$ is

$$
P = \frac{L}{1 + A}
$$

and that for $t \geq 0$ the population $P$ satisfies

$$
\frac{L}{1 + A} \leq P < L
$$

which is consistent with the graph in Figure 5.3.19. The horizontal asymptote in the graph at $P = L$ is confirmed by the limit

$$
\lim_{t \to +\infty} \frac{L}{1 + Ae^{-kLt}} = \frac{L}{1 + 0} = L
$$

Physically, $L$ represents the upper limit on the size of the population.

To investigate the remaining properties, we need the first and second derivatives of $P$ with respect to $t$. We leave it for you to confirm that

$$
\frac{dP}{dt} = kP(L - P) \tag{6}
$$

$$
\frac{d^2P}{dt^2} = k^2P(L - P)(L - 2P) \tag{7}
$$

Since $k > 0$, $P > 0$, and $L - P > 0$, it follows from (6) that $dP/dt > 0$ for all $t$. Thus, $P$ is always increasing and there are no stationary points, which is consistent with Figure 5.3.19.

Since $k^2 > 0$, $P > 0$, and $L - P > 0$, it follows from (7) that

$$\frac{d^2P}{dt^2} > 0 \quad \text{if} \quad L - 2P > 0$$

$$\frac{d^2P}{dt^2} < 0 \quad \text{if} \quad L - 2P < 0$$

Thus, the graph of $P$ versus $t$ is concave up if $P < L/2$, concave down if $P > L/2$, and has an inflection point where $P = L/2$, all of which is consistent with Figure 5.3.19.

Finally, we leave it as an exercise for you to confirm that the inflection point occurs at time

$$t = \frac{1}{Lk} \ln A = \frac{\ln A}{Lk} \tag{8}$$

by solving the equation

$$\frac{L}{2} = \frac{L}{1 + Ae^{-kLt}}$$

for $t$.  ◀

## EXERCISE SET 5.3   ⌁ Graphing Calculator   © CAS

In Exercises 1–10, give a complete graph of the polynomial, and label the coordinates of the stationary points and inflection points. Check your work with a graphing utility.

**1.** $x^2 - 2x - 3$   **2.** $1 + x - x^2$

**3.** $x^3 - 3x + 1$   **4.** $2x^3 - 3x^2 + 12x + 9$

**5.** $x^4 + 2x^3 - 1$   **6.** $x^4 - 2x^2 - 12$

**7.** $3x^5 - 5x^3$   **8.** $3x^4 + 4x^3$

**9.** $x(x - 1)^3$   **10.** $x^5 + 5x^4$

In Exercises 11–19, give a complete graph of the rational function, and label the coordinates of the stationary points and inflection points. Show the horizontal and vertical asymptotes, and label them with their equations. Check your work with a graphing utility.

**11.** $\dfrac{2x}{x - 3}$   **12.** $\dfrac{x}{x^2 - 1}$   **13.** $\dfrac{x^2}{x^2 - 1}$

**14.** $\dfrac{x^2 - 1}{x^2 + 1}$   **15.** $x^2 - \dfrac{1}{x}$   **16.** $\dfrac{2x^2 - 1}{x^2}$

**17.** $\dfrac{x^3 - 1}{x^3 + 1}$   **18.** $\dfrac{8}{4 - x^2}$   **19.** $\dfrac{x - 1}{x^2 - 4}$

In Exercises 20–22, the graph of the rational function crosses a horizontal asymptote. Give a complete graph of the function, and label the coordinates of the stationary points and inflection points. Show the horizontal and vertical asymptotes, and label the point(s) where the graph crosses a horizontal asymptote. Check your work with a graphing utility.

**20.** $\dfrac{3x^2 - 4x - 4}{x^2}$   **21.** $\dfrac{(x - 1)^2}{x^2}$   **22.** $2 + \dfrac{3}{x} - \dfrac{1}{x^3}$

⌁ **23.** In each part, match the functions with graphs I–VI without using a graphing utility, and then use a graphing utility to generate the graphs.

(a) $x^{1/3}$   (b) $x^{1/4}$   (c) $x^{1/5}$

(d) $x^{2/5}$   (e) $x^{4/3}$   (f) $x^{-1/3}$

I
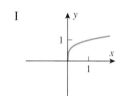

II
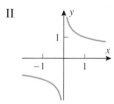

III
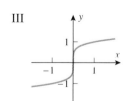

IV
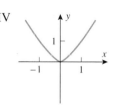

V
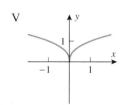

VI
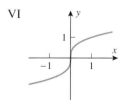

**24.** Sketch the general shape of the graph of $y = x^{1/n}$, and then explain in words what happens to the shape of the graph as $n$ increases if
(a) $n$ is a positive even integer
(b) $n$ is a positive odd integer.

In Exercises 25–32, give a complete graph of the function, and identify the location of all critical points and inflection points. Check your work with a graphing utility.

**25.** $\sqrt{x^2 - 1}$            **26.** $\sqrt[3]{x^2 - 4}$

**27.** $2x + 3x^{2/3}$       **28.** $4x - 3x^{4/3}$

**29.** $x\sqrt{3 - x}$         **30.** $4x^{1/3} - x^{4/3}$

**31.** $\dfrac{8(\sqrt{x} - 1)}{x}$       **32.** $\dfrac{1 + \sqrt{x}}{1 - \sqrt{x}}$

In Exercises 33–38, give a complete graph of the function, and identify the location of all relative extrema and inflection points. Check your work with a graphing utility.

**33.** $x + \sin x$         **34.** $x - \cos x$

**35.** $\sin x + \cos x$      **36.** $\sqrt{3}\cos x + \sin x$

**37.** $\sin^2 x, \quad 0 \le x \le 2\pi$

**38.** $x \tan x, \quad -\pi/2 < x < \pi/2$

In Exercises 39–44: (a) Find the limits of the function as $x \to +\infty$ and $x \to -\infty$. (b) Give a complete graph of the function, and identify the location of all relative extrema and inflection points. Check your work with a graphing utility.

**39.** $xe^x$         **40.** $xe^{-2x}$         **41.** $x^2 e^{-2x}$

**42.** $x^2 e^{2x}$       **43.** $xe^{x^2}$         **44.** $e^{-1/x^2}$

In Exercises 45–48: (a) Find the limits of the function as $x \to 0^+$ and $x \to +\infty$. (b) Give a complete graph of the function, and identify the location of all relative extrema and inflection points. Check your work with a graphing utility.

**45.** $x \ln x$     **46.** $x^2 \ln x$     **47.** $\dfrac{\ln x}{x^2}$     **48.** $\dfrac{\ln x}{\sqrt{x}}$

**49.** In each part: (i) Make a conjecture about the behavior of the graph in the vicinity of its $x$-intercepts. (ii) Make a rough sketch of the graph based on your conjecture and the limits of the polynomials as $x \to +\infty$ and $x \to -\infty$. (iii) Compare your sketch to the graph generated with a graphing utility.
(a) $y = x(x - 1)(x + 1)$    (b) $y = x^2(x - 1)^2(x + 1)^2$
(c) $y = x^2(x - 1)^2(x + 1)^3$   (d) $y = x(x - 1)^5(x + 1)^4$

**50.** Sketch the graph of $y = (x - a)^m(x - b)^n$ for the stated values of $m$ and $n$, assuming that $a \ne b$ (six graphs in total).
(a) $m = 1, \, n = 1, 2, 3$     (b) $m = 2, \, n = 2, 3$
(c) $m = 3, \, n = 3$

**51.** In each part, make a rough sketch of the graph using asymptotes and appropriate limits but no derivatives. Compare your sketch to that generated with a graphing utility.
(a) $y = \dfrac{3x^2 - 8}{x^2 - 4}$       (b) $y = \dfrac{x^2 + 2x}{x^2 - 1}$
(c) $y = \dfrac{2x - x^2}{x^2 + x - 2}$     (d) $y = \dfrac{x^2}{x^2 - x - 2}$

**52.** Sketch the graph of
$$y = \frac{1}{(x - a)(x - b)}$$
assuming that $a \ne b$.

**53.** Consider the family of curves $y = xe^{-bx}$ $(b > 0)$.
(a) Use a graphing utility to generate some members of this family.
(b) Discuss the effect of varying $b$ on the shape of the graph, and discuss the locations of the relative extrema and inflection points.

**54.** Consider the family of curves $y = e^{-bx^2}$ $(b > 0)$.
(a) Use a graphing utility to generate some members of this family.
(b) Discuss the effect of varying $b$ on the shape of the graph, and discuss the locations of the relative extrema and inflection points.

**55.** (a) Determine whether the following limits exist, and if so, find them:
$$\lim_{x \to +\infty} e^x \cos x, \quad \lim_{x \to -\infty} e^x \cos x$$
(b) Sketch the graphs of $y = e^x$, $y = e^{-x}$, and $y = e^x \cos x$ in the same coordinate system, and label any points of intersection.
(c) Use a graphing utility to generate some members of the family $y = e^{ax} \cos bx$ $(a > 0$ and $b > 0)$, and discuss the effect of varying $a$ and $b$ on the shape of the curve.

**56.** (**Oblique Asymptotes**) If a rational function $P(x)/Q(x)$ is such that the degree of the numerator exceeds the degree of the denominator by *one*, then the graph of $P(x)/Q(x)$ will have an *oblique asymptote*, that is, an asymptote that is neither vertical nor horizontal. To see why, we perform the division of $P(x)$ by $Q(x)$ to obtain
$$\frac{P(x)}{Q(x)} = (ax + b) + \frac{R(x)}{Q(x)}$$
where $ax + b$ is the quotient and $R(x)$ is the remainder. Use the fact that the degree of the remainder $R(x)$ is less than the degree of the divisor $Q(x)$ to help prove
$$\lim_{x \to +\infty} \left[ \frac{P(x)}{Q(x)} - (ax + b) \right] = 0$$
$$\lim_{x \to -\infty} \left[ \frac{P(x)}{Q(x)} - (ax + b) \right] = 0$$
As illustrated in the accompanying figure, these results tell us that the graph of the equation $y = P(x)/Q(x)$ "approaches" the line (an oblique asymptote) $y = ax + b$ as $x \to +\infty$ or $x \to -\infty$.

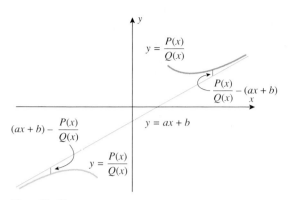

Figure Ex-56

In Exercises 57–61, sketch the graph of the rational function. Show all vertical, horizontal, and oblique asymptotes (see Exercise 56).

**57.** $\dfrac{x^2 - 2}{x}$  **58.** $\dfrac{x^2 - 2x - 3}{x + 2}$  **59.** $\dfrac{(x - 2)^3}{x^2}$

**60.** $\dfrac{4 - x^3}{x^2}$  **61.** $x + 1 - \dfrac{1}{x} - \dfrac{1}{x^2}$

**62.** Find all values of $x$ where the graph of

$$y = \frac{2x^3 - 3x + 4}{x^2}$$

crosses its oblique asymptote. [See Exercise 56.]

**63.** Let $f(x) = (x^3 + 1)/x$. Show that the graph of $y = f(x)$ approaches the curve $y = x^2$ "asymptotically" in the sense that

$$\lim_{x \to +\infty} [f(x) - x^2] = 0 \quad \text{and} \quad \lim_{x \to -\infty} [f(x) - x^2] = 0$$

Sketch the graph of $y = f(x)$ showing this asymptotic behavior.

**64.** Let $f(x) = (2 + 3x - x^3)/x$. Show that $y = f(x)$ approaches the curve $y = 3 - x^2$ asymptotically in the sense described in Exercise 63. Sketch the graph of $y = f(x)$ showing this asymptotic behavior.

**65.** A rectangular plot of land is to be fenced off so that the area enclosed will be 400 ft$^2$. Let $L$ be the length of fencing needed and $x$ the length of one side of the rectangle. Show

that $L = 2x + 800/x$ for $x > 0$, and sketch the graph of $L$ versus $x$ for $x > 0$.

**66.** A box with a square base and open top is to be made from sheet metal so that its volume is 500 in$^3$. Let $S$ be the area of the surface of the box and $x$ the length of a side of the square base. Show that $S = x^2 + 2000/x$ for $x > 0$, and sketch the graph of $S$ versus $x$ for $x > 0$.

**67.** The accompanying figure shows a computer-generated graph of the polynomial $y = 0.1x^5(x - 1)$ using a viewing window of $[-2, 2.5] \times [-1, 5]$. Show that the choice of the vertical scale caused the computer to miss important features of the graph. Find the features that were missed and make your own sketch of the graph that shows the missing features.

**68.** The accompanying figure shows a computer-generated graph of the polynomial $y = 0.1x^5(x + 1)^2$ using a viewing window of $[-2, 1.5] \times [-0.2, 0.2]$. Show that the choice of the vertical scale caused the computer to miss important features of the graph. Find the features that were missed and make your own sketch of the graph that shows the missing features.

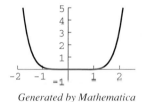

Generated by Mathematica

Figure Ex-67

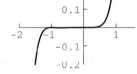

Generated by Mathematica

Figure Ex-68

**69.** Suppose that a population $P$ grows according to the logistic model given by Formula (5).
(a) At what rate is $P$ increasing at time $t = 0$?
(b) In words, describe how the rate of growth of $P$ varies with time.
(c) At what time is the population growing most rapidly?

**70.** Show that the inflection point of the logistic growth curve in Example 9 occurs at the time $t$ given by Formula (8).

## SUPPLEMENTARY EXERCISES

**1.** (a) If $x_1 < x_2$, what relationship must hold between $f(x_1)$ and $f(x_2)$ for $f$ to be increasing on an interval containing $x_1$ and $x$? Decreasing? Constant?
(b) What condition on $f'$ ensures that $f$ is increasing on an interval $[a, b]$? Decreasing? Constant?

**2.** (a) What condition on $f'$ ensures that $f$ is concave up on an open interval $I$? Concave down?
(b) What condition on $f''$ ensures that $f$ is concave up on an open interval $I$? Concave down?
(c) In words, what is an inflection point of $f$?

**3.** (a) Where on the graph of $y = f(x)$ would you expect $y$ to be increasing or decreasing most rapidly with respect to $x$?

(b) In words, what is a relative extremum?

(c) State a procedure for determining where the relative extrema of $f$ occur.

**4.** Determine whether the statement is true or false. If it is false, give an example that illustrates why.

(a) If $f$ has a relative maximum at $x_0$, then $f(x_0)$ is the largest value that $f(x)$ can have.

(b) If $f(x_0)$ is the largest value for $f$ on the interval $(a, b)$, then $f$ has a relative maximum at $x_0$.

(c) A function $f$ has a relative extremum at each of its critical points.

**5.** (a) According to the first derivative test, what conditions ensure that $f$ has a relative maximum at $x_0$? A relative minimum?

(b) According to the second derivative test, what conditions ensure that $f$ has a relative maximum at $x_0$? A relative minimum?

**6.** In each part, sketch a continuous curve $y = f(x)$ with the stated properties.

(a) $f(2) = 4$, $f'(2) = 1$, $f''(x) < 0$ for $x < 2$, $f''(x) > 0$ for $x > 2$

(b) $f(2) = 4$, $f''(x) > 0$ for $x < 2$, $f''(x) < 0$ for $x > 2$, and $\lim_{x \to 2^-} f'(x) = +\infty$, $\lim_{x \to 2^+} f'(x) = +\infty$

(c) $f(2) = 4$, $f''(x) < 0$ for $x \neq 2$, and $\lim_{x \to 2^-} f'(x) = 1$, $\lim_{x \to 2^+} f'(x) = -1$

**7.** In each part, find the location of all critical points, and use the first derivative test to classify them as relative maxima, relative minima, or neither.

(a) $f(x) = x^{1/3}(x - 7)^2$

(b) $f(x) = 2\sin x - \cos 2x$, $0 \leq x \leq 2\pi$

(c) $f(x) = 3x - (x - 1)^{3/2}$

**8.** In each part, find the location of all critical points, and use the second derivative test (where possible) to classify them as relative maxima, relative minima, or neither.

(a) $f(x) = x^{-1/2} + \frac{1}{9}x^{1/2}$

(b) $f(x) = x^2 + 8/x$

(c) $f(x) = \sin^2 x - \cos x$, $0 \leq x \leq 2\pi$

In Exercises 9–24, give a complete graph of $f$, and identify the limits as $x \to \pm\infty$, as well as locations of all relative extrema, inflection points, and asymptotes (as appropriate).

**9.** $f(x) = x^4 - 3x^3 + 3x^2 + 1$

**10.** $f(x) = x^5 - 4x^4 + 4x^3$

**11.** $f(x) = \tan(x^2 + 1)$

**12.** $f(x) = x - \cos x$

**13.** $f(x) = \dfrac{x^2}{x^2 + 2x + 5}$

**14.** $f(x) = \dfrac{25 - 9x^2}{x^3}$

**15.** $f(x) = \begin{cases} \frac{1}{2}x^2, & x \leq 0 \\ -x^2, & x > 0 \end{cases}$

**16.** $f(x) = (1 + x)^{2/3}(3 - x)^{1/3}$

**17.** $f(x) = x \ln x$

**18.** $f(x) = x^2 \ln x$

**19.** $f(x) = \dfrac{\ln x}{x^2}$

**20.** $f(x) = \ln(x^2 + 1)$

**21.** $f(x) = \dfrac{e^x}{x}$

**22.** $f(x) = xe^{-x}$

**23.** $f(x) = x^2 e^{1-x}$

**24.** $f(x) = x^3 e^{x-1}$

When using a graphing utility, important features of a graph may be missed if the viewing window is not chosen appropriately. This is illustrated in Exercises 25 and 26.

**25.** (a) Generate the graph of $f(x) = \frac{1}{3}x^3 - \frac{1}{400}x$ over the interval $[-5, 5]$, and make a conjecture about the location and nature of all critical points.

(b) Find the exact location of all critical points, and classify them as relative maxima, relative minima, or neither.

(c) Confirm the results in part (b) by graphing $f$ over an appropriate interval.

**26.** (a) Generate the graph of

$$f(x) = \tfrac{1}{5}x^5 - \tfrac{7}{8}x^4 + \tfrac{1}{3}x^3 + \tfrac{7}{2}x^2 - 6x$$

over the interval $[-5, 5]$, and make a conjecture about the location and nature of all critical points.

(b) Find the exact location of all critical points, and classify them as relative maxima, relative minima, or neither.

(c) Confirm the results in part (b) by graphing portions of $f$ over appropriate intervals. [*Note:* It will not be possible to find a single window in which all of the critical points are clearly visible.]

**27.** (a) Use a graphing utility to generate the graphs of $y = x$ and $y = (x^3 - 8)/(x^2 + 1)$ together over the interval $[-5, 5]$, and make a conjecture about the relationship between the two graphs.

(b) Use Exercise 56 of Section 5.3 to confirm your conjecture in part (a).

**28.** In parts (a)–(d), the complete graph of a polynomial with degree at most 6 is given. Find equations for polynomials that produce graphs with these shapes, and check your answers with a graphing utility.

(a)

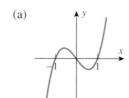

(b)

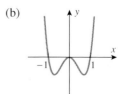

(c)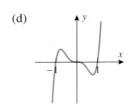

(d)

**29.** Find the equations of the tangent lines at all inflection points of the graph of

$$f(x) = x^4 - 6x^3 + 12x^2 - 8x + 3$$

**30.** Use implicit differentiation to show that a function defined implicitly by $\sin x + \cos y = 2y$ has a critical point whenever $\cos x = 0$. Then use either the first or second derivative test to classify these critical points as relative maxima or minima.

**31.** Let

$$f(x) = \frac{2x^3 + x^2 - 15x + 7}{(2x - 1)(3x^2 + x - 1)}$$

Graph $y = f(x)$, and find the equations of all horizontal and vertical asymptotes. Explain why there is no vertical asymptote at $x = \frac{1}{2}$, even though the denominator of $f$ is zero at that point.

**C** **32.** Let

$$f(x) = \frac{x^5 - x^4 - 3x^3 + 2x + 4}{x^7 - 2x^6 - 3x^5 + 6x^4 + 4x - 8}$$

(a) Use a CAS to factor the numerator and denominator of $f$, and use the results to determine the locations of all vertical asymptotes.

(b) Confirm that your answer is consistent with the graph of $f$.

**33.** (a) By inspection, find the largest and smallest possible values for $f(x) = e^{\sin x}$, and then confirm that your answers are consistent with the graph of $f$.

(b) Find the exact locations of the relative extrema.

(c) Estimate the locations of the inflection points in the interval $0 < x < 2\pi$ from the graph of $f''$.

**34.** For the general cubic polynomial

$$f(x) = ax^3 + bx^2 + cx + d \quad (a \neq 0)$$

find conditions on $a$, $b$, $c$, and $d$ to ensure that $f$ is always increasing or always decreasing on $(-\infty, +\infty)$.

**35.** In each part, approximate the coordinates $(x, y)$ of the relative extrema, and confirm that your answers are consistent with the graph of $f$.

(a) $f(x) = x^2 - \sin x$

(b) $f(x) = \sqrt{x^4 + 1} - \sqrt{x^2 + 1}$

(c) $f(x) = \dfrac{x}{x^2 - \sin x + 1}$

**36.** (a) Approximate to two decimal places the largest value of $k$ such that the function $f(x) = 1 + 2x + x^3 - x^4$ is one-to-one for $x \leq k$.

(b) For the value of $k$ found in part (a), find the domain and range of $f^{-1}$ and the value of $f^{-1}(-1)$ for the function $f(x) = 1 + 2x + x^3 - x^4$, $x \leq k$.

**37.** Consider the family of curves $y = xe^{-ax}$, $x \geq 0$, where $a$ is a positive constant.

(a) Use a graphing utility to graph some members of this family.

(b) Find the value of $y$ at $x = 0$ and the limiting value of $y$ as $x \to +\infty$; confirm that these values are consistent with your graphs.

(c) Find formulas for the coordinates of the relative extrema and inflection points, and confirm that these formulas are consistent with the graphs.

(d) How does increasing $a$ affect the graph?

**38.** Consider the family of curves $y = e^{-(x-a)/2b}$, where $a$ and $b$ are constants and $b > 0$.

(a) Use a graphing utility to graph some members of this family, first keeping $a$ fixed and varying $b$, and then keeping $b$ fixed and varying $a$.

(b) Find the value of $y$ at $x = a$ and the limiting values of $y$ as $x \to \pm\infty$; confirm that these values are consistent with your graphs.

(c) Find formulas for the coordinates of the relative extrema and inflection points, and confirm that these formulas are consistent with the graphs.

(d) If $a$ is kept fixed, how does increasing $b$ affect the shape of the graph?

(e) If $b$ is kept fixed, how does varying $a$ affect the graph?

**39.** Show that for successive positive integer values of $n$, the number $(1 + 1/n)^{n+1}$ is smaller than its predecessor. [*Hint:* Consider the function $f(x) = (x + 1) \ln(1 + 1/x)$.]

## EXPANDING THE CALCULUS HORIZON

# Functions from Data

*One of the most important procedures in applied science is using experimental data to discover relationships between variables. In this module we will discuss some mathematical techniques for doing this, and we will use these ideas to investigate principles of planetary motion and the cooling of liquids.*

### Fitting Curves to Data

Suppose that a scientist is looking for a relationship between two variables $x$ and $y$ and that measurements of corresponding values of these variables have produced a set of $n$ data points

$$(x_1, y_1), \quad (x_2, y_2), \quad (x_3, y_3), \ldots, \quad (x_n, y_n)$$

If the scientist uses the data in some way to obtain a relationship $y = f(x)$ between $x$ and $y$, then this equation is called a ***mathematical model*** for the data.

One way to obtain a mathematical model for a set of data is to look for a function $f$ whose graph passes through all of the data points; this is called an ***interpolating function***. Although interpolating functions are appropriate in certain situations, they do not adequately account for measurement errors in the data. For example, suppose that the relationship between $x$ and $y$ is known to be linear but that accuracy limitations in the measuring devices and random variations in experimental conditions produce a scatter plot such as that shown in Figure 1a. With the help of a computer, one can find a polynomial of degree 10 whose graph passes through all of the data points (Figure 1b). However, this polynomial model does not successfully convey the underlying linear relationship; a better approach is to look for a linear equation $y = mx + b$ whose graph more accurately describes the linear relationship, even if it does not pass through all (or any) of the data points (Figure 1c).

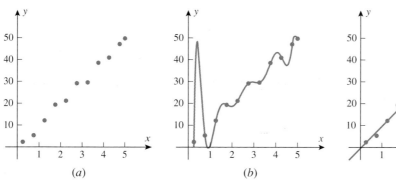

(a)                    (b)                    (c)

Figure 1

### Finding Mathematical Models

The most challenging part of finding a model $y = f(x)$ for experimental data is coming up with an appropriate form for the function $f$. Sometimes the form of the function will be suggested by the visual appearance of the scatter plot, and sometimes it will be dictated by a known physical law that relates $x$ and $y$. For example, Figure 1a strongly suggests that the relationship between $x$ and $y$ is linear, so in absence of additional information it would be natural to look for a linear model $y = mx + b$. In contrast, the scatter plot of U.S. population growth in Figure 2 strongly suggests some nonlinear relationship, so we must look for a nonlinear function for the model. The possibilities for nonlinear models are endless; however, there are theories in the study of population growth which suggest that in absence of environmental constraints, populations of people can be modeled over time by equations of the form $P = P_0 e^{kt}$, so in this case we might look for an equation of this form to model the data.

## Linear Models

The most important methods for finding linear models are based on the following idea: For any proposed linear model $y = mx + b$, draw a vertical connector from each data point $(x_i, y_i)$ to the line, and consider the differences $y_i - y$ (Figure 3). These differences, which are called **residuals**, may be viewed as "errors" that result when the line is used to model the data. Points above the line have positive errors, points below the line have negative errors, and points on the line have no error.

One way to choose a linear model is to look for a line $y = mx + b$ in which the sum of the residuals is zero, the logic being that this makes the positive and negative errors balance out. However, one can find examples where this procedure produces unacceptably poor models, so for reasons that we cannot discuss here the most common method for finding a linear model is to look for a line $y = mx + b$ in which the *sum of the squares* of the residuals is as small as possible. This is called the **least-squares line of best fit** or the **regression line**.

*Exercise 1*   One of the lines in Figure 4 is the regression line. Which one is it?

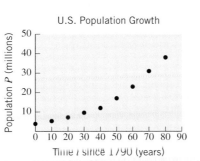

Figure 2

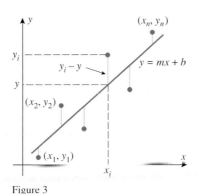

Figure 3

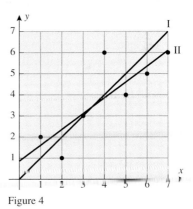

Figure 4

*Exercise 2*

(a) Most scientific calculators and CAS programs provide a method for finding regression lines. Read the documentation for your calculator or CAS to determine how to do this, and then find the regression line for the following $(x, y)$ data:

| $x$ | 1.0 | 1.5 | 2.0 | 2.5 | 3.0 | 3.5 | 4.0 |
|---|---|---|---|---|---|---|---|
| $y$ | 1.0 | 2.5 | 6.0 | 9.0 | 10.5 | 14.5 | 15.0 |

(b) Make a scatter plot of the data together with the regression line.

## How Good Is the Linear Model?

It is possible to compute a regression line, even in cases where the data have no apparent linear pattern. Thus, it is important to have some quantitative method of determining whether a linear model is appropriate for the data. The most common measure of linearity in data is called the **correlation coefficient**, which is usually denoted by the letter $r$. A detailed explanation of correlation coefficients and the formula used to compute them is outside the scope of this text. However, here are some of the basic ideas:

- The values of $r$ are in the interval $-1 \leq r \leq 1$, where $r$ has the same sign as the slope of the regression line.

- If $r = \pm 1$, then the data points all lie on a line, so a linear model is a perfect fit for the data.

- If $r = 0$, then the data points exhibit no linear tendency, so a linear model is inappropriate for the data.

The closer $r$ is to $\pm 1$, the more tightly the data points hug the regression line and the more appropriate the regression line is as a model; the closer $r$ is to 0, the more scattered the points and the less appropriate the regression line is as a model (Figure 5). Roughly stated, the value of $r^2$ is a measure of the percentage of data points that fall in a "tight linear band." Thus, $r = 0.5$ means that 25% of the points fall in a tight linear band, and $r = 0.9$ means that 81% of the points fall in a tight linear band. A precise explanation of what is meant by a "tight linear band" requires ideas from statistics.

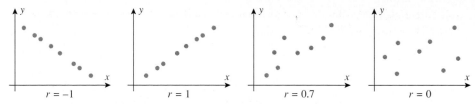

Figure 5

**Exercise 3**  If you have a scientific calculator, read the documentation to determine whether it produces the correlation coefficient when it computes a regression line. If you have a CAS, then the chances are that it will not automatically produce the correlation coefficient. However, on our website we have provided some CAS "miniprograms" that can be used to find regression lines with their associated correlation coefficients.

**Exercise 4**  Find the correlation coefficient for the data in Exercise 2.

**Exercise 5**

(a)  Table 1.1.1 of Chapter 1 gives the Indianapolis 500 qualifying speeds $S$ from 1975 to 1994. Take 1975 to be $t = 0$, and find the regression line and correlation coefficient for $S$ versus $t$.

(b)  Do you think that a linear model is reasonable for the data? Explain your reasoning.

(c)  Predict the qualifying speed for the year 2000.

(d)  What assumptions did you make in part (c)?

## Nonlinear Models

Three of the most important nonlinear models are

- **Exponential models** ($y = ae^{bx}$)
- **Logarithmic models** ($y = a + b \ln x$)
- **Power function models** ($y = ax^b$)

Many scientific calculators and computer programs can fit models of these types to data by the method of least squares. However, a useful alternative approach is to use logarithms to transform the original data into a form where linear models can be applied. This procedure, called **linearizing** the data, is based on the following idea:

- A set of $(x_i, y_i)$ data will have an exponential model if the transformed data $(x_i, \log y_i)$ have a linear model.
- A set of $(x_i, y_i)$ data will have a logarithmic model if the transformed data $(\log x_i, y_i)$ have a linear model.
- A set of $(x_i, y_i)$ data will have a power function model if the transformed data $(\log x_i, \log y_i)$ have a linear model.

The following exercise explains the reason for this.

### Exercise 6

(a) Suppose that $y = ae^{bx}$, and let $Y = \ln y$. Show that the graph of $Y$ versus $x$ is a line of slope $b$ and $Y$-intercept $\ln a$.

(b) Suppose that $y = a + b \ln x$, and let $X = \ln x$. Show that the graph of $y$ versus $X$ is a line of slope $b$ and $y$-intercept $a$.

(c) Suppose that $y = ax^b$, and let $Y = \ln y$ and $X = \ln x$. Show that the graph of $Y$ versus $X$ is a line of slope $b$ and $Y$-intercept $\ln a$.

(d) Show that in parts (a), (b), and (c) the statements remain true if the natural logarithm "ln" is replaced by the common logarithm "log".

### Exercise 7

(a) Find an exponential model $y = ae^{bx}$ for the following data by linearizing the data, finding the regression line for the linearized data, and then applying part (a) of Exercise 6 to find $a$ and $b$.

| $x$ | 0 | 1 | 2 | 3 | 4 | 5 | 6 | 7 |
|-----|-----|-----|-----|-----|-----|-----|-----|-----|
| $y$ | 3.9 | 5.3 | 7.2 | 9.6 | 12 | 17 | 23 | 31 |

(b) Make a scatter plot of the data together with the exponential model.

### Exercise 8

The table in Figure 6 shows the relationship between the time $T$ that it takes for each planet in our solar system to make one revolution around the Sun and the mean distance $d$ between the planet and the Sun during one revolution. The graph in Figure 6 is a plot of $\log T$ versus $\log d$.

(a) What type of model for $T$ as a function of $d$ is suggested by the graph?

(b) Find the regression line for the $(\log d, \log T)$ data.

(c) Use the appropriate part of Exercise 6 to express $T$ as a function of $d$.

(d) In part (c) you discovered Kepler's Third Law of Planetary Motion. Find some information about this law, and state the law in words.

| PLANET | MEAN DISTANCE $d$ FROM THE SUN | TIME $T$ FOR ONE REVOLUTION |
|--------|--------|--------|
| Mercury | 0.387 | 0.241 |
| Venus | 0.723 | 0.615 |
| Earth | 1.000 | 1.000 |
| Mars | 1.523 | 1.881 |
| Jupiter | 5.203 | 11.861 |
| Saturn | 9.541 | 29.457 |
| Uranus | 19.190 | 84.008 |
| Neptune | 30.086 | 164.784 |
| Pluto | 39.507 | 248.350 |

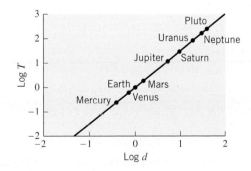

*Note:* Distances are measured in astronomical units (AU); 1 AU ≈ 93,000,000 mi. Time is measured in Earth years.

Figure 6

### Modeling Cooling

If a cup of hot coffee is left on a table to cool, then the graph of its temperature $T$ versus the elapsed time $t$ will have the general shape shown in Figure 7. The graph suggests that the coffee will cool quickly at first and then more and more slowly as its temperature approaches that of the room. To be more precise, it is shown in Physics that if the temperature of a liquid at time $t = 0$ is $T_0$ and if the room has a constant temperature of $T_a$, where $T_a < T_0$ (the room is cooler than the liquid), then the temperature $T$ of the liquid at time $t$ is given by

$$T = T_a + (T_0 - T_a)e^{kt}$$

where $k$ is a *negative* constant whose value depends on the physical characteristics of the liquid. This equation, called **Newton's Law of Cooling**, can also be written in the form

$$T - T_a = (T_0 - T_a)e^{kt}$$

which states that the difference between the temperature of the liquid and the temperature of the room has an exponential model.

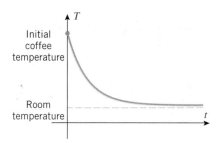

Figure 7

··········

*Exercise 9*    Table 1 shows temperature measurements of a cup of coffee at 1-minute intervals after it was placed in a room with a constant temperature of $27°C$.

(a) Find a model for the temperature $T$ as a function of the elapsed time $t$.

(b) Estimate the temperature of the coffee at the time it was placed in the room.

(c) Approximately how long will it take until the coffee temperature is within $5°C$ of the room temperature?

**Table 1**

| $t$ (min) | 1 | 2 | 3 | 4 | 5 | 6 | 7 | 8 | 9 | 10 |
|---|---|---|---|---|---|---|---|---|---|---|
| $T$ (°C) | 82.2 | 79.6 | 77.3 | 75.0 | 73.1 | 70.7 | 69.2 | 66.9 | 65.3 | 63.3 |

··········

*Module by Mary Ann Connors, USMA, West Point, and Howard Anton, Drexel University*

**Additional material for this module can be found on the World Wide Web at http://www.wiley.com/college/anton**

Isaac Barrow

# 6

# APPLICATIONS OF THE DERIVATIVE

*I*n this chapter we will study various applications of the derivative. For example, we will investigate problems that are concerned with finding the "best" way to perform a task—these are called *optimization problems*. Many optimization problems are concerned with time and cost. For example, if time is the main consideration in a problem, we might be interested in finding the quickest way to perform a task, and if cost is the main consideration, we might be interested in finding the most profitable way to perform the task. Mathematically, optimization problems can be reduced to finding the largest or smallest value of a function on some interval and determining where the largest or smallest value occurs; thus, part of our work in this chapter will focus on developing the mathematical tools for solving such problems. We will also use the derivative to study the motion of a particle moving along a line, and we will show how the derivative can be used to approximate solutions of equations.

## 6.1 ABSOLUTE MAXIMA AND MINIMA

*At the beginning of Section 5.2 we observed that if the graph of a function f is viewed as a two-dimensional mountain range (Figure 5.2.1), then the relative maxima and minima correspond to the tops of the hills and the bottoms of the valleys; that is, they are the high and low points in their immediate vicinity. In this section we will be concerned with the more encompassing problem of finding the highest and lowest points over the entire mountain range, that is, we will be looking for the top of the highest hill and the bottom of the deepest valley. In mathematical terms, we will be looking for the largest and smallest values of a function over an interval.*

............................................

**ABSOLUTE EXTREMA**

We will be concerned here with finding the largest and smallest values of a function over a finite or infinite interval $I$. We begin with some terminology.

---

**6.1.1   DEFINITION.**   A function $f$ is said to have an ***absolute maximum*** on an interval $I$ at the point $x_0$ if $f(x_0)$ is the largest value of $f$ on $I$; that is, $f(x_0) \geq f(x)$ for all $x$ in $I$. Similarly, $f$ is said to have an ***absolute minimum*** on $I$ at the point $x_0$ if $f(x_0)$ is the smallest value of $f$ on $I$; that is, $f(x_0) \leq f(x)$ for all $x$ in $I$. If $f$ has either an absolute maximum or absolute minimum on $I$ at $x_0$, then $f$ is said to have an ***absolute extremum*** on $I$ at $x_0$.

---

As illustrated in Figure 6.1.1, there is no guarantee that a function $f$ will have absolute extrema on a given interval.

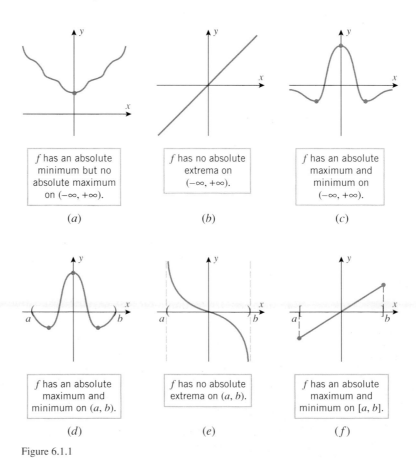

*f* has an absolute
minimum but no
absolute maximum
on $(-\infty, +\infty)$.

*(a)*

*f* has no absolute
extrema on
$(-\infty, +\infty)$.

*(b)*

*f* has an absolute
maximum and
minimum on
$(-\infty, +\infty)$.

*(c)*

*f* has an absolute
maximum and
minimum on $(a, b)$.

*(d)*

*f* has no absolute
extrema on $(a, b)$.

*(e)*

*f* has an absolute
maximum and
minimum on $[a, b]$.

*(f)*

Figure 6.1.1

EXISTENCE OF ABSOLUTE
EXTREMA

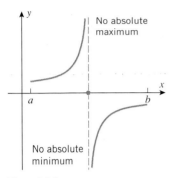

No absolute
maximum

No absolute
minimum

Figure 6.1.2

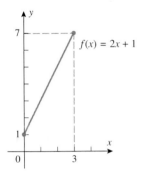

$f(x) = 2x + 1$

Figure 6.1.3

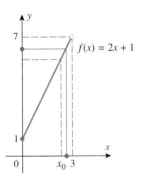

$f(x) = 2x + 1$

Figure 6.1.4

**FINDING ABSOLUTE EXTREMA ON
FINITE CLOSED INTERVALS**

The remainder of this section will focus on the following problem.

**6.1.2** PROBLEM.

(a) Determine whether a function $f$ has any absolute extrema on a given interval $I$.

(b) If there are absolute extrema, determine where they occur and what the absolute maximum and minimum values are.

Parts $(a)$–$(e)$ of Figure 6.1.1 show that a continuous function may or may not have relative maxima or minima on an infinite interval or on a finite open interval. However, the following theorem shows that a continuous function must have both an absolute maximum and an absolute minimum on every *finite closed* interval [see part $(f)$ of Figure 6.1.1].

**6.1.3** THEOREM (*Extreme-Value Theorem*).   *If a function $f$ is continuous on a finite closed interval $[a, b]$, then $f$ has both an absolute maximum and an absolute minimum on $[a, b]$.*

FOR THE READER.    Although the proof of this theorem is too difficult to include here, you should be able to convince yourself of its validity with a little experimentation—try graphing various continuous functions over the interval $[0, 1]$, and convince yourself that there is no way to avoid having a highest and lowest point on the graph. As a physical analogy, if you imagine the graph to be a roller coaster track starting at $x = 0$ and ending at $x = 1$, the roller coaster will have to pass through a highest point and a lowest point during the trip.

The hypotheses in the Extreme-Value Theorem are essential; for example, if $f$ is not continuous, then we could encounter a situation such as that in Figure 6.1.2, and if $f$ is continuous but the interval is not closed and finite, then we could encounter situations such as those in Figure 6.1.1. This is illustrated further in the following example.

### Example 1

The function $f(x) = 2x + 1$ is continuous, and hence is guaranteed to have both an absolute maximum and an absolute minimum on every finite closed interval and, in particular, on the interval $[0, 3]$. For this interval an absolute minimum occurs at $x = 0$ and an absolute maximum occurs at $x = 3$, at which points the absolute minimum and maximum values are $f(0) = 1$ and $f(3) = 7$ (Figure 6.1.3).

However, if we consider this same function on the half-open interval $[0, 3)$, then there is no longer an absolute maximum. To see why this so, observe that $f(3) = 7$ is no longer the absolute maximum because we have removed the point $x = 3$ from the interval. However, $f$ cannot have an absolute maximum in the interval at a point $x_0$ that is *less* than 3, because $f$ will have a larger value at any point in the interval to the right of $x_0$ (Figure 6.1.4). Thus, $f$ has no absolute maximum on the interval $[0, 3)$.   ◀

The Extreme-Value Theorem is an example of what mathematicians call an **existence theorem**. Such theorems state conditions under which something exists, in this case absolute extrema. However, knowing that something exists and finding it are two separate things, so we will now address the problem of finding the absolute extrema.

If $f$ is continuous on the finite closed interval $[a, b]$, then the absolute extrema of $f$ can occur either at the endpoints of the interval or inside on the open interval $(a, b)$. If the absolute extrema happen to fall inside, then the following theorem tells us that they must occur at critical points of $f$.

*✭ important ⟶ ▶*

> **6.1.4 THEOREM.** *If f is continuous on an open interval (a, b), and if f has an absolute extremum on this interval, then it must occur at a critical point of f.*

**Proof.** If $f$ has an absolute maximum on $(a, b)$ at $x_0$, then $f(x_0)$ is also a relative maximum for $f$; for if $f(x_0)$ is the largest value of $f$ on all of $(a, b)$, then $f(x_0)$ is certainly the largest value for $f$ in the immediate vicinity of $x_0$. Thus, $x_0$ is a critical point of $f$ by Theorem 5.2.2. The proof for absolute minima is similar. ∎

It follows from this theorem, that if $f$ is continuous on the finite closed interval $[a, b]$, then the absolute extrema occur either at the endpoints of the interval or at critical points inside the interval (Figure 6.1.5). Thus, we can use the following procedure to find the absolute extrema of a continuous function on a finite closed interval $[a, b]$.

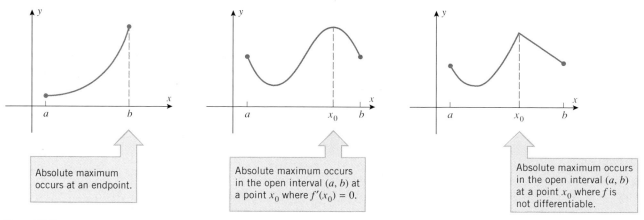

Absolute maximum occurs at an endpoint.

Absolute maximum occurs in the open interval $(a, b)$ at a point $x_0$ where $f'(x_0) = 0$.

Absolute maximum occurs in the open interval $(a, b)$ at a point $x_0$ where $f$ is not differentiable.

Figure 6.1.5

> **A Procedure for Finding the Absolute Extrema of a Continuous Function f on a Finite Closed Interval [a, b].**
>
> **Step 1.** Find the critical points of $f$ in $(a, b)$.
>
> **Step 2.** Evaluate $f$ at all the critical points and at the endpoints $a$ and $b$.
>
> **Step 3.** The largest of the values in Step 2 is the absolute maximum value of $f$ on $[a, b]$ and the smallest value is the absolute minimum.

## Example 2

Find the absolute maximum and minimum values of $f(x) = 2x^3 - 15x^2 + 36x$ on the interval $[1, 5]$, and determine where these values occur.

**Solution.** Since $f$ is differentiable, the absolute extrema must occur either at the endpoints of the interval $[1, 5]$ or at stationary points in the open interval $(1, 5)$. To find the stationary points, we must solve the equation $f'(x) = 0$, which can be written as

$$6x^2 - 30x + 36 = 6(x^2 - 5x + 6) = 6(x - 3)(x - 2) = 0$$

Thus, there are stationary points at $x = 2$ and $x = 3$. Evaluating $f$ at the endpoints and the stationary points yields

$$f(1) = 2(1)^3 - 15(1)^2 + 36(1) = 23$$
$$f(2) = 2(2)^3 - 15(2)^2 + 36(2) = 28$$
$$f(3) = 2(3)^3 - 15(3)^2 + 36(3) = 27$$
$$f(5) = 2(5)^3 - 15(5)^2 + 36(5) = 55$$

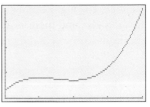

[1, 5] × [20, 55]
xScl = 1, yScl = 10

Figure 6.1.6

from which we conclude that an absolute minimum of $f(x) = 23$ occurs at $x = 1$ and an absolute maximum of $f(x) = 55$ occurs at $x = 5$. This is consistent with the graph of $f$ in Figure 6.1.6. ◀

### Example 3

Find the absolute extrema of $f(x) = 6x^{4/3} - 3x^{1/3}$ on the interval $[-1, 1]$, and determine where these values occur.

*Solution.* Differentiating we obtain

$$f'(x) = 8x^{1/3} - x^{-2/3} = x^{-2/3}(8x - 1) = \frac{8x - 1}{x^{2/3}}$$

**Table 6.1.1**

| $x$ | $-1$ | $0$ | $\frac{1}{8}$ | $1$ |
|---|---|---|---|---|
| $f(x)$ | $9$ | $0$ | $-\frac{9}{8}$ | $3$ |

Thus, $f'(x) = 0$ at $x = \frac{1}{8}$, and there is a point of nondifferentiability at $x = 0$. Evaluating $f$ at these critical points and the endpoints yields Table 6.1.1, from which we conclude that an absolute minimum of $f(x) = -\frac{9}{8}$ occurs at $x = \frac{1}{8}$, and an absolute maximum of $f(x) = 9$ occurs at $x = -1$. ◀

**ABSOLUTE EXTREMA ON INFINITE INTERVALS**

We observed earlier that a continuous function may or may not have absolute extrema on an infinite interval (see Figure 6.1.1). However, certain conclusions about the existence of absolute extrema of a continuous function $f$ on $(-\infty, +\infty)$ can be drawn from the behavior of $f(x)$ as $x \to -\infty$ and $x \to +\infty$ (Table 6.1.2).

**Table 6.1.2**

| | | | | |
|---|---|---|---|---|
| **LIMITS** | $\lim_{x \to -\infty} f(x) = +\infty$ <br> $\lim_{x \to +\infty} f(x) = +\infty$ | $\lim_{x \to -\infty} f(x) = -\infty$ <br> $\lim_{x \to +\infty} f(x) = -\infty$ | $\lim_{x \to -\infty} f(x) = -\infty$ <br> $\lim_{x \to +\infty} f(x) = +\infty$ | $\lim_{x \to -\infty} f(x) = +\infty$ <br> $\lim_{x \to +\infty} f(x) = -\infty$ |
| **CONCLUSION IF $f$ IS CONTINUOUS** | $f$ has an absolute minimum but no absolute maximum on $(-\infty, +\infty)$. | $f$ has an absolute maximum but no absolute minimum on $(-\infty, +\infty)$. | $f$ has neither an absolute maximum nor an absolute minimum on $(-\infty, +\infty)$. | $f$ has neither an absolute maximum nor an absolute minimum on $(-\infty, +\infty)$. |
| **GRAPH** | | | | |

### Example 4

What can you say about the existence of absolute extrema on $(-\infty, +\infty)$ for polynomials?

*Solution.* If $p(x)$ is a polynomial of odd degree, then

$$\lim_{x \to +\infty} p(x) \quad \text{and} \quad \lim_{x \to -\infty} p(x) \tag{1}$$

have opposite signs (one is $+\infty$ and the other is $-\infty$), so there are no absolute extrema. On the other hand, if $p(x)$ has even degree, then the limits in (1) have the same sign (both $+\infty$ or both $-\infty$). If the leading coefficient is positive, then both limits are $+\infty$, and there is an absolute minimum; if the leading coefficient is negative, then both limits are $-\infty$, and there is an absolute maximum. ◄

## Example 5

Determine by inspection whether $p(x) = 3x^4 + 4x^3$ has any absolute extrema. If so, find them, and state where they occur.

*Solution.* Since $p(x)$ has even degree and the leading coefficient is positive, $p(x) \to +\infty$ as $x \to \pm\infty$. Thus, there is an absolute minimum but no absolute maximum. From Theorem 6.1.4 [applied to the interval $(-\infty, +\infty)$], the absolute minimum must occur at a critical point of $p$. Since $p$ is differentiable, all critical points are stationary points, so we can find them by solving the equation $p'(x) = 0$. This equation is

$$12x^3 + 12x^2 = 12x^2(x + 1) = 0$$

from which we conclude that stationary points occur at $x = 0$ and $x = -1$. Evaluating $p$ at the stationary points yields

$$p(0) = 0 \quad \text{and} \quad p(-1) = -1$$

from which we conclude that $p$ has an absolute minimum of $p(x) = -1$, and this occurs at $x = -1$ (Figure 6.1.7). ◄

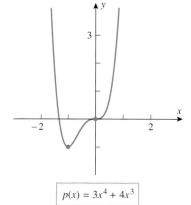

$p(x) = 3x^4 + 4x^3$

Figure 6.1.7

**ABSOLUTE EXTREMA ON OPEN INTERVALS**

We know that a continuous function may or may not have absolute extrema on an open interval. However, certain conclusions about the existence of absolute extrema of a continuous function $f$ on a finite open interval $(a, b)$ can be drawn from the behavior of $f(x)$ as $x \to a^+$ and as $x \to b^-$ (Table 6.1.3).

Table 6.1.3

| **LIMITS** | $\lim\limits_{x \to a^+} f(x) = +\infty$ $\lim\limits_{x \to b^-} f(x) = +\infty$ | $\lim\limits_{x \to a^+} f(x) = -\infty$ $\lim\limits_{x \to b^-} f(x) = -\infty$ | $\lim\limits_{x \to a^+} f(x) = -\infty$ $\lim\limits_{x \to b^-} f(x) = +\infty$ | $\lim\limits_{x \to a^+} f(x) = +\infty$ $\lim\limits_{x \to b^-} f(x) = -\infty$ |
|---|---|---|---|---|
| **CONCLUSION IF $f$ IS CONTINUOUS ON $(a, b)$** | $f$ has an absolute minimum but no absolute maximum on $(a, b)$. | $f$ has an absolute maximum but no absolute minimum on $(a, b)$. | $f$ has neither an absolute maximum nor an absolute minimum on $(a, b)$. | $f$ has neither an absolute maximum nor an absolute minimum on $(a, b)$. |
| **GRAPH** | | | | |

## Example 6

Determine whether the function

$$f(x) = \frac{1}{x^2 - x}$$

has any absolute extrema on the interval $(0, 1)$. If so, find them and state where they occur.

*Solution.*  Since $f$ is continuous on the interval $(0, 1)$ and

$$\lim_{x \to 0^+} f(x) = \lim_{x \to 0^+} \frac{1}{x^2 - x} = \lim_{x \to 0^+} \frac{1}{x(x-1)} = -\infty$$

$$\lim_{x \to 1^-} f(x) = \lim_{x \to 1^-} \frac{1}{x^2 - x} = \lim_{x \to 1^-} \frac{1}{x(x-1)} = -\infty$$

the function $f$ has an absolute maximum but no absolute minimum on the interval $(0, 1)$. By Theorem 6.1.4 the absolute maximum must occur at a critical point of $f$, so we need to look for stationary points or points of nondifferentiability in the interval $(0, 1)$. We have

$$f'(x) = -\frac{2x - 1}{\left(x^2 - x\right)^2}$$

so the only solution of the equation $f'(x) = 0$ is $x = \frac{1}{2}$. The denominator of $f'$ is zero if $x = 0$ or $x = 1$, but these critical points are of no concern here because they fall outside of the open interval $(0, 1)$. Thus, the absolute maximum occurs at $x = \frac{1}{2}$, and this absolute maximum is

$$f\left(\tfrac{1}{2}\right) = \frac{1}{\left(\frac{1}{2}\right)^2 - \frac{1}{2}} = -4$$

(Figure 6.1.8).  ◀

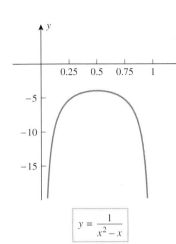

$$y = \frac{1}{x^2 - x}$$

Figure 6.1.8

**ABSOLUTE EXTREMA OF FUNCTIONS WITH ONE RELATIVE EXTREMUM**

If a continuous function has only one relative extremum on a finite or infinite interval $I$, then that relative extremum must of necessity also be an absolute extremum. To understand why this is so, suppose that $f$ has a relative maximum at a point $x_0$ on an interval $I$, and there are no other relative extrema of $f$ on $I$. If $f(x_0)$ is *not* the absolute maximum of $f$ on $I$, then the graph of $f$ has to make an upward turn somewhere on $I$ to rise above $f(x_0)$. However, this cannot happen because in the process of making an upward turn it would produce a second relative extremum on $I$ (Figure 6.1.9). Thus, $f(x_0)$ must be the absolute maximum as well as a relative maximum. This idea is captured in the following theorem, which we state without proof.

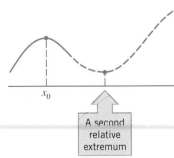

Figure 6.1.9

**6.1.5**  THEOREM.  *Suppose that $f$ is continuous and has exactly one relative extremum on an interval $I$, say at $x_0$.*

*(a)  If $f$ has a relative minimum at $x_0$, then $f(x_0)$ is the absolute minimum of $f$ on $I$.*

*(b)  If $f$ has a relative maximum at $x_0$, then $f(x_0)$ is the absolute maximum of $f$ on $I$.*

This theorem is often helpful in situations where other methods are difficult or tedious to apply.

## Example 7

Find all absolute extrema of the function $f(x) = x^3 - 3x^2 + 4$ on the interval

(a)  $(-\infty, +\infty)$      (b)  $(0, +\infty)$

*Solution (a).*  Because $f$ is a polynomial of odd degree, it follows from the discussion in Example 4 that there are no absolute extrema on the interval $(-\infty, +\infty)$.

*Solution (b).*  Since

$$\lim_{x \to +\infty} (x^3 - 3x^2 + 4) = +\infty$$

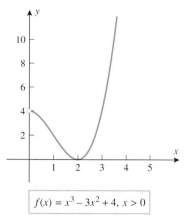

$f(x) = x^3 - 3x^2 + 4, \ x > 0$

Figure 6.1.10

we know that $f$ cannot have an absolute maximum on the interval $(0, +\infty)$. However, the limit

$$\lim_{x \to 0^+} (x^3 - 3x^2 + 4) = 4$$

is not infinite, so there is a possibility that $f$ may have an absolute minimum on this interval. In this case it would have to occur at a stationary point, which suggests that we look for solutions of the equation $f'(x) = 0$. But,

$$f'(x) = 3x^2 - 6x = 3x(x - 2)$$

so $f$ has stationary points at $x = 0$ and $x = 2$. However, $x = 0$ falls outside of the interval $(0, +\infty)$, so only the stationary point at $x = 2$ lies in the interval $(0, +\infty)$. Thus, Theorem 6.1.5 is applicable here. Since

$$f''(x) = 6x - 6$$

we have $f''(2) = 6 > 0$, so a relative minimum occurs at $x = 2$ by the second derivative test. Thus, $f(x)$ has an absolute minimum at $x = 2$, and this absolute minimum is $f(2) = 0$ (Figure 6.1.10). ◀

**ABSOLUTE EXTREMA AND PARAMETRIC CURVES**

Suppose that a curve $C$ is given parametrically by the equations

$$x = f(t), \quad y = g(t) \qquad (a \leq t \leq b)$$

where $f$ and $g$ are *continuous* on the finite closed interval $[a, b]$. It follows from the Extreme-Value Theorem that $f(t)$ and $g(t)$ have absolute maxima and absolute minima for $a \leq t \leq b$; this means that a particle moving along the curve cannot move away from the origin indefinitely—there must be a smallest and largest $x$-coordinate and a smallest and largest $y$-coordinate. Geometrically, the entire curve is contained within a box determined by these smallest and largest coordinates.

### Example 8

Suppose that the equations of motion for a paper airplane during its first 10 seconds of flight are

$$x = t - 3 \sin t, \quad y = 4 - 3 \cos t \qquad (0 \leq t \leq 10)$$

What are the highest and lowest points in the trajectory, and when is the airplane at those points?

*Solution.* The trajectory, pictured in Figure 6.1.11, is shown in more detail in Figure 1.7.2. We want to find the absolute maximum and minimum values of $y$ over the time interval $[0, 10]$ and the values of $t$ for which these absolute extrema occur. The absolute extrema

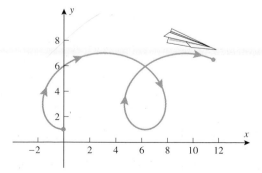

Figure 6.1.11

must occur either at the endpoints of the closed interval [0, 10] or at stationary points in the open interval (0, 10). To find the stationary points, we must solve the equation $dy/dt = 0$, which is

$$3 \sin t = 0$$

Thus, there are stationary points in the interval (0, 10) at $t = \pi, 2\pi$, and $3\pi$. Evaluating $y = 4 - 3\cos t$ at the endpoints and the stationary points yields

$$y = 4 - 3\cos 0 = 4 - 3 = 1$$
$$y = 4 - 3\cos \pi = 4 - (-3) = 7$$
$$y = 4 - 3\cos 2\pi = 4 - 3 = 1$$
$$y = 4 - 3\cos 3\pi = 4 - (-3) = 7$$
$$y = 4 - 3\cos 10 \approx 6.517$$

Thus, a high point of $y = 7$ is reached at times $t = \pi$ and $t = 3\pi$, and a low point of $y = 1$ is reached at times $t = 0$ and $t = 2\pi$. This is consistent with Figure 1.7.2. ◀

## EXERCISE SET 6.1 ⌇ Graphing Calculator ⃞C CAS

In Exercises 1–2, use the graph to find $x$-coordinates of the relative extrema and absolute extrema of $f$.

**1.**
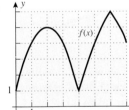

**2.**
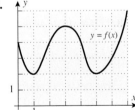

**3.** In each part, sketch the graph of a continuous function $f$ with the stated properties on the interval [0, 10].
  (a) $f$ has an absolute minimum at $x = 0$ and an absolute maximum at $x = 10$.
  (b) $f$ has an absolute minimum at $x = 2$ and an absolute maximum at $x = 7$.
  (c) $f$ has a relative minima at $x = 1$ and $x = 8$, has relative maxima at $x = 3$ and $x = 7$, has an absolute minimum at $x - 5$, and has an absolute maximum at $x = 10$.

**4.** In each part, sketch the graph of a continuous function $f$ with the stated properties on the interval $(-\infty, +\infty)$.
  (a) $f$ has no relative extrema or absolute extrema.
  (b) $f$ has an absolute minimum at $x = 0$ but no absolute maximum.
  (c) $f$ has an absolute maximum at $x = -5$ and an absolute minimum at $x = 5$.

In Exercises 5–14, find the absolute maximum and minimum values of $f$ on the given closed interval, and state where those values occur.

**5.** $f(x) = 4x^2 - 4x + 1$; $[0, 1]$

**6.** $f(x) = 8x - x^2$; $[0, 6]$

**7.** $f(x) = (x - 1)^3$; $[0, 4]$

**8.** $f(x) = 2x^3 - 3x^2 - 12x$; $[-2, 3]$

**9.** $f(x) = \dfrac{3x}{\sqrt{4x^2 + 1}}$; $[-1, 1]$

**10.** $f(x) = \left(x^2 + x\right)^{2/3}$; $[-2, 3]$

**11.** $f(x) = x - \tan x$; $[-\pi/4, \pi/4]$

**12.** $f(x) = \sin x - \cos x$; $[0, \pi]$

**13.** $f(x) = 1 + |9 - x^2|$; $[-5, 1]$

**14.** $f(x) = |6 - 4x|$; $[-3, 3]$

In Exercises 15–22, find the absolute maximum and minimum values of $f$, if any, on the given interval, and state where those values occur.

**15.** $f(x) = x^2 - 3x - 1$; $(-\infty, +\infty)$

**16.** $f(x) = 3 - 4x - 2x^2$; $(-\infty, +\infty)$

**17.** $f(x) = 4x^3 - 3x^4$; $(-\infty, +\infty)$

**18.** $f(x) = x^4 + 4x$; $(-\infty, +\infty)$

**19.** $f(x) = x^3 - 3x - 2$; $(-\infty, +\infty)$

**20.** $f(x) = x^3 - 9x + 1$; $(-\infty, +\infty)$

**21.** $f(x) = \dfrac{x^2}{x + 1}$; $(-5, -1)$   **22.** $f(x) = \dfrac{x + 3}{x - 3}$; $[-5, 5]$

In Exercises 23–34, use a graphing utility to estimate the absolute maximum and minimum values of $f$, if any, on the stated interval, and then use calculus methods to find the exact values.

**23.** $f(x) = \left(x^2 - 1\right)^2$; $(-\infty, +\infty)$

**24.** $f(x) = (x - 1)^2(x + 2)^2$; $(-\infty, +\infty)$

**25.** $f(x) = x^{2/3}(20 - x)$; $[-1, 20]$

**26.** $f(x) = \dfrac{x}{x^2 + 2}$; $[-1, 4]$

**27.** $f(x) = 1 + \dfrac{1}{x}$; $(0, +\infty)$

**28.** $f(x) = \dfrac{x}{x^2 + 1}$; $[0, +\infty)$

**29.** $f(x) = 2 \sec x - \tan x$; $[0, \pi/4]$

**30.** $f(x) = \sin^2 x + \cos x$; $[-\pi, \pi]$

**31.** $f(x) = x^3 e^{-2x}$; $[1, 4]$

**32.** $f(x) = \dfrac{\ln x}{x}$; $[1, e]$

**33.** $f(x) = \sin(\cos x)$; $[0, 2\pi]$

**34.** $f(x) = \cos(\sin x)$; $[0, \pi]$

**35.** Find the absolute maximum and minimum values of
$$f(x) = \begin{cases} 4x - 2, & x < 1 \\ (x - 2)(x - 3), & x \geq 1 \end{cases}$$
on $\left[\frac{1}{2}, \frac{7}{2}\right]$.

**36.** Let $f(x) = x^2 + px + q$. Find the values of $p$ and $q$ such that $f(1) = 3$ is an extreme value of $f$ on $[0, 2]$. Is this value a maximum or minimum?

If $f$ is a periodic function, then the locations of all absolute extrema on the interval $(-\infty, +\infty)$ can be obtained by finding the locations of the absolute extrema for one period and using the periodicity to locate the rest. Use this idea in Exercise 37 and 38 to find the absolute maximum and minimum values of the function, and state the $x$-values at which they occur.

**37.** $f(x) = 2 \sin 2x + \sin 4x$   **38.** $f(x) = 3 \cos \dfrac{x}{3} + 2 \cos \dfrac{x}{2}$

One way of proving that $f(x) \leq g(x)$ for all $x$ in a given interval is to show that $0 \leq g(x) - f(x)$ for all $x$ in the interval; and one way of proving the latter inequality is to show that the absolute minimum value of $g(x) - f(x)$ on the interval is nonnegative. Use this idea to prove the inequalities in Exercises 39 and 40.

**39.** Prove that $\sin x \leq x$ for all $x$ in the interval $[0, 2\pi]$.

**40.** Prove that $\ln x \leq x - 1$ on the interval $(0, +\infty)$.

**41.** What is the smallest possible slope for a tangent to the equation $y = x^3 - 3x^2 + 5x$?

**42.** (a) Show that
$$f(x) = \frac{64}{\sin x} + \frac{27}{\cos x}$$
has a minimum value, but no maximum value on the interval $(0, \pi/2)$.
(b) Find the minimum value.

**43.** Use a CAS to show that the absolute minimum value of
$$f(x) = x^2 + \frac{16x^2}{(8 - x)^2}, \quad x > 8$$
occurs at $x = 4(2 + \sqrt[3]{2})$ by using it to find $f'(x)$ and to solve the equation $f'(x) = 0$.

**44.** The concentration $C(t)$ of a drug in the bloodstream $t$ hours after it has been injected is commonly modeled by an equation of the form
$$C(t) = \frac{K(e^{-bt} - e^{-at})}{a - b}$$
where $K > 0$ and $a > b > 0$.
(a) At what time does the maximum concentration occur?
(b) Let $K = 1$ for simplicity, and use a graphing utility to check your result in part (a) by graphing $C(t)$ for various values of $a$ and $b$.

**45.** It can be proved that if $f$ is differentiable on $(a, b)$ and $L$ is a line that does not intersect the curve $y = f(x)$ over an interval $(a, b)$, then the points at which the curve is closest to or farthest from the line $L$, if any, occur at points where the tangent line to the curve is parallel to $L$ (see the accompanying figure). Use this result to find the points on the graph of $y = -x^2$ that are closest to and farthest from the line $y = 2 - x$ for $-1 \leq x \leq \frac{3}{2}$.

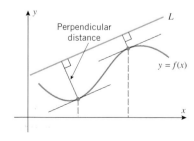

Figure Ex-45

**46.** Use the idea discussed in Exercise 45 to find the coordinates of all points on the graph of $y = x^3$ closest to and farthest from the line $y = \frac{4}{3}x - 1$ for $-1 \le x \le 1$.

**47.** Suppose that the equations of motion of a paper airplane during the first 12 seconds of flight are

$$x = t - 2\sin t, \quad y = 2 - 2\cos t \qquad (0 \le t \le 12)$$

What are the highest and lowest points in the trajectory, and when is the airplane at those points?

**48.** The accompanying figure shows the path of a fly whose equations of motion are

$$x = \frac{\cos t}{2 + \sin t}, \quad y = 3 + \sin(2t) - 2\sin^2 t \qquad (0 \le t \le 2\pi)$$

(a)  How high and low does it fly?
(b)  How far left and right of the origin does it fly?

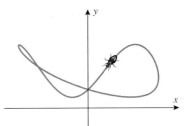

Figure Ex-48

**49.** Let $f(x) = ax^2 + bx + c$, where $a > 0$. Prove that $f(x) \ge 0$ for all $x$ if and only if $b^2 - 4ac \le 0$. [*Hint:* Find the minimum of $f(x)$.]

**50.** Prove Theorem 6.1.4 in the case where the extreme value is a minimum.

# 6.2 APPLIED MAXIMUM AND MINIMUM PROBLEMS

*In this section we will show how the methods discussed in the last section can be used to solve various applied optimization problems.*

**CLASSIFICATION OF OPTIMIZATION PROBLEMS**

The applied optimization problems that we will consider in this section fall into the following two categories:

- Problems that reduce to maximizing or minimizing a continuous function over a finite closed interval.

- Problems that reduce to maximizing or minimizing a continuous function over an infinite interval or a finite interval that is not closed.

For problems of the first type the Extreme-Value Theorem (6.1.3) guarantees that the problem has a solution, and we know that the solution can be obtained by examining the values of the function at the critical points and the endpoints. However, for problems of the second type there may or may not be a solution. Thus, part of the attack on such problems is to determine whether there actually is a solution. If the function is continuous and has exactly one relative extremum on the interval, then Theorem 6.1.5 guarantees the existence of a solution and provides a method for finding it. In cases where this theorem is not applicable some ingenuity may be required to solve the problem.

**PROBLEMS INVOLVING FINITE CLOSED INTERVALS**

## Example 1

Find the dimensions of a rectangle with perimeter 100 ft whose area is as large as possible.

*Solution.*  Let

$x = $ length of the rectangle (ft)

$y = $ width of the rectangle (ft)

$A = $ area of the rectangle (ft$^2$)

Then

$$A = xy \tag{1}$$

Since the perimeter of the rectangle is 100 ft, the variables $x$ and $y$ are related by the equation

$$2x + 2y = 100 \quad \text{or} \quad y = 50 - x \tag{2}$$

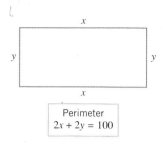

Figure 6.2.1

**Table 6.2.1**

| $x$ | 0 | 25 | 50 |
|---|---|---|---|
| $A$ | 0 | 625 | 0 |

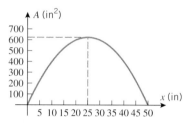

Figure 6.2.2

(See Figure 6.2.1.) Substituting (2) in (1) yields

$$A = x(50 - x) = 50x - x^2 \qquad (3)$$

Because $x$ represents a length it cannot be negative, and because the two sides of length $x$ cannot have a combined length exceeding the total perimeter of 100 ft, the variable $x$ must satisfy

$$0 \le x \le 50 \qquad (4)$$

Thus, we have reduced the problem to that of finding the value (or values) of $x$ in [0, 50], for which $A$ is maximum. Since $A$ is a polynomial in $x$, it is continuous on [0, 50], and so the maximum must occur at an endpoint of this interval or at a stationary point.

From (3) we obtain

$$\frac{dA}{dx} = 50 - 2x$$

Setting $dA/dx = 0$ we obtain

$$50 - 2x = 0$$

or $x = 25$. Thus, the maximum occurs at one of the points

$$x = 0, \quad x = 25, \quad x = 50$$

Substituting these values in (3) yields Table 6.2.1, which tells us that the maximum area of 625 ft$^2$ occurs at $x = 25$, which is consistent with the graph of (3) in Figure 6.2.2. From (2) the corresponding value of $y = 25$, so the rectangle of perimeter 100 ft with greatest area is a square with sides of length 25 ft. ◀

REMARK.   In this example we included $x = 0$ and $x = 50$ as possible values for $x$, even though both values lead to rectangles with two sides of length zero. Whether or not these values should be allowed will depend on our objective in the problem. If we view this purely as a mathematical problem, then there is nothing wrong with allowing sides of length zero. However, if we view this as an applied problem in which the rectangle will be formed from physical material, then these values should be excluded.

Example 1 illustrates the following five-step procedure that can be used for solving many applied maximum and minimum problems.

**Step 1.**   Draw an appropriate figure and label the quantities relevant to the problem.

**Step 2.**   Find a formula for the quantity to be maximized or minimized.

**Step 3.**   Using the conditions stated in the problem to eliminate variables, express the quantity to be maximized or minimized as a function of one variable.

**Step 4.**   Find the interval of possible values for this variable from the physical restrictions in the problem.

**Step 5.**   If applicable, use the techniques of the preceding section to obtain the maximum or minimum.

## Example 2

An open box is to be made from a 16-inch by 30-inch piece of cardboard by cutting out squares of equal size from the four corners and bending up the sides (Figure 6.2.3). What size should the squares be to obtain a box with largest possible volume?

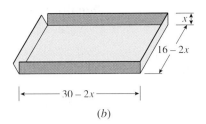

Figure 6.2.3

*Solution.* Let

    $x$ = length (in inches) of the sides of the squares to be cut out

    $V$ = volume (in cubic inches) of the resulting box

Because we are removing a square of side $x$ from each corner, the resulting box will have dimensions $16 - 2x$ by $30 - 2x$ by $x$ (Figure 6.2.3$b$). Since the volume of a box is the product of its dimensions, we have

$$V = (16 - 2x)(30 - 2x)x = 480x - 92x^2 + 4x^3 \tag{5}$$

The variable $x$ in this expression is subject to certain restrictions. Because $x$ represents a length it cannot be negative, and because the width of the cardboard is 16 inches we cannot cut out squares whose sides are more than 8 inches long. Thus, the variable $x$ in (5) must satisfy

$$0 \le x \le 8$$

and hence we have reduced our problem to finding the value (or values) of $x$ in the interval $[0, 8]$ for which (5) is maximum. From (5) we obtain

$$\frac{dV}{dx} = 480 - 184x + 12x^2 = 4(120 - 46x + 3x^2)$$

Setting $dV/dx = 0$ yields

$$120 - 46x + 3x^2 = 0$$

which can be solved by the quadratic formula to obtain the critical points

$$x = \tfrac{10}{3} \quad \text{and} \quad x = 12$$

Since $x = 12$ falls outside the interval $[0, 8]$, the maximum value of $V$ occurs either at the critical point $x = \tfrac{10}{3}$ or at one of the endpoints $x = 0$, $x = 8$. Substituting these values in (5) yields Table 6.2.2, which tells us that the greatest possible volume $V = \tfrac{19600}{27}$ in$^3 \approx 726$ in$^3$ occurs when we cut out squares whose sides have length $\tfrac{10}{3}$ inches. This is consistent with the graph of (5) shown in Figure 6.2.4. ◀

In Example 2 of Section 1.1 we used approximate graphical methods to solve a problem of piping oil from an offshore well to a point on the shore with minimal cost. We will now show how to solve that problem exactly using calculus.

**Table 6.2.2**

| $x$ | 0 | $\frac{10}{3}$ | 8 |
|---|---|---|---|
| $V$ | 0 | $\frac{19600}{27} \approx 726$ | 0 |

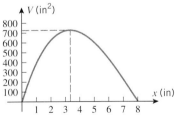

Figure 6.2.4

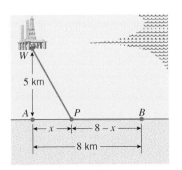

Figure 6.2.5

### Example 3

Figure 6.2.5 shows an offshore oil well located at a point $W$ that is 5 km from the closest point $A$ on a straight shoreline. Oil is to be piped from $W$ to a shore point $B$ that is 8 km from $A$ by piping it on a straight line under water from $W$ to some shore point $P$ between $A$ and $B$ and then on to $B$ via pipe along the shoreline. If the cost of laying pipe is \$1,000,000/km under water and \$500,000/km over land, where should the point $P$ be located to minimize the cost of laying the pipe?

***Solution.***  Let

$x =$ distance (in kilometers) between $A$ and $P$

$c =$ cost (in millions of dollars) for the entire pipeline

From Figure 6.2.5 the length of pipe under water is the distance between $W$ and $P$. By the Theorem of Pythagoras, that length is

$$\sqrt{x^2 + 25} \tag{6}$$

Also from Figure 6.2.5, the length of pipe over land is the distance between $P$ and $B$, which is

$$8 - x \tag{7}$$

From (6) and (7) it follows that the total cost $c$ (in millions of dollars) for the pipeline is

$$c = 1(\sqrt{x^2 + 25}) + \tfrac{1}{2}(8 - x) = \sqrt{x^2 + 25} + \tfrac{1}{2}(8 - x) \tag{8}$$

Because the distance between $A$ and $B$ is 8 km, the distance $x$ between $A$ and $P$ must satisfy

$$0 \le x \le 8$$

We have thus reduced our problem to finding the value (or values) of $x$ in the interval $[0, 8]$ for which (8) is a minimum. Since $c$ is a continuous function of $x$ on the closed interval $[0, 8]$, we can use the methods developed in the preceding section to find the minimum.

From (8) we obtain

$$\frac{dc}{dx} = \frac{x}{\sqrt{x^2 + 25}} - \frac{1}{2}$$

Setting $dc/dx = 0$ and solving for $x$ yields

$$\frac{x}{\sqrt{x^2 + 25}} = \frac{1}{2} \tag{9}$$

$$x^2 = \frac{1}{4}(x^2 + 25)$$

$$x = \pm \frac{5}{\sqrt{3}}$$

The number $-5/\sqrt{3}$ is not a solution of (9) and must be discarded, leaving $x = 5/\sqrt{3}$ as the only critical point. Since this point lies in the interval $[0, 8]$, the minimum must occur at one of the points

$$x = 0, \quad x = 5/\sqrt{3}, \quad x = 8$$

Substituting these values in (8) yields Table 6.2.3, which tells us that the least possible cost of the pipeline (to the nearest dollar) is $c = \$8,330,127$, and this occurs when the point $P$ is located at a distance of $5/\sqrt{3} \approx 2.89$ km from $A$. This is consistent with the graph in Figure 1.1.9c.  ◄

**Table 6.2.3**

| $x$ | 0 | $\frac{5}{\sqrt{3}}$ | 8 |
|---|---|---|---|
| $c$ | 9 | $\frac{10}{\sqrt{3}} + \left(4 - \frac{5}{2\sqrt{3}}\right) \approx 8.330127$ | $\sqrt{89} \approx 9.433981$ |

FOR THE READER.    If you have a CAS, use it to check all of the computations in this example. Specifically, differentiate $c$ with respect to $x$, solve the equation $dc/dx = 0$, and perform all of the numerical calculations.

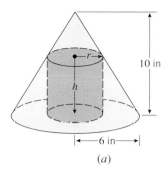

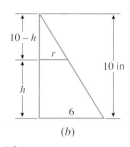

Figure 6.2.6

### Example 4

Find the radius and height of the right circular cylinder of largest volume that can be inscribed in a right circular cone with radius 6 inches and height 10 inches (Figure 6.2.6*a*).

*Solution.*  Let

$r$ = radius (in inches) of the cylinder

$h$ − height (in inches) of the cylinder

$V$ = volume (in cubic inches) of the cylinder

The formula for the volume of the inscribed cylinder is

$$V = \pi r^2 h \tag{10}$$

To eliminate one of the variables in (10) we need a relationship between $r$ and $h$. Using similar triangles (Figure 6.2.6*b*) we obtain

$$\frac{10 - h}{r} = \frac{10}{6} \quad \text{or} \quad h = 10 - \tfrac{5}{3}r \tag{11}$$

Substituting (11) into (10) we obtain

$$V = \pi r^2 \left(10 - \tfrac{5}{3}r\right) = 10\pi r^2 - \tfrac{5}{3}\pi r^3 \tag{12}$$

which expresses $V$ in terms of $r$ alone. Because $r$ represents a radius it cannot be negative, and because the radius of the inscribed cylinder cannot exceed the radius of the cone, the variable $r$ must satisfy

$$0 \le r \le 6$$

Thus, we have reduced the problem to that of finding the value (or values) of $r$ in $[0, 6]$ for which (12) is a maximum. Since $V$ is a continuous function of $r$ on $[0, 6]$, the methods developed in the preceding section apply.

From (12) we obtain

$$\frac{dV}{dr} = 20\pi r - 5\pi r^2 = 5\pi r(4 - r)$$

Setting $dV/dr = 0$ gives

$$5\pi r(4 - r) = 0$$

so $r = 0$ and $r = 4$ are critical points. Since these lie in the interval $[0, 6]$, the maximum must occur at one of the points

$$r = 0, \quad r = 4, \quad r = 6$$

#### Table 6.2.4

| $r$ | 0 | 4 | 6 |
|---|---|---|---|
| $V$ | 0 | $\frac{160}{3}\pi$ | 0 |

Substituting these values in (12) yields Table 6.2.4, which tells us that the maximum volume $V = \frac{160}{3}\pi \approx 168$ in$^3$ occurs when the inscribed cylinder has radius 4 in. When $r = 4$ it follows from (11) that $h = \frac{10}{3}$. Thus, the inscribed cylinder of largest volume has radius $r = 4$ in and height $h = \frac{10}{3}$ in.  ◄

........................................

**PROBLEMS INVOLVING INTERVALS THAT ARE NOT FINITE AND CLOSED**

### Example 5

A closed cylindrical can is to hold 1 liter (1000 cm$^3$) of liquid. How should we choose the height and radius to minimize the amount of material needed to manufacture the can?

*Solution.*  Let

$h$ = height (in cm) of the can

$r$ = radius (in cm) of the can

$S$ = surface area (in cm$^2$) of the can

Assuming there is no waste or overlap, the amount of material needed for manufacture will be the same as the surface area of the can. Since the can consists of two circular disks of radius $r$ and a rectangular sheet with dimensions $h$ by $2\pi r$ (Figure 6.2.7), the surface area will be

$$S = 2\pi r^2 + 2\pi r h \tag{13}$$

Since $S$ depends on two variables, $r$ and $h$, we will look for some condition in the problem that will allow us to express one of these variables in terms of the other. For this purpose, observe that the volume of the can is 1000 cm$^3$, so it follows from the formula $V = \pi r^2 h$ for the volume of a cylinder that

$$1000 = \pi r^2 h \quad \text{or} \quad h = \frac{1000}{\pi r^2} \tag{14–15}$$

Substituting (15) in (13) yields

$$S = 2\pi r^2 + \frac{2000}{r} \tag{16}$$

Thus, we have reduced the problem to finding a value of $r$ in the interval $(0, +\infty)$ for which $S$ is minimum, provided there actually is a minimum.* However, $S$ is a continuous function of $r$ on the interval $(0, +\infty)$ and

$$\lim_{r \to 0^+} \left( 2\pi r^2 + \frac{2000}{r} \right) = +\infty \quad \text{and} \quad \lim_{r \to +\infty} \left( 2\pi r^2 + \frac{2000}{r} \right) = +\infty$$

so the analysis in Table 6.1.3 implies that $S$ does have a minimum on the interval $(0, +\infty)$. Since this minimum must occur at a critical point, we calculate

$$\frac{dS}{dr} = 4\pi r - \frac{2000}{r^2} \tag{17}$$

Setting $dS/dr = 0$ gives

$$4\pi r - \frac{2000}{r^2} = 0 \quad \text{or} \quad r = \frac{10}{\sqrt[3]{2\pi}} \tag{18}$$

Since (18) is the only critical point in the interval $(0, +\infty)$, this value of $r$ yields the minimum value of $S$. From (15) the value of $h$ corresponding to this $r$ is

$$h = \frac{1000}{\pi(10/\sqrt[3]{2\pi})^2} = \frac{20}{\sqrt[3]{2\pi}} = 2r$$

It is not accidental here that the minimum occurs when the height of the can is equal to the diameter of its base (Exercise 27).

Area $2\pi r^2$          Area $2\pi r h$

Figure 6.2.7

*Second Solution.* The conclusion that a minimum occurs at the value of $r$ in (18) can be deduced from Theorem 6.1.5 and the second derivative test by noting that

---

*The value $r = 0$ must be excluded because a cylindrical can of radius 0 cm cannot have a volume of 1000 cm$^3$ [see (14)].

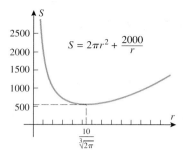

Figure 6.2.8

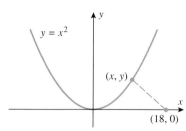

Figure 6.2.9

$$\frac{d^2S}{dr^2} = 4\pi + \frac{4000}{r^3}$$

is positive if $r > 0$ and hence is positive if $r = 10/\sqrt[3]{2\pi}$. This implies that a relative minimum, and therefore a minimum, occurs at the critical point $r = 10/\sqrt[3]{2\pi} \approx 5.4$.

*Third Solution.* The existence of a minimum is implied by the graph of $S$ versus $r$ in Figure 6.2.8. As shown in (18), this minimum occurs at $r = 10/\sqrt[3]{2\pi}$.   ◀

REMARK.    Note that $S$ has no maximum on $(0, +\infty)$. Thus, had we asked for the dimensions of the can requiring the maximum amount of material for its manufacture, there would have been no solution to the problem. Optimization problems with no solution are sometimes called *ill posed*.

## Example 6

Find a point on the curve $y = x^2$ that is closest to the point $(18, 0)$.

*Solution.*  The distance $L$ between $(18, 0)$ and an arbitrary point $(x, y)$ on the curve $y = x^2$ (Figure 6.2.9) is given by

$$L = \sqrt{(x - 18)^2 + (y - 0)^2}$$

Since $(x, y)$ lies on the curve, $x$ and $y$ satisfy $y = x^2$; thus,

$$L = \sqrt{(x - 18)^2 + x^4} \tag{19}$$

Because there are no restrictions on $x$, the problem reduces to finding a value of $x$ in $(-\infty, +\infty)$ for which (19) is minimum, provided such a value exists.

In problems of minimizing or maximizing a distance, there is a trick that is helpful for simplifying the computations. It is based on the observation that the distance and the square of the distance have their maximum or minimum at the same point (see Exercise 60). Thus, the minimum value of $L$ in (19) and the minimum value of

$$S = L^2 = (x - 18)^2 + x^4 \tag{20}$$

occur at the same $x$-value.

From (20),

$$\frac{dS}{dx} = 2(x - 18) + 4x^3 = 4x^3 + 2x - 36 \tag{21}$$

so that the critical points satisfy $4x^3 + 2x - 36 = 0$ or, equivalently,

$$2x^3 + x - 18 = 0 \tag{22}$$

To solve for $x$ we will begin by checking the divisors of $-18$ to see whether the polynomial on the left side has any integer roots (see Appendix F). These divisors are $\pm 1, \pm 2, \pm 3, \pm 6, \pm 9,$ and $\pm 18$. A check of these values shows that $x = 2$ is a root, so that $x - 2$ is a factor of the polynomial. After dividing the polynomial by this factor we can rewrite (22) as

$$(x - 2)(2x^2 + 4x + 9) = 0$$

Thus, the remaining solutions of (22) satisfy the quadratic equation

$$2x^2 + 4x + 9 = 0$$

But these solutions are imaginary numbers (use the quadratic formula), so that $x = 2$ is the only real solution of (22) and consequently the only critical point of $S$. To determine the nature of this critical point we will use the second derivative test. From (21),

$$\frac{d^2S}{dx^2} = 12x^2 + 2, \quad \text{so} \quad \left.\frac{d^2S}{dx^2}\right|_{x=2} = 50 > 0$$

which shows that a relative minimum occurs at $x = 2$. Since $x = 2$ is the only relative extremum for $L$, it follows from Theorem 6.1.5 that an absolute minimum value of $L$ also occurs at $x = 2$. Thus, the point on the curve $y = x^2$ closest to $(18, 0)$ is

$$(x, y) = (x, x^2) = (2, 4) \qquad \blacktriangleleft$$

...........................................

**AN APPLICATION TO ECONOMICS**

Three functions of importance to an economist or a manufacturer are

$C(x) = $ total cost of producing $x$ units of a product during some time period

$R(x) = $ total revenue from selling $x$ units of the product during the time period

$P(x) = $ total profit obtained by selling $x$ units of the product during the time period

These are called, respectively, the ***cost function***, ***revenue function***, and ***profit function***. If all units produced are sold, then these are related by

$$P(x) = R(x) - C(x) \tag{23}$$

$$[\text{profit}] = [\text{revenue}] - [\text{cost}]$$

The total cost $C(x)$ of producing $x$ units can be expressed as a sum

$$C(x) = a + M(x) \tag{24}$$

where $a$ is a constant, called ***overhead***, and $M(x)$ is a function representing ***manufacturing cost***. The overhead, which includes such fixed costs as rent and insurance, does not depend on $x$; it must be paid even if nothing is produced. On the other hand, the manufacturing cost $M(x)$, which includes such items as cost of materials and labor, depends on the number of items manufactured. It is shown in economics that with suitable simplifying assumptions, $M(x)$ can be expressed in the form

$$M(x) = bx + cx^2$$

where $b$ and $c$ are constants. Substituting this in (24) yields

$$C(x) = a + bx + cx^2 \tag{25}$$

If a manufacturing firm can sell all the items it produces for $p$ dollars apiece, then its total revenue $R(x)$ (in dollars) will be

$$R(x) = px \tag{26}$$

and its total profit $P(x)$ (in dollars) will be

$$P(x) = [\text{total revenue}] - [\text{total cost}] = R(x) - C(x) = px - C(x)$$

Thus, if the cost function is given by (25),

$$P(x) = px - (a + bx + cx^2) \tag{27}$$

Depending on such factors as number of employees, amount of machinery available, economic conditions, and competition, there will be some upper limit $l$ on the number of items a manufacturer is capable of producing and selling. Thus, during a fixed time period the variable $x$ in (27) will satisfy

$$0 \leq x \leq l$$

By determining the value or values of $x$ in $[0, l]$ that maximize (27), the firm can determine how many units of its product must be manufactured and sold to yield the greatest profit. This is illustrated in the following numerical example.

## Example 7

A liquid form of penicillin manufactured by a pharmaceutical firm is sold in bulk at a price of \$200 per unit. If the total production cost (in dollars) for $x$ units is

$$C(x) = 500{,}000 + 80x + 0.003x^2$$

and if the production capacity of the firm is at most 30,000 units in a specified time, how many units of penicillin must be manufactured and sold in that time to maximize the profit?

*Solution.*   Since the total revenue for selling $x$ units is $R(x) = 200x$, the profit $P(x)$ on $x$ units will be

$$P(x) = R(x) - C(x) = 200x - (500{,}000 + 80x + 0.003x^2) \tag{28}$$

Since the production capacity is at most 30,000 units, $x$ must lie in the interval $[0, 30{,}000]$. From (28)

$$\frac{dP}{dx} = 200 - (80 + 0.006x) = 120 - 0.006x$$

Setting $dP/dx = 0$ gives

$$120 - 0.006x = 0 \quad \text{or} \quad x = 20{,}000$$

Since this critical point lies in the interval $[0, 30{,}000]$, the maximum profit must occur at one of the points

$$x = 0, \quad x = 20{,}000, \quad \text{or} \quad x = 30{,}000$$

Substituting these values in (28) yields Table 6.2.5, which tells us that the maximum profit $P = \$700{,}000$ occurs when $x = 20{,}000$ units are manufactured and sold in the specified time.   ◄

**Table 6.2.5**

| $x$ | 0 | 20,000 | 30,000 |
|------|-----------|---------|---------|
| $P(x)$ | $-500{,}000$ | 700,000 | 400,000 |

**MARGINAL ANALYSIS**

Economists call $P'(x)$, $R'(x)$, and $C'(x)$ the **marginal profit**, **marginal revenue**, and **marginal cost**, respectively; and they interpret these quantities as the *additional* profit, revenue, and cost that result from producing and selling one additional unit of the product when the production and sales levels are at $x$ units. These interpretations follow from the local linear approximations of the profit, revenue, and cost functions. For example, it follows from Formula (7) of Section 3.6 that when the production and sales levels are at $x$ units the local linear approximation of the profit function is

$$P(x + \Delta x) \approx P(x) + P'(x)\Delta x$$

Thus, if $\Delta x = 1$ (one additional unit produced and sold), this formula implies

$$P(x + 1) \approx P(x) + P'(x)$$

and hence the *additional* profit that results from producing and selling one additional unit can be approximated as

$$P(x + 1) - P(x) \approx P'(x)$$

**A BASIC PRINCIPLE OF ECONOMICS**

It follows from (23) that $P'(x) = 0$ at those points where $C'(x) = R'(x)$, and this implies that the maximum profit must occur at a point where the marginal revenue is equal to the marginal cost; that is:

*The maximum profit occurs at a point where the cost of manufacturing and selling an additional unit of a product is exactly equal to the revenue generated by the additional unit.*

This is one of the basic principles of economics.

## EXERCISE SET 6.2

1. Express the number 10 as a sum of two nonnegative numbers whose product is as large as possible.

2. How should two nonnegative numbers be chosen so that their sum is 1 and the sum of their squares is
   (a) as large as possible
   (b) as small as possible?

3. Find a number in the closed interval $\left[\frac{1}{2}, \frac{3}{2}\right]$ such that the sum of the number and its reciprocal is
   (a) as small as possible
   (b) as large as possible.

4. A rectangular field is to be bounded by a fence on three sides and by a straight stream on the fourth side. Find the dimensions of the field with maximum area that can be enclosed with 1000 feet of fence.

5. A rectangular plot of land is to be fenced in using two kinds of fencing. Two opposite sides will use heavy-duty fencing selling for $3 a foot, while the remaining two sides will use standard fencing selling for $2 a foot. What are the dimensions of the rectangular plot of greatest area that can be fenced in at a cost of $6000?

6. A rectangle is to be inscribed in a right triangle having sides of length 6 in, 8 in, and 10 in. Find the dimensions of the rectangle with greatest area assuming the rectangle is positioned as in the accompanying figure.

7. Solve the problem in Exercise 6 assuming the rectangle is positioned as in the accompanying figure.

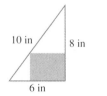 

Figure Ex-6          Figure Ex-7

8. A rectangle has its two lower corners on the $x$-axis and its two upper corners on the curve $y = 16 - x^2$. For all such rectangles, what are the dimensions of the one with largest area?

9. Find the dimensions of the rectangle with maximum area that can be inscribed in a circle of radius 10.

10. Find the dimensions of the rectangle of greatest area that can be inscribed in a semicircle of radius $R$ as shown in the accompanying figure.

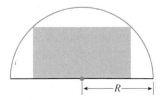

Figure Ex-10

11. A rectangular area of 3200 ft$^2$ is to be fenced off. Two opposite sides will use fencing costing $1 per foot and the remaining sides will use fencing costing $2 per foot. Find the dimensions of the rectangle of least cost.

12. Show that among all rectangles with perimeter $p$, the square has the maximum area.

13. Show that among all rectangles with area $A$, the square has the minimum perimeter.

14. A wire of length 12 in can be bent into a circle, bent into a square, or cut into two pieces to make both a circle and a square. How much wire should be used for the circle if the total area enclosed by the figure(s) is to be
    (a) a maximum           (b) a minimum?

15. Suppose that the number of bacteria in a culture at time $t$ is given by $N = 5000(25 + te^{-t/20})$.
    (a) Find the largest and smallest number of bacteria in the culture during the time interval $0 \le t \le 100$.
    (b) At what time during the time interval in part (a) is the number of bacteria decreasing most rapidly?

16. A church window consisting of a rectangle topped by a semicircle is to have a perimeter $p$. Find the radius of the semicircle if the area of the window is to be maximum.

17. A sheet of cardboard 12 in square is used to make an open box by cutting squares of equal size from the four corners and folding up the sides. What size squares should be cut to obtain a box with largest possible volume?

18. A square sheet of cardboard of side $k$ is used to make an open box by cutting squares of equal size from the four corners and folding up the sides. What size squares should be cut from the corners to obtain a box with largest possible volume?

19. An open box is to be made from a 3-ft by 8-ft rectangular piece of sheet metal by cutting out squares of equal size from the four corners and bending up the sides. Find the maximum volume that the box can have.

20. A closed rectangular container with a square base is to have a volume of 2250 in$^3$. The material for the top and bottom of the container will cost $2 per in$^2$, and the material for the sides will cost $3 per in$^2$. Find the dimensions of the container of least cost.

21. A closed rectangular container with a square base is to have a volume of 2000 cm$^3$. It costs twice as much per square centimeter for the top and bottom as it does for the sides. Find the dimensions of the container of least cost.

22. A container with square base, vertical sides, and open top is to be made from 1000 ft$^2$ of material. Find the dimensions of the container with greatest volume.

23. A rectangular container with two square sides and an open top is to have a volume of $V$ cubic units. Find the dimensions of the container with minimum surface area.

**24.** Find the dimensions of the right circular cylinder of largest volume that can be inscribed in a sphere of radius $R$.

**25.** Find the dimensions of the right circular cylinder of greatest surface area that can be inscribed in a sphere of radius $R$.

**26.** Show that the right circular cylinder of greatest volume that can be inscribed in a right circular cone has volume that is $\frac{4}{9}$ the volume of the cone (Figure Ex-26).

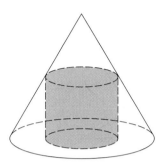

Figure Ex-26

**27.** A closed, cylindrical can is to have a volume of $V$ cubic units. Show that the can of minimum surface area is achieved when the height is equal to the diameter of the base.

**28.** A closed cylindrical can is to have a surface area of $S$ square units. Show that the can of maximum volume is achieved when the height is equal to the diameter of the base.

**29.** A cylindrical can, open at the top, is to hold 500 cm³ of liquid. Find the height and radius that minimize the amount of material needed to manufacture the can.

**30.** A soup can in the shape of a right circular cylinder of radius $r$ and height $h$ is to have a prescribed volume $V$. The top and bottom are cut from squares as shown in the accompanying figure. If the shaded corners are wasted, but there is no other waste, find the ratio $r/h$ for the can requiring the least material (including waste).

**31.** A box-shaped wire frame consists of two identical wire squares whose vertices are connected by four straight wires of equal length (Figure Ex-31). If the frame is to be made from a wire of length $L$, what should the dimensions be to obtain a box of greatest volume?

Figure Ex-30

Figure Ex-31

**32.** Suppose that the sum of the surface areas of a sphere and a cube is a constant.
  (a) Show that the sum of their volumes is smallest when the diameter of the sphere is equal to the length of an edge of the cube.
  (b) When will the sum of their volumes be greatest?

**33.** Find the height and radius of the cone of slant height $L$ whose volume is as large as possible.

**34.** A cone is made from a circular sheet of radius $R$ by cutting out a sector and gluing the cut edges of the remaining piece together (Figure Ex-34). What is the maximum volume attainable for the cone?

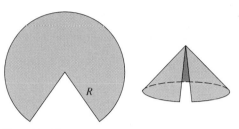

Figure Ex-34

**35.** A cone-shaped paper drinking cup is to hold 10 cm³ of water. Find the height and radius of the cup that will require the least amount of paper.

**36.** Find the dimensions of the isosceles triangle of least area that can be circumscribed about a circle of radius $R$.

**37.** Find the height and radius of the right circular cone with least volume that can be circumscribed about a sphere of radius $R$.

**38.** A trapezoid is inscribed in a semicircle of radius 2 so that one side is along the diameter (Figure Ex-38). Find the maximum possible area for the trapezoid. [*Hint:* Express the area of the trapezoid in terms of $\theta$.]

**39.** A drainage channel is to be made so that its cross section is a trapezoid with equally sloping sides (Figure Ex-39). If the sides and bottom all have a length of 5 ft, how should the angle $\theta$ ($0 \leq \theta \leq \pi/2$) be chosen to yield the greatest cross-sectional area?

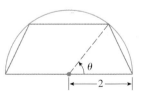

Figure Ex-38

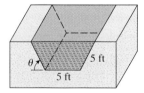

Figure Ex-39

**40.** A lamp is suspended above the center of a round table of radius $r$. How high above the table should the lamp be placed to achieve maximum illumination at the edge of the table? [Assume that the illumination $I$ is directly proportional to the cosine of the angle of incidence $\phi$ of the light rays and inversely proportional to the square of the distance $l$ from the light source (Figure Ex-40).]

**41.** A plank is used to reach over a fence 8 ft high to support a wall that is 1 ft behind the fence (Figure Ex-41). What is the length of the shortest plank that can be used? [*Hint:* Express the length of the plank in terms of the angle $\theta$ shown in the figure.]

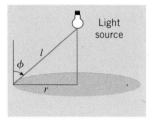

Figure Ex-40                    Figure Ex-41

**42.** A commercial cattle ranch currently allows 20 steers per acre of grazing land; on the average its steers weigh 2000 lb at market. Estimates by the Agriculture Department indicate that the average market weight per steer will be reduced by 50 lb for each additional steer added per acre of grazing land. How many steers per acre should be allowed in order for the ranch to get the largest possible total market weight for its cattle?

**43.** (a) A chemical manufacturer sells sulfuric acid in bulk at a price of $100 per unit. If the daily total production cost in dollars for $x$ units is

$$C(x) = 100{,}000 + 50x + 0.0025x^2$$

and if the daily production capacity is at most 7000 units, how many units of sulfuric acid must be manufactured and sold daily to maximize the profit?

(b) Would it benefit the manufacturer to expand the daily production capacity?

**44.** A firm determines that $x$ units of its product can be sold daily at $p$ dollars per unit, where

$$x = 1000 - p$$

The cost of producing $x$ units per day is

$$C(x) = 3000 + 20x$$

(a) Find the revenue function $R(x)$.

(b) Find the profit function $P(x)$.

(c) Assuming that the production capacity is at most 500 units per day, determine how many units the company must produce and sell each day to maximize the profit.

(d) Find the maximum profit.

(e) What price per unit must be charged to obtain the maximum profit?

**45.** In a certain chemical manufacturing process, the daily weight $y$ of defective chemical output depends on the total weight $x$ of all output according to the empirical formula

$$y = 0.01x + 0.00003x^2$$

where $x$ and $y$ are in pounds. If the profit is $100 per pound

of nondefective chemical produced and the loss is $20 per pound of defective chemical produced, how many pounds of chemical should be produced daily to maximize the total daily profit?

**46.** The cost $c$ (in dollars per hour) to run an ocean liner at a constant speed $v$ (in miles per hour) is given by $c = a + bv^n$, where $a$, $b$, and $n$ are positive constants with $n > 1$. Find the speed needed to make the cheapest 3000-mi run.

**47.** Two particles, $A$ and $B$, are in motion in the $xy$-plane. Their coordinates at each instant of time $t$ $(t \geq 0)$ are given by $x_A = t$, $y_A = 2t$, $x_B = 1 - t$, and $y_B = t$. Find the minimum distance between $A$ and $B$.

**48.** Follow the directions of Exercise 47, with $x_A = t$, $y_A = t^2$, $x_B = 2t$, and $y_B = 2$.

**49.** Prove that $(1, 0)$ is the closest point on the curve $x^2 + y^2 = 1$ to $(2, 0)$.

**50.** Find all points on the curve $y = \sqrt{x}$ for $0 \leq x \leq 3$ that are closest to, and at the greatest distance from, the point $(2, 0)$.

**51.** Find all points on the curve $x^2 - y^2 = 1$ closest to $(0, 2)$.

**52.** Find a point on the curve $x = 2y^2$ closest to $(0, 9)$.

**53.** Find the coordinates of the point $P$ on the curve

$$y = \frac{1}{x^2} \quad (x > 0)$$

where the segment of the tangent line at $P$ that is cut off by the coordinate axes has its shortest length.

**54.** Find the $x$-coordinate of the point $P$ on the parabola

$$y = 1 - x^2 \quad (0 < x \leq 1)$$

where the triangle that is enclosed by the tangent line at $P$ and the coordinate axes has the smallest area.

**55.** Where on the curve $y = \left(1 + x^2\right)^{-1}$ does the tangent line have the greatest slope?

**56.** A man is on the bank of a river that is 1 mile wide. He wants to travel to a town on the opposite bank, but 1 mile upstream. He intends to row on a straight line to some point $P$ on the opposite bank and then walk the remaining distance along the bank (Figure Ex-56). To what point should he row in order to reach his destination in the least time if

(a) he can walk 5 mi/h and row 3 mi/h

(b) he can walk 5 mi/h and row 4 mi/h?

**57.** A pipe of negligible diameter is to be carried horizontally around a corner from a hallway 8 ft wide into a hallway 4 ft wide (Figure Ex-57). What is the maximum length that the pipe can have? [An interesting discussion of this problem in the case where the diameter of the pipe is not neglected is given by Norman Miller in the *American Mathematical Monthly*, Vol. 56, 1949, pp. 177–179.]

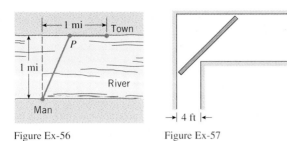

Figure Ex-56            Figure Ex-57

**58.** In an unknown physical quantity $x$ is measured $n$ times, the measurements $x_1, x_2, \ldots, x_n$ often vary because of uncontrollable factors such as temperature, atmospheric pressure, and so forth. Thus, a scientist is often faced with the problem of using $n$ different observed measurements to obtain an estimate $\bar{x}$ of an unknown quantity $x$. One method for making such an estimate is based on the *least squares principle*, which states that the estimate $\bar{x}$ should be chosen to minimize

$$s = (x_1 - \bar{x})^2 + (x_2 - \bar{x})^2 + \cdots + (x_n - \bar{x})^2$$

which is the sum of the squares of the deviations between the estimate $\bar{x}$ and the measured values. Show that the estimate resulting from the least squares principle is

$$\bar{x} = \frac{1}{n}(x_1 + x_2 + \cdots + x_n)$$

that is, $\bar{x}$ is the arithmetic average of the observed values.

**59.** Suppose that the intensity of a point light source is directly proportional to the strength of the source and inversely proportional to the square of the distance from the source. Two point light sources with strengths of $S$ and $8S$ are separated by a distance of 90 cm. Where on the line segment between the two sources is the intensity a minimum?

**60.** Prove: If $f(x) \geq 0$ on an interval $I$ and if $f(x)$ has a maximum value on $I$ at $x_0$, then $\sqrt{f(x)}$ also has a maximum value at $x_0$. Similarly for minimum values. [*Hint:* Use the fact that $\sqrt{x}$ is an increasing function on the interval $[0, +\infty)$.]

**61.** Fermat's* (biography on pp. 352–353) principle in optics states that light traveling from one point to another follows that path for which the total travel time is minimum. In a uniform medium, the paths of "minimum time" and "shortest distance" turn out to be the same, so that light, if unobstructed, travels along a straight line. Assume that we have a light source, a flat mirror, and an observer in a uniform medium. If a light ray leaves the source, bounces off the mirror, and travels on to the observer, then its path will consist of two line segments, as shown in Figure Ex-61. According to Fermat's principle, the path will be such that the total travel time $t$ is minimum or, since the medium is uniform, the path will be such that the total distance traveled from $A$ to $P$ to $B$ is as small as possible. Assuming the minimum

occurs when $dt/dx = 0$, show that the light ray will strike the mirror at the point $P$ where the "angle of incidence" $\theta_1$ equals the "angle of reflection" $\theta_2$.

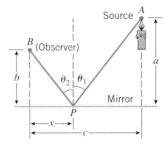

Figure Ex-61

**62.** Fermat's principle (Exercise 61) also explains why light rays traveling between air and water undergo bending (refraction). Imagine that we have two uniform media (such as air and water) and a light ray traveling from a source $A$ in one medium to an observer $B$ in the other medium (Figure Ex-62). It is known that light travels at a constant speed in a uniform medium, but more slowly in a dense medium (such as water) than in a thin medium (such as air). Consequently, the path of shortest time from $A$ to $B$ is not necessarily a straight line, but rather some broken line path $A$ to $P$ to $B$ allowing the light to take greatest advantage of its higher speed through the thin medium. Snell's[†] (biography on p. 353) law of refraction states that the path of the light ray will be such that

$$\frac{\sin \theta_1}{v_1} = \frac{\sin \theta_2}{v_2}$$

where $v_1$ is the speed of light in the first medium, $v_2$ is the speed of light in the second medium, and $\theta_1$ and $\theta_2$ are the angles shown in Figure Ex-62. Show that this follows from the assumption that the path of minimum time occurs when $dt/dx = 0$.

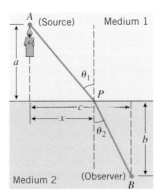

Figure Ex-62

**63.** A farmer wants to walk at a constant rate from her barn to a straight river, fill her pail, and carry it to her house in the least time.

(a) Explain how this problem relates to Fermat's principle and the light-reflection problem in Exercise 61.

(b) Use the result of Exercise 61 to describe geometrically the best path for the farmer to take.

(c) Use part (b) to determine where the farmer should fill her pail if her house and barn are located as in Figure Ex-63.

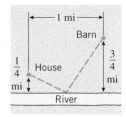

Figure Ex-63

## 6.3 RECTILINEAR MOTION (MOTION ALONG A LINE)

*In Section 1.5 we discussed the motion of a particle moving with constant velocity in one direction along a line, and in Section 3.1 we discussed the motion of a particle moving with variable velocity in one direction along a line. In this section we will investigate the more general situation in which a particle may move back and forth with variable velocity along a line. Some examples are a piston moving up and down in a cylinder, a buoy bobbing up and down in the waves, or an object attached to a vibrating spring.*

**TERMINOLOGY**

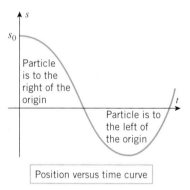

Position versus time curve

Figure 6.3.1

In this section we will assume that a particle representing some object is allowed to move in either direction along a coordinate line. This is called ***rectilinear motion***. The coordinate line might be an $x$-axis, a $y$-axis, or an axis that is inclined at some angle. To avoid being specific, we will denote the coordinate line as the $s$-axis. We will assume that units are chosen for measuring distance and time and that we begin observing the particle at time $t = 0$. As the particle moves along the $s$-axis, its coordinate is some function of the elapsed time $t$, say $s = s(t)$. We call $s(t)$ the ***position function*** of the particle, and we call the graph of $s$ versus $t$ the ***position versus time curve***.

Figure 6.3.1 shows a typical position versus time curve for a particle in rectilinear motion. We can tell from that graph that the coordinate of the particle at time $t = 0$ is $s_0$, and we can tell from the sign of $s$ when the particle is to the left or right of the origin as it moves along the coordinate line.

*PIERRE DE FERMAT (1601–1665). Fermat, the son of a successful French leather merchant, was a lawyer who practiced mathematics as a hobby. He received a Bachelor of Civil Laws degree from the University of Orleans in 1631 and subsequently held various government positions, including a post as councillor to the Toulouse parliament. Although he was apparently financially successful, confidential documents of that time suggest that his performance in office and as a lawyer was poor, perhaps because he devoted so much time to mathematics. Throughout his life, Fermat fought all his efforts to have his mathematical results published. He had the unfortunate habit of scribbling his work in the margins of books and often sent his results to friends without keeping copies for himself. As a result, he never received credit for many major achievements until his name was raised from obscurity in the mid-nineteenth century. It is now known that Fermat, simultaneously and independently of Descartes, developed analytic geometry. Unfortunately, Descartes and Fermat argued bitterly over various problems so that there was never any real cooperation between these two great geniuses.

Fermat solved many fundamental calculus problems. He obtained the first procedure to differentiating polynomials, and solved many important maximization, minimization, area, and tangent problems. His work served to inspire Isaac Newton. Fermat is best known for his work in number theory, the study of properties and relationships between whole numbers. He was the first mathematician to make substantial contributions to this field after the ancient Greek mathematician Diophantus. Unfortunately, none of Fermat's contemporaries appreciated his work in this area, a fact that eventually pushed Fermat into isolation and obscurity in later life. In addition to his work

## Example 1

Figure 6.3.2 shows the position versus time curve for a jackrabbit moving along an *s*-axis. In words, describe how the position of the rabbit changes with time.

*Solution.* The rabbit is at the point $s = -3$ at time $t = 0$. It moves in the positive direction until time $t = 4$, since $s$ is increasing. At time $t = 4$ the rabbit is at the point $s = 3$. At that time it turns around and travels in the negative direction until time $t = 7$, since $s$ is decreasing. At time $t = 7$ the rabbit is at the point $s = -1$, and it remains stationary at that point thereafter, since $s$ is constant for $t > 7$.  ◄

•••••••••••••••••••••••••••••••••••••
**INSTANTANEOUS VELOCITY**

In rectilinear motion, the rate at which the coordinate of a particle changes with time is called the *velocity* of the particle. More precisely, we make the following definition.

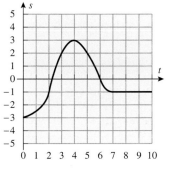

**6.3.1** DEFINITION. If $s(t)$ is the position function of a particle moving on a coordinate line, then the ***instantaneous velocity*** of the particle at time $t$ is defined by

$$v(t) = s'(t) = \frac{ds}{dt} \tag{1}$$

Geometrically, the instantaneous velocity at a given time is the slope of the tangent line to the position versus time curve at that time, and hence the sign of the velocity tells which way the particle is moving—a positive velocity means that $s$ is increasing with time, so the particle is moving in the positive direction; a negative velocity means that $s$ is decreasing with time, so the particle is moving in the negative direction (Figure 6.3.3). For example, in Figure 6.3.2 the rabbit is moving in the positive direction between times $t = 0$ and $t = 4$ and is moving in the negative direction between times $t = 4$ and $t = 7$.

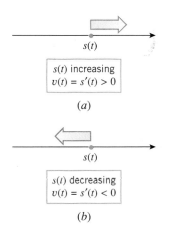

Figure 6.3.2

Recall from our discussion of uniform rectilinear motion in Section 1.5 that there is a distinction between the terms *speed* and *velocity*—speed describes how fast an object is moving without regard to direction, whereas velocity describes how fast it is moving and in what direction. Mathematically, we define the ***instantaneous speed*** of a particle to be the absolute value of its instantaneous velocity; that is,

••••••••••••••••••••••••••••••••••••••••••
**SPEED VERSUS VELOCITY**

$s(t)$

$s(t)$ increasing
$v(t) = s'(t) > 0$

(a)

$s(t)$

$s(t)$ decreasing
$v(t) = s'(t) < 0$

(b)

Figure 6.3.3

$$\left[ \begin{array}{c} \text{instantaneous} \\ \text{speed} \end{array} \right] = |v(t)| = \left| \frac{ds}{dt} \right| \tag{2}$$

in calculus and number theory, Fermat was one of the founders of probability theory and made major contributions to the theory of optics. Outside mathematics, Fermat was a classical scholar of some note, was fluent in French, Italian, Spanish, Latin, and Greek, and he composed a considerable amount of Latin poetry.

One of the great mysteries of mathematics is shrouded in Fermat's work in number theory. In the margin of a book by Diophantus, Fermat scribbled that for integer values of $n$ greater than 2, the equation $x^n + y^n = z^n$ has no nonzero integer solutions for $x$, $y$, and $z$. He stated, "I have discovered a truly marvelous proof of this, which however the margin is not large enough to contain." This result, which became known as "Fermat's last theorem," appeared to be true, but its proof evaded the greatest mathematical geniuses for 300 years until Professor Andrew Wiles of Princeton University presented a proof in June 1993 in a dramatic series of three lectures that drew international media attention (see *New York Times*, June 27, 1993). A prize of 100,000 German marks was offered in 1908 for the solution, but it is worthless today because of inflation.

† WILLEBRORD VAN ROIJEN SNELL (1591–1626). Dutch mathematician. Snell, who succeeded his father to the post of Professor of Mathematics at the University of Leiden in 1613, is most famous for the result of light refraction that bears his name. Although this phenomenon was studied as far back as the ancient Greek astronomer Ptolemy, until Snell's work the relationship was incorrectly thought to be $\theta_1/v_1 = \theta_2/v_2$. Snell's law was published by Descartes in 1638 without giving proper credit to Snell. Snell also discovered a method for determining distances by triangulation that founded the modern technique of mapmaking.

For example, if two particles on the same coordinate line are moving with velocities $v = 5$ m/s and $v = -5$ m/s, respectively, then the particles are moving in opposite directions, but they both have a speed of $|v| = 5$ m/s.

### Example 2

Let $s(t) = t^3 - 6t^2$ be the position function of a particle moving along an $s$-axis, where $s$ is in meters and $t$ is in seconds. Find the instantaneous velocity and speed, and show the graphs of position, velocity, and speed versus time.

*Solution.*   From (1) and (2), the instantaneous velocity and speed are given by

$$v(t) = \frac{ds}{dt} = 3t^2 - 12t \quad \text{and} \quad |v(t)| = |3t^2 - 12t|$$

The graphs of position, velocity, and speed versus time are shown in Figure 6.3.4. Observe that velocity and speed both have units of meters per second (m/s), since $s$ is in meters (m) and time is in seconds (s).   ◄

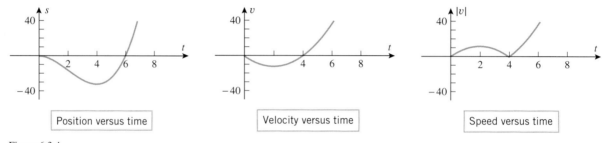

| Position versus time | Velocity versus time | Speed versus time |

Figure 6.3.4

The graphs in Figure 6.3.4 provide a wealth of visual information about the motion of the particle. For example, the position versus time curve tells us that the particle is to the left of the origin for $0 < t < 6$, is to the right of the origin for $t > 6$, and is at the origin at times $t = 0$ and $t = 6$. The velocity versus time curve tells us that the particle is moving in the negative direction if $0 < t < 4$, is moving in the positive direction if $t > 4$, and is momentarily stopped at times $t = 0$ and $t = 4$ (the velocity is zero at those times). The speed versus time curve tells us that the speed of the particle is increasing for $0 < t < 2$, decreasing for $2 < t < 4$, and increasing again for $t > 4$.

**ACCELERATION**

In rectilinear motion, the rate at which the velocity of a particle changes with time is called its *acceleration*. More precisely, we make the following definition.

---

**6.3.2**   DEFINITION.   If $s(t)$ is the position function of a particle moving on a coordinate line, then the instantaneous acceleration of the particle at time $t$ is defined by

$$a(t) = v'(t) = \frac{dv}{dt} \tag{3}$$

or alternatively, since $v(t) = s'(t)$,

$$a(t) = s''(t) = \frac{d^2s}{dt^2} \tag{4}$$

---

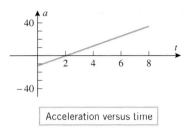

Acceleration versus time

Figure 6.3.5

**INTERPRETING THE SIGN OF ACCELERATION**

### Example 3

Let $s(t) = t^3 - 6t^2$ be the position function of a particle moving along an $s$-axis, where $s$ is in meters and $t$ is in seconds. Find the instantaneous acceleration $a(t)$, and show the graph of acceleration versus time.

*Solution.*  From Example 2, the instantaneous velocity of the particle is $v(t) = 3t^2 - 12t$, so the instantaneous acceleration is

$$a(t) = \frac{dv}{dt} = 6t - 12$$

and the acceleration versus time curve is the line shown in Figure 6.3.5. Note that in this example the acceleration has units of m/s$^2$, since $v$ is in meters per second (m/s) and time is in seconds (s).  ◄

We will say that a particle in rectilinear motion is **speeding up** when its instantaneous speed is increasing and is **slowing down** when its instantaneous speed is decreasing. In everyday language an object that is speeding up is said to be "accelerating" and an object that is slowing down is said to be "decelerating"; thus, one might expect that a particle in rectilinear motion will be speeding up when its instantaneous acceleration is positive and slowing down when it is negative. Although this is true for a particle moving in the positive direction, it is *not* true for a particle moving in the negative direction—a particle with negative velocity is speeding up when its acceleration is negative and slowing down when its acceleration is positive. This is because a positive acceleration implies an increasing velocity, and increasing a negative velocity decreases its absolute value; similarly, a negative acceleration implies a decreasing velocity, and decreasing a negative velocity increases its absolute value. In summary:

> **6.3.3**  INTERPRETING THE SIGN OF ACCELERATION.  *A particle in rectilinear motion is speeding up when its velocity and acceleration have the same sign and slowing down when they have opposite signs.*

FOR THE READER.     For a particle in rectilinear motion, what is happening when $v(t) = 0$? When $a(t) = 0$?

### Example 4

In Examples 2 and 3 we found the velocity versus time curve and the acceleration versus time curve for a particle with position function $s(t) = t^3 - 6t^2$. Use those curves to determine when the particle is speeding up and slowing down, and confirm that your results are consistent with the speed versus time curve obtained in Example 2.

*Solution.*  Over the time interval $0 < t < 2$ the velocity and acceleration are negative, so the particle is speeding up. This is consistent with the speed versus time curve, since the speed is increasing over this time interval. Over the time interval $2 < t < 4$ the velocity is negative and the acceleration is positive, so the particle is slowing down. This is also consistent with the speed versus time curve, since the speed is decreasing over this time interval. Finally, on the time interval $t > 4$ the velocity and acceleration are positive, so the particle is speeding up, which again is consistent with the speed versus time curve.  ◄

**ANALYZING THE POSITION VERSUS TIME CURVE**

The position versus time curve contains all of the significant information about the position and velocity of a particle in rectilinear motion:

- Where $s(t) > 0$, the particle is on the positive side of the $s$-axis.
- Where $s(t) < 0$, the particle is on the negative side of the $s$-axis.

- The slope of the tangent line at a point in time is the instantaneous velocity at that time.
- Where the tangent line has positive slope, the velocity is positive and the particle is moving in the positive direction.
- Where the tangent line has negative slope, the velocity is negative, and the particle is moving in the negative direction.
- Where the tangent line is horizontal, the velocity is zero, and the particle is momentarily stopped.

Information about the acceleration of a particle in rectilinear motion can also be deduced from the position versus time curve by examining its concavity. To see why this is so, observe that the position versus time curve will be concave up on intervals where $s''(t) > 0$, and it will be concave down on intervals where $s''(t) < 0$. But we know from (4) that $s''(t)$ is the instantaneous acceleration, so that on intervals where the position versus time curve is concave up the particle has a positive acceleration, and on intervals where it is concave down the particle has a negative acceleration.

Table 6.3.1 summarizes our observations about the position versus time curve.

**Table 6.3.1**

| POSITION VERSUS TIME CURVE | CHARACTERISTICS OF THE CURVE AT $t = t_0$ | BEHAVIOR OF THE PARTICLE AT TIME $t = t_0$ |
|---|---|---|
| | • $s(t_0) > 0$ <br> • Tangent line has positive slope. <br> • Curve is concave down. | • Particle is on the positive side of the origin. <br> • Particle is moving in the positive direction. <br> • Velocity is decreasing. <br> • Particle is slowing down. |
| | • $s(t_0) > 0$ <br> • Tangent line has negative slope. <br> • Curve is concave down. | • Particle is on the positive side of the origin. <br> • Particle is moving in the negative direction. <br> • Velocity is decreasing. <br> • Particle is speeding up. |
| | • $s(t_0) < 0$ <br> • Tangent line has negative slope. <br> • Curve is concave up. | • Particle is on the negative side of the origin. <br> • Particle is moving in the negative direction. <br> • Velocity is increasing. <br> • Particle is slowing down. |
| | • $s(t_0) > 0$ <br> • Tangent line has zero slope. <br> • Curve is concave down. | • Particle is on the positive side of the origin. <br> • Particle is momentarily stopped. <br> • Velocity is decreasing. |

## Example 5

Use the position versus time curve in Figure 6.3.2 to determine when the jackrabbit in Example 1 is speeding up and slowing down.

*Solution.* From $t = 0$ to $t = 2$, the acceleration and velocity are positive, so the rabbit is speeding up. From $t = 2$ to $t = 4$, the acceleration is negative and the velocity is positive, so the rabbit is slowing down. At $t = 4$, the velocity is zero, so the rabbit has momentarily stopped. From $t = 4$ to $t = 6$, the acceleration is negative and the velocity is negative, so the rabbit is speeding up. From $t = 6$ to $t = 7$, the acceleration is positive and the velocity is negative, so the rabbit is slowing down. Thereafter, the velocity is zero, so the rabbit has stopped. ◀

### Example 6

Suppose that the position function of a particle moving on a coordinate line is given by $s(t) = 2t^3 - 21t^2 + 60t + 3$. Analyze the motion of the particle for $t \geq 0$.

*Solution.* The velocity and acceleration at time $t$ are

$$v(t) = s'(t) = 6t^2 - 42t + 60 = 6(t-2)(t-5)$$
$$a(t) = v'(t) = 12t - 42 = 12\left(t - \tfrac{7}{2}\right)$$

At each instant we can determine the direction of motion from the sign of $v(t)$ and whether the particle is speeding up or slowing down from the signs of $v(t)$ and $a(t)$ together (Figures 6.3.6*a* and 6.3.6*b*). The motion of the particle is described schematically by the curved line in Figure 6.3.6*c*. At time $t = 0$ the particle is at the point $s(0) = 3$ moving right with velocity $v(0) = 60$ ft/s, but slowing down with acceleration $a(0) = -42$ ft/s$^2$. The particle continues moving right until time $t = 2$, when it stops at the point $s(2) = 55$, reverses

Analysis of the particle's direction

Figure 6.3.6*a*

Analysis of the particle's speed

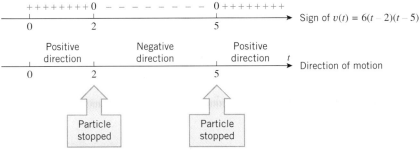

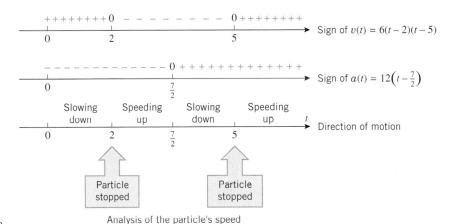

Figure 6.3.6*b*

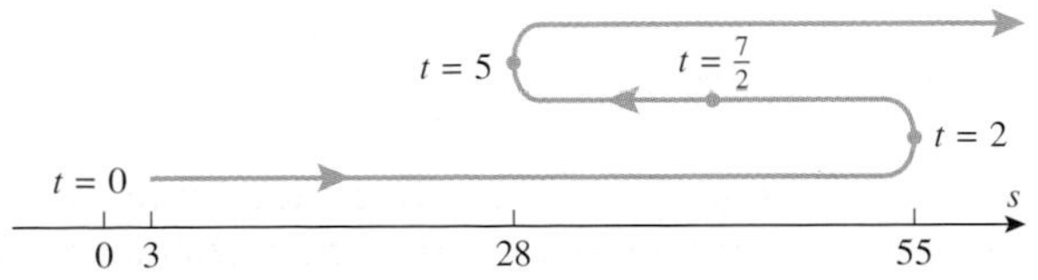

Figure 6.3.6*c*

direction, and begins to speed up with an acceleration of $a(2) = -18 \text{ ft/s}^2$. At time $t = \frac{7}{2}$ the particle begins to slow down, but continues moving left until time $t = 5$, when it stops at the point $s(5) = 28$, reverses direction again, and begins to speed up with acceleration $a(5) = 18 \text{ ft/s}^2$. The particle then continues moving right thereafter with increasing speed.

◀

REMARK. The curved line in Figure 6.3.6c is descriptive only. The actual path of the particle is back and forth on the coordinate line.

**FREE-FALL MOTION**

We will now discuss how some of the ideas in this section can be applied to the study of *free-fall motion*, which is the motion that occurs when an object near the Earth is imparted some initial vertical velocity (up or down), and thereafter moves on a vertical line. In modeling free-fall motion it is assumed that the only force acting on the object is the Earth's gravity and that the object stays sufficiently close to the Earth's surface so that the gravitational force is constant. In particular, air resistance and the gravitational pull of other celestial bodies are neglected.

In our study of free-fall motion, we will ignore the physical size of the object by treating it as a particle, and we will assume that the object moves along an $s$-axis whose origin is at the surface of the Earth and whose positive direction is up. With this convention, the $s$-coordinate of the particle is the height of the particle above the Earth's surface (Figure 6.3.7). The following result will be derived later using calculus and some basic principles of physics.

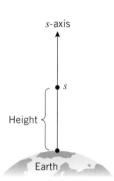

Figure 6.3.7

---

**6.3.4** THE FREE-FALL MODEL. Suppose that at time $t = 0$ an object at a height of $s_0$ above the Earth's surface is imparted an upward or downward velocity of $v_0$ and thereafter moves vertically subject only to the force of the Earth's gravity. If the positive direction of the $s$-axis is up, and if the origin is at the surface of the Earth, then at any time $t$ the height $s = s(t)$ of the object is given by the formula

$$s = s_0 + v_0 t - \tfrac{1}{2} g t^2 \tag{5}$$

where $g$ is a constant, called the ***acceleration due to gravity***. In this text we will use the following approximations for $g$, depending on the units of measurement:

$g = 9.8 \text{ m/s}^2$ [distance in meters and time in seconds]

$g = 32 \text{ ft/s}^2$ [distance in feet and time in seconds]

---

It follows from (5) that the instantaneous velocity and acceleration of an object in free-fall motion are

$$v = \frac{ds}{dt} = v_0 - gt \tag{6}$$

$$a = \frac{dv}{dt} = -g \tag{7}$$

REMARK. Because we have chosen the positive direction of the $s$-axis to be up, a positive velocity implies an upward motion and a negative velocity a downward motion. Thus, it makes sense that instantaneous acceleration $-g$ is negative, since an upward-moving object has positive velocity and negative acceleration, which implies that it is slowing down; and a downward-moving object has negative velocity and negative acceleration, which implies that it is speeding up. (It is a little confusing that the positive constant $g$ is called the *acceleration due to gravity* in 6.3.4, given that the instantaneous acceleration is actually the negative constant $-g$. This mismatch in terminology is caused by the upward orientation of the $s$-axis in Figure 6.3.7; had we chosen the positive direction to be down, then the instantaneous acceleration would have turned out to be $g$. However, our orientation has the advantage of allowing us to interpret $s$ as the height of the object.)

Nolan Ryan's rookie baseball card

### Example 7

Nolan Ryan, one of the fastest baseball pitchers of all time, was capable of throwing a baseball 150 ft/s (over 102 mi/h). Could Nolan Ryan have hit the 208-ft ceiling of the Houston Astrodome if he were capable of releasing a baseball upward at 100 ft/s from a height of 7 ft?

*Solution.* Taking $g = 32$ ft/s$^2$, $v_0 = 100$ ft/s, and $s_0 = 7$ ft in (5) and (6) yields the equations

$$s = 7 + 100t - 16t^2 \quad \text{and} \quad v = 100 - 32t \qquad (8\text{--}9)$$

whose graphs are shown in Figure 6.3.8. It is evident from the graph of $s$ versus $t$ that the maximum height of the baseball is less than 208 ft, so Ryan cannot hit the ceiling. However, let us go a step further and determine exactly how high he could throw the ball. The maximum height $s$ occurs at the stationary point obtained by solving the equation $ds/dt = 0$. However, $ds/dt = v$, which means that the maximum height occurs when $v = 0$, which from (9) can be expressed as

$$100 - 32t = 0 \qquad (10)$$

Solving this equation yields $t = 25/8$. To find the height $s$ at this time we substitute this value of $t$ in (8), from which we obtain

$$s = 7 + 100(25/8) - 16(25/8)^2 = 163.25 \text{ ft}$$

which is roughly 45 ft short of hitting the ceiling.  ◀

Figure 6.3.8

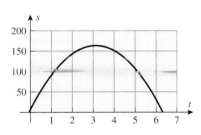
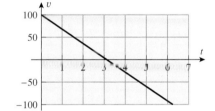

REMARK. Equation (10) can also be deduced by physical reasoning: The ball is moving up when the velocity is positive and moving down when the velocity is negative, so it makes sense that the velocity is zero when the ball reaches its peak.

## EXERCISE SET 6.3  ☒ Graphing Calculator  ⓒ CAS

1. The graphs of three position functions are shown in the accompanying figure. In each case determine the sign of the velocity and acceleration, then determine whether the particle is speeding up or slowing down.

2. The graphs of three velocity functions are shown in the accompanying figure. In each case determine the sign of the acceleration, then determine whether the particle is speeding up or slowing down.

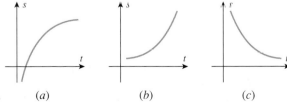

     (a)          (b)          (c)              (a)         (b)         (c)

Figure Ex-1                                Figure Ex-2

**3.** The position function of a particle moving on a horizontal $x$-axis is shown in the accompanying figure.
(a) Is the particle moving left or right at time $t_0$?
(b) Is the acceleration positive or negative at time $t_0$?
(c) Is the particle speeding up or slowing down at time $t_0$?
(d) Is the particle speeding up or slowing down at time $t_1$?

Figure Ex-3

**4.** For the graphs in the accompanying figure, match the position functions with their corresponding velocity functions.

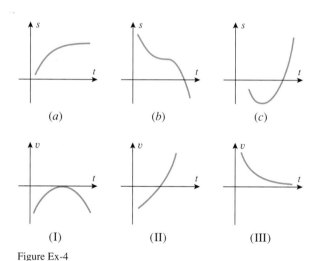

(a)     (b)     (c)

(I)     (II)     (III)

Figure Ex-4

**5.** Sketch a reasonable graph of $s$ versus $t$ for a mouse that is trapped in a narrow corridor (an $s$-axis with the positive direction to the right) and scurries back and forth as follows. It runs right with a constant speed of 1.2 m/s for awhile, then gradually slows down to 0.6 m/s, then quickly speeds up to 2.0 m/s, then gradually slows to a stop but immediately reverses direction and quickly speeds up to 1.2 m/s.

**6.** The accompanying figure shows the graph of $s$ versus $t$ for an ant that moves along a narrow vertical pipe (an $s$-axis with the positive direction up).
(a) When, if ever, is the ant above the origin?
(b) When, if ever, does the ant have velocity zero?
(c) When, if ever, is the ant moving down the pipe?

**7.** The accompanying figure shows the graph of velocity versus time for a particle moving along a coordinate line. Make a rough sketch of the graphs of speed versus time and acceleration versus time.

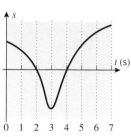

Figure Ex-6          Figure Ex-7

**8.** The accompanying figure shows the position versus time graph for an elevator that ascends 40 m from one stop to the next.
(a) Estimate the velocity when the elevator is halfway up.
(b) Sketch rough graphs of the velocity versus time curve and the acceleration versus time curve.

**9.** The accompanying figure shows the velocity versus time graph for a test run on the Grand Prix GTP. Using this graph, estimate
(a) the acceleration at 60 mi/h (in units of ft/s$^2$)
(b) the time at which the maximum acceleration occurs.
[Data from *Car and Driver Magazine*, October 1990.]

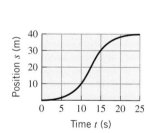     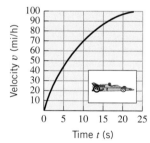

Figure Ex-8          Figure Ex-9

**10.** Let $s(t) = \sin(\pi t/4)$ be the position function of a particle moving along a coordinate line, where $s$ is in meters and $t$ is in seconds.
(a) Make a table showing the position, velocity, and acceleration to two decimal places at times $t = 1, 2, 3, 4$, and 5.
(b) At each of the times in part (a), determine whether the particle is stopped; if it is not, state its direction of motion.
(c) At each of the times in part (a), determine whether the particle is speeding up, slowing down, or neither.

In Exercises 11–14, the position function of a particle moving along a coordinate line is given, where $s$ is in feet and $t$ is in seconds.

(a) Find the velocity and acceleration functions.

(b) Find the position, velocity, speed, and acceleration at time $t = 1$.

(c) At what times is the particle stopped?

(d) When is the particle speeding up? Slowing down?

(e) Find the total distance traveled by the particle from time $t = 0$ to time $t = 5$.

**11.** $s(t) = t^3 - 6t^2, \quad t \geq 0$

**12.** $s(t) = t^4 - 4t + 2, \quad t \geq 0$

**13.** $s(t) = 3\cos(\pi t/2), \quad 0 \leq t \leq 5$

**14.** $s(t) = \dfrac{t}{t^2 + 4}, \quad t \geq 0$

**15.** Let $s(t) = t/(t^2 + 5)$ be the position function of a particle moving along a coordinate line, where $s$ is in meters and $t$ is in seconds. Use a graphing utility to generate the graphs of $s(t)$, $v(t)$, and $a(t)$ for $t \geq 0$, and use those graphs where needed.

(a) Use the appropriate graph to make a rough estimate of the time at which the particle first reverses the direction of its motion; and then find the time exactly.

(b) Find the exact position of the particle when it first reverses the direction of its motion.

(c) Use the appropriate graphs to make a rough estimate of the time intervals on which the particle is speeding up and on which it is slowing down; and then find those time intervals exactly.

**16.** Let $s(t) = t/e^t$ be the position function of a particle moving along a coordinate line, where $s$ is in meters and $t$ is in seconds. Use a graphing utility to generate the graphs of $s(t)$, $v(t)$, and $a(t)$ for $t \geq 0$, and use those graphs where needed.

(a) Use the appropriate graph to make a rough estimate of the time at which the particle first reverses the direction of its motion; and then find the time exactly.

(b) Find the exact position of the particle when it first reverses the direction of its motion.

(c) Use the appropriate graphs to make a rough estimate of the time intervals on which the particle is speeding up and on which it is slowing down; and then find those time intervals exactly.

In Exercises 17–22, the position function of a particle moving along a coordinate line is given. Use the method of Example 6 to analyze the motion of the particle for $t \geq 0$, and give a schematic picture of the motion (as in Figure 6.3.6).

**17.** $s = -3t + 2$

**18.** $s = t^3 - 6t^2 + 9t + 1$

**19.** $s = t^3 - 9t^2 + 24t$

**20.** $s = t + \dfrac{9}{t + 1}$

**21.** $s = \begin{cases} \cos t, & 0 \leq t \leq 2\pi \\ 1, & t > 2\pi \end{cases}$

**22.** $s = \sqrt{t}(4 - 4t + 2t^2)$

**23.** Let $s(t) = 5t^2 - 22t$ be the position function of a particle moving along a coordinate line, where $s$ is in feet and $t$ is in seconds.

(a) Find the maximum speed of the particle during the time interval $1 \leq t \leq 3$.

(b) When, during the time interval $1 \leq t \leq 3$, is the particle farthest from the origin? What is its position at that instant?

**24.** Let $s = 100/(t^2 + 12)$ be the position function of a particle moving along a coordinate line, where $s$ is in feet and $t$ is in seconds. Find the maximum speed of the particle for $t \geq 0$, and find the direction of motion of the particle when it has its maximum speed.

In Exercises 25–29, assume that the free-fall model applies and that the positive direction is up, so that Formulas (5), (6), and (7) can be used. In those problems stating that an object is "dropped" or "released from rest," you should interpret that to mean that the initial velocity of the object is zero. Take $g = 32$ ft/s$^2$ or $g = 9.8$ m/s$^2$, depending on the units.

**25.** A wrench is accidentally dropped at the top of an elevator shaft in a tall building.

(a) How many meters does the wrench fall in 1.5 s?

(b) What is the velocity of the wrench at that time?

(c) How long does it take for the wrench to reach a speed of 12 m/s?

(d) How long does it take for the wrench to fall 100 m?

**26.** In 1939, Joe Sprinz of the San Francisco Seals Baseball Club attempted to catch a ball dropped from a blimp at a height of 800 ft (for the purpose of breaking the record for catching a ball dropped from the greatest height set the preceding year by members of the Cleveland Indians).

(a) How long does it take for a ball to drop 800 ft?

(b) What is the velocity of a ball in miles per hour after an 800-ft drop (88 ft/s = 60 mi/h)?

[*Note:* As a practical matter, it is unrealistic to ignore wind resistance in this problem; however, even with the slowing effect of wind resistance, the impact of the ball slammed Sprinz's glove hand into his face, fractured his upper jaw in 12 places, broke five teeth, and knocked him unconscious. He dropped the ball!]

**27.** A projectile is launched upward from ground level with an initial speed of 60 m/s.

(a) How long does it take for the projectile to reach its highest point?

(b) How high does the projectile go?

(c) How long does it take for the projectile to drop back to the ground from its highest point?

(d) What is the speed of the projectile when it hits the ground?

**28.** (a) Use the results in Exercise 27 to make a conjecture about the relationship between the initial and final speeds of a projectile that is launched upward from ground level and returns to ground level.

(b) Prove your conjecture.

**29.** In Example 7, how fast would Nolan Ryan have to throw a ball upward from a height of 7 feet in order to hit the ceiling of the Astrodome?

**30.** The free-fall formulas (5) and (6) can be combined and rearranged in various useful ways. Derive the following variations of those formulas.

(a) $v^2 = v_0^2 - 2g(s - s_0)$   (b) $s = s_0 + \frac{1}{2}(v_0 + v)t$

(c) $s = s_0 + vt + \frac{1}{2}gt^2$

**31.** A rock, dropped from an unknown height, strikes the ground with a speed of 24 m/s. Use the formula in part (a) of Exercise 30 to find the unknown height.

**32.** A rock thrown downward with an unknown initial velocity from a height of 1000 ft reaches the ground in 5 s. Use the formula in part (c) of Exercise 30 to find the velocity of the rock when it hits the ground.

**33.** (a) A ball is thrown upward from a height $s_0$ with an initial velocity of $v_0$. Use the formula in part (a) of Exercise 30 to show that the maximum height of the ball is
$s_{max} = s_0 + v_0^2/2g$.

(b) Use this result to solve Exercise 29.

**34.** Let $s = t^3 - 6t^2 + 1$.

(a) Find $s$ and $v$ when $a = 0$.

(b) Find $s$ and $a$ when $v = 0$.

**35.** Let $s = \sqrt{2t^2 + 1}$ be the position function of a particle moving along a coordinate line.

(a) Use a graphing utility to generate the graph of $v$ versus $t$, and make a conjecture about the velocity of the particle as $t \to +\infty$.

(b) Check your conjecture by finding $\lim_{t \to +\infty} v$.

**36.** (a) Use the chain rule to show that for a particle in rectilinear motion $a = v(dv/ds)$.

(b) Let $s = \sqrt{3t + 7}, t \geq 0$. Find a formula for $v$ in terms of $s$ and use the equation in part (a) to find the acceleration when $s = 5$.

**37.** Suppose that the position function of two particles, $P_1$ and $P_2$, in motion along the same line are

$s_1 = \frac{1}{2}t^2 - t + 3$   and   $s_2 = -\frac{1}{4}t^2 + t + 1$

respectively, for $t \geq 0$.

(a) Prove that $P_1$ and $P_2$ do not collide.

(b) How close can $P_1$ and $P_2$ get to one another?

(c) During what intervals of time are they moving in opposite directions?

**38.** Let $s_A = 15t^2 + 10t + 20$ and $s_B = 5t^2 + 40t, t \geq 0$, be the position functions of cars $A$ and $B$ that are moving along parallel straight lanes of a highway.

(a) How far is car $A$ ahead of car $B$ when $t = 0$?

(b) At what instants of time are the cars next to one another?

(c) At what instant of time do they have the same velocity? Which car is ahead at this instant?

**39.** The accompanying figure shows the velocity versus distance graph for a 222 Remington Magnum 55 grain pointed soft point bullet.

(a) Use the graph to estimate the value of $dv/ds$ when the velocity is 2000 ft/s.

(b) Use the result in part (a) and the chain rule to approximate the acceleration when the velocity is 2000 ft/s. [*Hint:* See Exercise 36.]

[Data from the *Shooter's Bible*, No. 82, Stoeger Publishing Co., 1991.]

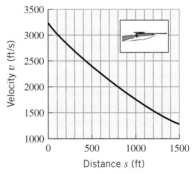

Figure Ex-39

**40.** Prove that a particle is speeding up if the velocity and acceleration have the same sign, and slowing down if they have opposite signs. [*Hint:* Let $r(t) = |v(t)| = \sqrt{v^2(t)}$, and find $r'(t)$.]

## 6.4  NEWTON'S METHOD

*In Section 2.4 we showed how to approximate the roots of an equation $f(x) = 0$ by using the Intermediate-Value Theorem and also by zooming in on the x-intercepts of $y = f(x)$ with a graphing utility. In this section we will study a technique, called Newton's Method, that is usually more efficient than either of those methods. Newton's Method is the technique used by many commercial and scientific computer programs for finding roots.*

**NEWTON'S METHOD**

In beginning algebra one learns that the solution of a first-degree equation $ax + b = 0$ is given by the formula $x = -b/a$, and the solutions of a second-degree equation

$$ax^2 + bx + c = 0$$

are given by the quadratic formula. Formulas also exist for the solutions of all third- and fourth-degree equations, although they are too complicated to be of practical use. In 1826 it was shown by the Norwegian mathematician Niels Henrik Abel[*] that it is impossible to construct a similar formula for the solutions of a *general* fifth-degree equation or higher. Thus, for a *specific* fifth-degree polynomial equation such as

$$x^5 - 9x^4 + 2x^3 - 5x^2 + 17x - 8 = 0$$

it may be difficult or impossible to find exact values for all of the solutions. Similar difficulties occur for trigonometric equations such as

$$x - \cos x = 0$$

as well as equations of other types. For such equations the solutions are generally approximated in some way, often by the method we will now discuss.

Suppose that we are trying to find a root $r$ of the equation $f(x) = 0$, and suppose that by some method we are able to obtain a rough initial estimate of $r$, say by generating the graph of $y = f(x)$ with a graphing utility and examining the x-intercepts. If we let $x_1$ denote our

---

[*] NIELS HENRIK ABEL (1802–1829). Norwegian mathematician. Abel was the son of a poor Lutheran minister and a remarkably beautiful mother from whom he inherited strikingly good looks. In his brief life of 26 years Abel lived in virtual poverty and suffered a succession of adversities; yet he managed to prove major results that altered the mathematical landscape forever. At the age of thirteen he was sent away from home to a school whose better days had long passed. By a stroke of luck the school had just hired a teacher named Bernt Michael Holmboe, who quickly discovered that Abel had extraordinary mathematical ability. Together, they studied the calculus texts of Euler and works of Newton and the later French mathematicians. By the time he graduated, Abel was familiar with most of the great mathematical literature. In 1820 his father died, leaving the family in dire financial straits. Abel was able to enter the University of Christiania in Oslo only because he was granted a free room and several professors supported him directly from their salaries. The University had no advanced courses in mathematics, so Abel took a preliminary degree in 1822 and then continued to study mathematics on his own. In 1824 he published at his own expense the proof that it is impossible to solve the general fifth-degree polynomial equation algebraically. With the hope that this landmark paper would lead to his recognition and acceptance by the European mathematical community, Abel sent the paper to the great German mathematician Gauss, who casually declared it to be a "monstrosity" and tossed it aside. However, in 1826 Abel's paper on the fifth-degree equation and other work was published in the first issue of a new journal, founded by his friend, Leopold Crelle. In the summer of 1826 he completed a landmark work on transcendental functions, which he submitted to the French Academy of Sciences in the hope of establishing himself as a major mathematician, for many young mathematicians had gained quick distinction by having their work accepted by the Academy. However, Abel waited in vain because the paper was either ignored or misplaced by one of the referees, and it did not surface again until two years after his death. That paper was later described by one major mathematician as "...the most important mathematical discovery that has been made in our century...." After submitting his paper, Abel returned to Norway, ill with tuberculosis and in heavy debt. While eking out a meager living as a tutor, he continued to produce great work and his fame spread. Soon great efforts were being made to secure a suitable mathematical position for him. Fearing that his great work had been lost by the Academy, he mailed a proof of the main results to Crelle in January of 1829. In April he suffered a violent hemorrhage and died. Two days later Crelle wrote to inform him that an appointment had been secured for him in Berlin and his days of poverty were over! Abel's great paper was finally published by the Academy twelve years after his death.

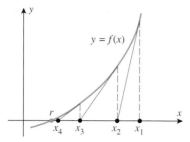

Figure 6.4.1

initial approximation to $r$, then we can generally improve on this approximation by moving along the tangent line to $y = f(x)$ at $x_1$ until we meet the $x$-axis at a point $x_2$ (Figure 6.4.1). Usually, $x_2$ will be closer to $r$ than $x_1$. To improve the approximation further, we can repeat the process by moving along the tangent line to $y = f(x)$ at $x_2$ until we meet the $x$-axis at a point $x_3$. Continuing in this way we can generate a succession of values $x_1, x_2, x_3, x_4, \ldots$ that will usually get closer and closer to $r$. This procedure for approximating $r$ is called *Newton's Method*.

To implement Newton's Method analytically, we must derive a formula that will tell us how to calculate each improved approximation from the preceding approximation. For this purpose, we note that the point-slope form of the tangent line to $y = f(x)$ at the initial approximation $x_1$ is

$$y - f(x_1) = f'(x_1)(x - x_1) \tag{1}$$

If $f'(x_1) \neq 0$, then this line is not parallel to the $x$-axis and consequently it crosses the $x$-axis at some point $(x_2, 0)$. Substituting the coordinates of this point in (1) yields

$$-f(x_1) = f'(x_1)(x_2 - x_1)$$

Solving for $x_2$ we obtain

$$x_2 = x_1 - \frac{f(x_1)}{f'(x_1)} \tag{2}$$

The next approximation can be obtained more easily. If we view $x_2$ as the starting approximation and $x_3$ the new approximation, we can simply apply (2) with $x_2$ in place of $x_1$ and $x_3$ in place of $x_2$. This yields

$$x_3 = x_2 - \frac{f(x_2)}{f'(x_2)} \tag{3}$$

provided $f'(x_2) \neq 0$. In general, if $x_n$ is the $n$th approximation, then it is evident from the pattern in (2) and (3) that the improved approximation $x_{n+1}$ is given by

> **Newton's Method**
>
> $$x_{n+1} = x_n - \frac{f(x_n)}{f'(x_n)}, \quad n = 1, 2, 3, \ldots \tag{4}$$

### Example 1

Use Newton's Method to approximate the real solutions of

$$x^3 - x - 1 = 0$$

*Solution.* Let $f(x) = x^3 - x - 1$, so $f'(x) = 3x^2 - 1$ and (4) becomes

$$x_{n+1} = x_n - \frac{x_n^3 - x_n - 1}{3x_n^2 - 1} \tag{5}$$

From the graph of $f$ in Figure 6.4.2, we see that the given equation has only one real solution. This solution lies between 1 and 2 because $f(1) = -1 < 0$ and $f(2) = 5 > 0$. We will use $x_1 = 1.5$ as our first approximation ($x_1 = 1$ or $x_1 = 2$ would also be reasonable choices).

Letting $n = 1$ in (5) and substituting $x_1 = 1.5$ yields

$$x_2 = 1.5 - \frac{(1.5)^3 - 1.5 - 1}{3(1.5)^2 - 1} = 1.34782609$$

(We used a calculator that displays nine digits.) Next, we let $n = 2$ in (5) and substitute $x_2 = 1.34782609$ to obtain

$$x_3 = 1.34782609 - \frac{(1.34782609)^3 - (1.34782609) - 1}{3(1.34782609)^2 - 1} = 1.32520040$$

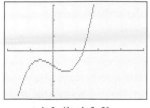

$[-2, 4] \times [-3, 3]$
$x\text{Scl} = 1, y\text{Scl} = 1$

$y = x^3 - x - 1$

Figure 6.4.2

If we continue this process until two identical approximations are generated in succession, we obtain

$$x_1 = 1.5$$
$$x_2 = 1.34782609$$
$$x_3 = 1.32520040$$
$$x_4 = 1.32471817$$
$$x_5 = 1.32471796$$
$$x_6 = 1.32471796$$

At this stage there is no need to continue further because we have reached the accuracy limit of our calculator, and all subsequent approximations that the calculator generates will be the same. Thus, the solution is approximately $x \approx 1.32471796$. ◀

### Example 2

It is evident from Figure 6.4.3 that if $x$ is in radians, then the equation

$$\cos x = x$$

has a solution between 0 and 1. Use Newton's Method to approximate it.

*Solution.*  Rewrite the equation as

$$x - \cos x = 0$$

and apply (4) with $f(x) = x - \cos x$. Since $f'(x) = 1 + \sin x$, (4) becomes

$$x_{n+1} = x_n - \frac{x_n - \cos x_n}{1 + \sin x_n} \qquad (6)$$

From Figure 6.4.3, the solution seems closer to $x = 1$ than $x = 0$, so we will use $x_1 = 1$ (radian) as our initial approximation. Letting $n = 1$ in (6) and substituting $x_1 = 1$ yields

$$x_2 = 1 - \frac{1 - \cos 1}{1 + \sin 1} = .750363868$$

Next, letting $n = 2$ in (6) and substituting this value of $x_2$ yields

$$x_3 = .750363868 - \frac{.750363868 - \cos(.750363868)}{1 + \sin(.750363868)} = .739112891$$

If we continue this process until two identical approximations are generated in succession, we obtain

$$x_1 = 1$$
$$x_2 = .750363868$$
$$x_3 = .739112891$$
$$x_4 = .739085133$$
$$x_5 = .739085133$$

Thus, to the accuracy limit of our calculator, the solution of the equation $\cos x = x$ is $x \approx .739085133$. ◀

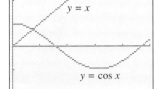

[0, 5] × [−2, 2]
xScl = 1, yScl = 1

Figure 6.4.3

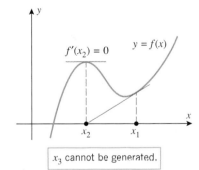

$x_3$ cannot be generated.

Figure 6.4.4

**SOME DIFFICULTIES WITH NEWTON'S METHOD**

When Newton's Method works, the approximations usually converge toward the solution with dramatic speed. However, there are situations in which the method fails. For example, if $f'(x_n) = 0$ for some $n$, then (4) involves a division by zero, making it impossible to generate $x_{n+1}$. However, this is to be expected because the tangent line to $y = f(x)$ is parallel to the $x$-axis where $f'(x_n) = 0$, and hence this tangent line does not cross the $x$-axis to generate the next approximation (Figure 6.4.4).

Newton's Method can fail for other reasons as well; sometimes it may overlook the root you are trying to find and converge to a different root, and sometimes it may fail to converge altogether. For example, consider the equation

$$x^{1/3} = 0$$

which has $x = 0$ as its only solution, and try to approximate this solution by Newton's Method with a starting value of $x_0 = 1$. Letting $f(x) = x^{1/3}$, Formula (4) becomes

$$x_{n+1} = x_n - \frac{(x_n)^{1/3}}{\frac{1}{3}(x_n)^{-2/3}} = x_n - 3x_n = -2x_n$$

Beginning with $x_1 = 1$, the successive values generated by this formula are

$$x_1 = 1, \quad x_2 = -2, \quad x_3 = 4, \quad x_4 = -8, \ldots$$

which obviously do not converge to $x = 0$. Figure 6.4.5 illustrates what is happening geometrically in this situation.

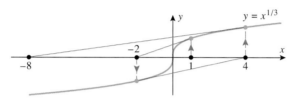

Figure 6.4.5

To learn more about the conditions under which Newton's Method converges and for a discussion of error questions, you should consult a book on numerical analysis. For a more in-depth discussion of Newton's Method and its relationship to contemporary studies of chaos and fractals, you may want to read the article, "Newton's Method and Fractal Patterns," by Phillip Straffin, which appears in *Applications of Calculus*, MAA Notes, Vol. 3, No. 29, 1993, published by the Mathematical Association of America.

**EXERCISE SET 6.4**   ☑ Graphing Calculator   ☐ CAS
· · · · · · · · · · · · · · · · · · · · · · · · · · · · · · · · · · · · · · · · · · · · · · · · · · · · · · · · · · · · · · · · · · · · · · · · · · · · · · · · · · · · · · · · · · · · · · ·

In this exercise set, use a calculator, and keep as many decimal places as it can display.

**1.** Approximate $\sqrt{2}$ by applying Newton's Method to the equation $x^2 - 2 = 0$.

**2.** Approximate $\sqrt{7}$ by applying Newton's Method to the equation $x^2 - 7 = 0$.

**3.** Approximate $\sqrt[3]{6}$ by applying Newton's Method to the equation $x^3 - 6 = 0$.

**4.** To what equation would you apply Newton's Method to approximate the $n$th root of $a$?

In Exercises 5–8, the equation has one real solution. Approximate it by Newton's Method.

**5.** $x^3 - x + 3 = 0$        **6.** $x^3 + x - 1 = 0$

**7.** $x^5 + x^4 - 5 = 0$        **8.** $x^5 - x + 1 = 0$

In Exercises 9–14, use a graphing utility to determine how many solutions the equation has, and then use Newton's Method to approximate the solution that satisfies the stated condition.

☑ **9.** $x^4 + x - 3 = 0; \ x < 0$

☑ **10.** $x^5 - 5x^3 - 2 = 0; \ x > 0$

☑ **11.** $2 \sin x = x; \ x > 0$        ☑ **12.** $\sin x = x^2; \ x > 0$

☑ **13.** $x - \tan x = 0; \ \pi/2 < x < 3\pi/2$

☑ **14.** $1 - e^x \cos x = 0; \ 0 < x < \pi$

In Exercises 15–18, use a graphing utility to determine the number of times the curves intersect; and then apply Newton's Method, where needed, to approximate the $x$-coordinates of all intersections.

**15.** $y = x^3$ and $y = \frac{1}{2}x - 1$

**16.** $y = e^{-x}$ and $y = \ln x$

**17.** $y = x^2$ and $y = \sqrt{2x + 1}$

**18.** $y = \frac{1}{8}x^3 + 1$ and $y = \cos 2x$

**19.** The *mechanic's rule* for approximating square roots states that $\sqrt{a} \approx x_{n+1}$, where

$$x_{n+1} = \frac{1}{2}\left(x_n + \frac{a}{x_n}\right), \quad n = 1, 2, 3, \dots$$

and $x_1$ is any positive approximation to $\sqrt{a}$.

(a) Apply Newton's Method to

$$f(x) = x^2 - a$$

to derive the mechanic's rule.

(b) Use the mechanic's rule to approximate $\sqrt{10}$.

**20.** Many calculators compute reciprocals using the approximation $1/a \approx x_{n+1}$, where

$$x_{n+1} = x_n(2 - ax_n), \quad n = 1, 2, 3, \dots$$

and $x_1$ is an initial approximation to $1/a$. This formula makes it possible to perform divisions using multiplications and subtractions, which is a faster procedure than dividing directly.

(a) Apply Newton's Method to

$$f(x) = (1/x) - a$$

to derive this approximation.

(b) Use the formula to approximate $\frac{1}{17}$.

**21.** Use Newton's Method to find the absolute minimum of

$$f(x) = \frac{1}{4}x^4 + x^2 + 5x$$

**22.** Use Newton's Method to find the absolute maximum of $f(x) = x \sin x$ on the interval $[0, \pi]$.

**23.** Use Newton's Method to find the coordinates of the point on the parabola $y = x^2$ that is closest to the point $(1, 0)$.

**24.** Use Newton's Method to find the dimensions of the rectangle of largest area that can be inscribed under the curve $y = \cos x$ for $0 \le x \le \pi/2$, as shown in the accompanying figure.

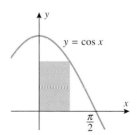

Figure Ex-24

**25.** (a) Show that on a circle of radius $r$, the central angle $\theta$ that subtends an arc whose length is 1.5 times the length $L$

of its chord satisfies the equation $\theta = 3\sin(\theta/2)$ (see the accompanying figure).

(b) Use Newton's Method to approximate $\theta$.

**26.** A *segment* of a circle is the region enclosed by an arc and its chord (see the accompanying figure). If $r$ is the radius of the circle and $\theta$ the angle subtended at the center of the circle, then it can be shown that the area $A$ of the segment is $A = \frac{1}{2}r^2(\theta - \sin\theta)$, where $\theta$ is in radians. Find the value of $\theta$ for which the area of the segment is one-fourth the area of the circle. Give $\theta$ to the nearest degree.

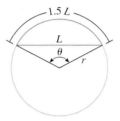

Figure Ex-25          Figure Ex-26

In Exercises 27 and 28, use Newton's Method to approximate all real values of $y$ satisfying the given equation for the indicated value of $x$.

**27.** $xy^4 + x^3y = 1; \ x = 1$

**28.** $xy - \cos\left(\frac{1}{2}xy\right) = 0; \ x = 2$

**29.** An *annuity* is a sequence of equal payments that are paid or received at regular time intervals. For example, you may want to deposit equal amounts at the end of each year into an interest-bearing account for the purpose of accumulating a lump sum at some future time. If, at the end of each year, interest of $i \times 100\%$ on the account balance for that year is added to the account, then the account is said to pay $i \times 100\%$ interest, *compounded annually*. It can be shown that if payments of $Q$ dollars are deposited at the end of each year into an account that pays $i \times 100\%$ compounded annually, then at the time when the $n$th payment and the accrued interest for the past year are deposited, the amount $S(n)$ in the account is given by the formula

$$S(n) = \frac{Q}{i}[(1 + i)^n - 1]$$

Suppose that you can invest \$5000 in an interest-bearing account at the end of each year, and your objective is to have \$250,000 on the 25th payment. What annual compound interest rate must the account pay for you to achieve your goal? [*Hint:* Show that the interest rate $i$ satisfies the equation $50i = (1 + i)^{25} - 1$, and solve it using Newton's Method.]

**30.** (a) Use a graphing utility to generate the graph of

$$f(x) = \frac{x}{x^2 + 1}$$

and use it to explain what happens if you apply Newton's Method with a starting value of $x_1 = 2$. Check your conclusion by computing $x_2, x_3, x_4,$ and $x_5$.

(b) Use the graph generated in part (a) to explain what happens if you apply Newton's Method with a starting value of $x_1 = 0.5$. Check your conclusion by computing $x_2$, $x_3$, $x_4$, and $x_5$.

**31.** (a) Apply Newton's Method to the function $f(x) = x^2 + 1$ with a starting value of $x_1 = 0.5$, and determine if the values of $x_2, \ldots, x_{10}$ appear to converge.

(b) Explain what is happening.

## 6.5 ROLLE'S THEOREM; MEAN-VALUE THEOREM

*In this section we will discuss a result called the Mean-Value Theorem. This theorem has so many important consequences that it is regarded as one of the major principles in calculus.*

**ROLLE'S THEOREM**

We will begin with a special case of the Mean-Value Theorem, called Rolle's Theorem, in honor of the mathematician Michel Rolle.[*] This theorem states the geometrically obvious fact that if the graph of a differentiable function crosses the $x$-axis at two points, $a$ and $b$, then somewhere between those points there must be at least one place where the tangent line is horizontal (Figure 6.5.1). The precise statement of the theorem is as follows:

---

**6.5.1** THEOREM (*Rolle's Theorem*).  *Let $f$ be differentiable on $(a, b)$ and continuous on $[a, b]$. If $f(a) = f(b) = 0$, then there is at least one point $c$ in $(a, b)$ where $f'(c) = 0$.*

---

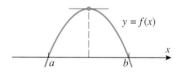

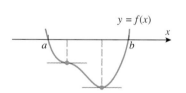

Figure 6.5.1

**Proof.**  Either $f(x)$ is equal to zero for all $x$ in $[a, b]$ or it is not. If it is, then $f'(x) = 0$ for all $x$ in $(a, b)$, since $f$ is constant on $(a, b)$. Thus, for any $c$ in $(a, b)$

$$f'(c) = 0$$

If $f(x)$ is not equal to zero for all $x$ in $[a, b]$, then there must be a point $x$ in $(a, b)$ where $f(x) > 0$ or $f(x) < 0$. We will consider the first case and leave the second as an exercise.

Since $f$ is continuous on $[a, b]$, it follows from the Extreme-Value Theorem (6.1.3) that $f$ has a maximum value at some point $c$ in $[a, b]$. Since $f(a) = f(b) = 0$ and $f(x) > 0$ at some point in $(a, b)$, the point $c$ cannot be an endpoint; it must lie in $(a, b)$. By hypothesis, $f$ is differentiable everywhere on $(a, b)$. In particular, it is differentiable at $c$ so that $f'(c) = 0$ by Theorem 6.1.4. ∎

---

[*] MICHEL ROLLE (1652-1719), French mathematician. Rolle, the son of a shopkeeper, received only an elementary education. He married early and as a young man struggled hard to support his family on the meager wages of a transcriber for notaries and attorneys. In spite of his financial problems and minimal education, Rolle studied algebra and Diophantine analysis (a branch of number theory) on his own. Rolle's fortune changed dramatically in 1682 when he published an elegant solution of a difficult, unsolved problem in Diophantine analysis. The public recognition of his achievement led to a patronage under minister Louvois, a job as an elementary mathematics teacher, and eventually to a short-term administrative post in the Ministry of War. In 1685 he joined the Académie des Sciences in a low-level position for which he received no regular salary until 1699. He stayed there until he died of apoplexy in 1719.

While Rolle's forté was always Diophantine analysis, his most important work was a book on the algebra of equations, called *Traité d'algèbre*, published in 1690. In that book Rolle firmly established the notation $\sqrt[n]{a}$ [earlier written as $\sqrt{\textcircled{n}\ a}$] for the $n$th root of $a$, and proved a polynomial version of the theorem that today bears his name. (Rolle's Theorem was named by Giusto Bellavitis in 1846.) Ironically, Rolle was one of the most vocal early antagonists of calculus. He strove intently to demonstrate that it gave erroneous results and was based on unsound reasoning. He quarreled so vigorously on the subject that the Académie des Sciences was forced to intervene on several occasions. Among his several achievements, Rolle helped advance the currently accepted size order for negative numbers. Descartes, for example, viewed $-2$ as smaller than $-5$. Rolle preceded most of his contemporaries by adopting the current convention in 1691.

### Example 1

The function $f(x) = \sin x$ has roots at $x = 0$ and $x = 2\pi$. Moreover, $f$ is continuous and differentiable everywhere, so it is differentiable on $(0, 2\pi)$ and continuous on $[0, 2\pi]$. Thus, Rolle's Theorem guarantees that there is at least one point $c$ in the interval $(0, 2\pi)$ where the tangent line to the graph of $y = \sin x$ is horizontal. Since $dy/dx = \cos x$, we can find $c$ by solving the equation $\cos c = 0$ on the interval $(0, 2\pi)$. This yields two values for $c$, namely $c_1 = \pi/2$ and $c_2 = 3\pi/2$ (Figure 6.5.2). ◀

REMARK.  In the preceding example, we were able to find the values of $c$ because the equation $f'(c) = 0$ was easy to solve. However, if this equation cannot be solved, then you will not be able to find values of $c$, even though you know they exist. This will rarely cause problems because usually one is more interested in knowing that the values of $c$ exist than in finding them.

The hypotheses in Rolle's Theorem are critical—if $f$ fails to be differentiable at even one point in the interval, then the theorem may fail. For example, the function $f(x) = |x| - 1$ has roots at $x = \pm 1$, yet there is no horizontal tangent line to the graph of $f$ over the interval $(-1, 1)$ (Figure 6.5.3).

Figure 6.5.2

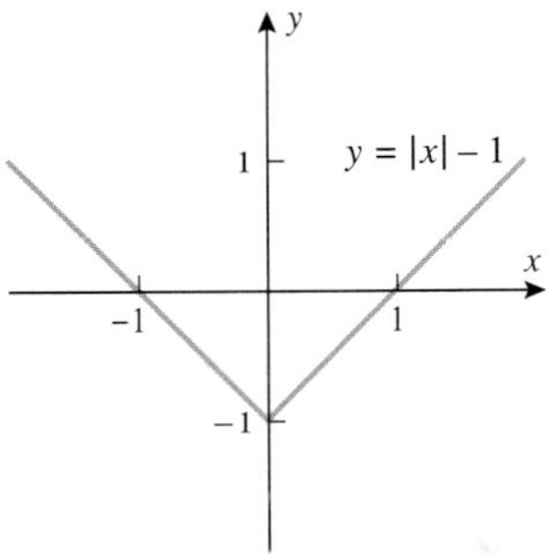

Figure 6.5.3

**THE MEAN-VALUE THEOREM**

Rolle's Theorem is a special case of the ***Mean-Value Theorem***, which states that between any two points $A$ and $B$ on the graph of a differentiable function, there must be at least one place where the tangent line to the curve is parallel to the secant line joining $A$ and $B$ (Figure 6.5.4).

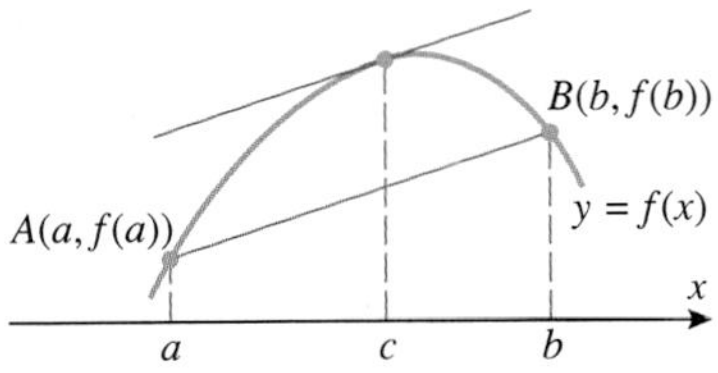

Noting that the slope of the secant line joining $A(a, f(a))$ and $B(b, f(b))$ is

$$\frac{f(b) - f(a)}{b - a}$$

and the slope of the tangent at $c$ is $f'(c)$, the Mean-Value Theorem can be stated precisely as follows.

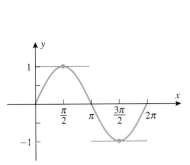

Figure 6.5.4

---

**6.5.2**  THEOREM (*Mean-Value Theorem*).  *Let $f$ be differentiable on $(a, b)$ and continuous on $[a, b]$. Then there is at least one point $c$ in $(a, b)$ where*

$$f'(c) = \frac{f(b) - f(a)}{b - a} \tag{1}$$

---

**VELOCITY INTERPRETATION OF THE MEAN-VALUE THEOREM**

There is a nice interpretation of the Mean-Value Theorem in the situation where $x = f(t)$ is the position versus time curve for a car moving along a straight road. In this case, the right side of (1) is the average velocity of the car over the time interval from $a \le t \le b$, and the left side is the instantaneous velocity at time $t = c$. Thus, the Mean-Value Theorem implies that at least once during the time interval the instantaneous velocity must equal the average velocity. This agrees with our real-world experience—if the average velocity for a trip is 40 mi/h, then sometime during the trip the speedometer has to read 40 mi/h.

### Example 2

You are driving on a straight highway on which the speed limit is 55 mi/h. At 8:05 A.M. a police car clocks your velocity at 50 mi/h and at 8:10 A.M. a second police car posted 5 mi down the road clocks your velocity at 55 mi/h. Explain why the police have a right to charge you with a speeding violation.

*Solution.* You traveled 5 mi in 5 min $\left(= \frac{1}{12} \text{ h}\right)$, so your average velocity was 60 mi/h. However, the Mean-Value Theorem guarantees the police that your instantaneous velocity was 60 mi/h at least once over the 5-mi section of highway. ◀

**PROOF OF THE MEAN-VALUE THEOREM**

*Motivation for the Proof of Theorem 6.5.2.* Figure 6.5.4 suggests that (1) will hold (i.e., the tangent line will be parallel to the secant line) at a point $c$ where the vertical distance between the curve and the secant line is maximum. Thus, to prove the Mean-Value Theorem it is natural to begin by looking for a formula for the vertical distance $v(x)$ between the curve $y = f(x)$ and the secant line joining $(a, f(a))$ and $(b, f(b))$.

*Proof of Theorem 6.5.2.* Since the two-point form of the equation of the secant line joining $(a, f(a))$ and $(b, f(b))$ is

$$y - f(a) = \frac{f(b) - f(a)}{b - a}(x - a)$$

or equivalently,

$$y = \frac{f(b) - f(a)}{b - a}(x - a) + f(a)$$

the difference $v(x)$ between the height of the graph of $f$ and the height of the secant line is

$$v(x) = f(x) - \left[\frac{f(b) - f(a)}{b - a}(x - a) + f(a)\right] \tag{2}$$

Since $f(x)$ is continuous on $[a, b]$ and differentiable on $(a, b)$, so is $v(x)$. Moreover,

$$v(a) = 0 \quad \text{and} \quad v(b) = 0$$

so that $v(x)$ satisfies the hypotheses of Rolle's Theorem on the interval $[a, b]$. Thus, there is a point $c$ in $(a, b)$ such that $v'(c) = 0$. But from Equation (2)

$$v'(x) = f'(x) - \frac{f(b) - f(a)}{b - a}$$

so

$$v'(c) = f'(c) - \frac{f(b) - f(a)}{b - a}$$

Thus, at the point $c$ in $(a, b)$, where $v'(c) = 0$, we have

$$f'(c) = \frac{f(b) - f(a)}{b - a}$$ ∎

### Example 3

(a)   Generate the graph of $f(x) = (x^3/4) + 1$ over the interval $[0, 2]$, and use it to determine the number of tangent lines to the graph of $f$ over the interval $(0, 2)$ that are parallel to the secant line joining the endpoints of the graph.

(b)   Show that $f$ satisfies the hypotheses of the Mean-Value Theorem on the interval $[0, 2]$, and find all values of $c$ in the interval $(0, 2)$ whose existence is guaranteed by the Mean-Value Theorem. Confirm that these values of $c$ are consistent with your graph in part (a).

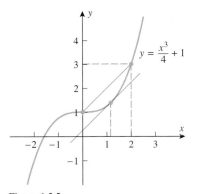

Figure 6.5.5

*Solution (a).* The graph of $f$ in Figure 6.5.5 suggests that there is only one tangent line over the interval $(0, 2)$ that is parallel to the secant line joining the endpoints.

*Solution (b).* The function $f$ is continuous and differentiable everywhere because it is a polynomial. In particular, $f$ is continuous on $[0, 2]$ and differentiable on $(0, 2)$, so the hypotheses of the Mean-Value Theorem are satisfied with $a = 0$ and $b = 2$. But

$$f(a) = f(0) = 1, \quad f(b) = f(2) = 3$$

$$f'(x) = \frac{3x^2}{4}, \qquad f'(c) = \frac{3c^2}{4}$$

so in this case Formula (1) becomes

$$\frac{3c^2}{4} = \frac{3 - 1}{2 - 0} \quad \text{or} \quad 3c^2 = 4$$

which has the two solutions $c = \pm 2/\sqrt{3} \approx \pm 1.15$. However, only the positive solution lies in the interval $[0, 2]$; this value of $c$ is consistent with Figure 6.5.5. ◀

**CONSEQUENCES OF THE MEAN-VALUE THEOREM**

We stated at the beginning of this section that the Mean-Value Theorem is the starting point for many important results in calculus. As an example of this, we will use it to prove Theorem 5.1.2, which was one of our fundamental tools for analyzing graphs of functions.

---

**5.1.2** THEOREM (*Revisited*). *Let $f$ be a function that is continuous on a closed interval $[a, b]$ and differentiable on the open interval $(a, b)$.*

*(a) If $f'(x) > 0$ for every value of $x$ in $(a, b)$, then $f$ is increasing on $[a, b]$.*

*(b) If $f'(x) < 0$ for every value of $x$ in $(a, b)$, then $f$ is decreasing on $[a, b]$.*

*(c) If $f'(x) = 0$ for every value of $x$ in $(a, b)$, then $f$ is constant on $[a, b]$.*

---

*Proof (a).* Suppose that $x_1$ and $x_2$ are points in $[a, b]$ such that $x_1 < x_2$. We must show that $f(x_1) < f(x_2)$. Because the hypotheses of the Mean-Value Theorem are satisfied on the entire interval $[a, b]$, they are satisfied on the subinterval $[x_1, x_2]$. Thus, there is some point $c$ in the open interval $(x_1, x_2)$ such that

$$f'(c) = \frac{f(x_2) - f(x_1)}{x_2 - x_1}$$

or equivalently,

$$f(x_2) - f(x_1) = f'(c)(x_2 - x_1) \qquad (3)$$

Since $c$ is in the open interval $(x_1, x_2)$, it follows that $a < c < b$; thus, $f'(c) > 0$. However, $x_2 - x_1 > 0$ since we assumed that $x_1 < x_2$. It follows from (3) that $f(x_2) - f(x_1) > 0$ or, equivalently, $f(x_1) < f(x_2)$, which is what we were to prove. The proofs of parts (*b*) and (*c*) are similar and are left as exercises. ∎

**THE CONSTANT DIFFERENCE THEOREM**

We know from our earliest study of derivatives that the derivative of a constant is zero. Part (*c*) of Theorem 5.1.2 is the converse of that result; that is, a function whose derivative is zero on an interval must be constant on that interval. If we apply this to the difference of two functions, we obtain the following useful theorem.

---

**6.5.3** THEOREM (*The Constant Difference Theorem*). *If $f$ and $g$ are continuous on a closed interval $[a, b]$, and if $f'(x) = g'(x)$ for all $x$ in the open interval $(a, b)$, then $f$ and $g$ differ by a constant on $[a, b]$; that is, there is a constant $k$ such that $f(x) - g(x) = k$ for all $x$ in $[a, b]$.*

---

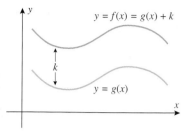

$y = f(x) = g(x) + k$

$k$

$y = g(x)$

If $f'(x) = g'(x)$ on an interval, then the graphs of $f$ and $g$ are vertical translations of one another.

Figure 6.5.6

**Proof.** Let $h(x) = f(x) - g(x)$. Then for every $x$ in $(a, b)$

$$h'(x) = f'(x) - g'(x) = 0$$

Thus, $h(x) = f(x) - g(x)$ is constant on $[a, b]$ by Theorem 5.1.2(c). ■

REMARK. This theorem remains true if the closed interval $[a, b]$ is replaced by a finite or infinite interval $(a, b)$, $[a, b)$, or $(a, b]$, provided $f$ and $g$ are differentiable on $(a, b)$ and continuous on the entire interval.

The Constant Difference Theorem has a simple geometric interpretation—it tells us that if $f$ and $g$ have the same derivative on an interval, then there is a constant $k$ such that $f(x) = g(x) + k$ for each $x$ in the interval; that is, the graphs of $f$ and $g$ can be obtained from one another by a vertical translation (Figure 6.5.6).

## EXERCISE SET 6.5   ⌢ Graphing Calculator   [C] CAS

In Exercises 1 and 2, use the graph of $f$ to find an interval $[a, b]$ on which Rolle's Theorem applies, and find all values of $c$ in that interval that satisfy the conclusion of the theorem.

**1.**
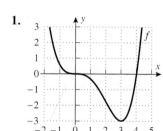

**2.**
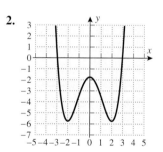

In Exercises 3–8, verify that the hypotheses of Rolle's Theorem are satisfied on the given interval, and find all values of $c$ in that interval that satisfy the conclusion of the theorem.

**3.** $f(x) = x^2 - 6x + 8$; $[2, 4]$

**4.** $f(x) = x^3 - 3x^2 + 2x$; $[0, 2]$

**5.** $f(x) = \cos x$; $[\pi/2, 3\pi/2]$

**6.** $f(x) = \dfrac{x^2 - 1}{x - 2}$; $[-1, 1]$

**7.** $f(x) = \frac{1}{2}x - \sqrt{x}$; $[0, 4]$

**8.** $f(x) = \dfrac{1}{x^2} - \dfrac{4}{3x} + \dfrac{1}{3}$; $[1, 3]$

**9.** Use the graph of $f$ in the accompanying figure to estimate all values of $c$ that satisfy the conclusion of the Mean-Value Theorem on the interval $[0, 8]$.

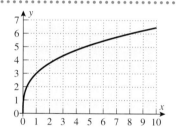

Figure Ex-9

**10.** Use the graph of $f$ in Exercise 9 to estimate all values of $c$ that satisfy the conclusion of the Mean-Value Theorem on the interval $[0, 4]$.

In Exercises 11–16, verify that the hypotheses of the Mean-Value Theorem are satisfied on the given interval, and find all values of $c$ in that interval that satisfy the conclusion of the theorem.

**11.** $f(x) = x^2 + x$; $[-4, 6]$

**12.** $f(x) = x^3 + x - 4$; $[-1, 2]$

**13.** $f(x) = \sqrt{x + 1}$; $[0, 3]$     **14.** $f(x) = x + \dfrac{1}{x}$; $[3, 4]$

**15.** $f(x) = \sqrt{25 - x^2}$; $[-5, 3]$

**16.** $f(x) = \dfrac{1}{x - 1}$; $[2, 5]$

⌢ **17.** (a) Find an interval $[a, b]$ on which

$$f(x) = x^4 + x^3 - x^2 + x - 2$$

satisfies the hypotheses of Rolle's Theorem.

(b) Generate the graph of $f'(x)$, and use it to make rough estimates of all values of $c$ in the interval obtained in part (a) that satisfy the conclusion of Rolle's Theorem.

(c) Use Newton's Method to improve on the rough estimates obtained in part (b).

**18.** Let $f(x) = x^3 + 4x$.

(a) Find the equation of the secant line through the points $(-2, f(-2))$ and $(1, f(1))$.

(b) Show that there is only one number $c$ in the interval $(-2, 1)$ that satisfies the conclusion of the Mean-Value Theorem for the secant line in part (a).

(c) Find the equation of the tangent line to the graph of $f$ at the point $(c, f(c))$.

(d) Use a graphing utility to generate the secant line in part (a) and the tangent line in part (c) in the same coordinate system, and confirm visually that the two lines seem parallel.

**19.** Let $f(x) = \tan x$.

(a) Show that there is no point $c$ in the interval $(0, \pi)$ such that $f'(c) = 0$, even though $f(0) = f(\pi) = 0$.

(b) Explain why the result in part (a) does not violate Rolle's Theorem.

**20.** Let $f(x) = x^{2/3}$, $a = -1$, and $b = 8$.

(a) Show that there is no point $c$ in $(a, b)$ such that

$$f'(c) = \frac{f(b) - f(a)}{b - a}$$

(b) Explain why the result in part (a) does not violate the Mean-Value Theorem.

**21.** (a) Show that if $f$ is differentiable on $(-\infty, +\infty)$, and if $y = f(x)$ and $y = f'(x)$ are graphed in the same coordinate system, then between any two $x$-intercepts of $f$ there is at least one $x$-intercept of $f'$.

(b) Give some examples that illustrate this.

**22.** Review Definitions 3.1.3 and 3.1.4 of average and instantaneous rate of change of $y$ with respect to $x$, and use the Mean-Value Theorem to show that if $f$ is differentiable on $(-\infty, +\infty)$, then in any interval $[x_0, x_1]$ there is at least one point where the instantaneous rate of change of $y$ with respect to $x$ is equal to the average rate of change over the interval.

In Exercises 23–25, use the result of Exercise 22.

**23.** An automobile travels 4 mi along a straight road in 5 min. Show that the speedometer read exactly 48 mi/h at least once during the trip.

**24.** At 11 A.M. on a certain morning the outside temperature was 76°F. At 11 P.M. that evening it had dropped to 52°F.

(a) Show that at some instant during this period the temperature was decreasing at the rate of 2°F/h.

(b) Suppose that you know that the temperature reached a high of 88°F sometime between 11 A.M. and 11 P.M.

Show that at some instant during this period the temperature was decreasing at a rate greater than 3°F/h.

**25.** Suppose that two runners in a 100-m dash finish in a tie. Show that they had the same velocity at least once during the race.

**26.** Use the fact that

$$\frac{d}{dx}(x^6 - 2x^2 + x) = 6x^5 - 4x + 1$$

to show that the equation $6x^5 - 4x + 1 = 0$ has at least one solution in the interval $(0, 1)$.

**27.** (a) Use the Constant Difference Theorem (6.5.3) to show that if $f'(x) = g'(x)$ for all $x$ in the interval $(-\infty, +\infty)$, and if $f$ and $g$ have the same value at any point $x_0$, then $f(x) = g(x)$ for all $x$ in $(-\infty, +\infty)$.

(b) Use the result in part (a) to prove the trigonometric identity $\sin^2 x + \cos^2 x = 1$.

**28.** (a) Use the Constant Difference Theorem (6.5.3) to show that if $f'(x) = g'(x)$ for all $x$ in $(-\infty, +\infty)$, and if $f(x_0) - g(x_0) = c$ at some point $x_0$, then

$$f(x) - g(x) = c$$

for all $x$ in $(-\infty, +\infty)$.

(b) Use the result in part (a) to show that the function

$$h(x) = (x - 1)^3 - (x^2 + 3)(x - 3)$$

is constant for all $x$ in $(-\infty, +\infty)$, and find the constant.

(c) Check the result in part (b) by multiplying out and simplifying the formula for $h(x)$.

**29.** (a) Use the Mean-Value Theorem to show that if $f$ is differentiable on an interval $I$, and if $|f'(x)| \leq M$ for all values of $x$ in $I$, then

$$|f(x) - f(y)| \leq M|x - y|$$

for all values of $x$ and $y$ in $I$.

(b) Use the result in part (a) to show that

$$|\sin x - \sin y| \leq |x - y|$$

for all real values of $x$ and $y$.

**30.** (a) Use the Mean-Value Theorem to show that if $f$ is differentiable on an open interval $I$, and if $|f'(x)| \geq M$ for all values of $x$ in $I$, then

$$|f(x) - f(y)| \geq M|x - y|$$

for all values of $x$ and $y$ in $I$.

(b) Use the result in part (a) to show that

$$|\tan x - \tan y| \geq |x - y|$$

for all values of $x$ and $y$ in the interval $(-\pi/2, \pi/2)$.

(c) Use the result in part (b) to show that

$$|\tan x + \tan y| \geq |x + y|$$

for all values of $x$ and $y$ in the interval $(-\pi/2, \pi/2)$.

**31.** (a) Use the Mean-Value Theorem to show that
$$\sqrt{y} - \sqrt{x} < \frac{y - x}{2\sqrt{x}}$$
if $0 < x < y$.

(b) Use the result in part (a) to show that if $x$ and $y$ are positive, then $\sqrt{xy} < \frac{1}{2}(x + y)$.

**32.** Show that if $f$ is differentiable on an open interval $I$ and $f'(x) \neq 0$ on $I$, the equation $f(x) = 0$ can have at most one real root in $I$.

**33.** Use the result in Exercise 32 to show the following:

(a) The equation $x^3 + 4x - 1 = 0$ has exactly one real root.

(b) If $b^2 - 3ac < 0$ and if $a \neq 0$, then the equation
$$ax^3 + bx^2 + cx + d = 0$$
has exactly one real root (possibly repeated).

**34.** Use the Mean-Value Theorem to prove that
$$1.71 < \sqrt{3} < 1.75$$
[*Hint:* Let $f(x) = \sqrt{x}$, $a = 3$, and $b = 4$ in the Mean-Value Theorem.]

**35.** (a) Show that if $f$ and $g$ are functions for which
$$f'(x) = g(x) \quad \text{and} \quad g'(x) = -f(x)$$
for all $x$, then $f^2(x) + g^2(x)$ is a constant.

(b) Give an example of functions $f$ and $g$ with this property.

**36.** (a) Show that if $f$ and $g$ are functions for which
$$f'(x) = g(x) \quad \text{and} \quad g'(x) = f(x)$$
for all $x$, then $f^2(x) - g^2(x)$ is a constant.

(b) Show that the function $f(x) = \frac{1}{2}(e^x + e^{-x})$ and the function $g(x) = \frac{1}{2}(e^x - e^{-x})$ have this property.

**37.** Let $g(x) = x^3 - 4x + 6$. Find $f(x)$ so that $f'(x) = g'(x)$ and $f(1) = 2$.

**38.** Let $f$ and $g$ be continuous on $[a, b]$ and differentiable on $(a, b)$. Prove: If $f(a) = g(a)$ and $f(b) = g(b)$, then there is a point $c$ in $(a, b)$ where $f'(c) = g'(c)$.

**39.** Illustrate the result in Exercise 38 by drawing an appropriate picture.

**40.** (a) Prove: If $f''(x) > 0$ for all $x$ in $(a, b)$, then $f'(x) = 0$ at most once in $(a, b)$.

(b) Give a geometric interpretation of the result in (a).

**41.** Prove part (*b*) of Theorem 5.1.2.

**42.** Prove part (*c*) of Theorem 5.1.2.

## SUPPLEMENTARY EXERCISES

**1.** (a) What inequality must $f(x)$ satisfy for the function $f$ to have an absolute maximum on an interval $I$ at $x_0$?

(b) What inequality must $f(x)$ satisfy for $f$ to have an absolute minimum on $I$ at $x_0$?

(c) What is the difference between an absolute extremum and a relative extremum?

**2.** According to the Extreme-Value Theorem, what conditions on a function $f$ and an interval $I$ guarantee that $f$ will have both an absolute maximum and an absolute minimum on $I$?

**3.** In each part, determine whether the statement is true or false, and justify your answer.

(a) If $f$ is differentiable on the open interval $(a, b)$, and if $f$ has an absolute extremum on that interval, then it must occur at a stationary point of $f$.

(b) If $f$ is continuous on the open interval $(a, b)$, and if $f$ is an absolute extremum on that interval, then it must occur at a stationary point of $f$.

**4.** Is it true or false that a particle in rectilinear motion is speeding up when its velocity is increasing and slowing down when its velocity is decreasing? Justify your answer.

**5.** Suppose that $f$ is continuous on the closed interval $[a, b]$ and differentiable on the open interval $(a, b)$, and suppose that $f(a) = f(b)$. Is it true or false that $f$ must have at least one stationary point in $(a, b)$? Justify your answer.

**6.** Draw an appropriate picture, and describe the basic idea of Newton's Method without using any formulas.

**7.** In each part, find the absolute minimum $m$ and the absolute maximum $M$ of $f$ on the given interval (if they exist), and state where the absolute extrema occur.

(a) $f(x) = 1/x$; $[-2, -1]$

(b) $f(x) = x^3 - x^4$; $\left[-1, \frac{3}{2}\right]$

(c) $f(x) = x^2(x - 2)^{1/3}$; $(0, 3]$

(d) $f(x) = e^x/x^2$; $(0, +\infty)$

**8.** In each part, find the absolute minimum $m$ and the absolute maximum $M$ of $f$ on the given interval (if they exist), and state where the absolute extrema occur.

(a) $f(x) = 2x/(x^2 + 3)$; $(0, 2]$

(b) $f(x) = 2x^5 - 5x^4 + 7$; $(-1, 3)$

(c) $f(x) = -|x^2 - 2x|$; $[1, 3]$

(d) $f(x) = x^x$; $[0, +\infty)$

**9.** Use Newton's Method to approximate the smallest positive solution of $\sin x + \cos x = 0$.

**10.** Use Newton's Method to approximate all three solutions of $x^3 - 4x + 1 = 0$.

**11.** In each part, determine whether all of the hypotheses of Rolle's Theorem are satisfied on the stated interval. If not, state which hypotheses fail; if so, find all values of $c$ guaranteed in the conclusion of the theorem.

(a) $f(x) = \sqrt{4 - x^2}$ on $[-2, 2]$

(b) $f(x) = x^{2/3} - 1$ on $[-1, 1]$

(c) $f(x) = \sin(x^2)$ on $[0, \sqrt{\pi}]$

**12.** In each part, determine whether all of the hypotheses of the Mean-Value Theorem are satisfied on the stated interval. If not, state which hypotheses fail; if so, find all values of $c$ guaranteed in the conclusion of the theorem.

(a) $f(x) = |x - 1|$ on $[-2, 2]$

(b) $f(x) = \dfrac{x + 1}{x - 1}$ on $[2, 3]$

(c) $f(x) = \begin{cases} 3 - x^2 & \text{if } x \leq 1 \\ 2/x & \text{if } x > 1 \end{cases}$ on $[0, 2]$

**13.** A church window consists of a blue semicircular section surmounting a clear rectangular section as shown in the accompanying figure. The blue glass lets through half as much light per unit area as the clear glass. Find the radius $r$ of the window that admits the most light if the perimeter of the entire window is to be $P$ feet.

**14.** Find the dimensions of the rectangle of maximum area that can be inscribed inside the ellipse $(x/4)^2 + (y/3)^2 = 1$ (see the accompanying figure).

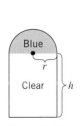

Figure Ex-13

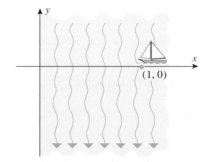

Figure Ex-14

**15.** (a) Can an object in rectilinear motion reverse direction if its acceleration is constant? Justify your answer using a velocity versus time curve.

(b) Can an object in rectilinear motion have increasing speed and decreasing acceleration? Justify your answer using a velocity versus time curve.

**16.** Suppose that the position function of a particle in rectilinear motion is given by the formula $s(t) = t/(t^2 + 5)$ for $t \geq 0$.

(a) Use a graphing utility to generate the position, velocity, and acceleration versus time curves.

(b) Use the appropriate graph to make a rough estimate of the time when the particle reverses direction, and then find that time exactly.

(c) Find the position, velocity, and acceleration at the instant when the particle reverses direction.

(d) Use the appropriate graphs to make rough estimates of the time intervals on which the particle is speeding up and the time intervals on which it is slowing down, and then find those time intervals exactly.

(e) When does the particle have its maximum and minimum velocities?

**17.** A basketball player, standing near the basket to grab a rebound, jumps 76.0 cm vertically.

(a) How much time does the player spend in the top 15.0 cm of the jump and how much time in the bottom 15.0 cm?

(b) In words, explain why basketball players seem to be suspended in air when they jump.

**18.** (a) Suppose that an object is released from rest from the top of a high building. Assuming that a free-fall model applies and that time is in seconds and distance is in meters, make a table that shows the distance traveled by the object and its speed to one decimal place at 1-second increments from $t = 0$ to $t = 4$.

(b) Confirm that doubling the elapsed time doubles the velocity, and explain why this happens.

(c) Confirm that doubling the elapsed time increases the distance traveled by a factor of 4, and explain why this happens.

**19.** Let

$$f(x) = \frac{x^3 + 2}{x^4 + 1}$$

(a) Generate the graph of $y = f(x)$, and use the graph to make rough estimates of the coordinates of the absolute extrema.

(b) Use a CAS to solve the equation $f'(x) = 0$ and then use it to make more accurate approximations of the coordinates in part (a).

**20.** As shown in the accompanying figure, suppose that a boat enters the river at the point $(1, 0)$ and maintains a heading toward the origin. As a result of the strong current, the boat follows the path

$$y = \frac{x^{10/3} - 1}{2x^{2/3}}$$

where $x$ and $y$ are in miles.

(a) Graph the path taken by the boat.

(b) Can the boat reach the origin? If not, discuss its fate and find how close it comes to the origin.

(c) What is the velocity of the boat in the $x$-direction at the instant when it is closest to the origin if the velocity in the $y$-direction is $-4$ mi/h at this instant?

Figure Ex-20

[c] **21.** Suppose that the position function of a particle in rectilinear motion is given by the formula

$$s(t) = \frac{t^2 + 1}{t^4 + 1}, \quad t \ge 0$$

(a) Use a CAS to find simplified formulas for the velocity $v(t)$ and the acceleration $a(t)$.

(b) Graph the position, velocity, and acceleration versus time curves.

(c) Use the appropriate graph to make a rough estimate of the time at which the particle is farthest from the origin and its distance from the origin at that time.

(d) Use the appropriate graph to make a rough estimate of the time interval during which the particle is moving in the positive direction.

(e) Use the appropriate graphs to make rough estimates of the time intervals during which the particle is speeding up and the time intervals during which it is slowing down.

(f) Use the appropriate graph to make a rough estimate of the maximum speed of the particle and the time at which the maximum speed occurs.

[c] **22.** Suppose that the number of individuals at time $t$ in a certain wildlife population is given by

$$N(t) = \frac{340}{1 + 9(0.77)^t}, \quad t \ge 0$$

where $t$ is in years. At approximately what instant of time is the size of the population increasing most rapidly?

**23.** According to *Kepler's law*, the planets in our solar system move in elliptical orbits around the Sun. If a planet's closest approach to the Sun occurs at time $t = 0$, then the distance $r$ from the center of the planet to the center of the Sun at some later time $t$ can be determined from the equation

$$r = a(1 - e\cos\phi)$$

where $a$ is the average distance between centers, $e$ is a positive constant that measures the "flatness" of the elliptical orbit, and $\phi$ is the solution of *Kepler's equation*

$$\frac{2\pi t}{T} = \phi - e\sin\phi$$

in which $T$ is the time it takes for one complete orbit of the planet. Estimate the distance from the Earth to the Sun when $t = 90$ days. [First find $\phi$ from Kepler's equation, and then use this value of $\phi$ to find the distance. Use $a = 150 \times 10^6$ km, $e = 0.0167$, and $T = 365$ days.]

**24.** Using the formulas in Exercise 23, find the distance from the planet Mars to the Sun when $t = 1$ year. For Mars use $a = 228 \times 10^6$ km, $e = 0.934$, and $T = 1.88$ years.

---

**EXPANDING THE CALCULUS HORIZON**

For additional material relating to this chapter, visit the Anton Website at http://www.wiley.com/college/anton

# 7

# INTEGRATION

Gottfried Leibniz

$\mathscr{T}$raditionally, that portion of calculus concerned with finding tangent lines and rates of change is called ***differential calculus*** and that portion concerned with finding areas is called ***integral calculus***. However, we will see in this chapter that the two problems are so closely related that the distinction between differential and integral calculus is often hard to discern.

In this chapter we will begin with an overview of the problem of finding areas—we will discuss what the term "area" means, and we will outline two approaches to defining and calculating areas. Following this overview, we will discuss the "Fundamental Theorem of Calculus", which is the theorem that relates the problems of finding tangent lines and areas, and we will discuss techniques for calculating areas. Finally, we will use the ideas in this chapter to continue our study of rectilinear motion and to reexamine the concept of a natural logarithm.

## 7.1 AN OVERVIEW OF THE AREA PROBLEM

*In this introductory section we will give an overview of the problem of defining and calculating areas of plane regions with curvilinear boundaries. All of the results in this section will be reexamined in more detail later in this chapter, so our purpose here is to introduce the fundamental concepts.*

**DEFINING AREA**

The main goal of this chapter is to study the following major problem of calculus:

> **7.1.1** THE AREA PROBLEM. Given a function $f$ that is continuous and nonnegative on an interval $[a, b]$, find the area between the graph of $f$ and the interval $[a, b]$ on the $x$-axis (Figure 7.1.1).

Area formulas for basic geometric figures, such as rectangles, polygons, and circles, date back to the earliest written records of mathematics. The first real advance beyond the elementary level of area computation was made by the Greek mathematician, Archimedes,[*] who devised an ingenious but cumbersome technique, called the *method of exhaustion*, for finding areas of regions bounded by parabolas, spirals, and various other curves.

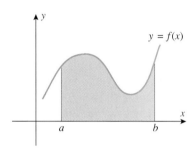

Figure 7.1.1

---

[*]ARCHIMEDES (287 B.C.–212 B.C.). Greek mathematician and scientist. Born in Syracuse, Sicily, Archimedes was the son of the astronomer Pheidias and possibly related to Heiron II, king of Syracuse. Most of the facts about his life come from the Roman biographer, Plutarch, who inserted a few tantalizing pages about him in the massive biography of the Roman soldier, Marcellus. In the words of one writer, "the account of Archimedes is slipped like a tissue-thin shaving of ham in a bull-choking sandwich."

Archimedes ranks with Newton and Gauss as one of the three greatest mathematicians who ever lived, and he is certainly the greatest mathematician of antiquity. His mathematical work is so modern in spirit and technique that it is barely distinguishable from that of a seventeenth-century mathematician, yet it was all done without benefit of algebra or a convenient number system. Among his mathematical achievements, Archimedes developed a general method (exhaustion) for finding areas and volumes, and he used the method to find areas bounded by parabolas and spirals and to find volumes of cylinders, paraboloids, and segments of spheres. He gave a procedure for approximating $\pi$ and bounded its value between $3\frac{10}{71}$ and $3\frac{1}{7}$. In spite of the limitations of the Greek numbering system, he devised methods for finding square roots and invented a method based on the Greek myriad (10,000) for representing numbers as large as 1 followed by 80 million billion zeros.

Of all his mathematical work, Archimedes was most proud of his discovery of the method for finding the volume of a sphere—he showed that the volume of a sphere is two-thirds the volume of the smallest cylinder that can contain it. At his request, the figure of a sphere and cylinder was engraved on his tombstone.

In addition to mathematics, Archimedes worked extensively in mechanics and hydrostatics. Nearly every schoolchild knows Archimedes as the absent-minded scientist who, on realizing that a floating object displaces its weight of liquid, leaped from his bath and ran naked through the streets of Syracuse shouting, "Eureka, Eureka!"—(meaning, "I have found it!"). Archimedes actually created the discipline of hydrostatics and used it to find equilibrium positions for various floating bodies. He laid down the fundamental postulates of mechanics, discovered the laws of levers, and calculated centers of gravity for various flat surfaces and solids. In the excitement of discovering the mathematical laws of the lever, he is said to have declared, "Give me a place to stand and I will move the earth."

Although Archimedes was apparently more interested in pure mathematics than its applications, he was an engineering genius. During the second Punic war, when Syracuse was attacked by the Roman fleet under the command of Marcellus, it was reported by Plutarch that Archimedes' military inventions held the fleet at bay for three years. He invented super catapults that showered the Romans with rocks weighing a quarter ton or more, and fearsome mechanical devices with iron "beaks and claws" that reached over the city walls, grasped the ships, and spun them against the rocks. After the first repulse, Marcellus called Archimedes a "geometrical Briareus (a hundred-armed mythological monster) who uses our ships like cups to ladle water from the sea."

Eventually the Roman army was victorious and contrary to Marcellus' specific orders the 75-year-old Archimedes was killed by a Roman soldier. According to one report of the incident, the soldier cast a shadow across the sand in which Archimedes was working on a mathematical problem. When the annoyed Archimedes yelled, "Don't disturb my circles," the soldier flew into a rage and cut the old man down.

With his death the Greek gift of mathematics passed into oblivion, not to be fully resurrected again until the sixteenth century. Unfortunately, there is no known accurate likeness or statue of this great man.

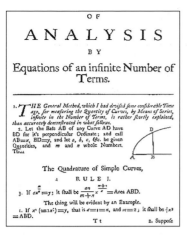

First page of the 1745 English translation of Newton's *De Analysi*

By the seventeenth century, several mathematicians had discovered how to obtain such areas more simply by calculating limits. However, the method of exhaustion and its successors lacked generality—for each different problem one had to devise special procedures. The major breakthrough in obtaining a general method for calculating areas was made independently by Newton and Leibniz, both of whom discovered that areas could be obtained by reversing the process of differentiation. This discovery, which is regarded as the beginning of calculus, was circulated by Newton in 1669 and published in 1711 in a paper entitled, *De Analysi per Aequationes Numero Terminorum Infinitas* (*On the Analysis by Means of Equations with Infinitely Many Terms*); and it was discovered by Leibniz around 1673 and stated in an unpublished manuscript dated November 11, 1675.

Before one can talk logically about methods for calculating areas, it is necessary to have a precise definition of what the term *area* means. To avoid a lot of mathematical formality, let us assume that the areas of geometric figures with straight boundaries, such as rectangles, triangles, and polygons, are defined and computed using the standard formulas for such figures. However, the problem of defining and computing areas of figures with *curvilinear* boundaries is more complicated and will require various limiting processes. For example, in the introductory section of this text we showed that the area of a circle could be viewed as a limit of areas of inscribed polygons (Figure 7 in the Introduction). Thus, once a definition is established for the area of a polygon, the area of a circle can be *defined* as a limit of areas of polygons.

## THE RECTANGLE METHOD FOR FINDING AREAS

There are two basic methods for finding the area of the region having the form shown in Figure 7.1.1—the *rectangle method* and the *antiderivative method*. The idea behind the rectangle method is as follows:

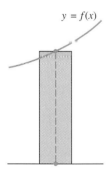

$y = f(x)$

- Divide the interval $[a, b]$ into $n$ equal subintervals, and over each subinterval construct a rectangle that extends from the $x$-axis to any point on the curve $y = f(x)$ that is above the subinterval; the particular point does not matter—it can be above the center, above an endpoint, or above any other point in the subinterval. In Figure 7.1.2 it is above the center.

- For each $n$, the total area of the rectangles can be viewed as an *approximation* to the exact area under the curve over the interval $[a, b]$. Moreover, it is evident intuitively that as $n$ increases these approximations will get better and better and will approach the exact area as a limit (Figure 7.1.3).

Figure 7.1.2

This procedure serves both as a mathematical definition and a method of computation—we can *define* the area under $y = f(x)$ over the interval $[a, b]$ as the limit of the areas of the approximating rectangles, and we can use the method itself to approximate this area.

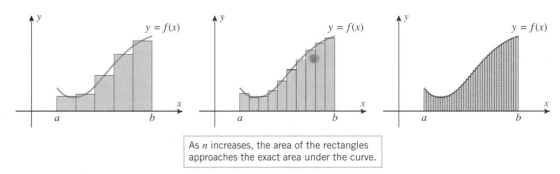

As $n$ increases, the area of the rectangles approaches the exact area under the curve.

Figure 7.1.3

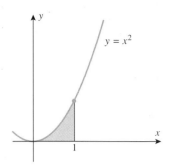

Figure 7.1.4

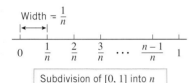

Subdivision of [0, 1] into $n$ subintervals of equal length

Figure 7.1.5

To illustrate this idea, we will use the rectangle method to approximate the area under the curve $y = x^2$ over the interval [0, 1] (Figure 7.1.4). We will begin by dividing the interval [0, 1] into $n$ equal subintervals, from which it follows that each subinterval has length $1/n$; the endpoints of the subintervals occur at

$$0, \quad \frac{1}{n}, \quad \frac{2}{n}, \quad \frac{3}{n}, \dots, \quad \frac{n-1}{n}, \quad 1$$

(Figure 7.1.5). We want to construct a rectangle over each of these intervals whose height is the value of the function $f(x) = x^2$ at any point in the interval. To be specific, let us use the right endpoints, in which case the heights of our rectangles will be

$$\left(\frac{1}{n}\right)^2, \quad \left(\frac{2}{n}\right)^2, \quad \left(\frac{3}{n}\right)^2, \dots, \quad 1$$

and since each rectangle has a base of width $1/n$, the total area $A_n$ of the $n$ rectangles will be

$$A_n = \left[\left(\frac{1}{n}\right)^2 + \left(\frac{2}{n}\right)^2 + \left(\frac{3}{n}\right)^2 + \cdots + 1^2\right]\left(\frac{1}{n}\right) \tag{1}$$

For example, if $n = 4$, then the total area of the four approximating rectangles would be

$$A_4 = \left[\left(\tfrac{1}{4}\right)^2 + \left(\tfrac{2}{4}\right)^2 + \left(\tfrac{3}{4}\right)^2 + 1^2\right]\left(\tfrac{1}{4}\right) = \tfrac{15}{32} = 0.46875$$

Table 7.1.1 shows the result of evaluating (1) on a computer for some increasingly large values of $n$. These computations suggest that the exact area is close to $\frac{1}{3}$.

**Table 7.1.1**

| $n$ | 4 | 10 | 100 | 1000 | 10,000 | 100,000 |
|---|---|---|---|---|---|---|
| $A_n$ | 0.468750 | 0.385000 | 0.338350 | 0.333834 | 0.333383 | 0.333338 |

FOR THE READER.    Use your calculating utility to confirm the value of $A_{10}$ given in Table 7.1.1.

**THE ANTIDERIVATIVE METHOD FOR FINDING AREAS**

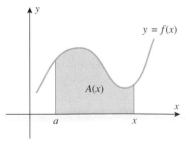

Figure 7.1.6

The antiderivative method for finding areas reflects the genius of Newton and Leibniz— they suggested that to find the area under the curve in Figure 7.1.1, one should first consider the more general problem of finding the area $A(x)$ under the curve from the point $a$ to an arbitrary point $x$ in the interval $[a, b]$ (Figure 7.1.6). Newton and Leibniz discovered independently that the *derivative* of the function $A(x)$ is easy to find, so that if one can figure out how to find $A(x)$ from $A'(x)$, then the area under the curve from $a$ to $b$ can be obtained by substituting $x = b$ in the area formula $A(x)$.

To illustrate how all of this works, let us begin with the problem of finding

$$A'(x) = \lim_{h \to 0} \frac{A(x + h) - A(x)}{h} \tag{2}$$

For simplicity, consider the case where $h > 0$. The numerator on the right side of (2) is the difference of two areas: the area between $a$ and $x + h$ minus the area between $a$ and $x$ (Figure 7.1.7a). If we let $c$ be the midpoint between $x$ and $x + h$, then this difference of areas can be approximated by the area of a rectangle with base $h$ and height $f(c)$ (Figure 7.1.7b). Thus,

$$\frac{A(x + h) - A(x)}{h} \approx \frac{f(c) \cdot h}{h} = f(c) \tag{3}$$

It seems plausible from Figure 7.1.7b that the error in approximation (3) will approach

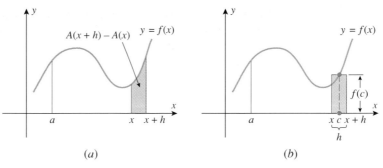

Figure 7.1.7

approach zero as $h \to 0$. If we accept this to be so, then it follows from (2) and (3) that

$$A'(x) = \lim_{h \to 0} \frac{A(x+h) - A(x)}{h} = \lim_{h \to 0} f(c) \tag{4}$$

Since $c$ is the midpoint between $x$ and $x + h$, it follows that $c \to x$ as $h \to 0$. But we have assumed $f$ to be a continuous function, so $f(c) \to f(x)$ as $c \to x$. Therefore,

$$\lim_{h \to 0} f(c) = f(x)$$

Thus, it follows from (4) that

$$A'(x) = f(x) \tag{5}$$

This is the result we were looking for; it tells us that *the derivative of the area function $A(x)$ is the function whose graph forms the upper boundary of the region.*

To illustrate how the antiderivative method works, let us apply it to the same problem we investigated with the rectangle method—finding the area under $y = x^2$ over the interval $[0, 1]$. The upper boundary of the region is the graph of $f(x) = x^2$, so it follows from (5) that the derivative of the area function is

$$A'(x) = x^2 \tag{6}$$

Thus, to find $A(x)$ we must look for a function whose derivative is $x^2$. This is called an ***antidifferentiation*** problem because we are trying to find $A(x)$ by "undoing" a differentiation. By simply guessing we see that

$$A(x) = \tfrac{1}{3}x^3$$

is one solution to (6). But this is not the only solution, since it follows from Theorem 6.5.3 that

$$A(x) = \tfrac{1}{3}x^3 + C \tag{7}$$

also satisfies (6) for any real value of $C$. We still have some work to do since this formula involves an unknown constant $C$ that must be determined. This is where the decision to solve the area problem for a general right-hand endpoint helps. If we consider the case where $x = 0$, then the interval $[0, x]$ reduces to a single point. If we agree that the area above a single point should be taken as zero, then it follows on substituting $x = 0$ in (7) that

$$A(0) = 0 + C = 0 \quad \text{or} \quad C = 0$$

so (7) simplifies to

$$A(x) = \tfrac{1}{3}x^3 \tag{8}$$

which is the formula for the area under $y = x^2$ over the interval $[0, x]$. For the area over the interval $[0, 1]$ we set $x = 1$ in (8), which yields $A(1) = \frac{1}{3}$ for the exact area under the curve. This confirms definitely what was suggested numerically in Table 7.1.1.

REMARK. Our success in finding the exact area under the curve $y = x^2$ hinged on our ability to guess at a function $A(x)$ whose derivative is $x^2$. Had we not been able to find such a function, then the antiderivative method would have failed and we would have been forced to rely on the rectangle method. Thus, whereas earlier in this text we were concerned with the process of differentiation, we will now also be concerned with the process of antidifferentiation.

## EXERCISE SET 7.1

In Exercises 1–4, use an appropriate formula from plane geometry to find the exact area between the graph of $f$ and the given interval; and then use the rectangle method to make a table of approximations $A_1, A_2, \ldots, A_{10}$ to the exact area, where $A_n$ is the approximation that results by dividing the interval into $n$ subintervals and constructing a rectangle over each subinterval whose height is the $y$-coordinate of the curve $y = f(x)$ at the right endpoint.

**1.** $f(x) = x$; $[0, 1]$

**2.** $f(x) = 4 - 2x$; $[0, 2]$

**3.** $f(x) = 6x + 2$; $[0, 2]$

**4.** $f(x) = \sqrt{1 - x^2}$; $[0, 1]$

**5.** Let $A(x) = x^2/2$. Confirm that $A'(x) = x$, and use the antiderivative method to find the exact area in Exercise 1.

**6.** Let $A(x) = 4x - x^2$. Confirm that $A'(x) = 4 - 2x$, and use the antiderivative method to find the exact area in Exercise 2.

**7.** Let $A(x) = 3x^2 + 2x$. Confirm that $A'(x) = 6x + 2$, and use the antiderivative method to find the exact area in Exercise 3.

**8.** Let $A(x) = \frac{1}{2}x\sqrt{1 - x^2} + \frac{1}{2}\sin^{-1} x$. Then confirm that $A'(x) = \sqrt{1 - x^2}$, and use the antiderivative method to find the exact area in Exercise 4.

**9.** Use the antiderivative method to find the exact area between the curve $y = e^x$ and the interval $[0, 1]$.

**10.** Use the antiderivative method to find the exact area between the curve $y = \sin x$ and the interval $[0, \pi]$.

## 7.2 THE INDEFINITE INTEGRAL; INTEGRAL CURVES AND DIRECTION FIELDS

*In the last section we saw that antidifferentiation plays an important role in finding exact areas. In this section we will develop some fundamental results about antidifferentiation that will ultimately lead us to systematic procedures for finding a function from its derivative.*

**7.2.1** DEFINITION. A function $F$ is called an **antiderivative** of a function $f$ on a given interval $I$ if $F'(x) = f(x)$ for all $x$ in the interval.

For example, the function $F(x) = \frac{1}{3}x^3$ is an antiderivative of $f(x) = x^2$ on the interval $(-\infty, +\infty)$ because for each $x$ in this interval

$$F'(x) = \frac{d}{dx}\left[\frac{1}{3}x^3\right] = x^2 = f(x)$$

However, this is not the only antiderivative of $F$ on this interval. If we add any constant $C$ to $\frac{1}{3}x^3$, then the function $F(x) = \frac{1}{3}x^3 + C$ is also an antiderivative of $f$ on $(-\infty, +\infty)$, since

$$F'(x) = \frac{d}{dx}\left[\frac{1}{3}x^3 + C\right] = x^2 + 0 = f(x)$$

In general, once any single antiderivative of a function is known, other antiderivatives can be obtained by adding constants to the known antiderivative. Thus,

$$\tfrac{1}{3}x^3, \quad \tfrac{1}{3}x^3 + 2, \quad \tfrac{1}{3}x^3 - 5, \quad \tfrac{1}{3}x^3 + \sqrt{2}$$

are all antiderivatives of $f(x) = x^2$.

WARNING.    Do not confuse derivatives and antiderivatives—the *derivative* of the function $f(x) = x^2$ is $f'(x) = 2x$, but the functions $F(x) = \tfrac{1}{3}x^3 + C$ are *antiderivatives* of $f$.

It is reasonable to ask if there are antiderivatives of a function $f$ that cannot be obtained by adding some constant to a known antiderivative $F$. The answer is *no*—once a single antiderivative of $f$ on an interval $I$ is known, all other antiderivatives on that interval are obtainable by adding constants to that antiderivative. This is so because Theorem 6.5.3 tells us that if two functions have the same derivative on an interval, then they differ by a constant on that interval. The following theorem summarizes these observations.

> **7.2.2    THEOREM.**    *If $F(x)$ is any antiderivative of $f(x)$ on an interval $I$, then for any constant $C$ the function $F(x) + C$ is also an antiderivative of $f(x)$ on that interval. Moreover, each antiderivative of $f(x)$ on the interval $I$ can be expressed in the form $F(x) + C$ by choosing the constant $C$ appropriately.*

**THE INDEFINITE INTEGRAL**

The process of finding antiderivatives is called **antidifferentiation** or **integration**. Thus, if

$$\frac{d}{dx}[F(x)] = f(x)$$

then integrating (or antidifferentiating) $f(x)$ produces the antiderivatives $F(x) + C$. We denote this by writing

$$\int f(x)\,dx = F(x) + C \tag{1}$$

For example, the antiderivatives of $f(x) = x^2$ are the functions $F(x) = \tfrac{1}{3}x^3 + C$, so

$$\int x^2\,dx = \tfrac{1}{3}x^3 + C$$

The "elongated s" that appears on the left side of (1) is called an **integral sign**[*] or an **indefinite integral**, the function $f(x)$ is called the **integrand**, and the constant $C$ is called the **constant of integration**. You should read Equation (1) as "the integral of $f(x)$ with respect to $x$ is equal to $F(x) + C$." The adjective "indefinite" emphasizes that the integration process does not produce a *definite* function, but rather a whole set of functions.

The $dx$ symbols in the differentiation and antidifferentiation operations

$$\frac{d}{dx}[\ \ ] \quad \text{and} \quad \int [\ \ ]\,dx$$

serve to identify the independent variable. If an independent variable other than $x$ is used, say $t$, then the notation must be adjusted appropriately. Thus,

$$\frac{d}{dt}[F(t)] = f(t) \quad \text{and} \quad \int f(t)\,dt = F(t) + C$$

are equivalent statements.

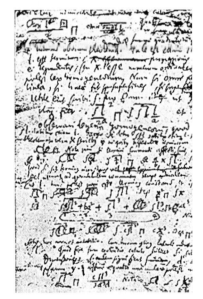

Extract from the manuscript of Leibniz dated October 29, 1675 in which the integral sign first appeared.

---

[*] This notation was devised by Leibniz. In his early papers Leibniz used the notation "omn." (an abbreviation for the Latin word "omnes") to denote integration. Then on October 29, 1675 he wrote, "It will be useful to write $\int$ for omn., thus $\int \ell$ for omn. $\ell$ . . . ." Two or three weeks later he refined the notation further and wrote $\int[\ \ ]\,dx$ rather than $\int$ alone. This notation is so useful and so powerful that its development by Leibniz must be regarded as a major milestone in the history of mathematics and science.

### Example 1

| DERIVATIVE FORMULA | EQUIVALENT INTEGRATION FORMULA |
|---|---|
| $\dfrac{d}{dx}[x^3] = 3x^2$ | $\displaystyle\int 3x^2\,dx = x^3 + C$ |
| $\dfrac{d}{dx}[\sqrt{x}] = \dfrac{1}{2\sqrt{x}}$ | $\displaystyle\int \dfrac{1}{2\sqrt{x}}\,dx = \sqrt{x} + C$ |
| $\dfrac{d}{dt}[\tan t] = \sec^2 t$ | $\displaystyle\int \sec^2 t\,dt = \tan t + C$ |
| $\dfrac{d}{du}[u^{3/2}] = \tfrac{3}{2}u^{1/2}$ | $\displaystyle\int \tfrac{3}{2}u^{1/2}\,du = u^{3/2} + C$ |

◀

For simplicity, the $dx$ is sometimes absorbed into the integrand. For example,

$$\int 1\,dx \quad \text{can be written as} \quad \int dx$$

$$\int \frac{1}{x^2}\,dx \quad \text{can be written as} \quad \int \frac{dx}{x^2}$$

**INTEGRATION FORMULAS**

Integration is essentially educated guesswork—given the derivative of a function $f$, one tries to guess what the function $f$ is. However, many basic integration formulas can be obtained directly from their companion differentiation formulas. Some of the most important ones are given in Table 7.2.1.

**Table 7.2.1**

| DIFFERENTIATION FORMULA | INTEGRATION FORMULA |
|---|---|
| 1. $\dfrac{d}{dx}[x] = 1$ | $\displaystyle\int dx = x + C$ |
| 2. $\dfrac{d}{dx}\left[\dfrac{x^{r+1}}{r+1}\right] = x^r \;\; (r \ne -1)$ | $\displaystyle\int x^r\,dx = \left[\dfrac{x^{r+1}}{r+1}\right] + C \;\; (r \ne -1)$ |
| 3. $\dfrac{d}{dx}[\sin x] = \cos x$ | $\displaystyle\int \cos x\,dx = \sin x + C$ |
| 4. $\dfrac{d}{dx}[-\cos x] = \sin x$ | $\displaystyle\int \sin x\,dx = -\cos x + C$ |
| 5. $\dfrac{d}{dx}[\tan x] = \sec^2 x$ | $\displaystyle\int \sec^2 x\,dx = \tan x + C$ |
| 6. $\dfrac{d}{dx}[-\cot x] = \csc^2 x$ | $\displaystyle\int \csc^2 x\,dx = -\cot x + C$ |
| 7. $\dfrac{d}{dx}[\sec x] = \sec x \tan x$ | $\displaystyle\int \sec x \tan x\,dx = \sec x + C$ |
| 8. $\dfrac{d}{dx}[-\csc x] = \csc x \cot x$ | $\displaystyle\int \csc x \cot x\,dx = -\csc x + C$ |
| 9. $\dfrac{d}{dx}[e^x] = e^x$ | $\displaystyle\int e^x\,dx = e^x + C$ |
| 10. $\dfrac{d}{dx}\left[\dfrac{b^x}{\ln b}\right] = b^x$ | $\displaystyle\int b^x\,dx = \dfrac{b^x}{\ln b} + C$ |
| 11. $\dfrac{d}{dx}[\ln|x|] = \dfrac{1}{x}$ | $\displaystyle\int \dfrac{dx}{x} = \ln|x| + C$ |

### Example 2

The second integration formula in this table will be easy to remember if you express it in words: *to integrate a power of $x$ (other than $-1$), add 1 to the power and divide by the new power*. Here are some examples:

$$\int x^2\, dx = \frac{x^3}{3} + C \qquad \boxed{r = 2}$$

$$\int x^3\, dx = \frac{x^4}{4} + C \qquad \boxed{r = 3}$$

$$\int \frac{1}{x^5}\, dx = \int x^{-5}\, dx = \frac{x^{-5+1}}{-5+1} + C = -\frac{1}{4x^4} + C \qquad \boxed{r = -5}$$

$$\int \sqrt{x}\, dx = \int x^{\frac{1}{2}}\, dx = \frac{x^{\frac{1}{2}+1}}{\frac{1}{2}+1} + C = \tfrac{2}{3}x^{\frac{3}{2}} + C = \tfrac{2}{3}(\sqrt{x})^3 + C \qquad \boxed{r = \tfrac{1}{2}}$$

$$\int x^{-1}\, dx = \int \frac{dx}{x} = \ln|x| + C \qquad \blacktriangleleft$$

**PROPERTIES OF THE INDEFINITE INTEGRAL**

If we differentiate an antiderivative of $f(x)$, we obtain $f(x)$ back again. Thus,

$$\frac{d}{dx}\left[\int f(x)\, dx\right] = f(x) \tag{2}$$

This result is helpful for proving the following basic properties of antiderivatives.

---

**7.2.3    THEOREM.**

(a)    *A constant factor can be moved through an integral sign; that is,*

$$\int cf(x)\, dx = c\int f(x)\, dx$$

(b)    *An antiderivative of a sum is the sum of the antiderivatives; that is,*

$$\int [f(x) + g(x)]\, dx = \int f(x)\, dx + \int g(x)\, dx$$

(c)    *An antiderivative of a difference is the difference of the antiderivatives; that is,*

$$\int [f(x) - g(x)]\, dx = \int f(x)\, dx - \int g(x)\, dx$$

---

*Proof.*    In each part we must show that the expression on the right side of the equation is an antiderivative of the integrand on the left side of the equation. This can be done using (2) as follows:

$$\frac{d}{dx}\left[c\int f(x)\, dx\right] = c\frac{d}{dx}\left[\int f(x)\, dx\right] = cf(x)$$

$$\frac{d}{dx}\left[\int f(x)\, dx + \int g(x)\, dx\right] = \frac{d}{dx}\left[\int f(x)\, dx\right] + \frac{d}{dx}\left[\int g(x)\, dx\right]$$
$$= f(x) + g(x)$$

$$\frac{d}{dx}\left[\int f(x)\, dx - \int g(x)\, dx\right] = \frac{d}{dx}\left[\int f(x)\, dx\right] - \frac{d}{dx}\left[\int g(x)\, dx\right]$$
$$= f(x) - g(x) \qquad \blacksquare$$

When applying Theorem 7.2.3, it is best to put in the constant of integration at the *very end* of the computations to obtain the simplest form of the answer. This is illustrated in the following example.

### Example 3

Evaluate

(a) $\int 4\cos x \, dx$     (b) $\int (x + x^2) \, dx$

*Solution (a).*

$$\int 4\cos x \, dx = 4\underbrace{\int \cos x \, dx}_{\text{Theorem 7.2.3(a)}} = 4\underbrace{(\sin x + C)}_{\text{Table 7.2.1}} = 4\sin x + 4C$$

Since $C$ is an arbitrary constant, so is $4C$. However, this latter form is unnecessarily complicated and can be avoided by deferring the insertion of the constant until the end of the computations; this procedure yields

$$\int 4\cos x \, dx = 4\int \cos x \, dx = 4\sin x + C$$

*Solution (b).*

$$\int (x + x^2) \, dx = \underbrace{\int x \, dx + \int x^2 \, dx}_{\text{Theorem 7.2.3(b)}} = \underbrace{\frac{x^2}{2} + \frac{x^3}{3}}_{\text{Table 7.2.1}} + C \qquad \blacktriangleleft$$

Parts (b) and (c) of Theorem 7.2.3 can be extended to more than two functions, which in combination with part (a) results in the following general formula:

$$\int [c_1 f_1(x) + c_2 f_2(x) + \cdots + c_n f_n(x)] \, dx$$
$$= c_1 \int f_1(x) \, dx + c_2 \int f_2(x) \, dx + \cdots + c_n \int f_n(x) \, dx \qquad (3)$$

### Example 4

$$\int (3x^6 - 2x^2 + 7x + 1) \, dx = 3\int x^6 \, dx - 2\int x^2 \, dx + 7\int x \, dx + \int 1 \, dx$$
$$= \frac{3x^7}{7} - \frac{2x^3}{3} + \frac{7x^2}{2} + x + C \qquad \blacktriangleleft$$

Sometimes it is useful to rewrite an integrand in a different form before performing the integration.

### Example 5

Evaluate

(a) $\int \dfrac{\cos x}{\sin^2 x} \, dx$     (b) $\int \dfrac{t^2 - 2t^4}{t^4} \, dt$

*Solution (a).*

$$\int \frac{\cos x}{\sin^2 x} \, dx = \int \frac{1}{\sin x}\frac{\cos x}{\sin x} \, dx = \underbrace{\int \csc x \cot x \, dx = -\csc x + C}_{\text{Formula 8 in Table 7.2.1}}$$

*Solution (b).*

$$\int \frac{t^2 - 2t^4}{t^4}\, dt = \int \left(\frac{1}{t^2} - 2\right) dt = \int (t^{-2} - 2)\, dt$$

$$= \frac{t^{-1}}{-1} - 2t + C = -\frac{1}{t} - 2t + C \qquad \blacktriangleleft$$

**INTEGRAL CURVES**

Graphs of antiderivatives of a function $f$ are called ***integral curves*** of $f$. We know from Theorem 7.2.2 that if $y = F(x)$ is any integral curve of $f(x)$, then all other integral curves are vertical translations of this curve, since they have equations of the form $y = F(x) + C$. For example, $y = \frac{1}{3}x^3$ is one integral curve for $f(x) = x^2$, so all the other integral curves have equations of the form $y = \frac{1}{3}x^3 + C$; conversely, the graph of any equation of this form is an integral curve (Figure 7.2.1).

In many problems one is interested in finding a function whose derivative satisfies specified conditions. The following example illustrates a geometric problem of this type.

### Example 6

Suppose that a point moves along some unknown curve $y = f(x)$ in the $xy$-plane in such a way that at each point $(x, y)$ on the curve, the tangent line has slope $x^2$. Find an equation for the curve given that it passes through the point $(2, 1)$.

*Solution.* We know that $dy/dx = x^2$, so

$$y = \int x^2\, dx = \frac{1}{3}x^3 + C$$

Since the curve passes through $(2, 1)$, a specific value for $C$ can be found by using the fact that $y = 1$ if $x = 2$. Substituting these values in the above equation yields

$$1 = \frac{1}{3}(2^3) + C \quad \text{or} \quad C = -\frac{5}{3}$$

so the curve is $y = \frac{1}{3}x^3 - \frac{5}{3}$. $\qquad \blacktriangleleft$

Observe that in this example the requirement that the unknown curve pass through the point $(2, 1)$ enabled us to determine a specific value for the constant of integration, thereby isolating the single integral curve $y = \frac{1}{3}x^3 - \frac{5}{3}$ from the family $y = \frac{1}{3}x^3 + C$ (Figure 7.2.2).

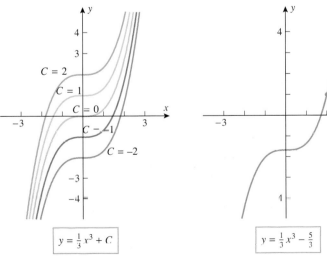

Figure 7.2.1

Figure 7.2.2

We will now consider another way of looking at integration that will be useful in our later work. Suppose that $f(x)$ is a known function and we are interested in finding a function $F(x)$ such that $y = F(x)$ satisfies the equation

$$\frac{dy}{dx} = f(x) \tag{4}$$

The solutions of this equation are the antiderivatives of $f(x)$, and we know that these can be obtained by integrating $f(x)$. For example, the solutions of the equation

$$\frac{dy}{dx} = x^2 \tag{5}$$

are

$$y = \int x^2 \, dx = \frac{x^3}{3} + C$$

Equation (4) is called a *differential equation* because it involves a derivative of an unknown function. Differential equations are different from the kinds of equations we have encountered so far in that the unknown is a *function* and not a *number* as in an equation such as $x^2 + 5x - 6 = 0$.

Sometimes we will not be interested in finding all of the solutions of (4), but rather we will want only the solution whose integral curve passes through a specified point $(x_0, y_0)$. For example, in Example 6 we solved (5) for the integral curve that passed through the point $(2, 1)$.

For simplicity, it is common in the study of differential equations to denote a solution of $dy/dx = f(x)$ as $y(x)$ rather than $F(x)$, as earlier. With this notation, the problem of finding a function $y(x)$ whose derivative is $f(x)$ and whose integral curve passes through the point $(x_0, y_0)$ is expressed as

$$\frac{dy}{dx} = f(x), \quad y(x_0) = y_0 \tag{6}$$

For reasons that will be explained later, this is called an *initial-value problem*, and the requirement that $y(x_0) = y_0$ is called the *initial condition* for the problem.

### Example 7

Solve the initial-value problem

$$\frac{dy}{dx} = \cos x, \quad y(0) = 1$$

*Solution.* The solution of the differential equation is

$$y = \int \cos x \, dx = \sin x + C \tag{7}$$

The initial condition $y(0) = 1$ implies that $y = 1$ if $x = 0$; substituting these values in (7) yields

$$1 = \sin(0) + C \quad \text{or} \quad C = 1$$

Thus, the solution of the initial-value problem is $y = \sin x + 1$. ◄

If we interpret $dy/dx$ as the slope of a tangent line, then at a point $(x, y)$ on an integral curve of the equation $dy/dx = f(x)$, the slope of the tangent line is $f(x)$. What is interesting about this is that the slopes of the tangent lines to the integral curves can be obtained without actually solving the differential equation. For example, if

$$\frac{dy}{dx} = \sqrt{x^2 + 1}$$

then we know without solving the equation that at the point where $x = 1$ the tangent line to an integral curve has slope $\sqrt{1^2 + 1} = \sqrt{2}$; and more generally, at a point where $x = a$, the tangent line to an integral curve has slope $\sqrt{a^2 + 1}$.

A geometric description of the integral curves of a differential equation $dy/dx = f(x)$ can be obtained by choosing a rectangular grid of points in the $xy$-plane, calculating the slopes of the tangent lines to the integral curves at the gridpoints, and drawing small portions of the tangent lines at those points. The resulting picture, which is called a **direction field** or **slope field** for the equation, shows the "direction" of the integral curves at the gridpoints. With sufficiently many gridpoints it is often possible to visualize the integral curves themselves; for example, Figure 7.2.3a shows a direction field for the differential equation $dy/dx = x^2$, and Figure 7.2.3b shows that same field with the integral curves imposed on it—the more gridpoints that are used, the more completely the direction field reveals the shape of the integral curves. However, the amount of computation can be considerable, so computers are usually used when direction fields with many gridpoints are needed.

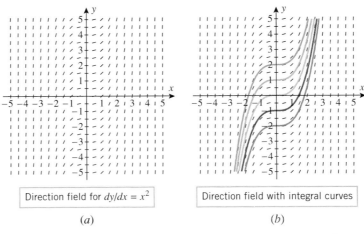

| Direction field for $dy/dx = x^2$ | Direction field with integral curves |
|---|---|
| *(a)* | *(b)* |

Figure 7.2.3

**EXERCISE SET 7.2**  📈 Graphing Calculator  C CAS
· · · · · · · · · · · · · · · · · · · · · · · · · · · · · · · · · · · · · · · · · · · · · · · · · · · · · · · · · · · · · · · · · · · · · · · · · · · ·

**1.** In each part, confirm that the formula is correct, and state a corresponding integration formula.

(a) $\dfrac{d}{dx}[\sqrt{1 + x^2}] = \dfrac{x}{\sqrt{1 + x^2}}$

(b) $\dfrac{d}{dx}[xe^x] = (x + 1)e^x$

**2.** In each part, confirm that the stated formula is correct by differentiating.

(a) $\displaystyle\int x \sin x \, dx = \sin x - x \cos x + C$

(b) $\displaystyle\int \dfrac{dx}{(1 - x^2)^{3/2}} = \dfrac{x}{\sqrt{1 - x^2}} + C$

In Exercises 3–6, find the derivative and state a corresponding integration formula.

**3.** $\dfrac{d}{dx}[\sqrt{x^3 + 5}]$

**4.** $\dfrac{d}{dx}\left[\dfrac{x}{x^2 + 3}\right]$

**5.** $\dfrac{d}{dx}[\sin(2\sqrt{x}\,)]$

**6.** $\dfrac{d}{dx}[\sin x - x \cos x]$

In Exercises 7 and 8, evaluate the integral by rewriting the integrand appropriately, if required, and then applying Formula 2 in Table 7.2.1.

**7.** (a) $\displaystyle\int x^8 \, dx$    (b) $\displaystyle\int x^{5/7} \, dx$    (c) $\displaystyle\int x^3 \sqrt{x} \, dx$

**8.** (a) $\displaystyle\int \sqrt[3]{x^2} \, dx$    (b) $\displaystyle\int \dfrac{1}{x^6} \, dx$    (c) $\displaystyle\int x^{-7/8} \, dx$

In Exercises 9–12, evaluate the integral by applying Theorem 7.2.3 and Formula 2 in Table 7.2.1 appropriately.

**9.** (a) $\displaystyle\int \dfrac{1}{2x^3} \, dx$    (b) $\displaystyle\int (u^3 - 2u + 7) \, du$

**10.** $\displaystyle\int (x^{2/3} - 4x^{-1/5} + 4) \, dx$

**11.** $\displaystyle\int (x^{-3} + \sqrt{x} - 3x^{1/4} + x^2)\, dx$

**12.** $\displaystyle\int \left( \frac{7}{y^{3/4}} - \sqrt[3]{y} + 4\sqrt{y} \right) dy$

In Exercises 13–30, evaluate the integral, and check your answer by differentiating.

**13.** $\displaystyle\int x(1 + x^3)\, dx$

**14.** $\displaystyle\int (2 + y^2)^2 \, dy$

**15.** $\displaystyle\int x^{1/3}(2 - x)^2 \, dx$

**16.** $\displaystyle\int (1 + x^2)(2 - x)\, dx$

**17.** $\displaystyle\int \frac{x^5 + 2x^2 - 1}{x^4}\, dx$

**18.** $\displaystyle\int \frac{1 - 2t^3}{t^3}\, dt$

**19.** $\displaystyle\int \left[ \frac{2}{x} + 3e^x \right] dx$

**20.** $\displaystyle\int \left[ \frac{1}{2t} - \sqrt{2}e^t \right] dt$

**21.** $\displaystyle\int [4 \sin x + 2 \cos x]\, dx$

**22.** $\displaystyle\int [4 \sec^2 x + \csc x \cot x]\, dx$

**23.** $\displaystyle\int \sec x(\sec x + \tan x)\, dx$

**24.** $\displaystyle\int \sec x(\tan x + \cos x)\, dx$

**25.** $\displaystyle\int \left[ \frac{1}{\theta} - 2e^\theta - \csc^2 \theta \right] d\theta$

**26.** $\displaystyle\int \frac{dy}{\csc y}$

**27.** $\displaystyle\int \frac{\sin x}{\cos^2 x}\, dx$

**28.** $\displaystyle\int \left[ \phi + \frac{2}{\sin^2 \phi} \right] d\phi$

**29.** $\displaystyle\int [1 + \sin^2 \theta \csc \theta]\, d\theta$

**30.** $\displaystyle\int \frac{\sin 2x}{\cos x}\, dx$

**31.** Evaluate the integral

$$\int \frac{1}{1 + \sin x}\, dx$$

by multiplying the numerator and denominator by an appropriate expression.

**c** **32.** For each of the integrals you evaluated in Exercises 13–31, use a CAS to check your answer. If the answer produced by the CAS does not match yours, show that the two answers are equivalent.

**33.** (a) Graph some representative integral curves of $f(x) = x$.
(b) Find an equation for the integral curve that passes through the point $(4, 7)$.

**34.** (a) Graph some representative integral curves of the function $f(x) = e^x/2$.
(b) Find an equation for the integral curve that passes through the point $(0, 1)$.

**35.** Use a graphing utility to generate some representative integral curves of the function $f(x) = 5x^4 - \sec^2 x$ over the interval $(-\pi/2, \pi/2)$.

**36.** Use a graphing utility to generate some representative integral curves of $f(x) = (x - 1)/x$ over the interval $(0, 5)$.

**37.** Suppose that a point moves along a curve $y = f(x)$ in the $xy$-plane in such a way that at each point $(x, y)$ on the curve

the tangent line has slope $-\sin x$. Find an equation for the curve, given that it passes through the point $(0, 2)$.

**38.** Suppose that a point moves along a curve $y = f(x)$ in the $xy$-plane in such a way that at each point $(x, y)$ on the curve the tangent line has slope $(x + 1)^2$. Find an equation for the curve, given that it passes through the point $(-2, 8)$.

In Exercises 39 and 40, solve the initial-value problems.

**39.** (a) $\dfrac{dy}{dx} = \sqrt[3]{x}, \ y(1) = 2$    (b) $\dfrac{dy}{dt} = \dfrac{1}{t}, \ y(-1) = 5$

(c) $\dfrac{dy}{dx} = \dfrac{x + 1}{\sqrt{x}}, \ y(1) = 0$

**40.** (a) $\dfrac{dy}{dx} = \dfrac{1}{(2x)^3}, \ y(1) = 0$

(b) $\dfrac{dy}{dt} = \sec^2 t - \sin t, \ y\left(\dfrac{\pi}{4}\right) = 1$

(c) $\dfrac{dy}{dx} = x^2 \sqrt{x^3}, \ y(0) = 0$

**41.** Find the general form of a function whose second derivative is $\sqrt{x}$. [*Hint:* Solve the equation $f''(x) = \sqrt{x}$ for $f(x)$ by integrating both sides twice.]

**42.** Find a function $f$ such that $f''(x) = x + \cos x$ and such that $f(0) = 1$ and $f'(0) = 2$. [*Hint:* Integrate both sides of the equation twice.]

In Exercises 43–45, find an equation of the curve that satisfies the given conditions.

**43.** At each point $(x, y)$ on the curve the slope is $2x + 1$; the curve passes through the point $(-3, 0)$.

**44.** At each point $(x, y)$ on the curve the slope equals the square of the distance between the point and the $y$-axis; the point $(-1, 2)$ is on the curve.

**45.** At each point $(x, y)$ on the curve, $y$ satisfies the condition $d^2y/dx^2 = 6x$; the line $y = 5 - 3x$ is tangent to the curve at the point where $x = 1$.

**46.** Suppose that a uniform metal rod 50 cm long is insulated laterally, and the temperatures at the exposed ends are maintained at $25°C$ and $85°C$, respectively. Assume that an $x$-axis is chosen as in the accompanying figure and that the temperature $T(x)$ at each point $x$ satisfies the equation

$$\frac{d^2 T}{dx^2} = 0$$

Find $T(x)$ for $0 \le x \le 50$.

Figure Ex-46

**47.** (a) Show that

$$F(x) = \tfrac{1}{6}(3x + 4)^2 \quad \text{and} \quad G(x) = \tfrac{3}{2}x^2 + 4x$$

differ by a constant by showing that they are antiderivatives of the same function.
(b) Find the constant $C$ such that $F(x) - G(x) = C$ by evaluating $F(x)$ and $G(x)$ at some point $x_0$.
(c) Check your answer in part (b) by simplifying the expression $F(x) - G(x)$ algebraically.

**48.** Follow the directions of Exercise 47 with

$$F(x) = \frac{x^2}{x^2 + 5} \quad \text{and} \quad G(x) = -\frac{5}{x^2 + 5}$$

In Exercises 49 and 50, use a trigonometric identity to help evaluate the integral.

**49.** $\displaystyle \int \tan^2 x \, dx$    **50.** $\displaystyle \int \cot^2 x \, dx$

**51.** Use the identities $\cos 2\theta = 1 - 2\sin^2 \theta = 2\cos^2 \theta - 1$ to help evaluate the integrals

(a) $\displaystyle \int \sin^2(x/2) \, dx$    (b) $\displaystyle \int \cos^2(x/2) \, dx$

**52.** Let $F$ and $G$ be the functions defined piecewise by

$$F(x) = \begin{cases} x, & x > 0 \\ -x, & x < 0 \end{cases} \quad \text{and} \quad G(x) = \begin{cases} x + 2, & x > 0 \\ -x + 3, & x < 0 \end{cases}$$

(a) Show that $F$ and $G$ have the same derivative.
(b) Show that $G(x) \neq F(x) + C$ for any constant $C$.
(c) Do parts (a) and (b) violate Theorem 7.2.2? Explain.

**53.** The speed of sound in air at $0°C$ (or 273 K on the Kelvin scale) is 1087 ft/s, but the speed $v$ increases as the temperature $T$ rises. Experimentation has shown that the rate of change of $v$ with respect to $T$ is

$$\frac{dv}{dT} = \frac{1087}{2\sqrt{273}} T^{-1/2}$$

where $v$ is in feet per second and $T$ is in kelvins (K). Find a formula that expresses $v$ as a function of $T$.

## 7.3 INTEGRATION BY SUBSTITUTION

*In this section we will study a technique, called* **substitution**, *that can often be used to transform complicated integration problems into simpler ones.*

*u*-SUBSTITUTION

The method of substitution can be motivated by examining the chain rule from the viewpoint of antidifferentiation. For this purpose, suppose that $F$ is an antiderivative of $f$ and that $g$ is a differentiable function. The chain rule implies that the derivative of $F(g(x))$ can be expressed as

$$\frac{d}{dx}[F(g(x))] = F'(g(x))g'(x)$$

which we can write in integral form as

$$\int F'(g(x))g'(x) \, dx = F(g(x)) + C \tag{1}$$

or since $F$ is an antiderivative of $f$,

$$\int f(g(x))g'(x) \, dx = F(g(x)) + C \tag{2}$$

For our purposes it will be useful to let $u = g(x)$ and to write $du/dx = g'(x)$ in the differential form $du = g'(x) \, dx$. With this notation (1) can be expressed as

$$\int f(u) \, du = F(u) + C \tag{3}$$

The process of evaluating an integral of form (2) by converting it into form (3) with the substitution

$$u = g(x) \quad \text{and} \quad du = g'(x) \, dx$$

is called the **method of u-substitution**. The following example illustrates how the method works.

### Example 1

Evaluate $\int (x^2 + 1)^{50} \cdot 2x \, dx$.

**Solution.** If we let $u = x^2 + 1$, then $du/dx = 2x$, which implies that $du = 2x \, dx$. Thus, the given integral can be written as

$$\int (x^2 + 1)^{50} \cdot 2x \, dx = \int u^{50} \, du = \frac{u^{51}}{51} + C = \frac{(x^2 + 1)^{51}}{51} + C \qquad \blacktriangleleft$$

It is important to realize that in the method of $u$-substitution you have control over the choice of $u$, but once you make that choice you have no control over the resulting expression for $du$. Thus, in the last example we *chose* $u = x^2 + 1$ but $du = 2x \, dx$ was *computed*. Fortunately, our choice of $u$, combined with the computed $du$, worked out perfectly to produce an integral involving $u$ that was easy to evaluate. However, in general, the method of $u$-substitution will fail if the chosen $u$ and the computed $du$ do not produce an integrand in which no expressions involving $x$ remain, or if you cannot evaluate the resulting integral. Thus, for example, the substitution $u = x^2 + 1$, $du = 2x \, dx$ will not work for the integral

$$\int (x^2 + 1)^{50} \cdot 2x \cos x \, dx$$

because this substitution results in the integral

$$\int u^{50} \cos x \, du$$

which still contains an expression involving $x$.

In general, there are no hard and fast rules for choosing $u$, and in some problems no choice of $u$ will work. In such cases other methods need to be used, some of which will be discussed later. Making appropriate choices for $u$ will come with experience, but you may find the following guidelines, combined with a mastery of the basic integrals in Table 7.2.1, helpful.

---

**Integration by Substitution**

**Step 1.** Make a choice for $u$, say $u = g(x)$.

**Step 2.** Compute $du/dx = g'(x)$.

**Step 3.** Make the substitution $u = g(x), du = g'(x) \, dx$.

At this stage, the *entire* integral must be in terms of $u$; no $x$'s should remain. If this is not the case, try a different choice of $u$.

**Step 4.** Evaluate the resulting integral, if possible.

**Step 5.** Replace $u$ by $g(x)$, so that the final answer is in terms of $x$.

---

### Example 2

The easiest substitutions occur when the integrand is the derivative of a known function, except for a constant added to or subtracted from the independent variable. For example,

$$\int \sin(x + 9) \, dx = \int \sin u \, du = -\cos u + C = -\cos(x + 9) + C$$

$$\underset{\substack{u = x + 9 \\ du = 1 \cdot dx = dx}}{}$$

$$\int (x - 8)^{23} \, dx = \int u^{23} \, du = \frac{u^{24}}{24} + C = \frac{(x - 8)^{24}}{24} + C \qquad \blacktriangleleft$$

$$\underset{\substack{u = x - 8 \\ du = 1 \cdot dx = dx}}{}$$

Another easy $u$-substitution occurs when the integrand is the derivative of a known function, except for a constant that multiplies or divides the independent variable. The following example illustrates two ways to evaluate such integrals.

### Example 3

Evaluate $\displaystyle\int \cos 5x \, dx$.

*Solution.*

$$\int \cos 5x \, dx = \int (\cos u) \cdot \frac{1}{5} du = \frac{1}{5} \int \cos u \, du = \frac{1}{5} \sin u + C = \frac{1}{5} \sin 5x + C$$

$$u = 5x$$
$$du = 5 \, dx \text{ or } dx = \tfrac{1}{5} \, du$$

*Alternative Solution.*   There is a variation of the preceding method that some people prefer. The substitution $u = 5x$ requires $du = 5 \, dx$. If there were a factor of 5 in the integrand, then we could group the 5 and $dx$ together to form the $du$ required by the substitution. Since there is no factor of 5, we will insert one and compensate by putting a factor of $\frac{1}{5}$ in front of the integral. The computations are as follows:

$$\int \cos 5x \, dx = \frac{1}{5} \int \cos 5x \cdot 5 \, dx = \frac{1}{5} \int \cos u \, du = \frac{1}{5} \sin u + C = \frac{1}{5} \sin 5x + C \quad \blacktriangleleft$$

$$u = 5x$$
$$du = 5 \, dx$$

### Example 4

Evaluate $\displaystyle\int \sin^2 x \cos x \, dx$.

*Solution.*   If we let $u = \sin x$, then

$$\frac{du}{dx} = \cos x, \quad \text{so} \quad du = \cos x \, dx$$

Thus,

$$\int \sin^2 x \cos x \, dx = \int u^2 \, du = \frac{u^3}{3} + C = \frac{\sin^3 x}{3} + C \quad \blacktriangleleft$$

### Example 5

Evaluate $\displaystyle\int \frac{e^{\sqrt{x}}}{\sqrt{x}} \, dx$.

*Solution.*   If we let $u = \sqrt{x}$, then

$$\frac{du}{dx} = \frac{1}{2\sqrt{x}}, \quad \text{so} \quad du = \frac{1}{2\sqrt{x}} \, dx \quad \text{or} \quad 2 \, du = \frac{1}{\sqrt{x}} \, dx$$

Thus,

$$\int \frac{e^{\sqrt{x}}}{\sqrt{x}} \, dx = \int 2e^u \, du = 2 \int e^u \, du = 2e^u + C = 2e^{\sqrt{x}} + C \quad \blacktriangleleft$$

### Example 6

$$\int \frac{dx}{\left(\frac{1}{3}x - 8\right)^5} = \int \frac{3 \, du}{u^5} = 3 \int u^{-5} \, du = -\frac{3}{4}u^{-4} + C = -\frac{3}{4}\left(\frac{1}{3}x - 8\right)^{-4} + C \quad \blacktriangleleft$$

$$u = \tfrac{1}{3}x - 8$$
$$du = \tfrac{1}{3} \, dx \text{ or } dx = 3 \, du$$

### Example 7

With the help of Theorem 7.2.3, a complicated integral can sometimes be computed by expressing it as a sum of simpler integrals. For example,

$$\int \left( \frac{1}{x} + \sec^2 \pi x \right) dx = \int \frac{dx}{x} + \int \sec^2 \pi x \, dx = \ln |x| + \int \sec^2 \pi x \, dx$$

$$= \ln |x| + \frac{1}{\pi} \int \sec^2 u \, du$$

$$u = \pi x$$
$$du = \pi \, dx \text{ or } dx = \frac{1}{\pi} \, du$$

$$= \ln |x| + \frac{1}{\pi} \tan u + C = \ln |x| + \frac{1}{\pi} \tan \pi x + C \qquad \blacktriangleleft$$

### Example 8

Evaluate $\int t^4 \sqrt[3]{3 - 5t^5} \, dt$.

*Solution.* After some possible false starts most readers would eventually hit on the following substitution:

$$\int t^4 \sqrt[3]{3 - 5t^5} \, dt = -\frac{1}{25} \int \sqrt[3]{u} \, du = -\frac{1}{25} \int u^{1/3} \, du$$

$$u = 3 - 5t^5$$
$$du = -25t^4 \, dt \text{ or } -\frac{1}{25} \, du = t^4 \, dt$$

$$= -\frac{1}{25} \frac{u^{4/3}}{4/3} + C = -\frac{3}{100} \left( 3 - 5t^5 \right)^{4/3} + C \qquad \blacktriangleleft$$

### Example 9

Evaluate $\int x^2 \sqrt{x - 1} \, dx$.

*Solution.* Let

$$u = x - 1 \quad \text{so that} \quad du = dx \qquad (4)$$

From the first equality in (4)

$$x^2 = (u + 1)^2 = u^2 + 2u + 1$$

so that

$$\int x^2 \sqrt{x - 1} \, dx = \int (u^2 + 2u + 1) \sqrt{u} \, du = \int (u^{5/2} + 2u^{3/2} + u^{1/2}) \, du$$

$$= \tfrac{2}{7} u^{7/2} + \tfrac{4}{5} u^{5/2} + \tfrac{2}{3} u^{3/2} + C$$

$$= \tfrac{2}{7} (x - 1)^{7/2} + \tfrac{4}{5} (x - 1)^{5/2} + \tfrac{2}{3} (x - 1)^{3/2} + C \qquad \blacktriangleleft$$

REMARK. Not every function can be integrated in terms of familiar functions using $u$-substitutions. For example, you will not find any $u$-substitution that will integrate

$$\int \sin(x^2) \, dx$$

in terms of functions encountered thus far in this text (try).

**INTEGRATION USING COMPUTER ALGEBRA SYSTEMS**

The advent of computer algebra systems has made it possible to evaluate many kinds of integrals that would be laborious to evaluate by hand. For example, *Mathematica*, *Maple*, and *Derive* all produce the following result in a matter of seconds:

$$\int \sqrt{2x - x^2}\, dx = \tfrac{1}{2}(x - 1)\sqrt{2x - x^2} - \tfrac{1}{2}\sin^{-1}(1 - x) + C$$

However, just as one would not want to rely on a calculator to compute $2 + 2$, so one would not want to use a CAS to integrate a simple function such as $f(x) = x^2$. Thus, even if you have a CAS, you will want to develop a reasonable level of competence in evaluating basic integrals. Moreover, the mathematical techniques that we will introduce for evaluating basic integrals are precisely the techniques that computer algebra systems use to evaluate more complicated integrals.

FOR THE READER. If you have a CAS, use it to calculate the integrals in the examples of this section. If your CAS produces a form of the answer that is different from the one in the text, then confirm algebraically that the two answers agree. Your CAS has various commands for simplifying answers. Explore the effect of using the CAS to simplify the expressions it produces for the integrals.

**EXERCISE SET 7.3** ⊠ Graphing Calculator ⊡ CAS

In Exercises 1–4, evaluate the integrals by making the indicated substitutions.

1. (a) $\displaystyle\int 2x\left(x^2 + 1\right)^{23} dx; \ u = x^2 + 1$

   (b) $\displaystyle\int \cos^3 x \sin x\, dx; \ u = \cos x$

   (c) $\displaystyle\int \frac{1}{\sqrt{x}} \sin \sqrt{x}\, dx; \ u = \sqrt{x}$

   (d) $\displaystyle\int \frac{3x\, dx}{\sqrt{4x^2 + 5}}; \ u = 4x^2 + 5$

   (e) $\displaystyle\int \frac{x^2}{x^3 - 4}\, dx; \ u = x^3 - 4$

2. (a) $\displaystyle\int \sec^2(4x + 1)\, dx; \ u = 4x + 1$

   (b) $\displaystyle\int y\sqrt{1 + 2y^2}\, dy; \ u = 1 + 2y^2$

   (c) $\displaystyle\int \sqrt{\sin \pi\theta} \cos \pi\theta\, d\theta; \ u = \sin \pi\theta$

   (d) $\displaystyle\int (2x + 7)(x^2 + 7x + 3)^{4/5}\, dx; \ u = x^2 + 7x + 3$

   (e) $\displaystyle\int \frac{e^x}{1 + e^x}\, dx; \ u = 1 + e^x$

3. (a) $\displaystyle\int \cot x \csc^2 x\, dx; \ u = \cot x$

   (b) $\displaystyle\int (1 + \sin t)^9 \cos t\, dt; \ u = 1 + \sin t$

   (c) $\displaystyle\int \frac{dx}{x \ln x}; \ u = \ln x$

   (d) $\displaystyle\int e^{-5x}\, dx; \ u = -5x$

   (e) $\displaystyle\int \frac{\sin 3\theta}{1 + \cos 3\theta}\, d\theta; \ u = 1 + \cos 3\theta$

4. (a) $\displaystyle\int x^2 \sqrt{1 + x}\, dx; \ u = 1 + x$

   (b) $\displaystyle\int [\csc(\sin x)]^2 \cos x\, dx; \ u = \sin x$

   (c) $\displaystyle\int e^{\tan x} \sec^2 x\, dx; \ u = \tan x$

   (d) $\displaystyle\int e^{2t}\sqrt{1 + e^{2t}}\, dt; \ u = 1 + e^{2t}$

   (e) $\displaystyle\int \frac{5x^4}{x^5 + 1}\, dx; \ u = x^5 + 1$

In Exercises 5–36, evaluate the integrals by making appropriate substitutions.

5. $\displaystyle\int e^{2x}\, dx$

6. $\displaystyle\int \frac{dx}{2x}$

7. $\displaystyle\int x\left(2 - x^2\right)^3 dx$

8. $\displaystyle\int (3x - 1)^5\, dx$

9. $\displaystyle\int \cos 8x\, dx$

10. $\displaystyle\int \sin 3x\, dx$

**11.** $\displaystyle\int \sec 4x \tan 4x \, dx$

**12.** $\displaystyle\int \sec^2 5x \, dx$

**13.** $\displaystyle\int t\sqrt{7t^2 + 12} \, dt$

**14.** $\displaystyle\int \frac{x}{\sqrt{4 - 5x^2}} \, dx$

**15.** $\displaystyle\int \frac{x^2}{\sqrt{x^3 + 1}} \, dx$

**16.** $\displaystyle\int \frac{1}{(1 - 3x)^2} \, dx$

**17.** $\displaystyle\int \frac{x}{(4x^2 + 1)^3} \, dx$

**18.** $\displaystyle\int x \cos(3x^2) \, dx$

**19.** $\displaystyle\int e^{\sin x} \cos x \, dx$

**20.** $\displaystyle\int x^3 e^{x^4} \, dx$

**21.** $\displaystyle\int x^2 e^{-2x^3} \, dx$

**22.** $\displaystyle\int \frac{e^x + e^{-x}}{e^x - e^{-x}} \, dx$

**23.** $\displaystyle\int \frac{\sin(5/x)}{x^2} \, dx$

**24.** $\displaystyle\int \frac{\sec^2(\sqrt{x})}{\sqrt{x}} \, dx$

**25.** $\displaystyle\int x^2 \sec^2(x^3) \, dx$

**26.** $\displaystyle\int \cos^3 2t \sin 2t \, dt$

**27.** $\displaystyle\int \frac{dx}{e^x}$

**28.** $\displaystyle\int \sqrt{e^x} \, dx$

**29.** $\displaystyle\int \sin^5 3t \cos 3t \, dt$

**30.** $\displaystyle\int \frac{\sin 2\theta}{(5 + \cos 2\theta)^3} \, d\theta$

**31.** $\displaystyle\int \cos 4\theta \sqrt{2 - \sin 4\theta} \, d\theta$

**32.** $\displaystyle\int \tan^3 5x \sec^2 5x \, dx$

**33.** $\displaystyle\int \sec^3 2x \tan 2x \, dx$

**34.** $\displaystyle\int [\sin(\sin \theta)] \cos \theta \, d\theta$

**35.** $\displaystyle\int \frac{e^{\sqrt{y}}}{\sqrt{y}} \, dy$

**36.** $\displaystyle\int \frac{dy}{\sqrt{y} e^{\sqrt{y}}}$

c **37.** For each of the integrals you evaluated in Exercises 5–36, use a CAS to check your answer. If the answer produced by the CAS does not match your own, show that the two answers are equivalent. [*Suggestion:* You may be able to obtain a match by applying the CAS "simplify" commands to the answer.]

In Exercises 38 and 39, evaluate the integrals assuming that $n$ is a positive integer and $b \neq 0$.

**38.** $\displaystyle\int \sqrt[n]{a + bx} \, dx \quad (b \neq 0)$

**39.** $\displaystyle\int \sin^n(a + bx) \cos(a + bx) \, dx$

c **40.** Use a CAS to check the answers you obtained in Exercises 38 and 39. If the answer produced by the CAS does not match yours, show that the two answers are equivalent. [*Suggestion: Mathematica* users may find it helpful to apply the Simplify command to the answer.]

In Exercises 41 and 42, evaluate the integrals by making the indicated substitutions.

**41.** $\displaystyle\int x\sqrt{x - 3} \, dx; \quad u = x - 3$

**42.** $\displaystyle\int \frac{y \, dy}{\sqrt{y + 1}}; \quad u = y + 1$

The integrals in Exercises 43–48 are a little trickier than those you have encountered thus far. To evaluate these integrals you will have to apply a trigonometric identity or modify the form of the integrand algebraically before making a substitution.

**43.** $\displaystyle\int \tan^2 3\theta \, d\theta$

**44.** $\displaystyle\int \sin^3 2\theta \, d\theta$

**45.** $\displaystyle\int \frac{t + 1}{t} \, dt$

**46.** $\displaystyle\int e^{2\ln x} \, dx$

**47.** $\displaystyle\int [\ln(e^x) + \ln(e^{-x})] \, dx$

**48.** $\displaystyle\int \cot x \, dx$

**49.** (a) Evaluate the integral $\int \sin x \cos x \, dx$ by two methods: first by letting $u = \sin x$, then by letting $u = \cos x$.

(b) Explain why the two apparently different answers obtained in part (a) are really equivalent.

**50.** (a) Evaluate $\int (5x - 1)^2 \, dx$ by two methods: first square and integrate, then let $u = 5x - 1$.

(b) Explain why the two apparently different answers obtained in part (a) are really equivalent.

In Exercises 51 and 52, solve the initial-value problems.

**51.** $\dfrac{dy}{dx} = \sqrt{3x + 1}; \quad y(1) = 5$

**52.** $\dfrac{dy}{dx} = 6 - 5\sin 2x; \quad y(0) = 3$

**53.** Find a function $f$ such that the slope of the tangent line at a point $(x, y)$ on the curve $y = f(x)$ is $\sqrt{3x + 1}$, and the curve passes through the point $(0, 1)$.

**54.** Use a graphing utility to generate some typical integral curves of $f(x) = x/(x^2 + 1)$ over the interval $(-5, 5)$.

**55.** Suppose that a population $p$ of frogs is estimated at the start of 1995 to be 100,000, and the growth model for the population assumes that the rate of growth (in thousands) after $t$ years will be $p'(t) = (4 + 0.15t)^{3/2}$. Estimate the projected population at the start of the year 2000.

**56.** Suppose that the radius $r$ of a spherical meteorite entering the Earth's atmosphere at time $t = 0$ decreases at a rate that is proportional to the square root of $t$, where $r$ is in meters and $t$ is in seconds. The initial radius is 10,000 m, and after 25 s the radius is 9000 m. Estimate the radius after 60 s.

## 7.4  SIGMA NOTATION

*In this section we will digress briefly from the main theme of this chapter to introduce a notation that can be used to write lengthy sums in a compact form. This material will be needed in many of the later chapters.*

**SIGMA NOTATION**

The notation we will discuss in this section is called *sigma notation* or *summation notation* because it uses the uppercase Greek letter $\Sigma$ (sigma) to denote various kinds of sums. To illustrate how this notation works, consider the sum

$$1^2 + 2^2 + 3^2 + 4^2 + 5^2$$

in which each term is of the form $k^2$, where $k$ is one of the integers from 1 to 5. In sigma notation this sum can be written as

$$\sum_{k=1}^{5} k^2$$

which is read "the summation of $k^2$, where $k$ runs from 1 to 5." The notation tells us to form the sum of the terms that result when we substitute successive integers for $k$ in the expression $k^2$, starting with $k = 1$ and ending with $k = 5$.

More generally, if $f(k)$ is a function of $k$, and if $m$ and $n$ are integers such that $m \leq n$, then

$$\sum_{k=m}^{n} f(k) \tag{1}$$

denotes the sum of the terms that result when we substitute successive integers for $k$, starting with $k = m$ and ending with $k = n$ (Figure 7.4.1).

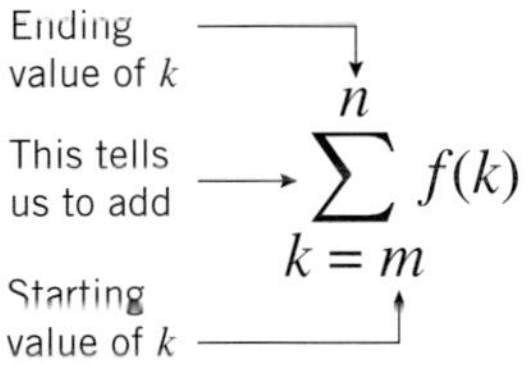

Ending value of $k$

This tells us to add → $\displaystyle\sum_{k=m}^{n} f(k)$

Starting value of $k$

Figure 7.4.1

**Example 1**

$$\sum_{k=4}^{8} k^3 = 4^3 + 5^3 + 6^3 + 7^3 + 8^3$$

$$\sum_{k=1}^{5} 2k = 2 \cdot 1 + 2 \cdot 2 + 2 \cdot 3 + 2 \cdot 4 + 2 \cdot 5 = 2 + 4 + 6 + 8 + 10$$

$$\sum_{k=0}^{5} (2k + 1) = 1 + 3 + 5 + 7 + 9 + 11$$

$$\sum_{k=0}^{5} (-1)^k (2k + 1) = 1 - 3 + 5 - 7 + 9 - 11$$

$$\sum_{k=-3}^{1} k^3 = (-3)^3 + (-2)^3 + (-1)^3 + 0^3 + 1^3 = -27 - 8 - 1 + 0 + 1$$

$$\sum_{k=1}^{3} k \sin\left(\frac{k\pi}{5}\right) = \sin\frac{\pi}{5} + 2 \sin\frac{2\pi}{5} + 3 \sin\frac{3\pi}{5} \qquad \blacktriangleleft$$

The numbers $m$ and $n$ in (1) are called, respectively, the *lower* and *upper limits of summation*; and the letter $k$ is called the *index of summation*. It is not essential to use $k$ as the index of summation; any letter not reserved for another purpose will do. For example,

$$\sum_{i=1}^{6} \frac{1}{i}, \quad \sum_{j=1}^{6} \frac{1}{j}, \quad \text{and} \quad \sum_{n=1}^{6} \frac{1}{n}$$

all denote the sum

$$1 + \frac{1}{2} + \frac{1}{3} + \frac{1}{4} + \frac{1}{5} + \frac{1}{6}$$

If the upper and lower limits of summation are the same, then the "sum" in (1) reduces to one term. For example,

$$\sum_{k=2}^{2} k^3 = 2^3 \quad \text{and} \quad \sum_{i=1}^{1} \frac{1}{i+2} = \frac{1}{1+2} = \frac{1}{3}$$

In the sums

$$\sum_{i=1}^{5} 2, \quad \sum_{k=3}^{6} 7, \quad \text{and} \quad \sum_{j=0}^{2} x^3$$

the expression to the right of the $\Sigma$ sign does not involve the index of summation. In such cases, we take all the terms in the sum to be the same, with one term for each allowable value of the summation index. Thus,

$$\sum_{i=1}^{5} 2 = 2 + 2 + 2 + 2 + 2$$

$$\sum_{k=3}^{6} 7 = 7 + 7 + 7 + 7$$

$$\sum_{j=0}^{2} x^3 = x^3 + x^3 + x^3$$

A sum can be written in more than one way with sigma notation by changing the limits of summation. For example, the sum of the first five positive even integers can be written in the following ways:

$$\sum_{k=1}^{5} 2k = 2 + 4 + 6 + 8 + 10$$

$$\sum_{k=0}^{4} (2k + 2) = 2 + 4 + 6 + 8 + 10$$

$$\sum_{k=2}^{6} (2k - 2) = 2 + 4 + 6 + 8 + 10$$

**CHANGING THE INDEX OF SUMMATION**

On occasion we will want to change the sigma notation for a given sum to a sigma notation with different limits of summation. The following example illustrates a method for doing this.

**Example 2**

Express

$$\sum_{k=3}^{7} 5^{k-2}$$

in sigma notation so that the lower limit of summation is 0 rather than 3.

*Solution.* If we define a new summation index $j$ by means of the formula

$$j = k - 3 \tag{2}$$

then $j$ runs from 0 up to 4 as $k$ runs from 3 up to 7. From (2), $k = j + 3$, so

$$\sum_{k=3}^{7} 5^{k-2} = \sum_{j=0}^{4} 5^{(j+3)-2} = \sum_{j=0}^{4} 5^{j+1}$$

As a check, the reader can verify that

$$\sum_{j=0}^{4} 5^{j+1} \quad \text{and} \quad \sum_{k=3}^{7} 5^{k-2}$$

both denote the sum $5 + 5^2 + 5^3 + 5^4 + 5^5$.   ◀

REMARK.   In the solution of Example 2 the summation index was changed from $k$ to $j$. If it is desirable to keep the same symbol for the summation index, we can change the $j$ back to $k$ *at the very end* and express the final result as

$$\sum_{k=0}^{4} 5^{k+1} \quad \text{instead of} \quad \sum_{j=0}^{4} 5^{j+1}$$

When we want to represent a general sum we will use letters with subscripts. For example, a general sum with five terms might be written as

$$a_1 + a_2 + a_3 + a_4 + a_5$$

or in sigma notation as

$$\sum_{k=1}^{5} a_k, \quad \sum_{j=1}^{5} a_j, \quad \text{or} \quad \sum_{m=1}^{5} a_m$$

A general sum with $n$ terms might be written as

$$b_1 + b_2 + \cdots + b_n$$

or in sigma notation as

$$\sum_{k=1}^{n} b_k, \quad \sum_{j=1}^{n} b_j, \quad \text{or} \quad \sum_{m=1}^{n} b_m$$

**PROPERTIES OF SIGMA NOTATION**

The following properties of sigma notation will help to manipulate sums:

**7.4.1   THEOREM.**

$$(a) \quad \sum_{k=1}^{n} ca_k = c \sum_{k=1}^{n} a_k$$

$$(b) \quad \sum_{k=1}^{n} (a_k + b_k) = \sum_{k=1}^{n} a_k + \sum_{k=1}^{n} b_k$$

$$(c) \quad \sum_{k=1}^{n} (a_k - b_k) = \sum_{k=1}^{n} a_k - \sum_{k=1}^{n} b_k$$

We will prove parts $(a)$ and $(b)$ and leave part $(c)$ as an exercise.

*Proof $(a)$.*

$$\sum_{k=1}^{n} ca_k = ca_1 + ca_2 + \cdots + ca_n = c(a_1 + a_2 + \cdots + a_n) = c \sum_{k=1}^{n} a_k$$

*Proof (b).*

$$\sum_{k=1}^{n}(a_k + b_k) = (a_1 + b_1) + (a_2 + b_2) + \cdots + (a_n + b_n)$$

$$= (a_1 + a_2 + \cdots + a_n) + (b_1 + b_2 + \cdots + b_n)$$

$$= \sum_{k=1}^{n} a_k + \sum_{k=1}^{n} b_k \qquad \blacksquare$$

REMARK.  Loosely phrased, this theorem states: *A constant factor can be moved through a sigma sign*; *sigma of a sum equals the sum of the sigmas*; *and sigma of a difference equals the difference of the sigmas.*

**SUMMATION FORMULAS**

The following formulas will be used in our later work.

---

**7.4.2  THEOREM.**

(a) $\displaystyle\sum_{k=1}^{n} k = 1 + 2 + 3 + \cdots + n = \frac{n(n+1)}{2}$

(b) $\displaystyle\sum_{k=1}^{n} k^2 = 1^2 + 2^2 + 3^2 + \cdots + n^2 = \frac{n(n+1)(2n+1)}{6}$

(c) $\displaystyle\sum_{k=1}^{n} k^3 = 1^3 + 2^3 + 3^3 + \cdots + n^3 = \left[\frac{n(n+1)}{2}\right]^2$

---

We will prove parts (*a*) and (*b*) and leave part (*c*) as an exercise.

*Proof (a).* If we write the terms of

$$\sum_{k=1}^{n} k = 1 + 2 + 3 + \cdots + (n-2) + (n-1) + n \qquad (3)$$

in the opposite order, we obtain

$$\sum_{k=1}^{n} k = n + (n-1) + (n-2) + \cdots + 3 + 2 + 1 \qquad (4)$$

Adding (3) and (4) term by term yields

$$2\sum_{k=1}^{n} k = \underbrace{(n+1) + (n+1) + (n+1) + \cdots + (n+1)}_{n\,\text{terms}} = n(n+1)$$

Thus,

$$\sum_{k=1}^{n} k = \frac{n(n+1)}{2}$$

*Proof (b).* This proof begins with a trick. Since

$$(k+1)^3 - k^3 = k^3 + 3k^2 + 3k + 1 - k^3 = 3k^2 + 3k + 1$$

we obtain

$$\sum_{k=1}^{n}[(k+1)^3 - k^3] = \sum_{k=1}^{n}(3k^2 + 3k + 1) \qquad (5)$$

Writing out the left side of (5) yields

$$[2^3 - 1^3] + [3^3 - 2^3] + [4^3 - 3^3] + \cdots + [(n+1)^3 - n^3] \qquad (6)$$

Observe that in (6) the $2^3$ in the first term cancels out the $-2^3$ in the second term, the $3^3$ in the second term cancels out the $-3^3$ in the third term, and so forth, so that the entire sum collapses like a folding telescope (hence, is called a ***telescoping sum***), leaving only $-1^3 + (n + 1)^3$. Thus, (5) can be rewritten as

$$-1 + (n + 1)^3 = \sum_{k=1}^{n}(3k^2 + 3k + 1) \tag{7}$$

or, from Theorem 7.4.1,

$$-1 + (n + 1)^3 = 3\sum_{k=1}^{n}k^2 + 3\sum_{k=1}^{n}k + \sum_{k=1}^{n}1 \tag{8}$$

But

$$\sum_{k=1}^{n}1 = \underbrace{1 + 1 + \cdots + 1}_{n\text{ terms}} = n$$

and by part $(a)$ of this theorem

$$\sum_{k=1}^{n}k = \frac{n(n + 1)}{2}$$

Thus, (8) can be written as

$$-1 + (n + 1)^3 = 3\sum_{k=1}^{n}k^2 + 3\frac{n(n + 1)}{2} + n$$

Therefore,

$$\sum_{k=1}^{n}k^2 = \frac{1}{3}\left[(n + 1)^3 - 3\frac{n(n + 1)}{2} - (n + 1)\right]$$

$$= \frac{n + 1}{6}[2(n + 1)^2 - 3n - 2]$$

$$= \frac{n + 1}{6}(2n^2 + n) = \frac{n(n + 1)(2n + 1)}{6}$$

$\blacksquare$

**Example 3**

Evaluate $\displaystyle\sum_{k=1}^{30}k(k + 1)$.

*Solution.*

$$\sum_{k=1}^{30}k(k + 1) = \sum_{k=1}^{30}(k^2 + k) = \sum_{k=1}^{30}k^2 + \sum_{k=1}^{30}k$$

$$= \frac{30(31)(61)}{6} + \frac{30(31)}{2} = 9920 \qquad \boxed{\text{Theorem } 7.4.2(a), (b)} \qquad \blacktriangleleft$$

REMARK.    In formulas such as

$$\sum_{k=1}^{n}k^2 = \frac{n(n + 1)(2n + 1)}{6}$$

or

$$1^2 + 2^2 + \cdots + n^2 = \frac{n(n + 1)(2n + 1)}{6}$$

the left side of the equality is said to express the sum in ***open form*** and the right side is said to express it in ***closed form***; the open form just indicates the terms to be added, while the closed form is an explicit formula for their sum.

### Example 4

Express $\displaystyle\sum_{k=1}^{n}(3+k)^2$ in closed form.

**Solution.**

$$\sum_{k=1}^{n}(3+k)^2 = \sum_{k=1}^{n}(9+6k+k^2) = \sum_{k=1}^{n}9 + 6\sum_{k=1}^{n}k + \sum_{k=1}^{n}k^2$$

$$= 9n + 6\frac{n(n+1)}{2} + \frac{n(n+1)(2n+1)}{6}$$

$$= \frac{1}{3}n^3 + \frac{7}{2}n^2 + \frac{73}{6}n \qquad\qquad \blacktriangleleft$$

FOR THE READER.   Your numerical calculating utility probably provides some way of evaluating sums that can be expressed in sigma notation. Check your documentation to find out how to do this, and then use your utility to confirm that the numerical result obtained in Example 3 is correct. If you have access to a CAS, then it provides some method for finding closed forms for sums such as those in Theorem 7.4.2. Use your CAS to confirm the formulas in that theorem, and then find closed forms for

$$\sum_{k=1}^{n}k^4 \quad \text{and} \quad \sum_{k=1}^{n}k^5$$

### EXERCISE SET 7.4   ⬚ Graphing Calculator   [c] CAS

**1.** Evaluate

(a) $\displaystyle\sum_{k=1}^{3}k^3$    (b) $\displaystyle\sum_{j=2}^{6}(3j-1)$    (c) $\displaystyle\sum_{i=-4}^{1}(i^2-i)$

(d) $\displaystyle\sum_{n=0}^{5}1$    (e) $\displaystyle\sum_{k=0}^{4}(-2)^k$    (f) $\displaystyle\sum_{n=1}^{6}\sin n\pi$.

**2.** Evaluate

(a) $\displaystyle\sum_{k=1}^{4}k\sin\frac{k\pi}{2}$    (b) $\displaystyle\sum_{j=0}^{5}(-1)^j$    (c) $\displaystyle\sum_{i=7}^{20}e^2$

(d) $\displaystyle\sum_{m=3}^{5}2^{m+1}$    (e) $\displaystyle\sum_{n=1}^{6}\ln n$    (f) $\displaystyle\sum_{k=0}^{10}\cos k\pi$.

In Exercises 3–12, write each expression in sigma notation, but do not evaluate.

**3.** $1 + 2 + 3 + \cdots + 10$

**4.** $3\cdot1 + 3\cdot2 + 3\cdot3 + \cdots + 3\cdot20$

**5.** $1\cdot2 + 2\cdot3 + 3\cdot4 + \cdots + 49\cdot50$

**6.** $1 + 2 + 2^2 + 2^3 + 2^4$

**7.** $2 + 4 + 6 + 8 + \cdots + 20$

**8.** $1 + 3 + 5 + 7 + \cdots + 15$

**9.** $1 - 3 + 5 - 7 + 9 - 11$

**10.** $1 - \dfrac{1}{2} + \dfrac{1}{3} - \dfrac{1}{4} + \dfrac{1}{5}$

**11.** $-1 + \dfrac{1}{2} - \dfrac{1}{3} + \dfrac{1}{4} - \dfrac{1}{5}$

**12.** $1 + \cos\dfrac{\pi}{7} + \cos\dfrac{2\pi}{7} + \cos\dfrac{3\pi}{7}$

**13.** (a) Express the sum of the even integers from 2 to 100 in sigma notation.

(b) Express the sum of the odd integers from 1 to 99 in sigma notation.

**14.** Express in sigma notation.

(a) $a_1 - a_2 + a_3 - a_4 + a_5$

(b) $-b_0 + b_1 - b_2 + b_3 - b_4 + b_5$

(c) $a_0 + a_1x + a_2x^2 + \cdots + a_nx^n$

(d) $a^5 + a^4b + a^3b^2 + a^2b^3 + ab^4 + b^5$

In Exercises 15–22, use Theorem 7.4.2 to evaluate the sums, and check your answers using the summation feature of a calculating utility.

**15.** $\displaystyle\sum_{k=1}^{100}k$    **16.** $\displaystyle\sum_{k=3}^{100}k$    **17.** $\displaystyle\sum_{k=1}^{20}k^2$

**18.** $\displaystyle\sum_{k=1}^{100}(7k+1)$    **19.** $\displaystyle\sum_{k=1}^{6}(4k^3-2k+1)$    **20.** $\displaystyle\sum_{k=4}^{20}k^2$

**21.** $\displaystyle\sum_{k=1}^{30} k(k-2)(k+2)$      **22.** $\displaystyle\sum_{k=1}^{6}(k-k^3)$

---

In Exercises 23–28, express the sums in closed form.

---

**23.** $\displaystyle\sum_{k=1}^{n}(4k-3)$    **24.** $\displaystyle\sum_{k=1}^{n-1} k^2$    **25.** $\displaystyle\sum_{k=1}^{n}\frac{3k}{n}$

**26.** $\displaystyle\sum_{k=1}^{n-1}\frac{k^2}{n}$    **27.** $\displaystyle\sum_{k=1}^{n-1}\frac{k^3}{n^2}$    **28.** $\displaystyle\sum_{k=1}^{n}\left(\frac{5}{n}-\frac{2k}{n}\right)$

[c] **29.** For each of the sums that you obtained in Exercises 23–28, use a CAS to check your answer. If the answer produced by the CAS does not match your own, show that the two answers are equivalent.

**30.** Let
$$S = \sum_{k=0}^{n} ar^k$$
Show that $S - rS = a - ar^{n+1}$ and hence that
$$\sum_{k=0}^{n} ar^k = \frac{a - ar^{n+1}}{1-r} \quad (r \neq 1)$$
(A sum of this form is called a **geometric sum**.)

**31.** In each part, rewrite the sum, if necessary, so that the lower limit is 0, and then use the formula derived in Exercise 30 to evaluate the sum. Check your answers using the summation feature of a calculating utility.

(a) $\displaystyle\sum_{k=1}^{20} 3^k$    (b) $\displaystyle\sum_{k=5}^{30} 2^k$    (c) $\displaystyle\sum_{k=0}^{100}(-1)^{k+1}\frac{1}{2^k}$

[c] **32.** In each part, make a conjecture about the limit by using a CAS to evaluate the sum for $n = 10$, 20, and 50; and then check your conjecture by using the formula in Exercise 30 to express the sum in closed form, and then finding the limit exactly.

(a) $\displaystyle\lim_{n\to+\infty}\sum_{k=0}^{n}\frac{1}{2^k}$    (b) $\displaystyle\lim_{n\to+\infty}\sum_{k=1}^{n}\left(\frac{3}{4}\right)^k$

---

In Exercises 33–36, express the function of $n$ in closed form, and then use L'Hôpital's rule to find the limit. [*Note:* L'Hôpital's rule was derived for functions of a real-valued variable $x$, whereas here the variable $n$ assumes only integer values. Thus, strictly speaking, L'Hôpital's rule cannot be used without justifying that it applies to functions of integer-valued variables. We will do this later in the text.]

---

**33.** $\displaystyle\lim_{n\to+\infty}\frac{1+2+3+\cdots+n}{n^2}$

**34.** $\displaystyle\lim_{n\to+\infty}\frac{1^2+2^2+3^2+\cdots+n^2}{n^3}$

**35.** $\displaystyle\lim_{n\to+\infty}\sum_{k=1}^{n}\frac{5k}{n^2}$    **36.** $\displaystyle\lim_{n\to+\infty}\sum_{k=1}^{n-1}\frac{2k^2}{n^3}$

**37.** Express $1 + 2 + 2^2 + 2^3 + 2^4 + 2^5$ in sigma notation with
(a) $j = 0$ as the lower limit of summation

(b) $j = 1$ as the lower limit of summation
(c) $j = 2$ as the lower limit of summation.

**38.** Express
$$\sum_{k=5}^{9} k2^{k+4}$$
in sigma notation with
(a) $k = 1$ as the lower limit of summation
(b) $k = 13$ as the upper limit of summation.

**39.** Change the limits of summation appropriately to simplify

(a) $\displaystyle\sum_{k=11}^{28}\sin\left(\frac{\pi}{k-10}\right)$    (b) $\displaystyle\sum_{k=6}^{12} e^{k-6}$

**40.** Show that the sum of the first $n$ consecutive positive odd integers is $n^2$.

**41.** The accompanying figure shows a square that is $n$ units by $n$ units that has been subdivided into a one-unit square and $n-1$ "$L$-shaped" regions. Use this figure to derive the result in Exercise 40.

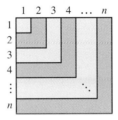

Figure Ex-41

**42.** Solve the equation $\displaystyle\sum_{k=1}^{n} k = 465$.

---

When part of each term of a sum cancels part of the next term, leaving only portions of the first and last terms at the end, the sum is said to **telescope**. In Exercises 43–46, evaluate the telescoping sum.

---

**43.** $\displaystyle\sum_{k=5}^{17}(3^k - 3^{k-1})$    **44.** $\displaystyle\sum_{k=1}^{50}\left(\frac{1}{k}-\frac{1}{k+1}\right)$

**45.** $\displaystyle\sum_{k=2}^{20}\left(\frac{1}{k^2}-\frac{1}{(k-1)^2}\right)$    **46.** $\displaystyle\sum_{k=1}^{100}(2^{k+1}-2^k)$

**47.** (a) Show that
$$\frac{1}{1\cdot 3}+\frac{1}{3\cdot 5}+\cdots+\frac{1}{(2n-1)(2n+1)}=\frac{n}{2n+1}$$
$$\left[Hint: \frac{1}{(2n-1)(2n+1)}=\frac{1}{2}\left(\frac{1}{2n-1}-\frac{1}{2n+1}\right).\right]$$

(b) Use the result in part (a) to find
$$\lim_{n\to+\infty}\sum_{k=1}^{n}\frac{1}{(2k-1)(2k+1)}$$

**48.** (a) Show that

$$\frac{1}{1 \cdot 2} + \frac{1}{2 \cdot 3} + \frac{1}{3 \cdot 4} + \cdots + \frac{1}{n(n+1)} = \frac{n}{n+1}$$

$$\left[ Hint: \frac{1}{n(n+1)} = \frac{1}{n} - \frac{1}{n+1}. \right]$$

(b) Use the result in part (a) to find

$$\lim_{n \to +\infty} \sum_{k=1}^{n} \frac{1}{k(k+1)}$$

**49.** By writing out the sums, determine whether the following are valid identities.

(a) $\displaystyle \int \left[ \sum_{i=1}^{n} f_i(x) \right] dx = \sum_{i=1}^{n} \left[ \int f_i(x)\, dx \right]$

(b) $\displaystyle \frac{d}{dx} \left[ \sum_{i=1}^{n} f_i(x) \right] = \sum_{i=1}^{n} \left[ \frac{d}{dx}[f_i(x)] \right]$

**50.** Which of the following are valid identities?

(a) $\displaystyle \sum_{i=1}^{n} a_i b_i = \sum_{i=1}^{n} a_i \sum_{i=1}^{n} b_i$

(b) $\displaystyle \sum_{i=1}^{n} \frac{a_i}{b_i} = \sum_{i=1}^{n} a_i \Bigg/ \sum_{i=1}^{n} b_i$

(c) $\displaystyle \sum_{i=1}^{n} a_i^2 = \left( \sum_{i=1}^{n} a_i \right)^2$

**51.** Let $\bar{x}$ denote the arithmetic average of the $n$ numbers $x_1, x_2, \ldots, x_n$. Use Theorem 7.4.1 to prove that

$$\sum_{i=1}^{n} (x_i - \bar{x}) = 0$$

**52.** Prove part $(c)$ of Theorem 7.4.1.

**53.** Prove part $(c)$ of Theorem 7.4.2. [*Hint:* Begin with the difference $(k+1)^4 - k^4$ and follow the steps used to prove part $(b)$ of the theorem.]

**54.** An artist wants to create a rough triangular design using uniform square tiles glued edge to edge. She places $n$ tiles in a row to form the base of the triangle and then makes each successive row two tiles shorter than the preceding row. Find a formula for the number of tiles used in the design. [*Hint:* Your answer will depend on whether $n$ is even or odd.]

**55.** An artist wants to create a sculpture by gluing together uniform spheres. She creates a rough rectangular base that has 50 spheres along one edge and 30 spheres along the other. She then creates successive layers by gluing spheres in the grooves of the preceding layer. How many spheres will there be in the sculpture?

## 7.5 THE DEFINITE INTEGRAL

*Recall from the informal discussion in Section 7.1 that if a function f is continuous and nonnegative on an interval [a, b], then the area under the graph of f over the interval [a, b] can be obtained by either the "rectangle method" or "the antiderivative method." In this section we will discuss the rectangle method in more detail, and we will introduce the concept of a "definite integral," which will link the concept of area to other important concepts such as length, volume, density, probability, and work.*

**A DEFINITION OF AREA**

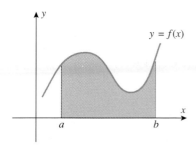

Figure 7.5.1

Our first goal in this section is to define formally what we mean by the area of a region $R$ that is bounded below by the $x$-axis, bounded on the sides by the vertical lines $x = a$ and $x = b$, and bounded above by the curve $y = f(x)$, where $f$ is continuous and nonnegative on the interval $[a, b]$ (Figure 7.5.1). We will start by defining the area of a rectangle to be the product of its length and width and defining the area of a region composed of finitely many rectangles to be the sum of the areas of those rectangles. To define the area of the region $R$, we will use these definitions and the rectangle method of Section 7.1. The basic idea is as follows (Figure 7.5.2):

- Divide the interval $[a, b]$ into $n$ equal subintervals.
- Over each subinterval construct a rectangle whose height is the value of $f$ at any point in the subinterval.
- The union of these rectangles forms a region $R_n$ whose area can be regarded as an approximation to the "area" $A$ of the region $R$.
- Repeat the process using more and more subdivisions.

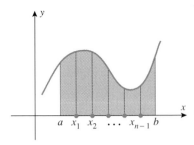

Figure 7.5.2

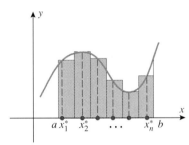

Figure 7.5.3

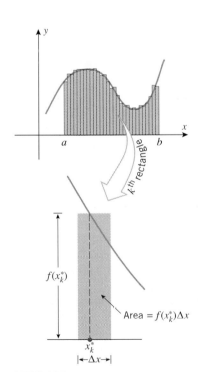

Figure 7.5.4

- Define the area of $R$ to be the limit of the areas of the approximating regions, $R_n$; that is,

$$A = \text{area}(R) = \lim_{n \to +\infty} [\text{area}(R_n)] \tag{1}$$

To make all of this more precise, it will be helpful to capture this procedure in mathematical notation. For this purpose, suppose that we divide the interval $[a, b]$ into $n$ subintervals by inserting $n - 1$ equally spaced points between $a$ and $b$, say

$$x_1, x_2, \ldots, x_{n-1}$$

(Figure 7.5.2). Each of these intervals has width $(b - a)/n$, which it is customary to denote by

$$\Delta x = \frac{b - a}{n}$$

In each subinterval we need to choose a point at which to evaluate the function $f$ to determine the height of a rectangle over that interval. If we denote those points by

$$x_1^*, x_2^*, \ldots, x_n^*$$

(Figure 7.5.3), then the areas of the rectangles constructed over these intervals will be

$$f(x_1^*)\Delta x, \quad f(x_2^*)\Delta x, \ldots, \quad f(x_n^*)\Delta x$$

(Figure 7.5.4), and the total area of the region $R_n$ will be

$$\text{area}(R_n) = f(x_1^*)\Delta x + f(x_2^*)\Delta x + \cdots + f(x_n^*)\Delta x$$

or in sigma notation,

$$\text{area}(R_n) = \sum_{k=1}^{n} f(x_k^*)\Delta x$$

With this notation (1) can be expressed as

$$A = \lim_{n \to +\infty} \sum_{k=1}^{n} f(x_k^*)\Delta x$$

which suggests the following definition of the area of the region $R$.

---

**7.5.1    DEFINITION (*Area Under a Curve*).**    If the function $f$ is continuous on $[a, b]$ and if $f(x) \geq 0$ for all $x$ in $[a, b]$, then the **area** under the curve $y = f(x)$ over the interval $[a, b]$ is defined by

$$A = \lim_{n \to +\infty} \sum_{k=1}^{n} f(x_k^*)\Delta x \tag{2}$$

---

REMARK.    Although this definition is satisfactory for our present purposes, there are some issues that would have to be resolved before it could be regarded as a rigorous mathematical definition. For example, we would have to prove that the limit actually exists and that its value does not depend on how the points $x_1^*, x_2^*, \ldots, x_n^*$ are chosen. It can be proved that this is true if $f$ is continuous on $[a, b]$, but the details are beyond the scope of this text.

The limit in Formula (2) is often difficult or impossible to find, so that when an *exact* area is needed the antiderivative method, which we will discuss in the next section, is the method of choice. However, if an *approximation* to the area will suffice, then instead of taking the limit we can approximate the area as

$$A \approx \sum_{k=1}^{n} f(x_k^*)\Delta x$$

where $n$ is sufficiently large to produce the required accuracy. For this purpose it is convenient to rewrite this sum as

$$\sum_{k=1}^{n} f(x_k^*)\Delta x = \Delta x \sum_{k=1}^{n} f(x_k^*) = \Delta x[f(x_1^*) + f(x_2^*) + \cdots + f(x_n^*)] \qquad (3)$$

where $\Delta x = (b - a)/n$. The calculation here involves only the sum of the values of the function at $n$ points, followed by a multiplication by $\Delta x$. The points $x_1^*, x_2^*, \ldots, x_n^*$ can be chosen arbitrarily in successive subintervals; however, the most common choices are at the left endpoints, the right endpoints, or the centers of the subintervals, in which cases Formula (3) is called the *left endpoint approximation*, the *right endpoint approximation*, or the *midpoint approximation* of the exact area (Figure 7.5.5).

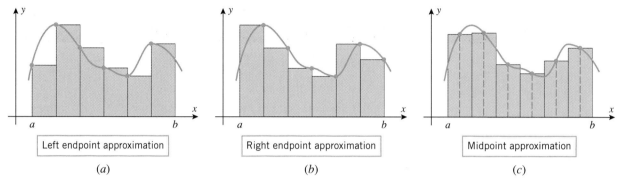

| Left endpoint approximation | Right endpoint approximation | Midpoint approximation |
| :---: | :---: | :---: |
| (a) | (b) | (c) |

Figure 7.5.5

## Example 1

Find the left endpoint, right endpoint, and midpoint approximations of the area under the curve $y = 9 - x^2$ over the interval $[0, 3]$ with $n = 10, n = 20$, and $n = 50$ (Figure 7.5.6).

*Solution.*   Details of the computations for the case $n = 10$ are shown to six decimal places in Table 7.5.1 and the results of all computations are given in Table 7.5.2.   ◀

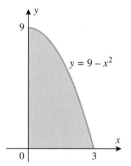

Figure 7.5.6

**Table 7.5.1**

$$n = 10, \; \Delta x = (b - a)/n = (3 - 0)/10 = 0.3$$

| | LEFT ENDPOINT APPROXIMATION | | RIGHT ENDPOINT APPROXIMATION | | MIDPOINT APPROXIMATION | |
| :---: | :---: | :---: | :---: | :---: | :---: | :---: |
| $k$ | $x_k^*$ | $9 - (x_k^*)^2$ | $x_k^*$ | $9 - (x_k^*)^2$ | $x_k^*$ | $9 - (x_k^*)^2$ |
| 1 | 0.0 | 9.000000 | 0.3 | 8.910000 | 0.15 | 8.977500 |
| 2 | 0.3 | 8.910000 | 0.6 | 8.640000 | 0.45 | 8.797500 |
| 3 | 0.6 | 8.640000 | 0.9 | 8.190000 | 0.75 | 8.437500 |
| 4 | 0.9 | 8.190000 | 1.2 | 7.560000 | 1.05 | 7.897500 |
| 5 | 1.2 | 7.560000 | 1.5 | 6.750000 | 1.35 | 7.177500 |
| 6 | 1.5 | 6.750000 | 1.8 | 5.760000 | 1.65 | 6.277500 |
| 7 | 1.8 | 5.760000 | 2.1 | 4.590000 | 1.95 | 5.197500 |
| 8 | 2.1 | 4.590000 | 2.4 | 3.240000 | 2.25 | 3.937500 |
| 9 | 2.4 | 3.240000 | 2.7 | 1.710000 | 2.55 | 2.497500 |
| 10 | 2.7 | 1.710000 | 3.0 | 0.000000 | 2.85 | 0.877500 |
| | | 64.350000 | | 55.350000 | | 60.075000 |
| $\Delta x \sum_{k=1}^{n} f(x_k^*)$ | | (.3)(64.350000) = 19.305000 | | (.3)(55.350000) = 16.605000 | | (.3)(60.075000) = 18.022500 |

**Table 7.5.2**

| $n$ | LEFT ENDPOINT APPROXIMATION | RIGHT ENDPOINT APPROXIMATION | MIDPOINT APPROXIMATION |
|---|---|---|---|
| 10 | 19.305000 | 16.605000 | 18.022500 |
| 20 | 18.663750 | 17.313750 | 18.005625 |
| 50 | 18.268200 | 17.728200 | 18.000900 |

REMARK. We will show in the next section that the exact area under $y = 9 - x^2$ over the interval $[0, 3]$ is 18 (i.e., 18 square units), so that in the preceding example the midpoint approximation is more accurate than either of the endpoint approximations. This can also be seen geometrically from the approximating rectangles: Since the graph of $y = 9 - x^2$ is decreasing over the interval $[0, 3]$, each left endpoint approximation overestimates the area, each right endpoint approximation underestimates the area, and each midpoint approximation falls between the overestimate and the underestimate (Figure 7.5.7). This is consistent with the values in Table 7.5.2. Later in the text we will investigate the error that results when an area is approximated by the midpoint rule.

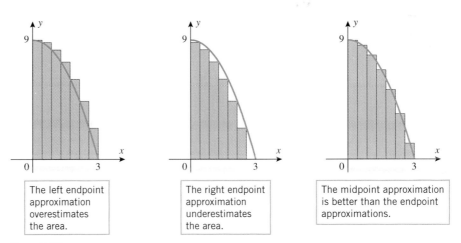

The left endpoint approximation overestimates the area.

The right endpoint approximation underestimates the area.

The midpoint approximation is better than the endpoint approximations.

Figure 7.5.7

**THE DEFINITE INTEGRAL OF A CONTINUOUS FUNCTION**

In Definition 7.5.1 we assumed that $f$ is continuous and nonnegative on the interval $[a, b]$. If $f$ is continuous and assumes both positive and negative values on $[a, b]$, then the limit

$$\lim_{n \to +\infty} \sum_{k=1}^{n} f(x_k^*)\Delta x \tag{4}$$

no longer represents the area between the curve $y = f(x)$ and the interval $[a, b]$; rather it represents a difference of areas—the area of the region that is above the interval $[a, b]$ and below the curve $y = f(x)$ minus the area of the region that is below the interval $[a, b]$ and above the curve $y = f(x)$. We call this the **net signed area** between the graph of $y = f(x)$ and the interval $[a, b]$. For example, in Figure 7.5.8a, the net signed area between the curve $y = f(x)$ and the interval $[a, b]$ is

$$(A_I + A_{III}) - A_{II} = [\text{area above } [a, b]] - [\text{area below } [a, b]]$$

To explain why the limit in (4) represents this net signed area, let us subdivide the interval $[a, b]$ in Figure 7.5.8a into $n$ equal subintervals and examine the terms in the sum

$$\sum_{k=1}^{n} f(x_k^*)\Delta x \tag{5}$$

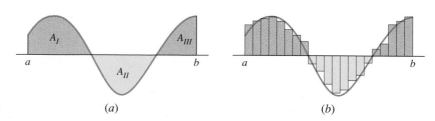

Figure 7.5.8          (*a*)          (*b*)

If $f(x_k^*)$ is positive, then the product $f(x_k^*)\Delta x$ represents the area of the rectangle with height $f(x_k^*)$ and base $\Delta x$ (the beige rectangles in Figure 7.5.8*b*). However, if $f(x_k^*)$ is negative, then the product $f(x_k^*)\Delta x$ is the *negative* of the area of the rectangle with height $|f(x_k^*)|$ and base $\Delta x$ (the green rectangles in Figure 7.5.8*b*). Thus, (5) represents the total area of the beige rectangles minus the total area of the green rectangles. As $n$ increases, the beige rectangles fill out the regions with areas $A_I$ and $A_{III}$ and the green rectangles fill out the region with area $A_{II}$, which explains why the limit in (4) represents the signed area between $y = f(x)$ and the interval $[a, b]$.

The limit in (4) is so important that there is some terminology and notation associated with it. We will denote this limit by the symbol

$$\int_a^b f(x)\,dx = \lim_{n \to +\infty} \sum_{k=1}^n f(x_k^*)\Delta x \tag{6}$$

which is called the **definite integral** of $f$ from $a$ to $b$. Geometrically, the definite integral represents the signed area between $y = f(x)$ and the interval $[a, b]$, and in the case where $f(x)$ is nonnegative on the interval $[a, b]$, the definite integral represents the area under the curve over the interval $[a, b]$. The numbers $a$ and $b$ are called the **lower limit of integration** and **upper limit of integration**, respectively, and $f(x)$ is called the **integrand**. The reason for the integral sign will become clear in the next section, where we will establish a link between the definite integral and the indefinite integral studied earlier.

In the simplest cases, definite integrals can be calculated using formulas from plane geometry to compute the signed areas.

### Example 2

Sketch the region whose area is represented by the definite integral, and evaluate the integral using an appropriate formula from geometry.

$$\text{(a)} \int_1^4 2\,dx \qquad \text{(b)} \int_{-1}^2 (x+2)\,dx \qquad \text{(c)} \int_0^1 \sqrt{1-x^2}\,dx$$

*Solution* (*a*). The graph of the integrand is the horizontal line $y = 2$, so the region is a rectangle of height 2 extending over the interval from 1 to 4 (Figure 7.5.9*a*). Thus,

$$\int_1^4 2\,dx = (\text{area of rectangle}) = 2(3) = 6$$

*Solution* (*b*). The graph of the integrand is the line $y = x + 2$, so the region is a trapezoid whose base extends from $x = -1$ to $x = 2$ (Figure 7.5.9*b*). Thus,

$$\int_{-1}^2 (x+2)\,dx = (\text{area of trapezoid}) = \tfrac{1}{2}(3)(1+4) = \tfrac{15}{2}$$

*Solution* (*c*). The graph of $y = \sqrt{1 - x^2}$ is the upper semicircle of radius 1, centered at the origin, so the region is the right quarter-circle extending from $x = 0$ to $x = 1$ (Figure 7.5.9*c*).

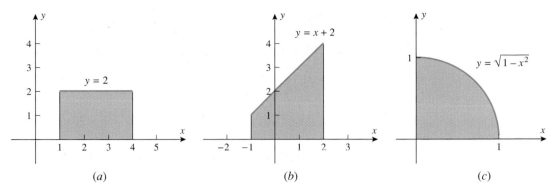

Figure 7.5.9

Thus,

$$\int_0^1 \sqrt{1 - x^2}\, dx = \text{(area of quarter-circle)} = \tfrac{1}{4}\pi(1^2) = \frac{\pi}{4}$$  ◀

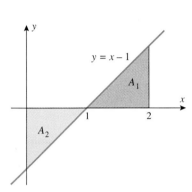

Figure 7.5.10

## Example 3

Evaluate

$$\text{(a) } \int_0^2 (x - 1)\, dx \qquad \text{(b) } \int_0^1 (x - 1)\, dx$$

***Solution.***  The graph of $y = x - 1$ is shown in Figure 7.5.10, and we leave it for you to verify that the shaded triangular regions both have area $\tfrac{1}{2}$. Over the interval $[0, 2]$ the net signed area is $A_1 - A_2 = \tfrac{1}{2} - \tfrac{1}{2} = 0$, and over the interval $[0, 1]$ the net signed area is $-A_2 = -\tfrac{1}{2}$. Thus,

$$\int_0^2 (x - 1)\, dx = 0 \quad \text{and} \quad \int_0^1 (x - 1)\, dx = -\tfrac{1}{2}$$  ◀

**THE RIEMANN INTEGRAL**

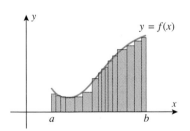

Figure 7.5.11

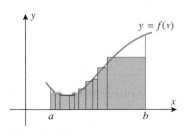

Figure 7.5.12

It is assumed in (6) that the function $f$ is continuous on the interval $[a, b]$ and that for each $n$ this interval is subdivided into $n$ subintervals of equal length to create bases for the approximating rectangles. Although equal lengths are useful for computations, this restriction is not essential. That is, the signed area between $y = f(x)$ and $[a, b]$ can be obtained using rectangles with different widths provided that successive subdivisions are constructed in such a way that the widths of the rectangles approach zero as $n$ increases (Figure 7.5.11). Thus, we must preclude the kind of situation that occurs in Figure 7.5.12 in which the right half of the interval is never subdivided. If this kind of subdivision were allowed, the error in the approximation would not approach zero as $n$ increased.

To provide for the added generality of unequal intervals, suppose that the interval $[a, b]$ is subdivided into $n$ subintervals whose widths are

$$\Delta x_1, \Delta x_2, \ldots, \Delta x_n$$

and let max $\Delta x_k$ denote the largest of the subinterval widths, which is read "the maximum of the $\Delta x_k$'s." The subintervals are said to form a ***partition*** of the interval $[a, b]$, and max $\Delta x_k$ is called the ***mesh size*** of the partition. For example, Figure 7.5.13 shows a partition of the interval $[0, 6]$ into four subintervals with a mesh size of 2.

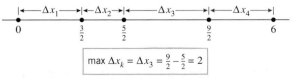

$$\max \Delta x_k = \Delta x_3 = \tfrac{9}{2} - \tfrac{5}{2} = 2$$

Figure 7.5.13

To generalize (6) so that it allows for unequal subinterval widths, we must replace the constant interval length $\Delta x$ by the variable interval length $\Delta x_k$, and we must replace $n \to +\infty$ by an expression to specify that the lengths of all the subintervals approach zero. We will use the expression $\max \Delta x_k \to 0$ for this purpose. With these modifications in notation (6) becomes

$$\int_a^b f(x)\,dx = \lim_{\max \Delta x_k \to 0} \sum_{k=1}^n f(x_k^*)\Delta x_k \tag{7}$$

The sum that appears in this expression is called a ***Riemann*** [*] ***sum***, and the limit is sometimes called the ***Riemann integral*** in honor of the German mathematician Bernhard Riemann who formulated many of the basic concepts of integration.

REMARK.    Some writers use the symbol $\|\Delta\|$ rather than $\max \Delta x_k$ for the mesh size of the partition, in which case (7) would be written as

$$\int_a^b f(x)\,dx = \lim_{\|\Delta\| \to 0} \sum_{k=1}^n f(x_k^*)\Delta x_k$$

**INTEGRABILITY**

Because the definite integral is defined as a limit, it is possible that the limit may not exist, in which case the definite integral would not exist. Thus, we make the following definition:

**7.5.2**   DEFINITION.   A function $f$ is said to be ***Riemann integrable*** or more simply ***integrable*** on a finite closed interval $[a, b]$ if the limit

$$\int_a^b f(x)\,dx = \lim_{\max \Delta x_k \to 0} \sum_{k=1}^n f(x_k^*)\Delta x_k$$

exists and does not depend on choice of the partitions or on the points $x_k^*$ in the subintervals.

At the end of this section we will discuss various conditions that ensure integrability, but for now suffice it to say that a function that is continuous on a finite closed interval $[a, b]$ is integrable on that interval.

---

[*] GEORG FRIEDRICH BERNHARD RIEMANN (1826–1866). German mathematician. Bernhard Riemann, as he is commonly known, was the son of a Protestant minister. He received his elementary education from his father and showed brilliance in arithmetic at an early age. In 1846 he enrolled at Göttingen University to study theology and philology, but he soon transferred to mathematics. He studied physics under W. E. Weber and mathematics under Karl Friedrich Gauss, whom some people consider to be the greatest mathematician who ever lived. In 1851 Riemann received his Ph.D. under Gauss, after which he remained at Göttingen to teach. In 1862, one month after his marriage, Riemann suffered an attack of pleuritis, and for the remainder of his life was an extremely sick man. He finally succumbed to tuberculosis in 1866 at age 39.

An interesting story surrounds Riemann's work in geometry. For his introductory lecture prior to becoming an associate professor, Riemann submitted three possible topics to Gauss. Gauss surprised Riemann by choosing the topic Riemann liked the least, the foundations of geometry. The lecture was like a scene from a movie. The old and failing Gauss, a giant in his day, watching intently as his brilliant and youthful protégé skillfully pieced together portions of the old man's own work into a complete and beautiful system. Gauss is said to have gasped with delight as the lecture neared its end, and on the way home he marveled at his student's brilliance. Gauss died shortly thereafter. The results presented by Riemann that day eventually evolved into a fundamental tool that Einstein used some 50 years later to develop relativity theory.

In addition to his work in geometry, Riemann made major contributions to the theory of complex functions and mathematical physics. The notion of the definite integral, as it is presented in most basic calculus courses, is due to him. Riemann's early death was a great loss to mathematics, for his mathematical work was brilliant and of fundamental importance.

It is assumed in Definition 7.5.2 that $[a, b]$ is a finite closed interval with $a < b$, and hence
the upper limit of integration in the definite integral is greater than the lower limit of integration. However, it will be convenient to extend this definition to allow for cases in which the
upper and lower limits of integration are equal or the lower limit of integration is greater than
the upper limit of integration. For this purpose we make the following special definitions.

---

**7.5.3** DEFINITION.

(a)  If $a$ is in the domain of $f$, we define

$$\int_a^a f(x)\, dx = 0$$

(b)  If $f$ is integrable on $[a, b]$, then we define

$$\int_b^a f(x)\, dx = -\int_a^b f(x)\, dx$$

---

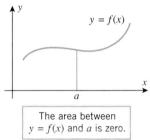

The area between
$y = f(x)$ and $a$ is zero.

Figure 7.5.14

REMARK.   Part (a) of this definition is consistent with the intuitive idea that the area
between a point on the $x$-axis and a curve $y = f(x)$ should be zero (Figure 7.5.14). Part
(b) of the definition is simply a useful convention; it states that interchanging the limits of
integration reverses the sign of the integral.

**Example 4**

(a) $\displaystyle \int_1^1 x^2\, dx = 0$

(b) $\displaystyle \int_1^0 \sqrt{1 - x^2}\, dx = -\int_0^1 \sqrt{1 - x^2}\, dx = -\frac{\pi}{4}$ ◄

Example 2(c)

Because definite integrals are defined as limits, they inherit many of the properties of
limits. For example, we know that constants can be moved through limit signs and that the
limit of a sum or difference is the sum or difference of the limits. Thus, you should not be
surprised by the following theorem, which we state without formal proof.

---

**7.5.4**  THEOREM.  *If $f$ and $g$ are integrable on $[a, b]$ and if $c$ is a constant, then $cf$,
$f + g$, and $f - g$ are integrable on $[a, b]$ and*

(a)  $\displaystyle \int_a^b cf(x)\, dx = c\int_a^b f(x)\, dx$

(b)  $\displaystyle \int_a^b [f(x) + g(x)]\, dx = \int_a^b f(x)\, dx + \int_a^b g(x)\, dx$

(c)  $\displaystyle \int_a^b [f(x) - g(x)]\, dx = \int_a^b f(x)\, dx - \int_a^b g(x)\, dx$

---

Part (*b*) of this theorem can be extended to more than two functions. More precisely,

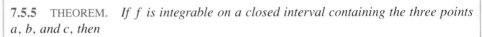

$$\int_a^b [f_1(x) + f_2(x) + \cdots + f_n(x)]\,dx$$

$$= \int_a^b f_1(x)\,dx + \int_a^b f_2(x)\,dx + \cdots + \int_a^b f_n(x)\,dx \qquad (8)$$

Some properties of definite integrals can be motivated by interpreting the integral as an area. For example, if $f$ is continuous and nonnegative on the interval $[a, b]$, and if $c$ is a point between $a$ and $b$, then the area under $y = f(x)$ over the interval $[a, b]$ can be split into two parts and expressed as the area under the graph from $a$ to $c$ plus the area under the graph from $c$ to $b$ (Figure 7.5.15), that is,

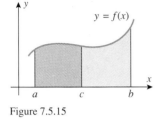

Figure 7.5.15

$$\int_a^b f(x)\,dx = \int_a^c f(x)\,dx + \int_c^b f(x)\,dx$$

This is a special case of the following theorem about definite integrals, which we state without proof.

---

**7.5.5** THEOREM. *If $f$ is integrable on a closed interval containing the three points $a$, $b$, and $c$, then*

$$\int_a^b f(x)\,dx = \int_a^c f(x)\,dx + \int_c^b f(x)\,dx \qquad (9)$$

*no matter how the points are ordered.*

---

The following theorem, which we state without formal proof, can also be motivated by interpreting definite integrals as areas.

---

**7.5.6** THEOREM.

(*a*) *If $f$ is integrable on $[a, b]$ and $f(x) \geq 0$ for all $x$ in $[a, b]$, then*

$$\int_a^b f(x)\,dx \geq 0$$

(*b*) *If $f$ and $g$ are integrable on $[a, b]$ and $f(x) \geq g(x)$ for all $x$ in $[a, b]$, then*

$$\int_a^b f(x)\,dx \geq \int_a^b g(x)\,dx$$

---

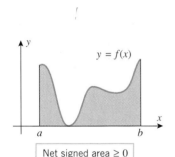

Net signed area ≥ 0

Figure 7.5.16

Geometrically, part (*a*) of this theorem states the obvious fact that if $f$ is nonnegative on $[a, b]$, then the net signed area between the graph of $f$ and the interval $[a, b]$ is also nonnegative (Figure 7.5.16). Part (*b*) has its simplest interpretation when $f$ and $g$ are nonnegative on $[a, b]$, in which case the theorem states that if the graph of $f$ does not go below the graph of $g$, then the area under the graph of $f$ is at least as large as the area under the graph of $g$ (Figure 7.5.17).

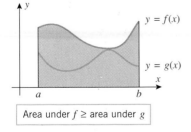

Area under $f$ ≥ area under $g$

Figure 7.5.17

REMARK. In words, part (*b*) of this theorem states that one can integrate both sides of the inequality $f(x) \geq g(x)$ without altering the sense of the inequality. We also note that in the case where $b > a$, both parts of the theorem remain true if $\geq$ is replaced by $\leq$, $>$, or $<$ throughout.

**Example 5**

Evaluate

$$\int_0^1 (5 - 3\sqrt{1 - x^2})\, dx$$

*Solution.* From parts (*a*) and (*c*) of Theorem 7.5.4 we can write

$$\int_0^1 (5 - 3\sqrt{1 - x^2})\, dx = \int_0^1 5\, dx - \int_0^1 3\sqrt{1 - x^2}\, dx = \int_0^1 5\, dx - 3\int_0^1 \sqrt{1 - x^2}\, dx$$

The first integral can be interpreted as the area of a rectangle of height 5 and base 1, so its value is 5, and from Example 2 the value of the second integral is $\pi/4$. Thus,

$$\int_0^1 (5 - 3\sqrt{1 - x^2})\, dx = 5 - 3\left(\frac{\pi}{4}\right) = 5 - \frac{3\pi}{4} \qquad \blacktriangleleft$$

**CONDITIONS FOR INTEGRABILITY**

The problem of determining precisely which functions are integrable is quite complex and beyond the scope of this text. However, there are a few basic results about integrability that are important to know; we begin with a definition.

> **7.5.7** DEFINITION. A function $f$ is said to be ***bounded*** on an interval $I$ if there is a positive number $M$ such that
>
> $$-M \leq f(x) \leq M$$
>
> for all $x$ in the interval $I$. Geometrically, this means that the graph of $f$ over the interval $I$ lies between the lines $y = -M$ and $y = M$.

For example, a continuous function $f$ is bounded on *every* finite closed interval because the Extreme-Value Theorem (6.1.3) implies that $f$ has an absolute maximum and an absolute minimum on the interval; hence, its graph will lie between the line $y = -M$ and $y = M$, provided we make $M$ large enough (Figure 7.5.18). In contrast, a function that has a vertical asymptote inside of an interval is not bounded on that interval because its graph over the interval cannot be made to lie between the lines $y = -M$ and $y = M$, no matter how large we make the value of $M$ (Figure 7.5.19).

The following theorem, which we state without proof, lists three of the most important facts about integrability.

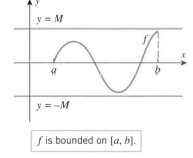

$f$ is bounded on $[a, b]$.

Figure 7.5.18

> **7.5.8** THEOREM. *Let $f$ be a function that is defined at all points in the finite closed interval $[a, b]$.*
> (*a*) *If $f$ is continuous on $[a, b]$, then $f$ is integrable on $[a, b]$.*
> (*b*) *If $f$ has finitely many points of discontinuity on $[a, b]$ but is bounded on $[a, b]$, then $f$ is integrable on $[a, b]$.*
> (*c*) *If $f$ is not bounded on $[a, b]$, then $f$ is not integrable on $[a, b]$.*

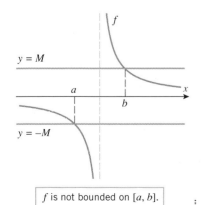

$f$ is not bounded on $[a, b]$.

Figure 7.5.19

FOR THE READER. Sketch the graph of a function over the interval $[0, 1]$ that has the properties stated in part (*b*) of this theorem.

**EXERCISE SET 7.5** ⊠ Graphing Calculator 　C CAS

1. (a) Use an appropriate geometric formula to find the exact area $A$ under the line $x + y = 4$ over the interval $[0, 4]$.
   (b) Sketch the rectangles for the left endpoint approximation to the area $A$ using $n = 4$ subintervals. Is that approximation greater than, less than, or equal to $A$? Explain your reasoning, and check your conclusion by calculating the left endpoint approximation.
   (c) Sketch the rectangles for the right endpoint approximation to the area $A$ using $n = 4$ subintervals. Is that approximation greater than, less than, or equal to $A$? Explain your reasoning, and check your conclusion by calculating the right endpoint approximation.
   (d) Sketch the rectangles for the midpoint approximation to the area $A$ using $n = 4$ subintervals. Is that approximation greater than, less than, or equal to $A$? Explain your reasoning, and check your conclusion by calculating the midpoint approximation.

2. Follow the directions of Exercise 1 for the area $A$ under the line $y = 3x$ over the interval $[2, 6]$.

3. Find the left endpoint, right endpoint, and midpoint approximations of the area under the curve $y = x^2 + 1$ over the interval $[0, 5]$ using $n = 5$ subintervals.

4. Find the left endpoint, right endpoint, and midpoint approximations of the area under the curve $y = x^3$ over the interval $[1, 6]$ using $n = 5$ subintervals.

5. Find the left endpoint, right endpoint, and midpoint approximations of the area under the curve $y = \cos x$ over the interval $[-\pi/2, \pi/2]$ using $n = 4$ subintervals.

6. Find the left endpoint, right endpoint, and midpoint approximations of the area under the curve $y = e^x$ over the interval $[0, 5]$ using $n = 5$ subintervals.

7. The accompanying figure shows five points on the graph of an unknown function $f$. Devise a strategy for using the known points to approximate the area $A$ under the graph of $y = f(x)$ over the interval $[1, 5]$. Describe your strategy, and use it to approximate $A$.

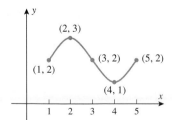

Figure Ex-7

8. (a) Use an appropriate geometric formula to find the exact area $A$ under the line $y = 3x + 1$ over the interval $[1, 5]$.
   (b) Show that the exact area is equal to the average value of the left endpoint and right endpoint approximations of $A$ obtained using $n = 4$ subintervals.
   (c) What is the explanation of the result in part (b)?

In Exercises 9–14, use a calculating utility to find the left endpoint, right endpoint, and midpoint approximations to the area under the curve $y = f(x)$ over the stated interval using $n = 10$ subintervals.

9. $y = 1/x$; $[1, 2]$ 　　　　10. $y = 1/x^2$; $[1, 3]$

11. $y = \sin x$; $[0, \pi/2]$ 　　12. $y = \sqrt{x}$; $[0, 4]$

13. $y = \ln x$; $[1, 2]$ 　　　　14. $y = e^x$; $[0, 1]$

15. If you have a programmable calculator, create a program for calculating the midpoint approximation of the area under a curve $y = f(x)$ over an interval $[a, b]$ using $n$ subintervals, and use the program to find midpoint approximations in Exercises 9–14 with
   (a) $n = 25$ 　　(b) $n = 50$ 　　(c) $n = 100$.

C 16. If you have a CAS, devise a procedure for using it to calculate the midpoint approximation of the area under a curve $y = f(x)$ over an interval $[a, b]$ using $n$ subintervals, and use the procedure to find the midpoint approximations in Exercises 9–14 with
   (a) $n = 25$ 　　(b) $n = 50$ 　　(c) $n = 100$.

In Exercises 17–20, sketch the region whose signed area is represented by the definite integral, and evaluate the integral using an appropriate formula from geometry, where needed.

17. (a) $\displaystyle\int_0^3 x\,dx$ 　　　　(b) $\displaystyle\int_{-2}^{-1} x\,dx$

　　(c) $\displaystyle\int_{-1}^4 x\,dx$ 　　　　(d) $\displaystyle\int_{-5}^5 x\,dx$

18. (a) $\displaystyle\int_0^2 \left(1 - \tfrac{1}{2}x\right) dx$ 　　(b) $\displaystyle\int_{-1}^1 \left(1 - \tfrac{1}{2}x\right) dx$

　　(c) $\displaystyle\int_2^3 \left(1 - \tfrac{1}{2}x\right) dx$ 　　(d) $\displaystyle\int_0^3 \left(1 - \tfrac{1}{2}x\right) dx$

19. (a) $\displaystyle\int_0^5 2\,dx$ 　　　　(b) $\displaystyle\int_0^\pi \cos x\,dx$

　　(c) $\displaystyle\int_{-1}^2 |2x - 3|\,dx$ 　　(d) $\displaystyle\int_{-1}^1 \sqrt{1 - x^2}\,dx$

20. (a) $\displaystyle\int_{-10}^{-5} 6\,dx$ 　　　(b) $\displaystyle\int_{-\pi/3}^{\pi/3} \sin x\,dx$

　　(c) $\displaystyle\int_0^3 |x - 2|\,dx$ 　　(d) $\displaystyle\int_0^2 \sqrt{4 - x^2}\,dx$

21. Use the areas shown in the accompanying figure to find
   (a) $\displaystyle\int_a^b f(x)\,dx$ 　　　(b) $\displaystyle\int_b^c f(x)\,dx$
   (c) $\displaystyle\int_a^c f(x)\,dx$ 　　　(d) $\displaystyle\int_a^d f(x)\,dx$.

Figure Ex-21

**22.** In each part, evaluate the integral, given that

$$f(x) = \begin{cases} 2x, & x \le 1 \\ 2, & x > 1 \end{cases}$$

(a) $\int_0^1 f(x)\,dx$     (b) $\int_{-1}^1 f(x)\,dx$

(c) $\int_1^{10} f(x)\,dx$     (d) $\int_{1/2}^5 f(x)\,dx$

**23.** Find $\int_{-1}^2 [f(x) + 2g(x)]\,dx$ if

$$\int_{-1}^2 f(x)\,dx = 5 \quad \text{and} \quad \int_{-1}^2 g(x)\,dx = -3$$

**24.** Find $\int_1^4 [3f(x) - g(x)]\,dx$ if

$$\int_1^4 f(x)\,dx = 2 \quad \text{and} \quad \int_1^4 g(x)\,dx = 10$$

**25.** Find $\int_1^5 f(x)\,dx$ if

$$\int_0^1 f(x)\,dx = -2 \quad \text{and} \quad \int_0^5 f(x)\,dx = 1$$

**26.** Find $\int_3^{-2} f(x)\,dx$ if

$$\int_{-2}^1 f(x)\,dx = 2 \quad \text{and} \quad \int_1^3 f(x)\,dx = -6$$

In Exercises 27 and 28, use Theorem 7.5.4 and appropriate formulas from geometry to evaluate the integrals.

**27.** (a) $\int_0^1 (x + 2\sqrt{1-x^2})\,dx$   (b) $\int_{-1}^3 (4 - 5x)\,dx$

**28.** (a) $\int_{-3}^0 (2 + \sqrt{9-x^2})\,dx$   (b) $\int_{-2}^2 (1 - 3|x|)\,dx$

In Exercises 29 and 30, use Theorem 7.5.6 to determine whether the value of the integral is positive or negative.

**29.** (a) $\int_2^3 \dfrac{\sqrt{x}}{1-x}\,dx$    (b) $\int_0^4 \dfrac{x^2}{3 - \cos x}\,dx$

**30.** (a) $\int_{-3}^{-1} \dfrac{x^4}{\sqrt{3-x}}\,dx$    (b) $\int_{-2}^2 \dfrac{x^3 - 9}{|x|+1}\,dx$

In Exercises 31 and 32, evaluate the integrals by completing the square and applying appropriate formulas from geometry.

**31.** $\int_0^{10} \sqrt{10x - x^2}\,dx$    **32.** $\int_0^3 \sqrt{6x - x^2}\,dx$

In Exercises 33 and 34, express the limits as definite integrals over the interval $[a, b]$. Do not try to evaluate the integrals.

**33.** (a) $\displaystyle\lim_{\max \Delta x_k \to 0} \sum_{k=1}^n 4x_k^*(1 - 3x_k^*)\Delta x_k$; $a = -3, b = 3$

   (b) $\displaystyle\lim_{\max \Delta x_k \to 0} \sum_{k=1}^n e^{x_k^*}\Delta x_k$; $a = 0, b = 1$

**34.** (a) $\displaystyle\lim_{\max \Delta x_k \to 0} \sum_{k=1}^n (x_k^*)^3 \Delta x_k$; $a = 1, b = 2$

   (b) $\displaystyle\lim_{\max \Delta x_k \to 0} \sum_{k=1}^n (\sin^2 x_k^*)\Delta x_k$; $a = 0, b = \pi/2$

In Exercises 35 and 36, evaluate the limit over the interval $[a, b]$ by expressing it as a definite integral and applying an appropriate formula from geometry.

**35.** $\displaystyle\lim_{\max \Delta x_k \to 0} \sum_{k=1}^n (3x_k^* + 1)\Delta x_k$; $a = 0, b = 1$

**36.** $\displaystyle\lim_{\max \Delta x_k \to 0} \sum_{k=1}^n \sqrt{4 - (x_k^*)^2}\,\Delta x_k$; $a = -2, b = 2$

In Exercises 37 and 38, use Formula (7) to express the integrals as limits of Riemann sums. Do not try to evaluate the integrals.

**37.** (a) $\int_1^2 2x\,dx$    (b) $\int_0^1 \dfrac{x}{x+1}\,dx$

**38.** (a) $\int_1^2 \ln x\,dx$    (b) $\int_{-\pi/2}^{\pi/2} (1 + \cos x)\,dx$

**39.** In this exercise you will find the area $A$ under the graph of $y = x$ over the interval $[1, 2]$ by calculating the limit of right endpoint approximations. For this particular problem, the area can be found much more easily using a formula from geometry, so our purpose here is not to provide a practical method for calculating the area, but rather to illustrate the idea that underlies the concept of a definite integral.

  (a) Suppose that the interval $[1, 2]$ is subdivided into $n$ equal subintervals of length $\Delta x = 1/n$ and that the points $x_1^*, x_2^*, \ldots, x_n^*$ are the right endpoints of the subintervals. Show that the right endpoint of the $k$th subinterval is

$$x_k^* = 1 + \frac{k}{n}$$

[*Suggestion:* Find $x_1^*, x_2^*,$ and $x_3^*$, and then look for the pattern.]

(b) Show that with $n$ subintervals the right endpoint approximation of the area $A$ is

$$\sum_{k=1}^{n} f(x_k^*)\Delta x = \sum_{k=1}^{n}\left[\left(1+\frac{k}{n}\right)\frac{1}{n}\right]$$

(c) Use Theorem 7.4.2 to show that the right endpoint approximation can be expressed as

$$\sum_{k=1}^{n} f(x_k^*)\Delta x = \frac{3}{2}+\frac{1}{2n}$$

(d) From (2), the area $A$ is

$$A = \lim_{n\to+\infty}\sum_{k=1}^{n} f(x_k^*)\Delta x$$

Find this limit, and check your answer by using a formula from geometry to calculate $A$.

40. Find the area $A$ in Exercise 39 as a limit of left endpoint approximations.

In Exercises 41–44, use the method of Exercise 39 to find the area under the curve $y = f(x)$ over the interval $[a, b]$ as a limit of right and left endpoint approximations.

41. $y = x^2$; $a = 0, b = 1$
42. $y = 4 - \frac{1}{4}x^2$; $a = 0, b = 3$
43. $y = x^3$; $a = 2, b = 6$
44. $y = 1 - x^3$; $a = -3, b = -1$

45. In each part, use Theorem 7.5.8 to determine whether the function $f$ is integrable on the interval $[-1, 1]$.
(a) $f(x) = e^x \cos x$

(b) $f(x) = \begin{cases} x/|x|, & x \neq 0 \\ 0, & x = 0 \end{cases}$

(c) $f(x) = \begin{cases} 1/x^2, & x \neq 0 \\ 0, & x = 0 \end{cases}$

(d) $f(x) = \begin{cases} \sin 1/x, & x \neq 0 \\ 0, & x = 0 \end{cases}$

46. It can be shown that every interval contains both rational and irrational numbers. Accepting this to be so, do you believe that the function

$$f(x) = \begin{cases} 1 & \text{if} \quad x \text{ is rational} \\ 0 & \text{if} \quad x \text{ is irrational} \end{cases}$$

is integrable on a closed interval $[a, b]$? Explain your reasoning.

47. It can be shown that the limit in Formula (7) has all of the limit properties stated in Theorem 2.2.2. Accepting this to be so, show that

(a) $\displaystyle\int_{a}^{b} cf(x)\,dx = c\int_{a}^{b} f(x)\,dx$

(b) $\displaystyle\int_{a}^{b} [f(x)+g(x)]\,dx = \int_{a}^{b} f(x)\,dx + \int_{a}^{b} g(x)\,dx$

48. Find the smallest and largest values that the Riemann sum

$$\sum_{k=1}^{3} f(x_k^*)\Delta x_k$$

can have on the interval $[0, 4]$ if $f(x) = x^2 - 3x + 4$ and $\Delta x_1 = 1, \Delta x_2 = 2, \Delta x_3 = 1$.

## 7.6 THE FUNDAMENTAL THEOREM OF CALCULUS

*In this section we will establish two basic relationships between definite and indefinite integrals that together constitute a result called the Fundamental Theorem of Calculus. One part of this theorem will relate the rectangle and antiderivative methods for calculating areas, and the second part will provide a powerful method for evaluating definite integrals using antiderivatives.*

**THE FUNDAMENTAL THEOREM OF CALCULUS**

To motivate the results we are looking for, let us begin by assuming that $f$ is nonnegative and continuous on the interval $[a, b]$, in which case the area $A$ under the graph of $f$ over the interval $[a, b]$ is represented by the definite integral

$$A = \int_{a}^{b} f(x)\,dx \tag{1}$$

(Figure 7.6.1).

Recall from our discussion of the antiderivative method in Section 7.1 that if $A(x)$ is the area under the graph of $f$ from $a$ to $x$ (Figure 7.6.2), then:

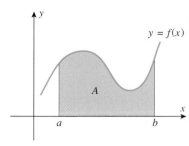

Figure 7.6.1

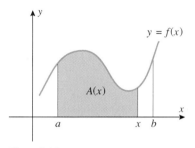

Figure 7.6.2

- $A'(x) = f(x)$
- $A(a) = 0$     The area under the curve from $a$ to $a$ is the area above the single point $a$, and hence is zero.
- $A(b) = A$     The area under the curve from $a$ to $b$ is $A$.

The formula $A'(x) = f(x)$ states that $A(x)$ is an antiderivative of $f(x)$, which implies that every other antiderivative of $f(x)$ can be obtained by adding a constant to $A(x)$. Accordingly, let

$$F(x) = A(x) + C$$

be any antiderivative of $f(x)$, and consider what happens when we subtract $F(a)$ from $F(b)$. We obtain

$$F(b) - F(a) = [A(b) + C] - [A(a) + C] = A(b) - A(a) = A - 0 = A$$

and hence (1) can be expressed as

$$\int_a^b f(x)\,dx = F(b) - F(a)$$

In words, this equation states that the definite integral can be evaluated by finding any antiderivative of the integrand and then subtracting the value of this antiderivative at the lower limit of integration from its value at the upper limit of integration. Although we derived this result subject to the assumption that $f$ is nonnegative on $[a, b]$, this assumption is not essential, as we will prove in the following theorem, which is the main tool used to evaluate definite integrals.

---

**7.6.1** THEOREM (*The Fundamental Theorem of Calculus, Part 1*). *If $f$ is continuous on $[a, b]$, and if $F$ is any antiderivative of $f$ on $[a, b]$, then*

$$\int_a^b f(x)\,dx = F(b) - F(a) \tag{2}$$

---

*Proof.* Let $x_1, x_2, \ldots, x_{n-1}$ be any points in $[a, b]$ such that

$$a < x_1 < x_2 < \cdots < x_{n-1} < b$$

These points divide $[a, b]$ into $n$ subintervals

$$[a, x_1], [x_1, x_2], \ldots, [x_{n-1}, b] \tag{3}$$

whose lengths, as usual, we denote by

$$\Delta x_1, \Delta x_2, \ldots, \Delta x_n$$

By hypothesis, $F'(x) = f(x)$ for all $x$ in $[a, b]$, so $F$ satisfies the hypotheses of the Mean-Value Theorem (6.5.2) on each subinterval in (3). Hence, we can find points $x_1^*, x_2^*, \ldots, x_n^*$ in the respective subintervals in (3) such that

$$F(x_1) - F(a) = F'(x_1^*)(x_1 - a) = f(x_1^*)\Delta x_1$$
$$F(x_2) - F(x_1) = F'(x_2^*)(x_2 - x_1) = f(x_2^*)\Delta x_2$$
$$F(x_3) - F(x_2) = F'(x_3^*)(x_3 - x_2) = f(x_3^*)\Delta x_3$$
$$\vdots \qquad\qquad \vdots \qquad\qquad \vdots$$
$$F(b) - F(x_{n-1}) = F'(x_n^*)(b - x_{n-1}) = f(x_n^*)\Delta x_n$$

Adding the preceding equations yields

$$F(b) - F(a) = \sum_{k=1}^{n} f(x_k^*)\Delta x_k \tag{4}$$

Let us now increase $n$ in such a way that max $\Delta x_k \to 0$. Since $f$ is assumed to be continuous, the right side of (4) approaches $\int_a^b f(x)\,dx$, by Theorem 7.5.8(a) and Formula (7) of Section 7.5. However, the left side of (4) is a constant that is independent of $n$; thus,

$$F(b) - F(a) = \lim_{\max \Delta x_k \to 0} \sum_{k=1}^n f(x_k^*)\Delta x_k = \int_a^b f(x)\,dx \qquad \blacksquare$$

It is standard to denote the difference $F(b) - F(a)$ as

$$F(x)\Big]_a^b = F(b) - F(a) \quad \text{or} \quad [F(x)]_a^b = F(b) - F(a)$$

For example, using the first of these notations we can express (2) as

$$\int_a^b f(x)\,dx = F(x)\Big]_a^b \tag{5}$$

### Example 1

Evaluate $\displaystyle\int_1^2 x\,dx$.

*Solution.* The function $F(x) = \frac{1}{2}x^2$ is an antiderivative of $f(x) = x$; thus, from (2)

$$\int_1^2 x\,dx = \frac{1}{2}x^2\Big]_1^2 = \frac{1}{2}(2)^2 - \frac{1}{2}(1)^2 = 2 - \frac{1}{2} = \frac{3}{2} \qquad \blacktriangleleft$$

### Example 2

In Example 1 of the last section we approximated the area under the graph of $y = 9 - x^2$ over the interval $[0, 3]$ using left endpoint, right endpoint, and midpoint approximations, all of which produced an approximation of roughly 18 (square units); and in the remark following that example we stated without proof that the exact area $A$ is 18 (square units). We can now confirm this using the Fundamental Theorem of Calculus as follows:

$$A = \int_0^3 (9 - x^2)\,dx = 9x - \frac{x^3}{3}\Big]_0^3 = \left(27 - \frac{27}{3}\right) - 0 = 18 \qquad \blacktriangleleft$$

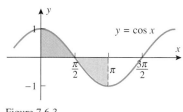

Figure 7.6.3

### Example 3

(a) Find the area under the curve $y = \cos x$ over the interval $[0, \pi/2]$ (Figure 7.6.3).

(b) Make a conjecture about the value of the integral

$$\int_0^\pi \cos x\,dx$$

and confirm your conjecture using the Fundamental Theorem of Calculus.

*Solution (a).* Since $\cos x \geq 0$ over the interval $[0, \pi/2]$, the area $A$ under the curve is

$$A = \int_0^{\pi/2} \cos x\,dx = \sin x\Big]_0^{\pi/2} = \sin\frac{\pi}{2} - \sin 0 = 1$$

*Solution (b).* The given integral can be interpreted as the signed area between the graph of $y = \cos x$ and the interval $[0, \pi]$. The graph in Figure 7.6.3 suggests that over the interval $[0, \pi]$ the portion of area above the $x$-axis is the same as the portion of area below the $x$-axis, so we conjecture that the signed area is zero; this implies that the value of the integral is zero. This is confirmed by the computations

$$\int_0^\pi \cos x\,dx = \sin x\Big]_0^\pi = \sin\pi - \sin 0 = 0 \qquad \blacktriangleleft$$

**THE RELATIONSHIP BETWEEN
DEFINITE AND INDEFINITE
INTEGRALS**

Observe that in the preceding examples we did not include a constant of integration in the antiderivatives. In general, when applying the Fundamental Theorem of Calculus there is no need to include a constant of integration because it will drop out anyhow. To see that this is so, let $F$ be any antiderivative of the integrand on $[a, b]$, and let $C$ be any constant; then

$$\int_a^b f(x)\, dx = F(x) + C \Big]_a^b = [F(b) + C] - [F(a) + C] = F(b) - F(a)$$

Thus, for purposes of evaluating a definite integral we can omit the constant of integration in

$$\int f(x)\, dx = F(x) + C$$

and express (5) as

$$\int_a^b f(x)\, dx = \left[ \int f(x)\, dx \right]_a^b \tag{6}$$

which relates the definite and indefinite integrals.

### Example 4

$$\int_1^9 \sqrt{x}\, dx = \int \sqrt{x}\, dx \Big]_1^9 = \int x^{1/2}\, dx \Big]_1^9 = \frac{2}{3}x^{3/2} \Big]_1^9 = \frac{2}{3}(27 - 1) = \frac{52}{3} \qquad \blacktriangleleft$$

REMARK. Usually, we will dispense with the step of displaying the indefinite integral explicitly and write the antiderivative immediately, as in our first three examples.

### Example 5

Table 7.2.1 will be helpful for the following computations.

$$\int_0^{\ln 3} 5e^x\, dx = 5\int_0^{\ln 3} e^x\, dx = 5e^x \Big|_0^{\ln 3} = 5(e^{\ln 3} - e^0) = 5(3 - 1) = 10$$

$$\int_1^2 \frac{1}{x}\, dx = \ln|x| \Big]_1^2 = \ln|2| - \ln|1| = \ln 2 - \ln 1 = \ln 2$$

$$\int_{-2}^{-1} \frac{1}{x}\, dx = \ln|x| \Big]_{-2}^{-1} = \ln|-1| - \ln|-2| = \ln 1 - \ln 2 = -\ln 2$$

$$\int_{-\pi/4}^{\pi/4} \sec x \tan x\, dx = \sec x \Big]_{-\pi/4}^{\pi/4} = \sec\left(\frac{\pi}{4}\right) - \sec\left(-\frac{\pi}{4}\right) = \frac{2}{\sqrt{2}} - \frac{2}{\sqrt{2}} = 0 \qquad \blacktriangleleft$$

WARNING. The requirement in the Fundamental Theorem of Calculus that $f$ be continuous on $[a, b]$ is important to keep in mind, for if you attempt to apply this theorem in cases where the integrand is not continuous on the interval of integration, then you may obtain erroneous results. For example, the function $f(x) = 1/x^2$ has a discontinuity at $x = 0$, so the Fundamental Theorem of Calculus cannot be used to integrate $f$ on any interval that contains $x = 0$. However, if we ignore this and blindly apply the theorem over the interval $[-1, 1]$, we obtain

$$\int_{-1}^1 \frac{1}{x^2}\, dx = -\frac{1}{x} \Big]_{-1}^1 = -[1 - (-1)] = -2$$

which is clearly erroneous because $f(x) = 1/x^2$ is a nonnegative function and hence cannot possibly produce a negative definite integral.

FOR THE READER.    If you have a CAS, read the documentation on evaluating definite integrals, and then check the results in the preceding examples.

The Fundamental Theorem of Calculus can be applied without modification to definite integrals in which the lower limit of integration is greater than or equal to the upper limit of integration.

**Example 6**

$$\int_1^1 x^2 \, dx = \frac{x^3}{3}\Bigg]_1^1 = \frac{1}{3} - \frac{1}{3} = 0$$

$$\int_4^0 x \, dx = \frac{x^2}{2}\Bigg]_4^0 = \left[\frac{0}{2} - \frac{16}{2}\right] = -8$$

The latter result is consistent with the result that would be obtained by first reversing the limits of integration in accordance with Definition 7.5.3(b):

$$\int_4^0 x \, dx = -\int_0^4 x \, dx = -\frac{x^2}{2}\Bigg]_0^4 = -\left[\frac{16}{2} - \frac{0}{2}\right] = -8 \qquad \blacktriangleleft$$

To integrate a continuous function that is defined piecewise on an interval $[a, b]$, split this interval into subintervals at the breakpoints of the function, and integrate separately over each subinterval in accordance with Theorem 7.5.5.

**Example 7**

Evaluate $\displaystyle\int_0^6 f(x) \, dx$ if

$$f(x) = \begin{cases} x^2, & x < 2 \\ 3x - 2, & x \geq 2 \end{cases}$$

*Solution.*  From Theorem 7.5.5

$$\int_0^6 f(x) \, dx = \int_0^2 f(x) \, dx + \int_2^6 f(x) \, dx = \int_0^2 x^2 \, dx + \int_2^6 (3x - 2) \, dx$$

$$= \frac{x^3}{3}\Bigg]_0^2 + \left[\frac{3x^2}{2} - 2x\right]_2^6 = \left(\frac{8}{3} - 0\right) + (42 - 2) = \frac{128}{3} \qquad \blacktriangleleft$$

**Example 8**

Evaluate $\displaystyle\int_{-1}^2 |x| \, dx$.

*Solution.*  Since $|x| = x$ when $x \geq 0$ and $|x| = -x$ when $x \leq 0$,

$$\int_{-1}^2 |x| \, dx = \int_{-1}^0 |x| \, dx + \int_0^2 |x| \, dx$$

$$= \int_{-1}^0 (-x) \, dx + \int_0^2 x \, dx$$

$$= -\frac{x^2}{2}\Bigg]_{-1}^0 + \frac{x^2}{2}\Bigg]_0^2 = \frac{1}{2} + 2 = \frac{5}{2} \qquad \blacktriangleleft$$

**DUMMY VARIABLES**

To evaluate a definite integral using the Fundamental Theorem of Calculus, one needs to be able to find an antiderivative of the integrand; thus, it is important to know what kinds of functions have antiderivatives. It is our next objective to show that all continuous functions have antiderivatives, but to do this we will need some preliminary results.

Formula (6) shows that there is a close relationship between the integrals

$$\int_a^b f(x)\,dx \quad \text{and} \quad \int f(x)\,dx$$

However, the definite and indefinite integrals differ in some important ways. For one thing, the two integrals are different kinds of objects—the definite integral is a *number* (the signed area between the graph of $y = f(x)$ and the interval $[a, b]$), whereas the indefinite integral is a *function*, or more accurately a set of functions [the antiderivatives of $f(x)$]. However, the two types of integrals also differ in the role played by the variable of integration. In an indefinite integral, the variable of integration is "passed through" to the antiderivative in the sense that integrating a function of $x$ produces a function of $x$, integrating a function of $t$ produces a function of $t$, and so forth. For example,

$$\int x^2\,dx = \frac{x^3}{3} + C \quad \text{and} \quad \int t^2\,dt = \frac{t^3}{3} + C$$

In contrast, the variable of integration in a definite integral is not passed through to the end result, since the end result is a number. Thus, integrating a function of $x$ over an interval and integrating the same function of $t$ over the same interval of integration produces the same value for the integral. For example,

$$\int_1^3 x^2\,dx = \frac{x^3}{3}\Bigg]_{x=1}^3 = \frac{27}{3} - \frac{1}{3} = \frac{26}{3} \quad \text{and} \quad \int_1^3 t^2\,dt = \frac{t^3}{3}\Bigg]_{t=1}^3 = \frac{27}{3} - \frac{1}{3} = \frac{26}{3}$$

However, this latter result should not be surprising, since the area under the graph of the curve $y = f(x)$ over an interval $[a, b]$ on the $x$-axis is the same as the area under the graph of the curve $y = f(t)$ over the interval $[a, b]$ on the $t$-axis (Figure 7.6.4).

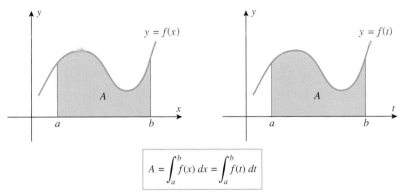

Figure 7.6.4

Because the variable of integration in a definite integral plays no role in the end result, it is often referred to as a ***dummy variable***. In summary:

> *Whenever you find it convenient to change the letter used for the variable of integration in a definite integral, you can do so without changing the value of the integral.*

**THE MEAN-VALUE THEOREM FOR INTEGRALS**

To reach our goal of showing that continuous functions have antiderivatives, we will need to develop a basic property of definite integrals, known as the *Mean-Value Theorem for Integrals*. In the next section we will use this theorem to extend the familiar idea of "average value" so that it applies to continuous functions, but here we will need it as a tool for developing other results.

Let $f$ be a continuous nonnegative function on $[a, b]$, and let $m$ and $M$ be the minimum and maximum values of $f(x)$ on this interval. Consider the rectangle of heights $m$ and $M$

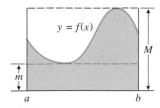

$y = f(x)$

$M$

$m$

$a$    $b$

Figure 7.6.5

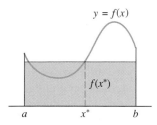

$y = f(x)$

$f(x^*)$

$a$    $x^*$    $b$

Figure 7.6.6

over the interval $[a, b]$ (Figure 7.6.5). It is clear geometrically from this figure that the area

$$A = \int_a^b f(x)\, dx$$

under $y = f(x)$ is at least as large as the area of the rectangle of height $m$ and no larger than the area of the rectangle of height $M$. It seems reasonable, therefore, that there is a rectangle over the interval $[a, b]$ of some appropriate height $f(x^*)$ between $m$ and $M$ whose area is precisely $A$; that is,

$$\int_a^b f(x)\, dx = f(x^*)(b - a)$$

(Figure 7.6.6). This is a special case of the following result.

---

**7.6.2    THEOREM (*The Mean-Value Theorem for Integrals*).**    *If $f$ is continuous on a closed interval $[a, b]$, then there is at least one number $x^*$ in $[a, b]$ such that*

$$\int_a^b f(x)\, dx = f(x^*)(b - a) \qquad (7)$$

---

*Proof.*    By the Extreme-Value Theorem (6.1.3), $f$ assumes a maximum value $M$ and a minimum value $m$ on $[a, b]$. Thus, for all $x$ in $[a, b]$,

$$m \le f(x) \le M$$

and from Theorem 7.5.6($b$)

$$\int_a^b m\, dx \le \int_a^b f(x)\, dx \le \int_a^b M\, dx$$

or

$$m(b - a) \le \int_a^b f(x)\, dx \le M(b - a) \qquad (8)$$

or

$$m \le \frac{1}{b - a} \int_a^b f(x)\, dx \le M$$

This implies that

$$\frac{1}{b - a} \int_a^b f(x)\, dx \qquad (9)$$

is a number between $m$ and $M$, and since $f(x)$ assumes the values $m$ and $M$ on $[a, b]$, it follows from the Intermediate-Value Theorem (2.4.8) that $f(x)$ must assume the value (9) at some point $x^*$ in $[a, b]$; that is,

$$\frac{1}{b - a} \int_a^b f(x)\, dx = f(x^*) \quad \text{or} \quad \int_a^b f(x)\, dx = f(x^*)(b - a) \qquad \blacksquare$$

**Example 9**

Since $f(x) = x^2$ is continuous on the interval $[1, 4]$, the Mean-Value Theorem for Integrals guarantees that there is a number $x^*$ in $[1, 4]$ such that

$$\int_1^4 x^2\, dx = f(x^*)(4 - 1) = (x^*)^2(4 - 1) = 3(x^*)^2$$

But

$$\int_1^4 x^2\, dx = \frac{x^3}{3} \bigg]_1^4 = 21$$

so that

$$3(x^*)^2 = 21 \quad \text{or} \quad (x^*)^2 = 7 \quad \text{or} \quad x^* = \pm\sqrt{7}$$

Thus, $x^* = \sqrt{7} \approx 2.65$ is the number in the interval $[1, 4]$ whose existence is guaranteed by the Mean-Value Theorem for Integrals. ◀

......................................

**PART 2 OF THE FUNDAMENTAL THEOREM OF CALCULUS**

In Section 7.1 we gave an informal argument to show that if $f$ is continuous and nonnegative on $[a, b]$, and if $A(x)$ is the area under the graph of $y = f(x)$ over the interval $[a, x]$ (Figure 7.6.2), then $A'(x) = f(x)$. But $A(x)$ can be expressed as the definite integral

$$A(x) = \int_a^x f(t)\, dt$$

(where we have used $t$ rather than $x$ as the variable of integration to avoid a conflict with the $x$ that appears as the upper limit of integration). Thus, the relationship $A'(x) = f(x)$ can be expressed as

$$\frac{d}{dx}\left[\int_a^x f(t)\, dt\right] = f(x)$$

This is a special case of the following more general result, which applies even if $f$ has negative values.

---

**7.6.3   THEOREM (*The Fundamental Theorem of Calculus, Part 2*).**   *If $f$ is continuous on an interval $I$, then $f$ has an antiderivative on $I$. In particular, if $a$ is any point in $I$, then the function $F$ defined by*

$$F(x) = \int_a^x f(t)\, dt$$

*is an antiderivative of $f$ on $I$; that is, $F'(x) = f(x)$ for each $x$ in $I$, or in an alternative notation*

$$\frac{d}{dx}\left[\int_a^x f(t)\, dt\right] = f(x) \tag{10}$$

---

*Proof.*   We will show first that $F(x)$ is defined at each point $x$ in the interval $I$. If $x > a$ and $x$ is in the interval $I$, then Theorem 7.5.8(a) applied to the interval $[a, x]$ and the continuity of $f$ on $I$ ensures that $F(x)$ is defined; and if $x$ is in the interval $I$ and $x \le a$, then Definition 7.5.3(b) combined with Theorem 7.5.8(a) ensures that $F(x)$ is defined. Thus, $F(x)$ is defined for all $x$ in $I$.

Next we will show that $F'(x) = f(x)$ for each $x$ in the interval $I$. If $x$ is not an endpoint of $I$, then it follows from the definition of a derivative that

$$F'(x) = \lim_{h \to 0} \frac{F(x + h) - F(x)}{h}$$

$$= \lim_{h \to 0} \frac{1}{h}\left[\int_a^{x+h} f(t)\, dt - \int_a^x f(t)\, dt\right]$$

$$= \lim_{h \to 0} \frac{1}{h}\left[\int_a^{x+h} f(t)\, dt + \int_x^a f(t)\, dt\right]$$

$$= \lim_{h \to 0} \frac{1}{h}\int_x^{x+h} f(t)\, dt \qquad \boxed{\text{Theorem 7.5.5}}$$

Applying the Mean-Value Theorem for Integrals (7.6.2) to the last expression, we obtain

$$F'(x) = \lim_{h \to 0} \frac{1}{h}[f(t^*) \cdot h] = \lim_{h \to 0} f(t^*) \tag{11}$$

where $t^*$ is some number between $x$ and $x + h$. Because $t^*$ is between $x$ and $x + h$, it follows that $t^* \to x$ as $h \to 0$. Thus, $f(t^*) \to f(x)$ as $h \to 0$, since $f$ is assumed continuous at $x$. Therefore, it follows from (11) that $F'(x) = f(x)$. If $x$ is an endpoint of the interval $I$, then the two-sided limits in the proof must be replaced by the appropriate one-sided limits, but otherwise the arguments are identical.   ■

In words, Formula (10) states:

*If a definite integral has a variable upper limit of integration and a continuous integrand, then the derivative of the integral with respect to its upper limit is equal to the integrand evaluated at the upper limit.*

### Example 10

Find

$$\frac{d}{dx}\left[\int_1^x t^3\,dt\right]$$

by applying Part 2 of the Fundamental Theorem of Calculus, and then confirm the result by performing the integration and then differentiating.

*Solution.*   The integrand is a continuous function, so from (10)

$$\frac{d}{dx}\left[\int_1^x t^3\,dt\right] = x^3$$

Alternatively, evaluating the integral and then differentiating yields

$$\int_1^x t^3\,dt = \frac{t^4}{4}\bigg]_{t=1}^{x} = \frac{x^4}{4} - \frac{1}{4}, \quad \frac{d}{dx}\left[\frac{x^4}{4} - \frac{1}{4}\right] = x^3$$

so the two methods for differentiating the integral agree.   ◄

### Example 11

Since

$$f(x) = \frac{\sin x}{x}$$

is continuous on any interval that does not contain the origin, it follows from (10) that on the interval $(0, +\infty)$ we have

$$\frac{d}{dx}\left[\int_1^x \frac{\sin t}{t}\,dt\right] = \frac{\sin x}{x}$$

Unlike the preceding example, there is no way to evaluate the integral in terms of familiar functions, so Formula (10) provides the only simple method for finding the derivative.
◄

**DIFFERENTIATION AND INTEGRATION ARE INVERSE PROCESSES**

The two parts of the Fundamental Theorem of Calculus, when taken together, tell us that differentiation and integration are inverse processes in the sense that each undoes the effect of the other. To see why this is so, note that Part 1 of the Fundamental Theorem of Calculus (7.6.1) implies that

$$\int_a^x f'(t)\,dt = f(x) - f(a)$$

which tells us that if the value of $f(a)$ is known, then function $f$ can be recovered from its derivative $f'$ by integrating. Conversely, Part 2 of the Fundamental Theorem of Calculus

(7.6.3) states that

$$\frac{d}{dx}\left[\int_a^x f(t)\,dt\right] = f(x)$$

which tells us that the function $f$ can be recovered from its integral by differentiating. Thus, differentiation and integration can be viewed as inverse processes.

It is common to treat parts 1 and 2 of the Fundamental Theorem of Calculus as a single theorem, and refer to it simply as the *Fundamental Theorem of Calculus*. This theorem ranks as one of the greatest discoveries in the history of science, and its formulation by Newton and Leibniz is generally regarded to be the "discovery of calculus."

## EXERCISE SET 7.6 ⬚ Graphing Calculator ⬚C CAS

1. In each part, use a definite integral to find the area of the region, and check your answer using an appropriate formula from geometry.

(a)          (b)          (c)

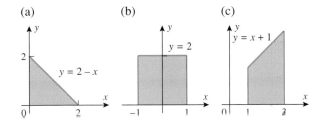

2. In each part, use a definite integral to find the area under the curve $y = f(x)$ over the stated interval, and check your answer using an appropriate formula from geometry.
   (a) $f(x) = x$; $[0, 5]$
   (b) $f(x) = 5$; $[3, 9]$
   (c) $f(x) = x + 3$; $[-1, 2]$

In Exercises 3–8, find the area under the curve $y = f(x)$ over the stated interval.

3. $f(x) = x^3$; $[2, 3]$

4. $f(x) = x^4$; $[-1, 1]$

5. $f(x) = \sqrt{x}$; $[1, 9]$

6. $f(x) = x^{-3/5}$; $[1, 4]$

7. $f(x) = e^x$; $[1, 3]$

8. $f(x) = \dfrac{1}{x}$; $[1, 5]$

In Exercises 9–24, evaluate the integrals using Part 1 of the Fundamental Theorem of Calculus.

9. $\displaystyle\int_{-3}^0 (x^2 - 4x + 7)\,dx$

10. $\displaystyle\int_{-1}^7 x(1 + x^3)\,dx$

11. $\displaystyle\int_1^3 \frac{1}{x^2}\,dx$

12. $\displaystyle\int_1^2 \frac{1}{x^6}\,dx$

13. $\displaystyle\int_4^9 2x\sqrt{x}\,dx$

14. $\displaystyle\int_1^8 (5x^{2/3} - 4x^{-2})\,dx$

15. $\displaystyle\int_{-\pi/2}^{\pi/2} \sin\theta\,d\theta$

16. $\displaystyle\int_0^{\pi/4} \sec^2\theta\,d\theta$

17. $\displaystyle\int_{-\pi/4}^{\pi/4} \cos x\,dx$

18. $\displaystyle\int_0^1 (x - \sec x \tan x)\,dx$

19. $\displaystyle\int_{\ln 2}^3 5e^x\,dx$

20. $\displaystyle\int_{1/2}^1 \frac{1}{2x}\,dx$

21. $\displaystyle\int_1^4 \left(\frac{3}{\sqrt{t}} - 5\sqrt{t} - t^{-3/2}\right)dt$

22. $\displaystyle\int_4^0 (4y^{-1/2} + 2y^{1/2} + y^{-5/2})\,dy$

23. $\displaystyle\int_{\pi/6}^{\pi/2} \left(x + \frac{2}{\sin^2 x}\right)dx$

24. $\displaystyle\int_1^2 (x^{-1} + \sqrt{2}e^x - \csc x \cot x)\,dx$

⬚C 25. For each of the integrals you evaluated in Exercises 9–24, use a CAS to check your answer. [*Note:* CAS programs have commands for evaluating definite integrals exactly or approximately. Use the exact evaluation here.]

⬚C 26. Use a CAS to evaluate the integral

$$\int_a^{4a} (a^{1/2} - x^{1/2})\,dx$$

and check the answer by hand.

In Exercises 27–29, use Theorem 7.5.5 to evaluate the given integrals.

27. (a) $\displaystyle\int_0^2 |2x - 3|\,dx$    (b) $\displaystyle\int_0^{3\pi/4} |\cos x|\,dx$

28. (a) $\displaystyle\int_{-1}^2 \sqrt{2 + |x|}\,dx$    (b) $\displaystyle\int_{-1}^1 |e^x - 1|\,dx$

**29.** $\displaystyle\int_{-2}^{3} f(x)\,dx$, where $f(x) = \begin{cases} -x, & x \geq 0 \\ x^2, & x < 0 \end{cases}$

[c] **30.** CAS programs provide methods for entering functions that are defined piecewise. Check your documentation to see how this is done, and then use the CAS to evaluate

$$\int_{0}^{4} f(x)\,dx, \quad \text{where} \quad f(x) = \begin{cases} \sqrt{x}, & 0 \leq x < 1 \\ 1/x^2, & x \geq 1 \end{cases}$$

Check the answer by hand.

In Exercises 31–33, use a calculating utility to find the midpoint approximation of the integral using $n = 20$ subintervals, and then find the exact value of the integral using Part 1 of the Fundamental Theorem of Calculus.

**31.** $\displaystyle\int_{1}^{3} \frac{1}{x^2}\,dx$     **32.** $\displaystyle\int_{0}^{\pi/2} \sin x\,dx$     **33.** $\displaystyle\int_{1}^{3} \frac{1}{x}\,dx$

[c] **34.** Compare the answers obtained by the midpoint rule in Exercises 31–33 to those obtained using the numerical (approximate) integration command of a CAS.

**35.** Find the area under the curve $y = x^2 + 1$ over the interval $[0, 3]$. Make a sketch of the region.

**36.** Find the area that is above the $x$-axis, but below the curve $y = (1 - x)(x - 2)$. Make a sketch of the region.

**37.** Find the area under the curve $y = 3\sin x$ over the interval $[0, 2\pi/3]$. Sketch the region.

**38.** Find the area below the interval $[-2, -1]$, but above the curve $y = x^3$. Make a sketch of the region.

**39.** Find the total area between the curve $y = x^2 - 3x - 10$ and the interval $[-3, 8]$. Make a sketch of the region. [*Hint:* Find the portion of area above the interval and the portion of area below the interval separately.]

**40.** (a) Use a graphing utility to generate the graph of

$$f(x) = \frac{1}{100}(x + 2)(x + 1)(x - 3)(x - 5)$$

and use the graph to make a conjecture about the sign of the integral

$$\int_{-2}^{5} f(x)\,dx$$

(b) Check your conjecture by evaluating the integral.

**41.** (a) Let $f$ be an odd function; that is, $f(-x) = -f(x)$. Invent a theorem that makes a statement about the value of an integral of the form

$$\int_{-a}^{a} f(x)\,dx$$

(b) Confirm that your theorem works for the integrals

$$\int_{-1}^{1} x^3\,dx \quad \text{and} \quad \int_{-\pi/2}^{\pi/2} \sin x\,dx$$

(c) Let $f$ be an even function; that is, $f(-x) = f(x)$. Invent a theorem that makes a statement about the rela-

tionship between the integrals

$$\int_{-a}^{a} f(x)\,dx \quad \text{and} \quad \int_{0}^{a} f(x)\,dx$$

(d) Confirm that your theorem works for the integrals

$$\int_{-1}^{1} x^2\,dx \quad \text{and} \quad \int_{-\pi/2}^{\pi/2} \cos x\,dx$$

[c] **42.** Use the theorem you invented in Exercise 41(a) to evaluate the integral

$$\int_{-5}^{5} \frac{x^7 - x^5 + x}{x^4 + x^2 + 7}\,dx$$

and check your answer with a CAS.

**43.** Define $F(x)$ by

$$F(x) = \int_{1}^{x} (t^3 + 1)\,dt$$

(a) Use Part 2 of the Fundamental Theorem of Calculus to find $F'(x)$.

(b) Check the result in part (a) by first integrating and then differentiating.

**44.** Define $F(x)$ by

$$F(x) = \int_{\pi/4}^{x} \cos 2t\,dt$$

(a) Use Part 2 of the Fundamental Theorem of Calculus to find $F'(x)$.

(b) Check the result in part (a) by first integrating and then differentiating.

In Exercises 45–48, use Part 2 of the Fundamental Theorem of Calculus to find the derivative.

**45.** (a) $\displaystyle\frac{d}{dx}\int_{1}^{x} \sin(\sqrt{t})\,dt$     (b) $\displaystyle\frac{d}{dx}\int_{0}^{x} e^{t^2}\,dt$

**46.** (a) $\displaystyle\frac{d}{dx}\int_{0}^{x} \frac{dt}{1 + \sqrt{t}}$     (b) $\displaystyle\frac{d}{dx}\int_{1}^{x} \ln t\,dt$

**47.** $\displaystyle\frac{d}{dx}\int_{x}^{0} \frac{t}{\cos t}\,dt$     [*Hint:* Use Definition 7.5.3(b).]

**48.** $\displaystyle\frac{d}{du}\int_{0}^{u} |x|\,dx$

**49.** Let $F(x) = \displaystyle\int_{2}^{x} \sqrt{3t^2 + 1}\,dt$. Find

  (a) $F(2)$     (b) $F'(2)$     (c) $F''(2)$

**50.** Let $F(x) = \displaystyle\int_{0}^{x} \frac{\cos t}{t^2 + 3}\,dt$. Find

  (a) $F(0)$     (b) $F'(0)$     (c) $F''(0)$

**51.** Let $F(x) = \displaystyle\int_{0}^{x} \frac{t - 3}{t^2 + 7}\,dt$ for $-\infty < x < +\infty$.

(a) Find the value of $x$ where $F$ attains its minimum value.

(b) Find intervals over which $F$ is only increasing or only decreasing.

(c) Find open intervals over which $F$ is only concave up or only concave down.

[c] **52.** Use the plotting and numerical integration commands of a CAS to generate the graph of the function $F$ in Exercise 51 over the interval $-20 \leq x \leq 20$, and confirm that the graph is consistent with the results obtained in that exercise.

**53.** (a) Over what open interval does the formula

$$F(x) = \int_1^x \frac{dt}{t}$$

represent an antiderivative of $f(x) = 1/x$?

(b) Find a point where the graph of $F$ crosses the $x$-axis.

**54.** (a) Over what open interval does the formula

$$F(x) = \int_1^x \frac{1}{t^2 - 9} \, dt$$

represent an antiderivative of

$$f(x) = \frac{1}{x^2 - 9}?$$

(b) Find a point where the graph of $F$ crosses the $x$-axis.

In Exercises 55 and 56, find all values of $x^*$ in the stated interval that satisfy Equation (7) in the Mean-Value Theorem for Integrals (7.6.2), and explain what these numbers represent.

**55.** (a) $f(x) = \sqrt{x}$; $[0, 9]$   (b) $f(x) = 1/x$; $[1, e]$

**56.** (a) $f(x) = \sin x$, $[-\pi, \pi]$   (b) $f(x) = 1/x^2$; $[1, 3]$

It was shown in the proof of the Mean-Value Theorem for Integrals that if $f$ is continuous on $[a, b]$, and if $m \leq f(x) \leq M$ on $[a, b]$, then

$$m(b - a) \leq \int_a^b f(x) \, dx \leq M(b - a)$$

[see (8)]. These inequalities make it possible to obtain bounds on the size of a definite integral from bounds on the size of its integrand. This is illustrated in Exercises 57–59.

**57.** Find the maximum and minimum values of $\sqrt{x^3 + 2}$ for $0 \leq x \leq 3$, and use these values to find bounds on the value of the integral

$$\int_0^3 \sqrt{x^3 + 2} \, dx$$

**58.** Find values of $m$ and $M$ such that $m \leq x \sin x \leq M$ for $0 \leq x \leq \pi$, and use these values to find bounds on the value of the integral

$$\int_0^\pi x \sin x \, dx$$

**59.** Show that

$$0 \leq \int_1^5 \ln x \, dx \leq 4 \ln 5$$

**60.** Prove:
(a) $[cF(x)]_a^b = c[F(x)]_a^b$
(b) $[F(x) + G(x)]_a^b = F(x)]_a^b + G(x)]_a^b$
(c) $[F(x) - G(x)]_a^b = F(x)]_a^b - G(x)]_a^b$.

# 7.7 RECTILINEAR MOTION REVISITED; AVERAGE VALUE

*In Section 6.3 we used the derivative to define the notions of instantaneous velocity and acceleration for a particle moving along a line. In this section we will resume the study of such motion using the tools of integration. We will also investigate the general problem of integrating a rate of change, and we will show how the definite integral can be used to define the average value of a continuous function. More applications of integration will be given in Chapter 8.*

**FINDING POSITION AND VELOCITY BY INTEGRATION**

Recall from Definitions 6.3.1 and 6.3.2 that if $s(t)$ is the position function of a particle moving on a coordinate line, then the instantaneous velocity and acceleration of the particle are given by the formulas

$$v(t) = s'(t) = \frac{ds}{dt} \quad \text{and} \quad a(t) = v'(t) = \frac{dv}{dt} = \frac{d^2s}{dt^2}$$

It follows from these formulas that $s(t)$ is an antiderivative of $v(t)$ and $v(t)$ is an antideriva-

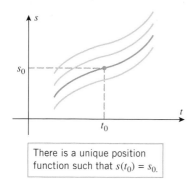

There is a unique position
function such that $s(t_0) = s_0$.

Figure 7.7.1

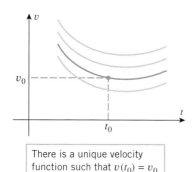

There is a unique velocity
function such that $v(t_0) = v_0$.

Figure 7.7.2

**UNIFORMLY ACCELERATED
MOTION**

tive of $a(t)$; that is,

$$s(t) = \int v(t)\, dt \qquad \text{and} \qquad v(t) = \int a(t)\, dt \qquad (1\text{-}2)$$

Thus, if the velocity of a particle is known, then its position function can be obtained from (1) by integration, provided there is sufficient additional information to determine the constant of integration. In particular, we can determine the constant of integration if we know the position $s_0$ of the particle at some time $t_0$, since this information determines a unique antiderivative $s(t)$ (Figure 7.7.1). Similarly, if the acceleration function of the particle is known, then its velocity function can be obtained from (2) by integration if we know the velocity $v_0$ of the particle at some time $t_0$ (Figure 7.7.2).

**Example 1**

Find the position function of a particle that is moving with velocity $v(t) = \cos \pi t$ along a coordinate line, assuming that the particle has coordinates $s = 4$ at time $t = 0$.

*Solution.* The position function is

$$s(t) = \int v(t)\, dt = \int \cos \pi t \, dt = \frac{1}{\pi} \sin \pi t + C$$

Since $s = 4$ when $t = 0$, it follows that

$$4 = s(0) = \frac{1}{\pi} \sin 0 + C = C$$

Thus,

$$s(t) = \frac{1}{\pi} \sin \pi t + 4 \qquad \blacktriangleleft$$

One of the most important cases of rectilinear motion occurs when a particle has constant acceleration. We call this *uniformly accelerated motion*.

We will show that if a particle moves with constant acceleration along an $s$-axis, and if the position and velocity of the particle are known at some point in time, say when $t = 0$, then it is possible to derive formulas for the position $s(t)$ and the velocity $v(t)$ at any time $t$. To see how this can be done, suppose that the particle has constant acceleration

$$a(t) = a \qquad (3)$$

and

$$s = s_0 \quad \text{when} \quad t = 0 \qquad (4)$$
$$v = v_0 \quad \text{when} \quad t = 0 \qquad (5)$$

where $s_0$ and $v_0$ are known. We call (4) and (5) the *initial conditions* for the motion.

With (3) as a starting point, we can integrate $a(t)$ to obtain $v(t)$, and we can integrate $v(t)$ to obtain $s(t)$, using an initial condition in each case to determine the constant of integration. The computations are as follows:

$$v(t) = \int a(t)\, dt = \int a \, dt = at + C_1 \qquad (6)$$

To determine the constant of integration $C_1$ we apply initial condition (5) to this equation to obtain

$$v_0 = v(0) = a \cdot 0 + C_1 = C_1$$

Substituting this in (6) and putting the constant term first yields

$$v(t) = v_0 + at$$

Since $v_0$ is constant, it follows that

$$s(t) = \int v(t)\, dt = \int (v_0 + at)\, dt = v_0 t + \tfrac{1}{2} at^2 + C_2 \qquad (7)$$

To determine the constant $C_2$ we apply initial condition (4) to this equation to obtain

$$s_0 = s(0) = v_0 \cdot 0 + \tfrac{1}{2}a \cdot 0 + C_2 = C_2$$

Substituting this in (7) and putting the constant term first yields

$$s(t) = s_0 + v_0 t + \tfrac{1}{2}at^2$$

In summary, we have the following result.

---

**7.7.1** UNIFORMLY ACCELERATED MOTION. *If a particle moves with constant acceleration a along an s-axis, and if the position and velocity at time $t = 0$ are $s_0$ and $v_0$, respectively, then the position and velocity functions of the particle are*

$$s(t) = s_0 + v_0 t + \tfrac{1}{2}at^2 \qquad (8)$$

$$v(t) = v_0 + at \qquad (9)$$

---

FOR THE READER. How can you tell from the velocity versus time curve whether a particle moving along a line has uniformly accelerated motion?

## Example 2

Suppose that an intergalactic spacecraft uses a sail and the "solar wind" to produce a constant acceleration of $0.032\ \text{m/s}^2$. Assuming that the spacecraft has a velocity of $10,000\ \text{m/s}$ when the sail is first raised, how far will the spacecraft travel in 1 hour, and what will its velocity be at that time?

*Solution.* In this problem the choice of a coordinate axis is at our discretion, so we will choose it to make the computations as simple as possible. Accordingly, let us introduce an $s$-axis whose positive direction is in the direction of motion, and let us take the origin to coincide with the position of the spacecraft at the time $t = 0$ when the sail is raised. Thus, the Formulas (8) and (9) for uniformly accelerated motion apply with

$$s_0 = s(0) = 0, \quad v_0 = v(0) = 10,000, \quad \text{and} \quad a = 0.032$$

Since 1 hour corresponds to $t = 3600$ s, it follows from (8) that in 1 hour the spacecraft travels a distance of

$$s(3600) = 10,000(3600) + \tfrac{1}{2}(0.032)(3600)^2 \approx 36,207,400\ \text{m}$$

and it follows from (9) that after 1 hour its velocity is

$$v(3600) = 10,000 + (0.032)(3600) \approx 10,115\ \text{m/s} \qquad \blacktriangleleft$$

## Example 3

A bus has stopped to pick up riders, and a woman is running at a constant velocity of 5 m/s to catch it. When she is 11 m behind the front door the bus pulls away with a constant acceleration of $1\ \text{m/s}^2$. From that point in time, how long will it take for the woman to reach the front door of the bus if she keeps running with a velocity of 5 m/s?

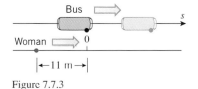

Figure 7.7.3

*Solution.* As shown in Figure 7.7.3, choose the $s$-axis so that the bus and the woman are moving in the positive direction, and the front door of the bus is at the origin at the time $t = 0$ when the bus begins to pull away. To catch the bus at some later time $t$, the woman will have to cover a distance $s_w(t)$ that is equal to 11 m plus the distance $s_b(t)$ traveled by the bus; that is, the woman will catch the bus when

$$s_w(t) = s_b(t) + 11 \qquad (10)$$

Since the woman has a constant velocity of 5 m/s, the distance she travels in $t$ seconds is $s_w(t) = 5t$. Thus, (10) can be written as

$$s_b(t) = 5t - 11 \qquad (11)$$

Since the bus has a constant acceleration of $a = 1 \text{ m/s}^2$, and since $s_0 = v_0 = 0$ at time $t = 0$ (why?), it follows from (8) that

$$s_b(t) = \tfrac{1}{2}t^2$$

Substituting this equation into (11) and reorganizing the terms yields the quadratic equation

$$\tfrac{1}{2}t^2 - 5t + 11 = 0 \quad \text{or} \quad t^2 - 10t + 22 = 0$$

Solving this equation for $t$ using the quadratic formula yields two solutions:

$$t = 5 - \sqrt{3} \approx 3.3 \quad \text{and} \quad t = 5 + \sqrt{3} \approx 6.7$$

(verify). Thus, the woman can reach the door at two different times, $t = 3.3$ s and $t = 6.7$ s. The reason that there are two solutions can be explained as follows: When the woman first reaches the door, she is running faster than the bus and can run past it if the driver does not see her. However, as the bus speeds up, it eventually catches up to her, and she has another chance to flag it down.    ◀

## THE FREE-FALL MODEL

In Section 6.3 we discussed the free-fall model of motion near the surface of the Earth with the promise that we would derive Formula (5) of that section later in the text; we will now show how to do this. As stated in 6.3.4 and illustrated in Figure 6.3.7, we will assume that the object moves on an $s$-axis whose origin is at the surface of the Earth and whose positive direction is up; and we will assume that the position and velocity of the object at time $t = 0$ are $s_0$ and $v_0$, respectively.

It is a fact of physics that a particle moving on a vertical line near the Earth's surface and subject only to the force of the Earth's gravity moves with constant acceleration. The magnitude of this constant, denoted by the letter $g$, is approximately 9.8 m/s$^2$ or 32 ft/s$^2$, depending on whether distance is measured in meters or feet.[*]

Recall that a particle is speeding up when its velocity and acceleration have the same sign and is slowing down when they have opposite signs. Thus, because we have chosen the positive direction to be up, it follows that the acceleration $a(t)$ of a particle in free fall is negative for all values of $t$. To see that this is so, observe that an upward-moving particle (positive velocity) is slowing down, so its acceleration must be negative; and a downward-moving particle (negative velocity) is speeding up, so its acceleration must also be negative. Thus, we conclude that

$$a(t) = -g$$

and hence it follows from (8) and (9) that the position and velocity functions of an object in free fall are

$$s(t) = s_0 + v_0 t - \tfrac{1}{2}gt^2 \tag{12}$$

$$v(t) = v_0 - gt \tag{13}$$

FOR THE READER.    Had we chosen the positive direction of the $s$-axis to be down, then the acceleration would have been $a(t) = g$ (why?). How would this have affected Formulas (12) and (13)?

### Example 4

A ball is thrown directly upward with an initial velocity of 49 m/s and is released from a point that is 8 m above the ground. Assuming that the free-fall model applies, how high will the ball travel?

---

[*]Strictly speaking, the constant $g$ varies with the latitude and the distance from the Earth's center. However, for motion at a fixed latitude and near the surface of the Earth, the assumption of a constant $g$ is satisfactory for many applications.

*Solution.*  Since distance is in meters, we take $g = 9.8$ m/s². Initially, we have $s_0 = 8$ and $v_0 = 49$, so from (12) and (13)

$$v(t) = -9.8t + 49$$

$$s(t) = -4.9t^2 + 49t + 8$$

The ball will rise until $v(t) = 0$, that is, until $-9.8t + 49 = 0$ or $t = 5$. At this instant the height above the ground will be

$$s(5) = -4.9(5)^2 + 49(5) + 8 = 130.5 \text{ m} \qquad \blacktriangleleft$$

### Example 5

A penny is released from rest near the top of the Empire State Building at a point that is 1250 ft above the ground (Figure 7.7.4). Assuming that the free-fall model applies, how long does it take for the penny to hit the ground, and what is its speed at the time of impact?

*Solution.*  Since distance is in feet, we take $g = 32$ ft/s². Initially, we have $s_0 = 1250$ and $v_0 = 0$, so from (12)

$$s(t) = -16t^2 + 1250 \tag{14}$$

Impact occurs when $s(t) = 0$. Solving this equation for $t$, we obtain

$$-16t^2 + 1250 = 0$$

$$t^2 = \frac{1250}{16} = \frac{625}{8}$$

$$t = \pm \frac{25}{\sqrt{8}} \approx \pm 8.8 \text{ s}$$

Since $t \geq 0$, we can discard the negative solution and conclude that it takes $25/\sqrt{8} \approx 8.8$ s for the penny to hit the ground. To obtain the velocity at the time of impact, we substitute $t = 25/\sqrt{8}$, $v_0 = 0$, and $g = 32$ in (13) to obtain

$$v\left(\frac{25}{\sqrt{8}}\right) = 0 - 32\left(\frac{25}{\sqrt{8}}\right) = -200\sqrt{2} \approx -282.8 \text{ ft/s}$$

Thus, the speed at the time of impact is

$$\left| v\left(\frac{25}{\sqrt{8}}\right) \right| = 200\sqrt{2} \approx 282.8 \text{ ft/s}$$

which is more than 192 mi/h!  $\qquad \blacktriangleleft$

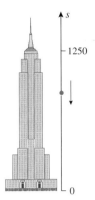

Figure 7.7.4

................................................

**INTEGRATING RATES OF CHANGE**

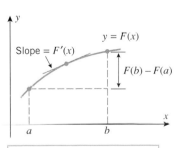

Integrating the slope of $y = F'(x)$ over the interval $[a, b]$ produces the change $F(b) - F(a)$ in the value of $F(x)$.

Figure 7.7.5

The Fundamental Theorem of Calculus

$$\int_a^b f(x)\,dx = F(b) - F(a) \tag{15}$$

has a useful interpretation that can be seen by rewriting it in a slightly different form. Since $F$ is an antiderivative of $f$ on the interval $[a, b]$, we can use the relationship $F'(x) = f(x)$ to rewrite (15) as

$$\int_a^b F'(x)\,dx = F(b) - F(a) \tag{16}$$

In this formula we can view $F'(x)$ as the rate of change of $F(x)$ with respect to $x$, and we can view $F(b) - F(a)$ as the *change* in the value of $F(x)$ as $x$ increases from $a$ to $b$ (Figure 7.7.5). Thus, we have the following useful principle.

**7.7.2**  INTEGRATING A RATE OF CHANGE.  Integrating the rate of change of $F(x)$ with respect to $x$ over an interval $[a, b]$ produces the change in the value of $F(x)$ that occurs as $x$ increases from $a$ to $b$.

Here are some examples of this idea:

- If $P(t)$ is a population (e.g., plants, animals, or people) at time $t$, then $P'(t)$ is the rate at which the population is changing at time $t$, and

$$\int_{t_1}^{t_2} P'(t)\, dt = P(t_2) - P(t_1)$$

is the change in the population between times $t_1$ and $t_2$.

- If $A(t)$ is the area of an oil spill at time $t$, then $A'(t)$ is the rate at which the area of the spill is changing at time $t$, and

$$\int_{t_1}^{t_2} A'(t)\, dt = A(t_2) - A(t_1)$$

is the change in the area of the spill between times $t_1$ and $t_2$.

- If $P'(x)$ is the marginal profit that results from producing and selling $x$ units of a product (see Section 6.2), then

$$\int_{x_1}^{x_2} P'(x)\, dx = P(x_2) - P(x_1)$$

is the change in the profit that results when the production level increases from $x_1$ units to $x_2$ units.

**DISPLACEMENT IN RECTILINEAR MOTION**

As another application of (16), suppose that $s(t)$ and $v(t)$ are the position and velocity functions of a particle moving on a coordinate line. Since $v(t)$ is the rate of change of $s(t)$ with respect to $t$, it follows from the principle in 7.7.2 that integrating $v(t)$ over an interval $[t_0, t_1]$ will produce the change in the value of $s(t)$ as $t$ increases from $t_0$ to $t_1$; that is,

$$\int_{t_0}^{t_1} v(t)\, dt = \int_{t_0}^{t_1} s'(t)\, dt = s(t_1) - s(t_0) \tag{17}$$

The expression $s(t_1) - s(t_0)$ in this formula is called the **displacement** or **change in position** of the particle over the time interval $[t_0, t_1]$. For a particle moving horizontally, the displacement is positive if the final position of the particle is to the right of its initial position, negative if it is to the left of its initial position, and zero if it coincides with the initial position (Figure 7.7.6).

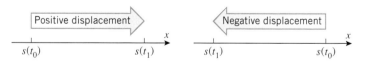

Figure 7.7.6

REMARK.   In physical problems it is important to associate the correct units with definite integrals. In general, the units for the definite integral

$$\int_a^b f(x)\, dx$$

will be units of $f(x)$ times units of $x$. This is because the definite integral is a limit of Riemann sums each of whose terms is a product of the form $f(x) \cdot \Delta x$. For example, if time is measured in seconds (s) and velocity is measured in meters per second (m/s), then integrating velocity over a time interval will produce a result whose units are in meters, since m/s $\times$ s $=$ m. Note that this is consistent with Formula (17), since displacement has units of length.

**DISTANCE TRAVELED IN RECTILINEAR MOTION**

In general, the displacement of a particle is not the same as the distance traveled by the particle. For example, a particle that travels 100 units in the positive direction and then 100 units in the negative direction travels a distance of 200 units but has a displacement of zero, since it returns to its starting point. The only case in which the displacement and the distance traveled are the same occurs when the particle moves in the positive direction without reversing the direction of its motion.

FOR THE READER.     What is the relationship between the displacement of a particle and the distance it travels if the particle moves in the negative direction without reversing the direction of motion?

From (17), integrating the velocity function of a particle over a time interval yields the displacement of a particle over that time interval. In contrast, to find the *total distance* traveled by the particle over the time interval (the distance traveled in the positive direction plus the distance traveled in the negative direction), we must integrate the *absolute value* of the velocity function; that is, we must integrate the speed:

$$\begin{bmatrix} \text{total distance} \\ \text{traveled during} \\ \text{time interval} \\ [t_0, t_1] \end{bmatrix} = \int_{t_0}^{t_1} |v(t)|\, dt \tag{18}$$

**Example 6**

A particle moves on a coordinate line so that its velocity at time $t$ is $v(t) = t^2 - 2t$ m/s.

(a)   Find the displacement of the particle during the time interval $0 \le t \le 3$.

(b)   Find the distance traveled by the particle during the time interval $0 \le t \le 3$.

*Solution (a).* From (17) the displacement is

$$\int_0^3 v(t)\, dt = \int_0^3 (t^2 - 2t)\, dt = \left[ \frac{t^3}{3} - t^2 \right]_0^3 = 0$$

Thus, the particle is at the same position at time $t = 3$ as at $t = 0$.

*Solution (b).* The velocity can be written as $v(t) = t^2 - 2t = t(t - 2)$, from which we see that $v(t) \le 0$ for $0 \le t \le 2$ and $v(t) \ge 0$ for $2 \le t \le 3$. Thus, it follows from (18) that the distance traveled is

$$\int_0^3 |v(t)|\, dt = \int_0^2 -v(t)\, dt + \int_2^3 v(t)\, dt$$

$$= \int_0^2 -(t^2 - 2t)\, dt + \int_2^3 (t^2 - 2t)\, dt$$

$$= -\left[ \frac{t^3}{3} - t^2 \right]_0^2 + \left[ \frac{t^3}{3} - t^2 \right]_2^3 = \frac{4}{3} + \frac{4}{3} = \frac{8}{3} \text{ m} \qquad \blacktriangleleft$$

**ANALYZING THE VELOCITY VERSUS TIME CURVE**

In Section 6.3 we showed how to use the position versus time curve to obtain information about the behavior of a particle moving on a coordinate line (Table 6.3.1). Similarly, there is valuable information that can be obtained from the *velocity versus time curve*. For example, the integral in (17) can be interpreted geometrically as the net signed area between the graph of $v(t)$ and the interval $[t_0, t_1]$, and it can be interpreted physically as the displacement of the particle over this interval. Thus, we have the following result.

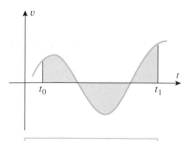

The net signed area is the displacement of the particle during the interval $[t_0, t_1]$.

Figure 7.7.7

### Example 7

Figure 7.7.8 shows three velocity versus time curves for a particle in rectilinear motion along a horizontal line. In each case, find the displacement of the particle over the time interval $0 \le t \le 4$, and explain what it tells you about the motion of the particle.

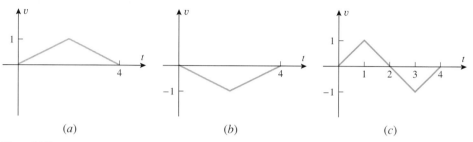

$(a)$              $(b)$              $(c)$

Figure 7.7.8

*Solution.*   In part $(a)$ of Figure 7.7.8 the net signed area under the curve is 2, so the particle is 2 units to the right of its starting point at the end of the time period. In part $(b)$ the net signed area under the curve is $-2$, so the particle is 2 units to the left of its starting point at the end of the time period. In part $(c)$ the net signed area under the curve is 0, so the particle is back at its starting point at the end of the time period.   ◄

Sometimes we will not want the net signed area between a curve $y = f(x)$ and an interval $[a, b]$, but rather the total area between the curve and the interval. This can be found by integrating $|f(x)|$ rather than $f(x)$ over the interval $[a, b]$.

### Example 8

Find the total area between the curve $y = 1 - x^2$ and the $x$-axis over the interval $[0, 2]$ (Figure 7.7.9).

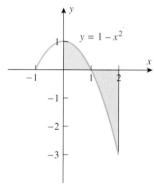

Figure 7.7.9

*Solution.*   The area $A$ is given by

$$A = \int_0^2 |1 - x^2|\, dx = \int_0^1 (1 - x^2)\, dx + \int_1^2 -(1 - x^2)\, dx$$

$$= \left[ x - \frac{x^3}{3} \right]_0^1 - \left[ x - \frac{x^3}{3} \right]_1^2$$

$$= \frac{2}{3} - \left( -\frac{4}{3} \right) = 2$$   ◄

From (18), integrating the speed $|v(t)|$ over a time interval $[t_0, t_1]$ produces the distance traveled by the particle during the time interval. However, we can also interpret the integral in (18) as the total area between the velocity versus time curve and the interval $[t_0, t_1]$ on the $t$-axis. Thus, we have the following result.

**Example 9**

For each of the velocity versus time curves in Figure 7.7.8 find the total distance traveled by the particle over the time interval $0 \leq t \leq 4$.

*Solution.*  In all three parts of Figure 7.7.8 the total area between the curve and the interval $[0, 4]$ is 2, so the particle travels a distance of 2 units during the time period in all three cases, even though the displacement is different in each case, as discussed in Example 7.  ◀

**AVERAGE VALUE OF A
CONTINUOUS FUNCTION**

In scientific work, numerical information is often summarized by computing some sort of *average* or *mean* value of the observed data. There are various kinds of averages, but the most common is the **arithmetic mean** or **arithmetic average**, which is formed by adding the data and dividing by the number of data points. Thus, the arithmetic average $\overline{a}$ of $n$ numbers $a_1, a_2, \ldots, a_n$ is

$$\overline{a} = \frac{1}{n}(a_1 + a_2 + \cdots + a_n) = \frac{1}{n}\sum_{k=1}^{n} a_k$$

In the case where the $a_k$'s are values of a function $f$, say,

$$a_1 = f(x_1), a_2 = f(x_2), \ldots, a_n = f(x_n)$$

then the arithmetic average $\overline{a}$ of these function values is

$$\overline{a} = \frac{1}{n}\sum_{k=1}^{n} f(x_k)$$

We will now show how to extend this concept so that we can compute not only the arithmetic average of finitely many function values but an average of *all* values of $f(x)$ as $x$ varies over a closed interval $[a, b]$. For this purpose recall the Mean-Value Theorem for Integrals (7.6.2), which states that if $f$ is continuous on the interval $[a, b]$, then there is at least one point $x^*$ in this interval such that

$$\int_{a}^{b} f(x)\,dx = f(x^*)(b - a)$$

The quantity

$$f(x^*) = \frac{1}{b - a}\int_{a}^{b} f(x)\,dx \tag{19}$$

will be our candidate for the average value of $f$ over the interval $[a, b]$. To explain what motivates this, divide the interval $[a, b]$ into $n$ subintervals of equal length

$$\Delta x = \frac{b - a}{n} \tag{20}$$

and choose arbitrary points $x_1^*, x_2^*, \ldots, x_n^*$ in successive subintervals. Then the arithmetic average of the numbers $f(x_1^*), f(x_2^*), \ldots, f(x_n^*)$ is

$$\text{ave} = \frac{1}{n}[f(x_1^*) + f(x_2^*) + \cdots + f(x_n^*)]$$

or from (20)

$$\text{ave} = \frac{1}{b - a}[f(x_1^*)\Delta x + f(x_2^*)\Delta x + \cdots + f(x_n^*)\Delta x] = \frac{1}{b - a}\sum_{k=1}^{n} f(x_k^*)\Delta x$$

Taking the limit as $n \to +\infty$ yields

$$\lim_{n \to +\infty} \frac{1}{b - a}\sum_{k=1}^{n} f(x_k^*)\Delta x = \frac{1}{b - a}\int_{a}^{b} f(x)\,dx$$

Since this equation describes what happens when we compute the average of "more and more" values of $f(x)$, we are led to the following definition.

**7.7.5**   DEFINITION.    If $f$ is continuous on $[a, b]$, then the *average value* (or *mean value*) of $f$ on $[a, b]$ is defined to be

$$f_{\text{ave}} = \frac{1}{b-a} \int_a^b f(x)\,dx \tag{21}$$

REMARK.    When $f$ is nonnegative on $[a, b]$, the quantity $f_{\text{ave}}$ has a simple geometric interpretation, which can be seen by writing (21) as

$$f_{\text{ave}} \cdot (b-a) = \int_a^b f(x)\,dx$$

The left side of this equation is the area of a rectangle with a height of $f_{\text{ave}}$ and base of length $b - a$, and the right side is the area under $y = f(x)$ over $[a, b]$. Thus, $f_{\text{ave}}$ is the height of a rectangle constructed over the interval $[a, b]$, whose area is the same as the area under the graph of $f$ over that interval (Figure 7.7.10). Note also that the Mean-Value Theorem, when expressed in form (21), ensures that there is always at least one point $x^*$ in $[a, b]$ at which the value of $f$ is equal to the average value of $f$ over the interval.

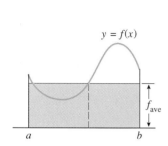

Figure 7.7.10

### Example 10

Find the average value of the function $f(x) = \sqrt{x}$ over the interval $[1, 4]$, and find all points in the interval at which the value of $f$ is the same as the average.

*Solution.*

$$f_{\text{ave}} = \frac{1}{b-a} \int_a^b f(x)\,dx = \frac{1}{4-1} \int_1^4 \sqrt{x}\,dx = \frac{1}{3} \left[ \frac{2x^{3/2}}{3} \right]_1^4$$

$$= \frac{1}{3} \left[ \frac{16}{3} - \frac{2}{3} \right] = \frac{14}{9} \approx 1.6$$

The $x$-values at which $f(x) = \sqrt{x}$ is the same as the average satisfy $\sqrt{x} = 14/9$, from which we obtain $x = 196/81 \approx 2.4$ (Figure 7.7.11).   ◀

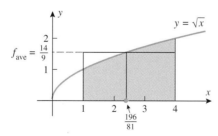

Figure 7.7.11

**AVERAGE VELOCITY REVISITED**

In Section 3.1 we considered the motion of a particle moving in the *positive direction* along a coordinate line, and we motivated the concept of instantaneous velocity in that special case by viewing it as the limit of average velocities over smaller and smaller time intervals. That discussion led us to conclude that the average velocity of the particle over a time interval could be interpreted as the slope of a secant line and the instantaneous velocity as the slope of a tangent line to the position versus time curve (Figure 3.1.5). We will now show that the same results are true in the more general case where the particle can move in either direction along the coordinate line.

For this purpose, suppose that $s(t)$ and $v(t)$ are the position and velocity functions of such a particle, and let us use Formula (21) to calculate the average velocity of the particle over a time interval $[t_0, t_1]$. This yields

$$v_{\text{ave}} = \frac{1}{t_1 - t_0} \int_{t_0}^{t_1} v(t)\, dt = \frac{1}{t_1 - t_0} \int_{t_0}^{t_1} s'(t)\, dt = \frac{s(t_1) - s(t_0)}{t_1 - t_0}$$

Thus, *the average velocity over a time interval is the displacement divided by the elapsed time.* Geometrically, this is the slope of the secant line shown in Figure 7.7.12. Moreover, if we allow $t_1$ to approach $t_0$, then the slopes of the secant lines approach the slope of the tangent line at $t_0$, which is the instantaneous velocity at that instant. Thus, the relationship between average and instantaneous velocity developed in Section 3.1 also applies to general rectilinear motion.

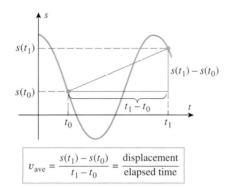

$$v_{\text{ave}} = \frac{s(t_1) - s(t_0)}{t_1 - t_0} = \frac{\text{displacement}}{\text{elapsed time}}$$

Figure 7.7.12

## EXERCISE SET 7.7    ⊵ Graphing Calculator    [c] CAS

1. (a) If $h'(t)$ is the rate of change of a child's height measured in inches per year, what does the integral $\int_0^{10} h'(t)\, dt$ represent, and what are its units?

   (b) If $r'(t)$ is the rate of change of the radius of a spherical balloon measured in centimeters per second, what does the integral $\int_1^2 r'(t)\, dt$ represent, and what are its units?

   (c) If $H(t)$ is the rate of change of the speed of sound with respect to temperature measured in ft/s per °F, what does the integral $\int_{32}^{100} H(t)\, dt$ represent, and what are its units?

   (d) If $v(t)$ is the velocity of a particle in rectilinear motion, measured in cm/h, what does the integral $\int_{t_1}^{t_2} v(t)\, dt$ represent, and what are its units?

2. (a) Suppose that sludge is emptied into a river at the rate of $V(t)$ gallons per minute, starting at time $t = 0$. Write an integral that represents the total volume of sludge that is emptied into the river during the first hour.

   (b) Suppose that the tangent line to a curve $y = f(x)$ has slope $m(x)$ at the point $x$. What does the integral $\int_{x_1}^{x_2} m(x)\, dx$ represent?

3. In each part, the velocity versus time curve is given for a particle moving along a line. Use the curve to find the displacement and the distance traveled by the particle over the time interval $0 \le t \le 3$.

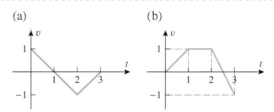

(a)     (b)

4. Sketch a velocity versus time curve for a particle that travels a distance of 5 units along a coordinate line during the time interval $0 \le t \le 10$ and has a displacement of 0 units.

5. The accompanying figure shows the acceleration versus time curve for a particle moving along a coordinate line. If the initial velocity of the particle is 20 m/s, estimate
   (a) the velocity at time $t = 4$ s.
   (b) the velocity at time $t = 6$ s.

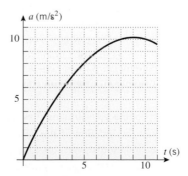

Figure Ex-5

**6.** Determine whether the particle in Exercise 5 is speeding up or slowing down at times $t = 4$ s and $t = 6$ s.

In Exercises 7–10, a particle moves along an $s$-axis. Use the given information to find the position function of the particle.

**7.** (a) $v(t) = t^3 - 2t^2 + 1$; $s(0) = 1$
  (b) $a(t) = 4\cos 2t$; $v(0) = -1$; $s(0) = -3$

**8.** (a) $v(t) = 1 + \sin t$; $s(0) = -3$
  (b) $a(t) = t^2 - 3t + 1$; $v(0) = 0$; $s(0) = 0$

**9.** (a) $v(t) = 2t - 3$; $s(1) = 5$
  (b) $a(t) = \cos t$; $v(\pi/2) = 2$; $s(\pi/2) = 0$

**10.** (a) $v(t) = t^{2/3}$; $s(8) = 0$
  (b) $a(t) = \sqrt{t}$; $v(4) = 1$; $s(4) = -5$

In Exercises 11–14, a particle moves with a velocity of $v(t)$ m/s along an $s$-axis. Find the displacement and the distance traveled by the particle during the given time interval.

**11.** (a) $v(t) = \sin t$; $0 \le t \le \pi/2$
  (b) $v(t) = \cos t$; $\pi/2 \le t \le 2\pi$

**12.** (a) $v(t) = 2t - 4$; $0 \le t \le 6$
  (b) $v(t) = |t - 3|$; $0 \le t \le 5$

**13.** (a) $v(t) = t^3 - 3t^2 + 2t$; $0 \le t \le 3$
  (b) $v(t) = e^t - 2$; $0 \le t \le 3$

**14.** (a) $v(t) = \frac{1}{2} - 1/t$; $1 \le t \le 3$
  (b) $v(t) = 3/\sqrt{t}$; $4 \le t \le 9$

In Exercises 15–18, a particle moves with acceleration $a(t)$ m/s² along an $s$-axis and has velocity $v_0$ m/s at time $t = 0$. Find the displacement and the distance traveled by the particle during the given time interval.

**15.** $a(t) = -2$; $v_0 = 3$; $1 \le t \le 4$

**16.** $a(t) = t - 2$; $v_0 = 0$; $1 \le t \le 5$

**17.** $a(t) = 1/\sqrt{5t + 1}$; $v_0 = 2$; $0 \le t \le 3$

**18.** $a(t) = \sin t$; $v_0 = 1$; $\pi/4 \le t \le \pi/2$

**19.** In each part use the given information to find the position, velocity, speed, and acceleration at time $t = 1$.
  (a) $v = \sin \frac{1}{2}\pi t$; $s = 0$ when $t = 0$
  (b) $a = -3t$; $s = 1$ and $v = 0$ when $t = 0$

**20.** The accompanying figure shows the velocity versus time curve over the time interval $1 \le t \le 5$ for a particle moving along a horizontal coordinate line.
  (a) What can you say about the sign of the acceleration over the time interval?
  (b) When is the particle speeding up? Slowing down?
  (c) What can you say about the location of the particle at time $t = 5$ relative to its location at time $t = 1$? Explain your reasoning.

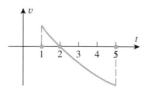

Figure Ex-20

In Exercises 21–24, sketch the curve and find the total area between the curve and the given interval on the $x$-axis.

**21.** $y = x^2 - 1$; $[0, 3]$    **22.** $y = \sin x$; $[0, 3\pi/2]$

**23.** $y = e^x - 1$; $[-1, 1]$    **24.** $y = \dfrac{x - 1}{x}$; $[\frac{1}{2}, 2]$

**25.** Suppose that the velocity function of a particle moving along an $s$-axis is $v(t) = 20t^2 - 100t + 50$ ft/s and that the particle is at the origin at time $t = 0$. Use a graphing utility to generate the graphs of $s(t)$, $v(t)$, and $a(t)$ for the first 6 s of motion.

**26.** Suppose that the acceleration function of a particle moving along an $s$-axis is $a(t) = 4t - 30$ m/s and that the position and velocity at time $t = 0$ are $s_0 = -5$ m and $v_0 = 3$ m/s. Use a graphing utility to generate the graphs of $s(t)$, $v(t)$, and $a(t)$ for the first 25 s of motion.

**27.** Let the velocity function for a particle that is at the origin initially and moves along an $s$-axis be $v(t) = 0.5 - te^{-t}$.
  (a) Generate the velocity versus time curve, and use it to make a conjecture about the sign of the displacement over the time interval $0 \le t \le 5$.
  (b) Use a CAS to find the displacement.

**28.** Let the velocity function for a particle that is at the origin initially and moves along an $s$-axis be $v(t) = t \ln(t + 0.1)$.
  (a) Generate the velocity versus time curve, and use it to make a conjecture about the sign of the displacement over the time interval $0 \le t \le 1$.
  (b) Use a CAS to find the displacement.

**29.** Suppose that at time $t = 0$ a particle is at the origin of an $x$-axis and has a velocity of $v_0 = 25$ cm/s. For the first 4 s thereafter it has no acceleration, and then it is acted on by a retarding force that produces a constant negative acceleration of $a = -10$ cm/s².
  (a) Sketch the acceleration versus time curve over the interval $0 \le t \le 12$.
  (b) Sketch the velocity versus time curve over the time interval $0 \le t \le 12$.
  (c) Find the $x$-coordinate of the particle at times $t = 8$ s and $t = 12$ s.
  (d) What is the maximum $x$-coordinate of the particle over the time interval $0 \le t \le 12$?

**30.** Formulas (8) and (9) for uniformly accelerated motion can be rearranged in various useful ways. For simplicity, let $s = s(t)$ and $v = v(t)$, and derive the following variations of those formulas.

(a) $a = \dfrac{v^2 - v_0^2}{2(s - s_0)}$     (b) $t = \dfrac{2(s - s_0)}{v_0 + v}$

(c) $s = s_0 + vt - \frac{1}{2}at^2$ [Note how this differs from (8).]

> Exercises 31–38 involve uniformly accelerated motion. In these exercises assume that the object is moving in the positive direction of a coordinate line, and apply Formulas (8) and (9) or those from Exercise 30, as appropriate. In some of these problems you will need the fact that 88 ft/s = 60 mi/h.

**31.** (a) An automobile traveling on a straight road decelerates uniformly from 55 mi/h to 25 mi/h in 30 s. Find its acceleration in ft/s$^2$.

(b) A bicycle rider traveling on a straight path accelerates uniformly from rest to 30 km/h in 1 min. Find his acceleration in km/s$^2$.

**32.** A car traveling 60 mi/h along a straight road decelerates at a constant rate of 10 ft/s$^2$.

(a) How long will it take until the speed is 45 mi/h?

(b) How far will the car travel before coming to a stop?

**33.** Spotting a police car, you hit the brakes on your new Porsche to reduce your speed from 90 mi/h to 60 mi/h at a constant rate over a distance of 200 ft.

(a) Find the acceleration in ft/s$^2$.

(b) How long does it take for you to reduce your speed to 55 mi/h?

(c) At the acceleration obtained in part (a), how long would it take for you to bring your Porsche to a complete stop from 90 mi/h?

**34.** A particle moving along a straight line is accelerating at a constant rate of 3 m/s$^2$. Find the initial velocity if the particle moves 40 m in the first 4 s.

**35.** A motorcycle, starting from rest, speeds up with a constant acceleration of 2.6 m/s$^2$. After it has traveled 120 m, it slows down with a constant acceleration of $-1.5$ m/s$^2$ until it attains a speed of 12 m/s. What is the distance traveled by the motorcycle at that point?

**36.** A sprinter in a 100-m race explodes out of the starting block with an acceleration of 4.0 m/s$^2$, which she sustains for 2.0 s. Her acceleration then drops to zero for the rest of race.

(a) What is her time for the race?

(b) Make a graph of her distance from the starting block versus time.

**37.** A car that has stopped at a toll booth leaves the booth with a constant acceleration of 2 ft/s$^2$. At the time the car leaves the booth it is 5000 ft behind a truck traveling with a constant velocity of 50 ft/s. How long will it take for the car to catch the truck, and how far will the car be from the toll booth at that time?

**38.** In the final sprint of a rowing race the challenger is rowing at a constant speed of 12 m/s. At the point where the leader is 100 m from the finish line and the challenger is 15 m behind, the leader is rowing at 8 m/s but starts accelerating at a constant 0.5 m/s$^2$. Who wins?

> In Exercises 39–48, assume that a free-fall model applies. Solve these exercises by applying Formulas (12) and (13) or, if appropriate, use those from Exercise 30 with $a = -g$. In these exercises take $g = 32$ ft/s$^2$ or $g = 9.8$ m/s$^2$, depending on the units.

**39.** A projectile is launched vertically upward from ground level with an initial velocity of 112 ft/s.

(a) Find the velocity at $t = 3$ s and $t = 5$ s.

(b) How high will the projectile rise?

(c) Find the speed of the projectile when it hits the ground.

**40.** A projectile fired downward from a height of 112 ft reaches the ground in 2 s. What is its initial velocity?

**41.** A projectile is fired vertically upward from ground level with an initial velocity of 16 ft/s.

(a) How long will it take for the projectile to hit the ground?

(b) How long will the projectile be moving upward?

**42.** A rock is dropped from the top of the Washington Monument, which is 555 ft high.

(a) How long will it take for the rock to hit the ground?

(b) What is the speed of the rock at impact?

**43.** A helicopter pilot drops a package when the helicopter is 200 ft above the ground and rising at a speed of 20 ft/s.

(a) How long will it take for the package to hit the ground?

(b) What will be its speed at impact?

**44.** A stone is thrown downward with an initial speed of 96 ft/s from a height of 112 ft.

(a) How long will it take for the stone to hit the ground?

(b) What will be its speed at impact?

**45.** A projectile is fired vertically upward with an initial velocity of 49 m/s from a tower 150 m high.

(a) How long will it take for the projectile to reach its maximum height?

(b) What is the maximum height?

(c) How long will it take for the projectile to pass its starting point on the way down?

(d) What is the velocity when it passes the starting point on the way down?

(e) How long will it take for the projectile to hit the ground?

(f) What will be its speed at impact?

**46.** A man drops a stone from a bridge. What is the height of the bridge if

(a) the stone hits the water 4 s later

(b) the sound of the splash reaches the man 4 s later? [Take 1080 ft/s as the speed of sound.]

**47.** In the final stages of a Moon landing, a lunar module fires its retrorockets and descends to a height of $h = 5$ m above the lunar surface (Figure Ex-47). At that point the retrorockets are cut off, and the module goes into free fall. Given that the Moon's gravity is 1/6 of the Earth's, find the speed of the module when it touches the lunar surface.

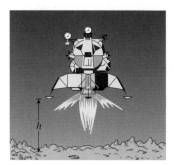

Figure Ex-47

**48.** Given that the Moon's gravity is 1/6 of the Earth's, how much faster would a projectile have to be launched upward from the surface of the Earth than from the surface of the Moon to reach a height of 1000 ft?

In Exercises 49–54, find the average value of the function over the given interval.

**49.** $f(x) = 3x$; $[1, 3]$        **50.** $f(x) = x^2$; $[-1, 2]$

**51.** $f(x) = \sin x$; $[0, \pi]$        **52.** $f(x) = \cos x$; $[0, \pi]$

**53.** $f(x) = 1/x$; $[1, e]$        **54.** $f(x) = e^x$; $[-1, \ln 5]$

**55.** (a) Find $f_{ave}$ of $f(x) = x^2$ over $[0, 2]$.
   (b) Find a point $x^*$ in $[0, 2]$ such that $f(x^*) = f_{ave}$.
   (c) Sketch the graph of $f(x) = x^2$ over $[0, 2]$ and construct a rectangle over the interval whose area is the same as the area under the graph of $f$ over the interval.

**56.** (a) Find $f_{ave}$ of $f(x) = 2x$ over $[0, 4]$.
   (b) Find a point $x^*$ in $[0, 4]$ such that $f(x^*) = f_{ave}$.
   (c) Sketch the graph of $f(x) = 2x$ over $[0, 4]$ and construct a rectangle over the interval whose area is the same as the area under the graph of $f$ over the interval.

**57.** (a) Suppose that the velocity function of a particle moving along a coordinate line is $v(t) = 3t^3 + 2$. Find the average velocity of the particle over the time interval $1 \le t \le 4$ by integrating.
   (b) Suppose that the position function of a particle moving along a coordinate line is $s(t) = 6t^2 + t$. Find the average velocity of the particle over the time interval $1 \le t \le 4$ algebraically.

**58.** (a) Suppose that the acceleration function of a particle moving along a coordinate line is $a(t) = t + 1$. Find the average acceleration of the particle over the time interval $0 \le t \le 5$ by integrating.
   (b) Suppose that the velocity function of a particle moving along a coordinate line is $v(t) = \cos t$. Find the average acceleration of the particle over the time interval $0 \le t \le \pi/4$ algebraically.

**59.** Water is run at a constant rate of 1 ft³/min to fill a cylindrical tank of radius 3 ft and height 5 ft. Assuming that the tank is empty initially, make a conjecture about the average weight of the water in the tank over the time period required to fill

it, and then check your conjecture by integrating. [Take the weight density of water to be 62.4 lb/ft³.]

**60.** (a) The temperature of a 10-m-long metal bar is 15°C at one end and 30°C at the other end. Assuming that the temperature increases linearly from the cooler end to the hotter end, what is the average temperature of the bar?
   (b) Explain why there must be a point on the bar where the temperature is the same as the average, and find it.

**61.** (a) Suppose that a reservoir supplies water to an industrial park at a constant rate of $r = 4$ gallons per minute (gal/min) between 8:30 A.M. and 9:00 A.M. How much water does the reservoir supply during that time period?
   (b) Suppose that one of the industrial plants increases its water consumption between 9:00 A.M. and 10:00 A.M. and that the rate at which the reservoir supplies water increases linearly, as shown in the accompanying figure. How much water does the reservoir supply during that 1-hour time period?
   (c) Suppose that from 10:00 A.M. to 12 noon the rate at which the reservoir supplies water is given by the formula $r(t) = 10 + \sqrt{t}$ gal/min, where $t = 0$ corresponds to 10:00 A.M. How much water does the reservoir supply during that 2-hour time period?

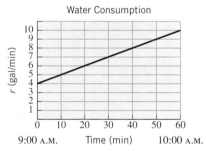

Water Consumption

Figure Ex-61

**62.** A traffic engineer monitors the rate at which cars enter the main highway during the afternoon rush hour. From her data she estimates that between 4:30 P.M. and 5:30 P.M. the rate $R(t)$ at which cars enter the highway is given by the formula $R(t) = 100(1 - 0.0001t^2)$ cars per minute, where $t = 0$ corresponds to 4:30 P.M.
   (a) When does the peak traffic flow into the highway occur?
   (b) Find the number of cars that enter the highway during the rush hour.

**63.** (a) Prove: If $f$ is continuous on $[a, b]$, then

$$\int_a^b [f(x) - f_{ave}] \, dx = 0$$

   (b) Does there exist a constant $c \ne f_{ave}$ such that

$$\int_a^b [f(x) - c] \, dx = 0?$$

## 7.8 EVALUATING DEFINITE INTEGRALS BY SUBSTITUTION

*In this section we will discuss two methods for evaluating definite integrals in which a substitution is required. We will also discuss methods for differentiating integrals whose limits of integration are functions.*

**TWO METHODS FOR MAKING SUBSTITUTIONS IN DEFINITE INTEGRALS**

Recall from Section 7.3 that indefinite integrals of the form

$$\int f(g(x))g'(x)\,dx$$

can sometimes be evaluated by making the $u$-substitution

$$u = g(x), \quad du = g'(x)\,dx \tag{1}$$

which converts the integral to the form

$$\int f(u)\,du$$

To apply this method to a definite integral of the form

$$\int_a^b f(g(x))g'(x)\,dx$$

we need to account for the effect that the substitution has on the $x$-limits of integration. There are two ways of doing this.

**Method 1**

First evaluate the indefinite integral

$$\int f(g(x))g'(x)\,dx$$

by substitution, and then use the relationship

$$\int_a^b f(g(x))g'(x)\,dx = \left[\int f(g(x))g'(x)\,dx\right]_a^b$$

to evaluate the definite integral. This procedure does not require any modification of the $x$-limits of integration.

**Method 2**

Make the substitution (1) directly in the definite integral, and then use the relationship $u = g(x)$ to replace the $x$-limits, $x = a$ and $x = b$, by corresponding $u$-limits, $u = g(a)$ and $u = g(b)$. This produces a new definite integral

$$\int_{g(a)}^{g(b)} f(u)\,du$$

that is expressed entirely in terms of $u$.

### Example 1

Use the two methods above to evaluate $\displaystyle\int_0^2 x(x^2 + 1)^3\,dx$.

*Solution by Method 1.*   If we let

$$u = x^2 + 1 \quad \text{so that} \quad du = 2x\,dx \tag{2}$$

then we obtain

$$\int x(x^2 + 1)^3\,dx = \frac{1}{2}\int u^3\,du = \frac{u^4}{8} + C = \frac{(x^2 + 1)^4}{8} + C$$

Thus,

$$\int_0^2 x(x^2+1)^3\,dx = \left[\int x(x^2+1)^3\,dx\right]_{x=0}^2 = \frac{(x^2+1)^4}{8}\Bigg]_{x=0}^2$$

$$= \frac{625}{8} - \frac{1}{8} = 78$$

*Solution by Method 2.* If we make the substitution $u = x^2 + 1$ in (2), then

$$u = 1 \quad \text{if} \quad x = 0$$
$$u = 5 \quad \text{if} \quad x = 2$$

Thus,

$$\int_0^2 x(x^2+1)^3\,dx = \frac{1}{2}\int_1^5 u^3\,du = \frac{u^4}{8}\Bigg]_{u=1}^5 = \frac{625}{8} - \frac{1}{8} = 78$$

which agrees with the result obtained by Method 1.   ◀

The following theorem states precise conditions under which Method 2 can be used. The proof is a straightforward application of the chain rule and the Fundamental Theorem of Calculus, but we will omit the details.

---

**7.8.1   THEOREM.**   *If $g'$ is continuous on $[a, b]$ and $f$ is continuous and has an anti-derivative on an interval containing the values of $g(x)$ for $a \le x \le b$, then*

$$\int_a^b f(g(x))g'(x)\,dx = \int_{g(a)}^{g(b)} f(u)\,du$$

---

The choice of methods for evaluating definite integrals by substitution is generally a matter of taste, but in the following examples we will use the second method, since the idea is new.

**Example 2**

Evaluate

$$\text{(a)} \int_0^{3/4} \frac{dx}{1-x} \qquad \text{(b)} \int_0^{\pi/8} \sin^5 2x \cos 2x\,dx$$

*Solution (a).* Let

$$u = 1 - x \quad \text{so that} \quad du = -dx$$

With this substitution we have

$$u = 1 \quad \text{if} \quad x = 0$$
$$u = \tfrac{1}{4} \quad \text{if} \quad x = \tfrac{3}{4}$$

Thus,

$$\int_0^{3/4} \frac{dx}{1-x} = -\int_1^{1/4} \frac{du}{u} = -\ln|u|\Bigg]_{u=1}^{1/4}$$

$$= -\left[\ln\left(\frac{1}{4}\right) - \ln(1)\right] = \ln 4$$

*Solution (b).* Let

$$u = \sin 2x \quad \text{so that} \quad du = 2\cos 2x\, dx \quad (\text{or } \tfrac{1}{2}\, du = \cos 2x\, dx)$$

With this substitution we have

$$u = \sin(0) = 0 \quad \text{if} \quad x = 0$$
$$u = \sin(\pi/4) = 1/\sqrt{2} \quad \text{if} \quad x = \pi/8$$

so

$$\int_0^{\pi/8} \sin^5 2x \cos 2x\, dx = \frac{1}{2} \int_0^{1/\sqrt{2}} u^5\, du = \frac{1}{2} \cdot \frac{u^6}{6} \bigg]_0^{1/\sqrt{2}}$$

$$= \frac{1}{2} \left[ \frac{1}{6(\sqrt{2})^6} - 0 \right] = \frac{1}{96} \quad \blacktriangleleft$$

## Example 3

In Example 8 of Section 4.4 we stated the following model for the temperature $T$ in degrees Fahrenheit ($^\circ$F) of a glass of lemonade $t$ hours after being placed in a room with a constant temperature of $70^\circ$F, given that the initial temperature of the lemonade was $40^\circ$F:

$$T = 70 - 30e^{-0.5t}$$

Find the average temperature $T_{\text{ave}}$ of the lemonade over the first 5 hours.

*Solution.* From Definition 7.7.5 the average value of $T$ over the time interval $[0, 5]$ is

$$T_{\text{ave}} = \frac{1}{5} \int_0^5 (70 - 30e^{-0.5t})\, dt \tag{3}$$

To evaluate this integral, we make the substitution

$$u = -0.5t \quad \text{so that} \quad du = -0.5\, dt \quad [\text{or } dt = -(1/0.5)\, du]$$

With this substitution we have

$$u = 0 \quad \text{if} \quad t = 0$$
$$u = -(0.5)5 = -2.5 \quad \text{if} \quad t = 5$$

Thus, (3) can be expressed as

$$T_{\text{ave}} = \frac{1}{5} \int_0^{-2.5} (70 - 30e^u) \left( -\frac{1}{0.5} \right) du = -\frac{1}{2.5} \int_0^{-2.5} (70 - 30e^u)\, du$$

$$= -\frac{1}{2.5} \left[ 70u - 30e^u \right]_{u=0}^{-2.5} = -\frac{1}{2.5} \left[ (-175 - 30e^{-2.5}) - (-30) \right]$$

$$= 58 + 12e^{-2.5} \approx 58.99^\circ\text{F} \quad \blacktriangleleft$$

REMARK.    Observe that the $u$-substitution in this example produced an integral in which the upper $u$-limit of integration was smaller than the lower $u$-limit of integration. In our computations we left the limits of integration in that order, but had we wanted to we could have reversed the order to put the larger limit on top and compensated by reversing the sign of the integral in accordance with Definition 7.5.3(b). The choice of procedures is a matter of taste, both produce the same result (verify).

FOR THE READER.    If you have a CAS, use it to evaluate the integral in the last example. See whether it makes any difference in the form of the answer if you express the exponent as $-t/2$ rather than $-0.5t$.

In Exercises 1 and 2, express the integral in terms of the variable $u$, but do not evaluate it.

**1.** (a) $\displaystyle\int_0^2 (x+1)^7 \, dx$; $u = x+1$

(b) $\displaystyle\int_{-1}^2 x\sqrt{8-x^2} \, dx$; $u = 8 - x^2$

(c) $\displaystyle\int_{-1}^1 \sin(\pi\theta) \, d\theta$; $u = \pi\theta$

(d) $\displaystyle\int_0^3 (x+2)(x-3)^{20} \, dx$; $u = x-3$

**2.** (a) $\displaystyle\int_0^1 e^{2x-1} \, dx$; $u = 2x-1$

(b) $\displaystyle\int_e^{e^2} \frac{\ln x}{x} \, dx$; $u = \ln x$

(c) $\displaystyle\int_0^{\pi/4} \tan^2 x \sec^2 x \, dx$; $u = \tan x$

(d) $\displaystyle\int_0^1 x^3\sqrt{x^2+3} \, dx$; $u = x^2+3$

In Exercises 3–12, evaluate the definite integral two ways: first by a $u$-substitution in the definite integral and then by a $u$-substitution in the corresponding indefinite integral.

**3.** $\displaystyle\int_0^1 (2x+1)^4 \, dx$

**4.** $\displaystyle\int_1^2 (4x-2)^3 \, dx$

**5.** $\displaystyle\int_{-1}^0 (1-2x)^3 \, dx$

**6.** $\displaystyle\int_1^2 (4-3x)^8 \, dx$

**7.** $\displaystyle\int_0^8 x\sqrt{1+x} \, dx$

**8.** $\displaystyle\int_{-5}^0 x\sqrt{4-x} \, dx$

**9.** $\displaystyle\int_0^{\pi/2} 4\sin(x/2) \, dx$

**10.** $\displaystyle\int_0^{\pi/6} 2\cos 3x \, dx$

**11.** $\displaystyle\int_{-\ln 3}^{\ln 3} \frac{e^x}{e^x+4} \, dx$

**12.** $\displaystyle\int_0^{\ln 5} e^x(3-4e^x) \, dx$

In Exercises 13–16, evaluate the definite integral by expressing it in terms of $u$ and evaluating the resulting integral using a formula from geometry.

**13.** $\displaystyle\int_0^{5/3} \sqrt{25-9x^2} \, dx$; $u = 3x$

**14.** $\displaystyle\int_0^2 x\sqrt{16-x^4} \, dx$; $u = x^2$

**15.** $\displaystyle\int_{\pi/3}^{\pi/2} \sin\theta\sqrt{1-4\cos^2\theta} \, d\theta$; $u = 2\cos\theta$

**16.** $\displaystyle\int_{e^{-6}}^{e^6} \frac{\sqrt{36-(\ln x)^2}}{x} \, dx$; $u = \ln x$

**17.** Find the area under the curve $y = \sin\pi x$ over the interval $[0, 1]$.

**18.** Find the area under the curve $y = 3\cos 2x$ over the interval $[0, \pi/8]$.

**19.** Find the area under the curve $y = 1/(x+5)^2$ over the interval $[3, 7]$.

**20.** Find the area under the curve $y = 1/(3x+1)^2$ over the interval $[0, 1]$.

**21.** Find the average value of $f(x) = e^{-2x}$ over the interval $[0, 4]$.

**22.** Find the average value of $f(x) = \sec^2 \pi x$ over the interval $\left[-\frac{1}{4}, \frac{1}{4}\right]$.

In Exercises 23–38, evaluate the integrals by any method.

**23.** $\displaystyle\int_0^1 \frac{dx}{\sqrt{3x+1}}$

**24.** $\displaystyle\int_1^2 \sqrt{5x-1} \, dx$

**25.** $\displaystyle\int_{-1}^1 \frac{x^2 \, dx}{\sqrt{x^3+9}}$

**26.** $\displaystyle\int_{-1}^0 6t^2(t^3+1)^{19} \, dt$

**27.** $\displaystyle\int_1^3 \frac{x+2}{\sqrt{x^2+4x+7}} \, dx$

**28.** $\displaystyle\int_1^2 \frac{dx}{x^2-6x+9}$

**29.** $\displaystyle\int_{-3\pi/4}^{\pi/4} \sin x \cos x \, dx$

**30.** $\displaystyle\int_0^{\pi/4} \sqrt{\tan x}\,\sec^2 x \, dx$

**31.** $\displaystyle\int_0^{\sqrt{\pi}} 5x\cos(x^2) \, dx$

**32.** $\displaystyle\int_{\pi^2}^{4\pi^2} \frac{1}{\sqrt{x}} \sin\sqrt{x} \, dx$

**33.** $\displaystyle\int_{\pi/12}^{\pi/9} \sec^2 3\theta \, d\theta$

**34.** $\displaystyle\int_0^{\pi/2} \sin^2 3\theta \cos 3\theta \, d\theta$

**35.** $\displaystyle\int_0^1 \frac{y^2 \, dy}{\sqrt{4-3y}}$

**36.** $\displaystyle\int_{-1}^4 \frac{x \, dx}{\sqrt{5+x}}$

**37.** $\displaystyle\int_0^e \frac{dx}{x+e}$

**38.** $\displaystyle\int_1^{\sqrt{2}} xe^{-x^2} \, dx$

☐ **39.** For each of the integrals you evaluated in Exercises 23–38, check your answer using a CAS.

☐ **40.** Use a CAS to find the exact value of the integral

$$\int_{-3}^1 \sqrt{3-2x-x^2} \, dx$$

and then confirm the result by hand calculation. [*Hint:* Complete the square.]

**41.** (a) Find $\displaystyle\int_0^1 f(3x+1) \, dx$ if $\displaystyle\int_1^4 f(x) \, dx = 5$.

(b) Find $\displaystyle\int_0^3 f(3x) \, dx$ if $\displaystyle\int_0^9 f(x) \, dx = 5$.

(c) Find $\displaystyle\int_{-2}^0 xf(x^2) \, dx$ if $\displaystyle\int_0^4 f(x) \, dx = 1$.

**42.** Given that $m$ and $n$ are positive integers, show that

$$\int_0^1 x^m(1-x)^n \, dx = \int_0^1 x^n(1-x)^m \, dx$$

by making a substitution. Do not attempt to evaluate the integrals.

**43.** Given that $n$ is a positive integer, show that

$$\int_0^{\pi/2} \sin^n x \, dx = \int_0^{\pi/2} \cos^n x \, dx$$

by using a trigonometric identity and making a substitution. Do not attempt to evaluate the integrals.

**44.** Given that $n$ is a positive integer, evaluate the integral

$$\int_0^1 x(1-x)^n \, dx$$

**45.** Suppose that at time $t = 0$ there are 750 bacteria in a growth medium and the bacteria population $y(t)$ grows at the rate $y'(t) = 802.137e^{1.528t}$ bacteria per hour. How many bacteria will there be in 12 hours?

**46.** Suppose that the value of a yacht in dollars after $t$ years of use is $V(t) = 275{,}000e^{-0.17t}$. What is the average value of the yacht over its first 10 years of use?

**47.** Suppose that a particle moving along a coordinate line has velocity $v(t) = 25 + 10e^{-0.05t}$ ft/s.
(a) What is the distance traveled by the particle from time $t = 0$ to time $t = 10$?
(b) Does the term $10e^{-0.05t}$ have much effect on the distance traveled by the particle over that time interval? Explain your reasoning.

**48.** Find a positive value of $k$ such that the area under the graph of $y = e^{2x}$ over the interval $[0, k]$ is 3 square units.

**49.** Electricity is supplied to homes in the form of *alternating current*, which means that the voltage has a sinusoidal waveform described by an equation of the form

$$V = V_p \sin(2\pi f t)$$

(see the accompanying figure). In this equation, $V_p$ is called the *peak voltage* or *amplitude* of the current, $f$ is called its *frequency*, and $1/f$ is called its *period*. The voltages $V$ and $V_p$ are measured in volts (V), the time $t$ is measured in seconds (s), and the frequency is measured in hertz (Hz) or sometimes in cycles per second. (A *cycle* is the electrical term for one period of the waveform.) Alternating current voltmeters read what is called the *rms* or *root-mean-square* value of $V$. By definition, this is the square root of the average value of $V^2$ over one period.
(a) Show that

$$V_{\text{rms}} = \frac{V_p}{\sqrt{2}}$$

[*Hint:* Compute the average over the cycle from $t = 0$ to $t = 1/f$, and use the identity $\sin^2 \theta = \frac{1}{2}(1 - \cos 2\theta)$ to help evaluate the integral.]

(b) In the United States, electrical outlets supply alternating current with an rms voltage of 120 V at a frequency of 60 Hz. What is the peak voltage at such an outlet?

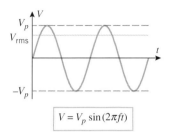

$$V = V_p \sin(2\pi f t)$$

Figure Ex-49

**50.** Show that if $f$ and $g$ are continuous functions, then

$$\int_0^t f(t-x)g(x)\, dx = \int_0^t f(x)g(t-x)\, dx$$

**51.** (a) Let $I = \displaystyle\int_0^a \frac{f(x)}{f(x) + f(a-x)}\, dx$. Show that $I = a/2$.

[*Hint:* Let $u = a - x$, and then express the integrand as the sum of two fractions.]

(b) Use the result of part (a) to find

$$\int_0^3 \frac{\sqrt{x}}{\sqrt{x} + \sqrt{3-x}}\, dx$$

(c) Use the result of part (a) to find

$$\int_0^{\pi/2} \frac{\sin x}{\sin x + \cos x}\, dx$$

**52.** Let $I = \displaystyle\int_{-1}^1 \frac{1}{1+x^2}\, dx$. Show that the substitution $x = 1/u$ results in

$$I = -\int_{-1}^1 \frac{1}{1+u^2}\, du = -I$$

so $2I = 0$, which implies that $I = 0$. However, this is impossible since the integrand of the given integral is positive over the interval of integration. Where is the error?

**53.** Find the limit

$$\lim_{n \to +\infty} \sum_{k=1}^n \frac{\sin(k\pi/n)}{n}$$

by evaluating an appropriate definite integral over the interval $[0, 1]$.

**C** **54.** Check your answer to Exercise 53 by evaluating the limit directly with a CAS.

**55.** (a) Prove that if $f$ is an odd function, then

$$\int_{-a}^a f(x)\, dx = 0$$

and give a geometric explanation of this result. [*Hint:* One way to prove that a quantity $q$ is zero is to show that $q = -q$.]

(b)  Prove that if $f$ is an even function, then

$$\int_{-a}^{a} f(x)\,dx = 2\int_{0}^{a} f(x)\,dx$$

and give a geometric explanation of this result. [*Hint:* Split the interval of integration from $-a$ to $a$ into two parts at 0.]

## 7.9 LOGARITHMIC FUNCTIONS FROM THE INTEGRAL POINT OF VIEW

*In Section 4.2 we discussed natural logarithms from the viewpoint of exponents; that is, we regarded $y = \ln x$ to mean that $e^y = x$. In this section we will show that $\ln x$ can also be expressed as an integral with a variable upper limit. This integral representation of $\ln x$ is important mathematically because it provides a convenient way of establishing properties such as differentiability and continuity. However, it is also important in applications because it provides a way of recognizing when integral solutions of problems can be expressed as natural logarithms.*

**THE LINK BETWEEN NATURAL LOGARITHMS AND INTEGRALS**

The connection between natural logarithms and integrals was made in the middle of the seventeenth century in the course of investigating areas under the curve $y = 1/t$. The problem being considered was to find values of $t_1, t_2, t_3, \ldots, t_n, \ldots$ for which the areas $A_1, A_2, A_3, \ldots, A_n, \ldots$ in Figure 7.9.1$a$ would be equal. Through the combined work of Isaac Newton, the Belgian Jesuit priest, Gregory of St. Vincent (1584–1667), and Gregory's student, Alfons A. de Sarasa (1618–1667), it was shown that by taking the points to be

$$t_1 = e, \quad t_2 = e^2, \quad t_3 = e^3, \ldots, \quad t_n = e^n, \ldots$$

each of the areas would be 1 (Figure 7.9.1$b$). Thus, in modern integral notation

$$\int_{1}^{e^n} \frac{1}{t}\,dt = n$$

which can be expressed as

$$\int_{1}^{e^n} \frac{1}{t}\,dt = \ln(e^n)$$

By comparing the upper limit of the integral and the expression inside the logarithm, it is a natural leap to the more general result

$$\int_{1}^{x} \frac{1}{t}\,dt = \ln x$$

which today we take as the formal definition of the natural logarithm.

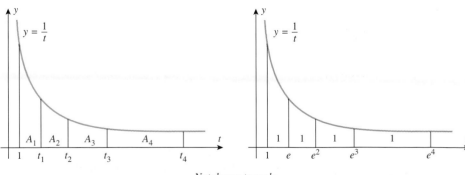

*Not drawn to scale*

Figure 7.9.1    $(a)$    $(b)$

> **7.9.1**   DEFINITION.   The **natural logarithm** of $x$ is denoted by $\ln x$ and is defined by the integral
>
> $$\ln x = \int_1^x \frac{1}{t}\,dt, \quad x > 0 \tag{1}$$

Geometrically, $\ln x$ is the area under the curve $y = 1/t$ from $t = 1$ to $t = x$ when $x > 1$, and $\ln x$ is the negative of the area under the curve $y = 1/t$ from $t = x$ to $t = 1$ when $0 < x < 1$ (Figure 7.9.2). If $x = 1$, then $\ln x = 0$, since the upper and lower limits in (1) are the same. All of this is consistent with the computer-generated graph of $y = \ln x$ in Figure 4.2.4.

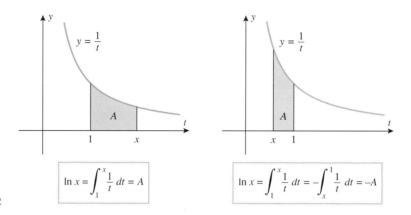

Figure 7.9.2

FOR THE READER.   Review Theorem 7.5.8, and then explain why $x$ is required to be positive in Definition 7.9.1.

---

**APPROXIMATING ln $x$ NUMERICALLY**

For specific values of $x$, the value of $\ln x$ can be approximated numerically by approximating the definite integral in (1), say by using the midpoint approximation that was discussed in Section 7.5.

**Example 1**

Approximate $\ln 2$ using the midpoint approximation with $n = 10$.

*Solution.*   From (1), the exact value of $\ln 2$ is represented by the integral

$$\ln 2 = \int_1^2 \frac{1}{t}\,dt$$

The midpoint rule is given in Formula (3) of Section 7.5. Expressed in terms of $t$, that formula is

$$\int_a^b f(t)\,dt \approx \Delta t \sum_{k=1}^{n} f(t_k^*)$$

where $\Delta t$ is the width of each subinterval and $t_1^*, t_2^*, \ldots, t_n^*$ are the midpoints. In this case we have 10 subintervals, so $\Delta t = (2-1)/10 = 0.1$. The computations to six decimal places are shown in Table 7.9.1. By comparison, a calculator set to display six decimal places gives $\ln 2 \approx 0.693147$, so the magnitude of the error in the midpoint approximation is about 0.000311. Greater accuracy in the midpoint approximation can be obtained by increasing $n$. For example, the midpoint approximation with $n = 100$ yields $\ln 2 \approx 0.693144$, which is correct to five decimal places.   ◀

**Table 7.9.1**

$n = 10$
$\Delta t = (b - a)/n = (2 - 1)/10 = 0.1$

| $k$ | $t_k^*$ | $1/t_k^*$ |
|-----|---------|-----------|
| 1  | 1.05 | 0.952381 |
| 2  | 1.15 | 0.869565 |
| 3  | 1.25 | 0.800000 |
| 4  | 1.35 | 0.740741 |
| 5  | 1.45 | 0.689655 |
| 6  | 1.55 | 0.645161 |
| 7  | 1.65 | 0.606061 |
| 8  | 1.75 | 0.571429 |
| 9  | 1.85 | 0.540541 |
| 10 | 1.95 | 0.512821 |
|    |      | 6.928355 |

$$\Delta t \sum_{k=1}^{n} f(t_k^*) = (0.1)(6.928355)$$
$$= 0.692836$$

Definition 7.9.1 is not only useful for approximating values of $\ln x$, but it is the key to establishing many of the fundamental properties of the natural logarithm. For example, in Section 4.4 we obtained the derivative

$$\frac{d}{dx}[\ln x] = \frac{1}{x} \quad (x > 0) \tag{2}$$

by *assuming* that $f(x) = \ln x$ is differentiable for $x > 0$. However, now that we have Definition 7.9.1 to work with, both the differentiability of $\ln x$ and Formula (2) follow immediately from Part 2 of the Fundamental Theorem of Calculus (7.6.3). Moreover, since differentiable functions are continuous, this also shows that $\ln x$ is continuous for $x > 0$.

Although it is not our objective to prove all of the properties of the functions we encounter, it is worthwhile to understand in principle how the differentiability and continuity of $\ln x$ can be used to establish differentiability and continuity of other important functions. For example, since the exponential function $e^x$ is the inverse of $\ln x$, it follows from Theorem 4.1.7, with $f(x) = \ln x$ and $f^{-1}(x) = e^x$, that $e^x$ is differentiable at any point $x$ where $f'(f^{-1}(x)) = 1/e^x \neq 0$. Since this holds for all $x$, it follows that $e^x$ is differentiable and hence continuous everywhere.

The differentiability $\ln x$ for $x > 0$ can be used to prove the differentiability of $\log_b x$ for $x > 0$ by using Formula (9) of Section 4.2 to express $\log_b x$ in terms of $\ln x$, and the differentiability of $e^x$ can be used to prove the differentiability of $b^x$ by expressing $b^x$ in terms of $e^x$ as $b^x = e^{x \ln b}$. We omit the details.

In Formulas (3), (4), and (5) of Section 4.2 we gave three limits for $e$, but at that time we did not have the mathematical tools to prove the existence of those limits; the following theorem does this.

---

**7.9.2**  THEOREM.

$(a)\ \lim_{x \to 0} (1 + x)^{1/x} = e \qquad (b)\ \lim_{x \to +\infty} \left(1 + \frac{1}{x}\right)^x = e \qquad (c)\ \lim_{x \to -\infty} \left(1 + \frac{1}{x}\right)^x = e$

---

*Proof.*  We will prove part $(a)$, and leave the proofs of the other parts for the exercises. Our proof will build on the differentiability of $\ln x$, and more specifically on the derivative of $\ln x$ at the point $x = 1$, namely

$$\frac{d}{dx}[\ln x]\bigg|_{x=1} = \frac{1}{x}\bigg|_{x=1} = 1$$

If we express this relationship using the definition of a derivative, we obtain

$$1 = \lim_{h \to 0} \frac{\ln(1 + h) - \ln 1}{h} = \lim_{h \to 0} \frac{\ln(1 + h)}{h} = \lim_{h \to 0} \ln(1 + h)^{1/h}$$

Thus, it follows that

$$e = e^{\lim_{h \to 0} \ln(1+h)^{1/h}}$$

which from the continuity of $e^x$ can be written as

$$e = \lim_{h \to 0} e^{\ln(1+h)^{1/h}} = \lim_{h \to 0} (1 + h)^{1/h}$$

Except for a difference in notation, this is what we wanted to prove.  ∎

The functions that we have dealt with thus far in this text are called *elementary functions*; they include polynomials, rational functions, power functions, exponential functions, logarithmic functions, trigonometric functions, and all other functions that can be obtained from these by addition, subtraction, multiplication, division, root extraction, composition, and by taking inverses.

However, there are many important functions that do not fall into this category. Such functions occur in many ways, but they commonly arise in the course of solving initial-value problems of the form

$$\frac{dy}{dx} = f(x), \quad y(x_0) = y_0 \tag{3}$$

Recall from Example 7 of Section 7.2 and the discussion preceding it that the basic method for solving (3) is to integrate $f(x)$, and then use the initial condition to determine the constant of integration. It can be proved that if $f$ is continuous, then (3) has a unique solution and that this procedure produces it. However, there is another approach: Instead of solving each initial-value problem individually, we can find a general formula for the solution of (3), and then apply that formula to solve specific problems. We will now show that

$$y(x) = y_0 + \int_{x_0}^{x} f(t)\, dt \tag{4}$$

is a formula for the solution of (3). To confirm that this is so we must show that $dy/dx = f(x)$ and that $y(x_0) = y_0$. The computations are as follows:

$$\frac{dy}{dx} = \frac{d}{dx}\left[ y_0 + \int_{x_0}^{x} f(t)\, dt \right] = 0 + f(x) = f(x)$$

$$y(x_0) = y_0 + \int_{x_0}^{x_0} f(t)\, dt = y_0 + 0 = y_0$$

### Example 2

In Example 7 of Section 7.2 we showed that the solution of the initial-value problem

$$\frac{dy}{dx} = \cos x, \quad y(0) = 1$$

is $y(x) = 1 + \sin x$. This initial-value problem can also be solved by applying Formula (4) with $f(x) = \cos x$, $x_0 = 0$, and $y_0 = 1$. This yields

$$y(x) = 1 + \int_{0}^{x} \cos t\, dt = 1 + \left[ \sin t \right]_{t=0}^{x} = 1 + \sin x \qquad \blacktriangleleft$$

In the last example we were able to perform the integration in Formula (4) and express the solution of the initial-value problem as an elementary function. However, sometimes this will not be possible, in which case the solution of the initial-value problem must be left in terms of an "unevaluated" integral. For example, from (4), the solution of the initial-value problem

$$\frac{dy}{dx} = e^{-x^2}, \quad y(0) = 1$$

is

$$y(x) = 1 + \int_{0}^{x} e^{-t^2}\, dt$$

However, it can be shown that there is no way to express the integral in this solution as an elementary function. Thus, we have encountered a *new* function, which we regard to be *defined* by the integral. A close relative of this function, known as the **error function**, plays an important role in probability and statistics, it is denoted by $\text{erf}(x)$ and is defined as

$$\text{erf}(x) = \frac{2}{\sqrt{\pi}} \int_{0}^{x} e^{-t^2}\, dt \tag{5}$$

Indeed, many of the most important functions in science and engineering are defined as integrals that have special names and notations associated with them. For example, the functions defined by

$$S(x) = \int_0^x \sin\left(\frac{\pi t^2}{2}\right) dt \quad \text{and} \quad C(x) = \int_0^x \cos\left(\frac{\pi t^2}{2}\right) dt \tag{6–7}$$

are called the ***Fresnel sine and cosine functions***, respectively, in honor of the French physicist Augustin Fresnel (1788–1827), who first encountered them in his study of diffraction of light waves.

**EVALUATING AND GRAPHING FUNCTIONS DEFINED BY INTEGRALS**

Computer programs evaluate functions defined by integrals by approximating the defining integral, say by the midpoint rule or some comparable method. For example, the following values of $S(1)$ and $C(1)$ were produced by a CAS that has a built-in algorithm for approximating definite integrals:

$$S(1) = \int_0^1 \sin\left(\frac{\pi t^2}{2}\right) dt \approx 0.438259, \qquad C(1) = \int_0^1 \cos\left(\frac{\pi t^2}{2}\right) dt \approx 0.779893$$

To generate graphs of functions defined by integrals, computer programs choose a set of $x$-values in the domain, approximate the integral for each of those values, and then plot the resulting points. Thus, there is a lot of computation involved in generating such graphs, since each plotted point requires the approximation of an integral. The graphs of the Fresnel functions in Figure 7.9.3 were generated in this way using a CAS.

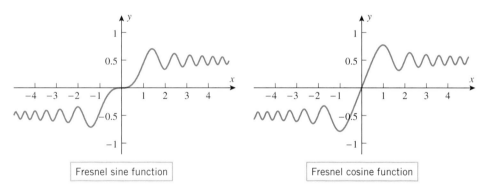

Figure 7.9.3

REMARK. Although it required a considerable amount of computation to generate the graphs of the Fresnel functions, the derivatives of $S(x)$ and $C(x)$ are easy to obtain using Part 2 of the Fundamental Theorem of Calculus (7.6.3); they are

$$S'(x) = \sin\left(\frac{\pi x^2}{2}\right) \quad \text{and} \quad C'(x) = \cos\left(\frac{\pi x^2}{2}\right) \tag{8–9}$$

These derivatives can be used to determine the locations of the relative extrema and inflection points and to investigate other properties of $S(x)$ and $C(x)$.

**INTEGRALS WITH FUNCTIONS AS LIMITS OF INTEGRATION**

Various applications can lead to integrals in which one or both of the limits of integration is a function of $x$. Some examples are

$$\int_x^1 \sqrt{\sin t}\, dt, \quad \int_{x^2}^{\sin x} \sqrt{t^3 + 1}\, dt, \quad \int_{\ln x}^{\pi} \frac{dt}{t^7 - 8}$$

We will complete this section by showing how to differentiate integrals of the form

$$\int_a^{g(x)} f(t)\, dt \tag{10}$$

where $a$ is constant. Derivatives of other kinds of integrals with functions as limits of integration will be discussed in the exercises.

To differentiate (10) we can view the integral as a composition $F(g(x))$, where

$$F(x) = \int_a^x f(t)\, dt$$

If we now apply the chain rule, we obtain

$$\frac{d}{dx}\left[\int_a^{g(x)} f(t)\, dt\right] = \frac{d}{dx}[F(g(x))] = F'(g(x))g'(x) = f(g(x))g'(x)$$

Theorem 7.6.3

Thus,

$$\frac{d}{dx}\left[\int_a^{g(x)} f(t)\, dt\right] = f(g(x))g'(x) \tag{11}$$

REMARK.   In words, *to differentiate an integral with a constant lower limit and a function as the upper limit, substitute the upper limit into the integrand, and multiply by the derivative of the upper limit.*

**Example 3**

$$\frac{d}{dx}\left[\int_1^{\sin x}(1 - t^2)\, dt\right] = (1 - \sin^2 x)\cos x = \cos^3 x$$ ◀

**EXERCISE SET 7.9**  ⬠ Graphing Calculator   [C] CAS
· · · · · · · · · · · · · · · · · · · · · · · · · · · · · · · · · · · · · · · · · · · · · · · · · · · · · · · · · · ·

1. Sketch the curve $y = 1/t$, and shade a region under the curve whose area is
   (a) $\ln 2$        (b) $-\ln 0.5$       (c) 2.

2. Sketch the curve $y = 1/t$, and shade two different regions under the curve whose area is $\ln 1.5$.

3. Given that $\ln a = 2$ and $\ln c = 5$, find
   (a) $\displaystyle\int_1^{ac} \frac{1}{t}\, dt$       (b) $\displaystyle\int_1^{1/c} \frac{1}{t}\, dt$
   (c) $\displaystyle\int_1^{a/c} \frac{1}{t}\, dt$       (d) $\displaystyle\int_1^{a^3} \frac{1}{t}\, dt$.

4. Given that $\ln a = 4$, find
   (a) $\displaystyle\int_1^{\sqrt{a}} \frac{1}{t}\, dt$       (b) $\displaystyle\int_1^{2a} \frac{1}{t}\, dt$
   (c) $\displaystyle\int_1^{2/a} \frac{1}{t}\, dt$       (d) $\displaystyle\int_2^{a} \frac{1}{t}\, dt$.

5. Approximate $\ln 5$ using the midpoint rule with $n = 10$, and estimate the magnitude of the error by comparing your answer to that produced directly by a calculating utility.

6. Approximate $\ln 3$ using the midpoint rule with $n = 20$, and estimate the magnitude of the error by comparing your answer to that produced directly by a calculating utility.

7. Simplify the expression and state the values of $x$ for which your simplification is valid.
   (a) $e^{-\ln x}$        (b) $e^{\ln x^2}$
   (c) $\ln\left(e^{-x^2}\right)$     (d) $\ln(1/e^x)$
   (e) $\exp(3\ln x)$     (f) $\ln(xe^x)$
   (g) $\ln\left(e^{x-\sqrt[3]{x}}\right)$    (h) $e^{x-\ln x}$

8. (a) Let $f(x) = e^{-2x}$. Find the simplest exact value of the function $f(\ln 3)$.
   (b) Let $f(x) = e^x + 3e^{-x}$. Find the simplest exact value of the function $f(\ln 2)$.

In Exercises 9 and 10, express the given quantity as a power of $e$.

9. (a) $3^\pi$              (b) $2^{\sqrt{2}}$

10. (a) $\pi^{-x}$         (b) $x^{2x}, \quad x > 0$

In Exercises 11 and 12, find the limits by making appropriate substitutions in the limits given in Theorem 7.9.2.

11. (a) $\displaystyle\lim_{x\to+\infty}\left(1 + \frac{1}{x}\right)^{2x}$       (b) $\displaystyle\lim_{x\to 0}(1 + 2x)^{1/x}$

**12. (a)** $\displaystyle\lim_{x \to +\infty} \left(1 + \frac{1}{3x}\right)^x$   **(b)** $\displaystyle\lim_{x \to 0} (1 + x)^{1/3x}$

---

In Exercises 13 and 14, find $g'(x)$ using Part 2 of the Fundamental Theorem of Calculus, and check your answer by evaluating the integral and then differentiating.

---

**13.** $g(x) = \displaystyle\int_1^x (t^2 - t)\,dt$   **14.** $g(x) = \displaystyle\int_\pi^x (1 - \cos t)\,dt$

---

In Exercises 15 and 16, find the derivative using Formula (11), and check your answer by evaluating the integral and then differentiating.

---

**15. (a)** $\displaystyle\frac{d}{dx}\int_1^{x^3} \frac{1}{t}\,dt$   **(b)** $\displaystyle\frac{d}{dx}\int_1^{\ln x} e^t\,dt$

**16. (a)** $\displaystyle\frac{d}{dx}\int_{-1}^{x^2} \sqrt{t+1}\,dt$   **(b)** $\displaystyle\frac{d}{dx}\int_\pi^{1/x} \sin t\,dt$

**17.** Let $F(x) = \displaystyle\int_0^x \frac{\cos t}{t^2 + 3}\,dt$. Find

  **(a)** $F(0)$   **(b)** $F'(0)$   **(c)** $F''(0)$.

**18.** Let $F(x) = \displaystyle\int_2^x \sqrt{3t^2 + 1}\,dt$. Find

  **(a)** $F(2)$   **(b)** $F'(2)$   **(c)** $F''(2)$.

**c 19. (a)** Use Formula (11) to find

$$\frac{d}{dx}\int_1^{x^2} t\sqrt{1 + t}\,dt$$

  **(b)** Use a CAS to evaluate the integral and differentiate the resulting function.

  **(c)** Use the simplification command of the CAS, if necessary, to confirm that answers in parts (a) and (b) are the same.

**20.** Show that

  **(a)** $\displaystyle\frac{d}{dx}\left[\int_x^a f(t)\,dt\right] = -f(x)$

  **(b)** $\displaystyle\frac{d}{dx}\left[\int_{g(x)}^a f(t)\,dt\right] = -f(g(x))g'(x)$.

---

In Exercises 21 and 22, use the results in Exercise 20 to find the derivative.

---

**21. (a)** $\displaystyle\frac{d}{dx}\int_x^1 \sin(t^2)\,dt$   **(b)** $\displaystyle\frac{d}{dx}\int_{\tan x}^3 \frac{t^2}{1 + t^2}\,dt$

**22. (a)** $\displaystyle\frac{d}{dx}\int_x^0 (t^2 + 1)^{40}\,dt$   **(b)** $\displaystyle\frac{d}{dx}\int_{1/x}^\pi \cos^3 t\,dt$

**23.** Find

$$\frac{d}{dx}\left[\int_{3x}^{x^2} \frac{t-1}{t^2 + 1}\,dt\right]$$

by writing

$$\int_{3x}^{x^2} \frac{t-1}{t^2 + 1}\,dt = \int_{3x}^0 \frac{t-1}{t^2 + 1}\,dt + \int_0^{x^2} \frac{t-1}{t^2 + 1}\,dt$$

**24.** Use Exercise 20(b) and the idea in Exercise 23 to show that

$$\frac{d}{dx}\int_{h(x)}^{g(x)} f(t)\,dt = f(g(x))g'(x) - f(h(x))h'(x)$$

**25.** Use the result obtained in Exercise 24 to perform the following differentiations:

  **(a)** $\displaystyle\frac{d}{dx}\int_{x^2}^{x^3} \sin^2 t\,dt$   **(b)** $\displaystyle\frac{d}{dx}\int_{-x}^x \frac{1}{1 + t}\,dt$.

**26.** Prove that the function

$$F(x) = \int_x^{3x} \frac{1}{t}\,dt$$

is constant on the interval $(0, +\infty)$ by using Exercise 24 to find $F'(x)$. What is that constant?

**27.** Let $F(x) = \int_0^x f(t)\,dt$, where $f$ is the function whose graph is shown in the accompanying figure.

  **(a)** Find $F(0)$, $F(3)$, $F(5)$, $F(7)$, and $F(10)$.

  **(b)** On what subintervals of the interval $[0, 10]$ is $F$ increasing? Decreasing?

  **(c)** Where does $F$ have its maximum value? Its minimum value?

  **(d)** Sketch the graph of $F$.

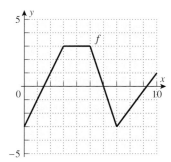

Figure Ex-27

**28.** Use the appropriate values found in part (a) of Exercise 27 to find the average value of $f$ over the interval $[0, 10]$.

---

In Exercises 29 and 30, express $F(x)$ in a piecewise form that does not involve an integral.

---

**29.** $F(x) = \displaystyle\int_{-1}^x |t|\,dt$

**30.** $F(x) = \displaystyle\int_0^x f(t)\,dt$, where $f(x) = \begin{cases} x, & 0 \le x \le 2 \\ 2, & x > 2 \end{cases}$

---

In Exercises 31–34, use Formula (4) to solve the initial-value problem.

**31.** $\dfrac{dy}{dx} = \sqrt[3]{x};\ y(1) = 2$

**32.** $\dfrac{dy}{dx} = \dfrac{x+1}{\sqrt{x}};\ y(1) = 0$

**33.** $\dfrac{dy}{dx} = \sec^2 x - \sin x;\ y(\pi/4) = 1$

**34.** $\dfrac{dy}{dx} = xe^{x^2};\ y(0) = 0$

**35.** Suppose that at time $t = 0$ there are $P_0$ individuals who have disease X, and suppose that a certain model for the spread of the disease predicts that the disease will spread at the rate of $r(t)$ individuals per day. Write a formula for the number of individuals who will have disease X after $x$ days.

**36.** Suppose that $v(t)$ is the velocity function of a particle moving along an $s$-axis. Write a formula for the coordinate of the particle at time $T$ if the particle is at the point $s_1$ at time $t = 1$.

**37.** The accompanying figure shows the graphs of $y = f(x)$ and $y = \int_0^x f(t)\,dt$. Determine which graph is which, and explain your reasoning.

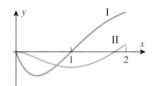

Figure Ex-37

**38.** (a) Make a conjecture about the value of the limit

$$\lim_{k \to 0} \int_1^x t^{k-1}\,dt \quad (x > 0)$$

(b) Check your conjecture by evaluating the integral, and then using L'Hôpital's rule to find the limit.

**39.** Let $F(x) = \int_0^x f(t)\,dt$, where $f$ is the function graphed in the accompanying figure.

(a) Where do the relative minima of $F$ occur?
(b) Where do the relative maxima of $F$ occur?
(c) Where does the absolute maximum of $F$ on the interval $[0, 5]$ occur?
(d) Where does the absolute minimum of $F$ on the interval $[0, 5]$ occur?
(e) Where is $F$ concave up? Concave down?
(f) Sketch the graph of $F$.

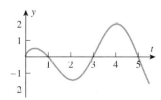

Figure Ex-39

c **40.** CAS programs have commands for working with most of the important nonelementary functions. Check your CAS

documentation for information about the error function erf($x$) [see Formula (5)], and then complete the following.

(a) Generate the graph of erf($x$).
(b) Use the graph to make a conjecture about the existence and location of any relative maxima and minima of erf($x$).
(c) Check your conjecture in part (b) using the derivative of erf($x$).
(d) Use the graph to make a conjecture about the existence and location of any inflection points of erf($x$).
(e) Check your conjecture in part (d) using the second derivative of erf($x$).
(f) Use the graph to make a conjecture about the existence of horizontal asymptotes of erf($x$).
(g) Check your conjecture in part (f) by using the CAS to find the limits of erf($x$) as $x \to \pm\infty$.

**41.** The Fresnel sine and cosine functions $S(x)$ and $C(x)$ were defined in Formulas (6) and (7) and graphed in Figure 7.9.3. Their derivatives were given in Formulas (8) and (9).

(a) At what points does $C(x)$ have relative minima? Relative maxima?
(b) Where do the inflection points of $C(x)$ occur?
(c) Confirm that your answers in parts (a) and (b) are consistent with the graph of $C(x)$.

**42.** Find the limit

$$\lim_{h \to 0} \frac{1}{h} \int_x^{x+h} \ln t\,dt$$

**43.** Find a function $f$ and a number $a$ such that

$$2 + \int_a^x f(t)\,dt = e^{3x}$$

**44.** (a) Give a geometric argument to show that

$$\frac{1}{x+1} < \int_x^{x+1} \frac{1}{t}\,dt < \frac{1}{x}, \quad x > 0$$

(b) Use the result in part (a) to prove that

$$\frac{1}{x+1} < \ln\left(1 + \frac{1}{x}\right) < \frac{1}{x}, \quad x > 0$$

(c) Use the result in part (b) to prove that

$$e^{\frac{x}{x+1}} < \left(1 + \frac{1}{x}\right)^x < e, \quad x > 0$$

and hence that

$$\lim_{x \to +\infty} \left(1 + \frac{1}{x}\right)^x = e$$

(d) Use the inequality in part (c) to prove that

$$\left(1 + \frac{1}{x}\right)^x < e < \left(1 + \frac{1}{x}\right)^{x+1}, \quad x > 0$$

**45.** Use a graphing utility to generate the graph of

$$y = \left(1 + \frac{1}{x}\right)^{x+1} - \left(1 + \frac{1}{x}\right)^x$$

in the window $[0, 100] \times [0, 0.2]$, and use that graph and part (d) of Exercise 44 to make a rough estimate of the error in the approximation

$$e \approx \left(1 + \frac{1}{50}\right)^{50}$$

46. Prove: If $f$ is continuous on an open interval $I$ and $a$ is any point in $I$, then

$$F(x) = \int_a^x f(t) \, dt$$

is continuous on $I$.

## SUPPLEMENTARY EXERCISES

1. Write a paragraph that describes the *rectangle method* for defining the area under a curve $y = f(x)$ over an interval $[a, b]$.

2. What is an *integral curve* of a function $f$? How are two integral curves of a function $f$ related?

3. The *definite integral* of $f$ over the interval $[a, b]$ is defined as the limit

$$\int_a^b f(x) \, dx = \lim_{\max \Delta x_k \to 0} \sum_{k=1}^n f(x_k^*) \Delta x_k$$

Explain what the various symbols on the right side of this equation mean.

4. State the two parts of the Fundamental Theorem of Calculus, and explain what is meant by the phrase "differentiation and integration are inverse processes."

5. Derive the formulas for the position and velocity functions of a particle that moves with uniformly accelerated motion along a coordinate line.

6. (a) Devise a procedure for finding upper and lower estimates of the area of the region in the accompanying figure (in cm²).
   (b) Use your procedure to find upper and lower estimates of the area.
   (c) Improve on the estimates you obtained in part (b).

Figure Ex-6

7. Suppose that

$$\int_0^1 f(x) \, dx = \frac{1}{2}, \quad \int_1^2 f(x) \, dx = \frac{1}{4},$$

$$\int_0^3 f(x) \, dx = -1, \quad \int_0^1 g(x) \, dx = 2$$

In each part, use this information to evaluate the given inte-

gral, if possible. If there is not enough information to evaluate the integral, then say so.

(a) $\displaystyle\int_0^2 f(x) \, dx$  (b) $\displaystyle\int_1^3 f(x) \, dx$  (c) $\displaystyle\int_2^3 5f(x) \, dx$

(d) $\displaystyle\int_1^0 g(x) \, dx$  (e) $\displaystyle\int_0^1 g(2x) \, dx$  (f) $\displaystyle\int_0^1 [g(x)]^2 \, dx$

8. In each part, use the information in Exercise 7 to evaluate the given integral. If there is not enough information to evaluate the integral, then say so.

(a) $\displaystyle\int_0^1 [f(x) + g(x)] \, dx$  (b) $\displaystyle\int_0^1 f(x) g(x) \, dx$

(c) $\displaystyle\int_0^1 \frac{f(x)}{g(x)} \, dx$  (d) $\displaystyle\int_0^1 [4g(x) - 3f(x)] \, dx$

9. In each part, evaluate the integral. Where appropriate, you may use a geometric formula.

(a) $\displaystyle\int_{-1}^1 1 + \sqrt{1 - x^2} \, dx$

(b) $\displaystyle\int_0^3 (x\sqrt{x^2 + 1} - \sqrt{9 - x^2}) \, dx$

(c) $\displaystyle\int_0^1 x\sqrt{1 - x^4} \, dx$

10. Evaluate the integral $\int_0^1 |2x - 1| \, dx$, and sketch the region whose area it represents.

11. One of the numbers $\pi$, $\pi/2$, $35\pi/128$, $1 - \pi$ is the correct value of the integral

$$\int_0^\pi \sin^8 x \, dx$$

Use the accompanying graph of $y = \sin^8 x$ and a logical process of elimination to find the correct value. [Do not attempt to evaluate the integral.]

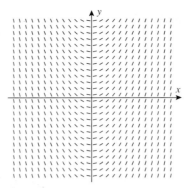

Figure Ex-11

**12.** Evaluate

$$\int \frac{e^{2x}}{e^x + 3} \, dx$$

[*Hint:* Divide $e^x + 3$ into $e^{2x}$.]

**13.** Give a convincing geometric argument to show that

$$\int_1^e \ln x \, dx + \int_0^1 e^x \, dx = e$$

**14.** In each part, find the limit by interpreting it as a limit of Riemann sums in which the interval $[0, 1]$ is divided into $n$ subintervals of equal length.

(a) $\displaystyle\lim_{n \to +\infty} \frac{\sqrt{1} + \sqrt{2} + \sqrt{3} + \cdots + \sqrt{n}}{n^{3/2}}$

(b) $\displaystyle\lim_{n \to +\infty} \frac{1^4 + 2^4 + 3^4 + \cdots + n^4}{n^5}$

(c) $\displaystyle\lim_{n \to +\infty} \frac{e^{1/n} + e^{2/n} + e^{3/n} + \cdots + e^{n/n}}{n}$

**15.** (a) Divide the interval $[1, 2]$ into 5 subintervals of equal length, and use appropriate Riemann sums to show that

$$0.2\left[\tfrac{1}{1.2} + \tfrac{1}{1.4} + \tfrac{1}{1.6} + \tfrac{1}{1.8} + \tfrac{1}{2.0}\right] < \ln 2$$

$$< 0.2\left[\tfrac{1}{1.0} + \tfrac{1}{1.2} + \tfrac{1}{1.4} + \tfrac{1}{1.6} + \tfrac{1}{1.8}\right]$$

(b) Show that if the interval $[1, 2]$ is divided into $n$ subintervals of equal length, then

$$\sum_{k=1}^n \frac{1}{n+k} < \ln 2 < \sum_{k=0}^{n-1} \frac{1}{n+k}$$

(c) Show that the difference between the two sums in part (b) is $1/2n$, and use this result to show that the sums in part (a) approximate $\ln 2$ with an error of at most 0.1.

(d) How large must $n$ be to ensure that the sums in part (b) approximate $\ln 2$ to three decimal places?

**16.** The accompanying figure shows the direction field for a differential equation $dy/dx = f(x)$. Which of the following functions is most likely to be $f(x)$?

$$\sqrt{x}, \quad \sin x, \quad x^4, \quad x$$

Explain your reasoning.

Figure Ex-16

**17.** In each part, confirm the stated equality.

(a) $1 \cdot 2 + 2 \cdot 3 + \cdots + n(n+1) = \tfrac{1}{3}n(n+1)(n+2)$

(b) $\displaystyle\lim_{n \to +\infty} \sum_{k=1}^{n-1} \left(\frac{9}{n} - \frac{k}{n^2}\right) = \frac{17}{2}$

(c) $\displaystyle\sum_{i=1}^3 \left(\sum_{j=1}^2 (i + j)\right) = 21$

**18.** Express

$$\sum_{k=4}^{18} k(k - 3)$$

in sigma notation with

(a) $k = 0$ as the lower limit of summation

(b) $k = 5$ as the lower limit of summation.

**19.** (a) Show that the substitutions $u = \sec x$ and $u = \tan x$ produce different values for the integral

$$\int \sec^2 x \tan x \, dx$$

(b) Explain why both are correct.

**20.** Use the two substitutions in Exercise 19 to evaluate the definite integral

$$\int_0^{\pi/4} \sec^2 x \tan x \, dx$$

and confirm that they produce the same result.

**21.** Evaluate the integral

$$\int \sqrt{1 + x^{-2/3}} \, dx$$

by making the substitution $u = 1 + x^{2/3}$.

**22.** (a) Express Formula 8 of Section 7.5 in sigma notation.

(b) If $c_1, c_2, \ldots, c_n$ are constants and $f_1, f_2, \ldots, f_n$ are integrable functions on $[a, b]$, do you think it is always true that

$$\int_a^b \left(\sum_{k=1}^n c_k f_k(x)\right) dx = \sum_{k=1}^n \left[c_k \int_a^b f_k(x) \, dx\right]?$$

Explain your reasoning.

**23.** Find an integral formula for the antiderivative of $1/(1+x^2)$ on the interval $(-\infty, +\infty)$ whose value at $x = 1$ is (a) 0 and (b) 2.

**c 24.** Let $F(x) = \int_0^x \dfrac{t-3}{t^2+7}\, dt$.

   (a) Find the intervals on which $F$ is increasing. Decreasing.

   (b) Find the open intervals on which $F$ is concave up. Concave down.

   (c) Find the $x$-values, if any, at which the function $F$ has absolute extrema.

   (d) Use a CAS to graph $F$, and confirm that the results in parts (a), (b), and (c) are consistent with the graph.

**25.** Prove that the function
$$F(x) = \int_0^x \frac{1}{1+t^2}\, dt + \int_0^{1/x} \frac{1}{1+t^2}\, dt$$
is constant on the interval $(0, +\infty)$.

**26.** What is the natural domain of the function
$$F(x) = \int_1^x \frac{1}{t^2-9}\, dt?$$
Explain your reasoning.

**27.** In each part, determine the values of $x$ for which $F(x)$ is positive, negative, or zero without performing the integration; explain your reasoning.

   (a) $F(x) = \displaystyle\int_1^x \frac{t^4}{t^2+3}\, dt$    (b) $F(x) = \displaystyle\int_{-1}^x \sqrt{4-t^2}\, dt$

**28.** Find a formula (defined piecewise) for the upper boundary of the trapezoid shown in the accompanying figure, and then integrate that function to derive the formula for the area of the trapezoid given on the inside front cover of this text.

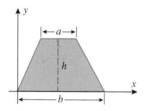

Figure Ex-28

**29.** An engineer studying the power consumption of a manufacturing plant determines that the plant's daily rate of electricity usage in kilowatts per hour (kW/h) can be reasonably modeled by the formula
$$R(t) = 2000e^{-t/48} + 500\sin\left(\tfrac{\pi}{12}t\right) \quad (0 \le t \le 24)$$

   (a) How many kilowatts of electricity does the plant use in a 24-hour period?

   (b) Find the average rate of electricity usage over the first 8 hours of operation.

   (c) Generate the graph of $R(t)$ over the first 8-hour period, and use it to make a rough estimate of the maximum rate of electricity usage during that period and when it occurs.

   (d) Determine the maximum rate of electricity usage during the first 8-hour period to two decimal places.

**30.** Suppose that a tumor grows at the rate of $r(t) = t/7$ grams (g) per week. When, during the second 26 weeks of growth, is the weight of the tumor the same as its average weight during that period?

**31.** The velocity of a particle moving along an $s$-axis is measured at 5-s intervals for 40 s, and the velocity function is modeled by a smooth curve drawn through the data points, as shown in the accompanying figure.

   (a) Does the particle have constant acceleration? Explain your reasoning.

   (b) Is there any 15-s time interval during which the acceleration is constant? Explain your reasoning.

   (c) Estimate the average velocity of the particle over the 40-s time period.

   (d) Estimate the distance traveled by the particle from time $t = 0$ to time $t = 40$.

   (e) Is the particle ever slowing down during the 40-s time period? Explain your reasoning.

   (f) Is there sufficient information for you to determine the $s$-coordinate of the particle at time $t = 10$? If so, find it. If not, explain what additional information you need.

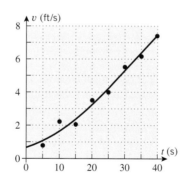

Figure Ex-31

**32.** Suppose that a particle moves along the $x$-axis so that its $x$-coordinate at time $t$ is given by $x = ae^{kt} + be^{-kt}$.

   (a) Show that the acceleration is proportional to $x$.

   (b) Assuming that the velocity of the particle at time $t = 0$ is $v_0$, find a formula for the acceleration function in terms of $a$, $b$, $x$, and $v_0$.

In Exercises 33–42, evaluate the integrals by hand, and check your answers with a CAS if you have one.

**33.** $\displaystyle\int \frac{\cos 3x}{\sqrt{5+2\sin 3x}}\, dx$     **34.** $\displaystyle\int \frac{\sqrt{3+\sqrt{x}}}{\sqrt{x}}\, dx$

**35.** $\displaystyle\int \frac{x^2}{(ax^3+b)^2}\, dx$     **36.** $\displaystyle\int x\sec^2(ax^2)\, dx$

**37.** $\displaystyle\int [\ln(e^x) + \ln(e^{-x})]\,dx$

**38.** $\displaystyle\int_{-2}^{-1} \left( u^{-4} + 3u^{-2} - \frac{1}{u^5} \right) du$

**39.** $\displaystyle\int_{e}^{e^2} \frac{dx}{x\ln x}$

**40.** $\displaystyle\int_{0}^{1} \frac{dx}{\sqrt{e^x}}$

**41.** $\displaystyle\int_{0}^{\ln\sqrt{2}} \frac{1 + \cos(e^{-2x})}{e^{2x}}\,dx$

**42.** $\displaystyle\int_{0}^{1} \sin^2(\pi x)\cos(\pi x)\,dx$

**C** **43.** Use a CAS to approximate the area of the region in the first quadrant that lies below the curve $y = x + x^2 - x^3$ and above the $x$-axis.

**C** **44.** In each part, use a CAS to solve the initial-value problem.

(a) $\dfrac{dy}{dx} = x^2 \cos 3x; \ \ y(\pi/2) = -1$

(b) $\dfrac{dy}{dx} = \dfrac{x^3}{(4 + x^2)^{3/2}}; \ \ y(0) = -2$

**C** **45.** In each part, use a CAS, where needed, to solve for $k$.

(a) $\displaystyle\int_{1}^{k} (x^3 - 2x - 1)\,dx = 0, \quad k > 1$

(b) $\displaystyle\int_{0}^{k} (x^2 + \sin 2x)\,dx = 3, \quad k \geq 0$

**C** **46.** Use a CAS to approximate the largest and smallest values of the integral

$$\int_{-1}^{x} \frac{t}{\sqrt{2 + t^3}}\,dt$$

for $1 \leq x \leq 3$

**C** **47.** The function $J_0$ defined by

$$J_0(x) = \frac{1}{\pi}\int_{0}^{\pi} \cos(x\sin t)\,dt$$

is called the **Bessel function of order zero**.

(a) Use a CAS to graph the equation $y = J_0(x)$ over the interval $0 \leq x \leq 8$.

(b) Find $J_0(1)$.

(c) Find the smallest positive zero of $J_0(x)$.

**C** **48.** Let $A$ be the area under the curve $y = x^2$ over the interval $[0, 1]$.

(a) Find $A$ by using Part 1 of the Fundamental Theorem of Calculus.

(b) Find $A$ by computing the limit of the left endpoint approximations by hand, and then find the limit using a CAS.

(c) Find $A$ by computing the limit of the right endpoint approximations by hand, and then find the limit using a CAS.

**C** **49.** In number theory, $\pi(n)$ denotes the number of prime numbers that are less than or equal to the positive integer $n$. For example, it can be shown with the help of a computer that $\pi(100,000) = 9592$; that is, there are 9592 prime numbers that are less than or equal to 100,000. There are two useful approximations to $\pi(n)$ that are appropriate for large values of $n$:

$$\pi(n) \approx \frac{n}{\ln n} \quad \text{and} \quad \pi(n) \approx \int_{2}^{n} \frac{1}{\ln t}\,dt$$

Use a CAS to determine which of these approximations produces the better estimate of $\pi(100,000)$.

---

### EXPANDING THE CALCULUS HORIZON

# Blammo the Human Cannonball

*Blammo the Human Cannonball will be fired from a cannon and hopes to land in a small net at the opposite end of the circus arena. Your job as Blammo's manager is to do the mathematical calculations that will allow Blammo to perform his death-defying act safely. The methods that you will use are from the field of **ballistics** (the study of projectile motion).*

■ **The Problem**

Blammo's cannon has a **muzzle velocity** of 35 m/s, which means that Blammo will leave the muzzle with that velocity. The muzzle opening will be 5 m above the ground, and Blammo's objective is to land in a net that is also 5 m above the ground and that extends a distance of 10 m between 90 m and 100 m from the cannon opening (Figure 1). Your mathematical problem is to determine the ***elevation angle*** $\alpha$ of the cannon (the angle from the horizontal to the cannon barrel) that will make Blammo land in the net.

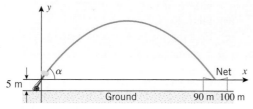

Figure 1

## Modeling Assumptions

Blammo's trajectory will be determined by his initial velocity, the elevation angle of the cannon, and the forces that act on him after he leaves the muzzle. We will assume that the only force acting on Blammo after he leaves the muzzle is the downward force of the Earth's gravity. In particular, we will ignore the effect of air resistance. It will be convenient to introduce the $xy$-coordinate system shown in Figure 1 and to assume that Blammo is at the origin at time $t = 0$. We will also assume that Blammo's motion can be decomposed into two independent components, a horizontal component parallel to the $x$-axis and a vertical component parallel to the $y$-axis. We will analyze the horizontal and vertical components of Blammo's motion separately, and then we will combine the information to obtain a complete picture of his trajectory.

## Blammo's Equations of Motion

We will denote the position and velocity functions for Blammo's horizontal component of motion by $x(t)$ and $v_x(t)$, and we will denote the position and velocity functions for his vertical component of motion by $y(t)$ and $v_y(t)$.

Since the only force acting on Blammo after he leaves the muzzle is the downward force of the Earth's gravity, there are no horizontal forces to alter his initial horizontal velocity $v_x(0)$. Thus, Blammo will have a constant velocity of $v_x(0)$ in the $x$-direction; this implies that

$$x(t) = v_x(0)t \tag{1}$$

In the $y$-direction Blammo is acted on only by the downward force of the Earth's gravity. Thus, his motion in this direction is governed by the free-fall model; hence, from (12) in Section 7.7 his vertical position function is

$$y(t) = y(0) + v_y(0)t - \tfrac{1}{2}gt^2$$

Taking $g = 9.8$ m/s$^2$, and using the fact that $y(0) = 0$, this equation can be written as

$$y(t) = v_y(0)t - 4.9t^2 \tag{2}$$

..........

***Exercise 1***    At time $t = 0$ Blammo's velocity is 35 m/s, and this velocity is directed at an angle $\alpha$ with the horizontal. It is a fact of physics that the initial velocity components $v_x(0)$ and $v_y(0)$ can be obtained geometrically from the muzzle velocity and the angle of elevation using the triangle shown in Figure 2. We will justify this later in the text, but for now use this fact to show that Equations (1) and (2) can be expressed as

$$x(t) = (35 \cos \alpha)t$$
$$y(t) = (35 \sin \alpha)t - 4.9t^2$$

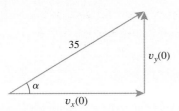

Figure 2

..........
## Exercise 2

(a) Use the result in Exercise 1 to find the velocity functions $v_x(t)$ and $v_y(t)$ in terms of the elevation angle $\alpha$.

(b) Find the time $t$ at which Blammo is at his maximum height above the $x$-axis, and show that this maximum height (in meters) is

$$y_{\text{max}} = 62.5 \sin^2 \alpha$$

..........
## Exercise 3
The equations obtained in Exercise 1 can be viewed as parametric equations for Blammo's trajectory. Show, by eliminating the parameter $t$, that if $0 < \alpha < \pi/2$, then Blammo's trajectory is given by the equation

$$y = (\tan \alpha)x - \frac{0.004}{\cos^2 \alpha} x^2$$

Explain why Blammo's trajectory is a parabola.

## Finding the Elevation Angle

Define Blammo's *horizontal range* $R$ to be the horizontal distance he travels until he returns to the height of the muzzle opening ($y = 0$). Your objective is to find elevation angles that will make the horizontal range fall between 90 m and 100 m, thereby ensuring that Blammo lands in the net.

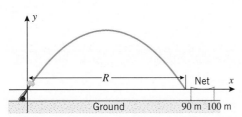

Figure 3

..........
## Exercise 4
Use a graphing utility and either the parametric equations obtained in Exercise 1 or the single equation obtained in Exercise 3 to generate Blammo's trajectories, taking elevation angles at increments of $10°$ from $15°$ to $85°$. In each case, determine visually whether Blammo lands in the net.

..........
## Exercise 5
Find the time required for Blammo to return to his starting height ($y = 0$), and use that result to show that Blammo's range $R$ is given by the formula

$$R = 125 \sin 2\alpha$$

·············
*Exercise 6*

(a) Use the result in Exercise 5 to find two elevation angles that will allow Blammo to hit the midpoint of the net 95 m away.

(b) The tent is 55 m high. Explain why the larger elevation angle cannot be used.

·············
*Exercise 7*   How much can the smaller elevation angle in Exercise 6 vary and still have Blammo hit the net between 90 m and 100 m?

## Blammo's Shark Trick

Blammo is to be fired from 5 m above ground level with a muzzle velocity of 35 m/s over a flaming wall that is 20 m high and past a 5-m-high shark pool (Figure 4). To make the feat impressive, the pool will be made as long as possible. Your job as Blammo's manager is to determine the length of the pool, how far to place the cannon from the wall, and what elevation angle to use to ensure that Blammo clears the pool.

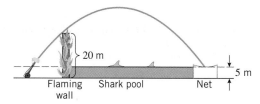

| | | |
|---|---|---|
| 20 m | | 5 m |
| Flaming wall | Shark pool | Net |

Figure 4

·············
*Exercise 8*   Prepare a written presentation of the problem and your solution of it that is at an appropriate level for an engineer, physicist, or mathematician to read. Your presentation should contain the following elements: an explanation of all notation, a list and description of all formulas that will be used, a diagram that shows the orientation of any coordinate systems that will be used, a description of any assumptions you make to solve the problem, graphs that you think will enhance the presentation, and a clear step-by-step explanation of your solution.

·······················································································

*Module by*: John Rickert, *Rose-Hulman Institute of Technology*
            Howard Anton, *Drexel University*

# APPENDIX A

# Real Numbers, Intervals, and Inequalities

**REAL NUMBERS**

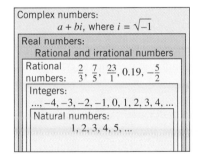

Figure A.1

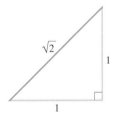

Figure A.2

**COMPLEX NUMBERS**

Figure A.1 describes the various categories of numbers that we will encounter in this text. The simplest numbers are the **natural numbers**

$$1, \quad 2, \quad 3, \quad 4, \quad 5, \ldots$$

These are a subset of the **integers**

$$\ldots, \quad -4, \quad -3, \quad -2, \quad -1, \quad 0, \quad 1, \quad 2, \quad 3, \quad 4, \ldots$$

and these in turn are a subset of the **rational numbers**, which are the numbers formed by taking ratios of integers (avoiding division by 0). Some examples are

$$\tfrac{2}{3}, \quad \tfrac{7}{5}, \quad 23 = \tfrac{23}{1}, \quad 0.19 = \tfrac{19}{100}, \quad -\tfrac{5}{2} = \tfrac{-5}{2} = \tfrac{5}{-2}$$

The early Greeks believed that every measurable quantity had to be a rational number. However, this idea was overturned in the fifth century B.C. by Hippasus of Metapontum[*] who demonstrated the existence of **irrational numbers**, that is, numbers that cannot be expressed as the ratio of two integers. Using geometric methods, he showed that the length of the hypotenuse of the triangle in Figure A.2 could not be expressed as a ratio of integers, thereby proving that $\sqrt{2}$ is an irrational number. Some other examples of irrational numbers are

$$\sqrt{3}, \quad \sqrt{5}, \quad 1 + \sqrt{2}, \quad \sqrt[3]{7}, \quad \pi, \quad \cos 19°$$

The rational and irrational numbers together comprise what is called the **real number system**, and both the rational and irrational numbers are called **real numbers**.

Because the square of a real number cannot be negative, the equation

$$x^2 = -1$$

has no solutions in the real number system. In the eighteenth century mathematicians remedied this problem by inventing a new number, which they denoted by

$$i = \sqrt{-1}$$

and which they defined to have the property $i^2 = -1$. This, in turn, led to the development

---

[*] HIPPASUS OF METAPONTUM (circa 500 B.C.). A Greek Pythagorean philosopher. According to legend, Hippasus made his discovery at sea and was thrown overboard by fanatic Pythagoreans because his result contradicted their doctrine. The discovery of Hippasus is one of the most fundamental in the entire history of science.

of the **complex numbers**, which are numbers of the form

$$a + bi$$

where $a$ and $b$ are real numbers. Some examples are

$$2 + 3i \qquad 3 - 4i \qquad 6i \qquad \tfrac{2}{3}$$

$$[a = 2, b = 3] \qquad [a = 3, b = -4] \qquad [a = 0, b = 6] \qquad [a = \tfrac{2}{3}, b = 0]$$

Observe that every real number $a$ is also a complex number because it can be written as

$$a = a + 0i$$

Thus, the real numbers are a subset of the complex numbers. Those complex numbers that are not real numbers are called **imaginary numbers**. Although we will be concerned primarily with real numbers in this text, imaginary numbers will arise in the course of solving equations. For example, the solutions of the quadratic equation

$$ax^2 + bx + c = 0$$

which are given by the **quadratic formula**

$$x = \frac{-b \pm \sqrt{b^2 - 4ac}}{2a}$$

are imaginary if the quantity $b^2 - 4ac$ is negative.

**DIVISION BY ZERO**

Division by zero is not allowed in numerical computations because it leads to mathematical inconsistencies. For example, if $1/0$ were assigned some numerical value, say $1/0 = p$, then it would follow that $0 \cdot p = 1$, which is incorrect.

**DECIMAL REPRESENTATION OF REAL NUMBERS**

Rational and irrational numbers can be distinguished by their decimal representations. Rational numbers have decimals that are **repeating**, by which we mean that at some point in the decimal some fixed block of numbers begins to repeat indefinitely. For example,

$$\tfrac{4}{3} = 1.333\ldots, \qquad \tfrac{3}{11} = .272727\ldots, \qquad \tfrac{1}{2} = .50000\ldots, \qquad \tfrac{5}{7} = .714285714285714285\ldots$$

$$\underbrace{\quad}_{\text{3 repeats}} \qquad \underbrace{\quad}_{\text{27 repeats}} \qquad \underbrace{\quad}_{\text{0 repeats}} \qquad \underbrace{\quad}_{\text{714285 repeats}}$$

Decimals in which zero repeats from some point on are called **terminating decimals**. For brevity, it is usual to omit the repetitive zeros in terminating decimals and for other repeating decimals to write the repeating digits only once but with a bar over them to indicate the repetition. For example,

$$\tfrac{1}{2} = .5, \qquad \tfrac{12}{4} = 3, \qquad \tfrac{8}{25} = .32, \qquad \tfrac{4}{3} = 1.\overline{3}, \qquad \tfrac{3}{11} = .\overline{27}, \qquad \tfrac{5}{7} = .\overline{714285}$$

Irrational numbers have nonrepeating decimals, so we can be certain that the decimals

$$\sqrt{2} = 1.414213562373095\ldots \quad \text{and} \quad \pi = 3.141592653589793\ldots$$

do not repeat from some point on. Moreover, if we stop the decimal expansion of an irrational number at some point, we get only an approximation to the number, never an exact value. For example, even if we compute $\pi$ to 1000 decimal places, as in Figure A.3, we still have only an approximation.

```
3.141592653589793238462643383279502884197169
399375105820974944592307816406286208998628034
825342117067982148086513282306647093844609550
582231725359408128481117450284102701938521105
559644622948954930381964428810975665933446128
475648233786783165271201909145648566923460348
610454326648213393607260249141273724587006606
315588174881520920962829254091715364367892590
360011330530548820466521384146951941511609433
057270365759591953092186117381932611793105118
548074462379962749567351885752724891227938183
011949129833673362440656643086021394946395224
737190702179860943702770539217176293176752384
674818467669405132000568127145263560827785771
342757789609173637178721468440901224953430146
549585371050792279689258923542019956112129021
960864034418159813629774771309960518707211349
999998372978049951059731732816096318595024459
455346908302642522308253344685035261931188171
010003137838752886587533208381420617177669147
303598253490428755468731159562863882353787593
751957781857780532171226806613001927876611195
909216420198
```

Figure A.3

**REMARK.** Beginning mathematics students are sometimes taught to approximate $\pi$ by $\tfrac{22}{7}$. Keep in mind, however, that this is only an approximation, since

$$\tfrac{22}{7} = 3.\overline{142857}$$

is a rational number whose decimal representation begins to differ from $\pi$ in the third decimal place.

**COORDINATE LINES**

Figure A.4

In 1637 René Descartes[*] published a philosophical work called *Discourse on the Method of Rightly Conducting the Reason*. In the back of that book was an appendix that the British philosopher John Stuart Mill described as "the greatest single step ever made in the progress of the exact sciences." In that appendix René Descartes linked together algebra and geometry, thereby creating a new subject called *analytic geometry*; it gave a way of describing algebraic formulas by geometric curves and, conversely, geometric curves by algebraic formulas.

The key step in analytic geometry is to establish a correspondence between real numbers and points on a line. To do this, choose any point on the line as a reference point, and call it the *origin*; and then arbitrarily choose one of the two directions along the line to be the *positive direction*, and let the other be the *negative direction*. It is usual to mark the positive direction with an arrowhead, as in Figure A.4, and to take the positive direction to the right when the line is horizontal. Next, choose a convenient unit of measure, and represent each positive number $r$ by the point that is $r$ units from the origin in the positive direction, each negative number $-r$ by the point that is $r$ units from the origin in the negative direction from the origin, and 0 by the origin itself (Figure A.5). The number associated with a point $P$ is called the *coordinate* of $P$, and the line is called a *coordinate line*, a *real number line*, or a *real line*.

Figure A.5

**INEQUALITY NOTATION**

The real numbers can be ordered by size as follows: If $b - a$ is positive, then we write either $a < b$ (read "$a$ is less than $b$") or $b > a$ (read "$b$ is greater than $a$"). We write $a \leq b$ to mean $a < b$ or $a = b$, and we write $a < b < c$ to mean that $a < b$ and $b < c$. As one traverses a coordinate line in the positive direction, the real numbers increase in size, so on a horizontal coordinate line the inequality $a < b$ implies that $a$ is to left of $b$, and the inequalities $a < b < c$ imply that $a$ is to the left of $c$, and $b$ lies between $a$ and $c$. The meaning of such symbols as

$$a \leq b < c, \quad a \leq b \leq c, \quad \text{and} \quad a < b < c < d$$

should be clear. For example, you should be able to confirm that all of the following are true statements:

$$3 < 8, \quad -7 < 1.5, \quad -12 \leq -\pi, \quad 5 \leq 5, \quad 0 \leq 2 \leq 4,$$
$$8 \geq 3, \quad 1.5 > -7, \quad -\pi > -12, \quad 5 \geq 5, \quad 3 > 0 > -1 > -3$$

**REVIEW OF SETS**

In the following discussion we will be concerned with certain sets of real numbers, so it will be helpful to review the basic ideas about sets. Recall that a *set* is a collection of objects, called *elements* or *members* of the set. In this text we will be concerned primarily with sets whose members are numbers or points that lie on a line, a plane, or in three-dimensional

[*] RENÉ DESCARTES (1596–1650). Descartes, a French aristocrat, was the son of a government official. He graduated from the University of Poitiers with a law degree at age 20. After a brief probe into the pleasures of Paris he became a military engineer, first for the Dutch Prince of Nassau and then for the German Duke of Bavaria. It was during his service as a soldier that Descartes began to pursue mathematics seriously and develop his analytic geometry. After the wars, he returned to Paris where he stalked the city as an eccentric, wearing a sword in his belt and a plumed hat. He lived in leisure, seldom arose before 11 A.M., and dabbled in the study of human physiology, philosophy, glaciers, meteors, and rainbows. He eventually moved to Holland, where he published his *Discourse on the Method*, and finally to Sweden where he died while serving as tutor to Queen Christina. Descartes is regarded as a genius of the first magnitude. In addition to major contributions in mathematics and philosophy, he is considered, along with William Harvey, to be a founder of modern physiology.

space. We will denote sets by capital letters and elements by lowercase letters. To indicate that $a$ is a member of the set $A$ we will write $a \in A$ (read "$a$ belongs to $A$"), and to indicate that $a$ is not a member of the set $A$ we will write $a \notin A$ (read "$a$ does not belong to $A$"). For example, if $A$ is the set of positive integers, then $5 \in A$, but $-5 \notin A$. Sometimes sets arise that have no members (e.g., the set of odd integers that are divisible by 2). A set with no members is called an ***empty set*** or a ***null set*** and is denoted by the symbol $\varnothing$.

Some sets can be described by listing their members between braces. The order in which the members are listed does not matter, so, for example, the set $A$ of positive integers that are less than 6 can be expressed as

$$A = \{1, 2, 3, 4, 5\} \quad \text{or} \quad A = \{2, 3, 1, 5, 4\}$$

We can also write $A$ in *set-builder notation* as

$$A = \{x : x \text{ is an integer and } 0 < x < 6\}$$

which is read "$A$ is the set of all $x$ such that $x$ is an integer and $0 < x < 6$." In general, to express a set $S$ in set-builder notation we write $S = \{x : \underline{\hspace{1cm}}\}$ in which the line is replaced by a property that uniquely defines the set $S$.

## INTERVALS

In calculus we will be concerned with sets of real numbers, called ***intervals***, that correspond to line segments on a coordinate line. For example, if $a < b$, then the ***open interval*** from $a$ to $b$, denoted by $(a, b)$, is the line segment extending from $a$ to $b$, *excluding* the endpoints; and the ***closed interval*** from $a$ to $b$, denoted by $[a, b]$, is the line segment extending from $a$ to $b$, *including* the endpoints (Figure A.6). These sets can be expressed in set-builder notation as

$$(a, b) = \{x : a < x < b\} \qquad \boxed{\text{The open interval from } a \text{ to } b}$$

$$[a, b] = \{x : a \le x \le b\} \qquad \boxed{\text{The closed interval from } a \text{ to } b}$$

The open interval $(a, b)$

$a$ $b$

$a$ $b$

The closed interval $[a, b]$

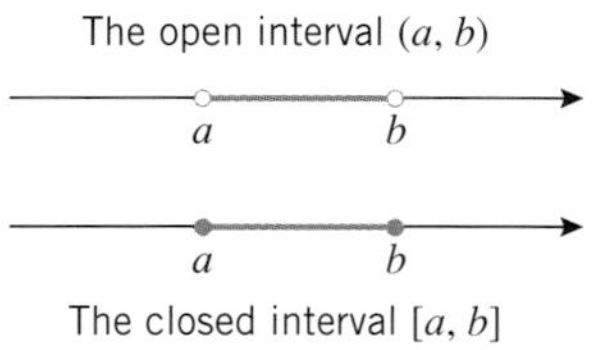

Figure A.6

**REMARK.**    Observe that in this notation and in the corresponding Figure A.6, parentheses and open dots mark endpoints that are excluded from the interval, whereas brackets and closed dots mark endpoints that are included in the interval. Observe also, that in set-builder notation for the intervals, it is understood that $x$ is a real number, even though it is not stated explicitly.

As shown in Table 1, an interval can include one endpoint and not the other; such intervals are called ***half-open*** (or sometimes ***half-closed***). Moreover, the table also shows that it is possible for an interval to extend indefinitely in one or both directions. To indicate that an interval extends indefinitely in the positive direction we write $+\infty$ (read "positive infinity") in place of a right endpoint, and to indicate that an interval extends indefinitely in the negative direction we write $-\infty$ (read "negative infinity") in place of a left endpoint. Intervals that extend between two real numbers are called ***finite intervals***, whereas intervals that extend indefinitely in one or both directions are called ***infinite intervals***.

**REMARK.**    By convention, infinite intervals of the form $[a, +\infty)$ or $(-\infty, b]$ are considered to be closed because they contain their endpoint, and intervals of the form $(a, +\infty)$ and $(-\infty, b)$ are considered to be open because they do not include their endpoint. The interval $(-\infty, +\infty)$, which is the set of all real numbers, has no endpoints and can be regarded as either open or closed, as convenient. This set is often denoted by the special symbol $\mathbb{R}$. To distinguish verbally between the open interval $(0, +\infty) = \{x : x > 0\}$ and the closed interval $[0, +\infty) = \{x : x \ge 0\}$, we will call $x$ ***positive*** if $x > 0$ and ***nonnegative*** if $x \ge 0$. Thus, a positive number must be nonnegative, but a nonnegative number need not be positive, since it might possibly be 0.

**Table 1**

| INTERVAL NOTATION | SET NOTATION | GEOMETRIC PICTURE | CLASSIFICATION |
|---|---|---|---|
| $(a, b)$ | $\{x : a < x < b\}$ | | Finite; open |
| $[a, b]$ | $\{x : a \leq x \leq b\}$ | | Finite; closed |
| $[a, b)$ | $\{x : a \leq x < b\}$ | | Finite; half-open |
| $(a, b]$ | $\{x : a < x \leq b\}$ | | Finite; half-open |
| $(-\infty, b]$ | $\{x : x \leq b\}$ | | Infinite; closed |
| $(-\infty, b)$ | $\{x : x < b\}$ | | Infinite; open |
| $[a, +\infty)$ | $\{x : x \geq a\}$ | | Infinite; closed |
| $(a, +\infty)$ | $\{x : x > a\}$ | | Infinite; open |
| $(-\infty, +\infty)$ | $\mathbb{R}$ | | Infinite; open and closed |

**UNIONS AND INTERSECTIONS OF INTERVALS**

If $A$ and $B$ are sets, then the ***union*** of $A$ and $B$ (denoted by $A \cup B$) is the set whose members belong to $A$ or $B$ (or both), and the ***intersection*** of $A$ and $B$ (denoted by $A \cap B$) is the set whose members belong to both $A$ and $B$. For example,

$$\{x : 0 < x < 5\} \cup \{x : 1 < x < 7\} = \{x : 0 < x < 7\}$$

$$\{x : x < 1\} \cap \{x : x \geq 0\} = \{x : 0 \leq x < 1\}$$

$$\{x : x < 0\} \cap \{x : x > 0\} = \varnothing$$

or in interval notation,

$$(0, 5) \cup (1, 7) = (0, 7)$$

$$(-\infty, 1) \cap [0, +\infty) = [0, 1)$$

$$(-\infty, 0) \cap (0, +\infty) = \varnothing$$

**ALGEBRAIC PROPERTIES OF INEQUALITIES**

The following algebraic properties of inequalities will be used frequently in this text. We omit the proofs.

> **A.1**   THEOREM (***Properties of Inequalities***).    *Let $a$, $b$, $c$, and $d$ be real numbers.*
> (a)   *If $a < b$ and $b < c$, then $a < c$.*
> (b)   *If $a < b$, then $a + c < b + c$ and $a - c < b - c$.*
> (c)   *If $a < b$, then $ac < bc$ when $c$ is positive and $ac > bc$ when $c$ is negative.*
> (d)   *If $a < b$ and $c < d$, then $a + c < b + d$.*
> (e)   *If $a$ and $b$ are both positive or both negative and $a < b$, then $1/a > 1/b$.*

If we call the direction of an inequality its *sense*, then these properties can be paraphrased as follows:

(b)   *The sense of an inequality is unchanged if the same number is added to or subtracted from both sides.*

(c)   *The sense of an inequality is unchanged if both sides are multiplied by the same positive number, but the sense is reversed if both sides are multiplied by the same negative number.*

(d)   *Inequalities with the same sense can be added.*

(e)   *If both sides of an inequality have the same sign, then the sense of the inequality is reversed by taking the reciprocal of each side.*

REMARK.     These properties remain true if the symbols $<$ and $>$ are replaced by $\leq$ and $\geq$ in Theorem A.1.

## Example 1

| STARTING INEQUALITY | OPERATION | RESULTING INEQUALITY |
|---|---|---|
| $-2 < 6$ | Add 7 to both sides. | $5 < 13$ |
| $-2 < 6$ | Subtract 8 from both sides. | $-10 < -2$ |
| $-2 < 6$ | Multiply both sides by 3. | $-6 < 18$ |
| $-2 < 6$ | Multiply both sides by $-3$. | $6 > -18$ |
| $3 < 7$ | Multiply both sides by 4. | $12 < 28$ |
| $3 < 7$ | Multiply both sides by $-4$. | $-12 > -28$ |
| $3 < 7$ | Take reciprocals of both sides. | $\frac{1}{3} > \frac{1}{7}$ |
| $-8 < -6$ | Take reciprocals of both sides. | $-\frac{1}{8} > -\frac{1}{6}$ |
| $4 < 5, -7 < 8$ | Add corresponding sides. | $-3 < 13$ |

◀

**SOLVING INEQUALITIES**

A **solution** of an inequality in an unknown $x$ is a value for $x$ that makes the inequality a true statement. For example, $x = 1$ is a solution of the inequality $x < 5$, but $x = 7$ is not. The set of all solutions of an inequality is called its **solution set**. It can be shown that if one does not multiply both sides of an inequality by zero or an expression involving an unknown, then the operations in Theorem A.1 will not change the solution set of the inequality. The process of finding the solution set of an inequality is called **solving** the inequality.

## Example 2

Solve $3 + 7x \leq 2x - 9$.

*Solution.*   We will use the operations of Theorem A.1 to isolate $x$ on one side of the inequality.

$$3 + 7x \leq 2x - 9 \quad \text{Given.}$$

$$7x \leq 2x - 12 \quad \text{We subtracted 3 from both sides.}$$

$$5x \leq -12 \quad \text{We subtracted } 2x \text{ from both sides.}$$

$$x \leq -\tfrac{12}{5} \quad \text{We multiplied both sides by } \tfrac{1}{5}.$$

Because we have not multiplied by any expressions involving the unknown $x$, the last inequality has the same solution set as the first. Thus, the solution set is the interval $\left(-\infty, -\frac{12}{5}\right]$ shown in Figure A.7.     ◀

$-\frac{12}{5}$

Figure A.7

## Example 3

Solve $7 \leq 2 - 5x < 9$.

*Solution.*   The given inequality is actually a combination of the two inequalities

$$7 \leq 2 - 5x \quad \text{and} \quad 2 - 5x < 9$$

We could solve the two inequalities separately, then determine the values of $x$ that satisfy both by taking the intersection of the two solution sets. However, it is possible to work with the combined inequalities in this problem:

$$7 \leq 2 - 5x < 9 \qquad \text{Given.}$$

$$5 \leq -5x < 7 \qquad \text{We subtracted 2 from each member.}$$

$$-1 \geq x > -\frac{7}{5} \qquad \text{We multiplied by } -\frac{1}{5} \text{ and reversed the sense of the inequalities.}$$

$$-\frac{7}{5} < x \leq -1 \qquad \text{For clarity, we rewrote the inequalities with the smaller number on the left.}$$

Thus, the solution set is the interval $\left(-\frac{7}{5}, -1\right]$ shown in Figure A.8. ◄

### Example 4

Solve $x^2 - 3x > 10$.

*Solution.* By subtracting 10 from both sides, the inequality can be rewritten as

$$x^2 - 3x - 10 > 0$$

Factoring the left side yields

$$(x + 2)(x - 5) > 0$$

The values of $x$ for which $x + 2 = 0$ or $x - 5 = 0$ are $x = -2$ and $x = 5$. These points divide the coordinate line into three open intervals,

$$(-\infty, -2), \quad (-2, 5), \quad (5, +\infty)$$

on each of which the product $(x + 2)(x - 5)$ has constant sign. To determine those signs we will choose an *arbitrary* point in each interval at which we will determine the sign; these are called **test points**. As shown in Figure A.9, we will use $-3$, $0$, and $6$ as our test points. The results can be organized as follows:

| INTERVAL | TEST POINT | SIGN OF $(x + 2)(x - 5)$ AT THE TEST POINT |
|---|---|---|
| $(-\infty, -2)$ | $-3$ | $(-)(-) = +$ |
| $(-2, 5)$ | $0$ | $(+)(-) = -$ |
| $(5, +\infty)$ | $6$ | $(+)(+) = +$ |

The pattern of signs in the intervals is shown on the number line in the middle of Figure A.9. We deduce that the solution set is $(-\infty, -2) \cup (5, +\infty)$, which is shown at the bottom of Figure A.9. ◄

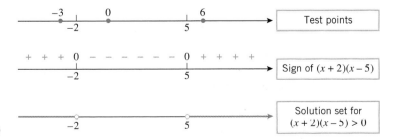

**Figure A.9**

### Example 5

Solve $\dfrac{2x - 5}{x - 2} < 1$.

*Solution.* We could start by multiplying both sides by $x - 2$ to eliminate the fraction. However, this would require us to consider the cases $x - 2 > 0$ and $x - 2 < 0$ separately because the sense of the inequality would be reversed in the second case, but not the first. The following approach is simpler:

$$\frac{2x-5}{x-2} < 1 \qquad \boxed{\text{Given.}}$$

$$\frac{2x-5}{x-2} - 1 < 0 \qquad \boxed{\text{We subtracted 1 from both sides to obtain a 0 on the right.}}$$

$$\frac{(2x-5)-(x-2)}{x-2} < 0 \qquad \boxed{\text{We combined terms.}}$$

$$\frac{x-3}{x-2} < 0 \qquad \boxed{\text{We simplified.}}$$

The quantity $x-3$ is zero if $x = 3$, and the quantity $x - 2$ is zero if $x = 2$. These points divide the coordinate line into three open intervals,

$$(-\infty, 2), \quad (2, 3), \quad (3, +\infty)$$

on each of which the quotient $(x-3)/(x-2)$ has constant sign. Using 0, 2.5, and 4 as test points (Figure A.10), we obtain the following results:

| INTERVAL | TEST POINT | SIGN OF $(x-3)(x-2)$ AT THE TEST POINT |
|---|---|---|
| $(-\infty, 2)$ | 0 | $(-)/(-) = +$ |
| $(2, 3)$ | 2.5 | $(-)/(+) = -$ |
| $(3, +\infty)$ | 4 | $(+)/(+) = +$ |

The signs of the quotient are shown in the middle of Figure A.10. From the figure we see that the solution set consists of all real values of $x$ such that $2 < x < 3$. This is the interval $(2, 3)$ shown at the bottom of Figure A.10. ◀

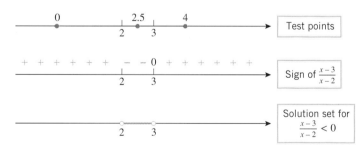

Figure A.10

**EXERCISE SET A**

**1.** Among the terms *integer*, *rational*, and *irrational*, which ones apply to the given number?
(a) $-\frac{3}{4}$  (b) 0  (c) $\frac{24}{8}$
(d) 0.25  (e) $-\sqrt{16}$  (f) $2^{1/2}$
(g) 0.020202…  (h) 7.000…

**2.** Which of the terms *integer*, *rational*, and *irrational* apply to the given number?
(a) 0.31311311131111…  (b) 0.729999…
(c) 0.376237623762…  (d) $17\frac{4}{5}$

**3.** The repeating decimal 0.137137137… can be expressed as a ratio of integers by writing

$$x = 0.137137137\ldots$$
$$1000x = 137.137137137\ldots$$

and subtracting to obtain $999x = 137$ or $x = \frac{137}{999}$. Use this idea, where needed, to express the following decimals as ratios of integers.
(a) 0.123123123…  (b) 12.7777…
(c) 38.07818181…  (d) 0.4296000…

**4.** Show that the repeating decimal 0.99999... represents the number 1. Since 1.000... is also a decimal representation of 1, this problem shows that a real number can have two different decimal representations. [*Hint:* Use the technique of Exercise 3.]

**5.** The Rhind Papyrus, which is a fragment of Egyptian mathematical writing from about 1650 B.C., is one of the oldest known examples of written mathematics. It is stated in the papyrus that the area $A$ of a circle is related to its diameter $D$ by

$$A = \left(\tfrac{8}{9}D\right)^2$$

(a) What approximation to $\pi$ were the Egyptians using?

(b) Use a calculating utility to determine if this approximation is better or worse than the approximation of $\frac{22}{7}$.

**6.** The following are all famous approximations to $\pi$:

$$\frac{333}{106} \quad \text{Adrian Athoniszoon, c. 1583}$$

$$\frac{355}{113} \quad \text{Tsu Chung-Chi and others}$$

$$\frac{63}{25}\left(\frac{17 + 15\sqrt{5}}{7 + 15\sqrt{5}}\right) \quad \text{Ramanujan}$$

$$\frac{22}{7} \quad \text{Archimedes}$$

$$\frac{223}{71} \quad \text{Archimedes}$$

(a) Use a calculating utility to order these approximations according to size.

(b) Which of these approximations is closest to but larger than $\pi$?

(c) Which of these approximations is closest to but smaller than $\pi$?

(d) Which of these approximations is most accurate?

**7.** In each line of the table in the accompanying figure, check the blocks, if any, that describe a valid relationship between the real numbers $a$ and $b$. The first line is already completed as an illustration.

| $a$ | $b$ | $a < b$ | $a \le b$ | $a > b$ | $a \ge b$ | $a = b$ |
|---|---|---|---|---|---|---|
| 1 | 6 | ✓ | ✓ | | | |
| 6 | 1 | | | | | |
| −3 | 5 | | | | | |
| 5 | −3 | | | | | |
| −4 | −4 | | | | | |
| 0.25 | $\frac{1}{3}$ | | | | | |
| $-\frac{1}{4}$ | $-\frac{3}{4}$ | | | | | |

Figure Ex-7

**8.** In each line of the table in the accompanying figure, check the blocks, if any, that describe a valid relationship between the real numbers $a$, $b$, and $c$.

| $a$ | $b$ | $c$ | $a < b < c$ | $a \le b \le c$ | $a < b \le c$ | $a \le b < c$ |
|---|---|---|---|---|---|---|
| −1 | 0 | 2 | | | | |
| 2 | 4 | −3 | | | | |
| $\frac{1}{2}$ | $\frac{1}{2}$ | $\frac{3}{4}$ | | | | |
| −5 | −5 | −5 | | | | |
| 0.75 | 1.25 | 1.25 | | | | |

Figure Ex 8

**9.** Which of the following are always correct if $a \le b$?

(a) $a - 3 \le b - 3$      (b) $-a \le -b$

(c) $3 - a \le 3 - b$      (d) $6a \le 6b$

(e) $a^2 \le ab$      (f) $a^3 \le a^2 b$

**10.** Which of the following are always correct if $a \le b$ and $c \le d$?

(a) $a + 2c \le b + 2d$      (b) $a - 2c \le b - 2d$

(c) $a - 2c \ge b - 2d$

**11.** For what values of $a$ are the following inequalities valid?

(a) $a \le a$      (b) $a < a$

**12.** If $a \le b$ and $b \le a$, what can you say about $a$ and $b$?

**13.** (a) If $a < b$ is true, does it follow that $a \le b$ must also be true?

(b) If $a \le b$ is true, does it follow that $a < b$ must also be true?

**14.** In each part, list the elements in the set.

(a) $\{x : x^2 - 5x = 0\}$

(b) $\{x : x$ is an integer satisfying $-2 < x < 3\}$

**15.** In each part, express the set in the notation $\{x : \underline{\hspace{1cm}}\}$.

(a) $\{1, 3, 5, 7, 9, \ldots\}$

(b) the set of even integers

(c) the set of irrational numbers

(d) $\{7, 8, 9, 10\}$

**16.** Let $A = \{1, 2, 3\}$. Which of the following sets are equal to $A$?

(a) $\{0, 1, 2, 3\}$      (b) $\{3, 2, 1\}$

(c) $\{x : (x - 3)(x^2 - 3x + 2) = 0\}$

**17.** In the accompanying figure, let

$S = $ the set of points inside the square

$T = $ the set of points inside the triangle

$C = $ the set of points inside the circle

and let $a$, $b$, and $c$ be the points shown. Answer the following as true or false.

(a) $T \subset C$

(b) $T \subset S$

(c) $a \notin T$

(d) $a \notin S$

(e) $b \in T$ and $b \in C$

(f) $a \in C$ or $a \in T$

(g) $c \in T$ and $c \notin C$

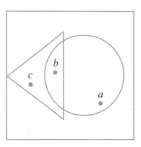

Figure Ex-17

18. List all subsets of
   (a) $\{a_1, a_2, a_3\}$            (b) $\varnothing$.

19. In each part, sketch on a coordinate line all values of $x$ that satisfy the stated condition.
   (a) $x \leq 4$           (b) $x \geq -3$          (c) $-1 \leq x \leq 7$
   (d) $x^2 = 9$           (e) $x^2 \leq 9$          (f) $x^2 \geq 9$

20. In parts (a)–(d), sketch on a coordinate line all values of $x$, if any, that satisfy the stated conditions.
   (a) $x > 4$      and      $x \leq 8$
   (b) $x \leq 2$      or      $x \geq 5$
   (c) $x > -2$      and      $x \geq 3$
   (d) $x \leq 5$      and      $x > 7$

21. Express in interval notation.
   (a) $\{x : x^2 \leq 4\}$            (b) $\{x : x^2 > 4\}$

22. In each part, sketch the set on a coordinate line.
   (a) $[-3, 2] \cup [1, 4]$          (b) $[4, 6] \cup [8, 11]$
   (c) $(-4, 0) \cup (-5, 1)$         (d) $[2, 4) \cup (4, 7)$
   (e) $(-2, 4) \cap (0, 5]$          (f) $[1, 2.3) \cup (1.4, \sqrt{2})$
   (g) $(-\infty, -1) \cup (-3, +\infty)$   (h) $(-\infty, 5) \cap [0, +\infty)$

In Exercises 23–44, solve the inequality and sketch the solution on a coordinate line.

23. $3x - 2 < 8$

24. $\frac{1}{5}x + 6 \geq 14$

25. $4 + 5x \leq 3x - 7$

26. $2x - 1 > 11x + 9$

27. $3 \leq 4 - 2x < 7$

28. $-2 \geq 3 - 8x \geq -11$

29. $\frac{x}{x - 3} < 4$

30. $\frac{x}{8 - x} \geq -2$

31. $\frac{3x + 1}{x - 2} < 1$

32. $\frac{\frac{1}{2}x - 3}{4 + x} > 1$

33. $\frac{4}{2 - x} \leq 1$

34. $\frac{3}{x - 5} \leq 2$

35. $x^2 > 9$

36. $x^2 \leq 5$

37. $(x - 4)(x + 2) > 0$

38. $(x - 3)(x + 4) < 0$

39. $x^2 - 9x + 20 \leq 0$

40. $2 - 3x + x^2 \geq 0$

41. $\frac{2}{x} < \frac{3}{x - 4}$

42. $\frac{1}{x + 1} \geq \frac{3}{x - 2}$

43. $x^3 - x^2 - x - 2 > 0$

44. $x^3 - 3x + 2 \leq 0$

In Exercises 45 and 46, find all values of $x$ for which the given expression yields a real number.

45. $\sqrt{x^2 + x - 6}$

46. $\sqrt{\dfrac{x + 2}{x - 1}}$

47. Fahrenheit and Celsius temperatures are related by the formula $C = \frac{5}{9}(F - 32)$. If the temperature in degrees Celsius ranges over the interval $25 \leq C \leq 40$ on a certain day, what is the temperature range in degrees Fahrenheit that day?

48. Every integer is either even or odd. The even integers are those that are divisible by 2, so $n$ is even if and only if $n = 2k$ for some integer $k$. Each odd integer is one unit larger than an even integer, so $n$ is odd if and only if $n = 2k + 1$ for some integer $k$. Show:
   (a) If $n$ is even, then so is $n^2$
   (b) If $n$ is odd, then so is $n^2$.

49. Prove the following results about sums of rational and irrational numbers:
   (a) rational + rational = rational
   (b) rational + irrational = irrational.

50. Prove the following results about products of rational and irrational numbers:
   (a) rational · rational = rational
   (b) rational · irrational = irrational    (provided the rational factor is nonzero).

51. Show that the sum or product of two irrational numbers can be rational or irrational.

52. Classify the following as rational or irrational and justify your conclusion.
   (a) $3 + \pi$              (b) $\frac{3}{4}\sqrt{2}$
   (c) $\sqrt{8}\sqrt{2}$          (d) $\sqrt{\pi}$
   (See Exercises 49 and 50.)

53. Prove: The average of two rational numbers is a rational number, but the average of two irrational numbers can be rational or irrational.

54. Can a rational number satisfy $10^x = 3$?

55. Solve: $8x^3 - 4x^2 - 2x + 1 < 0$.

56. Solve: $12x^3 - 20x^2 \geq -11x + 2$.

57. Prove: If $a, b, c,$ and $d$ are positive numbers such that $a < b$ and $c < d$, then $ac < bd$. (This result gives conditions under which inequalities can be "multiplied together.")

58. Is the number represented by the decimal

   $0.101001000100001000001 \ldots$

   rational or irrational? Explain your reasoning.

# APPENDIX B

## Absolute Value

**B.1** DEFINITION. The *absolute value* or *magnitude* of a real number $a$ is denoted by $|a|$ and is defined by

$$|a| = \begin{cases} a & \text{if} \quad a \geq 0 \\ -a & \text{if} \quad a < 0 \end{cases}$$

### Example 1

$$|5| = 5 \qquad \left|-\tfrac{4}{7}\right| = -\left(-\tfrac{4}{7}\right) = \tfrac{4}{7} \qquad |0| = 0 \qquad\qquad \blacktriangleleft$$

$\boxed{\text{Since } 5 > 0}$ $\boxed{\text{Since } -\tfrac{4}{7} < 0}$ $\boxed{\text{Since } 0 \geq 0}$

Note that the effect of taking the absolute value of a number is to strip away the minus sign if the number is negative and to leave the number unchanged if it is nonnegative.

### Example 2

Solve $|x - 3| = 4$.

*Solution.* Depending on whether $x - 3$ is positive or negative, the equation $|x - 3| = 4$ can be written as

$$x - 3 = 4 \quad \text{or} \quad x - 3 = -4$$

Solving these two equations gives $x = 7$ and $x = -1$. $\qquad \blacktriangleleft$

### Example 3

Solve $|3x - 2| = |5x + 4|$.

*Solution.* Because two numbers with the same absolute value are either equal or differ in sign, the given equation will be satisfied if either

$$3x - 2 = 5x + 4 \quad \text{or} \quad 3x - 2 = -(5x + 4)$$

Solving the first equation yields $x = -3$ and solving the second yields $x = -\tfrac{1}{4}$; thus, the given equation has the solutions $x = -3$ and $x = -\tfrac{1}{4}$. $\qquad \blacktriangleleft$

Recall from algebra that a number is called a *square root* of $a$ if its square is $a$. Recall also that every positive real number has two square roots, one positive and one negative; the positive square root is denoted by $\sqrt{a}$ and the negative square root by $-\sqrt{a}$. For example, the positive square root of 9 is $\sqrt{9} = 3$, and the negative square root of 9 is $-\sqrt{9} = -3$.

REMARK.    Readers who may have been taught to write $\sqrt{9} = \pm 3$ should stop doing so, since it is incorrect.

It is a common error to write $\sqrt{a^2} = a$. Although this equality is correct when $a$ is nonnegative, it is false for negative $a$. For example, if $a = -4$, then

$$\sqrt{a^2} = \sqrt{(-4)^2} = \sqrt{16} = 4 \neq a$$

A result that is correct for all $a$ is given in the following theorem.

---

**B.2**    THEOREM.    *For any real number $a$,*

$$\sqrt{a^2} = |a|$$

---

*Proof.*    Since $a^2 = (+a)^2 = (-a)^2$, the numbers $+a$ and $-a$ are square roots of $a^2$. If $a \geq 0$, then $+a$ is the nonnegative square root of $a^2$, and if $a < 0$, then $-a$ is the nonnegative square root of $a^2$. Since $\sqrt{a^2}$ denotes the nonnegative square root of $a^2$, it follows that

$$\sqrt{a^2} = +a \quad \text{if} \quad a \geq 0$$
$$\sqrt{a^2} = -a \quad \text{if} \quad a < 0$$

That is, $\sqrt{a^2} = |a|$.    ■

**PROPERTIES OF ABSOLUTE VALUE**

---

**B.3**    THEOREM.    *If $a$ and $b$ are real numbers, then*

(a)    $|-a| = |a|$        A number and its negative have the same absolute value.
(b)    $|ab| = |a||b|$      The absolute value of a product is the product of the absolute values.
(c)    $|a/b| = |a|/|b|$    The absolute value of a ratio is the ratio of the absolute values.

---

We will prove parts $(a)$ and $(b)$ only.

*Proof $(a)$.*    From Theorem B.2,

$$|-a| = \sqrt{(-a)^2} = \sqrt{a^2} = |a|$$

*Proof $(b)$.*    From Theorem B.2 and a basic property of square roots,

$$|ab| = \sqrt{(ab)^2} = \sqrt{a^2 b^2} = \sqrt{a^2}\sqrt{b^2} = |a||b|$$    ■

REMARK.    In part $(c)$ of Theorem B.3 we did not explicitly state that $b \neq 0$, but this must be so since division by zero is not allowed. Whenever divisions occur in this text, it will be assumed that the denominator is not zero, even if we do not mention it explicitly.

The result in part $(b)$ of Theorem B.3 can be extended to three or more factors. More precisely, for any $n$ real numbers, $a_1, a_2, \ldots, a_n$, it follows that

$$|a_1 a_2 \cdots a_n| = |a_1||a_2| \cdots |a_n| \tag{1}$$

In the special case where $a_1, a_2, \ldots, a_n$ have the same value, $a$, it follows from (1) that

$$|a^n| = |a|^n \tag{2}$$

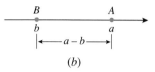

(a)

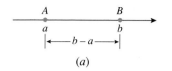

(b)

Figure B.1

The notion of absolute value arises naturally in distance problems. For example, suppose that $A$ and $B$ are points on a coordinate line that have coordinates $a$ and $b$, respectively. Depending on the relative positions of the points, the distance $d$ between them will be $b - a$ or $a - b$ (Figure B.1). In either case, the distance can be written as $d = |b - a|$, so we have the following result.

> **B.4** THEOREM (*Distance Formula*). *If $A$ and $B$ are points on a coordinate line with coordinates $a$ and $b$, respectively, then the distance $d$ between $A$ and $B$ is $d = |b - a|$.*

This theorem provides useful geometric interpretations of some common mathematical expressions:

| EXPRESSION | GEOMETRIC INTERPRETATION ON A COORDINATE LINE |
|---|---|
| $|x - a|$ | The distance between $x$ and $a$ |
| $|x + a|$ | The distance between $x$ and $-a$ (since $|x + a| = |x - (-a)|$) |
| $|x|$ | The distance between $x$ and the origin (since $|x| = |x - 0|$) |

### INEQUALITIES WITH ABSOLUTE VALUES

Inequalities of the form $|x - a| < k$ and $|x - a| > k$ arise so often that we have summarized the key facts about them in Table 1.

**Table 1**

| INEQUALITY ($k > 0$) | GEOMETRIC INTERPRETATION | FIGURE | ALTERNATIVE FORMS OF THE INEQUALITY |
|---|---|---|---|
| $|x - a| < k$ | $x$ is within $k$ units of $a$. | 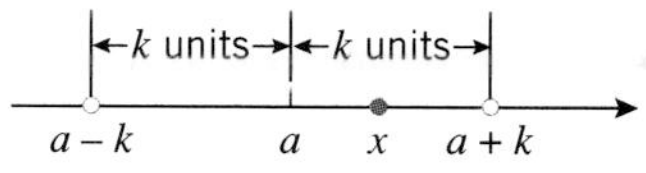 $\leftarrow k$ units$\rightarrow$$\leftarrow k$ units$\rightarrow$ $a - k$ $\quad a \quad x \quad a + k$ | $-k < x - a < k$ $a - k < x < a + k$ |
| $|x - a| > k$ | $x$ is more than $k$ units away from $a$. | $\leftarrow k$ units$\rightarrow$$\leftarrow k$ units$\rightarrow$ $a - k$ $\quad a \quad a + k \quad x$ | $x - a < -k$ or $x - a > k$ $x < a - k$ or $x > a + k$ |

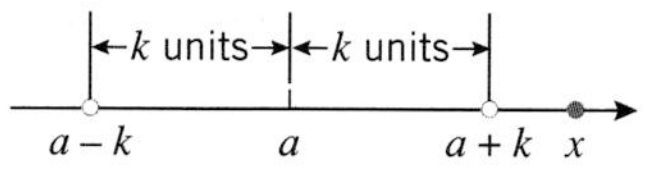

REMARK. The statements in this table remain true if $<$ is replaced by $\leq$ and $>$ by $\geq$, and if the open dots are replaced by closed dots in the illustrations.

### Example 4

Solve

(a) $|x - 3| < 4$    (b) $|x + 4| \geq 2$    (c) $\dfrac{1}{|2x - 3|} > 5$

*Solution (a).* The inequality $|x - 3| < 4$ can be rewritten as

$$-4 < x - 3 < 4$$

Adding 3 throughout yields

$$-1 < x < 7$$

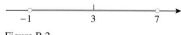

Figure B.2

which can be written in interval notation as $(-1, 7)$. Observe that this solution set consists of all $x$ that are within 4 units of 3 on a number line (Figure B.2), which is consistent with Table 1.

*Solution (b).* The inequality $|x + 4| \geq 2$ will be satisfied if

$$x + 4 \leq -2 \quad \text{or} \quad x + 4 \geq 2$$

Solving for $x$ in the two cases yields

$$x \leq -6 \quad \text{or} \quad x \geq -2$$

which can be expressed in interval notation as

$$(-\infty, -6] \cup [-2, +\infty)$$

Observe that the solution set consists of all $x$ that are at least 2 units away from $-4$ on a number line (Figure B.3), which is consistent with Table 1 and the remark that follows it.

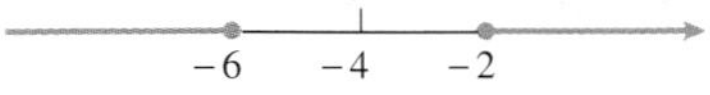

Figure B.3

*Solution (c).* Observe first that $x = \frac{3}{2}$ results in a division by zero, so this value of $x$ cannot be in the solution set. Putting this aside for the moment, we will begin by taking reciprocals on both sides and reversing the sense of the inequality in accordance with Theorem A.1(*d*) of Appendix A; then we will use Theorem B.3 to rewrite the inequality $1/|2x - 3| > 5$ in a more familiar form:

$$|2x - 3| < \tfrac{1}{5}$$

$$|2||x - \tfrac{3}{2}| < \tfrac{1}{5} \qquad \boxed{\text{Theorem B.3(}b\text{)}}$$

$$|x - \tfrac{3}{2}| < \tfrac{1}{10} \qquad \boxed{\text{We multiplied both sides by } 1/|2| = 1/2.}$$

$$-\tfrac{1}{10} < x - \tfrac{3}{2} < \tfrac{1}{10} \qquad \boxed{\text{Table 1}}$$

$$\tfrac{7}{5} < x < \tfrac{8}{5} \qquad \boxed{\text{We added 3/2 throughout.}}$$

As noted earlier, we must eliminate $x = \frac{3}{2}$ to avoid a division by zero, so the solution set is

$$\tfrac{7}{5} < x < \tfrac{3}{2} \quad \text{or} \quad \tfrac{3}{2} < x < \tfrac{8}{5}$$

which can be expressed in interval notation as $\left(\frac{7}{5}, \frac{3}{2}\right) \cup \left(\frac{3}{2}, \frac{8}{5}\right)$. (See Figure B.4.)  ◄

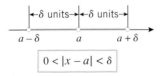

Figure B.4

## AN INEQUALITY FROM CALCULUS

One of the most important inequalities in calculus is

$$0 < |x - a| < \delta \tag{3}$$

where $\delta$ (Greek "delta") is a positive real number. This is equivalent to the two inequalities

$$0 < |x - a| \quad \text{and} \quad |x - a| < \delta$$

the first of which is satisfied by all $x$ except $x = a$, and the second of which is satisfied by all $x$ that are within $\delta$ units of $a$ on a coordinate line. Combining these two restrictions, we conclude that the solution set of (3) consists of all $x$ in the interval $(a - \delta, a + \delta)$ except $x = a$ (Figure B.5). Stated another way, the solution set of (3) is

$$(a - \delta, a) \cup (a, a + \delta) \tag{4}$$

It is *not* generally true that $|a + b| = |a| + |b|$. For example, if $a = 1$ and $b = -1$, then $|a + b| = 0$, whereas $|a| + |b| = 2$. It is true, however, that *the absolute value of a sum is always less than or equal to the sum of the absolute values.* This is the content of the following useful theorem, called the **triangle inequality**.

Figure B.5

## THE TRIANGLE INEQUALITY

---

**B.5**    THEOREM (*Triangle Inequality*).    *If a and b are any real numbers, then*

$$|a + b| \leq |a| + |b| \tag{5}$$

---

*Proof.* Observe first that $a$ satisfies the inequality

$$-|a| \leq a \leq |a|$$

because either $a = |a|$ or $a = -|a|$, depending on the sign of $a$. The corresponding inequality for $b$ is

$$-|b| \le b \le |b|$$

Adding the two inequalities we obtain

$$-(|a| + |b|) \le a + b \le (|a| + |b|) \tag{6}$$

Let us now consider the cases $a + b \ge 0$ and $a + b < 0$ separately. In the first case, $a + b = |a + b|$, so the right-hand inequality in (6) yields the triangle inequality (5). In the second case, $a + b = -|a + b|$, so the left-hand inequality in (6) can be written as

$$-(|a| + |b|) \le -|a + b|$$

which yields the triangle inequality (5) on multiplying by $-1$. ∎

REMARK. The name "triangle inequality" arises from a geometric interpretation of the inequality that can be made when $a$ and $b$ are complex numbers. A more detailed explanation is outside the scope of this text.

## EXERCISE SET B

1. Compute $|x|$ if
   (a) $x = 7$
   (b) $x = -\sqrt{2}$
   (c) $x = k^2$
   (d) $x = -k^2$.

2. Rewrite $\sqrt{(x-6)^2}$ without using a square root or absolute value sign.

In Exercises 3–10, find all values of $x$ for which the given statement is true.

3. $|x - 3| = 3 - x$
4. $|x + 2| = x + 2$
5. $|x^2 + 9| = x^2 + 9$
6. $|x^2 + 5x| = x^2 + 5x$
7. $|3x^2 + 2x| = x|3x + 2|$
8. $|6 - 2x| = 2|x - 3|$
9. $\sqrt{(x+5)^2} = x + 5$
10. $\sqrt{(3x-2)^2} = 2 - 3x$

11. Verify $\sqrt{a^2} = |a|$ for $a = 7$ and $a = -7$.

12. Verify the inequalities $-|a| \le a \le |a|$ for $a = 2$ and for $a = -5$.

13. Let $A$ and $B$ be points with coordinates $a$ and $b$. In each part find the distance between $A$ and $B$.
    (a) $a = 9$, $b = 7$
    (b) $a = 2$, $b = 3$
    (c) $a = -8$, $b = 6$
    (d) $a = \sqrt{2}$, $b = -3$
    (e) $a = -11$, $b = -4$
    (f) $a = 0$, $b = -5$

14. Is the equality $\sqrt{a^4} = a^2$ valid for all values of $a$? Explain.

15. Let $A$ and $B$ be points with coordinates $a$ and $b$. In each part, use the given information to find $b$.
    (a) $a = -3$, $B$ is to the left of $A$, and $|b - a| = 6$.
    (b) $a = -2$, $B$ is to the right of $A$, and $|b - a| = 9$.
    (c) $a = 5$, $|b - a| = 7$, and $b > 0$.

16. Let $E$ and $F$ be points with coordinates $e$ and $f$. In each part, determine whether $E$ is to the left or to the right of $F$ on a coordinate line.
    (a) $f - e = 4$
    (b) $e - f = 4$
    (c) $f - e = -6$
    (d) $e - f = -7$

In Exercises 17–24, solve for $x$.

17. $|6x - 2| = 7$
18. $|3 + 2x| = 11$
19. $|6x - 7| = |3 + 2x|$
20. $|4x + 5| = |8x - 3|$
21. $|9x| - 11 = x$
22. $2x - 7 = |x + 1|$
23. $\left|\dfrac{x+5}{2-x}\right| = 6$
24. $\left|\dfrac{x-3}{x+4}\right| = 5$

In Exercises 25–36, solve for $x$ and express the solution in terms of intervals.

25. $|x + 6| < 3$
26. $|7 - x| \le 5$
27. $|2x - 3| \le 6$
28. $|3x + 1| < 4$
29. $|x + 2| > 1$
30. $|\frac{1}{2}x - 1| \ge 2$
31. $|5 - 2x| \ge 4$
32. $|7x + 1| > 3$
33. $\dfrac{1}{|x-1|} < 2$
34. $\dfrac{1}{|3x+1|} \ge 5$
35. $\dfrac{3}{|2x-1|} \ge 4$
36. $\dfrac{2}{|x+3|} < 1$

37. For which values of $x$ is $\sqrt{(x^2 - 5x + 6)^2} = x^2 - 5x + 6$?

38. Solve $3 \le |x - 2| \le 7$ for $x$.

39. Solve $|x - 3|^2 - 4|x - 3| = 12$ for $x$. [Hint: Begin by letting $u = |x - 3|$.]

40. Verify the triangle inequality $|a + b| \le |a| + |b|$ (Theorem B.5) for
    (a) $a = 3$, $b = 4$
    (b) $a = -2$, $b = 6$
    (c) $a = -7$, $b = -8$
    (d) $a = -4$, $b = 4$.

41. Prove: $|a - b| \le |a| + |b|$.

42. Prove: $|a| - |b| \le |a - b|$.

43. Prove: $\big| |a| - |b| \big| \le |a - b|$. [Hint: Use Exercise 42.]

# APPENDIX C

## Coordinate Planes and Lines

**RECTANGULAR COORDINATE SYSTEMS**

Just as points on a coordinate line can be associated with real numbers, so points in a plane can be associated with pairs of real numbers by introducing a ***rectangular coordinate system*** (also called a ***Cartesian coordinate system***). A rectangular coordinate system consists of two perpendicular coordinate lines, called ***coordinate axes***, that intersect at their origins. Usually, but not always, one axis is horizontal with its positive direction to the right, and the other is vertical with its positive direction up. The intersection of the axes is called the ***origin*** of the coordinate system.

It is common to call the horizontal axis the ***x-axis*** and the vertical axis the ***y-axis***, in which case the plane and the axes together are referred to as the ***xy-plane*** (Figure C.1). Although labeling the axes with the letters $x$ and $y$ is common, other letters may be more appropriate in specific applications. Figure C.2 shows a $uv$-plane and a $ts$-plane—the first letter in the name of the plane always refers to the horizontal axis and the second to the vertical axis.

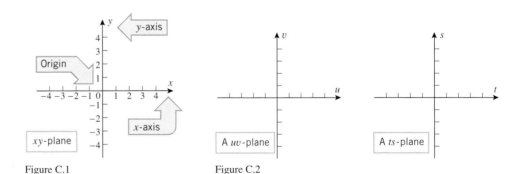

Figure C.1          Figure C.2

**COORDINATES**

Every point $P$ in a coordinate plane can be associated with a unique ordered pair of real numbers by drawing two lines through $P$, one perpendicular to the $x$-axis and the other perpendicular to the $y$-axis (Figure C.3). If the first line intersects the $x$-axis at the point with coordinate $a$ and the second line intersects the $y$-axis at the point with coordinate $b$, then we associate the ordered pair of real numbers $(a, b)$ with the point $P$. The number $a$ is called the ***x-coordinate*** or ***abscissa*** of $P$ and the number $b$ is called the ***y-coordinate*** or ***ordinate*** of $P$. We will say that $P$ has ***coordinates*** $(a, b)$ and write $P(a, b)$ when we want to emphasize that the coordinates of $P$ are $(a, b)$. We can also reverse the above procedure and find the point $P$ associated with the coordinates $(a, b)$ by locating the intersection of the dashed

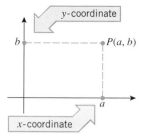

Figure C.3

lines in Figure C.3. Because of this one-to-one correspondence between coordinates and points, we will sometimes blur the distinction between points and ordered pairs of numbers by talking about the *point* $(a, b)$.

REMARK.    Recall that the symbol $(a, b)$ also denotes the open interval between $a$ and $b$; the appropriate interpretation will usually be clear from the context.

In a rectangular coordinate system the coordinate axes divide the plane into four regions called **quadrants**. These are numbered counterclockwise with roman numerals as shown in Figure C.4. As indicated in that figure, it is easy to determine the quadrant in which a given point lies from the signs of its coordinates: a point with two positive coordinates $(+, +)$ lies in Quadrant I, a point with a negative $x$-coordinate and a positive $y$-coordinate $(-, +)$ lies in Quadrant II, and so forth. Points with a zero $x$-coordinate lie on the $y$-axis and points with a zero $y$-coordinate lie on the $x$-axis.

To **plot** a point $P(a, b)$ means to locate the point with coordinates $(a, b)$ in a coordinate plane. For example, in Figure C.5 we have plotted the points

$$P(2, 5), \quad Q(-4, 3), \quad R(-5, -2), \quad \text{and} \quad S(4, -3)$$

Observe how the signs of the coordinates identify the quadrants in which the points lie.

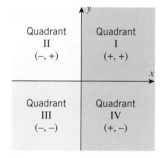

Figure C.4

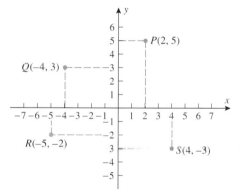

Figure C.5

**GRAPHS**

The correspondence between points in a plane and ordered pairs of real numbers makes it possible to visualize algebraic equations as geometric curves, and, conversely, to represent geometric curves by algebraic equations. To understand how this is done, suppose that we have an $xy$-coordinate system and an equation involving two variables $x$ and $y$, say

$$6x - 4y = 10, \quad y = \sqrt{x}, \quad x = y^3 + 1, \quad \text{or} \quad x^2 + y^2 = 1$$

We define a **solution** of such an equation to be any ordered pair of real numbers $(a, b)$ whose coordinates satisfy the equation when we substitute $x = a$ and $y = b$. For example, the ordered pair $(3, 2)$ is a solution of the equation $6x - 4y = 10$, since the equation is satisfied by $x = 3$ and $y = 2$ (verify). However, the ordered pair $(2, 0)$ is not a solution of this equation, since the equation is not satisfied by $x = 2$ and $y = 0$ (verify).

The following definition makes the association between equations in $x$ and $y$ and curves in the $xy$-plane.

> **C.1**    DEFINITION.    The set of all solutions of an equation in $x$ and $y$ is called the **solution set** of the equation, and the set of all points in the $xy$-plane whose coordinates are members of the solution set is called the **graph** of the equation.

One of the main themes in calculus is to identify the exact shape of a graph. Point plotting is one approach to obtaining a graph, but this method has limitations, as discussed in the following example.

### Example 1

Sketch the graph of $y = x^2$.

*Solution.*   The solution set of the equation has infinitely many members, since we can substitute an arbitrary value for $x$ into the right side of $y = x^2$ and compute the associated $y$ to obtain a point $(x, y)$ in the solution set. The fact that the solution set has infinitely many members means that we cannot obtain the *entire* graph of $y = x^2$ by point plotting. However, we can obtain an *approximation* to the graph by plotting some sample members of the solution set and connecting them with a smooth curve, as in Figure C.6. The problem with this method is that we cannot be sure how the graph behaves *between* the plotted points. For example, the curves in Figure C.7 also pass through the plotted points and hence are legitimate candidates for the graph in the absence of additional information. Moreover, even if we use a graphing calculator or a computer program to generate the graph, as in Figure C.8, we have the same problem because graphing technology uses point-plotting algorithms to generate graphs. Indeed, in Section 1.3 of the text we see examples where graphing technology can be fooled into producing grossly inaccurate graphs.   ◀

| $x$ | $y = x^2$ | $(x, y)$ |
|---|---|---|
| 0 | 0 | $(0, 0)$ |
| 1 | 1 | $(1, 1)$ |
| 2 | 4 | $(2, 4)$ |
| 3 | 9 | $(3, 9)$ |
| −1 | 1 | $(−1, 1)$ |
| −2 | 4 | $(−2, 4)$ |
| −3 | 9 | $(−3, 9)$ |

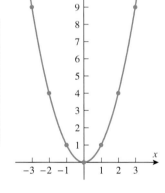

Figure C.6

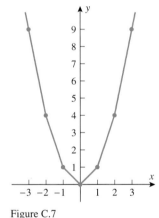

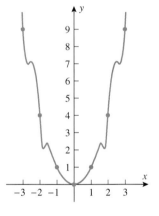

Figure C.7

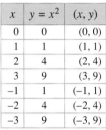

Figure C.8

In spite of its limitations, point plotting by hand or with the help of graphing technology can be useful, so here are two more examples.

### Example 2

Sketch the graph of $y = \sqrt{x}$.

*Solution.*   If $x < 0$, then $\sqrt{x}$ is an imaginary number. Thus, we can only plot points for which $x \geq 0$, since points in the $xy$-plane have real coordinates. Figure C.9 shows the graph obtained by point plotting and a graph obtained with a graphing calculator.   ◀

### Example 3

Sketch the graph of $y^2 - 2y - x = 0$.

*Solution.*   To calculate coordinates of points on the graph of an equation in $x$ and $y$, it is desirable to have $y$ expressed in terms of $x$ or of $x$ in terms of $y$. In this case it is easier to

| $x$ | $y = \sqrt{x}$ | $(x, y)$ |
|---|---|---|
| 0 | 0 | $(0, 0)$ |
| 1 | 1 | $(1, 1)$ |
| 2 | $\sqrt{2}$ | $(2, \sqrt{2}) \approx (2, 1.4)$ |
| 3 | $\sqrt{3}$ | $(3, \sqrt{3}) \approx (3, 1.7)$ |
| 4 | 2 | $(4, 2)$ |

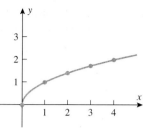
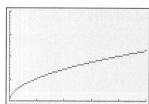

Figure C.9

express $x$ in terms of $y$, so we rewrite the equation as

$$x = y^2 - 2y$$

Members of the solution set can be obtained from this equation by substituting arbitrary values for $y$ in the right side and computing the associated values of $x$ (Figure C.10).  ◀

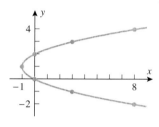

| $y$ | $x = y^2 - 2y$ | $(x, y)$ |
|---|---|---|
| $-2$ | 8 | $(8, -2)$ |
| $-1$ | 3 | $(3, -1)$ |
| 0 | 0 | $(0, 0)$ |
| 1 | $-1$ | $(-1, 1)$ |
| 2 | 0 | $(0, 2)$ |
| 3 | 3 | $(3, 3)$ |
| 4 | 8 | $(8, 4)$ |

Figure C.10

REMARK.  Most graphing calculators and computer graphing programs require that $y$ be expressed in terms of $x$ to generate a graph in the $xy$-plane. In Section 1.7 we discuss a method for circumventing this restriction.

**Example 4**

Sketch the graph of $y = 1/x$.

*Solution.*  Because $1/x$ is undefined at $x = 0$, we can only plot points for which $x \neq 0$. This forces a break, called a *discontinuity*, in the graph at $x = 0$ (Figure C.11).  ◀

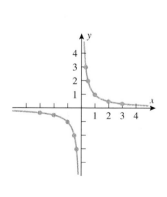

| $x$ | $y = 1/x$ | $(x, y)$ |
|---|---|---|
| $\frac{1}{3}$ | 3 | $\left(\frac{1}{3}, 3\right)$ |
| $\frac{1}{2}$ | 2 | $\left(\frac{1}{2}, 2\right)$ |
| 1 | 1 | $(1, 1)$ |
| 2 | $\frac{1}{2}$ | $\left(2, \frac{1}{2}\right)$ |
| 3 | $\frac{1}{3}$ | $\left(3, \frac{1}{3}\right)$ |
| $-\frac{1}{3}$ | $-3$ | $\left(-\frac{1}{3}, -3\right)$ |
| $-\frac{1}{2}$ | $-2$ | $\left(-\frac{1}{2}, -2\right)$ |
| $-1$ | $-1$ | $(-1, -1)$ |
| $-2$ | $-\frac{1}{2}$ | $\left(-2, -\frac{1}{2}\right)$ |
| $-3$ | $-\frac{1}{3}$ | $\left(-3, -\frac{1}{3}\right)$ |

Figure C.11

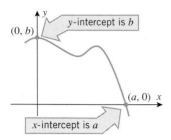

Figure C.12

Points where a graph intersects the coordinate axes are of special interest in many problems. As illustrated in Figure C.12, intersections of a graph with the $x$-axis have the form $(a, 0)$ and intersections with the $y$-axis have the form $(0, b)$. The number $a$ is called an **$x$-intercept** of the graph and the number $b$ a **$y$-intercept**.

## Example 5

Find all intercepts of

(a) $3x + 2y = 6$     (b) $x = y^2 - 2y$     (c) $y = 1/x$

**Solution (a).** To find the $x$-intercepts we set $y = 0$ and solve for $x$:

$$3x = 6 \quad \text{or} \quad x = 2$$

To find the $y$-intercepts we set $x = 0$ and solve for $y$:

$$2y = 6 \quad \text{or} \quad y = 3$$

As we will see later, the graph of $3x + 2y = 6$ is the line shown in Figure C.13.

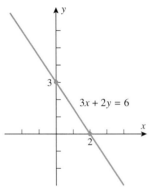

Figure C.13

**Solution (b).** To find the $x$-intercepts, set $y = 0$ and solve for $x$:

$$x = 0$$

Thus, $x = 0$ is the only $x$-intercept. To find the $y$-intercepts, set $x = 0$ and solve for $y$:

$$y^2 - 2y = 0$$
$$y(y - 2) = 0$$

So the $y$-intercepts are $y = 0$ and $y = 2$. The graph is shown in Figure C.10.

**Solution (c).** To find the $x$-intercepts, set $y = 0$:

$$\frac{1}{x} = 0$$

This equation has no solutions (why?), so there are no $x$-intercepts. To find $y$-intercepts we would set $x = 0$ and solve for $y$. But, substituting $x = 0$ leads to a division by zero, which is not allowed, so there are no $y$-intercepts either. The graph of the equation is shown in Figure C.11.   ◄

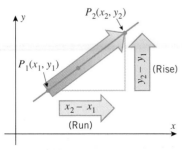

Figure C.14

To obtain equations of lines we will first need to discuss the concept of *slope*, which is a numerical measure of the "steepness" of a line.

Consider a particle moving left to right along a *nonvertical* line from a point $P_1(x_1, y_1)$ to a point $P_2(x_2, y_2)$. As shown in Figure C.14, the particle moves $y_2 - y_1$ units in the $y$-direction as it travels $x_2 - x_1$ units in the positive $x$-direction. The vertical change $y_2 - y_1$ is called the **rise**, and the horizontal change $x_2 - x_1$ the **run**. The ratio of the rise over the run can be used to measure the steepness of the line, which leads us to the following definition.

**C.2**   DEFINITION.   If $P_1(x_1, y_1)$ and $P_2(x_2, y_2)$ are points on a nonvertical line, then the **slope** $m$ of the line is defined by

$$m = \frac{\text{rise}}{\text{run}} = \frac{y_2 - y_1}{x_2 - x_1} \tag{1}$$

REMARK.   Observe that this definition does not apply to vertical lines. For such lines we have $x_2 = x_1$ (a zero run), which means that the formula for $m$ involves a division by zero. For this reason, the slope of a vertical line is **undefined**, which is sometimes described informally by stating that a vertical line has **infinite slope**.

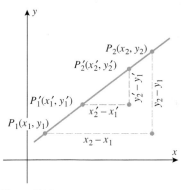

Figure C.15

When calculating the slope of a nonvertical line from Formula (1), it does not matter which two points on the line you use for the calculation, as long as they are distinct. This can be proved using Figure C.15 and similar triangles to show that

$$m = \frac{y_2 - y_1}{x_2 - x_1} = \frac{y_2' - y_1'}{x_2' - x_1'}$$

Moreover, once you choose two points to use for the calculation, it does not matter which one you call $P_1$ and which one you call $P_2$ because reversing the points reverses the sign of both the numerator and denominator of (1) and hence has no effect on the ratio.

### Example 6

In each part find the slope of the line through

(a)   the points $(6, 2)$ and $(9, 8)$
(b)   the points $(2, 9)$ and $(4, 3)$
(c)   the points $(-2, 7)$ and $(5, 7)$.

*Solution.*

(a) $m = \dfrac{8 - 2}{9 - 6} = \dfrac{6}{3} = 2$     (b) $m = \dfrac{3 - 9}{4 - 2} = \dfrac{-6}{2} = -3$     (c) $m = \dfrac{7 - 7}{5 - (-2)} = 0$     ◄

### Example 7

Figure C.16 shows the three lines determined by the points in Example 6 and explains the significance of their slopes.     ◄

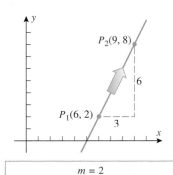

$m = 2$
Traveling left to right, a point on the line rises two units for each unit it moves in the positive $x$-direction.

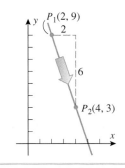

$m = -3$
Traveling left to right, a point on the line falls three units for each unit it moves in the positive $x$-direction.

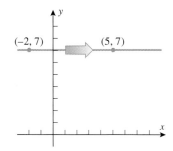

$m = 0$
Traveling left to right, a point on the line neither rises nor falls.

Figure C.16

As illustrated in this example, the slope of a line can be positive, negative, or zero. A positive slope means that the line is inclined upward to the right, a negative slope means that the line is inclined downward to the right, and a zero slope means that the line is horizontal.

An undefined slope means that the line is vertical. Figure C.17 shows various lines through the origin with their slopes.

The following theorem shows how slopes can be used to tell whether two lines are parallel or perpendicular.

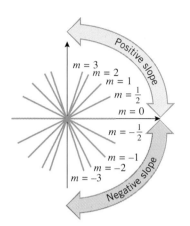

Figure C.17

---

**C.3**   THEOREM.

(a)   *Two nonvertical lines with slopes $m_1$ and $m_2$ are parallel if and only if they have the same slope, that is,*

$$m_1 = m_2$$

(b)   *Two nonvertical lines with slopes $m_1$ and $m_2$ are perpendicular if and only if the product of their slopes is $-1$, that is,*

$$m_1 m_2 = -1$$

*This relationship can also be expressed as $m_1 = -1/m_2$ or $m_2 = -1/m_1$, which states that nonvertical lines are perpendicular if and only if their slopes are negative reciprocals of one another.*

---

A complete proof of this theorem is a little tedious, but it is not hard to motivate the results informally. Let us start with part (a).

Suppose that $L_1$ and $L_2$ are nonvertical parallel lines with slopes $m_1$ and $m_2$, respectively. If the lines are parallel to the $x$-axis, then $m_1 = m_2 = 0$, and we are done. If they are not parallel to the $x$-axis, then both lines intersect the $x$-axis; and for simplicity assume that they are oriented as in Figure C.18a. On each line choose the point whose run relative to the point of intersection with the $x$-axis is 1. On line $L_1$ the corresponding rise will be $m_1$ and on $L_2$ it will be $m_2$. However, because the lines are parallel, the shaded triangles in the figure must be congruent (verify), so $m_1 = m_2$. Conversely, the condition $m_1 = m_2$ can be used to show that the shaded triangles are congruent, from which it follows that the lines make the same angle with the $x$-axis and hence are parallel (verify).

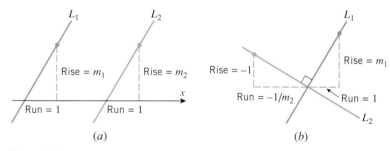

Figure C.18

Now suppose that $L_1$ and $L_2$ are nonvertical perpendicular lines with slopes $m_1$ and $m_2$, respectively; and for simplicity assume that they are oriented as in Figure C.18b. On line $L_1$ choose the point whose run relative to the point of intersection of the lines is 1, in which case the corresponding rise will be $m_1$; and on line $L_2$ choose the point whose rise relative to the point of intersection is $-1$, in which case the corresponding run will be $-1/m_2$. Because the lines are perpendicular, the shaded triangles in the figure must be congruent (verify), and hence the ratios of corresponding sides of the triangles must be equal. Taking into account that for line $L_2$ the vertical side of the triangle has length 1 and the horizontal side has length $-1/m_2$ (since $m_2$ is negative), the congruence of the triangles implies that

$m_1/1 = (-1/m_2)/1$ or $m_1m_2 = -1$. Conversely, the condition $m_1 = -1/m_2$ can be used to show that the shaded triangles are congruent, from which it can be deduced that the lines are perpendicular (verify).

### Example 8

Use slopes to show that the points $A(1, 3)$, $B(3, 7)$, and $C(7, 5)$ are vertices of a right triangle.

*Solution.* We will show that the line through $A$ and $B$ is perpendicular to the line through $B$ and $C$. The slopes of these lines are

$$m_1 = \frac{7 - 3}{3 - 1} = 2 \quad \text{and} \quad m_2 = \frac{5 - 7}{7 - 3} = -\frac{1}{2}$$

Slope of the line<br>through $A$ and $B$      Slope of the line<br>through $B$ and $C$

Since $m_1m_2 = -1$, the line through $A$ and $B$ is perpendicular to the line through $B$ and $C$; thus, $ABC$ is a right triangle (Figure C.19). ◀

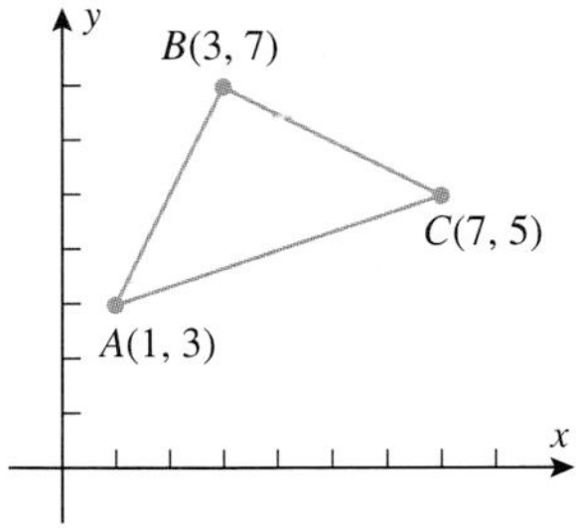

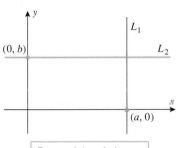

Figure C.19

---

**LINES PARALLEL TO THE COORDINATE AXES**

We now turn to the problem of finding equations of lines that satisfy specified conditions. The simplest cases are lines parallel to the coordinate axes. A line parallel to the $y$-axis intersects the $x$-axis at some point $(a, 0)$. This line consists precisely of those points whose $x$-coordinate is equal to $a$ (Figure C.20). Similarly, a line parallel to the $x$-axis intersects the $y$-axis at some point $(0, b)$. This line consists precisely of those points whose $y$-coordinate is equal to $b$ (Figure C.20). Thus, we have the following theorem.

Every point on $L_1$ has an $x$-coordinate of $a$ and every point on $L_2$ has a $y$-coordinate of $b$.

Figure C.20

> **C.4**   THEOREM.   *The vertical line through $(a, 0)$ and the horizontal line through $(0, b)$ are represented, respectively, by the equations*
>
> $$x = a \quad \text{and} \quad y = b$$

### Example 9

The graph of $x = -5$ is the vertical line through $(-5, 0)$, and the graph of $y = 7$ is the horizontal line through $(0, 7)$ (Figure C.21). ◀

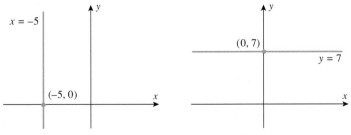

Figure C.21

---

**LINES DETERMINED BY POINT AND SLOPE**

There are infinitely many lines that pass through any given point in the plane. However, if we specify the slope of the line in addition to a point on it, then the point and the slope together determine a unique line (Figure C.22).

Let us now consider how to find an equation of a nonvertical line $L$ that passes through a point $P_1(x_1, y_1)$ and has slope $m$. If $P(x, y)$ is any point on $L$, different from $P_1$, then the

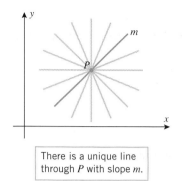

There is a unique line through $P$ with slope $m$.

Figure C.22

slope $m$ can be obtained from the points $P(x, y)$ and $P_1(x_1, y_1)$; this gives

$$m = \frac{y - y_1}{x - x_1}$$

which can be rewritten as

$$y - y_1 = m(x - x_1) \tag{2}$$

With the possible exception of $(x_1, y_1)$, we have shown that every point on $L$ satisfies (2). But $x = x_1$, $y = y_1$ satisfies (2), so that all points on $L$ satisfy (2). We leave it as an exercise to show that every point satisfying (2) lies on $L$.

In summary, we have the following theorem.

---

**C.5**   THEOREM.   *The line passing through $P_1(x_1, y_1)$ and having slope $m$ is given by the equation*

$$y - y_1 = m(x - x_1) \tag{3}$$

*This is called the **point-slope form** of the line.*

---

### Example 10

Find the point-slope form of the line through $(4, -3)$ with slope 5.

*Solution.*   Substituting the values $x_1 = 4$, $y_1 = -3$, and $m = 5$ in (3) yields the point-slope form $y + 3 = 5(x - 4)$.   ◀

**LINES DETERMINED BY SLOPE AND $y$-INTERCEPT**

A nonvertical line crosses the $y$-axis at some point $(0, b)$. If we use this point in the point-slope form of its equation, we obtain

$$y - b = m(x - 0)$$

which we can rewrite as $y = mx + b$. To summarize:

---

**C.6**   THEOREM.   *The line with $y$-intercept $b$ and slope $m$ is given by the equation*

$$y = mx + b \tag{4}$$

*This is called the **slope-intercept form** of the line.*

---

$$y = mx + b$$

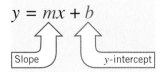

Figure C.23

REMARK.   Note that $y$ is alone on one side of Equation (4). When the equation of a line is written in this way the slope of the line and its $y$-intercept can be determined by inspection of the equation—the slope is the coefficient of $x$ and the $y$-intercept is the constant term (Figure C.23).

### Example 11

| EQUATION | SLOPE | $y$-INTERCEPT |
|---|---|---|
| $y = 3x + 7$ | $m = 3$ | $b = 7$ |
| $y = -x + \frac{1}{2}$ | $m = -1$ | $b = \frac{1}{2}$ |
| $y = x$ | $m = 1$ | $b = 0$ |
| $y = \sqrt{2}x - 8$ | $m = \sqrt{2}$ | $b = -8$ |
| $y = 2$ | $m = 0$ | $b = 2$ |

◀

**Example 12**

Find the slope-intercept form of the equation of the line that satisfies the stated conditions:

(a)   slope is $-9$; crosses the $y$-axis at $(0, -4)$

(b)   slope is 1; passes through the origin

(c)   passes through $(5, -1)$; perpendicular to $y = 3x + 4$

(d)   passes through $(3, 4)$ and $(2, -5)$.

*Solution (a).* From the given conditions we have $m = -9$ and $b = -4$, so (4) yields $y = -9x - 4$.

*Solution (b).* From the given conditions $m = 1$ and the line passes through $(0, 0)$, so $b = 0$. Thus, it follows from (4) that $y = x + 0$ or $y = x$.

*Solution (c).* The given line has slope 3, so the line to be determined will have slope $m = -\frac{1}{3}$. Substituting this slope and the given point in the point-slope form (3) and then simplifying yields

$$y - (-1) = -\tfrac{1}{3}(x - 5)$$
$$y = -\tfrac{1}{3}x + \tfrac{2}{3}$$

*Solution (d).* We will first find the point-slope form, then solve for $y$ in terms of $x$ to obtain the slope-intercept form. From the given points the slope of the line is

$$m = \frac{-5 - 4}{2 - 3} = 9$$

We can use either of the given points for $(x_1, y_1)$ in (3). We will use $(3, 4)$. This yields the point-slope form

$$y - 4 = 9(x - 3)$$

Solving for $y$ in terms of $x$ yields the slope-intercept form

$$y = 9x - 23$$

We leave it for the reader to show that the same equation results if $(2, -5)$ rather than $(3, 4)$ is used for $(x_1, y_1)$ in (3).   ◀

**THE GENERAL EQUATION OF A LINE**

An equation that is expressible in the form

$$Ax + By + C = 0 \tag{5}$$

where $A$, $B$, and $C$ are constants and $A$ and $B$ are not both zero, is called a ***first-degree equation*** in $x$ and $y$. For example,

$$4x + 6y - 5 = 0$$

is a first-degree equation in $x$ and $y$ since it has form (5) with

$$A = 4, \quad B = 6, \quad C = -5$$

In fact, all the equations of lines studied in this section are first-degree equations in $x$ and $y$.

   The following theorem states that the first-degree equations in $x$ and $y$ are precisely the equations whose graphs in the $xy$-plane are straight lines.

> **C.7**   THEOREM.   *Every first-degree equation in x and y has a straight line as its graph and, conversely, every straight line can be represented by a first-degree equation in x and y.*

Because of this theorem, (5) is sometimes called the ***general equation*** of a line or a ***linear equation*** in $x$ and $y$.

### Example 13

Graph the equation $3x - 4y + 12 = 0$.

*Solution.* Since this is a linear equation in $x$ and $y$, its graph is a straight line. Thus, to sketch the graph we need only plot any two points on the graph and draw the line through them. It is particularly convenient to plot the points where the line crosses the coordinate axes. These points are $(0, 3)$ and $(-4, 0)$ (verify), so the graph is the line in Figure C.24. ◀

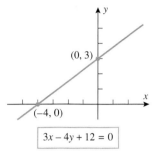

(0, 3)

(−4, 0)

$3x - 4y + 12 = 0$

Figure C.24

### Example 14

Find the slope of the line in Example 13.

*Solution.* Solving the equation for $y$ yields

$$y = \tfrac{3}{4}x + 3$$

which is the slope-intercept form of the line. Thus, the slope is $m = \tfrac{3}{4}$. ◀

---

### EXERCISE SET C

. . . . . . . . . . . . . . . . . . . . . . . . . . . . . . . . . . . . . . . . . . . . . . . . . . . . . . . .

**1.** Draw the rectangle, three of whose vertices are $(6, 1)$, $(-4, 1)$, and $(6, 7)$, and find the coordinates of the fourth vertex.

**2.** Draw the triangle whose vertices are $(-3, 2)$, $(5, 2)$, and $(4, 3)$, and find its area.

> In Exercises 3 and 4, draw a rectangular coordinate system and sketch the set of points whose coordinates $(x, y)$ satisfy the given conditions.

**3.** (a) $x = 2$     (b) $y = -3$     (c) $x \geq 0$
    (d) $y = x$     (e) $y \geq x$     (f) $|x| \geq 1$

**4.** (a) $x = 0$              (b) $y = 0$
    (c) $y < 0$           (d) $x \geq 1$ and $y \leq 2$
    (e) $x = 3$           (f) $|x| = 5$

> In Exercises 5–12, sketch the graph of the equation. (A calculating utility will be helpful in some of these problems.)

**5.** $y = 4 - x^2$            **6.** $y = 1 + x^2$

**7.** $y = \sqrt{x - 4}$        **8.** $y = -\sqrt{x + 1}$

**9.** $x^2 - x + y = 0$     **10.** $x = y^3 - y^2$

**11.** $x^2 y = 2$            **12.** $xy = -1$

**13.** Find the slope of the line through
    (a) $(-1, 2)$ and $(3, 4)$     (b) $(5, 3)$ and $(7, 1)$
    (c) $(4, \sqrt{2})$ and $(-3, \sqrt{2})$     (d) $(-2, -6)$ and $(-2, 12)$.

**14.** Find the slopes of the sides of the triangle with vertices $(-1, 2)$, $(6, 5)$, and $(2, 7)$.

**15.** Use slopes to determine whether the given points lie on the same line.
    (a) $(1, 1)$, $(-2, -5)$, and $(0, -1)$
    (b) $(-2, 4)$, $(0, 2)$, and $(1, 5)$

**16.** Draw the line through $(4, 2)$ with slope
    (a) $m = 3$     (b) $m = -2$     (c) $m = -\tfrac{3}{4}$.

**17.** Draw the line through $(-1, -2)$ with slope
    (a) $m = \tfrac{3}{5}$     (b) $m = -1$     (c) $m = \sqrt{2}$.

**18.** An equilateral triangle has one vertex at the origin, another on the $x$-axis, and the third in the first quadrant. Find the slopes of its sides.

**19.** List the lines in the accompanying figure in the order of increasing slope.

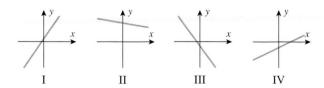

I         II         III         IV

**20.** List the lines in the accompanying figure in the order of increasing slope.

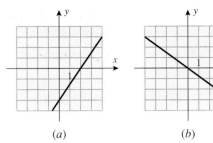

I      II      III      IV

**21.** A particle, initially at $(1, 2)$, moves along a line of slope $m = 3$ to a new position $(x, y)$.
(a) Find $y$ if $x = 5$.      (b) Find $x$ if $y = -2$.

**22.** A particle, initially at $(7, 5)$, moves along a line of slope $m = -2$ to a new position $(x, y)$.
(a) Find $y$ if $x = 9$.      (b) Find $x$ if $y = 12$.

**23.** Let the point $(3, k)$ lie on the line of slope $m = 5$ through $(-2, 4)$; find $k$.

**24.** Given that the point $(k, 4)$ is on the line through $(1, 5)$ and $(2, -3)$, find $k$.

**25.** Find $x$ if the slope of the line through $(1, 2)$ and $(x, 0)$ is the negative of the slope of the line through $(4, 5)$ and $(x, 0)$.

**26.** Find $x$ and $y$ if the line through $(0, 0)$ and $(x, y)$ has slope $\frac{1}{2}$, and the line through $(x, y)$ and $(7, 5)$ has slope 2.

**27.** Use slopes to show that $(3, -1)$, $(6, 4)$, $(-3, 2)$, and $(-6, -3)$ are vertices of a parallelogram.

**28.** Use slopes to show that $(3, 1)$, $(6, 3)$, and $(2, 9)$ are vertices of a right triangle.

**29.** Graph the equations
(a) $2x + 5y = 15$      (b) $x = 3$
(c) $y = -2$      (d) $y = 2x - 7$.

**30.** Graph the equations
(a) $\dfrac{x}{3} - \dfrac{y}{4} = 1$      (b) $x = -8$
(c) $y = 0$      (d) $x = 3y + 2$.

**31.** Graph the equations
(a) $y = 2x - 1$      (b) $y = 3$
(c) $y = -2x$.

**32.** Graph the equations
(a) $y = 2 - 3x$      (b) $y = \frac{1}{4}x$
(c) $y = -\sqrt{3}$.

**33.** Find the slope and $y$-intercept of
(a) $y = 3x + 2$      (b) $y = 3 - \frac{1}{4}x$
(c) $3x + 5y = 8$      (d) $y = 1$
(e) $\dfrac{x}{a} + \dfrac{y}{b} = 1$.

**34.** Find the slope and $y$-intercept of
(a) $y = -4x + 2$      (b) $x = 3y + 2$
(c) $\dfrac{x}{2} + \dfrac{y}{3} = 1$      (d) $y - 3 = 0$
(e) $a_0 x + a_1 y = 0$    $(a_1 \neq 0)$.

In Exercises 35 and 36, use the graph to find the equation of the line in slope-intercept form.

**35.**

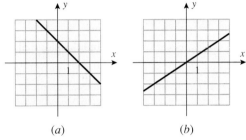

(a)          (b)

Figure Ex-35

**36.**

(a)          (b)

Figure Ex-36

In Exercises 37–48, find the slope-intercept form of the line satisfying the given conditions.

**37.** Slope $= -2$, $y$-intercept $= 4$.

**38.** $m = 5$, $b = -3$.

**39.** The line is parallel to $y = 4x - 2$ and its $y$-intercept is 7.

**40.** The line is parallel to $3x + 2y = 5$ and passes through $(-1, 2)$.

**41.** The line is perpendicular to $y = 5x + 9$ and its $y$-intercept is 6.

**42.** The line is perpendicular to $x - 4y = 7$ and passes through $(3, -4)$.

**43.** The line passes through $(2, 4)$ and $(1, -7)$.

**44.** The line passes through $(-3, 6)$ and $(-2, 1)$.

**45.** The $y$-intercept is 2 and the $x$-intercept is $-4$.

**46.** The $y$-intercept is $b$ and the $x$-intercept is $a$.

**47.** The line is perpendicular to the $y$-axis and passes through $(-4, 1)$.

**48.** The line is parallel to $y = -5$ and passes through $(-1, -8)$.

**49.** In each part, classify the lines as parallel, perpendicular, or neither.
(a) $y = 4x - 7$ and $y = 4x + 9$
(b) $y = 2x - 3$ and $y = 7 - \frac{1}{2}x$
(c) $5x - 3y + 6 = 0$ and $10x - 6y + 7 = 0$
(d) $Ax + By + C = 0$ and $Bx - Ay + D = 0$
(e) $y - 2 = 4(x - 3)$ and $y - 7 = \frac{1}{4}(x - 3)$

**50.** In each part, classify the lines as parallel, perpendicular, or neither.
(a) $y = -5x + 1$ and $y = 3 - 5x$

(b) $y - 1 = 2(x - 3)$ and $y - 4 = -\frac{1}{2}(x + 7)$

(c) $4x + 5y + 7 = 0$ and $5x - 4y + 9 = 0$

(d) $Ax + By + C = 0$ and $Ax + By + D = 0$

(e) $y = \frac{1}{2}x$ and $x = \frac{1}{2}y$

**51.** For what value of $k$ will the line $3x + ky = 4$
   (a) have slope 2
   (b) have $y$-intercept 5
   (c) pass through the point $(-2, 4)$
   (d) be parallel to the line $2x - 5y = 1$
   (e) be perpendicular to the line $4x + 3y = 2$?

**52.** Sketch the graph of $y^2 = 3x$ and explain how this graph is related to the graphs of $y = \sqrt{3x}$ and $y = -\sqrt{3x}$.

**53.** Sketch the graph of $(x - y)(x + y) = 0$ and explain how it is related to the graphs of $x - y = 0$ and $x + y = 0$.

**54.** Graph $F = \frac{9}{5}C + 32$ in a $CF$-coordinate system.

**55.** Graph $u = 3v^2$ in a $uv$-coordinate system.

**56.** Graph $Y = 4X + 5$ in a $YX$-coordinate system.

**57.** A point moves in the $xy$-plane in such a way that at any time $t$ its coordinates are given by $x = 5t + 2$ and $y = t - 3$. By expressing $y$ in terms of $x$, show that the point moves along a straight line.

**58.** A point moves in the $xy$-plane in such a way that at any time $t$ its coordinates are given by $x = 1 + 3t^2$ and $y = 2 - t^2$. By expressing $y$ in terms of $x$, show that the point moves along a straight-line path and specify the values of $x$ for which the equation is valid.

**59.** Find the area of the triangle formed by the coordinate axes and the line through $(1, 4)$ and $(2, 1)$.

**60.** Draw the graph of $4x^2 - 9y^2 = 0$.

**61.** In each part, name an appropriate coordinate system for graphing the equation [e.g., an $\alpha\beta$-coordinate system in part (a)], and state whether the graph of the equation is a line in that coordinate system.
   (a) $3\alpha - 2\beta = 5$
   (b) $A = 2000(1 + 0.06t)$
   (c) $A = \pi r^2$
   (d) $E = mc^2$       ($c$ constant)
   (e) $V = C(1 - rt)$     ($r$ and $C$ constant)
   (f) $V = \frac{1}{3}\pi r^2 h$     ($r$ constant)
   (g) $V = \frac{1}{3}\pi r^2 h$     ($h$ constant)

# APPENDIX D

## Distance, Circles, and Quadratic Equations

**DISTANCE BETWEEN TWO POINTS IN THE PLANE**

Suppose that we are interested in finding the distance $d$ between two points $P_1(x_1, y_1)$ and $P_2(x_2, y_2)$ in the $xy$-plane. If, as in Figure D.1, we form a right triangle with $P_1$ and $P_2$ as vertices, then it follows from Theorem B.4 in Appendix B that the sides of that triangle have lengths $|x_2 - x_1|$ and $|y_2 - y_1|$. Thus, it follows from the Theorem of Pythagoras that

$$d = \sqrt{|x_2 - x_1|^2 + |y_2 - y_1|^2} = \sqrt{(x_2 - x_1)^2 + (y_2 - y_1)^2}$$

and hence we have the following result.

---

**D.1** THEOREM. *The distance $d$ between two points $P_1(x_1, y_1)$ and $P_2(x_2, y_2)$ in a coordinate plane is given by*

$$d = \sqrt{(x_2 - x_1)^2 + (y_2 - y_1)^2} \tag{1}$$

---

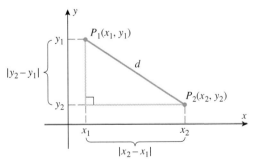

Figure D.1

REMARK. To apply Formula (1) the scales on the coordinate axes must be the same; otherwise, we would not have been able to use the Theorem of Pythagoras in the derivation. Moreover, when using Formula (1) it does not matter which point is labeled $P_1$ and which one is labeled $P_2$, since reversing the points changes the signs of $x_2 - x_1$ and $y_2 - y_1$; this has no effect on the value of $d$ because these quantities are squared in the formula. When it is important to emphasize the points, the distance between $P_1$ and $P_2$ is denoted by $d(P_1, P_2)$ or $d(P_2, P_1)$.

### Example 1

Find the distance between the points $(-2, 3)$ and $(1, 7)$.

*Solution.* If we let $(x_1, y_1)$ be $(-2, 3)$ and let $(x_2, y_2)$ be $(1, 7)$, then (1) yields

$$d = \sqrt{[1 - (-2)]^2 + [7 - 3]^2} = \sqrt{3^2 + 4^2} = \sqrt{25} = 5 \qquad \blacktriangleleft$$

### Example 2

It can be shown that the converse of the Theorem of Pythagoras is true; that is, if the sides of a triangle satisfy the relationship $a^2 + b^2 = c^2$, then the triangle must be a right triangle. Use this result to show that the points $A(4, 6)$, $B(1, -3)$, and $C(7, 5)$ are vertices of a right triangle.

*Solution.* The points and the triangle are shown in Figure D.2. From (1), the lengths of the sides of the triangles are

$$d(A, B) = \sqrt{(1 - 4)^2 + (-3 - 6)^2} = \sqrt{9 + 81} = \sqrt{90}$$

$$d(A, C) = \sqrt{(7 - 4)^2 + (5 - 6)^2} = \sqrt{9 + 1} = \sqrt{10}$$

$$d(B, C) = \sqrt{(7 - 1)^2 + [5 - (-3)]^2} = \sqrt{36 + 64} = \sqrt{100} = 10$$

Since

$$[d(A, B)]^2 + [d(A, C)]^2 = [d(B, C)]^2$$

it follows that $\triangle ABC$ is a right triangle with hypotenuse $BC$. $\qquad \blacktriangleleft$

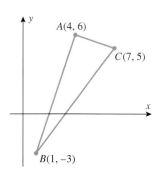

Figure D.2

····················

**THE MIDPOINT FORMULA**

It is often necessary to find the coordinates of the midpoint of a line segment joining two points in the plane. To derive the midpoint formula, we will start with two points on a coordinate line. If we assume that the points have coordinates $a$ and $b$ and that $a \le b$, then, as shown in Figure D.3, the distance between $a$ and $b$ is $b - a$, and the coordinate of the midpoint between $a$ and $b$ is

$$a + \tfrac{1}{2}(b - a) = \tfrac{1}{2}a + \tfrac{1}{2}b = \tfrac{1}{2}(a + b)$$

which is the arithmetic average of $a$ and $b$. Had the points been labeled with $b \le a$, the same formula would have resulted (verify). Therefore, *the midpoint of two points on a coordinate line is the arithmetic average of their coordinates, regardless of their relative positions.*

If we now let $P_1(x_1, y_1)$ and $P_2(x_2, y_2)$ be any two points in the plane and $M(x, y)$ the midpoint of the line segment joining them (Figure D.4), then it can be shown using similar triangles that $x$ is the midpoint of $x_1$ and $x_2$ on the $x$-axis and $y$ is the midpoint of $y_1$ and $y_2$ on the $y$-axis, so

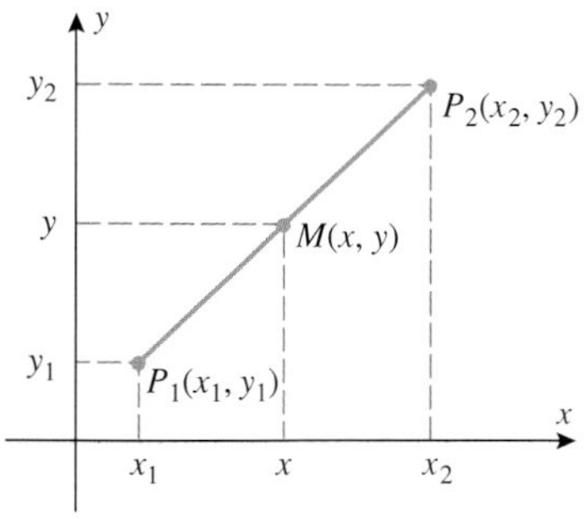

$$x = \tfrac{1}{2}(x_1 + x_2) \quad \text{and} \quad y = \tfrac{1}{2}(y_1 + y_2)$$

Thus, we have the following result.

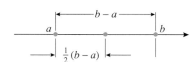

Figure D.3

---

**D.2** THEOREM (*The Midpoint Formula*).    *The midpoint of the line segment joining two points $(x_1, y_1)$ and $(x_2, y_2)$ in a coordinate plane is*

$$\left(\tfrac{1}{2}(x_1 + x_2), \tfrac{1}{2}(y_1 + y_2)\right) \qquad\qquad (2)$$

---

Figure D.4

### Example 3

Find the midpoint of the line segment joining $(3, -4)$ and $(7, 2)$.

*Solution.* From (2) the midpoint is

$$\left(\tfrac{1}{2}(3 + 7), \tfrac{1}{2}(-4 + 2)\right) = (5, -1) \qquad \blacktriangleleft$$

**CIRCLES**

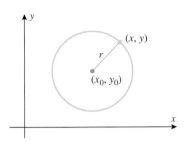

Figure D.5

If $(x_0, y_0)$ is a fixed point in the plane, then the circle of radius $r$ centered at $(x_0, y_0)$ is the set of all points in the plane whose distance from $(x_0, y_0)$ is $r$ (Figure D.5). Thus, a point $(x, y)$ will lie on this circle if and only if

$$\sqrt{(x - x_0)^2 + (y - y_0)^2} = r$$

or equivalently,

$$(x - x_0)^2 + (y - y_0)^2 = r^2 \tag{3}$$

This is called the **standard form of the equation of a circle**.

### Example 4

Find an equation for the circle of radius 4 centered at $(-5, 3)$.

*Solution.* From (3) with $x_0 = -5$, $y_0 = 3$, and $r = 4$ we obtain

$$(x + 5)^2 + (y - 3)^2 = 16$$

If desired, this equation can be written in an expanded form by squaring the terms and then simplifying:

$$(x^2 + 10x + 25) + (y^2 - 6y + 9) - 16 = 0$$
$$x^2 + y^2 + 10x - 6y + 18 = 0 \qquad \blacktriangleleft$$

### Example 5

Find an equation for the circle with center $(1, -2)$ that passes through $(4, 2)$.

*Solution.* The radius $r$ of the circle is the distance between $(4, 2)$ and $(1, -2)$, so

$$r = \sqrt{(1 - 4)^2 + (-2 - 2)^2} = 5$$

We now know the center and radius, so we can use (3) to obtain the equation

$$(x - 1)^2 + (y + 2)^2 = 25 \quad \text{or} \quad x^2 + y^2 - 2x + 4y - 20 = 0 \qquad \blacktriangleleft$$

**FINDING THE CENTER AND RADIUS OF A CIRCLE**

When you encounter an equation of form (3), you will know immediately that its graph is a circle; its center and radius can then be found from the constants that appear in the equation:

$$\underbrace{(x - x_0)^2}_{x\text{-coordinate of the center is } x_0} + \underbrace{(y - y_0)^2}_{y\text{-coordinate of the center is } y_0} = \underbrace{r^2}_{\text{radius squared}}$$

### Example 6

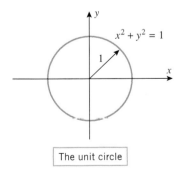

The unit circle

Figure D.6

| EQUATION OF A CIRCLE | CENTER $(x_0, y_0)$ | RADIUS $r$ |
|---|---|---|
| $(x - 2)^2 + (y - 5)^2 = 9$ | $(2, 5)$ | 3 |
| $(x + 7)^2 + (y + 1)^2 = 16$ | $(-7, -1)$ | 4 |
| $x^2 + y^2 = 25$ | $(0, 0)$ | 5 |
| $(x - 4)^2 + y^2 = 5$ | $(4, 0)$ | $\sqrt{5}$ |

$\blacktriangleleft$

The circle $x^2 + y^2 = 1$, which is centered at the origin and has radius 1, is of special importance; it is called the **unit circle** (Figure D.6).

**OTHER FORMS FOR THE EQUATION OF A CIRCLE**

An alternative version of Equation (3) can be obtained by squaring the terms and simplifying. This yields an equation of the form

$$x^2 + y^2 + dx + ey + f = 0 \tag{4}$$

where $d$, $e$, and $f$ are constants. (See the final equations in Examples 4 and 5.)

Still another version of the equation of a circle can be obtained by multiplying both sides of (4) by a nonzero constant $A$. This yields an equation of the form

$$Ax^2 + Ay^2 + Dx + Ey + F = 0 \tag{5}$$

where $A$, $D$, $E$, and $F$ are constants and $A \neq 0$.

If the equation of a circle is given by (4) or (5), then the center and radius can be found by first rewriting the equation in standard form, then reading off the center and radius from that equation. The following example shows how to do this using the technique of ***completing the square***. However, in preparation for the example, recall that completing the square is a method for rewriting an expression of the form

$$x^2 + bx$$

as a difference of two squares. The procedure is to take half the coefficient of $x$, square it, and then add and subtract that result from the original expression to obtain

$$x^2 + bx = x^2 + bx + (b/2)^2 - (b/2)^2 = [x + (b/2)]^2 - (b/2)^2$$

### Example 7

Find the center and radius of the circle with equation

(a) $x^2 + y^2 - 8x + 2y + 8 = 0$    (b) $2x^2 + 2y^2 + 24x - 81 = 0$

*Solution* (*a*). First, group the $x$-terms, group the $y$-terms, and take the constant to the right side:

$$(x^2 - 8x) + (y^2 + 2y) = -8$$

Next we want to add the appropriate constant within each set of parentheses to complete the square, and subtract the same constant outside the parentheses to maintain equality. The appropriate constant is obtained by taking half the coefficient of the first-degree term and squaring it. This yields

$$(x^2 - 8x + 16) - 16 + (y^2 + 2y + 1) - 1 = -8$$

from which we obtain

$$(x - 4)^2 + (y + 1)^2 = -8 + 16 + 1 \quad \text{or} \quad (x - 4)^2 + (y + 1)^2 = 9$$

Thus from (3), the circle has center $(4, -1)$ and radius 3.

*Solution* (*b*). The given equation is of form (5). We will first divide through by 2 (the coefficient of the squared terms) to reduce the equation to form (4). Then we will proceed as in part (a) of this example. The computations are as follows:

$$x^2 + y^2 + 12x - \tfrac{81}{2} = 0 \qquad \boxed{\text{We divided through by 2.}}$$

$$(x^2 + 12x) + y^2 = \tfrac{81}{2}$$

$$(x^2 + 12x + 36) + y^2 = \tfrac{81}{2} + 36 \qquad \boxed{\text{We completed the square.}}$$

$$(x + 6)^2 + y^2 = \tfrac{153}{2}$$

From (3), the circle has center $(-6, 0)$ and radius $\sqrt{\tfrac{153}{2}}$.  ◀

**DEGENERATE CASES OF A CIRCLE**

There is no guarantee that an equation of form (5) represents a circle. For example, suppose that we divide both sides of (5) by $A$, then complete the squares to obtain

$$(x - x_0)^2 + (y - y_0)^2 = k$$

Depending on the value of $k$, the following situations occur:

- $(k > 0)$   The graph is a circle with center $(x_0, y_0)$ and radius $\sqrt{k}$.
- $(k = 0)$   The only solution of the equation is $x = x_0$, $y = y_0$, so the graph is the single point $(x_0, y_0)$.
- $(k < 0)$   The equation has no real solutions and consequently no graph.

**Example 8**

Describe the graphs of

$$(a)\ (x - 1)^2 + (y + 4)^2 = -9 \qquad (b)\ (x - 1)^2 + (y + 4)^2 = 0$$

*Solution (a).* There are no real values of $x$ and $y$ that will make the left side of the equation negative. Thus, the solution set of the equation is empty, and the equation has no graph.

*Solution (b).* The only values of $x$ and $y$ that will make the left side of the equation 0 are $x = 1$, $y = -4$. Thus, the graph of the equation is the single point $(1, -4)$.   ◀

The following theorem summarizes our observations.

---

**D.3**   THEOREM.   *An equation of the form*

$$Ax^2 + Ay^2 + Dx + Ey + F = 0 \tag{6}$$

*where $A \neq 0$, represents a circle, or a point, or else has no graph.*

---

REMARK.   The last two cases in Theorem D.3 are called ***degenerate cases***. In spite of the fact that these degenerate cases can occur, (6) is often called the ***general equation of a circle***.

**THE GRAPH of $y = ax^2 + bx + c$**

An equation of the form

$$y = ax^2 + bx + c \quad (a \neq 0) \tag{7}$$

is called a ***quadratic equation in x***. Depending on whether $a$ is positive or negative, the graph, which is called a ***parabola***, has one of the two forms shown in Figure D.7. In both cases the parabola is symmetric about a vertical line parallel to the $y$-axis. This line of symmetry cuts the parabola at a point called the ***vertex***. The vertex is the low point on the curve if $a > 0$ and the high point if $a < 0$.

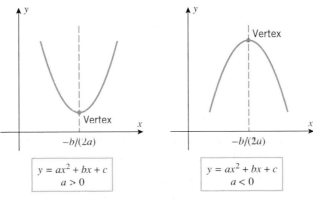

Figure D.7

| $x$ | $y = x^2 - 2x - 2$ |
|-----|-----|
| $-1$ | $1$ |
| $0$ | $-2$ |
| $1$ | $-3$ |
| $2$ | $-2$ |
| $3$ | $1$ |

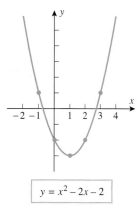

$y = x^2 - 2x - 2$

Figure D.8

| $x$ | $y = -x^2 + 4x - 5$ |
|-----|-----|
| $0$ | $-5$ |
| $1$ | $-2$ |
| $2$ | $-1$ |
| $3$ | $-2$ |
| $4$ | $-5$ |

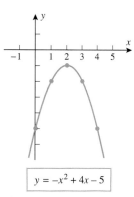

$y = -x^2 + 4x - 5$

Figure D.9

In the exercises (Exercise 78) we will help the reader show that the $x$-coordinate of the vertex is given by the formula

$$x = -\frac{b}{2a} \tag{8}$$

With the aid of this formula, a reasonably accurate graph of a quadratic equation in $x$ can be obtained by plotting the vertex and two points on each side of it.

### Example 9

Sketch the graph of

    (a) $y = x^2 - 2x - 2$    (b) $y = -x^2 + 4x - 5$

*Solution (a).* The equation is of form (7) with $a = 1$, $b = -2$, and $c = -2$, so by (8) the $x$-coordinate of the vertex is

$$x = -\frac{b}{2a} = 1$$

Using this value and two additional values on each side, we obtain Figure D.8.

*Solution (b).* The equation is of form (7) with $a = -1$, $b = 4$, and $c = -5$, so by (8) the $x$-coordinate of the vertex is

$$x = -\frac{b}{2a} = 2$$

Using this value and two additional values on each side, we obtain the table and graph in Figure D.9.    ◀

Quite often the intercepts of a parabola $y = ax^2 + bx + c$ are important to know. The $y$-intercept, $y = c$, results immediately by setting $x = 0$. However, in order to obtain the $x$-intercepts, if any, we must set $y = 0$ and then solve the resulting quadratic equation $ax^2 + bx + c = 0$.

### Example 10

Solve the inequality

$$x^2 - 2x - 2 > 0$$

*Solution.* Because the left side of the inequality does not have readily discernible factors, the test-point method illustrated in Example 4 of Appendix A is not convenient to use. Instead, we will give a graphical solution. The given inequality is satisfied for those values of $x$ where the graph of $y = x^2 - 2x - 2$ is above the $x$-axis. From Figure D.8 those are the values of $x$ to the left of the smaller intercept or to the right of the larger intercept. To find these intercepts we set $y = 0$ to obtain

$$x^2 - 2x - 2 = 0$$

Solving by the quadratic formula gives

$$x = \frac{-b \pm \sqrt{b^2 - 4ac}}{2a} = \frac{2 \pm \sqrt{12}}{2} = 1 \pm \sqrt{3}$$

Thus, the $x$-intercepts are

$$x = 1 + \sqrt{3} \approx 2.7 \quad \text{and} \quad x = 1 - \sqrt{3} \approx -0.7$$

and the solution set of the inequality is

$$(-\infty, 1 - \sqrt{3}) \cup (1 + \sqrt{3}, +\infty)$$  ◄

REMARK.   Note that the decimal approximations of the intercepts calculated in the preceding example agree with the graph in Figure D.8. Observe, however, that we used the exact values of the intercepts to express the solution. The choice of exact versus approximate values is often a matter of judgment that depends on the purpose for which the values are to be used. Numerical approximations often provide a sense of size that exact values do not, but they can introduce severe errors if not used with care.

### Example 11

From Figure D.9 we see that the parabola $y = -x^2 + 4x - 5$ has no $x$-intercepts. This can also be seen algebraically by solving for the $x$-intercepts. Setting $y = 0$ and solving the resulting equation

$$-x^2 + 4x - 5 = 0$$

by the quadratic formula yields

$$y = \frac{-4 \pm \sqrt{16 - 20}}{-2} = 2 \pm i$$

Because the solutions are complex numbers, there are no (real) $x$-intercepts.  ◄

### Example 12

A ball is thrown straight up from the surface of the Earth at time $t = 0$ s with an initial velocity of 24.5 m/s. If air resistance is ignored, it can be shown that the distance $s$ (in meters) of the ball above the ground after $t$ seconds is given by

$$s = 24.5t - 4.9t^2 \tag{9}$$

(a)   Graph $s$ versus $t$, making the $t$-axis horizontal and the $s$-axis vertical.

(b)   How high does the ball rise above the ground?

*Solution (a).*  Equation (9) is of form (7) with $a = -4.9$, $b = 24.5$, and $c = 0$, so by (8) the $t$-coordinate of the vertex is

$$t = -\frac{b}{2a} = -\frac{24.5}{2(-4.9)} = 2.5 \text{ s}$$

and consequently the $s$-coordinate of the vertex is

$$s = 24.5(2.5) - 4.9(2.5)^2 = 30.625 \text{ m}$$

The factored form of (9) is

$$s = 4.9t(5 - t)$$

so the graph has $t$-intercepts $t = 0$ and $t = 5$. From the vertex and the intercepts we obtain the graph shown in Figure D.10.

*Solution (b).*  From the $s$-coordinate of the vertex we deduce that the ball rises 30.625 m above the ground.  ◄

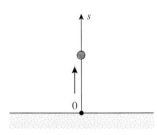

Earth surface

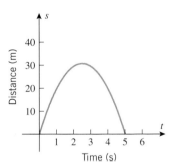

Figure D.10

..........................................

**THE GRAPH of x = ay² + by + c**

If $x$ and $y$ are interchanged in (7), the resulting equation,

$$x = ay^2 + by + c$$

is called a ***quadratic equation in y***. The graph of such an equation is a parabola with its line of symmetry parallel to the $x$-axis and its vertex at the point with $y$-coordinate $y = -b/(2a)$ (Figure D.11). Some problems relating to such equations appear in the exercises.

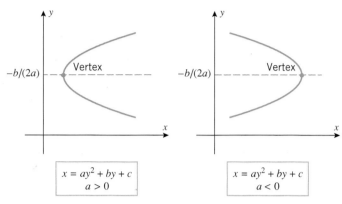

Figure D.11

## EXERCISE SET D

**1.** Where in this section did we use the fact that the same scale was used on both coordinate axes?

In Exercises 2–5, find
(a) the distance between $A$ and $B$
(b) the midpoint of the line segment joining $A$ and $B$.

**2.** $A(2, 5)$, $B(-1, 1)$

**3.** $A(7, 1)$, $B(1, 9)$

**4.** $A(2, 0)$, $B(-3, 6)$

**5.** $A(-2, -6)$, $B(-7, -4)$

In Exercises 6–10, use the distance formula to solve the given problem.

**6.** Prove that $(1, 1)$, $(-2, -8)$, and $(4, 10)$ lie on a straight line.

**7.** Prove that the triangle with vertices $(5, -2)$, $(6, 5)$, $(2, 2)$ is isosceles.

**8.** Prove that $(1, 3)$, $(4, 2)$, and $(-2, -6)$ are vertices of a right triangle and then specify the vertex at which the right angle occurs.

**9.** Prove that $(0, -2)$, $(-4, 8)$, and $(3, 1)$ lie on a circle with center $(-2, 3)$.

**10.** Prove that for all values of $t$ the point $(t, 2t - 6)$ is equidistant from $(0, 4)$ and $(8, 0)$.

**11.** Find $k$, given that $(2, k)$ is equidistant from $(3, 7)$ and $(9, 1)$.

**12.** Find $x$ and $y$ if $(4, -5)$ is the midpoint of the line segment joining $(-3, 2)$ and $(x, y)$.

In Exercises 13 and 14, find an equation of the given line.

**13.** The line is the perpendicular bisector of the line segment joining $(2, 8)$ and $(-4, 6)$.

**14.** The line is the perpendicular bisector of the line segment joining $(5, -1)$ and $(4, 8)$.

**15.** Find the point on the line $4x - 2y + 3 = 0$ that is equidistant from $(3, 3)$ and $(7, -3)$. [*Hint:* First find an equation of the line that is the perpendicular bisector of the line segment joining $(3, 3)$ and $(7, -3)$.]

**16.** Find the distance from the point $(3, -2)$ to the line
(a) $y = 4$ (b) $x = -1$.

**17.** Find the distance from $(2, 1)$ to the line $4x - 3y + 10 = 0$. [*Hint:* Find the foot of the perpendicular dropped from the point to the line.]

**18.** Find the distance from $(8, 4)$ to the line $5x + 12y - 36 = 0$. [*Hint:* See the hint in Exercise 17.]

**19.** Use the method described in Exercise 17 to prove that the distance $d$ from $(x_0, y_0)$ to the line $Ax + By + C = 0$ is

$$d = \frac{|Ax_0 + By_0 + C|}{\sqrt{A^2 + B^2}}$$

**20.** Use the formula in Exercise 19 to solve Exercise 17.

**21.** Use the formula in Exercise 19 to solve Exercise 18.

**22.** Prove: For any triangle, the perpendicular bisectors of the sides meet at a point. [*Hint:* Position the triangle with one vertex on the $y$-axis and the opposite side on the $x$-axis, so that the vertices are $(0, a)$, $(b, 0)$, and $(c, 0)$.]

In Exercises 23 and 24, find the center and radius of each circle.

**23.** (a) $x^2 + y^2 = 25$
(b) $(x-1)^2 + (y-4)^2 = 16$
(c) $(x+1)^2 + (y+3)^2 = 5$
(d) $x^2 + (y+2)^2 = 1$

**24.** (a) $x^2 + y^2 = 9$
(b) $(x-3)^2 + (y-5)^2 = 36$
(c) $(x+4)^2 + (y+1)^2 = 8$
(d) $(x+1)^2 + y^2 = 1$

In Exercises 25–32, find the standard equation of the circle satisfying the given conditions.

**25.** Center $(3, -2)$; radius $= 4$.

**26.** Center $(1, 0)$; diameter $= \sqrt{8}$.

**27.** Center $(-4, 8)$; circle is tangent to the $x$-axis.

**28.** Center $(5, 8)$; circle is tangent to the $y$-axis.

**29.** Center $(-3, -4)$; circle passes through the origin.

**30.** Center $(4, -5)$; circle passes through $(1, 3)$.

**31.** A diameter has endpoints $(2, 0)$ and $(0, 2)$.

**32.** A diameter has endpoints $(6, 1)$ and $(-2, 3)$.

In Exercises 33–44, determine whether the equation represents a circle, a point, or no graph. If the equation represents a circle, find the center and radius.

**33.** $x^2 + y^2 - 2x - 4y - 11 = 0$

**34.** $x^2 + y^2 + 8x + 8 = 0$

**35.** $2x^2 + 2y^2 + 4x - 4y = 0$

**36.** $6x^2 + 6y^2 - 6x + 6y = 3$

**37.** $x^2 + y^2 + 2x + 2y + 2 = 0$

**38.** $x^2 + y^2 - 4x - 6y + 13 = 0$

**39.** $9x^2 + 9y^2 = 1$

**40.** $(x^2/4) + (y^2/4) = 1$

**41.** $x^2 + y^2 + 10y + 26 = 0$

**42.** $x^2 + y^2 - 10x - 2y + 29 = 0$

**43.** $16x^2 + 16y^2 + 40x + 16y - 7 = 0$

**44.** $4x^2 + 4y^2 - 16x - 24y = 9$

**45.** Find an equation of
(a) the bottom half of the circle $x^2 + y^2 = 16$
(b) the top half of the circle $x^2 + y^2 + 2x - 4y + 1 = 0$.

**46.** Find an equation of
(a) the right half of the circle $x^2 + y^2 = 9$
(b) the left half of the circle $x^2 + y^2 - 4x + 3 = 0$.

**47.** Graph
(a) $y = \sqrt{25 - x^2}$
(b) $y = \sqrt{5 + 4x - x^2}$.

**48.** Graph
(a) $x = -\sqrt{4 - y^2}$
(b) $x = 3 + \sqrt{4 - y^2}$.

**49.** Find an equation of the line that is tangent to the circle
$$x^2 + y^2 = 25$$
at the point $(3, 4)$ on the circle.

**50.** Find an equation of the line that is tangent to the circle at the point $P$ on the circle
(a) $x^2 + y^2 + 2x - 9$;  $P(2, -1)$
(b) $x^2 + y^2 - 6x + 4y = 13$;  $P(4, 3)$.

**51.** For the circle $x^2 + y^2 = 20$ and the point $P(-1, 2)$:
(a) Is $P$ inside, outside, or on the circle?
(b) Find the largest and smallest distances between $P$ and points on the circle.

**52.** Follow the directions of Exercise 51 for the circle
$$x^2 + y^2 - 2y - 4 = 0$$
and the point $P\left(3, \frac{5}{2}\right)$.

**53.** Referring to the accompanying figure, find the coordinates of the points $T$ and $T'$, where the lines $L$ and $L'$ are tangent to the circle of radius 1 with center at the origin.

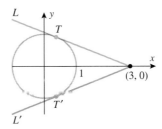

Figure Ex-53

**54.** A point $(x, y)$ moves so that its distance to $(2, 0)$ is $\sqrt{2}$ times its distance to $(0, 1)$.
(a) Show that the point moves along a circle.
(b) Find the center and radius.

**55.** A point $(x, y)$ moves so that the sum of the squares of its distances from $(4, 1)$ and $(2, -5)$ is 45.
(a) Show that the point moves along a circle.
(b) Find the center and radius.

**56.** Find all values of $c$ for which the system of equations
$$\begin{cases} x^2 - y^2 = 0 \\ (x - c)^2 + y^2 = 1 \end{cases}$$
has 0, 1, 2, 3, or 4 solutions. [*Hint:* Sketch a graph.]

In Exercises 57–70, graph the parabola and label the coordinates of the vertex and the intersections with the coordinate axes.

**57.** $y = x^2 + 2$

**58.** $y = x^2 - 3$

**59.** $y = x^2 + 2x - 3$

**60.** $y = x^2 - 3x - 4$

**61.** $y = -x^2 + 4x + 5$

**62.** $y = -x^2 + x$

**63.** $y = (x - 2)^2$

**64.** $y = (3 + x)^2$

**65.** $x^2 - 2x + y = 0$

**66.** $x^2 + 8x + 8y = 0$

**67.** $y = 3x^2 - 2x + 1$

**68.** $y = x^2 + x + 2$

**69.** $x = -y^2 + 2y + 2$

**70.** $x = y^2 - 4y + 5$

**71.** Find an equation of
(a) the right half of the parabola $y = 3 - x^2$
(b) the left half of the parabola $y = x^2 - 2x$.

**72.** Find an equation of
(a) the upper half of the parabola $x = y^2 - 5$
(b) the lower half of the parabola $x = y^2 - y - 2$.

**73.** Graph
(a) $y = \sqrt{x + 5}$

(b) $x = -\sqrt{4 - y}$.

**74.** Graph
(a) $y = 1 + \sqrt{4 - x}$

(b) $x = 3 + \sqrt{y}$.

**75.** If a ball is thrown straight up with an initial velocity of 32 ft/s, then after $t$ seconds the distance $s$ above its starting height, in feet, is given by $s = 32t - 16t^2$.
(a) Graph this equation in a $ts$-coordinate system ($t$-axis horizontal).
(b) At what time $t$ will the ball be at its highest point, and how high will it rise?

**76.** A rectangular field is to be enclosed with 500 ft of fencing along three sides and by a straight stream on the fourth side. Let $x$ be the length of each side perpendicular to the stream, and let $y$ be the length of the side parallel to the stream.
(a) Express $y$ in terms of $x$.
(b) Express the area $A$ of the field in terms of $x$.
(c) What is the largest area that can be enclosed?

**77.** A rectangular plot of land is to be enclosed using two kinds of fencing. Two opposite sides will have heavy-duty fencing costing $3/ft, while the other two sides will have standard fencing costing $2/ft. A total of $600 is available for the fencing. Let $x$ be the length of each side with the heavy-duty fencing, and let $y$ be the length of each side with the standard fencing.
(a) Express $y$ in terms of $x$.
(b) Find a formula for the area $A$ of the rectangular plot in terms of $x$.
(c) What is the largest area that can be enclosed?

**78.** (a) By completing the square, show that the quadratic equation $y = ax^2 + bx + c$ can be rewritten as

$$y = a\left(x + \frac{b}{2a}\right)^2 + \left(c - \frac{b^2}{4a}\right)$$

if $a \neq 0$.

(b) Use the result in part (a) to show that the graph of the quadratic equation $y = ax^2 + bx + c$ has its high point at $x = -b/(2a)$ if $a < 0$ and its low point there if $a > 0$.

In Exercises 79 and 80, solve the given inequality.

**79.** (a) $2x^2 + 5x - 1 < 0$

(b) $x^2 - 2x + 3 > 0$

**80.** (a) $x^2 + x - 1 > 0$

(b) $x^2 - 4x + 6 < 0$

**81.** At time $t = 0$ a ball is thrown straight up from a height of 5 ft above the ground. After $t$ seconds its distance $s$, in feet, above the ground is given by $s = 5 + 40t - 16t^2$.
(a) Find the maximum height of the ball above the ground.
(b) Find, to the nearest tenth of a second, the time when the ball strikes the ground.
(c) Find, to the nearest tenth of a second, how long the ball will be more than 12 ft above the ground.

**82.** Find all values of $x$ at which points on the parabola $y = x^2$ lie below the line $y = x + 3$.

# Trigonometry Review

## TRIGONOMETRIC FUNCTIONS AND IDENTITIES

**ANGLES**

Angles in the plane can be generated by rotating a ray about its endpoint. The starting position of the ray is called the **initial side** of the angle, the final position is called the **terminal side** of the angle, and the point at which the initial and terminal sides meet is called the **vertex** of the angle. We allow for the possibility that the ray may make more than one complete revolution. Angles are considered to be **positive** if generated counterclockwise and **negative** if generated clockwise (Figure E.1).

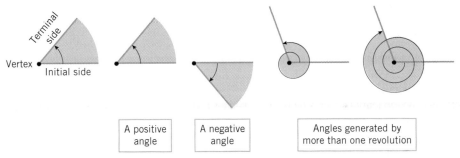

Figure E.1

There are two standard measurement systems for describing the size of an angle: **degree measure** and **radian measure**. In degree measure, one degree (written $1°$) is the measure of an angle generated by $1/360$ of one revolution. Thus, there are $360°$ in an angle of one revolution, $180°$ in an angle of one-half revolution, $90°$ in an angle of one-quarter revolution (a *right angle*), and so forth. Degrees are divided into sixty equal parts, called **minutes**, and minutes are divided into sixty equal parts, called **seconds**. Thus, one minute (written $1'$) is $1/60$ of a degree, and one second (written $1''$) is $1/60$ of a minute. Smaller subdivisions of a degree are expressed as fractions of a second.

In radian measure, angles are measured by the length of the arc that the angle subtends on a circle of radius 1 when the vertex is at the center. One unit of arc on a circle of radius 1 is called one **radian** (written 1 radian or 1 rad) (Figure E.2), and hence the entire circumference of a circle of radius 1 is $2\pi$ radians. It follows that an angle of $360°$ subtends an arc of $2\pi$ radians, an angle of $180°$ subtends an arc of $\pi$ radians, an angle of $90°$ subtends an arc of $\pi/2$ radians, and so forth. Figure E.3 and Table 1 show the relationship between degree measure and radian measure for some important positive angles.

REMARK. Observe that in Table 1, angles in degrees are designated by the degree symbol, but angles in radians have no units specified. This is standard practice—when no units are specified for an angle, it is understood that the units are radians.

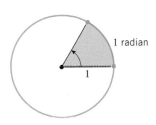

1 radian

1

Figure E.2

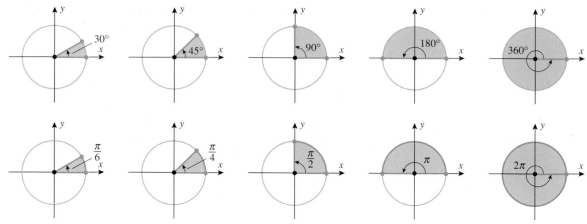

Figure E.3

**Table 1**

| DEGREES | 30° | 45° | 60° | 90° | 120° | 135° | 150° | 180° | 270° | 360° |
|---|---|---|---|---|---|---|---|---|---|---|
| RADIANS | $\dfrac{\pi}{6}$ | $\dfrac{\pi}{4}$ | $\dfrac{\pi}{3}$ | $\dfrac{\pi}{2}$ | $\dfrac{2\pi}{3}$ | $\dfrac{3\pi}{4}$ | $\dfrac{5\pi}{6}$ | $\pi$ | $\dfrac{3\pi}{2}$ | $2\pi$ |

From the fact that $\pi$ radians corresponds to $180°$, we obtain the following formulas, which are useful for converting from degrees to radians and conversely.

$$1° = \frac{\pi}{180}\text{rad} \approx 0.01745 \text{ rad} \tag{1}$$

$$1 \text{ rad} = \left(\frac{180}{\pi}\right)^{\circ} \approx 57° \, 17' \, 44.8'' \tag{2}$$

**Example 1**

(a) Express $146°$ in radians.     (b) Express 3 radians in degrees.

*Solution (a).* From (1), degrees can be converted to radians by multiplying by a conversion factor of $\pi/180$. Thus,

$$146° = \left(\frac{\pi}{180} \cdot 146\right) \text{rad} = \frac{73\pi}{90} \text{ rad} \approx 2.5482 \text{ rad}$$

*Solution (b).* From (2), radians can be converted to degrees by multiplying by a conversion factor of $180/\pi$. Thus,

$$3 \text{ rad} = \left(3 \cdot \frac{180}{\pi}\right)^{\circ} = \left(\frac{540}{\pi}\right)^{\circ} \approx 171.9° \qquad \blacktriangleleft$$

**RELATIONSHIPS BETWEEN ARC LENGTH, ANGLE, RADIUS, AND AREA**

There is a theorem from plane geometry which states that for two concentric circles, the ratio of the arc lengths subtended by a central angle is equal to the ratio of the corresponding radii (Figure E.4). In particular, if $s$ is the arc length subtended on a circle of radius $r$ by a central angle of $\theta$ radians, then by comparison with the arc length subtended by that angle on a circle of radius 1 we obtain

$$\frac{s}{\theta} = \frac{r}{1}$$

from which we obtain the following relationships between the central angle $\theta$, the radius $r$, and the subtended arc length $s$ when $\theta$ is in radians (Figure E.5):

$$\theta = s/r \qquad \text{and} \qquad s = r\theta \tag{3–4}$$

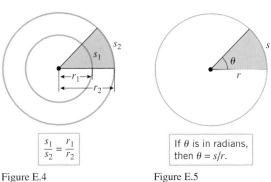

$$\frac{s_1}{s_2} = \frac{r_1}{r_2}$$

If $\theta$ is in radians, then $\theta = s/r$.

Figure E.4                    Figure E.5

The shaded region in Figure E.5 is called a **sector**. It is a theorem from plane geometry that the ratio of the area $A$ of this sector to the area of the entire circle is the same as the ratio of the central angle of the sector to the central angle of the entire circle; thus, if the angles are in radians, we have

$$\frac{A}{\pi r^2} = \frac{\theta}{2\pi}$$

Solving for $A$ yields the following formula for the area of a sector in terms of the radius $r$ and the angle $\theta$ in radians:

$$A = \tfrac{1}{2}r^2\theta \tag{5}$$

## TRIGONOMETRIC FUNCTIONS FOR RIGHT TRIANGLES

The **sine**, **cosine**, **tangent**, **cosecant**, **secant**, and **cotangent** of a positive acute angle $\theta$ can be defined as ratios of the sides of a right triangle. Using the notation from Figure E.6, these definitions take the following form:

$$
\begin{array}{ll}
\sin\theta = \dfrac{\text{side opposite }\theta}{\text{hypotenuse}} = \dfrac{y}{r}, & \csc\theta = \dfrac{\text{hypotenuse}}{\text{side opposite }\theta} = \dfrac{r}{y} \\[2ex]
\cos\theta = \dfrac{\text{side adjacent to }\theta}{\text{hypotenuse}} = \dfrac{x}{r}, & \sec\theta = \dfrac{\text{hypotenuse}}{\text{side adjacent to }\theta} = \dfrac{r}{x} \\[2ex]
\tan\theta = \dfrac{\text{side opposite }\theta}{\text{side adjacent to }\theta} = \dfrac{y}{x}, & \cot\theta = \dfrac{\text{side adjacent to }\theta}{\text{side opposite }\theta} = \dfrac{x}{y}
\end{array}
\tag{6}
$$

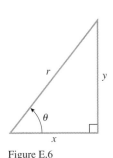

Figure E.6

We will call sin, cos, tan, csc, sec, and cot the **trigonometric functions**. Because similar triangles have proportional sides, the values of the trigonometric functions depend only on the size of $\theta$ and not on the particular right triangle used to compute the ratios. Moreover, in these definitions it does not matter whether $\theta$ is measured in degrees or radians.

### Example 2

Recall from geometry that the two legs of a $45°\!-\!45°\!-\!90°$ triangle are of equal size and that the hypotenuse of a $30°\!-\!60°\!-\!90°$ triangle is twice the shorter leg, where the shorter leg is opposite the $30°$ angle. These facts and the Theorem of Pythagoras yield Figure E.7. From that figure we obtain the results in Table 2.

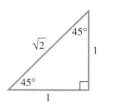
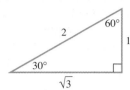

Figure E.7

**Table 2**

| | | |
|---|---|---|
| $\sin 45° = 1/\sqrt{2},$ | $\cos 45° = 1/\sqrt{2},$ | $\tan 45° = 1$ |
| $\csc 45° = \sqrt{2},$ | $\sec 45° = \sqrt{2},$ | $\cot 45° = 1$ |
| $\sin 30° = 1/2,$ | $\cos 30° = \sqrt{3}/2,$ | $\tan 30° = 1/\sqrt{3}$ |
| $\csc 30° = 2,$ | $\sec 30° = 2/\sqrt{3},$ | $\cot 30° = \sqrt{3}$ |
| $\sin 60° = \sqrt{3}/2,$ | $\cos 60° = 1/2,$ | $\tan 60° = \sqrt{3}$ |
| $\csc 60° = 2/\sqrt{3},$ | $\sec 60° = 2,$ | $\cot 60° = 1/\sqrt{3}$ |

◄

**ANGLES IN RECTANGULAR COORDINATE SYSTEMS**

Because the angles of a right triangle are between $0°$ and $90°$, the formulas in (6) are not directly applicable to negative angles or to angles greater than $90°$. To extend the trigonometric functions to include these cases, it will be convenient to consider angles in rectangular coordinate systems. An angle is said to be in *standard position* in an $xy$-coordinate system if its vertex is at the origin and its initial side is on the positive $x$-axis (Figure E.8).

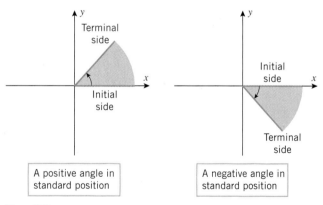

Figure E.8

To define the trigonometric functions of an angle $\theta$ in standard position, construct a circle of radius $r$, centered at the origin, and let $P(x, y)$ be the intersection of the terminal side of $\theta$ with this circle (Figure E.9). We make the following definition.

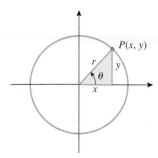

Figure E.9

**E.1** DEFINITION.

$$\sin \theta = \frac{y}{r}, \quad \cos \theta = \frac{x}{r}, \quad \tan \theta = \frac{y}{x}$$

$$\csc \theta = \frac{r}{y}, \quad \sec \theta = \frac{r}{x}, \quad \cot \theta = \frac{x}{y}$$

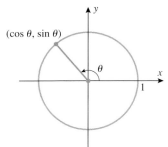

Figure E.10

Note that the formulas in this definition agree with those in (6), so there is no conflict with the earlier definition of the trigonometric functions for triangles. However, this definition applies to all angles (except for cases where a zero denominator occurs).

In the special case where $r = 1$, we have $\sin\theta = y$ and $\cos\theta = x$, so the terminal side of the angle $\theta$ intersects the unit circle at the point $(\cos\theta, \sin\theta)$ (Figure E.10). It follows from Definition E.1 that the remaining trigonometric functions of $\theta$ are expressible as (verify)

$$\tan\theta = \frac{\sin\theta}{\cos\theta}, \quad \cot\theta = \frac{\cos\theta}{\sin\theta} = \frac{1}{\tan\theta}, \quad \sec\theta = \frac{1}{\cos\theta}, \quad \csc\theta = \frac{1}{\sin\theta} \quad (7\text{--}10)$$

These observations suggest the following procedure for evaluating the trigonometric functions of common angles:

- Construct the angle $\theta$ in standard position in an $xy$-coordinate system.
- Find the coordinates of the intersection of the terminal side of the angle and the unit circle; the $x$- and $y$-coordinates of this intersection are the values of $\cos\theta$ and $\sin\theta$, respectively.
- Use Formulas (7) through (10) to find the values of the remaining trigonometric functions from the values of $\cos\theta$ and $\sin\theta$.

### Example 3

Evaluate the trigonometric functions of $\theta = 150°$.

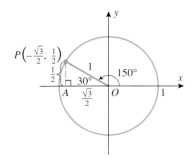

Figure E.11

*Solution.* Construct a unit circle and place the angle $\theta = 150°$ in standard position (Figure E.11). Since $\angle AOP$ is $30°$ and $\triangle OAP$ is a $30°$–$60°$–$90°$ triangle, the leg $AP$ has length $\frac{1}{2}$ (half the hypotenuse) and the leg $OA$ has length $\sqrt{3}/2$ by the Theorem of Pythagoras. Thus, the coordinates of $P$ are $(-\sqrt{3}/2, 1/2)$, from which we obtain

$$\sin 150° = \frac{1}{2}, \quad \cos 150° = -\frac{\sqrt{3}}{2}, \quad \tan 150° = \frac{\sin 150°}{\cos 150°} = \frac{1/2}{-\sqrt{3}/2} = -\frac{1}{\sqrt{3}}$$

$$\csc 150° = \frac{1}{\sin 150°} = 2, \quad \sec 150° = \frac{1}{\cos 150°} = -\frac{2}{\sqrt{3}}$$

$$\cot 150° = \frac{1}{\tan 150°} = -\sqrt{3} \qquad \blacktriangleleft$$

### Example 4

Evaluate the trigonometric functions of $\theta = 5\pi/6$.

*Solution.* Since $5\pi/6 = 150°$, this problem is equivalent to that of Example 3. From that example we obtain

$$\sin\frac{5\pi}{6} = \frac{1}{2}, \quad \cos\frac{5\pi}{6} = -\frac{\sqrt{3}}{2}, \quad \tan\frac{5\pi}{6} = -\frac{1}{\sqrt{3}}$$

$$\csc\frac{5\pi}{6} = 2, \quad \sec\frac{5\pi}{6} = -\frac{2}{\sqrt{3}}, \quad \cot\frac{5\pi}{6} = -\sqrt{3} \qquad \blacktriangleleft$$

### Example 5

Evaluate the trigonometric functions of $\theta = -\pi/2$.

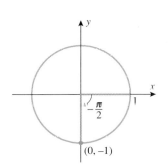

Figure E.12

*Solution.* As shown in Figure E.12, the terminal side of $\theta = -\pi/2$ intersects the unit circle at the point $(0, -1)$, so

$$\sin(-\pi/2) = -1, \quad \cos(-\pi/2) = 0$$

and from Formulas (7) through (10),

$$\tan(-\pi/2) = \frac{\sin(-\pi/2)}{\cos(-\pi/2)} = \frac{-1}{0} \quad \text{(undefined)}$$

$$\cot(-\pi/2) = \frac{\cos(-\pi/2)}{\sin(-\pi/2)} = \frac{0}{-1} = 0$$

$$\sec(-\pi/2) = \frac{1}{\cos(-\pi/2)} = \frac{1}{0} \quad \text{(undefined)}$$

$$\csc(-\pi/2) = \frac{1}{\sin(-\pi/2)} = \frac{1}{-1} = -1 \qquad \blacktriangleleft$$

The reader should be able to obtain all of the results in Table 3 by the methods illustrated in the last three examples. The dashes indicate quantities that are undefined.

**Table 3**

| | $\theta = 0$ (0°) | $\pi/6$ (30°) | $\pi/4$ (45°) | $\pi/3$ (60°) | $\pi/2$ (90°) | $2\pi/3$ (120°) | $3\pi/4$ (135°) | $5\pi/6$ (150°) | $\pi$ (180°) | $3\pi/2$ (270°) | $2\pi$ (360°) |
|---|---|---|---|---|---|---|---|---|---|---|---|
| $\sin\theta$ | 0 | $1/2$ | $1/\sqrt{2}$ | $\sqrt{3}/2$ | 1 | $\sqrt{3}/2$ | $1/\sqrt{2}$ | $1/2$ | 0 | $-1$ | 0 |
| $\cos\theta$ | 1 | $\sqrt{3}/2$ | $1/\sqrt{2}$ | $1/2$ | 0 | $-1/2$ | $-1/\sqrt{2}$ | $-\sqrt{3}/2$ | $-1$ | 0 | 1 |
| $\tan\theta$ | 0 | $1/\sqrt{3}$ | 1 | $\sqrt{3}$ | — | $-\sqrt{3}$ | $-1$ | $-1/\sqrt{3}$ | 0 | — | 0 |
| $\csc\theta$ | — | 2 | $\sqrt{2}$ | $2/\sqrt{3}$ | 1 | $2/\sqrt{3}$ | $\sqrt{2}$ | 2 | — | $-1$ | — |
| $\sec\theta$ | 1 | $2/\sqrt{3}$ | $\sqrt{2}$ | 2 | — | $-2$ | $-\sqrt{2}$ | $-2/\sqrt{3}$ | $-1$ | — | 1 |
| $\cot\theta$ | — | $\sqrt{3}$ | 1 | $1/\sqrt{3}$ | 0 | $-1/\sqrt{3}$ | $-1$ | $-\sqrt{3}$ | — | 0 | — |

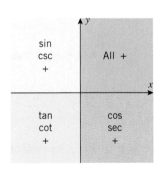

Figure E.13

REMARK.  It is only in special cases that exact values for trigonometric functions can be obtained; usually, a calculating utility or a computer program will be required.

The signs of the trigonometric functions of an angle are determined by the quadrant in which the terminal side of the angle falls. For example, if the terminal side falls in the first quadrant, then $x$ and $y$ are positive in Definition E.1, so all of the trigonometric functions have positive values. If the terminal side falls in the second quadrant, then $x$ is negative and $y$ is positive, so sin and csc are positive, but all other trigonometric functions are negative. The diagram in Figure E.13 shows which trigonometric functions are positive in the various quadrants. The reader will find it instructive to check that the results in Table 3 are consistent with Figure E.13.

**TRIGONOMETRIC IDENTITIES**

A *trigonometric identity* is an equation involving trigonometric functions that is true for all angles for which both sides of the equation are defined. One of the most important identities in trigonometry can be derived by applying the Theorem of Pythagoras to the triangle in Figure E.9 to obtain

$$x^2 + y^2 = r^2$$

Dividing both sides by $r^2$ and using the definitions of $\sin\theta$ and $\cos\theta$ (Definition E.1), we obtain the following fundamental result:

$$\sin^2\theta + \cos^2\theta = 1 \tag{11}$$

The following identities can be obtained from (11) by dividing through by $\cos^2\theta$ and $\sin^2\theta$, respectively, then applying Formulas (7) through (10):

$$\tan^2\theta + 1 = \sec^2\theta \tag{12}$$

$$1 + \cot^2\theta = \csc^2\theta \tag{13}$$

If $(x, y)$ is a point on the unit circle, then the points $(-x, y)$, $(-x, -y)$, and $(x, -y)$ also lie on the unit circle (why?), and the four points form corners of a rectangle with sides parallel to the coordinate axes (Figure E.14$a$). The $x$- and $y$-coordinates of each corner represent the cosine and sine of an angle in standard position whose terminal side passes through the corner; hence we obtain the identities in parts $(b)$, $(c)$, and $(d)$ of Figure E.14 for sine and cosine. Dividing those identities leads to identities for the tangent. In summary:

$$\sin(\pi - \theta) = \sin\theta, \qquad \sin(\pi + \theta) = -\sin\theta, \qquad \sin(-\theta) = -\sin\theta \tag{14–16}$$
$$\cos(\pi - \theta) = -\cos\theta, \qquad \cos(\pi + \theta) = -\cos\theta, \qquad \cos(-\theta) = \cos\theta \tag{17–19}$$
$$\tan(\pi - \theta) = -\tan\theta, \qquad \tan(\pi + \theta) = \tan\theta, \qquad \tan(-\theta) = -\tan\theta \tag{20–22}$$

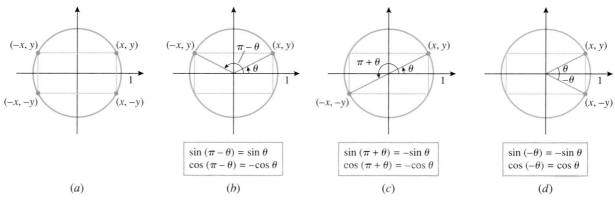

$(a)$        $(b)$        $(c)$        $(d)$

Figure E.14

Two angles in standard position that have the same terminal side must have the same values for their trigonometric functions since their terminal sides intersect the unit circle at the same point. In particular, two angles whose radian measures differ by a multiple of $2\pi$ have the same terminal side and hence have the same values for their trigonometric functions. This yields the identities

$$\sin\theta = \sin(\theta + 2\pi) = \sin(\theta - 2\pi) \tag{23}$$
$$\cos\theta = \cos(\theta + 2\pi) = \cos(\theta - 2\pi) \tag{24}$$

and more generally,

$$\sin\theta = \sin(\theta \pm 2n\pi), \quad n = 0, 1, 2, \ldots \tag{25}$$
$$\cos\theta = \cos(\theta \pm 2n\pi), \quad n = 0, 1, 2, \ldots \tag{26}$$

Identities (20) through (22) imply that

$$\tan\theta = \tan(\theta + \pi) \qquad \text{and} \qquad \tan\theta = \tan(\theta - \pi) \tag{27–28}$$

Identity (27) is just (21) with the terms in the sum reversed, and identity (28) follows from (20) and (22) (verify). These two identities state that adding or subtracting $\pi$ from an angle does not affect the value of the tangent of the angle. It follows that the same is true for any multiple of $\pi$; thus,

$$\tan\theta = \tan(\theta \pm n\pi), \quad n = 0, 1, 2, \ldots \tag{29}$$

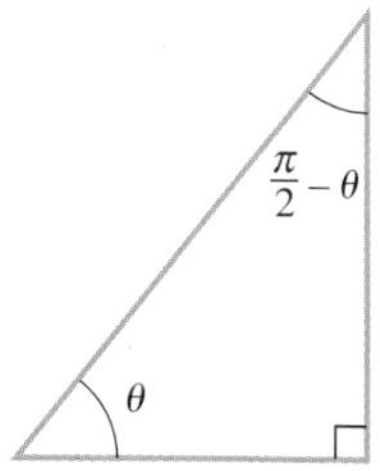

Figure E.15 shows complementary angles $\theta$ and $(\pi/2) - \theta$ of a right triangle. It follows from (6) that

$$\sin\theta = \frac{\text{side opposite } \theta}{\text{hypotenuse}} = \frac{\text{side adjacent to } (\pi/2) - \theta}{\text{hypotenuse}} = \cos\left(\frac{\pi}{2} - \theta\right)$$

$$\cos\theta = \frac{\text{side adjacent to } \theta}{\text{hypotenuse}} = \frac{\text{side opposite } (\pi/2) - \theta}{\text{hypotenuse}} = \sin\left(\frac{\pi}{2} - \theta\right)$$

which yields the identities

$$\sin\left(\frac{\pi}{2} - \theta\right) = \cos\theta, \quad \cos\left(\frac{\pi}{2} - \theta\right) = \sin\theta, \quad \tan\left(\frac{\pi}{2} - \theta\right) = \cot\theta \tag{30–32}$$

where the third identity results from dividing the first two. These identities are also valid for angles that are not acute and for negative angles as well.

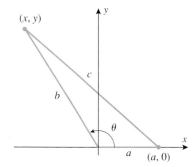

Figure E.15

**THE LAW OF COSINES**

The next theorem, called the *law of cosines*, generalizes the Theorem of Pythagoras. This result is important in its own right and is also the starting point for some important trigonometric identities.

---

**E.2** THEOREM (*Law of Cosines*). *If the sides of a triangle have lengths $a$, $b$, and $c$, and if $\theta$ is the angle between the sides with lengths $a$ and $b$, then*

$$c^2 = a^2 + b^2 - 2ab\cos\theta$$

---

*Proof.* Introduce a coordinate system so that $\theta$ is in standard position and the side of length $a$ falls along the positive $x$-axis. As shown in Figure E.16, the side of length $a$ extends from the origin to $(a, 0)$ and the side of length $b$ extends from the origin to some point $(x, y)$. From the definition of $\sin\theta$ and $\cos\theta$ we have $\sin\theta = y/b$ and $\cos\theta = x/b$, so

$$y = b\sin\theta, \quad x = b\cos\theta \tag{33}$$

From the distance formula in Theorem D.1 of Appendix D, we obtain

$$c^2 = (x - a)^2 + (y - 0)^2$$

so that, from (33),

$$c^2 = (b\cos\theta - a)^2 + b^2\sin^2\theta$$

$$= a^2 + b^2(\cos^2\theta + \sin^2\theta) - 2ab\cos\theta$$

$$= a^2 + b^2 - 2ab\cos\theta$$

Figure E.16

which completes the proof. ∎

We will now show how the law of cosines can be used to obtain the following identities, called the *addition formulas* for sine and cosine:

$$\sin(\alpha + \beta) = \sin\alpha\cos\beta + \cos\alpha\sin\beta \tag{34}$$

$$\cos(\alpha + \beta) = \cos\alpha\cos\beta - \sin\alpha\sin\beta \tag{35}$$

$$\sin(\alpha - \beta) = \sin\alpha\cos\beta - \cos\alpha\sin\beta \tag{36}$$

$$\cos(\alpha - \beta) = \cos\alpha\cos\beta + \sin\alpha\sin\beta \tag{37}$$

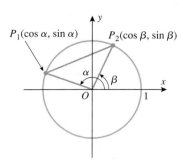

Figure E.17

We will derive (37) first. In our derivation we will assume that $0 \leq \beta < \alpha < 2\pi$ (Figure E.17). As shown in the figure, the terminal sides of $\alpha$ and $\beta$ intersect the unit circle at the points $P_1(\cos\alpha, \sin\alpha)$ and $P_2(\cos\beta, \sin\beta)$. If we denote the lengths of the sides of triangle $OP_1P_2$ by $OP_1$, $P_1P_2$, and $OP_2$, then $OP_1 = OP_2 = 1$ and, from the distance formula in Theorem D.1 of Appendix D,

$$(P_1P_2)^2 = (\cos\beta - \cos\alpha)^2 + (\sin\beta - \sin\alpha)^2$$
$$= (\sin^2\alpha + \cos^2\alpha) + (\sin^2\beta + \cos^2\beta) - 2(\cos\alpha\cos\beta + \sin\alpha\sin\beta)$$
$$= 2 - 2(\cos\alpha\cos\beta + \sin\alpha\sin\beta)$$

But angle $P_2OP_1 = \alpha - \beta$, so that the law of cosines yields

$$(P_1P_2)^2 = (OP_1)^2 + (OP_2)^2 - 2(OP_1)(OP_2)\cos(\alpha - \beta)$$
$$= 2 - 2\cos(\alpha - \beta)$$

Equating the two expressions for $(P_1P_2)^2$ and simplifying, we obtain

$$\cos(\alpha - \beta) = \cos\alpha\cos\beta + \sin\alpha\sin\beta$$

which completes the derivation of (37).

We can use (31) and (37) to derive (36) as follows:

$$\sin(\alpha - \beta) = \cos\left[\frac{\pi}{2} - (\alpha - \beta)\right] = \cos\left[\left(\frac{\pi}{2} - \alpha\right) - (-\beta)\right]$$
$$= \cos\left(\frac{\pi}{2} - \alpha\right)\cos(-\beta) + \sin\left(\frac{\pi}{2} - \alpha\right)\sin(-\beta)$$
$$= \cos\left(\frac{\pi}{2} - \alpha\right)\cos\beta - \sin\left(\frac{\pi}{2} - \alpha\right)\sin\beta$$
$$= \sin\alpha\cos\beta - \cos\alpha\sin\beta$$

Identities (34) and (35) can be obtained from (36) and (37) by substituting $-\beta$ for $\beta$ and using the identities

$$\sin(-\beta) = -\sin\beta, \quad \cos(-\beta) = \cos\beta$$

We leave it for the reader to derive the identities

$$\tan(\alpha + \beta) = \frac{\tan\alpha + \tan\beta}{1 - \tan\alpha\tan\beta} \qquad \tan(\alpha - \beta) = \frac{\tan\alpha - \tan\beta}{1 + \tan\alpha\tan\beta} \qquad (38\text{–}39)$$

Identity (38) can be obtained by dividing (34) by (35) and then simplifying. Identity (39) can be obtained from (38) by substituting $-\beta$ for $\beta$ and simplifying.

In the special case where $\alpha = \beta$, identities (34), (35), and (38) yield the ***double-angle formulas***

$$\sin 2\alpha = 2\sin\alpha\cos\alpha \tag{40}$$

$$\cos 2\alpha = \cos^2\alpha - \sin^2\alpha \tag{41}$$

$$\tan 2\alpha = \frac{2\tan\alpha}{1 - \tan^2\alpha} \tag{42}$$

By using the identity $\sin^2\alpha + \cos^2\alpha = 1$, (41) can be rewritten in the alternative forms

$$\cos 2\alpha = 2\cos^2\alpha - 1 \qquad \text{and} \qquad \cos 2\alpha = 1 - 2\sin^2\alpha \qquad (43\text{–}44)$$

If we replace $\alpha$ by $\alpha/2$ in (43) and (44) and use some algebra, we obtain the ***half-angle formulas***

$$\cos^2\frac{\alpha}{2} = \frac{1 + \cos\alpha}{2} \qquad \text{and} \qquad \sin^2\frac{\alpha}{2} = \frac{1 - \cos\alpha}{2} \qquad (45\text{–}46)$$

We leave it for the exercises to derive the following ***product-to-sum formulas*** from (34) through (37):

$$\sin\alpha\cos\beta = \frac{1}{2}[\sin(\alpha-\beta)+\sin(\alpha+\beta)] \tag{47}$$

$$\sin\alpha\sin\beta = \frac{1}{2}[\cos(\alpha-\beta)-\cos(\alpha+\beta)] \tag{48}$$

$$\cos\alpha\cos\beta = \frac{1}{2}[\cos(\alpha-\beta)+\cos(\alpha+\beta)] \tag{49}$$

We also leave it for the exercises to derive the following ***sum-to-product formulas***:

$$\sin\alpha + \sin\beta = 2\sin\frac{\alpha+\beta}{2}\cos\frac{\alpha-\beta}{2} \tag{50}$$

$$\sin\alpha - \sin\beta = 2\cos\frac{\alpha+\beta}{2}\sin\frac{\alpha-\beta}{2} \tag{51}$$

$$\cos\alpha + \cos\beta = 2\cos\frac{\alpha+\beta}{2}\cos\frac{\alpha-\beta}{2} \tag{52}$$

$$\cos\alpha - \cos\beta = -2\sin\frac{\alpha+\beta}{2}\sin\frac{\alpha-\beta}{2} \tag{53}$$

**FINDING AN ANGLE FROM THE VALUE OF ITS TRIGONOMETRIC FUNCTIONS**

There are numerous situations in which it is necessary to find an unknown angle from a known value of one of its trigonometric functions. The following example illustrates a method for doing this.

**Example 6**
Find $\theta$ if $\sin\theta = \frac{1}{2}$.

*Solution.*   We begin by looking for positive angles that satisfy the equation. Because $\sin\theta$ is positive, the angle $\theta$ must terminate in the first or second quadrant. If it terminates in the first quadrant, then the hypotenuse of $\triangle OAP$ in Figure E.18$a$ is double the leg $AP$, so

$$\theta = 30° = \frac{\pi}{6}\text{ radians}$$

If $\theta$ terminates in the second quadrant (Figure E.18$b$), then the hypotenuse of $\triangle OAP$ is double the leg $AP$, so $\angle AOP = 30°$, which implies that

$$\theta = 180° - 30° = 150° = \frac{5\pi}{6}\text{ radians}$$

Now that we have found these two solutions, all other solutions are obtained by adding or subtracting multiples of $360°$ ($2\pi$ radians) to them. Thus, the entire set of solutions is given by the formulas

$$\theta = 30° \pm n\cdot 360°, \quad n = 0, 1, 2, \ldots$$

and

$$\theta = 150° \pm n\cdot 360°, \quad n = 0, 1, 2, \ldots$$

or in radian measure,

$$\theta = \frac{\pi}{6} \pm n\cdot 2\pi, \quad n = 0, 1, 2, \ldots$$

and

$$\theta = \frac{5\pi}{6} \pm n\cdot 2\pi, \quad n = 0, 1, 2, \ldots \quad \blacktriangleleft$$

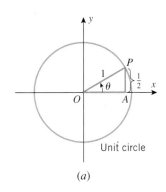

Unit circle

(a)

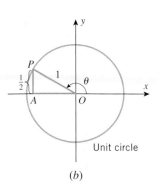

Unit circle

(b)

Figure E.18

## EXERCISE SET E

In Exercises 1 and 2, express the angles in radians.

**1.** (a) $75°$   (b) $390°$   (c) $20°$   (d) $138°$

**2.** (a) $420°$   (b) $15°$   (c) $225°$   (d) $165°$

In Exercises 3 and 4, express the angles in degrees.

**3.** (a) $\pi/15$   (b) $1.5$   (c) $8\pi/5$   (d) $3\pi$

**4.** (a) $\pi/10$   (b) $2$   (c) $2\pi/5$   (d) $7\pi/6$

In Exercises 5 and 6, find the exact values of all six trigonometric functions of $\theta$.

**5.** (a)          (b)          (c)

**6.** (a)          (b)          (c)

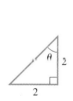  

In Exercises 7–12, the angle $\theta$ is an acute angle of a right triangle. Solve the problems by drawing an appropriate right triangle. Do *not* use a calculator.

**7.** Find $\sin\theta$ and $\cos\theta$ given that $\tan\theta = 3$.

**8.** Find $\sin\theta$ and $\tan\theta$ given that $\cos\theta = \frac{2}{3}$.

**9.** Find $\tan\theta$ and $\csc\theta$ given that $\sec\theta = \frac{5}{2}$.

**10.** Find $\cot\theta$ and $\sec\theta$ given that $\csc\theta = 4$.

**11.** Find the length of the side adjacent to $\theta$ given that the hypotenuse has length 6 and $\cos\theta = 0.3$.

**12.** Find the length of the hypotenuse given that the side opposite $\theta$ has length 2.4 and $\sin\theta = 0.8$.

In Exercises 13 and 14, the value of an angle $\theta$ is given. Find the values of all six trigonometric functions of $\theta$ without using a calculator.

**13.** (a) $225°$   (b) $-210°$   (c) $5\pi/3$   (d) $-3\pi/2$

**14.** (a) $330°$   (b) $-120°$   (c) $9\pi/4$   (d) $-3\pi$

In Exercises 15 and 16, use the information to find the exact values of the remaining five trigonometric functions of $\theta$.

**15.** (a) $\cos\theta = \frac{3}{5}$, $0 < \theta < \pi/2$

(b) $\cos\theta = \frac{3}{5}$, $-\pi/2 < \theta < 0$

(c) $\tan\theta = -1/\sqrt{3}$, $\pi/2 < \theta < \pi$

(d) $\tan\theta = -1/\sqrt{3}$, $-\pi/2 < \theta < 0$

(e) $\csc\theta = \sqrt{2}$, $0 < \theta < \pi/2$

(f) $\csc\theta = \sqrt{2}$, $\pi/2 < \theta < \pi$

**16.** (a) $\sin\theta = \frac{1}{4}$, $0 < \theta < \pi/2$

(b) $\sin\theta = \frac{1}{4}$, $\pi/2 < \theta < \pi$

(c) $\cot\theta = \frac{1}{3}$, $0 < \theta < \pi/2$

(d) $\cot\theta = \frac{1}{3}$, $\pi < \theta < 3\pi/2$

(e) $\sec\theta = -\frac{5}{2}$, $\pi/2 < \theta < \pi$

(f) $\sec\theta = -\frac{5}{2}$, $\pi < \theta < 3\pi/2$

In Exercises 17 and 18, use a calculating utility to find $x$ to four decimal places.

**17.** (a)          (b)

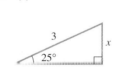 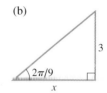

**18.** (a)          (b)

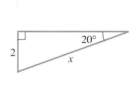 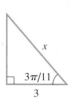

**19.** In each part, let $\theta$ be an acute angle of a right triangle. Express the remaining five trigonometric functions in terms of $a$.

(a) $\sin\theta = a/3$   (b) $\tan\theta = a/5$   (c) $\sec\theta = a$

In Exercises 20–27, find all values of $\theta$ (in radians) that satisfy the given equation. Do not use a calculator.

**20.** (a) $\cos\theta = -1/\sqrt{2}$          (b) $\sin\theta = -1/\sqrt{2}$

**21.** (a) $\tan\theta = -1$          (b) $\cos\theta = \frac{1}{2}$

**22.** (a) $\sin\theta = -\frac{1}{2}$          (b) $\tan\theta = \sqrt{3}$

**23.** (a) $\tan\theta = 1/\sqrt{3}$          (b) $\sin\theta = -\sqrt{3}/2$

**24.** (a) $\sin\theta = -1$          (b) $\cos\theta = -1$

**25.** (a) $\cot\theta = -1$      (b) $\cot\theta = \sqrt{3}$

**26.** (a) $\sec\theta = -2$      (b) $\csc\theta = -2$

**27.** (a) $\csc\theta = 2/\sqrt{3}$      (b) $\sec\theta = 2/\sqrt{3}$

> In Exercises 28 and 29, find the values of all six trigonometric functions of $\theta$.

**28.**

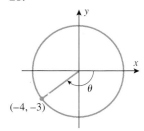

**29.**

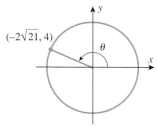

$(-2\sqrt{21}, 4)$

**30.** Find all values of $\theta$ (in radians) such that
(a) $\sin\theta = 1$    (b) $\cos\theta = 1$    (c) $\tan\theta = 1$
(d) $\csc\theta = 1$    (e) $\sec\theta = 1$    (f) $\cot\theta = 1$.

**31.** Find all values of $\theta$ (in radians) such that
(a) $\sin\theta = 0$    (b) $\cos\theta = 0$    (c) $\tan\theta = 0$
(d) $\csc\theta$ is undefined    (e) $\sec\theta$ is undefined
(f) $\cot\theta$ is undefined.

**32.** How could you use a ruler and protractor to approximate $\sin 17°$ and $\cos 17°$?

**33.** Find the length of the circular arc on a circle of radius 4 cm subtended by an angle of
(a) $\pi/6$      (b) $150°$.

**34.** Find the radius of a circular sector that has an angle of $\pi/3$ and a circular arc length of 7 units.

**35.** A point $P$ moving counterclockwise on a circle of radius 5 cm traverses an arc length of 2 cm. What is the angle swept out by a radius from the center to $P$?

**36.** Find a formula for the area $A$ of a circular sector in terms of its radius $r$ and arc length $s$.

**37.** As shown in the accompanying figure, a right circular cone is made from a circular piece of paper of radius $R$ by cutting out a sector of angle $\theta$ radians and gluing the cut edges of the remaining piece together. Find
(a) the radius $r$ of the base of the cone in terms of $R$ and $\theta$
(b) the height $h$ of the cone in terms of $R$ and $\theta$.

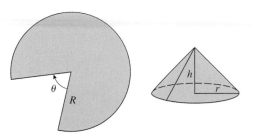

Figure Ex-37

**38.** As shown in the accompanying figure, let $r$ and $L$ be the radius of the base and the slant height of a right circular cone. Show that the lateral surface area, $S$, of the cone is $S = \pi rL$. [*Hint:* As shown in the figure in Exercise 37, the lateral surface of the cone becomes a circular sector when cut along a line from the vertex to the base and flattened.]

Figure Ex-38

**39.** Two sides of a triangle have lengths of 3 cm and 7 cm and meet at an angle of $60°$. Find the area of the triangle.

**40.** Let $ABC$ be a triangle whose angles at $A$ and $B$ are $30°$ and $45°$. If the side opposite the angle $B$ has length 9, find the lengths of the remaining sides and the size of the angle $C$.

**41.** A 10-foot ladder leans against a house and makes an angle of $67°$ with level ground. How far is the top of the ladder above the ground? Express your answer to the nearest tenth of a foot.

**42.** From a point 120 feet on level ground from a building, the angle of elevation to the top of the building is $76°$. Find the height of the building. Express your answer to the nearest foot.

**43.** An observer on level ground is at a distance $d$ from a building. The angles of elevation to the bottom of the windows on the second and third floors are $\alpha$ and $\beta$, respectively. Find the distance $h$ between the bottoms of the windows in terms of $\alpha$, $\beta$, and $d$.

**44.** From a point on level ground, the angle of elevation to the top of a tower is $\alpha$. From a point that is $d$ units closer to the tower, the angle of elevation is $\beta$. Find the height $h$ of the tower in terms of $\alpha$, $\beta$, and $d$.

> In Exercises 45 and 46, do *not* use a calculator.

**45.** If $\cos\theta = \frac{2}{3}$ and $0 < \theta < \pi/2$, find
(a) $\sin 2\theta$      (b) $\cos 2\theta$.

**46.** If $\tan\alpha = \frac{3}{4}$ and $\tan\beta = 2$, where $0 < \alpha < \pi/2$ and $0 < \beta < \pi/2$, find
(a) $\sin(\alpha - \beta)$      (b) $\cos(\alpha + \beta)$.

**47.** Express $\sin 3\theta$ and $\cos 3\theta$ in terms of $\sin\theta$ and $\cos\theta$.

> In Exercises 48–58, derive the given identities.

**48.** $\dfrac{\cos\theta \sec\theta}{1 + \tan^2\theta} = \cos^2\theta$

**49.** $\dfrac{\cos\theta \tan\theta + \sin\theta}{\tan\theta} = 2\cos\theta$

**50.** $2\csc 2\theta = \sec\theta \csc\theta$      **51.** $\tan\theta + \cot\theta = 2\csc 2\theta$

**52.** $\dfrac{\sin 2\theta}{\sin \theta} - \dfrac{\cos 2\theta}{\cos \theta} = \sec \theta$

**53.** $\dfrac{\sin \theta + \cos 2\theta - 1}{\cos \theta - \sin 2\theta} = \tan \theta$

**54.** $\sin 3\theta + \sin \theta = 2 \sin 2\theta \cos \theta$

**55.** $\sin 3\theta - \sin \theta = 2 \cos 2\theta \sin \theta$

**56.** $\tan \dfrac{\theta}{2} = \dfrac{1 - \cos \theta}{\sin \theta}$     **57.** $\tan \dfrac{\theta}{2} = \dfrac{\sin \theta}{1 + \cos \theta}$

**58.** $\cos \left(\dfrac{\pi}{3} + \theta\right) + \cos \left(\dfrac{\pi}{3} - \theta\right) = \cos \theta$

Exercises 59 and 60 refer to an arbitrary triangle $ABC$ in which the side of length $a$ is opposite angle $A$, the side of length $b$ is opposite angle $B$, and the side of length $c$ is opposite angle $C$.

**59.** Prove: The area of a triangle $ABC$ can be written as

$$\text{area} = \tfrac{1}{2} bc \sin A$$

Find two other similar formulas for the area.

**60.** Prove the *law of sines*: In any triangle, the ratios of the sides to the sines of the opposite angles are equal; that is,

$$\frac{a}{\sin A} = \frac{b}{\sin B} = \frac{c}{\sin C}$$

**61.** Use identities (34) through (37) to express each of the following in terms of $\sin \theta$ or $\cos \theta$.

(a) $\sin \left(\dfrac{\pi}{2} + \theta\right)$     (b) $\cos \left(\dfrac{\pi}{2} + \theta\right)$

(c) $\sin \left(\dfrac{3\pi}{2} - \theta\right)$     (d) $\cos \left(\dfrac{3\pi}{2} + \theta\right)$

**62.** Derive identities (38) and (39).

**63.** Derive identity

(a) (47)     (b) (48)     (c) (49).

**64.** If $A = \alpha + \beta$ and $B = \alpha - \beta$, then $\alpha = \tfrac{1}{2}(A + B)$ and $\beta = \tfrac{1}{2}(A - B)$ (verify). Use this result and identities (47) through (49) to derive identity

(a) (50)     (b) (52)     (c) (53).

**65.** Substitute $-\beta$ for $\beta$ in identity (50) to derive identity (51).

**66.** (a) Express $3 \sin \alpha + 5 \cos \alpha$ in the form

$$C \sin(\alpha + \phi)$$

(b) Show that a sum of the form

$$A \sin \alpha + B \cos \alpha$$

can be rewritten in the form $C \sin(\alpha + \phi)$.

**67.** Show that the length of the diagonal of the parallelogram in the accompanying figure is

$$d = \sqrt{a^2 + b^2 + 2ab \cos \theta}$$

Figure Ex-67

# APPENDIX F

## Solving Polynomial Equations

In the subsection of Section 5.3 entitled A Brief Review of Polynomials, we reviewed some of the basic ideas and terminology concerning polynomials. We will assume in this appendix that you have read that material, and we will also assume that you know how to divide polynomials using long division and synthetic division. If you need to review those techniques, refer to an algebra book.

**THE REMAINDER THEOREM**

When two positive integers are divided, the numerator can be expressed as the quotient plus the remainder over the divisor, where the remainder is less than the divisor. For example,

$$\tfrac{17}{5} = 3 + \tfrac{2}{5}$$

If we multiply this equation through by 5, we obtain

$$17 = 5 \cdot 3 + 2$$

which states that the *numerator is the divisor times the quotient plus the remainder*.

The following theorem, which we state without proof, is an analogous result for division of polynomials.

---

**F.1    THEOREM.**    *If $p(x)$ and $s(x)$ are polynomials, and if $s(x)$ is not the zero polynomial, then $p(x)$ can be expressed as*

$$p(x) = s(x)q(x) + r(x)$$

*where $q(x)$ and $r(x)$ are the quotient and remainder that result when $p(x)$ is divided by $s(x)$, and the degree of $r(x)$ is less than the degree of $s(x)$.*

---

In the special case where $p(x)$ is divided by a first-degree polynomial of the form $x - c$, the degree of the remainder must be 0, since it is less than the degree of $x - c$, which is 1. This implies that the remainder is a constant, say $r$. Thus, Theorem F.1 implies that

$$p(x) = (x - c)q(x) + r$$

and this in turn implies that $p(c) = r$. In summary, we have the following theorem.

---

**F.2    THEOREM (*Remainder Theorem*).**    *If a polynomial $p(x)$ is divided by $x - c$, then the remainder is $p(c)$.*

---

### Example 1

According to the Remainder Theorem, the remainder on dividing

$$p(x) = 2x^3 + 3x^2 - 4x - 3$$

by $x + 4$ should be

$$p(-4) = 2(-4)^3 + 3(-4)^2 - 4(-4) - 3 = -67$$

Show that this is so.

*Solution.*  By long division

$$
\begin{array}{r}
2x^2 - 5x + 16 \\
x + 4\overline{)2x^3 + 3x^2 - 4x - 3} \\
\underline{2x^3 + 8x^2\phantom{xxxxxxxx}} \\
-5x^2 - 4x\phantom{xxxx} \\
\underline{-5x^2 - 20x\phantom{xxxx}} \\
16x - 3 \\
\underline{16x + 64} \\
-67
\end{array}
$$

which shows that the remainder is $-67$.

*Alternative Solution.*  Because we are dividing by an expression of the form $x - c$ (where $c = -4$), we can use synthetic division rather than long division. The computations are

$$
\begin{array}{r|rrrr}
-4 & 2 & 3 & -4 & -3 \\
   &   & -8 & 20 & -64 \\
\hline
   & 2 & -5 & 16 & -67
\end{array}
$$

which again shows that the remainder is $-67$.  ◀

**THE FACTOR THEOREM**

To *factor* a polynomial $p(x)$ is to write it as a product of lower-degree polynomials, called *factors* of $p(x)$. For $s(x)$ to be a factor of $p(x)$ there must be no remainder when $p(x)$ is divided by $s(x)$. For example, if $p(x)$ can be factored as

$$p(x) = s(x)q(x) \tag{1}$$

then

$$\frac{p(x)}{s(x)} = q(x) \tag{2}$$

so dividing $p(x)$ by $s(x)$ produces a quotient $q(x)$ with no remainder. Conversely, (2) implies (1), so $s(x)$ is a factor of $p(x)$ if there is no remainder when $p(x)$ is divided by $s(x)$.

In the special case where $x - c$ is a factor of $p(x)$, the polynomial $p(x)$ can be expressed as

$$p(x) = (x - c)q(x)$$

which implies that $p(c) = 0$. Conversely, if $p(c) = 0$, then the Remainder Theorem implies that $x - c$ is a factor of $p(x)$, since the remainder is 0 when $p(x)$ is divided by $x - c$. These results are summarized in the following theorem.

> **F.3**  THEOREM (*Factor Theorem*).  *A polynomial $p(x)$ has a factor $x - c$ if and only if $p(c) = 0$.*

It follows from this theorem that the statements below say the same thing in different ways:

- $x - c$ is a factor of $p(x)$.
- $p(c) = 0$.
- $c$ is a zero of $p(x)$.
- $c$ is a root of the equation $p(x) = 0$.
- $c$ is a solution of the equation $p(x) = 0$.
- $c$ is an $x$-intercept of $y = p(x)$.

### Example 2

Confirm that $x - 1$ is a factor of

$$p(x) = x^3 - 3x^2 - 13x + 15$$

by dividing $x - 1$ into $p(x)$ and checking that the remainder is zero.

*Solution.* By long division

$$
\begin{array}{r}
x^2 - 2x - 15 \\
x - 1 \overline{\smash{)}x^3 - 3x^2 - 13x + 15} \\
\underline{x^3 - x^2} \\
-2x^2 - 13x \\
\underline{-2x^2 + 2x} \\
-15x + 15 \\
\underline{-15x + 15} \\
0
\end{array}
$$

which shows that the remainder is zero.

*Alternative Solution.* Because we are dividing by an expression of the form $x - c$, we can use synthetic division rather than long division. The computations are

$$
\begin{array}{r|rrrr}
1\rfloor & 1 & -3 & -13 & 15 \\
& & 1 & -2 & -15 \\
\hline
& 1 & -2 & -15 & 0
\end{array}
$$

which again confirms that the remainder is zero.  ◀

**USING ONE FACTOR TO FIND OTHER FACTORS**

If $x - c$ is a factor of $p(x)$, and if $q(x) = p(x)/(x - c)$, then

$$p(x) = (x - c)q(x) \tag{3}$$

so that additional linear factors of $p(x)$ can be obtained by factoring the quotient $q(x)$.

### Example 3

Factor

$$p(x) = x^3 - 3x^2 - 13x + 15 \tag{4}$$

completely into linear factors.

*Solution.* We showed in Example 2 that $x - 1$ is a factor of $p(x)$ and we also showed that $p(x)/(x - 1) = x^2 - 2x - 15$. Thus,

$$x^3 - 3x^2 - 13x + 15 = (x - 1)(x^2 - 2x - 15)$$

Factoring $x^2 - 2x - 15$ by inspection yields

$$x^3 - 3x^2 - 13x + 15 = (x - 1)(x - 5)(x + 3)$$

which is the complete linear factorization of $p(x)$.  ◀

A general quadratic equation $ax^2 + bx + c = 0$ can be solved by using the quadratic formula to express the solutions of the equation in terms of the coefficients. Versions of this formula were known since Babylonian times, and by the seventeenth century formulas had been obtained for solving general cubic and quartic equations. However, attempts to find formulas for the solutions of general fifth-degree equations and higher proved fruitless. The reason for this became clear in 1829 when the French mathematician Evariste Galois (1811–1832) proved that it is impossible to express the solutions of a general fifth-degree equation or higher in terms of its coefficients using algebraic operations.

Today, we have powerful computer programs for finding the zeros of specific polynomials. For example, it takes only seconds for a computer algebra system, such as *Mathematica*, *Maple*, or *Derive*, to show that the zeros of the polynomial

$$p(x) = 10x^4 - 23x^3 - 10x^2 + 29x + 6 \tag{5}$$

are

$$x = -1, \quad x = -\tfrac{1}{5}, \quad x = \tfrac{3}{2}, \quad \text{and} \quad x = 2 \tag{6}$$

The algorithms that these programs use to find the integer and rational zeros of a polynomial, if any, are based on the following theorem, which is proved in advanced algebra courses.

---

**F.4** THEOREM. *Suppose that*

$$p(x) = c_n x^n + c_{n-1} x^{n-1} + \cdots + c_1 x + c_0$$

*is a polynomial with integer coefficients.*

(a) *If $r$ is an integer zero of $p(x)$, then $r$ must be a divisor of the constant term $c_0$.*

(b) *If $r = a/b$ is a rational zero of $p(x)$ in which all common factors of $a$ and $b$ have been canceled, then $a$ must be a divisor of the constant term $c_0$, and $b$ must be a divisor of the leading coefficient $c_n$.*

---

For example, in (5) the constant term is 6 (which has divisors $\pm 1, \pm 2, \pm 3$, and $\pm 6$) and the leading coefficient is 10 (which has divisors $\pm 1, \pm 2, \pm 5$, and $\pm 10$). Thus, the only possible integer zeros of $p(x)$ are

$$\pm 1, \quad \pm 2, \quad \pm 3, \quad \pm 6$$

and the only possible noninteger rational zeros are

$$\pm \tfrac{1}{2}, \quad \pm \tfrac{1}{5}, \quad \pm \tfrac{1}{10}, \quad \pm \tfrac{2}{5}, \quad \pm \tfrac{3}{2}, \quad \pm \tfrac{3}{5}, \quad \pm \tfrac{3}{10}, \quad \pm \tfrac{6}{5}$$

Using a computer, it is a simple matter to evaluate $p(x)$ at each of the numbers in these lists to show that its only zeros are the numbers in (6).

## Example 4

Solve the equation $x^3 + 3x^2 - 7x - 21 = 0$.

*Solution.* The solutions of the equation are the zeros of the polynomial

$$p(x) = x^3 + 3x^2 - 7x - 21$$

We will look for integer zeros first. All such zeros must divide the constant term, so the only possibilities are $\pm 1, \pm 3, \pm 7$, and $\pm 21$. Substituting these values into $p(x)$ (or using the method of Exercise 6) shows that $x = -3$ is an integer zero. This tells us that $x + 3$ is a factor of $p(x)$ and that $p(x)$ can be written as

$$x^3 + 3x^2 - 7x - 21 = (x + 3)q(x)$$

where $q(x)$ is the quotient that results when $x^3 + 3x^2 - 7x - 21$ is divided by $x + 3$. We leave it for you to perform the division and show that $q(x) = x^2 - 7$; hence,

$$x^3 + 3x^2 - 7x - 21 = (x + 3)(x^2 - 7) = (x + 1)(x + \sqrt{7})(x - \sqrt{7})$$

which tells us that the solutions of the given equation are $x = 3$, $x = \sqrt{7} \approx 2.65$, and $x = -\sqrt{7} \approx -2.65$.   ◄

## EXERCISE SET F   ⌇ Graphing Calculator   [c] CAS

In Exercises 1 and 2, find the quotient $q(x)$ and the remainder $r(x)$ that result when $p(x)$ is divided by $s(x)$.

**1.** (a) $p(x) = x^4 + 3x^3 - 5x + 10$;  $s(x) = x^2 - x + 2$
   (b) $p(x) = 6x^4 + 10x^2 + 5$;  $s(x) = 3x^2 - 1$
   (c) $p(x) = x^5 + x^3 + 1$;  $s(x) = x^2 + x$

**2.** (a) $p(x) = 2x^4 - 3x^3 + 5x^2 + 2x + 7$; $s(x) = x^2 - x + 1$
   (b) $p(x) = 2x^5 + 5x^4 - 4x^3 + 8x^2 + 1$; $s(x) = 2x^2 - x + 1$
   (c) $p(x) = 5x^6 + 4x^2 + 5$; $s(x) = x^3 + 1$

In Exercises 3 and 4, use synthetic division to find the quotient $q(x)$ and the remainder $r$ that result when $p(x)$ is divided by $s(x)$.

**3.** (a) $p(x) = 3x^3 - 4x - 1$;  $s(x) = x - 2$
   (b) $p(x) = x^4 - 5x^2 + 4$;  $s(x) = x + 5$
   (c) $p(x) = x^5 - 1$;  $s(x) = x - 1$

**4.** (a) $p(x) = 2x^3 - x^2 - 2x + 1$;  $s(x) = x - 1$
   (b) $p(x) = 2x^4 + 3x^3 - 17x^2 - 27x - 9$;  $s(x) = x + 4$
   (c) $p(x) = x^7 + 1$;  $s(x) = x - 1$

**5.** Let $p(x) = 2x^4 + x^3 - 3x^2 + x - 4$. Use synthetic division and the Remainder Theorem to find $p(0)$, $p(1)$, $p(-3)$, and $p(7)$.

**6.** Let $p(x)$ be the polynomial in Example 4. Use synthetic division and the Remainder Theorem to evaluate $p(x)$ at $x = \pm 1, \pm 3, \pm 7,$ and $\pm 21$.

**7.** Let $p(x) = x^3 + 4x^2 + x - 6$. Find a polynomial $q(x)$ and a constant $r$ such that
   (a) $p(x) = (x - 2)q(x) + r$
   (b) $p(x) = (x + 1)q(x) + r$.

**8.** Let $p(x) = x^5 - 1$. Find a polynomial $q(x)$ and a constant $r$ such that
   (a) $p(x) = (x + 1)q(x) + r$
   (b) $p(x) = (x - 1)q(x) + r$.

**9.** In each part, make a list of all possible candidates for the rational zeros of $p(x)$.
   (a) $p(x) = x^7 + 3x^3 - x + 24$
   (b) $p(x) = 3x^4 - 2x^2 + 7x - 10$
   (c) $p(x) = x^{35} - 17$

**10.** Find all integer zeros of

$$p(x) = x^6 + 5x^5 - 16x^4 - 15x^3 - 12x^2 - 38x - 21$$

In Exercises 11–15, factor the polynomials completely.

**11.** $p(x) = x^3 - 2x^2 - x + 2$

**12.** $p(x) = 3x^3 + x^2 - 12x - 4$

**13.** $p(x) = x^4 + 10x^3 + 36x^2 + 54x + 27$

**14.** $p(x) = 2x^4 + x^3 - 19x^2 + 9$

**15.** $p(x) = x^5 + 4x^4 - 4x^3 - 34x^2 - 45x - 18$

[c] **16.** For each of the factorizations that you obtained in Exercises 11–15, check your answer using a CAS.

In Exercises 17–21, find all solutions of the equations.

**17.** $x^3 + 3x^2 + 4x + 12 = 0$

**18.** $2x^3 - 5x^2 - 10x + 3 = 0$

**19.** $3x^4 + 14x^3 + 14x^2 - 8x - 8 = 0$

**20.** $2x^4 - x^3 - 14x^2 - 5x + 6 = 0$

**21.** $x^5 - 2x^4 - 6x^3 + 5x^2 + 8x + 12 = 0$

[c] **22.** For each of the equations you solved in Exercises 17–21, check your answer using a CAS.

**23.** Find all values of $k$ for which $x - 1$ is a factor of the polynomial $p(x) = k^2 x^3 - 7kx + 10$.

**24.** Is $x + 3$ a factor of $x^7 + 2187$? Justify your answer.

[c] **25.** A 3-cm-thick slice is cut from a cube, leaving a volume of 196 cm³. Use a CAS to find the length of a side of the original cube.

**26.** (a) Show that there is no rational number that exceeds its cube by 1.
   (b) Does there exist a real number that exceeds its cube by 1? Justify your answer.

**27.** Use the Factor Theorem to show each of the following.
   (a) $x - y$ is a factor of $x^n - y^n$ for all positive integer values of $n$.
   (b) $x + y$ is a factor of $x^n - y^n$ for all positive even integer values of $n$.
   (c) $x + y$ is a factor of $x^n + y^n$ for all positive odd integer values of $n$.

# APPENDIX G

## Selected Proofs

An extensive excursion into proofs of limit theorems would be too time consuming to undertake, so we have selected a few proofs of results from Section 2.2 that illustrate some of the basic ideas.

---

**G.1 THEOREM.** *Let $k$ be a constant, and suppose that $\lim\limits_{x \to a} f(x) = L_1$ and that $\lim\limits_{x \to a} g(x) = L_2$. Then*

*(a)* $\lim\limits_{x \to a} k = k$

*(b)* $\lim\limits_{x \to a} [f(x) + g(x)] = \lim\limits_{x \to a} f(x) + \lim\limits_{x \to a} g(x) = L_1 + L_2$

*(c)* $\lim\limits_{x \to a} [f(x)g(x)] = \lim\limits_{x \to a} f(x) \lim\limits_{x \to a} g(x) = L_1 L_2$

---

*Proof (a).* We will apply Definition 2.3.3 with $f(x) = k$ and $L = k$. Thus, given $\epsilon > 0$, we must find a number $\delta > 0$ such that

$$|k - k| < \epsilon \quad \text{if} \quad 0 < |x - a| < \delta$$

or equivalently,

$$0 < \epsilon \quad \text{if} \quad 0 < |x - a| < \delta$$

But the condition on the left side of this statement is *always* true, no matter how $\delta$ is chosen. Thus, any positive value for $\delta$ will suffice.

*Proof (b).* We must show that given $\epsilon > 0$ we can find a number $\delta > 0$ such that

$$|(f(x) + g(x)) - (L_1 + L_2)| < \epsilon \quad \text{if} \quad 0 < |x - a| < \delta \tag{1}$$

However, from the limits of $f$ and $g$ in the hypothesis of the theorem we can find numbers $\delta_1$ and $\delta_2$ such that

$$|f(x) - L_1| < \epsilon/2 \quad \text{if} \quad 0 < |x - a| < \delta_1$$

$$|g(x) - L_2| < \epsilon/2 \quad \text{if} \quad 0 < |x - a| < \delta_2$$

Moreover, the inequalities on the left sides of these statements *both* hold if we replace $\delta_1$ and $\delta_2$ by any positive number $\delta$ that is less than both $\delta_1$ and $\delta_2$. Thus, for any such $\delta$ it follows that

$$|f(x) - L_1| + |g(x) - L_2| < \epsilon \quad \text{if} \quad 0 < |x - a| < \delta \tag{2}$$

However, it follows from the triangle inequality [Theorem 1.2.2(d)] that

$$|(f(x) + g(x)) - (L_1 + L_2)| = |(f(x) - L_1) + (g(x) - L_2)|$$
$$\leq |f(x) - L_1| + |g(x) - L_2|$$

so that (1) follows from (2).

***Proof (c).*** We must show that given $\epsilon > 0$ we can find a number $\delta > 0$ such that

$$|f(x)g(x) - L_1L_2| < \epsilon \quad \text{if} \quad 0 < |x - a| < \delta \tag{3}$$

To find $\delta$ it will be helpful to express (3) in a different form. If we rewrite $f(x)$ and $g(x)$ as

$$f(x) = L_1 + (f(x) - L_1) \quad \text{and} \quad g(x) = L_2 + (g(x) - L_2)$$

then the inequality on the left side of (3) can be expressed as (verify)

$$|L_1(g(x) - L_2) + L_2(f(x) - L_1) + (f(x) - L_1)(g(x) - L_2)| < \epsilon \tag{4}$$

Since

$$\lim_{x \to a} f(x) = L_1 \quad \text{and} \quad \lim_{x \to a} g(x) = L_2$$

we can find positive numbers $\delta_1, \delta_2, \delta_3,$ and $\delta_4$ such that

$$
\begin{array}{ll}
|f(x) - L_1| < \sqrt{\epsilon/3} & \text{if} \quad 0 < |x - a| < \delta_1 \\[2mm]
|f(x) - L_1| < \dfrac{\epsilon}{3(1 + |L_2|)} & \text{if} \quad 0 < |x - a| < \delta_2 \\[3mm]
|g(x) - L_2| < \sqrt{\epsilon/3} & \text{if} \quad 0 < |x - a| < \delta_3 \\[2mm]
|g(x) - L_2| < \dfrac{\epsilon}{3(1 + |L_1|)} & \text{if} \quad 0 < |x - a| < \delta_4
\end{array}
\tag{5}
$$

Moreover, the inequalities on the left sides of these four statements *all* hold if we replace $\delta_1, \delta_2, \delta_3,$ and $\delta_4$ by any number $\delta$ that is smaller than $\delta_1, \delta_2, \delta_3,$ and $\delta_4$. Thus, for any such $\delta$ it follows with the help of the triangle inequality that

$$
\begin{aligned}
|L_1(g(x) &- L_2) + L_2(f(x) - L_1) + (f(x) - L_1)(g(x) - L_2)| \\[2mm]
&\leq |L_1(g(x) - L_2)| + |L_2(f(x) - L_1)| + |(f(x) - L_1)(g(x) - L_2)| \\[2mm]
&= |L_1||g(x) - L_2| + |L_2||f(x) - L_1| + |f(x) - L_1||g(x) - L_2| \\[2mm]
&< |L_1|\frac{\epsilon}{3(1 + |L_1|)} + |L_2|\frac{\epsilon}{3(1 + |L_2|)} + \sqrt{\epsilon/3}\sqrt{\epsilon/3} \quad \boxed{\text{From (5)}} \\[2mm]
&= \frac{\epsilon}{3}\frac{|L_1|}{1 + |L_1|} + \frac{\epsilon}{3}\frac{|L_2|}{1 + |L_2|} + \frac{\epsilon}{3} \\[2mm]
&< \frac{\epsilon}{3} + \frac{\epsilon}{3} + \frac{\epsilon}{3} = \epsilon \quad \boxed{\text{Since } \dfrac{|L_1|}{1 + |L_1|} < 1 \text{ and } \dfrac{|L_2|}{1 + |L_2|} < 1}
\end{aligned}
$$

which shows that (4) holds for the $\delta$ selected.  ∎

REMARK.    Do not be alarmed if the proof of part (c) seems difficult; it takes some experience with proofs of this type to develop a feel for choosing the right $\delta$. Your initial goal should be to understand the ideas and the computations.

Next, we will prove Theorem 2.4.5 for two-sided limits.

**PROOF OF A BASIC CONTINUITY PROPERTY**

> **G.2**    THEOREM (***Theorem 2.4.5***).    *If* $\lim\limits_{x \to c} g(x) = L$ *and if the function* $f$ *is continuous at* $L$, *then* $\lim\limits_{x \to c} f(g(x)) = f(L)$; *that is,* $\lim\limits_{x \to c} f(g(x)) = f(\lim\limits_{x \to c} g(x))$.

***Proof.***    We must show that given $\epsilon > 0$, we can find a number $\delta > 0$ such that

$$|f(g(x)) - f(L)| < \epsilon \quad \text{if} \quad 0 < |x - c| < \delta \tag{6}$$

Since $f$ is continuous at $L$, we have

$$\lim_{u \to L} f(u) = f(L)$$

and hence we can find a number $\delta_1 > 0$ such that

$$|f(u) - f(L)| < \epsilon \quad \text{if} \quad |u - L| < \delta_1$$

In particular, if $u = g(x)$, then

$$|f(g(x)) - f(L)| < \epsilon \quad \text{if} \quad |g(x) - L| < \delta_1 \tag{7}$$

But $\lim_{x \to c} g(x) = L$, and hence there is a number $\delta > 0$ such that

$$|g(x) - L| < \delta_1 \quad \text{if} \quad 0 < |x - c| < \delta \tag{8}$$

Thus, if $x$ satisfies the condition on the right side of statement (8), then it follows that $g(x)$ satisfies the condition on the right side of statement (7), and this implies that the condition on the left side of statement (6) is satisfied, completing the proof. ∎

**PROOF OF THE CHAIN RULE**

Next, we will prove the chain rule (Theorem 3.5.2), but first we need a preliminary result.

---

**G.3**   THEOREM.   *If $f$ is differentiable at $x$ and if $y = f(x)$, then*

$$\Delta y = f'(x)\Delta x + \epsilon \Delta x$$

*where $\epsilon \to 0$ as $\Delta x \to 0$ and $\epsilon = 0$ if $\Delta x = 0$.*

---

*Proof.*   Define

$$\epsilon = \begin{cases} \dfrac{f(x + \Delta x) - f(x)}{\Delta x} - f'(x) & \text{if } \Delta x \neq 0 \\ \\ 0 & \text{if } \Delta x = 0 \end{cases} \tag{9}$$

If $\Delta x \neq 0$, it follows from (9) that

$$\epsilon \Delta x = [f(x + \Delta x) - f(x)] - f'(x)\Delta x \tag{10}$$

But

$$\Delta y = f(x + \Delta x) - f(x) \tag{11}$$

so (10) can be written as

$$\epsilon \Delta x = \Delta y - f'(x)\Delta x$$

or

$$\Delta y = f'(x)\Delta x + \epsilon \Delta x \tag{12}$$

If $\Delta x = 0$, then (12) still holds (why?), so (12) is valid for all values of $\Delta x$. It remains to show that $\epsilon \to 0$ as $\Delta x \to 0$. But this follows from the assumption that $f$ is differentiable at $x$, since

$$\lim_{\Delta x \to 0} \epsilon = \lim_{\Delta x \to 0} \left[ \frac{f(x + \Delta x) - f(x)}{\Delta x} - f'(x) \right] = f'(x) - f'(x) = 0 \qquad \blacksquare$$

We are now ready to prove the chain rule.

---

**G.4**   THEOREM (*Theorem 3.5.2*).   *If $g$ is differentiable at the point $x$ and $f$ is differentiable at the point $g(x)$, then the composition $f \circ g$ is differentiable at the point $x$. Moreover, if $y = f(g(x))$ and $u = g(x)$, then*

$$\frac{dy}{dx} = \frac{dy}{du} \cdot \frac{du}{dx}$$

---

*Proof.* Since $g$ is differentiable at $x$ and $u = g(x)$, it follows from Theorem G.3 that

$$\Delta u = g'(x)\Delta x + \epsilon_1 \Delta x \tag{13}$$

where $\epsilon_1 \to 0$ as $\Delta x \to 0$. And since $y = f(u)$ is differentiable at $u = g(x)$, it follows from Theorem G.3 that

$$\Delta y = f'(u)\Delta u + \epsilon_2 \Delta u \tag{14}$$

where $\epsilon_2 \to 0$ as $\Delta u \to 0$.

Factoring out the $\Delta u$ in (14) and then substituting (13) yields

$$\Delta y = [f'(u) + \epsilon_2][g'(x)\Delta x + \epsilon_1 \Delta x]$$

or

$$\Delta y = [f'(u) + \epsilon_2][g'(x) + \epsilon_1]\Delta x$$

or if $\Delta x \neq 0$,

$$\frac{\Delta y}{\Delta x} = [f'(u) + \epsilon_2][g'(x) + \epsilon_1] \tag{15}$$

But (13) implies that $\Delta u \to 0$ as $\Delta x \to 0$, and hence $\epsilon_1 \to 0$ and $\epsilon_2 \to 0$ as $\Delta x \to 0$. Thus, from (15)

$$\lim_{\Delta x \to 0} \frac{\Delta y}{\Delta x} = f'(u)g'(x)$$

or

$$\frac{dy}{dx} = f'(u)g'(x) = \frac{dy}{du} \cdot \frac{du}{dx}$$ ∎

**PROOF THAT RELATIVE EXTREMA OCCUR AT CRITICAL POINTS**

In this subsection we will prove Theorem 5.2.2, which states that the relative extrema of a function occur at critical points.

> **G.5**   THEOREM (*Theorem 5.2.2*).   *If a function $f$ has any relative extrema, then they occur either at points where $f'(x) = 0$ or at points where $f$ is not differentiable.*

*Proof.*   There are two possibilities—either $f$ is differentiable at a point $x_0$ or it is not. If it is not, then $x_0$ is a critical point for $f$ and we are done. If $f$ is differentiable at $x_0$, then we must show that $f'(x_0) = 0$. We will do this by showing that $f'(x_0) \geq 0$ and $f'(x_0) \leq 0$, from which it follows that $f'(x_0) = 0$. From the definition of a derivative we have

$$f'(x_0) = \lim_{h \to 0} \frac{f(x_0 + h) - f(x_0)}{h}$$

so that

$$f'(x_0) = \lim_{h \to 0^+} \frac{f(x_0 + h) - f(x_0)}{h} \tag{16}$$

and

$$f'(x_0) = \lim_{h \to 0^-} \frac{f(x_0 + h) - f(x_0)}{h} \tag{17}$$

Because $f$ has a relative maximum at $x_0$, there is an open interval $(a, b)$ containing $x_0$ in which $f(x) \leq f(x_0)$ for all $x$ in $(a, b)$.

Assume that $h$ is sufficiently small so that $x_0 + h$ lies in the interval $(a, b)$. Thus,

$$f(x_0 + h) \leq f(x_0) \quad \text{or equivalently,} \quad f(x_0 + h) - f(x_0) \leq 0$$

Thus, if $h$ is negative,

$$\frac{f(x_0 + h) - f(x_0)}{h} \geq 0 \tag{18}$$

and if $h$ is positive,

$$\frac{f(x_0 + h) - f(x_0)}{h} \leq 0 \tag{19}$$

But an expression that never assumes negative values cannot approach a negative limit and an expression that never assumes positive values cannot approach a positive limit, so that

$$f'(x_0) = \lim_{h \to 0^-} \frac{f(x_0 + h) - f(x_0)}{h} \geq 0 \qquad \boxed{\text{From (17) and (18)}}$$

and

$$f'(x_0) = \lim_{h \to 0^+} \frac{f(x_0 + h) - f(x_0)}{h} \leq 0 \qquad \boxed{\text{From (16) and (19)}}$$

Since $f'(x_0) \geq 0$ and $f'(x_0) \leq 0$, it must be that $f'(x_0) = 0$. ∎

## EXERCISE SET G

In Exercises 1 and 2, use the appropriate parts of Theorem G.1 to prove the stated result.

**1.** If $\lim_{x \to a} f(x) = L_1$, and $\lim_{x \to a} g(x) = L_2$, then

$$\lim_{x \to a} [f(x) - g(x)] = L_1 - L_2$$

**2.** If $\lim_{x \to a} f(x) = L$ and $k$ is a constant, then

$$\lim_{x \to a} [kf(x)] = kL$$

In Exercises 3–6, use Definitions 2.3.4 and 2.3.5 to prove the stated results.

**3.** For any constant $k$, $\lim_{x \to +\infty} k = k$.

**4.** For any constant $k$, $\lim_{x \to -\infty} k = k$.

**5.** If $\lim_{x \to -\infty} f(x) = L_1$ and $\lim_{x \to -\infty} g(x) = L_2$, then
$\lim_{x \to -\infty} [f(x) + g(x)] = L_1 + L_2$.

**6.** If $\lim_{x \to +\infty} f(x) = L_1$ and $\lim_{x \to +\infty} g(x) = L_2$, then
$\lim_{x \to +\infty} [f(x) + g(x)] = L_1 + L_2$.

In Exercises 7 and 8, use Definitions 2.3.6 and 2.3.7 for the proofs.

**7.** Suppose that $\lim_{x \to a} f(x) = +\infty$ and $\lim_{x \to a} g(x) = +\infty$.
   (a) Prove: $\lim_{x \to a} [f(x) + g(x)] = +\infty$.
   (b) Is it true that $\lim_{x \to a} [f(x) - g(x)] = 0$?

**8.** Suppose that $\lim_{x \to a} f(x) = -\infty$ and $\lim_{x \to a} g(x) = +\infty$.
   (a) Prove: $\lim_{x \to a} [f(x) - g(x)] = -\infty$.
   (b) Is it true that $\lim_{x \to a} [f(x) + g(x)] = 0$?

**9.** Prove: $\lim_{x \to a} f(x) = L$ if and only if

$$\lim_{x \to a} [f(x) - L] = 0$$

**10.** Prove: If $\lim_{x \to a} f(x) = L$, then $\lim_{x \to a} |f(x)| = |L|$.

# ANSWERS TO ODD-NUMBERED EXERCISES

......................................................

▶ **Exercise Set for Introduction (Page 12)**

1. **(a)** $\frac{41}{333}$ **(b)** $\frac{115}{9}$ **(c)** $\frac{20943}{550}$ **(d)** $\frac{537}{1250}$

3. **(a)** $\frac{223}{71} < \frac{333}{106} < \frac{63}{25}\left(\frac{17+15\sqrt{5}}{7+15\sqrt{5}}\right) < \frac{355}{113} < \frac{22}{7}$ **(b)** $\frac{63}{25}\left(\frac{17+15\sqrt{5}}{7+15\sqrt{5}}\right)$ **(c)** $\frac{333}{106}$ **(d)** $\frac{63}{25}\left(\frac{17+15\sqrt{5}}{7+15\sqrt{5}}\right)$ 5. 3.1416 (Machin); 3.0418

7. **(a)** $\frac{7}{11} = 0.636363\ldots = \frac{6}{10} + \frac{3}{100} + \frac{6}{1000} + \frac{3}{10000} + \frac{6}{100000} + \frac{3}{1000000} + \cdots$

 **(b)** $\frac{8}{33} = 0.242424\ldots = \frac{2}{10} + \frac{4}{100} + \frac{2}{1000} + \frac{4}{10000} + \frac{2}{100000} + \frac{4}{1000000} + \cdots$

 **(c)** $\frac{5}{12} = 0.416666\ldots = \frac{4}{10} + \frac{1}{100} + \frac{6}{1000} + \frac{6}{10000} + \frac{6}{100000} + \frac{6}{1000000} + \cdots$ 9. **(a)** 2.6458 **(b)** 7.0711

▶ **Exercise Set 1.1 (Page 22)**

1. **(a)** 1943 **(b)** 1960; 4200 **(c)** no, you need the year's population **(d)** war, marketing

 **(e)** news of health risk, social pressure, antismoking campaigns, increased taxation

3. **(a)** $-2.9, -2.0, 2.35, 2.9$ **(b)** none **(c)** 0 **(d)** $-1.75 \le x \le 2.15$ **(e)** $y_{max} = 2.8$ at $x = -2.6$; $y_{min} = -2.2$ at $x = 1.2$

5. **(a)** 2, 4 **(b)** none **(c)** $x \le 2; 4 \le x$ **(d)** $y_{min} = -1$; no maximum

7. **(a)** no; war, pestilence, flood, earthquakes **(b)** decreases for 8 hours, takes a jump upward, and repeats

9. **(a)** $L = x + 2000/x$ **(b)** $x > 0$; $x$ must be smaller than the width of the building, which was not given.

 **(c)**  **(d)** 89.44 11. **(a)** $r \approx 3.4, h \approx 13.8$

 **(b)** taller

 **(c)** $r \approx 3.1, h \approx 16.0, C \approx 4.76$

▶ **Exercise Set 1.2 (Page 33)**

1. **(a)** $-2$; 10; 10; 25; 4; $27t^2 - 2$ **(b)** 0; 4; $-4$; 6; $2\sqrt{2}$; $f(3t) = 1/3t$ for $t > 1$ and $f(3t) = 6t$ for $t \le 1$

3. **(a)** $x \ne 3$ **(b)** $x \le -\sqrt{3}, x \ge \sqrt{3}$ **(c)** $(-\infty, +\infty)$ **(d)** $x \ne 0$ **(e)** $x \ne \left(2n + \frac{1}{2}\right)\pi, n = 0, \pm 1, \pm 2, \ldots$

5. **(a)** $x \le 3$ **(b)** $-2 \le x \le 2$ **(c)** $x \ge 0$ **(d)** all $x$ **(e)** all $x$ 7. **(a)** yes **(b)** yes **(c)** no **(d)** no

9. $h = L(1 - \cos\theta)$ 11. 13.

 **(a)** $f(x) = \begin{cases} 2x + 1, & x < 0 \\ 4x + 1, & x \ge 0 \end{cases}$ 15. **(a)** $V = (8 - 2x)(15 - 2x)x$

 **(b)** $-\infty < x < \infty, -\infty < V < \infty$

 **(b)** $g(x) = \begin{cases} 1 - 2x, & x < 0 \\ 1, & 0 \le x < 1 \\ 2x - 1, & x \ge 1 \end{cases}$ **(c)** $0 < x < 4$

17. **(i)** $x = 1, -2$ **(ii)** $g(x) = x + 1$, all $x$ 19. **(a)** $25°F$ **(b)** $2°F$ **(c)** $-15°F$ 21. $5°F$ 23. $D(t) = 1000 - 20t$

▶ **Exercise Set 1.3 (Page 45)**

1. (e)    3. (b), (c)    5. $[-3, 3] \times [0, 5]$    9. $[-5, 14] \times [-60, 40]$    11. $[-0.1, 0.1] \times [-3, 3]$

13. $[-1000, 1050] \times [-1500000, 10000]$    15. $[-2, 2] \times [-20, 20]$    19. (a) $f(x) = \sqrt{16 - x^2}$    35. $4.6455, -0.6455$

(b) $f(x) = -\sqrt{16 - x^2}$

(e) no

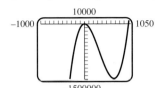

▶ **Exercise Set 1.4 (Page 57)**

1. (a) 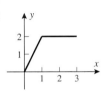    (b)     (c)     (d)

3. (a) 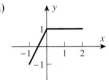    (b) 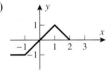    (c) 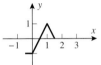    (d)

5. 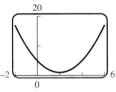    7. 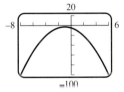    9. 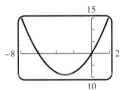    11.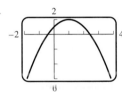

13.     15. 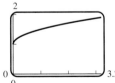    17. 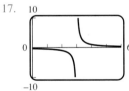    19.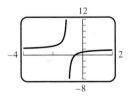

21. 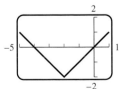    23. 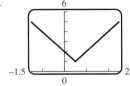    25. 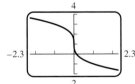    27.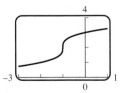

29. (a) 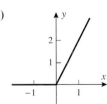    (b) $y = \begin{cases} 0, & x \le 0 \\ 2x, & x > 0 \end{cases}$    31. $x^2 + 2x + 1$, all $x$;    33. $3\sqrt{x - 1}, x \ge 1$;    35. (a) 3

$2x - x^2 - 1$, all $x$;    $\sqrt{x - 1}, x \ge 1$;    (b) 9

$2x^3 + 2x$, all $x$;    $2x - 2, x > 1$;    (c) 2

$2x/(x^2 + 1)$, all $x$    $2, x > 1$    (d) 2

37. (a) $x^4 + 1$    (b) $x^2 + 4x + 3$    (c) $x^2 + 4x + 5$    (d) $\dfrac{1}{x^2} + 1$    (e) $x^2 + 2xh + h^2 + 1$    (f) $x^2 + 1$    (g) $x + 1$    (h) $9x^2 + 1$

39. $2x^2 - 2x + 1$, all $x$; $4x^2 + 2x$, all $x$    41. $1 - x, x \le 1$; $\sqrt{1 - x^2}, |x| \le 1$    43. $\dfrac{1}{1 - 2x}, x \ne \dfrac{1}{2}, 1; -\dfrac{1}{2x} - \dfrac{1}{2}, x \ne 0, 1$

45. $x^{-6} + 1$    47. (a) $g(x) = \sqrt{x}, h(x) = x + 2$    (b) $g(x) = |x|, h(x) = x^2 - 3x + 5$

49. (a) $g(x) = x^2, h(x) = \sin x$   (b) $g(x) = 3/x, h(x) = 5 + \cos x$

51. (a) $f(x) = x^3, g(x) = 1 + \sin x, h(x) = x^2$   (b) $f(x) = \sqrt{x}, g(x) = 1 - x, h(x) = \sqrt[3]{x}$

53. 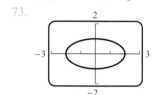    55. 

57. $\pm 2$    59. $6x + 3h$    61. $-\dfrac{1}{x(x+h)}$    63. (a) origin
    (b) $x$-axis
    (c) $y$-axis
    (d) none

65. (a)

| $x$ | $-3$ | $-2$ | $-1$ | $0$ | $1$ | $2$ | $3$ |
|---|---|---|---|---|---|---|---|
| $f(x)$ | $1$ | $-5$ | $-1$ | $0$ | $-1$ | $-5$ | $1$ |

(b)

| $x$ | $-3$ | $-2$ | $-1$ | $0$ | $1$ | $2$ | $3$ |
|---|---|---|---|---|---|---|---|
| $f(x)$ | $1$ | $5$ | $-1$ | $0$ | $1$ | $-5$ | $-1$ |

67. (a) even  (b) odd  (c) odd  (d) neither    69. (a) even  (b) odd  (c) even  (d) neither  (e) odd  (f) even

71. (a) $y$-axis  (b) origin  (c) $x$-axis, $y$-axis, origin

73.    77. (a)    (b)    79. yes; $f(x) = x^k, g(x) = x^n$

---

► **Exercise Set 1.5 (Page 71)**

1. (a) $-\frac{3}{2}, -\frac{1}{18}, \frac{2}{3}$  (b) yes    3. $\text{III} < \text{II} < \text{IV} < \text{I}$

5. (a) The slopes are equal; the points lie on the same line.

(b) The slopes $-1, 3, \frac{1}{3}$ are not equal; the points do not lie on a line.    7. (a) 14   (b) $-\frac{1}{3}$

9. $\frac{13}{7}$    11. (a) $153°$  (b) $45°$  (c) $117°$  (d) $89°$    13. (a) $60°$  (b) $117°$    15. $y = -2x + 4$    17. $y = 4x + 7$

19. $y = -\frac{1}{5}x + 6$    21. $y = 11x - 18$    23. (a) parallel  (b) perpendicular  (c) parallel  (d) perpendicular  (e) neither

25. (a) $y = \frac{3}{2}x - 3$  (b) $y = -\frac{3}{4}x$    27. (a) $\frac{9}{10}$ ft/s  (b) $-4$  (c) $-2.2$  (d) $\frac{80}{9}$ s

29. (a) $-\frac{4}{3}$ ft/s$^2$  (b) $v = -\frac{4}{3}t + \frac{13}{3}$  (c) $v = \frac{13}{3}$ ft/s    31. (b) $-\frac{9}{10}$ cm/s  (c) $\frac{9}{10}$ cm/s

33. (a) 0 mi/h  (b) 48 mi/h  (c) 240 mi

35. (a)    (b) $v = \begin{cases} 10t & \text{if} \quad 0 \le t \le 10 \\ 100 & \text{if} \quad 10 \le t \le 100 \\ 600 - 5t & \text{if} \quad 100 \le t \le 120 \end{cases}$    37. (a) $y = x/9$    (c) 26.11 in
    (b) 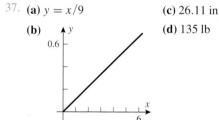    (d) 135 lb

39. $y = 1.2x + 2$    41. (a) $T_C = \frac{5}{9}(T_F - 32)$  (b) $\frac{5}{9}$  (c) $-40°$ (F or C)  (d) $37°$    43. (a) $p = 0.098h + 1$  (b) 10.20 m

45. (a) $r = -0.0125t + 0.8$  (b) 64 days

47. (a) $C_1 = 2x, C_2 = 25 + (x/4)$  (b) $x = 15$    49. (a) $H \approx 181$
    (c) The Universe would be even older.

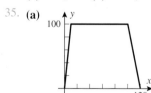

▶ **Exercise Set 1.6 (Page 89)**

1. **(a)** $y = 3x + b$      **(c)**
   **(b)** $y = 3x + 6$

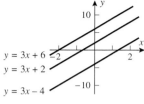

$y = 3x + 6$
$y = 3x + 2$
$y = 3x - 4$

3. **(a)** $y = mx + 2$      **(c)**
   **(b)** $y = -x + 2$

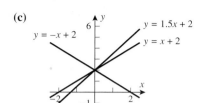

$y = -x + 2$
$y = 1.5x + 2$
$y = x + 2$

5. **(a)** slope: $-1$                    **(b)** $y$-intercept: $y = -1$

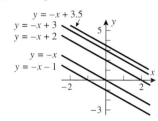

$y = -x + 3.5$
$y = -x + 3$
$y = -x + 2$
$y = -x$
$y = -x - 1$

$y = -1.5x - 1$
$y = -x - 1$
$y = -1$
$y = 2.5x - 1$
$y = 2x - 1$

**(c)** pass through $(-4, 2)$                **(d)** $x$-intercept: $x = 1$

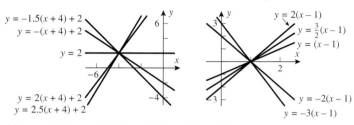

$y = -1.5(x + 4) + 2$
$y = -(x + 4) + 2$
$y = 2$
$y = 2(x + 4) + 2$
$y = 2.5(x + 4) + 2$

$y = 2(x - 1)$
$y = \frac{3}{2}(x - 1)$
$y = (x - 1)$
$y = -2(x - 1)$
$y = -3(x - 1)$

7. $y = \pm\dfrac{9 - x_0 x}{\sqrt{9 - x_0^2}}$

9.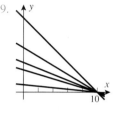

11. **(a)** VI    **(b)** IV    **(c)** III    **(d)** V    **(e)** I    **(f)** II

13. **(a)**

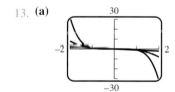
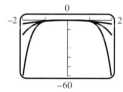

**(b)**

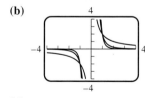
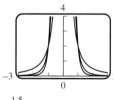

**(c)**

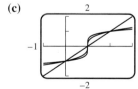
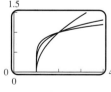

15. **(a)**

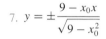

$y = 3x^2$   $y = 2x^2$
$y = x^3$
$y = -x^2$
$y = -3x^2$   $y = -2x^2$

**(b)**

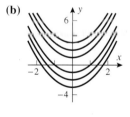

**(c)**

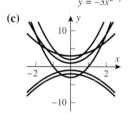

17. **(a)**

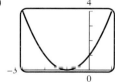 

**(b)**

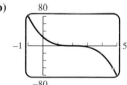 

**(c)**

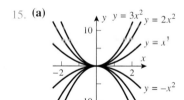

**(d)**

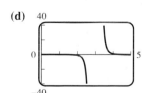
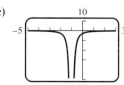

**19. (a)** 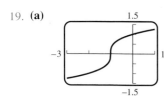 **(b)** 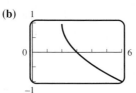 **(c)** 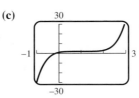 **(d)**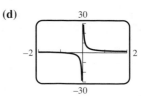

**21.** 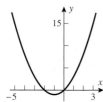  **23.** $t = 0.445\sqrt{d}$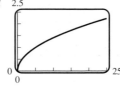

**25. (a)** newton-meters (N·m)  **27. (a)** $k = 0.000045$ N·m²  **29. (a)** II; $y = 1, x = -1, 2$

**(b)** 20 N·m  **(b)** 0.000005 N  **(b)** I; $y = 0, x = -2, 3$

**(c)**

| V (L) | 0.25 | 0.5 | 1.0 | 1.5 | 2.0 |
|---|---|---|---|---|---|
| P (N/m²) | $80 \times 10^3$ | $40 \times 10^3$ | $20 \times 10^3$ | $13.3 \times 10^3$ | $10 \times 10^3$ |

**(c)** 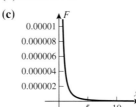  **(c)** IV; $y = 2$

**(d)** III; $y = 0, x = -2$

**(d)**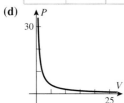

**(d)** The force becomes infinite; the force tends to zero.

**31.** Order the six trigonometric functions as sin, cos, tan, cot, sec, csc:  **(a)** pos, pos, pos, pos, pos, pos

**(b)** neg, zero, undefined, zero, undefined, neg   **(c)** pos, neg, neg, neg, neg, pos   **(d)** neg, pos, neg, neg, pos, neg

**(e)** neg, neg, pos, pos, neg, neg   **(f)** neg, pos, neg, neg, pos, neg

**33. (a)** use $\sin(\pi - x) = \sin x$; 0.588   **(b)** use $\cos(-x) = \cos x$; 0.924   **(c)** use $\sin(2\pi + x) = \sin x$; 0.588

**(d)** use $\cos(\pi - x) = -\cos x$; $-0.924$   **(e)** use $\sin 2x = 2\sin x\sqrt{1 - \sin^2 x}$; 0.951.   **(f)** use $\cos^2 x = 1 - \sin^2 x$; 0.654

**35. (a)** $-a$   **(b)** $b$   **(c)** $-c$   **(d)** $\pm\sqrt{1 - a^2}$   **(e)** $-b$   **(f)** $-a$   **(g)** $\pm 2b\sqrt{1 - b^2}$   **(h)** $2b^2 - 1$

**(i)** $1/b$   **(j)** $-1/a$   **(k)** $1/c$   **(l)** $(1 - b)/2$   **37.** 80,936 km

**39.** The second quarter revolves twice (720°) about its own center.   **41. (a)** $y = 3\sin(x/2)$   **(b)** $y = 4\cos 2x$   **(c)** $y = -5\sin 4x$

**43. (a)** $y = \sin[x + (\pi/2)]$   **(b)** $y = 3 + 3\sin(2x/9)$   **(c)** $y = 1 + 2\sin\left[2\left(x - \dfrac{\pi}{4}\right)\right]$

**45. (a)** $3, \pi/2, 0$   **(b)** $2, 2, 0$   **(c)** $1, 4\pi, 0$

**47. (b)** $A = \sqrt{A_1^2 + A_2^2}, \theta = \tan^{-1}(A_2/A_1)$   **(c)** $x = \dfrac{5\sqrt{13}}{2}\sin\left(2\pi t + \tan^{-1}\dfrac{1}{2\sqrt{3}}\right)$

▶ **Exercise Set 1.7 (Page 100)**

**1. (a)** 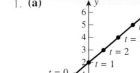  **(c)**

| t | 0 | 1 | 2 | 3 | 4 | 5 |
|---|---|---|---|---|---|---|
| x | −1 | 0 | 1 | 2 | 3 | 4 |
| y | 1 | 2 | 3 | 4 | 5 | 6 |

**3.**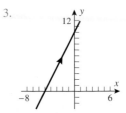

5.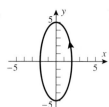

7.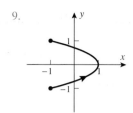

9.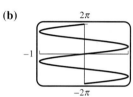

11.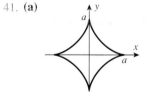

13. $x = 5\cos t$, $y = -5\sin t$, $0 \leq t \leq 2\pi$    15. $x = 2$, $y = t$    17. $x = t^2$, $y = t$, $-1 \leq t \leq 1$

19. (a) IV    21. (a)
(b) II
(c) V
(d) VI
(e) III
(f) I

(b)

| $t$ | 0 | 1 | 2 | 3 | 4 | 5 |
|-----|---|---|---|---|---|---|
| $x$ | 0 | 5.5 | 8 | 4.5 | -8 | -32.5 |
| $y$ | 1 | 1.5 | 3 | 5.5 | 9 | 13.5 |

(c) $t = 0, 2\sqrt{3}$
(d) $0 < t < 2\sqrt{2}$
(e) 2

23. (a) 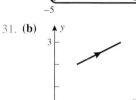    (b)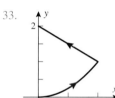

25. (a) $\dfrac{x - x_0}{x_1 - x_0} = \dfrac{y - y_0}{y_1 - y_0}$
(c) $x = 1 + t$, $y = -2 + 6t$
(d) $x = 2 - t$, $y = 4 - 6t$

27. (b) $\frac{1}{2}$    (c) $\frac{3}{4}$

31. (b)    33.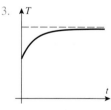

35. (a) $x = 4\cos t$,
      $y = 3\sin t$
  (b) $x = -1 + 4\cos t$,
      $y = 2 + 3\sin t$

37. (a) $x = 400\sqrt{2}t$,
      $y = 400\sqrt{2}t - 4.9t^2$
  (b) 16,326.53 m
  (c) 65,306.12 m

39. (a) ellipses with fixed center, varying axes of symmetry
   (b) (assume $a \neq 0$, $b \neq 0$) ellipses with varying center, fixed axes of symmetry
   (c) circles of radius 1 with centers on line $y = x - 1$

41. (a)

---

▶ **Chapter 1 Supplementary Exercises (Page 103)**

1. 1940–1945    3.    5. $C = 5x^2 + (64/x)$    7.    9. (a) $V = (6 - 2x)(5 - x)x$
   (b) $0 < x < 3$
   (c) 3.57 ft × 3.79 ft × 1.21 ft

11. no solution    13. $1/(2 - x^2)$    15.

| $x$ | -4 | -3 | -2 | -1 | 0 | 1 | 2 | 3 | 4 |
|-----|----|----|----|----|---|---|---|---|---|
| $f(x)$ | 0 | -1 | 2 | 1 | 3 | -2 | -3 | 4 | -4 |
| $g(x)$ | 3 | 2 | 1 | -3 | -1 | -4 | 4 | -2 | 0 |
| $(f \circ g)(x)$ | 4 | -3 | -2 | -1 | 1 | 0 | -4 | 2 | 3 |
| $(g \circ f)(x)$ | -1 | -3 | 4 | -4 | -2 | 1 | 2 | 0 | 3 |

17. (a) odd
   (b) even
   (c) neither
   (d) even

19. (b) 295.72 ft    21. $C$: (2.0944, 1.9132); $D$: (4.1888, 1.2284); $B$: (-2.0944, -1.9132); $A$: (-4.1888, -1.2284)

23. (a) circles of radius 1 centered on the parabola $y = x^2$    (b) parabolas that open up with vertices on the line $y = x/2$

25.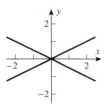

27.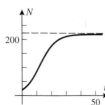

29. $d = \sqrt{(x-1)^2 + (\sqrt{x}-2)^2}$;
    $x = 1.358094$

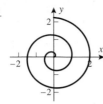

31. 0.48 ft

33. **(a)** 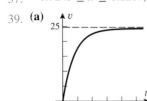   **(b)** about 10 years   **(c)** 220 sheep

35. **(a)** 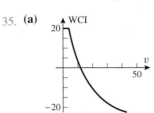   **(b)** $3°F, -11°F, -18°F, -22°F$   **(c)** $v = 35, 19, 12, 7$ mi/h

37. $-0.7245 \le x \le 1.2207$; $-1.0551 \le y \le 1.4902$

39. **(a)** 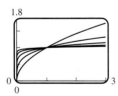   **(c)** For large $t$ the velocity approaches $c$.
    **(d)** No, but it comes arbitrarily close.
    **(e)** 3.013 s

---

▶ **Chapter 1 Horizon Module (Page 106)**

1. **(a)** $c = 0.5$: $0.25, 6.25 \times 10^{-2}, 3.91 \times 10^{-3}, 1.53 \times 10^{-5}, 2.32 \times 10^{-10}, 5.42 \times 10^{-20}, 2.94 \times 10^{-39}, 8.64 \times 10^{-78}, 7.46 \times 10^{-155},$
   $5.56 \times 10^{-309}$; $c = 1$: $1, 1, 1, 1, 1, 1, 1, 1, 1, 1$; $c = 2$: $4, 16, 256, 65536, \approx 4.29 \times 10^9, \approx 1.84 \times 10^{19}, \approx 3.40 \times 10^{38},$
   $\approx 1.16 \times 10^{77}, \approx 1.34 \times 10^{154}, \approx 1.80 \times 10^{308}$

2. $2, 2.25, 2.2361111, 2.23606798, 2.23606798, \ldots$    3. **(a)** $\frac{1}{2}, \frac{1}{4}, \frac{1}{8}, \frac{1}{16}, \frac{1}{32}, \frac{1}{64}$   **(b)** $y_n = 1/2^n$

4. **(a)** $y_{n+1} = 1.05 y_n$   **(b)** $y_1 = \$1050, y_2 = \$1102.50, y_3 = \$1157.62, y_4 = \$1215.51, y_5 = \$1276.28$
   **(c)** $y_{n+1} = 1.05 y_n (n \ge 1)$   **(d)** $y_n = (1.05)^n 1000$; $y_{15} = \$2078.93$

5. **(a)** $x^{1/2}, x^{1/4}, x^{1/8}, x^{1/16}, x^{1/32}$

   **(b)** They tend to the horizontal line $y = 1$ with a hole at $x = 0$.

   1.8

   0
   0          3

6. **(a)** $\frac{1}{2}, \frac{2}{3}, \frac{3}{5}, \frac{5}{8}, \frac{8}{13}, \frac{13}{21}, \frac{21}{34}, \frac{34}{55}, \frac{55}{89}, \frac{89}{144}$

   **(b)** $1, 2, 3, 5, 8, 13, 21, 34, 55, 89$;
   each numerator is the sum of the previous two

   **(c)** $\frac{144}{233}, \frac{233}{377}, \frac{377}{610}, \frac{610}{987}, \frac{987}{1597}, \frac{1597}{2584}, \frac{2584}{4181}, \frac{4181}{6765}, \frac{6765}{10946}, \frac{10946}{17711}$

   **(d)** $F_0 = 1, F_1 = 1, F_n = F_{n-1} + F_{n-2}$ for $n \ge 2$

   **(e)** the positive solution

7. **(a)** $y_1 = cr, y_2 = cy_1 = cr^2, y_3 = cr^3, y_4 = cr^4$   **(b)** $y_n = cr^n$
   **(c)** If $r = 1$, then $y_n = c$ for all $n$; if $r < 1$, then $y_n$ tends to zero; if $r > 1$, then $y_n$ gets ever larger.

9. **(a)** $0.261, 0.559, 0.715, 0.591, 0.701$   **(b)** It appears to approach a value somewhere near 0.65.

---

▶ **Exercise Set 2.1 (Page 124)**

1. **(a)** $-1$   **(b)** 3   **(c)** does not exist   **(d)** 1   **(e)** $-1$   **(f)** 3    3. **(a)** 1   **(b)** 1   **(c)** 1   **(d)** 1   **(e)** $-\infty$   **(f)** $+\infty$

5. **(a)** 0   **(b)** 0   **(c)** 0   **(d)** 3   **(e)** $+\infty$   **(f)** $+\infty$    7. **(a)** $-\infty$   **(b)** $+\infty$   **(c)** does not exist   **(d)** undefined   **(e)** 2   **(f)** 0

9. **(a)** $-\infty$   **(b)** $-\infty$   **(c)** $-\infty$   **(d)** 1   **(e)** 1   **(f)** 2    11. **(a)** 0   **(b)** 0   **(c)** 0   **(d)** 0   **(e)** does not exist   **(f)** does not exist

13. for all $x_0 \neq -4$    15. **(a)** At $x = 3$ the one-sided limits fail to exist.
   **(b)** At $x = -2$ the two-sided limit exists but is not equal to $F(-2)$.   **(c)** At $x = 3$ the limit fails to exist.

17. **(a)** $\frac{1}{3}$   **(b)** $+\infty$   **(c)** $-\infty$    19. **(a)** 3   **(b)** does not exist    21. **(a)** $y = 2$   **(b)** $y = 20.086$   **(c)** no horizontal asymptote

23. **(a)** $\lim\limits_{x \to 0^+} \dfrac{\sin x}{x}$   **(b)** $\lim\limits_{x \to 0^+} \dfrac{x-1}{x+1}$   **(c)** $\lim\limits_{x \to 0^-} (1+2x)^{1/x}$   25. **(a)** $f(x) = \begin{cases} 1+(1/x), & x < 0 \\ -1+(1/x), & x \geq 0 \end{cases}$   **(b)** yes; $f(x) = (\sin x)/x$

29. **(a)** catastrophic subtraction when the $x$-interval is small (the size depending on the calculating utility).   **(c)** no.

▶ Exercise Set 2.2 **(Page 137)**

1. **(a)** $-6$   **(b)** 13   **(c)** $-8$   **(d)** 16   **(e)** 2   **(f)** $-\frac{1}{2}$   **(g)** The denominator tends to zero but the numerator does not.

**(h)** The denominator tends to zero but the numerator does not.   3. **(a)** 7   **(b)** $-3$   **(c)** $\pi$   **(d)** $-6$   **(e)** 36   **(f)** $-\infty$   5. 0

7. 8   9. 4   11. $-\frac{4}{5}$   13. $\frac{3}{2}$   15. 0   17. 0   19. $-\sqrt{5}$   21. $1/\sqrt{6}$   23. $\sqrt{3}$   25. $+\infty$   27. does not exist

29. $-\infty$   31. $+\infty$   33. does not exist   35. $+\infty$   37. $-\infty$   39. $-\frac{1}{7}$   41. 6   43. $+\infty$   45. $+\infty$   47. $-\infty$

49. **(a)** 2   51. **(a)** 3   **(b)**   53. **(a)** Theorem 2.2.2($a$) does not apply.   55. $\frac{1}{4}$   57. 0

**(b)** 2   **(b)** $\lim\limits_{x \to 0^+} \left( \dfrac{1}{x} - \dfrac{1}{x^2} \right) = \lim\limits_{x \to 0^+} \left( \dfrac{x-1}{x^2} \right) = -\infty$

**(c)** 2

59. $a/2$   61. $\lim\limits_{x \to +\infty} p(x) = (-1)^n \infty$ and $\lim\limits_{x \to -\infty} p(x) = +\infty$

63. For $m > n$, the limits are both zero; for $m = n$, the limits are equal to the leading coefficient of $p$; for $n > m$, the limits are $\pm\infty$.

65. The left and/or right limits could be $\pm\infty$; or the limit could exist and equal any preassigned real number.

▶ Exercise Set 2.3 **(Page 145)**

1. **(a)** $|x| < 0.1$   **(b)** $|x - 3| < 0.0025$   **(c)** $|x - 4| < 0.000125$   3. **(a)** $x_1 = 3.8025, x_2 = 4.2025$   **(b)** $\delta = 0.1975$

5. $\delta = 0.05$   7. $\delta = \frac{1}{700}$   9. $\delta = 0.05$   11. $\delta = \frac{1}{9000}$   13. $\delta = 1$   15. $\delta = \frac{1}{3}\epsilon$   17. $\delta = \frac{1}{2}\epsilon$   19. $\delta = \epsilon$

21. $\delta = \min\left(1, \frac{1}{6}\epsilon\right)$   23. $\delta = \min\left( \dfrac{1}{6}, \dfrac{\epsilon}{18} \right)$   25. $\delta = 2\epsilon$   27. $\delta = \epsilon$   29. **(a)** $\sqrt{10}$   **(b)** 99   **(c)** $-10$   **(d)** $-101$

31. **(a)** $-\sqrt{\dfrac{1-\epsilon}{\epsilon}}; \sqrt{\dfrac{1-\epsilon}{\epsilon}}$   **(b)** $\sqrt{\dfrac{1-\epsilon}{\epsilon}}$   **(c)** $-\sqrt{\dfrac{1-\epsilon}{\epsilon}}$   33. 10   35. 999   37. 203   39. 57.5   41. $\dfrac{1}{\sqrt{\epsilon}}$

43. $-2 - \dfrac{1}{\epsilon}$   45. $\dfrac{1}{\epsilon} - 1$   47. $-\dfrac{5}{2} - \dfrac{11}{2\epsilon}$   49. **(a)** $|x| < \frac{1}{10}$   **(b)** $|x - 1| < \frac{1}{1000}$   **(c)** $|x - 3| < \frac{1}{10\sqrt{10}}$   **(d)** $|x| < \frac{1}{10}$

51. $\delta = 1/\sqrt{M}$   53. $\delta = 1/M$   55. $\delta = 1/(-M)^{1/4}$   57. $\delta = \epsilon$   59. $\delta = \epsilon^2$   61. $\delta = \epsilon$

63. **(a)** $\delta = -1/M$   65. **(a)** $N = M - 1$   67. $\delta = \min\left(2, \frac{1}{8}\epsilon\right)$   69. $\delta = 0.0442$

**(b)** $\delta = 1/M$   **(b)** $N = M - 1$

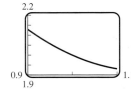

▶ Exercise Set 2.4 **(Page 156)**

1. **(a)** not continuous, $x = 2$   **(b)** not continuous, $x = 2$   **(c)** not continuous, $x = 2$   **(d)** continuous   **(e)** continuous   **(f)** continuous

3. **(a)** not continuous, $x = 1, 3$   **(b)** continuous   **(c)** not continuous, $x = 1$   **(d)** continuous   **(e)** not continuous, $x = 3$

**(f)** continuous   5. **(a)** 3   **(b)** 3

7. **(a)**   **(b)**   **(c)**   **(d)**

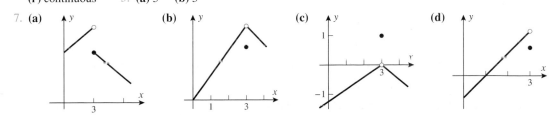

**9. (a)**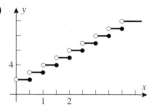

**(b)** One second could cost you one dollar.     **11.** none     **13.** none

**15.** $f$ is not defined at $x = \pm 4$.     **17.** $f$ is not defined at $x = \pm 3$.     **19.** none     **21.** none     **23. (a)** $k = 5$     **(b)** $k = \frac{4}{3}$

**25. (a)** 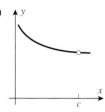     **(b)**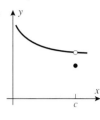

**27. (a)** $x = 0$, not removable
     **(b)** $x = -3$, removable
     **(c)** $x = 2$, removable;
          $x = -2$, not removable

**29. (a)** $x = \frac{1}{2}$, not removable;
     at $x = -3$, removable

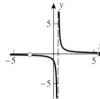

**(b)** $(2x - 1)(x + 3)$

**33. (a)** $f(x) = k$ for $x \neq c$, $f(c) = 0$; $g(x) = l$ for $x \neq c$, $g(c) = 0$. If $k = -l$, then $f + g$ is continuous; otherwise it is not.
     **(b)** $f(x) = k$ for $x \neq c$, $f(c) = 1$; $g(x) = l \neq 0$ for $x \neq c$, $g(c) = 1$. If $kl = 1$, then $fg$ is continuous; otherwise it is not.
**37.** $f(x) = 1$ for $0 \leq x < 1$, $f(x) = -1$ for $1 \leq x \leq 2$     **43.** $x = -1.25$, $x = 0.75$
**45.** $x = -1.605$, $x = 1.375$     **47.** $x = 2.24$     **49.** $x = 4.847$ cm

---

▶ **Exercise Set 2.5 (Page 163)**

**1.** none     **3.** $x = n\pi$, $n = 0, \pm 1, \pm 2, \ldots$     **5.** $x = n\pi$, $n = 0, \pm 1, \pm 2, \ldots$     **7.** none
**9.** $2n\pi + (\pi/6)$, $2n\pi + (5\pi/6)$, $n = 0, \pm 1, \pm 2, \ldots$
**11. (a)** $\sin x$, $x^3 + 7x + 1$     **(b)** $|x|$, $\sin x$     **(c)** $x^3$, $\cos x$, $x + 1$     **(d)** $\sqrt{x}$, $3 + x$, $\sin x$, $2x$     **(e)** $\sin x$, $\sin x$     **(f)** $x^5 - 2x^3 + 1$, $\cos x$
**13.** 1     **15.** $-\sqrt{3}/2$     **17.** 3     **19.** $-1$     **21.** 0     **23.** $\frac{7}{3}$     **25.** 1     **27.** 2     **29.** 0     **31.** $-\frac{25}{49}$     **33.** does not exist
**35.** 3     **37.** $k = \frac{1}{2}$     **39. (a)** 1     **(b)** 0     **(c)** 1     **41.** $-\pi$     **43.** $-x \leq x \cos(50\pi/x) \leq x$
**45.** $\lim\limits_{x \to 0} f(x) = 1$ by the Squeezing Theorem.     **47.** $g(x) = -\dfrac{1}{x}$, $h(x) = \dfrac{1}{x}$;
          $\lim\limits_{x \to +\infty} \dfrac{\sin x}{x} = 0$ by the Squeezing Theorem.

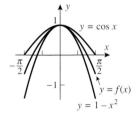

**51. (a)** 0.17365     **53. (a)** 0.08749     **55. (b)** 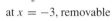     **(c)** 0.739     **57. (a)** symmetry about the equatorial plane
     **(b)** 0.17453          **(b)** 0.08727

▶ Chapter 2 Supplementary Exercises **(Page 165)**

1. **(a)** 1   **(b)** does not exist   **(c)** does not exist   **(d)** 1   **(e)** 3   **(f)** 0   **(g)** 0   **(h)** 2   **(i)** $\frac{1}{2}$

5. **(a)** $\frac{1}{4}$   **(b)** 4   7. **(a)** 0.405   17. **(a)** 1.449   **(b)** $x = 0, \pm 1.896$

19. **(a)** $\sqrt{5}$, does not exist, $\sqrt{10}$, $\sqrt{10}$, does not exist, $+\infty$, does not exist   **(b)** 5, 10, 0, 0, 10, $-\infty$, $+\infty$

21. $a/b$   23. does not exist   25. 0   27. $3 - k$   31. 2.71828   33. 0.54030   35. 0.49996   37. 0.07747

39. **(b)**   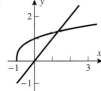   **(d)** 1, 1.26, 1.31, 1.322, 1.324, 1.3246, 1.3247   41. $x = \sqrt[5]{x + 2}$; 1.267168

---

▶ Exercise Set 3.1 **(Page 175)**

1. **(a)** $\frac{7}{2}$   **(d)**
   **(b)** 3
   **(c)** $x_0$

3. **(a)** $-\frac{1}{6}$   **(d)** 
   **(b)** $-\frac{1}{4}$
   **(c)** $-1/x_0^2$

5. **(a)** $2x_0$   7. **(a)** $1/(2\sqrt{x_0})$
   **(b)** 4   **(b)** $\frac{1}{2}$

9. **(a)** 4 m/s   11. **(a)** $t_0$   **(b)** 0   **(c)** speeding up   **(d)** slowing down   13. straight line with slope equal to the velocity

15. **(a)** 72°F at about 4:30 P.M.   17. **(a)** first year   **(d)** 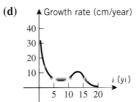   19. **(a)** 320,000 ft
   **(b)** 4°F/h   **(b)** 6 cm/year   **(b)** 8000 ft/s
   **(c)** $-7$°F/h at about 9 P.M.   **(c)** 10 cm/year at about age 14   **(c)** 45 ft/s
   **(d)** 24,000 ft/s

21. **(a)** 720 ft/min   **(b)** 192 ft/min

---

▶ Exercise Set 3.2 **(Page 186)**

1. 2, 0, $-2$, $-1$   3. **(b)** 3   5.   7. $y = 5x - 16$   9. $6x$; $y = 18x - 27$   11. $3x^2$; $y = 0$
   **(c)** 3
   13. $\dfrac{1}{2\sqrt{x + 1}}$; $y = \frac{1}{6}x + \frac{5}{3}$   15. $-1/x^2$
   17. $2ax$   19. $-1/2x^{3/2}$   21. $8t + 1$

23. **(a)** D   25. **(a)**   **(b)**   **(c)**   27. **(a)** $x^2$, 3   29. 8
   **(b)** F   **(b)** $\sqrt{x}$, 1
   **(c)** B
   **(d)** C
   **(e)** A
   **(f)** E

31. $y = -2x + 1$      33. **(h)**

| $h$ | 0.5 | 0.1 | 0.01 | 0.001 | 0.0001 | 0.00001 |
|---|---|---|---|---|---|---|
| $[f(1 + h) - f(1)]/h$ | 1.6569 | 1.4355 | 1.3911 | 1.3868 | 1.3863 | 1.3863 |

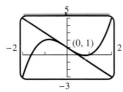

35. **(a)** dollars per foot  **(b)** the price per additional foot  **(c)** positive  **(d)** $1000

37. **(a)** $F \approx 200$ lb, $dF/d\theta \approx 60$ lb/rad  **(b)** $\mu \approx 0.3$  39. **(a)** $T \approx 120°$F, $dT/dt \approx -4.5°$F/min  **(b)** $k = -0.1$

45. $f(1) = 0$, $f'(1) = 5$

---

► **Exercise Set 3.3 (Page 197)**

1. $28x^6$  3. $24x^7 + 2$  5. $0$  7. $-\frac{1}{3}(7x^6 + 2)$  9. $3ax^2 + 2bx + c$  11. $24x^{-9} + (1/\sqrt{x})$  13. $-3x^{-4} - 7x^{-8}$

15. $18x^2 - \frac{3}{2}x + 12$  17. $-15x^{-2} - 14x^{-3} + 48x^{-4} + 32x^{-5}$  19. $12x(3x^2 + 1)$  21. $-\frac{5}{4}$  23. $\dfrac{3}{(2t+1)^2}$  25. $\frac{7}{16}$

27. $-29$  29. $32t$  31. $3\pi r^2$  33. **(a)** $4\pi r^2$  **(b)** $100\pi$  35. **(a)** $-\frac{37}{4}$  **(b)** $-\frac{23}{16}$

37. **(a)** $10$  **(b)** $19$  **(c)** $9$  **(d)** $-1$  39. $y = 5x + 17$  41. **(a)** $42x - 10$  **(b)** $24$  **(c)** $2/x^3$  **(d)** $700x^3 - 96x$

43. **(a)** $-210x^{-8} + 60x^2$  45. **(a)** $0$  49. $F'(x) = xf''(x) + 2f'(x)$  51. $\left(1, \frac{5}{6}\right) \left(2, \frac{2}{3}\right)$ 1.5  53. $0$

  **(b)** $-6x^{-4}$  **(b)** $112$

  **(c)** $6a$  **(c)** $360$

55. $0$  57. $y = 3x^2 - x - 2$  59. $x = \frac{1}{2}$  61. $2 \pm \sqrt{3}$  63. $-2x_0$  67. $-\dfrac{2GmM}{r^3}$  69. $f'(x) > 0$ for all $x \neq 0$

73. **(a)** $2(1 + x^{-1})(x^{-3} + 7) + (2x + 1)(-x^{-2})(x^{-3} + 7) + (2x + 1)(1 + x^{-1})(-3x^{-4})$  **(b)** $3(7x^6 + 2)(x^7 + 2x - 3)^2$

75. not differentiable at $x = 1$  77. $a = 6, b = -3$  79. **(a)** $x = \frac{2}{3}$  **(b)** $x = \pm 2$

83. **(a)** $n(n-1)(n-2)\cdots 1$  **(b)** $0$  **(c)** $a_n n(n-1)(n-2)\cdots 1$

85. **(b)** $f$ and all its derivatives up to $f^{(n-1)}(x)$ are continuous on $(a, b)$.

---

► **Exercise Set 3.4 (Page 202)**

1. $-2\sin x - 3\cos x$  3. $\dfrac{x\cos x - \sin x}{x^2}$  5. $x^3 \cos x + (3x^2 + 5)\sin x$  7. $\sec x \tan x - \sqrt{2}\sec^2 x$  9. $\sec^3 x + \sec x \tan^2 x$

11. $-\csc^3 x - \csc x \cot^2 x$  13. $-\dfrac{\csc x}{1 + \csc x}$  15. $0$  17. $\dfrac{1}{(1 + x\tan x)^2}$  19. $-x\cos x - 2\sin x$  21. $-x\sin x + 5\cos x$

23. $-4\sin x \cos x$  27. **(a)** $y = x$  **(b)** $y = 2x - (\pi/2) + 1$  **(c)** $y = 2x + (\pi/2) - 1$

29. **(a)** $x = \pm \pi/2, \pm 3\pi/2$  **(b)** $x = -3\pi/2, \pi/2$  **(c)** no horizontal tangent line  **(d)** $x = \pm 2\pi, \pm \pi, 0$  31. $0.087$ ft/degree

33. $1.75$ m/degree  35. **(a)** $-\cos x$  **(b)** $\cos x$

37. **(a)** all $x$  **(b)** all $x$  **(c)** $x \neq (\pi/2) + n\pi, n = 0, \pm 1, \pm 2, \dots$

  **(d)** $x \neq n\pi, n = 0, \pm 1, \pm 2, \dots$  **(e)** $x \neq (\pi/2) + n\pi, n = 0, \pm 1, \pm 2, \dots$  **(f)** $x \neq n\pi, n = 0, \pm 1, \pm 2, \dots$

  **(g)** $x \neq (2n + 1)\pi, n = 0, \pm 1, \pm 2, \dots$  **(h)** $x \neq n\pi/2, n = 0, \pm 1, \pm 2, \dots$  **(i)** all $x$  39. $3, 7, 11, \dots$  41. $\sec^2 y$

---

► **Exercise Set 3.5 (Page 208)**

1. $37(x^3 + 2x)^{36}(3x^2 + 2)$  3. $-2\left(x^3 - \dfrac{7}{x}\right)^{-3}\left(3x^2 + \dfrac{7}{x^2}\right)$  5. $\dfrac{24(1 - 3x)}{(3x^2 - 2x + 1)^4}$  7. $\dfrac{3}{4\sqrt{x}\sqrt{4 + 3\sqrt{x}}}$  9. $3x^2\cos(x^3)$

11. $8x\sec^2(4x^2)$  13. $-20\cos^4 x \sin x$  15. $-\dfrac{2}{x^3}\cos\left(\dfrac{1}{x^2}\right)$  17. $28x^6 \sec^2(x^7)\tan(x^7)$  19. $-\dfrac{5\sin(5x)}{2\sqrt{\cos(5x)}}$

21. $-3\left[x + \csc(x^3 + 3)\right]^{-4}\left[1 - 3x^2\csc(x^3 + 3)\cot(x^3 + 3)\right]$  23. $\dfrac{x(10 - 3x^2)}{\sqrt{5 - x^2}}$  25. $10x^3\sin 5x \cos 5x + 3x^2\sin^2 5x$

27. $-x^3\sec\left(\dfrac{1}{x}\right)\tan\left(\dfrac{1}{x}\right) + 5x^4\sec\left(\dfrac{1}{x}\right)$  29. $\sin(\cos x)\sin x$  31. $-6\cos^2(\sin 2x)\sin(\sin 2x)\cos 2x$

33. $12(5x + 8)^{13}(x^3 + 7x)^{11}(3x^2 + 7) + 65(x^3 + 7x)^{12}(5x + 8)^{12}$  35. $\dfrac{33(x - 5)^2}{(2x + 1)^4}$  37. $-\dfrac{2(2x + 3)^2(52x^2 + 96x + 3)}{(4x^2 - 1)^9}$

39. $5\left[x\sin 2x + \tan^4(x^7)\right]^4\left[2x\cos 2x + \sin 2x + 28x^6\tan^3(x^7)\sec^2(x^7)\right]$  41. $-25x\cos(5x) - 10\sin(5x) - 2\cos(2x)$

43. $4(1 - x)^{-3}$  45. $y = -x$  47. $y = -1$  49. $3\cot^2\theta\csc^2\theta$  51. $\pi(b - a)\sin 2\pi\omega$

53. **(a)** 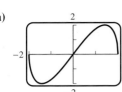 **(c)** $\dfrac{4 - 2x^2}{\sqrt{4 - x^2}}$ 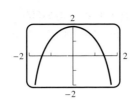 **(d)** $y - \sqrt{3} = \dfrac{2}{\sqrt{3}}(x - 1)$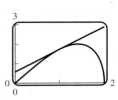

55. **(c)** $f = 1/T$  **(d)** amplitude $= 0.6$ cm, $T = 2\pi/15$ seconds per oscillation, $f = 15/(2\pi)$ oscillations per second

57. **(a)** $10 \text{ lb/in}^2, -2 \text{ lb/in}^2/\text{mi}$  **(b)** $-0.6 \text{ lb/in}^2/\text{s}$   59. $\begin{cases} \cos x, & 0 < x < \pi \\ -\cos x, & -\pi < x < 0 \end{cases}$   61. **(a)** $-\dfrac{1}{x}\cos\dfrac{1}{x} + \sin\dfrac{1}{x}$

63. **(a)** 21  **(b)** $-36$   65. 6   67. $1/2x$   69. $\frac{2}{3}x$   73. $f'(g(h(x)))g'(h(x))h'(x)$

---

▶ **Exercise Set 3.6 (Page 217)**

1. **(a)** 4, 5  **(b)** 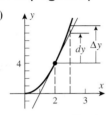   3. **(a)** 0.5, 1  **(b)** 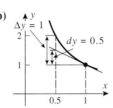   5. $3x^2\,dx, 3x^2\,\Delta x + 3x(\Delta x)^2 + (\Delta x)^3$

7. $(2x - 2)\,dx, 2x\,\Delta x + (\Delta x)^2 - 2\,\Delta x$   9. **(a)** $(12x^2 - 14x)\,dx$  **(b)** $(-x\sin x + \cos x)\,dx$

11. **(a)** $\dfrac{2 - 3x}{2\sqrt{1 - x}}\,dx$   13. **(a)** $f(x) \approx 1 + 3(x - 1)$   15. **(a)** $1 + \frac{1}{2}x, 0.95, 1.05$  **(b)**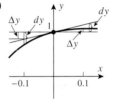
  **(b)** $-17(1 + x)^{-18}\,dx$      **(b)** $f(1 + \Delta x) \approx 1 + 3\,\Delta x$
           **(c)** 1.06

25. **(a)** 0.0174533  **(b)** $x_0 = 45°$  **(c)** 0.694765   27. 83.16   29. 8.0625   31. 8.9944   33. 0.1   35. 0.8573

37. $|x| < 0.1692$      39. $|x| < 0.6316$      41. 0.0225   43. 0.0048   45. **(a)** $\pm 2 \text{ ft}^2$
                                                                **(b)** side: $\pm 1\%$; area: $\pm 2\%$

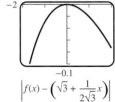

$\left| f(x) - \left(\sqrt{3} + \dfrac{1}{2\sqrt{3}}x\right) \right|$              $|f(x) - x|$

47. **(a)** opposite: $\pm 0.151$ in; adjacent: $\pm 0.087$ in  **(b)** opposite: $\pm 3.0\%$; adjacent: $\pm 1.0\%$   49. $\pm 10\%$   51. $\pm 0.017 \text{ cm}^2$

53. $\pm 6\%$   55. $\pm 0.5\%$   57. $0.236 \text{ cm}^3$   59. **(a)** $\alpha = 1.5 \times 10^{-5}/°\text{C}$  **(b)** 180.1 cm long

---

▶ **Chapter 3 Supplementary Exercises (Page 219)**

5. $x = -\frac{7}{2}, 2, -\frac{1}{2}$   7. $x = 1, -\frac{1}{15}$   9. **(a)** $x = -2, -1, 1, 3$  **(b)** $(-\infty, -2), (-1, 1), (3, +\infty)$  **(c)** $(-2, -1), (1, 3)$  **(d)** 4

11. $y = -16x, y = -145x/4$   13. $x = n\pi \pm (\pi/4), n = 0, \pm 1, \pm 2, \ldots$

17. **(a)** $-0.5, 1, 0.5$  **(b)** $\pi/4, 1, \pi/2$  **(c)** $3, -1, 0$   19. **(a)** 2000 gal/min  **(b)** 2500 gal/min

21. **(a)** between 139.48 m and 144.55 m  **(b)** $|d\phi| \le 0.98°$   23. **(a)** 3.6  **(b)** $-0.777778$   25. 2.772589

27. 58.75 ft/s   29. $\pm 0.535428$

---

▶ **Chapter 3 Horizon Module (Page 221)**

2. **(a)** $\{(x, y) : 0 \le x^2 + y^2 \le 2l_1\}$  **(b)** $\{(x, y) : l_1 - l_2 \le x^2 + y^2 \le l_1 + l_2\}$  **(c)** $\{(x, y) : l_2 - l_1 \le x^2 + y^2 \le l_1 + l_2\}$

3. $\left(\dfrac{\sqrt{2}+3\sqrt{6}}{4}, \dfrac{7\sqrt{2}+3\sqrt{6}}{4}\right)$　　5.

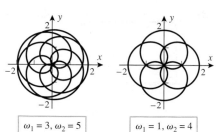 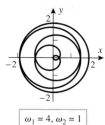

$\boxed{\omega_1 = 3,\ \omega_2 = 5}$　　$\boxed{\omega_1 = 1,\ \omega_2 = 4}$　　$\boxed{\omega_1 = 4,\ \omega_2 = 1}$

6. $x = 2\cos t,\ y = 2\sin t$, a circle of radius 2　　7. **(e)** $\pm 27.2660°$

9. **(a)** $d\theta_1/dt = \frac{1}{3},\ d\theta_2/dt = 0$　**(b)** $d\theta_1/dt = 0,\ d\theta_2/dt = -\frac{1}{3}$

---

► Exercise Set 4.1 **(Page 233)**

1. **(a)** yes　　3. **(a)** yes　5. **(a)** yes　**(d)** yes　　7. **(a)** no　　9. **(b)** $[-2, 2], [-8, 8]$　**(c)**
**(b)** no　　　**(b)** no　　**(b)** yes　**(e)** no　　　**(b)** no
**(c)** yes　　　　　　　**(c)** no　**(f)** no　　　**(c)** yes
**(d)** no

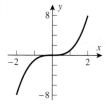

11. **(a)** no　**(b)** yes　**(c)** yes　　13. $x^{1/5}$　　15. $\frac{1}{7}(x+6)$　　17. $\sqrt[3]{(x+5)/3}$　　19. $(x^3+1)/2$　　21. $-\sqrt{3/x}$

23. $\begin{cases} (5/2) - x, & x > 1/2 \\ 1/x, & 0 < x \le 1/2 \end{cases}$　　25. $x^{1/4} - 2$ for $x \ge 16$　　27. $\frac{1}{2}(3 - x^2)$ for $x \le 0$　　29. $\frac{1}{10}(1 + \sqrt{1 - 20x})$ for $x \le -4$

31. **(a)** $y = (6.214 \times 10^{-4})x$　　　　33. **(b)**　　　　**(c)** No, because $f(g(x)) = x$ for $x > 1$
**(b)** $x = \dfrac{10^4}{6.214}\,y$　　　　　　　　　　　　　but the domain of $g$ is $x \ge 0$.
**(c)** how many meters in $y$ miles

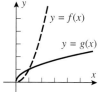

$y = f(x)$
$y = g(x)$

35. **(b)** symmetric about the line $y = x$　　37. **(b)** $1 - (\sqrt{3}/3)$　　39. 10

41.　　　　　　　　　43. 3　　　　　47.　　　　　　　　49. $\frac{88}{7}$

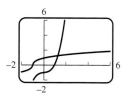 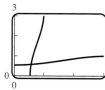 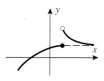

---

► Exercise Set 4.2 **(Page 243)**

1. **(a)** $-4$　**(b)** 4　**(c)** $\frac{1}{4}$　　3. **(a)** 2.9690　**(b)** 0.0341　　5. **(a)** 4　**(b)** $-5$　**(c)** 1　**(d)** $\frac{1}{2}$　　7. **(a)** 1.3655　**(b)** $-0.3011$

9. **(a)** $2r + \dfrac{s}{2} + \dfrac{t}{2}$　**(b)** $s - 3r - t$　　11. **(a)** $1 + \log x + \frac{1}{2}\log(x - 3)$　**(b)** $2\ln|x| + 3\ln\sin x - \frac{1}{2}\ln(x^2 + 1)$　　13. $\log\frac{256}{3}$

15. $\ln\dfrac{\sqrt[3]{x}(x+1)^2}{\cos x}$　　17. 0.01　　19. $e^2$　　21. 4　　23. $10^5$　　25. $\sqrt{3/2}$　　27. $-\dfrac{\ln 3}{2\ln 5}$　　29. $\frac{1}{3}\ln\frac{7}{2}$　　31. $-2$

33. $0, -\ln 2$　　35. **(a)**　　　　　**(b)**　　　　　　37. $2.8777, -0.3174$　　39. 3

  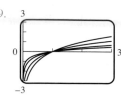

41. $x = 3.6541, y = 1.2958$    43. **(a)** no    **(d)** 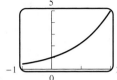    45. $\log \frac{1}{2} < 0$, so $3 \log \frac{1}{2} < 2 \log \frac{1}{2}$

**(b)** $y = 2^{x/4}$

**(c)** $y = 2^{-x}$

47. 201 days    49. **(a)** 7.4, basic    **(b)** 4.2, acidic    **(c)** 6.4, acidic    **(d)** 5.9, acidic

51. **(a)** 140 dB, damage    **(b)** 120 dB, damage    **(c)** 80 dB, no damage    **(d)** 75 dB, no damage

53. $\approx 200$    55. **(a)** $\approx 5 \times 10^{16}$ J    **(b)** $\approx 0.67$    57. $e^{-2}$

---

▶ **Exercise Set 4.3 (Page 253)**

1. $\frac{2}{3}(2x - 5)^{-2/3}$    3. $\frac{9}{2(x+2)^2}\left[\frac{x-1}{x+2}\right]^{1/2}$    5. $\frac{1}{3}x^2(5x^2+1)^{-5/3}(25x^2+9)$    7. $-\frac{15[\sin(3/x)]^{3/2}\cos(3/x)}{2x^2}$

9. **(a)** $\frac{2 - 3x^2 - y}{x}$    **(b)** $-\frac{1}{x^2} - 2x$    11. $-\frac{x}{y}$    13. $\frac{1 - 2xy - 3y^3}{x^2 + 9xy^2}$    15. $-\frac{y^2}{x^2}$    17. $\frac{1 - 2xy^2\cos(x^2y^2)}{2x^2y\cos(x^2y^2)}$

19. $\frac{1 - 3y^2\tan^2(xy^2+y)\sec^2(xy^2+y)}{3(2xy+1)\tan^2(xy^2+y)\sec^2(xy^2+y)}$    21. $-\frac{21}{16y^3}$    23. $\frac{2y}{x^2}$    25. $\frac{\sin y}{(1 + \cos y)^3}$    27. $-1, +1$    29. $-0.1312$

31. $-\frac{9}{13}$    35. **(a)**    **(b)** $\pm 1.1547$    37. $\frac{2t^3 + 3a^2}{2a^3 - 6at}$    39. $-\frac{b^2\lambda}{a^2\omega}$    41. $a = \frac{1}{4}, b = \frac{5}{4}$

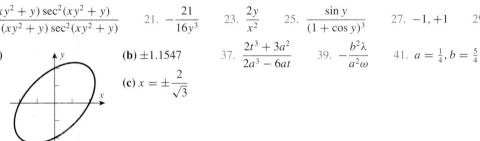

**(c)** $x = \pm\frac{2}{\sqrt{3}}$

43. $y = (\sqrt{3}/3)x, \; y = -(\sqrt{3}/3)x$    45. $-\frac{2y^3 + 3t^2y}{(6ty^2 + t^3)\cos t}$    47. $\frac{dy}{dt} = \frac{3\cos 3x - y^2}{2xy}\frac{dx}{dt}$    49. $-1, \frac{2}{3}$

53. $\frac{1}{15y^2 + 1}$    55. $\frac{1}{10y^4 + 3y^2}$

---

▶ **Exercise Set 4.4 (Page 260)**

1. $\frac{1}{x}$    3. $\frac{2\ln x}{x}$    5. $\frac{\sec^2 x}{\tan x}$    7. $\frac{1 - x^2}{x(1 + x^2)}$    9. $\frac{3x^2 - 14x}{x^3 - 7x^2 - 3}$    11. $\frac{1}{2x\sqrt{\ln x}}$    13. $-\frac{1}{x}\sin(\ln x)$

15. $3x^2\log_2(3 - 2x) - \frac{2x^3}{(\ln 2)(3 - 2x)}$    17. $\frac{2x(1 + \log x) - x/(\ln 10)}{(1 + \log x)^2}$    19. $7e^{7x}$    21. $x^2e^x(x + 3)$    23. $\frac{4}{(e^x + e^{-x})^2}$

25. $(x\sec^2 x + \tan x)e^{x\tan x}$    27. $(1 - 3e^{3x})e^{x - e^{3x}}$    29. $\frac{x - 1}{e^x - x}$    31. $-\frac{y}{x(y + 1)}$    33. $-\tan x + \frac{3x}{4 - 3x^2}$

35. $x\sqrt[3]{1 + x^2}\left[\frac{1}{x} + \frac{2x}{3(1 + x^2)}\right]$    37. $\frac{(x^2 - 8)^{1/3}\sqrt{x^3 + 1}}{x^6 - 7x + 5}\left[\frac{2x}{3(x^2 - 8)} + \frac{3x^2}{2(x^3 + 1)} - \frac{6x^5 - 7}{x^6 - 7x + 5}\right]$    39. $2^x\ln 2$

41. $\pi^{\sin x}(\ln \pi)\cos x$    43. $(x^3 - 2x)^{\ln x}\left[\frac{3x^2 - 2}{x^3 - 2x}\ln x + \frac{1}{x}\ln(x^3 - 2x)\right]$    45. $(\ln x)^{\tan x}\left[\frac{\tan x}{x \ln x} + (\sec^2 x)\ln(\ln x)\right]$

49. **(a)** $k^n e^{kx}$    **(b)** $(-1)^n k^n e^{-kx}$    51. $-\frac{1}{\sqrt{2\pi}\sigma^3}(x - \mu)\exp\left[-\frac{1}{2}\left(\frac{x - \mu}{\sigma}\right)^2\right]$    53. **(a)** $-\frac{1}{x(\ln x)^2}$    **(b)** $-\frac{\ln 2}{x(\ln x)^2}$

55. $-\frac{qk_0}{2T^2}\exp\left[-\frac{q(T - T_0)}{2T_0 T}\right]$    57. $ex^{e-1}$    59. **(a)** 1    61. **(b)** 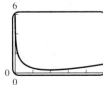    **(d)** must take the value zero in between

**(b)** $\ln 10$    **(e)** $x = 2$

**63. (b)**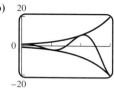

**65. (a)** 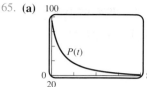  **(b)** The population tends to 19.  **(c)** The *rate* tends to zero.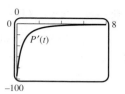

---

▶ **Exercise Set 4.5 (Page 267)**

1. **(a)** $-\pi/2$  **(b)** $\pi$  **(c)** $-\pi/4$  **(d)** $0$    3. $1/2, -\sqrt{3}, -1/\sqrt{3}, 2, -2/\sqrt{3}$    5. $\frac{4}{5}, \frac{3}{5}, \frac{3}{4}, \frac{5}{3}, \frac{5}{4}$    7. **(a)** $\pi/7$  **(b)** $0$  **(c)** $2\pi/7$

**(d)** $201\pi - 630$    9. **(a)** $0 \le x \le \pi$  **(b)** $-1 \le x \le 1$  **(c)** $-\pi/2 < x < \pi/2$  **(d)** $-\infty < x < +\infty$    11. $\frac{24}{25}$

13. **(a)** $\dfrac{1}{\sqrt{1+x^2}}$  **(c)** $\dfrac{\sqrt{x^2-1}}{x}$    15. **(a)** 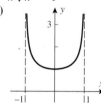   **(b)** domain of $\cot^{-1} x$ is $(-\infty, +\infty)$, range is $(0, \pi)$; domain of $\csc^{-1} x$ is $(-\infty, -1] \cup [1, +\infty)$, range is $[-\pi/2, 0) \cup (0, \pi/2]$.

**(b)** $\dfrac{1}{x}$  **(d)** $\sqrt{x^2-1}$

17. **(a)** $55.0°$  **(b)** $33.6°$  **(c)** $25.8°$    19. **(a)** $x = 3.6964$ rad  **(b)** $\theta = -76.7°$

21. **(a)** $\dfrac{1}{\sqrt{9-x^2}}$  **(b)** $-\dfrac{2}{\sqrt{1-(2x+1)^2}}$    23. **(a)** $\dfrac{7}{x\sqrt{x^{14}-1}}$  **(b)** $-\dfrac{1}{\sqrt{e^{2x}-1}}$    25. **(a)** $-\dfrac{1}{|x|\sqrt{x^2-1}}$  **(b)** $\begin{cases} 1, & \sin x > 0 \\ -1, & \sin x < 0 \end{cases}$

27. **(a)** $\dfrac{e^x}{x\sqrt{x^2-1}} + e^x \sec^{-1} x$  **(b)** $\dfrac{3x^2(\sin^{-1} x)^2}{\sqrt{1-x^2}} + 2x(\sin^{-1} x)^3$    29. $\dfrac{(3x^2 + \tan^{-1} y)(1+y^2)}{(1+y^2)e^y - x}$

31. **(a)** 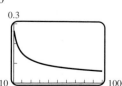    33. **(b)** 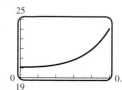    35. **(a)** 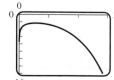  **(b)**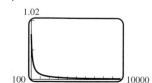

37. **(b)** $23°$    39. $32°$ or $58°$; $32°$    41. $29°$

---

▶ **Exercise Set 4.6 (Page 274)**

1. **(b)** $A = x^2$  **(c)** $\dfrac{dA}{dt} = 2x \dfrac{dx}{dt}$  **(d)** $12$ ft$^2$/min    3. **(a)** $\dfrac{dV}{dt} = \pi\left(r^2 \dfrac{dh}{dt} + 2rh \dfrac{dr}{dt}\right)$  **(b)** $-20\pi$ in$^3$/s; decreasing

5. **(a)** $\dfrac{d\theta}{dt} = \dfrac{\cos^2 \theta}{x^2}\left(x\dfrac{dy}{dt} - y\dfrac{dx}{dt}\right)$  **(b)** $-\frac{5}{16}$ rad/s; decreasing    7. $\dfrac{4\pi}{15}$ in$^2$/min    9. $\dfrac{1}{\sqrt{\pi}}$ mi/h    11. $4860\pi$ cm$^3$/min

13. $\frac{5}{6}$ ft/s    15. $\dfrac{125}{\sqrt{61}}$ ft/s    17. $704$ ft/s    19. **(a)** $500$ mi, $1716$ mi  **(b)** $1354$ mi; $27.7$ mi/min    21. $\dfrac{9}{20\pi}$ ft/min

23. $125\pi$ ft$^3$/min    25. $250$ mi/h    27. $\dfrac{36\sqrt{69}}{25}$ ft/min    29. $\dfrac{8\pi}{5}$ km/s    31. $600\sqrt{7}$ mi/h    33. **(a)** $-\dfrac{60}{7}$ units per second

**(b)** falling    35. $-4$ units per second    37. $e^2$    39. $4.5$ cm/s; away    43. $\dfrac{20}{9\pi}$ cm/s

---

▶ **Exercise Set 4.7 (Page 284)**

1. **(a)** $\frac{2}{3}$  **(b)** $\frac{2}{3}$    3. $1$    5. $1$    7. $1$    9. $-1$    11. $0$    13. $-\infty$    15. $0$    17. $2$    19. $0$

21. $\pi$    23. $-\frac{5}{3}$    25. $e^{-3}$    27. $e^2$    29. $e^{2/\pi}$    31. $0$    33. $\frac{1}{2}$    35. $+\infty$    39. **(b)** $2$

41. $0$    43. $e^3$    45. no horizontal asymptote    47. $y = 1$

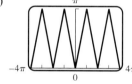

49. **(a)** 0 **(b)** $+\infty$ **(c)** 0 **(d)** $-\infty$ **(f)** $+\infty$ **(g)** $-\infty$ 51. 1 53. does not exist

55. $Vt/L$ 57. **(c)** $1024\left(\sqrt[1024]{0.3}-1\right) = -1.20327; \ 1024\left(\sqrt[1024]{2}-1\right) = 0.69338$

59. $k=-1, l=\pm 2\sqrt{2}$ 61. does not exist

---

▶ **Chapter 4 Supplementary Exercises (Page 286)**

3. **(a)** $0/0, \infty/\infty$ **(b)** no 5. **(a)** $\frac{1}{2}(x+1)^{1/3}$ **(b)** none **(c)** $\frac{1}{2}\ln(x-1)$ **(d)** $\frac{x+2}{x-1}$

7. **(a)** $y+1=2(x-1)$ **(b)** $y=1$ 9. $15x+2$ 11. **(a)** $+\infty$ **(b)** $\frac{1}{2}$ **(c)** $\ln a$

15. $-b - \dfrac{a}{\sqrt{2}}$ cm/s 17. **(a)**

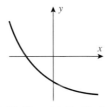

**(b)** $y = e^{-x/2}\sin 2x$ intersects $y = e^{-x/2}$ at $x = \pi/4$
and $y = -e^{-x/2}$ at $x = -\pi/4, 3\pi/4$

19. **(a)** $\dfrac{1}{x}(1+x)^{(1/x)-1} - \dfrac{(1+x)^{1/x}}{x^2}\ln(1+x)$ **(b)** $e^x\left[x^{e^x-1} + x^{e^x}\ln x\right]$ **(c)** $3x^2$ **(d)** $\dfrac{abe^{-x}}{(1+be^{-x})^2}$

**(e)** $\dfrac{6x^{4/3}y^{1/3} - 3x^{1/3}y - 2y^{4/3}}{2x^{4/3} + 3xy^{1/3}}$ **(f)** $\dfrac{5x+3}{6x(x+1)} - \cot x - \tan x$ 21. **(b)** $x = 3.654$ 23. $e^{1/e}$

---

▶ **Exercise Set 5.1 (Page 296)**

1. **(a)** $f' > 0, f'' > 0$ **(b)** $f' > 0, f'' < 0$ **(c)** $f' < 0, f'' > 0$ **(d)** $f' < 0, f'' < 0$

3. A: $dy/dx < 0, \ d^2y/dx^2 > 0$
B: $dy/dx > 0, \ d^2y/dx^2 < 0$
C: $dy/dx < 0, \ d^2y/dx^2 < 0$

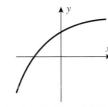

5. $x = -1, 0, 1, 2$ 7. **(a)** $[4, 6]$ **(b)** $[1, 4], [6, 7]$ **(c)** $(1, 2), (3, 5)$ **(d)** $(2, 3), (5, 7)$ **(e)** $x = 2, 3, 5$

9. **(a)** $\left[\frac{5}{2}, +\infty\right)$ **(b)** $\left(-\infty, \frac{5}{2}\right)$ **(c)** $(-\infty, +\infty)$ **(d)** none **(e)** none

11. **(a)** $(-\infty, +\infty)$ **(b)** none **(c)** $(-2, +\infty)$ **(d)** $(-\infty, -2)$ **(e)** $-2$

13. **(a)** $[1, +\infty)$ **(b)** $(-\infty, 1]$ **(c)** $(-\infty, 0), \left(\frac{2}{3}, +\infty\right)$ **(d)** $\left(0, \frac{2}{3}\right)$ **(e)** $0, \frac{2}{3}$

15. **(a)** $[0, +\infty)$ **(b)** $(-\infty, 0]$ **(c)** $(-\sqrt{2/3}, \sqrt{2/3})$ **(d)** $(-\infty, -\sqrt{2/3}), (\sqrt{2/3}, +\infty)$ **(e)** $-\sqrt{2/3}, \sqrt{2/3}$

17. **(a)** $(-\infty, +\infty)$ **(b)** none **(c)** $(-\infty, -2)$ **(d)** $(-2, +\infty)$ **(e)** $-2$

19. **(a)** $[-1, +\infty)$ **(b)** $(-\infty, -1]$ **(c)** $(-\infty, 0), (2, +\infty)$ **(d)** $(0, 2)$ **(e)** $0, 2$

21. **(a)** $(-\infty, 0]$ **(b)** $[0, +\infty)$ **(c)** $(-\infty, -1), (1, +\infty)$ **(d)** $(-1, 1)$ **(e)** $-1, 1$

23. **(a)** $[0, +\infty)$ **(b)** $(-\infty, 0]$ **(c)** $(-1, 1)$ **(d)** $(-\infty, -1), (1, +\infty)$ **(e)** $-1, 1$

25. **(a)** $[\pi, 2\pi]$
**(b)** $[0, \pi]$
**(c)** $(\pi/2, 3\pi/2)$
**(d)** $(0, \pi/2), (3\pi/2, 2\pi)$
**(e)** $\pi/2, 3\pi/2$

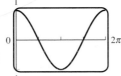

27. **(a)** $(-\pi/2, \pi/2)$
**(b)** none
**(c)** $(0, \pi/2)$
**(d)** $(-\pi/2, 0)$
**(e)** $0$

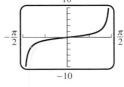

29. **(a)** $[0, \pi/4], [3\pi/4, \pi]$
**(h)** $[\pi/4, 3\pi/4]$
**(c)** $(\pi/2, \pi)$
**(d)** $(0, \pi/2)$
**(e)** $\pi/2$

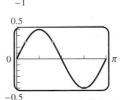

31. **(a)**

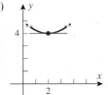

**(b)**

**(c)**

33. **(a)** $(a, 0)$     35.   2.5 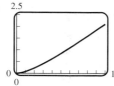     37. $x \geq \sin x$   4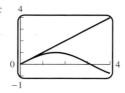
   **(b)** none

39. points of inflection at $x = -2, 2$; 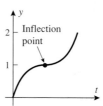     41. $-0.175, 0$     43. $-2.45, 0.65, 2.75$
   concave up on $(-5, -2)$, $(2, 5)$;
   concave down on $(-2, 2)$;
   increasing on $[-3.5829, 0.2513]$ and $[3.3316, 5]$;
   decreasing on $[-5, -3.5829]$, $[0.2513, 3.3316]$

47. **(a)** true     49. **(c)** 1     53. the eighth day  1000 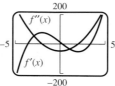     55.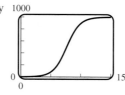

---

▶ **Exercise Set 5.2 (Page 304)**

1. **(a)** 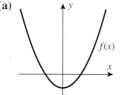     **(b)** 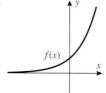     **(c)** 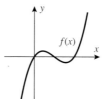     **(d)**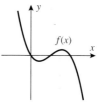

5. **(b)** nothing     **(c)** $f$ has a relative minimum at $x = 1$, $g$ has no relative extremum at $x = 1$.

7. **(a)** $x = -3, 1$ (stationary points)     **(b)** $x = 0, \pm\sqrt{3}$ (stationary points)     9. **(a)** $x = \pm\sqrt{2}$ (stationary points)     **(b)** no critical points

11. **(a)** $x = -1$ (stationary point)     **(b)** $n\pi/3, n = 0, \pm1, \pm2, \ldots$ (stationary points)     13. **(a)** $x = 2$     **(b)** $x = 0$     **(c)** $x = 1, 3$

15. **(a)** $x = 0$, relative max; $x = \pm\sqrt{5}$, relative min     **(b)** $x = 0$, relative min     17. relative max of 5 at $x = -2$

19. relative min of 0 at $x = \pi$, relative max of 1 at $x = \pi/2, 3\pi/2$     21. no relative extrema

23. relative min of 0 at $x = 1$, relative max of $\frac{4}{27}$ at $x = \frac{1}{3}$     25. relative min of 0 at $x = 0$, relative max of 1 at $x = 1, -1$

27. relative min of 0 at $x = 0$     29. relative min of 0 at $x = 0$     31. relative min of 0 at $x = 0$

33. relative min of 0 at $x = 2, -2$, relative max of 4 at $x = 0$

35. relative min of 0 at $x = \pi/2, \pi, 3\pi/2$;  1 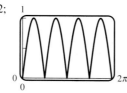     37. relative min of 0 at $x = \pi/2, 3\pi/2$;  1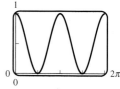
   relative max of 1 at $x = \pi/4$,                            relative max of 1 at $x = \pi$
   $3\pi/4, 5\pi/4, 7\pi/4$

39. relative min of $-1/e$ at $x = 1/e$   2.5 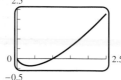     41. relative min of 0 at $x = 0$;   0.14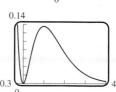
                                                       relative max of $1/e^2$ at $x = 1$

43. relative minima at $x = -3.58, 3.33$;   250 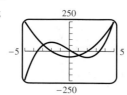     45. relative max at $x = 0.255$
   relative max at $x = 0.25$

47. relative min at $x = -1.20$;
    relative max at $x = 1.80$

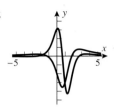

49. **(a)** 54
    **(b)** 9

51. **(b)**

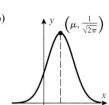

$\left(\mu, \frac{1}{\sqrt{2\pi}}\right)$

53. $f(x) = -2x^3 + 3x^2$

55. **(a)**

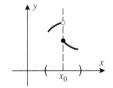

$f(x_0)$ is not an extreme value

**(b)**

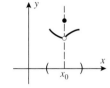

$f(x_0)$ is a relative maximum

**(c)**

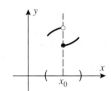

$f(x_0)$ is a relative minimum

▶ **Exercise Set 5.3** (Page 319)

1.

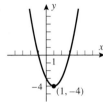

3.

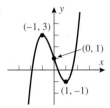

5.

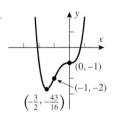

7. $\left(-\frac{1}{\sqrt{2}}, \frac{7\sqrt{2}}{8}\right)$

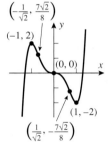

9.

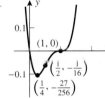

11.

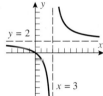

13.

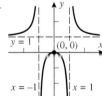

15.

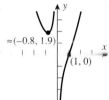

17.

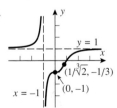

19.

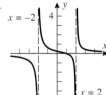

21.

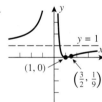

23. **(a)** VI
    **(b)** I
    **(c)** III
    **(d)** V
    **(e)** IV
    **(f)** II

25.

27.

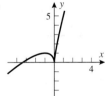

29.

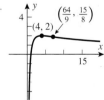

31.

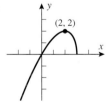

33.

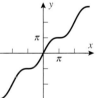

35.

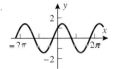

37.

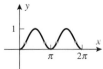

39. **(a)** $+\infty, 0$
    **(b)**

41. **(a)** $0, +\infty$
    **(b)**

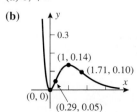

**43. (a)** $+\infty$, $-\infty$

**(b)**

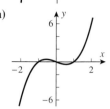

**45. (a)** $0$; $+\infty$

**(b)**

**47. (a)** $-\infty$; $0$

**(b)**

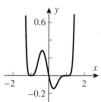

**49. (a)**

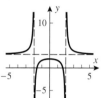

**(b)**

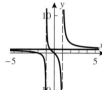

**(c)**

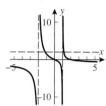

**(d)**

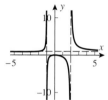

**51. (a)**

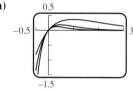

**(b)**

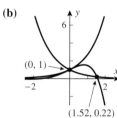

**(c)**

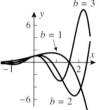

**(d)**

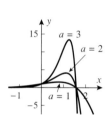

**53. (a)**

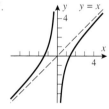

**55. (a)** The limit does not exist.

**(b)**

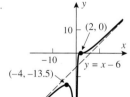

**(c)**

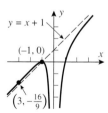

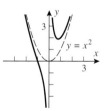

**57.**

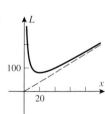

**59.**

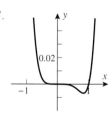

**61.**

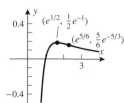

$y = x + 1$

$(-1, 0)$

$\left(3, -\frac{16}{9}\right)$

**63.**

$y = x^2$

**65.**

**67.**

**69. (a)** $\dfrac{kL^2A}{(1+A)^2}$

**(c)** $t = \dfrac{1}{Lk}\ln A$

---

► **Chapter 5 Supplementary Exercises (Page 321)**

**7. (a)** relative max at $x = 1$, relative min at $x = 7$, neither at $x = 0$

**(b)** relative max at $x = \pi/2$, $3\pi/2$;

relative min at $x = 7\pi/6$, $11\pi/6$

**(c)** relative max at $x = 5$

**9.** $\lim\limits_{x \to -\infty} f(x) = +\infty$, $\lim\limits_{x \to +\infty} f(x) = +\infty$;

relative min at $x = 0$;

points of inflection at $x = \frac{1}{2}$, $1$;

no asymptotes

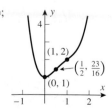

$(1, 2)$

$\left(\frac{1}{2}, \frac{23}{16}\right)$

$(0, 1)$

11. $\lim\limits_{x \to \pm\infty} f(x)$ does not exist;
critical point at $x = 0$;
relative min at $x = 0$;
point of inflection when $1 + 4x^2 \tan(x^2 + 1) = 0$;
vertical asymptotes at $x = \pm\sqrt{\pi(n + \frac{1}{2}) - 1}$, $n = 0, 1, 2, \ldots$

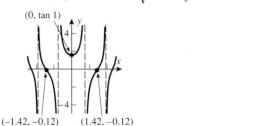

(0, tan 1)

(−1.42, −0.12)     (1.42, −0.12)

13. critical points at $x = -5, 0$;
relative max at $x = -5$, relative min at $x = 0$;
points of inflection at $x = -7.26, -1.44, 1.20$;
horizontal asymptote $y = 1$ as $x \to \pm\infty$

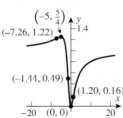

$\left(-5, \frac{5}{4}\right)$
(−7.26, 1.22)
(−1.44, 0.49)
(1.20, 0.16)
(0, 0)

15. $\lim\limits_{x \to -\infty} f(x) = +\infty$, $\lim\limits_{x \to +\infty} f(x) = -\infty$;
critical point at $x = 0$;
no extrema;
inflection point at $x = 0$ ($f$ changes concavity);
no asymptotes

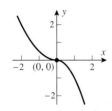

(0, 0)

17. $\lim\limits_{x \to +\infty} f(x) = +\infty$, $\lim\limits_{x \to 0^+} f(x) = 0$, $\lim\limits_{x \to 0^+} f'(x) = -\infty$;
critical point at $x = 1/e$;
relative min at $x = 1/e$;
no points of inflection;
no asymptotes

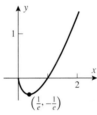

$\left(\frac{1}{e}, -\frac{1}{e}\right)$

19. critical point at $x = e^{1/2}$;
relative max at $x = e^{1/2}$;
point of inflection at $x = e^{5/6}$;
horizontal asymptote $y = 0$ as $x \to +\infty$

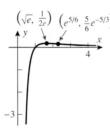

$\left(\sqrt{e}, \frac{1}{2e}\right)$  $\left(e^{5/6}, \frac{5}{6}e^{-5/3}\right)$

21. $\lim\limits_{x \to +\infty} f(x) = +\infty$;
critical point at $x = 1$;
relative min at $x = 1$;
no points of inflection;
vertical asymptote $x = 0$;
horizontal asymptote $y = 0$ for $x \to -\infty$

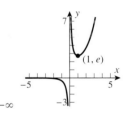

$(1, e)$

23. critical points at $x = 0, 2$;
relative min at $x = 0$, relative max at $x = 2$;
points of inflection at $x = 2 \pm \sqrt{2}$;
horizontal asymptote $y = 0$ as $x \to +\infty$;
$\lim\limits_{x \to -\infty} f(x) = +\infty$

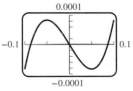

$\left(2, \frac{4}{e}\right)$
$(2 + \sqrt{2}, 1.04)$
$(2 - \sqrt{2}, 0.59)$
(0, 0)

25. (a)

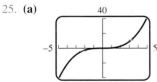

(b) relative max at $x = -\frac{1}{20}$, relative min at $x = \frac{1}{20}$
(c) The finer details can be seen when graphing over a much smaller $x$-window.

27. (a)

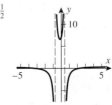

29. $y = 2x$, $y = 3$

31. $f(x) = \dfrac{x^2 + x - 7}{3x^2 + x - 1}$, $x \neq \frac{1}{2}$

33. **(a)** $\sin x = 1$, $\sin x = -1$

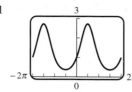

**(b)** relative maxima at $x = 2n\pi + \pi/2$, $y = e$;
relative minima at $x = 2n\pi - \pi/2$, $y = 1/e$, $n = 0, \pm 1, \pm 2, \ldots$

**(c)** when $\sin x = \dfrac{-1 + \sqrt{5}}{2}$

35. **(a)** relative min $-0.232466$ at $x = 0.450184$

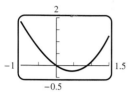

**(b)** relative max $0$ at $x = 0$;
relative min $-0.107587$ at $x = \pm 0.674841$

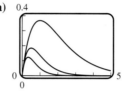

**(c)** relative max $0.876839$ at $x = 0.886352$;
relative min $-0.355977$ at $x = -1.244155$

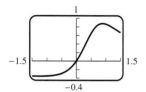

37. **(a)**

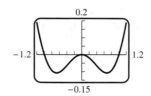

**(b)** $y = 0$ at $x = 0$;
$\lim\limits_{x \to +\infty} y = 0$

**(c)** relative max at $x = 1/a$;
inflection point at $x = 2/a$

**(d)** The maximum and the inflection point move toward the origin.

---

▶ **Chapter 5 Horizon Module (Page 324)**

1. Line II is the regression line.

2. **(a)** $y = 5.035714286x - 4.232142857$

    **(b)**

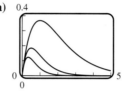

4. $r = 0.9907002406$

5. **(a)** $S = 2.155239850t + 190.3600714$;
$r = 0.9569426456$

    **(b)** Yes, because $r$ is close to 1.

    **(c)** 244.241068 mi/h

    **(d)** It is assumed that the line will still give a good estimate in the year 2000.

7. **(a)** $y = 3.923208367 + e^{0.2934589528x}$

    **(b)**

8. **(a)** an exponential model

    **(b)** $\log T = 1.719666407 \times 10^{-4} + 1.499661719 \log d$

    **(c)** $T = 1.000396046\, d^{1.499661719}$

    **(d)** The squares of the periods of revolution of the planets are proportional to the cubes of their mean distances.

9. **(a)** $T = 27 + 57.8\, e^{-0.046t}$    **(b)** $T_0 = 84.9°\text{C}$    **(c)** 53.19 min

---

▶ **Exercise Set 6.1 (Page 337)**

1. relative maxima at $x = 2, 6$; absolute max at $x = 6$; relative and absolute min at $x = 4$

3. **(a)**          **(b)**          **(c)**

5. maximum value 1 at $x = 0, 1$; minimum value 0 at $x = \frac{1}{2}$    7. maximum value 27 at $x = 4$, minimum value $-1$ at $x = 0$

9. maximum value $3/\sqrt{5}$ at $x = 1$, minimum value $-3/\sqrt{5}$ at $x = -1$

11. maximum value $1 - (\pi/4)$ at $x = -\pi/4$, minimum value $(\pi/4) - 1$ at $x = \pi/4$

13. maximum value 17 at $x = -5$, minimum value 1 at $x = -3$    15. minimum value $f\left(\frac{3}{2}\right) = -\frac{13}{4}$, no maximum

17. maximum value $f(1) = 1$, no minimum    19. no maximum or minimum    21. maximum value $f(-2) = -4$, no minimum

23. minimum value 0 for $x = \pm 1$, no maximum    25. maximum value 48 at $x = 8$, minimum value 0 at $x = 0, 20$

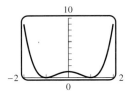          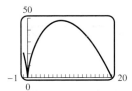

27. no maximum or minimum    29. maximum value 2 at $x = 0$, minimum value $\sqrt{3}$ at $x = \pi/6$

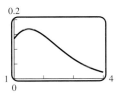          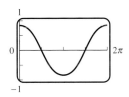

31. maximum value $\frac{27}{8}e^{-3}$ at $x = \frac{3}{2}$,    33. maximum value $\sin(1) \approx 0.84147$,
    minimum value $64/e^8$ at $x = 4$         minimum value $-\sin(1) \approx -0.84147$

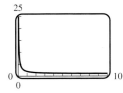          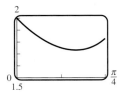

35. maximum value 2, minimum value $-\frac{1}{4}$

37. maximum value $3\sqrt{3}/2$ at $x = (\pi/6) + n\pi$, minimum value $-3\sqrt{3}/2$ at $x = (5\pi/6) + n\pi$, $n = 0, \pm 1, \pm 2, \ldots$

41. $f'(1) = 2$    45. $\left(\frac{1}{2}, -\frac{1}{4}\right)$ is closest, $(-1, -1)$ is farthest    47. maximum $y = 4$ at $t = \pi, 3\pi$; minimum $y = 0$ at $t = 0, 2\pi$

---

▶ **Exercise Set 6.2 (Page 348)**

1. $5, 5$    3. **(a)** 1    **(b)** $\frac{1}{2}$    5. 500 ft × 750 ft    7. 5 in × $\frac{12}{5}$ in    9. $10\sqrt{2}$ in × $10\sqrt{2}$ in

11. 80 ft ($1 fencing), 40 ft ($2 fencing)    15. **(a)** maximum $N = 161{,}788$, minimum $N = 125{,}000$    **(b)** 40    17. 2 in square

19. $\frac{200}{27}$ ft$^3$    21. base 10 cm square, height 20 cm    23. ends $\sqrt[3]{3V/4}$ units square, height $\frac{4}{3}\sqrt[3]{3V/4}$

25. height $= 2\sqrt{(5 - \sqrt{5})/10}\,R$, radius $= \sqrt{(5 + \sqrt{5})/10}\,R$    29. height $=$ radius $= \sqrt[3]{500/\pi}$    31. $L/12$ by $L/12$ by $L/12$

33. height $= L/\sqrt{3}$, radius $= \sqrt{2/3}\,L$    35. radius $= \sqrt[6]{450/\pi^2}$, height $= \frac{30}{\pi}\sqrt[3]{\pi^2/450}$ cm    37. height $= 4R$, radius $= \sqrt{2}R$

39. $\pi/3$    41. $5\sqrt{5}$ ft    43. **(a)** 7000    **(b)** yes    45. 13,722 lb    47. $1/\sqrt{5}$    51. $(-\sqrt{2}, 1), (\sqrt{2}, 1)$    53. $\left(\sqrt{2}, \frac{1}{2}\right)$

55. $\left(-1/\sqrt{3}, \frac{3}{4}\right)$    57. $4(1 + 2^{2/3})^{3/2}$ ft    59. 30 cm from the weaker source    63. **(c)** $\frac{1}{4}$ mile downstream from the house

---

▶ **Exercise Set 6.3 (Page 359)**

1. **(a)** positive, negative, slowing down    3. **(a)** left    5.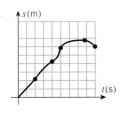
   **(b)** positive, positive, speeding up         **(b)** negative
   **(c)** negative, positive, slowing down         **(c)** speeding up
                                                     **(d)** slowing down

7.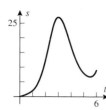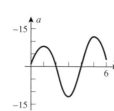

9. **(a)** 6.7 ft/s$^2$
   **(b)** $t = 0$ s

11. **(a)** $v(t) = 3t^2 - 12t$, $a(t) = 6t - 12$
    **(b)** $s(1) = -5$ ft, $v(1) = -9$ ft/s,
    $|v(1)| = 9$ ft/s, $a(1) = -6$ ft/s$^2$
    **(c)** 0, 4
    **(d)** speeding up for $0 < t < 2$ and $4 < t$,
    slowing down for $2 < t < 4$
    **(e)** 39 ft

13. **(a)** $v(t) = -(3\pi/2)\sin(\pi t/2)$, $a(t) = -(3\pi^2/4)\cos(\pi t/2)$
    **(b)** $s(1) = 0$ ft, $v(1) = -3\pi/2$ ft/s, $|v(1)| = 3\pi/2$ ft/s, $a(1) = 0$ ft/s$^2$  **(c)** 0, 2, 4
    **(d)** speeding up for $0 < t < 1$, $2 < t < 3$, and $4 < t < 5$; slowing down for $1 < t < 2$ and $3 < t < 4$  **(e)** 15 ft

15. **(a)** $\sqrt{5}$  **(b)** $\sqrt{5}/10$  **(c)** speeding up for $\sqrt{5} < t < \sqrt{15}$,
    slowing down for $0 < t < \sqrt{5}$ and $\sqrt{15} < t$

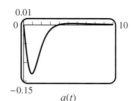

17.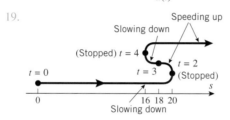

Constant speed

19.

21.

23. **(a)** 12 ft/s  **(b)** $t = 2.2$, $s = -24.2$  25. **(a)** 11.025 m  **(b)** $-14.7$ m/s  **(c)** 1.2245 s  **(d)** $t = 4.5175$ s

27. **(a)** 6.12 s  29. 113.42 ft/s  31. 29.39 m  33. **(b)** 113.42 ft/s  35. **(a)**  **(b)** $\sqrt{2}$
**(b)** 183.67 m
**(c)** 6.12 s
**(d)** 60 m/s

37. **(b)** $\frac{2}{3}$ unit  **(c)** $0 \le t < 1$ and $t > 2$  39. **(a)** $-1.25$ ft/s/ft  **(b)** $-2500$ ft/s$^2$

---

▶ **Exercise Set 6.4 (Page 366)**

1. 1.414213562  3. 1.817120593  5. $-1.671699882$  7. 1.224439550
9. $-1.452626879$  11. 1.895494267  13. 4.493409458

15. $-1.165373043$  17. $-0.474626618, 1.395336994$  19. **(b)** 3.162277660  21. $-4.098859132$

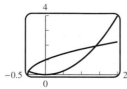

23. $(0.589754512, 0.347810385)$     25. **(b)** $171°$     27. $-1.220744085, 0.724491959$     29. $5.3362\%$

---

▶ Exercise Set 6.5 **(Page 372)**

1. $[0, 4], c = 3$     3. $c = 3$     5. $c = \pi$     7. $c = 1$     9. $c = 1.54$     11. $1$     13. $\frac{5}{4}$     15. $-\sqrt{5}$

17. **(a)** $[-2, 1]$     **(b)** $c = -1.29$                     **(c)** $-1.2885843$     37. $f(x) = x^3 - 4x + 5$     39.

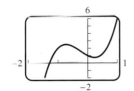     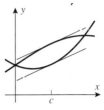

---

▶ Chapter 6 Supplementary Exercises **(Page 374)**

3. **(a)** true     **(b)** false     5. true

7. **(a)** $M = -\frac{1}{2}$ at $x = -2$; $m = -1$ at $x = -1$     **(b)** $M = \frac{27}{256}$ at $x = \frac{3}{4}$; $m = -2$ at $x = -1$

 **(c)** $m \approx -1.9356$ at $x = \frac{12}{7}$; $M = 9$ at $x = 3$     **(d)** $m = e^2/4$ at $x = 2$     9. $2.3561945$

11. **(a)** yes, $c = 0$     **(b)** no     **(c)** yes, $c = \sqrt{\pi}/2$     13. $r = 2P/(8 + 3\pi)$ ft     15. **(a)** yes     **(b)** yes

17. **(a)** $0.3501$ s, $0.0820$ s     19.                     **(b)** minimum: $(-2.111985, -0.355116)$;

maximum: $(0.372591, 2.012931)$

21. **(a)** $v = -2\dfrac{t(t^4 + 2t^2 - 1)}{(t^4 + 1)^2}$, $a = 2\dfrac{3t^8 + 10t^6 - 12t^4 - 6t^2 + 1}{(t^4 + 1)^3}$     **(c)** $t = 0.64, s = 1.2$     **(d)** $0 \le t \le 0.64$ s

 **(e)** speeding up when $0 \le t < 0.36$ and $0.64 < t < 1.1$, otherwise slowing down

 **(f)** maximum speed $= 1.05$ m/s when $t = 1.10$ s     23. $149.988 \times 10^6$ km

---

▶ Exercise Set 7.1 **(Page 382)**

1. $A = 1/2$                     3. $A = 16$                     5. $\frac{1}{2}$     7. $16$     9. $e - 1$

| $n$ | 1 | 2 | 3 | 4 | 5 |
|---|---|---|---|---|---|
| $A_n$ | 1.0000 | 0.7500 | 0.6666 | 0.6250 | 0.6000 |

| $n$ | 1 | 2 | 3 | 4 | 5 |
|---|---|---|---|---|---|
| $A_n$ | 28.0000 | 22.0000 | 20.0000 | 19.0000 | 18.4000 |

| $n$ | 6 | 7 | 8 | 9 | 10 |
|---|---|---|---|---|---|
| $A_n$ | 0.5833 | 0.5714 | 0.5625 | 0.5556 | 0.5500 |

| $n$ | 6 | 7 | 8 | 9 | 10 |
|---|---|---|---|---|---|
| $A_n$ | 18.0000 | 17.7143 | 17.5000 | 17.3333 | 17.2000 |

---

▶ Exercise Set 7.2 **(Page 389)**

1. **(a)** $\displaystyle\int \frac{x}{\sqrt{1 + x^2}}\, dx = \sqrt{1 + x^2} + C$     **(b)** $\displaystyle\int (x + 1)e^x\, dx = xe^x + C$

3. $\dfrac{d}{dx}\left[\sqrt{x^3 + 5}\right] = \dfrac{3x^2}{2\sqrt{x^3 + 5}}$,  so  $\displaystyle\int \frac{3x^2}{2\sqrt{x^3 + 5}}\, dx = \sqrt{x^3 + 5} + C.$

5. $\dfrac{d}{dx}\left[\sin(2\sqrt{x})\right] = \dfrac{\cos(2\sqrt{x})}{\sqrt{x}}$,  so  $\displaystyle\int \frac{\cos(2\sqrt{x})}{\sqrt{x}}\, dx = \sin(2\sqrt{x}) + C.$

7. **(a)** $(x^9/9) + C$     **(b)** $\frac{7}{12}x^{12/7} + C$     **(c)** $\frac{2}{9}x^{9/2} + C$     9. **(a)** $-\frac{1}{4}x^{-4} + C$     **(b)** $(u^4/4) - u^2 + 7u + C$

11. $-\frac{1}{2}x^{-2} + \frac{2}{3}x^{3/2} - \frac{12}{5}x^{5/4} + \frac{1}{3}x^3 + C$     13. $(x^2/2) + (x^5/5) + C$     15. $3x^{4/3} - \frac{12}{7}x^{7/3} + \frac{3}{10}x^{10/3} + C$     17. $\dfrac{x^2}{2} - \dfrac{2}{x} + \dfrac{1}{3x^3} + C$

19. $2\ln x + 3e^x + C$     21. $-4\cos x + 2\sin x + C$     23. $\tan x + \sec x + C$     25. $\ln\theta - 2e^\theta + \cot\theta + C$     27. $\sec x + C$

29. $\theta - \cos\theta + C$    31. $\tan x - \sec x + C$    33. **(a)**    **(b)** $f(x) = (x^2/2) + 5$    35.

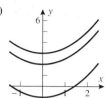

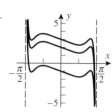

37. $f(x) = \cos x + 1$    39. **(a)** $y(x) = \frac{3}{4}x^{4/3} + \frac{5}{4}$   **(b)** $y(t) = \ln|t| + 5$   **(c)** $y(x) = \frac{2}{3}x^{3/2} + 2x^{1/2} - \frac{8}{3}$

41. $f(x) = \frac{4}{15}x^{5/2} + C_1 x + C_2$    43. $y = x^2 + x - 6$    45. $y = x^3 - 6x + 7$    47. **(b)** $F(0) - G(0) = \frac{8}{3}$    49. $\tan x - x + C$

51. **(a)** $\frac{1}{2}(x - \sin x) + C$   **(b)** $\frac{1}{2}(x + \sin x) + C$    53. $v = \dfrac{1087}{\sqrt{273}}T^{1/2}$ ft/s

---

▶ **Exercise Set 7.3 (Page 395)**

1. **(a)** $\dfrac{(x^2 + 1)^{24}}{24} + C$   **(b)** $-\dfrac{\cos^4 x}{4} + C$   **(c)** $-2\cos\sqrt{x} + C$   **(d)** $\frac{3}{4}\sqrt{4x^2 + 5} + C$   **(e)** $\frac{1}{3}\ln(x^3 - 4) + C$

3. **(a)** $-\frac{1}{2}\cot^2 x + C$   **(b)** $\frac{1}{10}(1 + \sin t)^{10} + C$   **(c)** $\ln|\ln x| + C$   **(d)** $-\frac{1}{5}e^{-5x} + C$   **(e)** $-\frac{1}{3}\ln|(1 + \cos 3\theta)| + C$

5. $\frac{1}{2}e^{2x} + C$   7. $-\dfrac{(2 - x^2)^4}{8} + C$   9. $\frac{1}{8}\sin 8x + C$   11. $\frac{1}{4}\sec 4x + C$   13. $\frac{1}{21}(7t^2 + 12)^{3/2} + C$   15. $\frac{2}{3}\sqrt{x^3 + 1} + C$

17. $-\frac{1}{16}(4x^2 + 1)^{-2} + C$   19. $e^{\sin x} + C$   21. $-\frac{1}{6}e^{-2x^3} + C$   23. $\frac{1}{5}\cos(5/x) + C$   25. $\frac{1}{3}\tan(x^3) + C$   27. $-e^{-x} + C$

29. $\frac{1}{18}\sin^6 3t + C$   31. $-\frac{1}{6}(2 - \sin 4\theta)^{3/2} + C$   33. $\frac{1}{6}\sec^3 2x + C$   35. $2e^{\sqrt{y}} + C$   39. $\dfrac{1}{b(n+1)}\sin^{n+1}(a + bx) + C$

41. $\frac{2}{5}(x - 3)^{5/2} + 2(x - 3)^{3/2} + C$   43. $\frac{1}{3}(\tan 3\theta - 3\theta) + C$   45. $t + \ln|t| + C$   47. $\displaystyle\int [\ln(e^x) + \ln(e^{-x})]\,dx = C$

49. **(a)** with $u = \sin x$, $\frac{1}{2}\sin^2 x + C_1$; with $u = \cos x$, $-\frac{1}{2}\cos^2 x + C_2$   **(b)** because they differ by a constant

51. $y(x) = \frac{2}{9}(3x + 1)^{3/2} + \frac{29}{9}$    53. $f(x) = \frac{2}{9}(3x + 1)^{3/2} + \frac{7}{9}$    55. 100,416

---

▶ **Exercise Set 7.4 (Page 402)**

1. **(a)** 36   **(b)** 55   **(c)** 40   **(d)** 6   **(e)** 11   **(f)** 0   3. $\displaystyle\sum_{k=1}^{10} k$   5. $\displaystyle\sum_{k=1}^{49} k(k + 1)$   7. $\displaystyle\sum_{k=1}^{10} 2k$   9. $\displaystyle\sum_{k=1}^{6}(-1)^{k+1}(2k - 1)$

11. $\displaystyle\sum_{k=1}^{5}(-1)^k \frac{1}{k}$   13. **(a)** $\displaystyle\sum_{1}^{50} 2k$   **(b)** $\displaystyle\sum_{1}^{50}(2k - 1)$   15. 5050   17. 2870   19. 1728   21. 214,365   23. $2n^2 - n$

25. $\frac{3}{2}(n+1)$   27. $\frac{1}{4}(n-1)^2$   31. **(a)** $\displaystyle\sum_{k=0}^{19} 3^{k+1} = \frac{3}{2}(3^{20}-1)$   **(b)** $\displaystyle\sum_{k=0}^{25} 2^{k+5} = 2^{31}-2^5$   **(c)** $\displaystyle\sum_{k=0}^{100}(-1)\left(\frac{-1}{2}\right)^k = -\frac{2}{3}\left(1 + \frac{1}{2^{101}}\right)$

33. $\dfrac{n + 1}{2n}$; $\frac{1}{2}$   35. $\dfrac{5(n + 1)}{2n}$; $\frac{5}{2}$   37. **(a)** $\displaystyle\sum_{j=0}^{5} 2^j$   **(b)** $\displaystyle\sum_{j=1}^{6} 2^{j-1}$   **(c)** $\displaystyle\sum_{j=2}^{7} 2^{j-2}$   39. **(a)** $\displaystyle\sum_{k=1}^{18}\sin\left(\frac{\pi}{k}\right)$   **(b)** $\displaystyle\sum_{k=0}^{6} e^k = \dfrac{e^7 - 1}{e - 1}$

43. $3^{17} - 3^4$   45. $-\frac{399}{400}$   47. **(b)** $\frac{1}{2}$   49. Both identities are valid.   55. 18,755

---

▶ **Exercise Set 7.5 (Page 414)**

1. **(a)** 8   **(b)** greater than $A$   **(c)** less than $A$   **(d)** equal to $A$

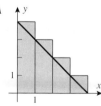

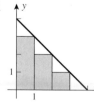

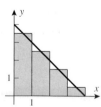

3. 35, 60, 46.25   5. $\dfrac{(1 + \sqrt{2})\pi}{4} \approx 1.896$, $\dfrac{(1 + \sqrt{2})\pi}{4} \approx 1.896$, $\dfrac{\pi\sqrt{2}\cos(\pi/8)}{2} \approx 2.052$

7. left endpoints: $A \approx (2 + 3 + 2 + 1)(1) = 8$; right endpoints: $A \approx (3 + 2 + 1 + 2)(1) = 8$

9. 0.718771403, 0.668771403, 0.692835360    11. 0.919403170, 1.07648280, 1.001028825

13. 0.351220577, 0.420535296, 0.386502483

15. **(a)** 0.693097198, 0.666154270, 1.000164512, 5.336963538, 0.386327689, 1.718167282

    **(b)** 0.693134682, 0.666538346, 1.000041125, 5.334644416, 0.386302694, 1.718253191

    **(c)** 0.693144056, 0.666634573, 1.000010281, 5.333803776, 0.3862964444, 1.718274669

17. **(a)** $A = \frac{9}{2}$     **(b)** $-A = -\frac{3}{2}$     **(c)** $-A_1 + A_2 = \frac{15}{2}$     **(d)** $-A_1 + A_2 = 0$

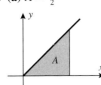  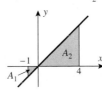 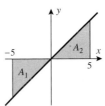

19. **(a)** $A = 10$     **(b)** $A_1 - A_2 = 0$ by symmetry     **(c)** $A_1 + A_2 = \frac{13}{2}$     **(d)** $\pi/2$     21. **(a)** 0.8

    **(b)** $-2.6$

    **(c)** $-1.8$

    **(d)** $-0.3$

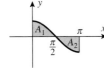 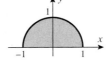 

23. $-1$     25. 3     27. **(a)** $(1 + \pi)/2$     **(b)** $-4$     29. **(a)** negative     **(b)** positive     31. $25\pi/2$

33. **(a)** $\displaystyle\int_{-3}^{3} 4x(1 - 3x)\, dx$     **(b)** $\displaystyle\int_{0}^{1} e^x\, dx$     35. $\frac{5}{2}$

37. **(a)** $\displaystyle\lim_{\max \Delta x_k \to 0} \sum_{k=1}^{n} 2x_k^* \Delta x_k; a = 1, b = 2$     **(b)** $\displaystyle\lim_{\max \Delta x_k \to 0} \sum_{k=1}^{n} \frac{x_k^*}{x_k^* + 1} \Delta x_k; a = 0, b = 1$

39. **(d)** $\frac{3}{2}$     41. $\frac{1}{3}$     43. 320     45. **(a)** yes     **(b)** yes     **(c)** no     **(d)** yes

---

▶ **Exercise Set 7.6 (Page 425)**

1. **(a)** $\displaystyle\int_{0}^{2} (2 - x)\, dx = 2$     **(b)** $\displaystyle\int_{-1}^{1} 2\, dx = 4$     **(c)** $\displaystyle\int_{1}^{3} (x + 1)\, dx = 6$     3. $\frac{65}{4}$     5. $\frac{52}{3}$     7. $e^3 - e$     9. 48     11. $\frac{2}{3}$

13. $\frac{844}{5}$     15. 0     17. $\sqrt{2}$     19. $5e^3 - 10$     21. $-\frac{55}{3}$     23. $\frac{\pi^2}{9} + 2\sqrt{3}$     27. **(a)** $\frac{5}{2}$     **(b)** $2 - \frac{\sqrt{2}}{2}$     29. $-\frac{11}{6}$

31. $0.665867079; \frac{2}{3}$     33. $1.098242635; \ln 3 \approx 1.098612289$     35. 12     37. $\frac{9}{2}$

39. $A_1 = \frac{23}{6}, A_2 = \frac{343}{6}, A_3 = \frac{243}{6}, A = \frac{203}{2}$     41. **(a)** The integral is zero.     43. **(a)** $x^3 + 1$

    **(c)** $\displaystyle\int_{-a}^{a} f(x)\, dx = 2 \int_{0}^{a} f(x)\, dx$

45. **(a)** $\sin\sqrt{x}$     **(b)** $e^{x^2}$     47. $-\dfrac{x}{\cos x}$     49. **(a)** 0     **(b)** $\sqrt{13}$     **(c)** $6/\sqrt{13}$

51. **(a)** $x = 3$     **(b)** increasing on $[3, +\infty)$, decreasing on $(-\infty, 3]$     **(c)** concave up on $(-1, 7)$, concave down on $(-\infty, -1)$ and $(7, +\infty)$

53. **(a)** $(0, +\infty)$     **(b)** $x = 1$     55. **(a)** $x^* = 4$     **(b)** $x^* = e - 1$     57. $3\sqrt{2} \le \displaystyle\int_{0}^{3} \sqrt{x^3 + 2}\, dx \le 3\sqrt{29}$

---

▶ **Exercise Set 7.7 (Page 437)**

1. **(a)** the increase in height in inches, during the first 10 years

    **(b)** the change in the radius in cm, during the time interval $t = 1$ to $t = 2$ seconds

    **(c)** the change in the speed of sound in ft/s, during an increase in temperature from $t = 32°$F to $t = 100°$F

    **(d)** the displacement of the particle in cm, during the time interval $t = t_1$ to $t = t_2$ seconds

3. **(a)** displacement $= -\frac{1}{2}$; distance $= \frac{1}{2}$     **(b)** displacement $= 5$; distance $= \frac{5}{2}$

5. **(a)** 31.3 m/s     **(b)** 55.15 m/s     7. **(a)** $\frac{1}{4}t^4 - \frac{2}{3}t^3 + t + 1$     **(b)** $-\cos 2t - t - 2$     9. **(a)** $t^2 - 3t + 7$     **(b)** $-\cos t + t - (\pi/2)$

11. **(a)** displacement $= 1$; distance $= 1$　　**(b)** displacement $= -1$; distance $= 3$

13. **(a)** displacement $= \frac{9}{4}$; distance $= \frac{11}{4}$　　**(b)** displacement $= e^3 - 7$; distance $= e^3 - 9 + 4\ln 2$

15. displacement $= -6$; distance $= \frac{13}{2}$　　17. displacement $= \frac{204}{25}$; distance $= \frac{204}{25}$

19. **(a)** $s = 2/\pi, v = 1, |v| = 1, a = 0$　　**(b)** $s = \frac{1}{2}, v = -\frac{3}{2}, |v| = \frac{3}{2}, a = -3$　　21. $\frac{22}{3}$　　23. $(1/e) + e - 2$

25. **(a)** 　　**(b)** 　　**(c)** 　　27. **(a)** The displacement is always positive.

29. **(a)** $a(t) = \begin{cases} 0, & t < 4 \\ -10, & t > 4 \end{cases}$　　**(b)** $v(t) = \begin{cases} 25, & t < 4 \\ 65 - 10t, & t > 4 \end{cases}$　　**(c)** $x(t) = \begin{cases} 25t, & t < 4 \\ 65t - 5t^2 - 80, & t > 4 \end{cases}$　　**(d)** $x(6.5) = 131.25$

so $x(8) = 120, x(12) = -20.$

31. **(a)** $-\frac{22}{15}$ ft/s$^2$　　**(b)** $\frac{1}{7200}$ km/s$^2$　　33. **(a)** $-\frac{121}{5}$ ft/s$^2$　　**(b)** $\frac{70}{33}$ s　　**(c)** $\frac{60}{11}$ s

35. 280 m　　37. 100 s; 10,000 ft　　39. **(a)** $-48$ ft/s　　**(b)** 196 ft　　**(c)** 112 ft/s

41. **(a)** 1 s　　**(b)** $\frac{1}{2}$ s　　43. **(a)** $(5 + 5\sqrt{33})/8$ s　　**(b)** $20\sqrt{33}$ ft/s

45. **(a)** 5 s　　**(b)** 272.5 m　　**(c)** 10 s　　**(d)** $-49$ m/s　　**(e)** 12.46 s　　**(f)** 73.1 m/s　　47. 4.04 m/s　　49. 6　　51. $2/\pi$　　53. $\dfrac{1}{e-1}$

55. **(a)** $\frac{4}{3}$　　**(c)**　　57. **(a)** $\frac{263}{4}$　　59. $1404\pi$ lb　　61. **(a)** 120 gal　　63. **(b)** no

　　**(b)** $2/\sqrt{3}$　　　　　**(b)** 31　　　　　**(b)** 420 gal

　　　　　　　**(c)** 2076.36 gal

---

▶ Exercise Set 7.8 **(Page 444)**

1. **(a)** $\displaystyle\int_1^3 u^7\, du$　　**(b)** $-\dfrac{1}{2}\displaystyle\int_7^4 u^{1/2}\,du$　　**(c)** $\dfrac{1}{\pi}\displaystyle\int_{-\pi}^{\pi}\sin u\, du$　　**(d)** $\displaystyle\int_{-3}^0 (u+5)u^{20}du$　　3. $\frac{121}{5}$　　5. 10　　7. $\frac{1192}{15}$

9. $8 - (4\sqrt{2})$　　11. $\ln\frac{21}{13}$　　13. $\frac{25}{12}\pi$　　15. $\pi/8$　　17. $2/\pi$　　19. $\frac{1}{24}$　　21. $\dfrac{1 - e^{-8}}{8}$　　23. $\frac{2}{3}$　　25. $\frac{2}{3}(\sqrt{10} - 2\sqrt{2})$

27. $2(\sqrt{7} - \sqrt{3})$　　29. 0　　31. 0　　33. $(\sqrt{3} - 1)/3$　　35. $\frac{106}{405}$　　37. $\ln 2$　　41. **(a)** $\frac{5}{3}$　　**(b)** $\frac{5}{3}$　　**(c)** $-\frac{1}{2}$

45. 48,233,525,650　　47. **(a)** 328.69 ft　　**(b)** yes　　49. **(b)** 169.7 V　　51. **(b)** $\frac{3}{2}$　　**(c)** $\pi/4$　　53. $2/\pi$

---

▶ Exercise Set 7.9 **(Page 451)**

1. **(a)**　　**(b)**　　**(c)**　　3. **(a)** 7

　　　　　　　　　　　　　　　　**(b)** $-5$

　　　　　　　　　　　　　　　　**(c)** $-3$

　　　　　　　　　　　　　　　　**(d)** 6

5. 1.603210678; magnitude of error is $< 0.0063$　　7. **(a)** $x^{-1}, x > 0$　　**(b)** $x^2, x \neq 0$　　**(c)** $-x^2, -\infty < x < +\infty$

　　**(d)** $-x, -\infty < x < +\infty$　　**(e)** $x^3, x > 0$　　**(f)** $\ln x + x, x > 0$　　**(g)** $x - \sqrt[3]{x}, -\infty < x < +\infty$　　**(h)** $\dfrac{e^x}{x}, x > 0$

9. **(a)** $e^{\pi \ln 3}$ **(b)** $e^{\sqrt{2}\ln 2}$ 11. **(a)** $e^2$ **(b)** $e^2$ 13. $x^2 - x$ 15. **(a)** $3/x$ **(b)** 1 17. **(a)** 0 **(b)** $\frac{1}{3}$ **(c)** 0

19. **(a)** $2x^3\sqrt{1+x^2}$ **(b)** $-\frac{2}{3}(x^2+1)^{3/2} + \frac{2}{5}(x^2+1)^{5/2} - \frac{4\sqrt{2}}{15}$ 21. **(a)** $-\sin x^2$ **(b)** $-\tan^2 x$

23. $-3\dfrac{3x-1}{9x^2+1} + 2x\dfrac{x^2-1}{x^4+1}$ 25. **(a)** $3x^2\sin^2(x^3) - 2x\sin^2(x^2)$ **(b)** $\dfrac{2}{1-x^2}$

27. **(a)** $F(0) = 0$, $F(3) = 0$, $F(5) = 6$, $F(7) = 6$, $F(10) = 3$

   **(b)** increasing on $\left[\frac{3}{2}, 6\right]$ and $\left[\frac{37}{4}, 10\right]$, decreasing on $\left[0, \frac{3}{2}\right]$ and $\left[6, \frac{37}{4}\right]$

   **(c)** maximum $\frac{15}{2}$ at $x = 6$, minimum $-\frac{9}{4}$ at $x = \frac{3}{2}$

**(d)**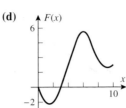

29. $F(x) = \begin{cases} (1-x^2)/2, & x < 0 \\ (1+x^2)/2, & x \geq 0 \end{cases}$

31. $y(x) = \frac{5}{4} + \frac{3}{4}x^{4/3}$ 33. $y(x) = \tan x + \cos x - (\sqrt{2}/2)$ 35. $P(x) = P_0 + \displaystyle\int_0^x r(t)\, dt$ individuals

37. I is the derivative of II. 39. **(a)** $t = 3$

**(f)**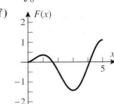

   **(b)** $t = 1$

   **(c)** $t = 5$

   **(d)** $t = 3$

   **(e)** $F$ is concave up on $\left(0, \frac{1}{2}\right)$ and $(2, 4)$,

   concave down on $\left(\frac{1}{2}, 2\right)$ and $(4, 5)$.

41. **(a)** relative maxima at $x = \pm\sqrt{4k+1}$, $k = 0, 1, \ldots$; 43. $f(x) = 3e^{3x}$, $a = \frac{1}{3}\ln 2$ 45. 0.06

   relative minima at $x = \pm\sqrt{4k-1}$, $k = 1, 2, \ldots$

   **(b)** $x = \pm\sqrt{2k}$, $k = 1, 2, \ldots$, and at $x = 0$

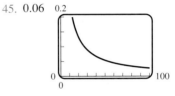

---

▶ **Chapter 7 Supplementary Exercises (Page 454)**

5. $s(t) = \frac{1}{2}at^2 + v_0 t + s_0$, $v(t) = u(t) + v_0$

7. **(a)** $\frac{3}{4}$ **(b)** $-\frac{3}{2}$ **(c)** $-\frac{35}{4}$ **(d)** $-2$ **(e)** not enough information **(f)** not enough information

9. **(a)** $2 + (\pi/2)$ **(b)** $\frac{1}{3}(10^{3/2} - 1) - \frac{9\pi}{4}$ **(c)** $\pi/8$ 11. $35\pi/128$ 15. **(d)** $n \geq 1000$

21. $(x^{2/3} + 1)^{3/2} + C$ 23. **(a)** $\displaystyle\int_1^x \frac{1}{1+t^2}\, dt$ **(b)** $\displaystyle\int_{\tan[(\pi/4)-2]}^x \frac{1}{1+t^2}\, dt$

27. **(a)** $F(x)$ is 0 if $x = 1$, positive if $x > 1$, and negative if $x < 1$.

   **(b)** $F(x)$ is 0 if $x = -1$, positive if $-1 < x \leq 2$, and negative if $-2 \leq x < -1$.

29. **(a)** 37,773.06 kW **(c)** **(d)** 2285.32 kW/h 31. **(a)** no **(d)** 141.5 ft

   **(b)** 2200.32 kW/h **(b)** $25 < t < 40$ **(e)** no

   **(c)** 3.54 ft/s **(f)** no

33. $\frac{1}{3}\sqrt{5 + 2\sin 3x} + C$ 35. $-\dfrac{1}{3a^2x^3 + 3ab} + C$ 37. $C$ 39. $\ln 2$ 41. $\frac{3}{8} + \frac{1}{2}\left(\sin 1 - \sin\frac{1}{4}\right)$ 43. 1.007514

45. **(a)** $k = 2.073948$ 47. **(a)** **(b)** 0.7651976866 49. The integral is better.

   **(b)** $k = 1.837992$

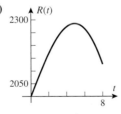

   **(c)** $x = 2.404826$

▶ **Chapter 7 Horizon Module (Page 457)**

2. **(a)** $v_x(t) = 35\cos\alpha$, $v_y(t) = 35\sin\alpha - 9.8t$    4.    65    5. $t = \dfrac{35\sin\alpha}{4.9}$ s

   **(b)** $t = \dfrac{35\sin\alpha}{9.8}$

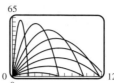

6. **(a)** $\alpha = 0.4316565575$, $1.139139769$ rad $\approx 24.73°$, $65.27°$

   **(b)** $y(t) < 50$ is required; but $y(1.139) \approx 51.56$ m, so his height would be 56.56 m.

7. $0.4019 < \alpha < 0.4636$ rad or $23.03° < \alpha < 26.57°$

▶ **Exercise Set A (Page A8)**

1. **(a)** rational    **(e)** integer, rational    3. **(a)** $\frac{41}{333}$    5. **(a)** $\frac{256}{81}$    7.

   **(b)** integer, rational    **(f)** irrational    **(b)** $\frac{115}{9}$    **(b)** worse

   **(c)** integer, rational    **(g)** rational    **(c)** $\frac{20943}{550}$

   **(d)** rational    **(h)** integer, rational    **(d)** $\frac{537}{1250}$

| Line | 2 | 3 | 4 | 5 | 6 | 7 |
|---|---|---|---|---|---|---|
| Blocks | 3, 4 | 1, 2 | 3, 4 | 2, 4, 5 | 1, 2 | 3, 4 |

9. (a), (d), (f)    11. **(a)** all values    **(b)** none    13. **(a)** yes    **(b)** no

15. **(a)** $\{x : x$ is a positive odd integer$\}$    **(b)** $\{x : x$ is an even integer$\}$    **(c)** $\{x : x$ is irrational$\}$    **(d)** $\{x : x$ is an integer and $7 \le x \le 10\}$

17. **(a)** false    **(b)** true    **(c)** true    **(d)** false    **(e)** true    **(f)** true    **(g)** true

19. **(a)** ———•——→  4    **(b)** ←•——•——→  −3    **(c)** ——•———•——→  −1  7

   **(d)** •————•——→  −3  3    **(e)** •———•———→  −3  3    **(f)** •————•——→  −3  3

21. **(a)** $[-2, 2]$    **(b)** $(-\infty, -2) \cup (2, +\infty)$    23. $\left(-\infty, \frac{10}{3}\right)$ ———○→ $\frac{10}{3}$    25. $\left(-\infty, -\frac{11}{2}\right]$ ———•→ $-\frac{11}{2}$

27. $\left(-\frac{3}{2}, \frac{1}{2}\right]$ ○———•→ $-\frac{3}{2}$ $\frac{1}{2}$    29. $(-\infty, 3) \cup (4, +\infty)$ ○———○→ 3  4    31. $\left(-\frac{3}{2}, 2\right)$ ○———○→ $-\frac{3}{2}$ 2

33. $(-\infty, -2] \cup (2, +\infty)$ •———○→ −2  2    35. $(-\infty, -3) \cup (3, +\infty)$ ○———○→ −3  3

37. $(-\infty, -2) \cup (4, +\infty)$ ○———○→ −2  4    39. $[4, 5]$ •———•→ 4  5    41. $(-8, 0) \cup (4, +\infty)$ ○——○——○→ −8  0  4

43. $(2, +\infty)$ ○———→ 2    45. $(-\infty, -3) \cup [2, +\infty)$    47. $77 \le F \le 104$    55. $\left(-\infty, -\frac{1}{2}\right)$

▶ **Exercise Set B (Page A15)**

1. **(a)** 7    **(b)** $\sqrt{2}$    **(c)** $k^2$    **(d)** $k^2$    3. $x \le 3$    5. all real $x$    7. $x \ge 0$ or $x = -\frac{2}{3}$    9. $x \ge -5$

13. **(a)** 2    **(b)** 1    **(c)** 14    **(d)** $3 + \sqrt{2}$    **(e)** 7    **(f)** 5    15. **(a)** −9    **(b)** 7    **(c)** 12    17. $-\frac{5}{6}, \frac{3}{2}$    19. $\frac{1}{2}, \frac{5}{2}$

21. $-\frac{11}{10}, \frac{11}{8}$    23. $1, \frac{17}{5}$    25. $(-9, -3)$    27. $\left[-\frac{3}{2}, \frac{9}{2}\right]$    29. $(-\infty, -3) \cup (-1, +\infty)$    31. $\left(-\infty, \frac{1}{2}\right] \cup \left[\frac{9}{2}, +\infty\right)$

33. $\left(-\infty, \frac{1}{2}\right) \cup \left(\frac{3}{2}, +\infty\right)$    35. $\left[\frac{1}{8}, \frac{1}{2}\right) \cup \left(\frac{1}{2}, \frac{7}{8}\right]$    37. $x \in (-\infty, 2] \cup [3, +\infty)$    39. $-3, 9$

▶ **Exercise Set 26C (Page A26)**

1. $(-4, 7)$

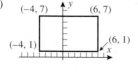

3. **(a)**

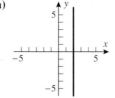

**(b)**

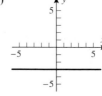

**(c)**

**(d)** 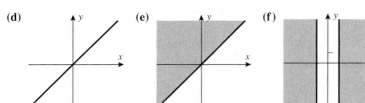 **(e)** **(f)**

5. 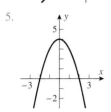 7. 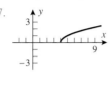 9. 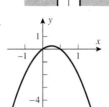 11. 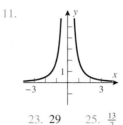 13. **(a)** $\frac{1}{2}$

**(b)** $-1$

**(c)** $0$

**(d)** not defined

15. **(a)** yes    17. 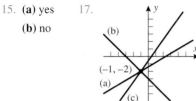    19. III < II < IV < I    21. **(a)** 14    23. 29    25. $\frac{13}{7}$

**(b)** no    **(b)** $-\frac{1}{3}$

29. **(a)** 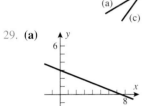    **(b)**    **(c)**    **(d)**

31. **(a)** 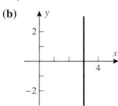    **(b)** 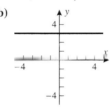    **(c)**    33.

| | (a) | (b) | (c) | (d) | (e) |
|---|---|---|---|---|---|
| Slope | 3 | –1/4 | –3/5 | 0 | –b/a |
| y intercept | 2 | 3 | 8/3 | 1 | b |

35. **(a)** $y = \frac{3}{2}x - 3$   **(b)** $y = -\frac{3}{4}x$    37. $y = -2x + 4$    39. $y = 4x + 7$    41. $y = -\frac{1}{5}x + 6$    43. $y = 11x - 18$

45. $y = \frac{1}{2}x + 2$    47. $y = 1$    49. **(a)** parallel    **(b)** perpendicular    **(c)** parallel    **(d)** perpendicular    **(e)** neither

51. **(a)** $-\frac{3}{2}$    53. the union of the graphs of $x - y = 0$ and $x + y = 0$    55. 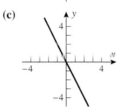    59. $\frac{49}{6}$    61. **(a)** yes

**(b)** $\frac{4}{5}$    **(b)** yes

**(c)** $\frac{5}{2}$    **(c)** no

**(d)** $-\frac{15}{2}$    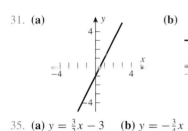    **(d)** yes

**(e)** $-4$    **(e)** yes

**(f)** yes

**(g)** no

▶ **Exercise Set D (Page A36)**

1. in the proof of Theorem D.1    3. **(a)** 10    **(b)** (4, 5)    5. **(a)** $\sqrt{29}$    **(b)** $\left(-\frac{9}{2}, -5\right)$    11. 0    13. $y = -3x + 4$

15. $\left(\frac{29}{8}, -\frac{23}{4}\right)$    17. 3    21. 4    23. **(a)** (0, 0); 5    **(b)** (1, 4); 4    **(c)** (−1, −3); $\sqrt{5}$    **(d)** (0, −2); 1

25. $(x - 3)^2 + (y + 2)^2 = 16$    27. $(x + 4)^2 + (y - 8)^2 = 64$    29. $(x + 3)^2 + (y + 4)^2 = 25$    31. $(x - 1)^2 + (y - 1)^2 = 2$

33. circle; center (1, 2), radius 4    35. circle; center (−1, 1), radius $\sqrt{2}$    37. the point (−1, −1)    39. circle; center (0, 0), radius $\frac{1}{3}$

41. no graph    43. circle; center $\left(-\frac{5}{4}, -\frac{1}{2}\right)$, radius $\frac{3}{2}$    45. **(a)** $y = -\sqrt{16 - x^2}$    **(b)** $y = 2 + \sqrt{3 - 2x - x^2}$

**47. (a)** 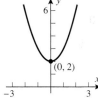 **(b)** 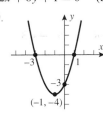   **49.** $y = -\frac{3}{4}x + \frac{25}{4}$   **51. (a)** inside   **53.** $(1/3, \pm\sqrt{8}/3)$

**(b)** largest $3\sqrt{5}$, smallest $\sqrt{5}$

**55. (a)** equation: $2x^2 + 2y^2 - 12x + 8y + 1 = 0$   **(b)** center $(3, -2)$, radius $5/\sqrt{2}$

**57.** 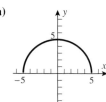   **59.** 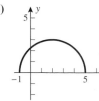   **61.** 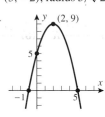   **63.**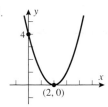

**65.** 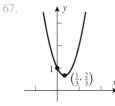   **67.** 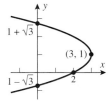   **69.** 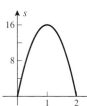   **71. (a)** $x = \sqrt{3 - y}$

**(b)** $x = 1 - \sqrt{y + 1}$

**73. (a)** 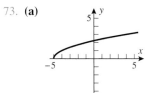  **(b)** 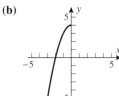   **75. (a)**   **77. (a)** $y = 150 - \frac{3}{2}x$

**(b)** $A = 150x - \frac{3}{2}x^2$

**(c)** $3750$ ft$^2$

**79. (a)** $(-5 - \sqrt{33})/4 < x < (-5 + \sqrt{33})/4$   **(b)** $-\infty < x < +\infty$   **81. (a)** 30 ft   **(b)** 2.6 s   **(c)** 2.1 s

---

▶ **Exercise Set E (Page A49)**

**1. (a)** $\frac{5}{12}\pi$   **3. (a)** $12°$   **5.**

| | $\sin\theta$ | $\cos\theta$ | $\tan\theta$ | $\csc\theta$ | $\sec\theta$ | $\cot\theta$ |
|---|---|---|---|---|---|---|
| (a) | $\sqrt{21}/5$ | $2/5$ | $\sqrt{21}/2$ | $5/\sqrt{21}$ | $5/2$ | $2/\sqrt{21}$ |
| (b) | $3/4$ | $\sqrt{7}/4$ | $3/\sqrt{7}$ | $4/3$ | $4/\sqrt{7}$ | $\sqrt{7}/3$ |
| (c) | $3/\sqrt{10}$ | $1/\sqrt{10}$ | $3$ | $\sqrt{10}/3$ | $\sqrt{10}$ | $1/3$ |

**7.** $\sin\theta = 3/\sqrt{10}, \cos\theta = 1/\sqrt{10}$

**(b)** $\frac{13}{6}\pi$   **(b)** $(270/\pi)°$

**(c)** $\frac{1}{9}\pi$   **(c)** $288°$

**(d)** $\frac{23}{30}\pi$   **(d)** $540°$

**9.** $\tan\theta = \sqrt{21}/2, \csc\theta = 5/\sqrt{21}$   **11.** 1.8   **13.**

| | $\theta$ | $\sin\theta$ | $\cos\theta$ | $\tan\theta$ | $\csc\theta$ | $\sec\theta$ | $\cot\theta$ |
|---|---|---|---|---|---|---|---|
| (a) | $225°$ | $-1/\sqrt{2}$ | $-1/\sqrt{2}$ | $1$ | $-\sqrt{2}$ | $-\sqrt{2}$ | $1$ |
| (b) | $-210°$ | $1/2$ | $-\sqrt{3}/2$ | $-1/\sqrt{3}$ | $2$ | $-2/\sqrt{3}$ | $-\sqrt{3}$ |
| (c) | $5\pi/3$ | $-\sqrt{3}/2$ | $1/2$ | $-\sqrt{3}$ | $-2/\sqrt{3}$ | $2$ | $-1/\sqrt{3}$ |
| (d) | $-3\pi/2$ | $1$ | $0$ | — | $1$ | — | $0$ |

**15.**

| | $\sin\theta$ | $\cos\theta$ | $\tan\theta$ | $\csc\theta$ | $\sec\theta$ | $\cot\theta$ |
|---|---|---|---|---|---|---|
| (a) | $4/5$ | $3/5$ | $4/3$ | $5/4$ | $5/3$ | $3/4$ |
| (b) | $-4/5$ | $3/5$ | $-4/3$ | $-5/4$ | $5/3$ | $-3/4$ |
| (c) | $1/2$ | $-\sqrt{3}/2$ | $-1/\sqrt{3}$ | $2$ | $-2/\sqrt{3}$ | $-\sqrt{3}$ |
| (d) | $-1/2$ | $\sqrt{3}/2$ | $-1/\sqrt{3}$ | $-2$ | $2/\sqrt{3}$ | $-\sqrt{3}$ |
| (e) | $1/\sqrt{2}$ | $1/\sqrt{2}$ | $1$ | $\sqrt{2}$ | $\sqrt{2}$ | $1$ |
| (f) | $1/\sqrt{2}$ | $-1/\sqrt{2}$ | $-1$ | $\sqrt{2}$ | $-\sqrt{2}$ | $-1$ |

**17. (a)** 1.2679   **(b)** 3.5753

**19.**

| | $\sin\theta$ | $\cos\theta$ | $\tan\theta$ | $\csc\theta$ | $\sec\theta$ | $\cot\theta$ |
|---|---|---|---|---|---|---|
| (a) | $a/3$ | $\sqrt{9 - a^2}/3$ | $a/\sqrt{9 - a^2}$ | $3/a$ | $3/\sqrt{9 - a^2}$ | $\sqrt{9 - a^2}/a$ |
| (b) | $a/\sqrt{a^2 + 25}$ | $5/\sqrt{a^2 + 25}$ | $a/5$ | $\sqrt{a^2 + 25}/a$ | $\sqrt{a^2 + 25}/5$ | $5/a$ |
| (c) | $\sqrt{a^2 - 1}/a$ | $1/a$ | $\sqrt{a^2 - 1}$ | $a/\sqrt{a^2 - 1}$ | $a$ | $1/\sqrt{a^2 - 1}$ |

21. **(a)** $3\pi/4 \pm n\pi, n = 0, 1, 2, \ldots$    **(b)** $\pi/3 \pm 2n\pi$ and $5\pi/3 \pm 2n\pi, n = 0, 1, 2, \ldots$

23. **(a)** $\pi/6 \pm n\pi, n = 0, 1, 2, \ldots$    **(b)** $4\pi/3 \pm 2n\pi$ and $5\pi/3 \pm 2n\pi, n = 0, 1, 2, \ldots$

25. **(a)** $3\pi/4 \pm n\pi, n = 0, 1, 2, \ldots$    **(b)** $\pi/6 \pm n\pi, n = 0, 1, 2, \ldots$

27. **(a)** $\pi/3 \pm 2n\pi$ and $2\pi/3 \pm 2n\pi, n = 0, 1, 2, \ldots$    **(b)** $\pi/6 \pm 2n\pi$ and $11\pi/6 \pm 2n\pi, n = 0, 1, 2, \ldots$

29. $\sin\theta = 2/5, \cos\theta = -\sqrt{21}/5, \tan\theta = -2/\sqrt{21}, \csc\theta = 5/2, \sec\theta = -5/\sqrt{21}, \cot\theta = -\sqrt{21}/2$

31. **(a)** $\theta = \pm n\pi, n = 0, 1, 2, \ldots$    **(b)** $\theta = \pi/2 \pm n\pi, n = 0, 1, 2, \ldots$    **(c)** $\theta = \pm n\pi, n = 0, 1, 2, \ldots$

    **(d)** $\theta = \pm n\pi, n = 0, 1, 2, \ldots$    **(e)** $\theta = \pi/2 \pm n\pi, n = 0, 1, 2, \ldots$    **(f)** $\theta = \pm n\pi, n = 0, 1, 2, \ldots$

33. **(a)** $2\pi/3$ cm   **(b)** $10\pi/3$ cm    35. $\frac{2}{5}$    37. **(a)** $\dfrac{2\pi - \theta}{2\pi}R$   **(b)** $\dfrac{\sqrt{4\pi\theta - \theta^2}}{2\pi}R$    39. $\frac{21}{4}\sqrt{3}$    41. 9.2 ft

43. $h = d(\tan\beta - \tan\alpha)$    45. **(a)** $4\sqrt{5}/9$   **(b)** $-\frac{1}{9}$    47. $\sin 3\theta = 3\sin\theta\cos^2\theta - \sin^3\theta, \cos 3\theta = \cos^3\theta - 3\sin^2\theta\cos\theta$

61. **(a)** $\cos\theta$   **(b)** $-\sin\theta$   **(c)** $-\cos\theta$   **(d)** $\sin\theta$

---

▶ Exercise Set F **(Page A56)**

1. **(a)** $x^2 + 4x + 2, -11x + 6$    3. **(a)** $3x^2 + 6x + 8, 15$    5.

    **(b)** $2x^2 + 4, 9$     **(b)** $x^3 - 5x^2 + 20x - 100, 504$

    **(c)** $x^3 - x^2 + 2x - 2, 2x + 1$     **(c)** $x^4 + x^3 + x^2 + x + 1, 0$

| $x$ | 0 | 1 | $-3$ | 7 |
|---|---|---|---|---|
| $p(x)$ | $-4$ | $-3$ | 101 | 5001 |

7. **(a)** $x^2 + 6x + 13, 20$

    **(b)** $x^2 + 3x - 2, -4$

9. **(a)** $\pm 1, \pm 2, \pm 3, \pm 4, \pm 6, \pm 8, \pm 12, \pm 24$    **(b)** $\pm 1, \pm 2, \pm 5, \pm 10, \pm\frac{1}{3}, \pm\frac{2}{3}, \pm\frac{5}{3}, \pm\frac{10}{3}$    **(c)** $\pm 1, \pm 17$    11. $(x+1)(x-1)(x-2)$

13. $(x+3)^3(x+1)$    15. $(x+3)(x+2)(x+1)^2(x-3)$    17. $-3$    19. $-2, -\frac{2}{3}$    21. $-2, 2, 3$    23. 2, 5    25. 7 cm

# PHOTO CREDITS

················································

**Introduction**
Page 1 (top): Granger Collection. Page 1 (bottom): Ed Honowitz/Tony Stone Images/New York, Inc. Page 2 (top): Courtesy Chris Brislawn, Los Alamos National Laboratory. Page 2 (bottom): Courtesy Dr. Adrian Maudsley, UCSF/NASA. Page 3 (top): Image generated from simulations carried out by the Aerospace Engineering and Mechanics Group at the Army HPC Research Center, Minneapolis, MN. Page 3 (center): Courtesy NOAA. Page 3 (just above bottom): Scott Camazine/Photo Researchers. Page 3 (bottom): Courtesy Applied Chaos Laboratory, Georgia Institute of Technology. Pages 10–11: Corbis-Bettmann.

**Chapter 1**
Page 15 (signature): Courtesy Smithsonian Institution. Page 15 (portrait): Corbis-Bettmann. Page 15 (bottom): Stephen Johnson/Tony Stone Images/New York, Inc. Page 17 (left): from *Visual Display of Quantitative Information*, ©1983, Edward R. Tufte. Published by Graphics Press, Cheshire, CT. Page 17 (right): from U.S. Department of Health and Human Services. Page 100: Corbis-Bettmann.

**Chapter 2**
Page 111 (signature): Courtesy Smithsonian Institution. Page 111 (portrait): Corbis-Bettmann. Page 111 (bottom): Barry Blackman/Tony Stone Images/New York, Inc.

**Chapter 3**
Page 169 (signature): Corbis-Bettmann. Page 169 (portrait): Courtesy the David Eugene Smith Collection, Columbia University. Page 169 (bottom): J. W. Burkey/Tony Stone Images/New York, Inc.

**Chapter 4**
Page 225 (signature): Courtesy New York Public Library. Page 225 (portrait): Corbis-Bettmann. Page 225 (bottom): Ken Fisher/Tony Stone Images/New York, Inc. Page 242: Bob Gruen/Star File.

**Chapter 5**
Page 289 (signature): Courtesy of the New York Public Library. Page 289 (portrait): Corbis-Bettmann. Page 289 (bottom): David Chambers/Tony Stone Images/New York, Inc.

**Chapter 6**
Page 329 (signature): Courtesy The British Library. Page 329 (portrait): Corbis-Bettmann. Page 329 (bottom): Donald Johnston/Tony Stone Images/New York, Inc. Page 359: UPI/Corbis-Bettmann.

**Chapter 7**
Page 377 (signature): Courtesy New York Public Library. Page 377 (portrait): Granger Collection. Page 377 (bottom): Cyber image/Tony Stone Images/New York, Inc. Page 379: Courtesy New York Public Library. Page 383: Reproduced from C. I. Gerhardt's *Briefwechsel von G.W. Leibniz mit Mathematikern* (1899). Page 457: J. Yulsman/The Image Bank.

# INDEX

## RATIONAL FUNCTIONS CONTAINING POWERS OF $a + bu$ IN THE DENOMINATOR

60. $\int \dfrac{u\,du}{a + bu} = \dfrac{1}{b^2}[bu - a\ln|a + bu|] + C$

61. $\int \dfrac{u^2\,du}{a + bu} = \dfrac{1}{b^3}\left[\dfrac{1}{2}(a + bu)^2 - 2a(a + bu) + a^2\ln|a + bu|\right] + C$

62. $\int \dfrac{u\,du}{(a + bu)^2} = \dfrac{1}{b^2}\left[\dfrac{a}{a + bu} + \ln|a + bu|\right] + C$

63. $\int \dfrac{u^2\,du}{(a + bu)^2} = \dfrac{1}{b^3}\left[bu - \dfrac{a^2}{a + bu} - 2a\ln|a + bu|\right] + C$

64. $\int \dfrac{u\,du}{(a + bu)^3} = \dfrac{1}{b^2}\left[\dfrac{a}{2(a + bu)^2} - \dfrac{1}{a + bu}\right] + C$

65. $\int \dfrac{du}{u(a + bu)} = \dfrac{1}{a}\ln\left|\dfrac{u}{a + bu}\right| + C$

66. $\int \dfrac{du}{u^2(a + bu)} = -\dfrac{1}{au} + \dfrac{b}{a^2}\ln\left|\dfrac{a + bu}{u}\right| + C$

67. $\int \dfrac{du}{u(a + bu)^2} = \dfrac{1}{a(a + bu)} + \dfrac{1}{a^2}\ln\left|\dfrac{u}{a + bu}\right| + C$

## RATIONAL FUNCTIONS CONTAINING $a^2 \pm u^2$ IN THE DENOMINATOR ($a > 0$)

68. $\int \dfrac{du}{a^2 + u^2} = \dfrac{1}{a}\tan^{-1}\dfrac{u}{a} + C$

69. $\int \dfrac{du}{a^2 - u^2} = \dfrac{1}{2a}\ln\left|\dfrac{u + a}{u - a}\right| + C$

70. $\int \dfrac{du}{u^2 - a^2} = \dfrac{1}{2a}\ln\left|\dfrac{u - a}{u + a}\right| + C$

71. $\int \dfrac{bu + c}{a^2 + u^2}\,du = \dfrac{b}{2}\ln(a^2 + u^2) + \dfrac{c}{a}\tan^{-1}\dfrac{u}{a} + C$

## INTEGRALS OF $\sqrt{a^2 + u^2}$, $\sqrt{a^2 - u^2}$, $\sqrt{u^2 - a^2}$ AND THEIR RECIPROCALS ($a > 0$)

72. $\int \sqrt{u^2 + a^2}\,du = \dfrac{u}{2}\sqrt{u^2 + a^2} + \dfrac{a^2}{2}\ln(u + \sqrt{u^2 + a^2}) + C$

73. $\int \sqrt{u^2 - a^2}\,du = \dfrac{u}{2}\sqrt{u^2 - a^2} - \dfrac{a^2}{2}\ln|u + \sqrt{u^2 - a^2}| + C$

74. $\int \sqrt{a^2 - u^2}\,du = \dfrac{u}{2}\sqrt{a^2 - u^2} + \dfrac{a^2}{2}\sin^{-1}\dfrac{u}{a} + C$

75. $\int \dfrac{du}{\sqrt{u^2 + a^2}} = \ln(u + \sqrt{u^2 + a^2}) + C$

76. $\int \dfrac{du}{\sqrt{u^2 - a^2}} = \ln|u + \sqrt{u^2 - a^2}| + C$

77. $\int \dfrac{du}{\sqrt{a^2 - u^2}} = \sin^{-1}\dfrac{u}{a} + C$

## POWERS OF $u$ MULTIPLYING OR DIVIDING $\sqrt{a^2 - u^2}$ OR ITS RECIPROCAL

78. $\int u^2\sqrt{a^2 - u^2}\,du = \dfrac{u}{8}(2u^2 - a^2)\sqrt{a^2 - u^2} + \dfrac{a^4}{8}\sin^{-1}\dfrac{u}{a} + C$

79. $\int \dfrac{\sqrt{a^2 - u^2}\,du}{u} = \sqrt{a^2 - u^2} - a\ln\left|\dfrac{a + \sqrt{a^2 - u^2}}{u}\right| + C$

80. $\int \dfrac{\sqrt{a^2 - u^2}\,du}{u^2} = -\dfrac{\sqrt{a^2 - u^2}}{u} - \sin^{-1}\dfrac{u}{a} + C$

81. $\int \dfrac{u^2\,du}{\sqrt{a^2 - u^2}} = -\dfrac{u}{2}\sqrt{a^2 - u^2} + \dfrac{a^2}{2}\sin^{-1}\dfrac{u}{a} + C$

82. $\int \dfrac{du}{u\sqrt{a^2 - u^2}} = -\dfrac{1}{a}\ln\left|\dfrac{a + \sqrt{a^2 - u^2}}{u}\right| + C$

83. $\int \dfrac{du}{u^2\sqrt{a^2 - u^2}} = -\dfrac{\sqrt{a^2 - u^2}}{a^2 u} + C$

## POWERS OF $u$ MULTIPLYING OR DIVIDING $\sqrt{u^2 \pm a^2}$ OR THEIR RECIPROCALS

84. $\int u\sqrt{u^2 + a^2}\,du = \dfrac{1}{3}(u^2 + a^2)^{3/2} + C$

85. $\int u\sqrt{u^2 - a^2}\,du = \dfrac{1}{3}(u^2 - a^2)^{3/2} + C$

86. $\int \dfrac{du}{u\sqrt{u^2 + a^2}} = -\dfrac{1}{a}\ln\left|\dfrac{a + \sqrt{u^2 + a^2}}{u}\right| + C$

87. $\int \dfrac{du}{u\sqrt{u^2 - a^2}} = \dfrac{1}{a}\sec^{-1}\left|\dfrac{u}{a}\right| + C$

88. $\int \dfrac{\sqrt{u^2 - a^2}\,du}{u} = \sqrt{u^2 - a^2} - a\sec^{-1}\left|\dfrac{u}{a}\right| + C$

89. $\int \dfrac{\sqrt{u^2 + a^2}\,du}{u} = \sqrt{u^2 + a^2} - a\ln\left|\dfrac{a + \sqrt{u^2 + a^2}}{u}\right| + C$

90. $\int \dfrac{du}{u^2\sqrt{u^2 \pm a^2}} = \mp\dfrac{\sqrt{u^2 \pm a^2}}{a^2 u} + C$

91. $\int u^2\sqrt{u^2 + a^2}\,du = \dfrac{u}{8}(2u^2 + a^2)\sqrt{u^2 + a^2} - \dfrac{a^4}{8}\ln(u + \sqrt{u^2 + a^2}) + C$

92. $\int u^2\sqrt{u^2 - a^2}\,du = \dfrac{u}{8}(2u^2 - a^2)\sqrt{u^2 - a^2} - \dfrac{a^4}{8}\ln|u + \sqrt{u^2 - a^2}| + C$

93. $\int \dfrac{\sqrt{u^2 + a^2}}{u^2}\,du = -\dfrac{\sqrt{u^2 + a^2}}{u} + \ln(u + \sqrt{u^2 + a^2}) + C$

94. $\int \dfrac{\sqrt{u^2 - a^2}}{u^2}\,du = -\dfrac{\sqrt{u^2 - a^2}}{u} + \ln|u + \sqrt{u^2 - a^2}| + C$

95. $\int \dfrac{u^2}{\sqrt{u^2 + a^2}}\,du = \dfrac{u}{2}\sqrt{u^2 + a^2} - \dfrac{a^2}{2}\ln(u + \sqrt{u^2 + a^2}) + C$

96. $\int \dfrac{u^2}{\sqrt{u^2 - a^2}}\,du = \dfrac{u}{2}\sqrt{u^2 - a^2} + \dfrac{a^2}{2}\ln|u + \sqrt{u^2 - a^2}| + C$

## INTEGRALS CONTAINING $(a^2 + u^2)^{3/2}$, $(a^2 - u^2)^{3/2}$, $(u^2 - a^2)^{3/2}$ ($a > 0$)

97. $\int \dfrac{du}{(a^2 - u^2)^{3/2}} = \dfrac{u}{a^2\sqrt{a^2 - u^2}} + C$

98. $\int \dfrac{du}{(u^2 \pm a^2)^{3/2}} = \pm\dfrac{u}{a^2\sqrt{u^2 \pm a^2}} + C$

99. $\int (a^2 - u^2)^{3/2}\,du = -\dfrac{u}{8}(2u^2 - 5a^2)\sqrt{a^2 - u^2} + \dfrac{3a^4}{8}\sin^{-1}\dfrac{u}{a} + C$

100. $\int (u^2 + a^2)^{3/2}\,du = \dfrac{u}{8}(2u^2 + 5a^2)\sqrt{u^2 + u^2} + \dfrac{3a^4}{8}\ln(u + \sqrt{u^2 + a^2}) + C$

101. $\int (u^2 - a^2)^{3/2}\,du = \dfrac{u}{8}(2u^2 - 5a^2)\sqrt{u^2 - a^2} + \dfrac{3a^4}{8}\ln|u + \sqrt{u^2 - a^2}| + C$

102. $\displaystyle\int u\sqrt{a + bu}\,du = \frac{2}{15b^2}(3bu - 2a)(a + bu)^{3/2} + C$

103. $\displaystyle\int u^2\sqrt{a + bu}\,du = \frac{2}{105b^3}(15b^2u^2 - 12abu + 8a^2)(a + bu)^{3/2} + C$

104. $\displaystyle\int u^n\sqrt{a + bu}\,du = \frac{2u^n(a + bu)^{3/2}}{b(2n + 3)} - \frac{2an}{b(2n + 3)}\int u^{n-1}\sqrt{a + bu}\,du$

105. $\displaystyle\int \frac{u\,du}{\sqrt{a + bu}} = \frac{2}{3b^2}(bu - 2a)\sqrt{a + bu} + C$

106. $\displaystyle\int \frac{u^2\,du}{\sqrt{a + bu}} = \frac{2}{15b^3}(3b^2u^2 - 4abu + 8a^2)\sqrt{a + bu} + C$

107. $\displaystyle\int \frac{u^n\,du}{\sqrt{a + bu}} = \frac{2u^n\sqrt{a + bu}}{b(2n + 1)} - \frac{2an}{b(2n + 1)}\int \frac{u^{n-1}\,du}{\sqrt{a + bu}}$

108. $\displaystyle\int \frac{du}{u\sqrt{a + bu}} = \begin{cases} \dfrac{1}{\sqrt{a}}\ln\left|\dfrac{\sqrt{a + bu} - \sqrt{a}}{\sqrt{a + bu} + \sqrt{a}}\right| + C & (a > 0) \\[3mm] \dfrac{2}{\sqrt{-a}}\tan^{-1}\sqrt{\dfrac{a + bu}{-a}} + C & (a < 0) \end{cases}$

109. $\displaystyle\int \frac{du}{u^n\sqrt{a + bu}} = -\frac{\sqrt{a + bu}}{a(n - 1)u^{n-1}} - \frac{b(2n - 3)}{2a(n - 1)}\int \frac{du}{u^{n-1}\sqrt{a + bu}}$

110. $\displaystyle\int \frac{\sqrt{a + bu}\,du}{u} = 2\sqrt{a + bu} + a\int \frac{du}{u\sqrt{a + bu}}$

111. $\displaystyle\int \frac{\sqrt{a + bu}\,du}{u^n} = -\frac{(a + bu)^{3/2}}{a(n - 1)u^{n-1}} - \frac{b(2n - 5)}{2a(n - 1)}\int \frac{\sqrt{a + bu}\,du}{u^{n-1}}$

112. $\displaystyle\int \sqrt{2au - u^2}\,du = \frac{u - a}{2}\sqrt{2au - u^2} + \frac{a^2}{2}\sin^{-1}\left(\frac{u - a}{a}\right) + C$

113. $\displaystyle\int u\sqrt{2au - u^2}\,du = \frac{2u^2 - au - 3a^2}{6}\sqrt{2au - u^2} + \frac{a^3}{2}\sin^{-1}\left(\frac{u - a}{a}\right) + C$

114. $\displaystyle\int \frac{\sqrt{2au - u^2}\,du}{u} = \sqrt{2au - u^2} + a\sin^{-1}\left(\frac{u - a}{a}\right) + C$

115. $\displaystyle\int \frac{\sqrt{2au - u^2}\,du}{u^2} = -\frac{2\sqrt{2au - u^2}}{u} - \sin^{-1}\left(\frac{u - a}{a}\right) + C$

116. $\displaystyle\int \frac{du}{\sqrt{2au - u^2}} = \sin^{-1}\left(\frac{u - a}{a}\right) + C$

117. $\displaystyle\int \frac{du}{u\sqrt{2au - u^2}} = -\frac{\sqrt{2au - u^2}}{au} + C$

118. $\displaystyle\int \frac{u\,du}{\sqrt{2au - u^2}} = -\sqrt{2au - u^2} + a\sin^{-1}\left(\frac{u - a}{a}\right) + C$

119. $\displaystyle\int \frac{u^2\,du}{\sqrt{2au - u^2}} = -\frac{(u + 3a)}{2}\sqrt{2au - u^2} + \frac{3a^2}{2}\sin^{-1}\left(\frac{u - a}{a}\right) + C$

120. $\displaystyle\int \frac{du}{(2au - u^2)^{3/2}} = \frac{u - a}{a^2\sqrt{2au - u^2}} + C$

121. $\displaystyle\int \frac{u\,du}{(2au - u^2)^{3/2}} = \frac{u}{a\sqrt{2au - u^2}} + C$

122. $\displaystyle\int_0^{\pi/2}\sin^n u\,du = \int_0^{\pi/2}\cos^n u\,du = \begin{cases} \dfrac{1\cdot 3\cdot 5\cdot\,\cdots\,\cdot(n - 1)}{2\cdot 4\cdot 6\cdot\,\cdots\,\cdot n}\cdot\dfrac{\pi}{2} & \left(\begin{array}{l} n \text{ an even} \\ \text{integer and} \\ n \geq 2 \end{array}\right) \\[5mm] \dfrac{2\cdot 4\cdot 6\cdot\,\cdots\,\cdot(n - 1)}{3\cdot 5\cdot 7\cdot\,\cdots\,\cdot n} & \left(\begin{array}{l} n \text{ an odd} \\ \text{integer and} \\ n \geq 3 \end{array}\right) \end{cases}$

# TRIGONOMETRIC IDENTITIES

## PYTHAGOREAN IDENTITIES

$\sin^2\theta + \cos^2\theta = 1 \qquad \tan^2\theta + 1 = \sec^2\theta \qquad 1 + \cot^2\theta = \csc^2\theta$

## SIGN IDENTITIES

$\sin(-\theta) = -\sin\theta \qquad \cos(-\theta) = \cos\theta \qquad \tan(-\theta) = -\tan\theta$

$\csc(-\theta) = -\csc\theta \qquad \sec(-\theta) = \sec\theta \qquad \cot(-\theta) = -\cot\theta$

## COMPLEMENT IDENTITIES

$\sin\left(\dfrac{\pi}{2} - \theta\right) = \cos\theta \qquad \cos\left(\dfrac{\pi}{2} - \theta\right) = \sin\theta \qquad \tan\left(\dfrac{\pi}{2} - \theta\right) = \cot\theta$

$\csc\left(\dfrac{\pi}{2} - \theta\right) = \sec\theta \qquad \sec\left(\dfrac{\pi}{2} - \theta\right) = \csc\theta \qquad \cot\left(\dfrac{\pi}{2} - \theta\right) = \tan\theta$

## SUPPLEMENT IDENTITIES

$\sin(\pi - \theta) = \sin\theta \qquad \cos(\pi - \theta) = -\cos\theta \qquad \tan(\pi - \theta) = -\tan\theta$

$\csc(\pi - \theta) = \csc\theta \qquad \sec(\pi - \theta) = -\sec\theta \qquad \cot(\pi - \theta) = -\cot\theta$

$\sin(\pi + \theta) = -\sin\theta \qquad \cos(\pi + \theta) = -\cos\theta \qquad \tan(\pi + \theta) = \tan\theta$

$\csc(\pi + \theta) = -\csc\theta \qquad \sec(\pi + \theta) = -\sec\theta \qquad \cot(\pi + \theta) = \cot\theta$

## ADDITION FORMULAS

$\sin(\alpha + \beta) = \sin\alpha\cos\beta + \cos\alpha\sin\beta$

$\sin(\alpha - \beta) = \sin\alpha\cos\beta - \cos\alpha\sin\beta$

$\cos(\alpha + \beta) = \cos\alpha\cos\beta - \sin\alpha\sin\beta$

$\cos(\alpha - \beta) = \cos\alpha\cos\beta + \sin\alpha\sin\beta$

$\tan(\alpha + \beta) = \dfrac{\tan\alpha + \tan\beta}{1 - \tan\alpha\tan\beta}$

$\tan(\alpha - \beta) = \dfrac{\tan\alpha - \tan\beta}{1 + \tan\alpha\tan\beta}$

## DOUBLE-ANGLE FORMULAS

$\sin 2\alpha = 2\sin\alpha\cos\alpha \qquad\qquad \cos 2\alpha = 2\cos^2\alpha - 1$

$\cos 2\alpha = \cos^2\alpha - \sin^2\alpha \qquad\qquad \cos 2\alpha = 1 - 2\sin^2\alpha$

## HALF-ANGLE FORMULAS

$\sin^2\dfrac{\alpha}{2} = \dfrac{1 - \cos\alpha}{2} \qquad\qquad \cos^2\dfrac{\alpha}{2} = \dfrac{1 + \cos\alpha}{2}$